T0130976

Queueing Networks
and Markov Chains

Queueing Networks and Markov Chains

Modeling and Performance Evaluation with Computer Science Applications

Second Edition

Gunter Bolch

Stefan Greiner

Hermann de Meer

Kishor S. Trivedi

WILEY-
INTERSCIENCE

A JOHN WILEY & SONS, INC., PUBLICATION

Library of Congress Cataloging-in-Publication Data:

Queueing networks and Markov chains : modeling and performance evaluation with computer science
    applications / Gunter Bolch ... [et al.].—2nd rev. and enlarged ed.
        p. cm.
    "A Wiley-Interscience publication."
    Includes bibliographical references and index.
    ISBN-13 978-0-471-56525-3 (acid-free paper)
    ISBN-10 0-471-56525-3 (acid-free paper)
    1. Markov processes.   2. Queuing theory.   I. Bolch, Gunter.

    QA76.9E94Q48 2006
    004.2'401519233—dc22   200506965

Printed in the United States of America

Contents

2   Markov Chains   51

3   Steady-State Solutions of Markov Chains   123

4 Steady-State Aggregation/Disaggregation Methods **185**

5 Transient Solution of Markov Chains **209**

6 Single Station Queueing Systems **241**

8   Algorithms for Product-Form Networks   369

9   Approximation Algorithms for Product-Form Networks   421

12 Performance Analysis Tools   657

13 Applications   703

Preface to the Second Edition

Nearly eight years have passed since the publication of the first edition of this book. In this second edition, we have thoroughly revised all the chapters. Many examples and problems are updated, and many new examples and problems have been added. A significant addition is a new chapter on simulation methods and applications. Application to current topics such as wireless system performance, Internet performance, J2EE applications, and Kanban systems performance are added. New material on non-Markovian and fluid stochastic Petri nets, along with solution techniques for Markov regenerative processes, is added. Topics that are covered briefly include self-similarity, large deviation theory, and diffusion approximation. The topic of hierarchical and fixed-point iterative models is also covered briefly. Our collective research experience and the application of these methods in practice for the past 30 years (at the time of writing) have been distilled in these chapters as much as possible. We hope that the book will be of use as a classroom textbook as well as of use for practicing engineers. Researchers will also find valuable information here.

We wish to thank many of our current students and former postdoctoral associates:

- Dr. Jörg Barner, for his contribution of the methodological background section in Chapter 1. He again supported us a lot in laying out the chapters and producing and improving figures and plots and with intensive proofreading.

- Pawan Choudhary, who helped considerably with the simulation chapter, Dr. Dharmaraja Selvamuthu helped with the section on SHARPE, and Dr. Hairong Sun helped in reading several chapters.

- Felix Engelhard, who was responsible for the new or extended sections on distributions, parameter estimation, Petri nets, and non-Markovian systems. He also did a thorough proofreading.

- Patrick Wüchner, for preparing the sections on matrix-analytic and matrix-geometric methods as well as the MMPP and MAP sections, and also for intensive proofreading.

- Dr. Michael Frank, who wrote and extended several sections: batch system and networks, summation method, and Kanban systems.

- Lassaad Essafi, who wrote the application section on differentiated services in the Internet.

Thanks are also due to Dr. Samuel Kounev and Prof. Alejandro Buchmann for allowing us to use their paper "Performance Modelling and Evaluation of Large Scale J2EE Applications" to produce the J2EE section, which is a shortened and adapted version of their paper.

Our special thanks are due to Prof. Helena Szczerbicka for her invaluable contribution to Chapter 11 on simulation and to modeling methodology section of Chapter 1. Her overall help with the second edition is also appreciated.

We also thank Val Moliere, George Telecki, Emily Simmons, and Whitney A. Lesch from John Wiley & Sons for their patience and encouragement.

The support from the Euro-NGI (Design and Engineering of the Next Generation Internet) Network of Excellence, European Commission grant IST-507613, is acknowledged.

Finally, a Web page has been set up for further information regarding the second edition. The URL is http://www.net.fmi.uni-passau.de/QNMC2/

Gunter Bolch, Stefan Greiner, Hermann de Meer, Kishor S. Trivedi

Erlangen, Passau, Durham, August 2005

Preface to the First Edition

Queueing networks and Markov chains are commonly used for the performance and reliability evaluation of computer, communication, and manufacturing systems. Although there are quite a few books on the individual topics of queueing networks and Markov chains, we have found none that covers both of these topics. The purpose of this book, therefore, is to offer a detailed treatment of queueing systems, queueing networks, and continuous and discrete-time Markov chains.

In addition to introducing the basics of these subjects, we have endeavored to:

- Provide some in-depth numerical solution algorithms.

- Incorporate a rich set of examples that demonstrate the application of the different paradigms and corresponding algorithms.

- Discuss stochastic Petri nets as a high-level description language, thereby facilitating automatic generation and solution of voluminous Markov chains.

- Treat in some detail approximation methods that will handle large models.

- Describe and apply four software packages throughout the text.

- Provide problems as exercises.

This book easily lends itself to a course on performance evaluation in the computer science and computer engineering curricula. It can also be used for a course on stochastic models in mathematics, operations research and industrial engineering departments. Because it incorporates a rich and comprehensive set of numerical solution methods comparatively presented, the text may also well serve practitioners in various fields of applications as a reference book for algorithms.

With sincere appreciation to our friends, colleagues, and students who so ably and patiently supported our manuscript project, we wish to publicly acknowledge:

- Jörg Barner and Stephan Kösters, for their painstaking work in keying the text and in laying out the figures and plots.

- Peter Bazan, who assisted both with the programming of many examples and comprehensive proofreading.

- Hana Ševčíková, who lent a hand in solving many of the examples and contributed with proofreading.

- János Sztrik, for his comprehensive proofreading.

- Doris Ehrenreich, who wrote the first version of the section on communication systems.

- Markus Decker, who prepared the first draft of the mixed queueing networks section.

- Those who read parts of the manuscript and provided many useful comments, including: Khalid Begain, Oliver Düsterhöft, Ricardo Fricks, Swapna Gokhale, Thomas Hahn, Christophe Hirel, Graham Horton, Steve Hunter, Demetres Kouvatsos, Yue Ma, Raymond Marie, Varsha Mainkar, Victor Nicola, Cheul Woo Ro, Helena Szczerbicka, Lorrie Tomek, Bernd Wolfinger, Katinka Wolter, Martin Zaddach, and Henry Zang.

Gunter Bolch and Stefan Greiner are grateful to Fridolin Hofmann, and Hermann de Meer is grateful to Bernd Wolfinger, for their support in providing the necessary freedom from distracting obligations.

Thanks are also due to Teubner B.G. Publishing House for allowing us to borrow sections from the book entitled **Leistungsbewertung von Rechensystemen** (originally in German) by one of the coauthors, Gunter Bolch. In the present book, these sections are integrated in Chapters 1 and 7 through 10.

We also thank Andrew Smith, Lisa Van Horn, and Mary Lynn of John Wiley & Sons for their patience and encouragement.

The financial support from the SFB (Collaborative Research Centre) 182 ("Multiprocessor and Network Configurations") of the DFG (Deutsche Forschungsgemeinschaft) is acknowledged.

Finally, a Web page has been set up for further information regarding the book. The URL is `http://www4.cs.fau.de/QNMC/`

GUNTER BOLCH, STEFAN GREINER, HERMANN DE MEER, KISHOR S. TRIVEDI

Erlangen, June 1998

1

Introduction

1.1 MOTIVATION

Information processing system designers need methods for the quantification of system design factors such as performance and reliability. Modern computer, communication, and production line systems process complex workloads with random service demands. Probabilistic and statistical methods are commonly employed for the purpose of performance and reliability evaluation. The purpose of this book is to explore major probabilistic modeling techniques for the performance analysis of information processing systems. Statistical methods are also of great importance but we refer the reader to other sources [Jain91, Triv01] for this topic. Although we concentrate on performance analysis, we occasionally consider reliability, availability, and combined performance and reliability analysis. Performance measures that are commonly of interest include throughput, resource utilization, loss probability, and delay (or response time).

The most direct method for performance evaluation is based on actual measurement of the system under study. However, during the design phase, the system is not available for such experiments, and yet performance of a given design needs to be predicted to verify that it meets design requirements and to carry out necessary trade-offs. Hence, abstract models are necessary for performance prediction of designs. The most popular models are based on discrete-event simulation (DES). DES can be applied to almost all problems of interest, and system details to the desired degree can be captured in such simulation models. Furthermore, many software packages are available that facilitate the construction and execution of DES models.

The principal drawback of DES models, however, is the time taken to run such models for large, realistic systems, particularly when results with high accuracy (i.e., narrow confidence intervals) are desired. A cost-effective alternative to DES models, analytic models can provide relatively quick answers to "what if" questions and can provide more insight into the system being studied. However, analytic models are often plagued by unrealistic assumptions that need to be made in order to make them tractable. Recent advances in stochastic models and numerical solution techniques, availability of software packages, and easy access to workstations with large computational capabilities have extended the capabilities of analytic models to more complex systems.

Analytical models can be broadly classified into state-space models and non-state-space models. Most commonly used state-space models are Markov chains. First introduced by A. A. Markov in 1907, Markov chains have been in use in performance analysis since around 1950. In the past decade, considerable advances have been made in the numerical solution techniques, methods of automated state-space generation, and the availability of software packages. These advances have resulted in extensive use of Markov chains in performance and reliability analysis. A Markov chain consists of a set of states and a set of labeled transitions between the states. A state of the Markov chain can model various conditions of interest in the system being studied. These could be the number of jobs of various types waiting to use each resource, the number of resources of each type that have failed, the number of concurrent tasks of a given job being executed, and so on. After a sojourn in a state, the Markov chain will make a transition to another state. Such transitions are labeled with either probabilities of transition (in case of discrete-time Markov chains) or rates of transition (in case of continuous-time Markov chains).

Long run (steady-state) dynamics of Markov chains can be studied using a system of linear equations with one equation for each state. Transient (or time dependent) behavior of a continuous-time Markov chain gives rise to a system of first-order, linear, ordinary differential equations. Solution of these equations results in state probabilities of the Markov chain from which desired performance measures can be easily obtained. The number of states in a Markov chain of a complex system can become very large, and, hence, automated generation and efficient numerical solution methods for underlying equations are desired. A number of concise notations (based on queueing networks and stochastic Petri nets) have evolved, and software packages that automatically generate the underlying state space of the Markov chain are now available. These packages also carry out efficient solution of steady-state and transient behavior of Markov chains. In spite of these advances, there is a continuing need to be able to deal with larger Markov chains and much research is being devoted to this topic.

If the Markov chain has nice structure, it is often possible to avoid the generation and solution of the underlying (large) state space. For a class of queueing networks, known as product-form queueing networks (PFQN), it is possible to derive steady-state performance measures without resorting to the underlying state space. Such models are therefore called non-state-space models. Other examples of non-state-space models are directed acyclic task precedence graphs [SaTr87] and fault-trees [STP96]. Other examples of methods exploiting Markov chains with "nice" structure are matrix-geometric methods [Neut81] (see Section 3.2).

Relatively large PFQN can be solved by means of a small number of simpler equations. However, practical queueing networks can often get so large that approximate methods are needed to solve such PFQN. Furthermore, many practical queueing networks (so-called non-product-form queueing networks, NPFQN) do not satisfy restrictions implied by product form. In such cases, it is often possible to obtain accurate approximations using variations of algorithms used for PFQNs. Other approximation techniques using hierarchical and fixed-point iterative methods are also used.

The flowchart shown in Fig. 1.1 gives the organization of this book. After a brief treatment on methodological background (Section 1.2), Section 1.3 covers the basics of probability and statistics. In Chapter 2, Markov chains basics are presented together with generation methods for them. Exact steady-state solution techniques for Markov chains are given in Chapter 3 and their aggregation/disaggregation counterpart in Chapter 4. These aggregation/disaggregation solution techniques are useful for practical Markov chain models with very large state spaces. Transient solution techniques for Markov chains are introduced in Chapter 5.

Chapter 6 deals with the description and computation of performance measures for single-station queueing systems in steady state. A general description of queueing networks is given in Chapter 7. Exact solution methods for PFQN are described in detail in Chapter 8 while approximate solution techniques for PFQN are described in Chapter 9. Solution algorithms for different types of NPFQN (such as networks with priorities, nonexponential service times, blocking, or parallel processing) are presented in Chapter 10.

Since there are many practical problems that may not be analytically tractable, discrete-event simulation is commonly used in this situations. We introduce the basics of DES in Chapter 11. For the practical use of modeling techniques described in this book, software packages (tools) are needed. Chapter 12 is devoted to the introduction of a queueing network tool, a stochastic Petri net tool, a tool based on Markov chains and a toolkit with many model types, and the facility for hierarchical modeling is also introduced. Each tool is described in some detail together with a simple example. Throughout the book we have provided many example applications of different algorithms introduced in the book. Finally, Chapter 13 is devoted to several large real-life applications of the modeling techniques presented in the book.

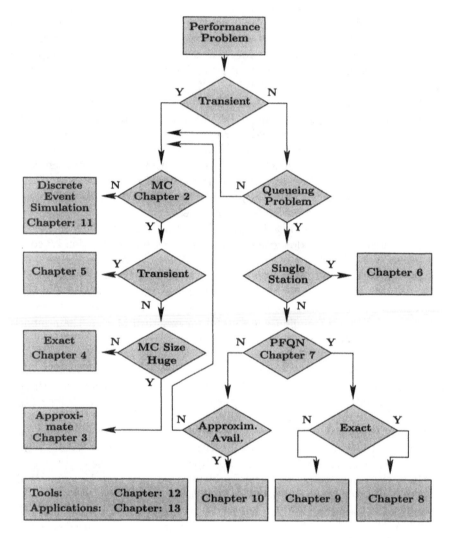

Fig. 1.1 Flowchart describing how to find the appropriate chapter for a given performance problem.

1.2 METHODOLOGICAL BACKGROUND

The focus of this book is the application of stochastic and probabilistic methods to obtain conclusions about performance and reliability properties of a wide range of systems. In general, a system can be regarded as a collection of components which are organized and interact in order to fulfill a common task [IEEE90].

Reactive systems and nondeterminism: The real-world systems treated in this book usually contain at least one digital component which controls the operation of other analog or digital components, and the whole system *reacts* to stimuli triggered by its *environment.* As an example, consider a computer communication network in which components like routers, switches, hubs, and communication lines fulfill the common task of transferring data packets between the various computers connected to the network. If the system of interest is the communication network only, the connected computers can be regarded as its environment which triggers the network by sending and receiving data packets. The behavior of the systems studied in this book can be characterized as *nondeterministic* since the stimulation by the environment is usually unpredictable. In case of a communication system like the Internet, the *workload* depends largely on the number of active users. When exactly a specific user will start to access information on the WWW via a browser is usually not known in advance. Another source of nondeterminism is the potential failure of one or several system components, which in most cases leads to an altered behavior of the complete system.

Modeling vs. Measurement: In contrast to the *empirical* methods of measurement, i.e., the collection of output data during the observation of an executing system, the *deductive* methods of model-based performance evaluation have the advantage to be applicable in situations when the system of interest is not yet existing. Deductive methods can thus be applied during the early design phases of the system development process in order to ensure that the final product meets its performance and reliability requirements. Although the material presented in this book is restricted to modeling approaches, it should be noticed that measurement as a supplementary technique can be employed to validate that the conclusions obtained by model-based performance evaluation can be translated into *useful* statements about the real-world system.

Another possible scenario for the application of modeling is the situation in which measurements on an existing system would either be too dangerous or too expensive. New policies, decision rules, or information flows can be explored without disrupting the ongoing operation of the real system. Moreover, new hardware architectures, scheduling algorithms, routing protocols, or reconfiguration strategies can be tested without committing resources for their acquisition/implementation. Also, the behavior of an existing system under a variety of anticipated workloads and environments can be evaluated very cost-effectively in advance by model-based approaches.

1.2.1 Problem Formulation

Before a meaningful model-based evaluation can commence, one should carefully consider what performance metric is of interest besides the nature of the system. This initial step is indispensable since it determines what is the appropriate formalism to be used. Most of the formalisms presented in the following chapters are suitable for the evaluation of specific metrics but inappropriate for the derivation of others. In general, it is important to consider the crucial aspects of the application domain with respect to the metrics to be evaluated before starting with the formalization process. Here, the *application context* strongly determines the kind of information that is meaningful in a concrete modeling exercise.

As an illustrative example, consider the power outage problem of computer systems. For a given hardware configuration, there is no ideal way to represent it without taking into consideration the software applications which run on the hardware and which of course have to be reflected in the model. In a real-time context, such as flight control, even the shortest power failure might have catastrophic implications for the system being controlled. Therefore, an appropriate reliability model of the flight control computer system has to be very sensitive to such a (hopefully) rare event of short duration. In contrast, the total number of jobs processed or the work accomplished by the computer hardware during the duration of a flight is probably a less important performance measure for such a safety-critical system. If the same hardware configuration is used in a transaction processing system, however, short outages are less significant for the proper system operation but the throughput is of predominant importance. As a consequence thereof, it is not useful to represent effects of short interruptions in the model, since they are of less importance in this application context.

Another important aspect to consider at the beginning of a model-based evaluation is how a reactive real-world system — as the core object of the study — is triggered by its environment. The stimulation of the system by its environment has to be captured in such a way during formalization so it reflects the conditions given in the real world as accurately as possible. Otherwise, the measures obtained during the evaluation process cannot be meaningfully retransformed into statements about the specific scenario in the application domain. In the context of stochastic modeling, the expression of the environment's influence on the system in the model is usually referred to as *workload modeling*. A frequently applied technique is the characterization of the arriving workload, e.g., the parts which enter a production line or the arriving data packets in a communication system, as a stochastic *arrival process*. Various arrival processes which are suitable in specific real-world scenarios can be defined (see Section 6.8).

The following four categories of system properties which are within the scope of the methods presented in this book can be identified:

Performance Properties: They are the oldest targets of performance evaluation and have been calculated already for non-computing systems like telephone switching centers [Erla17] or patient flows in hospitals [Jack54] using closed-form descriptions from applied probability theory. Typical properties to be evaluated are the mean throughput of served customers, the mean waiting, or response time and the utilization of the various system resources. The IEEE standard glossary of software engineering terminology [IEEE90] contains the following definition:

Definition 1.1   Performance: The degree to which a system or component accomplishes its designated functions within given constraints, such as speed, accuracy, or memory usage.

Reliability and Availability: Requirements of these types can be evaluated quantitatively if the system description contains information about the failure and repair behavior of the system components. In some cases it is also necessary to specify the conditions under which a new user cannot get access to the service offered by the operational system. The information about the failure behavior of system components is usually based on heuristics which are reflected in the parameters of probability distributions. In [IEEE90], software reliability is defined as:

Definition 1.2   Reliability: The probability that the software will not cause the failure of the system for a specified time under specified conditions.

System reliability is a measure for the continuity of correct service, whereas availability measures for a system refer to its readiness for correct service, as stated by the following definition from [IEEE90]:

Definition 1.3   Availability: The ability of a system to perform its required function at a stated instant or over a stated period of time. It is usually expressed as the availability ratio, i.e., the proportion of time that the service is actually available for use by the Customers within the agreed service hours.

Note that reliability and availability are related yet distinct system properties: a system which — during a mission time of 100 days — fails on average every two minutes but becomes operational again after a few milliseconds is not very reliable but nevertheless highly available.

Dependability and Performability: These terms and the definitions for them originated from the area of *dependable* and *fault tolerant* computing. The following definition for dependability is taken from [ALRL04]:

Definition 1.4   Dependability: The dependability of a computer system is the ability to deliver a service that can justifiably be trusted. The service delivered by a system is its behavior as it is perceived by its user(s); a user is another system (physical, human) that interacts with the former at the service interface.

This is a rather general definition which comprises the five attributes availability, reliability, *maintainability* — the systems ability to undergo modifications or repairs, *integrity* — the absence of improper system alterations and *safety* as a measure for the continuous delivery of service free from occurrences of catastrophic failures. The term *performability* was coined by J.F. MEYER [Meye78] as a measure to assess a system's ability to perform when performance degrades as a consequence of faults:

Definition 1.5   Performability: The probability that the system reaches an accomplishment level y over a utilization interval $(0, t)$. That is, the probability that the system does a certain amount of useful work over a mission time t.

Subsequently, many other measures are included under performance as we shall see in Section 2.2. Informally, the performability refers to performance in the presence of failures/repair/recovery of components and the system. Performability is of special interest for *gracefully degrading systems* [Beau77]. In Section 2.2, a framework based on *Markov reward models* (MRMs) is presented which provides recipes for a selection of the right model type and the definition of an appropriate performance measure.

1.2.2   The Modeling Process

The first step of a model-based performance evaluation consists of the formalization process, during which the modeler generates a *formal* description of the real-world system. Figure 1.2 illustrates the basic idea: Starting from an informal system description, e.g. in natural language, which includes structural and functional information as well as the desired performance and reliability requirements, the modeler creates a formal model of the real-world system using a specific *conceptualization*. A conceptualization is an *abstract, simplified* view of the reference reality which is represented for some purpose. Two kinds of conceptualizations for the purpose of performance evaluation are presented in detail in this book: If the system is to be represented as a queueing network, the modeler applies a "routed job flow" *modeling paradigm* in which the real-world system is conceptualized as a set of service stations which are connected by edges through which independent entities "flow" through the network and sojourn in the queues and servers of the service stations (see Chapter 7). In an alternative Markov chain conceptualization a "state-transition" modeling paradigm is applied in which the possible trajectories through the system's global state space are represented as a graph whose directed arcs represent the transitions between subsequent system states (see Chapter 2). The main difference between the two conceptualizations is that the queueing network formalism is oriented more towards the *structure* of the real-world system, whereas in the Markov chain formalization the emphasis is put on the description of the system *behavior* on the underlying state-space level. A Markov chain can be regarded to "mimic" the behavior of the executing real-world system, whereas the service stations and jobs of a queueing network establish a one-to-one correspondence to the components of the real-world system. As indicated in Fig. 1.2, a Markov chain serves as the underlying *semantic* model of the high-level queueing network model.

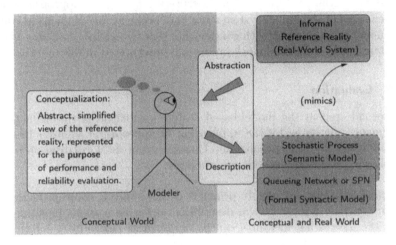

Fig. 1.2 Formalization of a real-world system.

During the formalization process the following abstractions with respect to the real-world system are applied:

- In both conceptualizations the behavior of the real-world system is regarded to evolve in a *discrete-event* fashion, even if the real-world system contains components which exhibit continuous behavior, such as the movements of a conveyor belt of a production line.

- The application of the queueing network formalism abstracts away from all *synchronization* mechanisms which may be present in the real-world system. If the representation of these synchronization mechanisms is crucial in order to obtain useful results from the evaluation, the modeler can resort to variants of stochastic Petri nets as an alternative description technique (see Section 2.3 and Section 2.3.6) in which almost arbitrary synchronization patterns can be captured.

- The *core abstractions* applied during the formalization process are the association of system *activity durations* with *random variables* and the inclusion of *branching probabilities* to represent alternative system evolutions. Both abstractions *resolve* the nondeterminism inherent in the real-world system and turn the formal queueing network or Markov chain prototype into an "executable" specification [Wing01]. For these, at any moment during their operation each possible future evolution has a well-defined probability to occur. Depending on which kind of random variables are used to represent the durations of the system activities either a discrete-time interpretation using a DTMC or a continuous-time interpretation of the system behavior based on a CTMC is achieved. It should be noted that for systems with *asynchronously* evolving components the continuous-time interpretation is more appropriate since changes of the global system state may occur at any moment in continuous time. Systems with components that evolve in a *lock-step* fashion triggered by a global clock are usually interpreted in discrete-time.

1.2.3   Evaluation

The second step in the model-based system evaluation is the deduction of performance measures by the application of appropriate solution methods. Depending on the conceptualization chosen during the formalization process the following solution methods are available:

Analytical Solutions: The core principle of the analytic solution methods is to represent the formal system description either as a single equation from which the interesting measures can be obtained as closed-form solutions, or as a set of system equations from which exact or approximate measures can be calculated by appropriate algorithms from numerical mathematics.

1. *Closed-form solutions* are available if the system can be described as a simple queueing system (see Chapter 6) or for simple product-form queueing networks (PFQN) [ChMa83] (see Section 7.3) or for structured small CTMCs. For these kind of formalizations equations can be derived from which the mean number of jobs in the service stations can be calculated as a *closed-form solution,* i.e., the solutions can be expressed analytically in terms of a bounded number of well-known operations. Also from certain types of Markov chains with regular structure (see Section 3.1), closed-form representations like the well-known Erlang-B and Erlang-C formulae [Erla17] can be derived. The measures can either be computed by ad-hoc programming or with the help of computer algebra packages such as Mathematica [Mat05]. A big advantage of the closed-form solutions is their moderate computational complexity which enables a fast calculation of performance measures even for larger system descriptions.

2. *Numerical solutions*: Many types of equations which can be derived from a formal system description do not possess a closed-form solution, e.g., in the case of complex systems of integro-differential equations. In these cases, approximate solutions can be obtained by the appli-

cation of algorithms from numerical mathematics, many of which are implemented in computer algebra packages [Mat05] or are integrated in performance analysis tools such as SHARPE [HSZT00], SPNP [HTT00], or TimeNET [ZFGH00] (see Chapter 12). The formal system descriptions can be either given as a queueing network, stochastic Petri net or another high-level modeling formalism, from which a state-space representation is generated manually or by the application of state-space generation algorithms. Depending on the stochastic information present in the high-level description, various types of system state equations which mimic the dynamics of the modeled system can be derived and solved by appropriate algorithms. The numerical solution of Markov models is discussed in Chapters 3 – 5, numerical solution methods for queueing networks can be found in Chapters 7 – 10. In comparison to closed-form solution approaches, numerical solution methods usually have a higher computational complexity.

Simulation Solutions: For many types of models no analytic solution method is feasible, because either a theory for the derivation of proper system equations is not known, or the computational complexity of an applicable numerical solution algorithm is too high. In this situation, solutions can be obtained by the application of discrete-event simulation (DES), which is described in detail in Chapter 11. Instead of solving system equations which have been derived from the formal model, the DES algorithm "executes" the model and collects the information about the observed behavior for the subsequent derivation of performance measures. In order to increase the quality of the results, the simulation outputs collected during multiple "executions" of the model are collected and from which the interesting measures are calculated by statistical methods. All the formalizations presented in this book, i.e., queueing networks, stochastic Petri nets, or Markov chains can serve as input for a DES, which is the most flexible and generally applicable solution method. Since the underlying state space does not have to be generated, simulation is not affected by the state-space explosion problem. Thus, simulation can also be employed for the analysis of complex models for which the numerical approaches would fail because of an exuberant number of system states.

Hybrid solutions: There exists a number of approaches in which different modeling formalisms and solution methods are combined in oder to exploit their complementing strengths. Examples of hybrid solution methods are mixed simulation and analytical/numerical approaches, or the combination of fault trees, reliability block diagrams, or reliability graphs, and Markov models [STP96]. Also product-form queueing networks and stochastic Petri nets or non-product-form networks and their solution methods can be combined. More generally, this approach can be characterized as intermingling of *state-space*-based and *non-state-space*-based methods [STP96]. A combination of analytic and simulative solutions of connected sub-models may be employed to combine the benefits of both solution methods [Sarg94, ShSa83]. More criteria for a choice between simulation and analytical/numerical solutions are discussed in Chapter 11.

Largeness Tolerance: Many high-level specification techniques, queueing systems, generalized stochastic Petri nets (GSPNs), and stochastic reward nets (SRNs), as the most prominent representatives, have been suggested in the literature to automate the model generation [HaTr93]. GSPNs/SRNs that are covered in more detail in Section 2.3, can be characterized as *tolerating* largeness of the underlying computational models and providing effective means for generating large state spaces.

Largeness Avoidance: Another way to deal with large models is *to avoid* the creation of such models from the beginning. The major largeness-avoidance technique we discuss in this book is that of product-form queueing networks. The main idea is, the structure of the underlying CTMC allows for an efficient solution that obviates the need for generation, storage, and solution of the large state space. The second method of avoiding largeness is to separate the originally single large problem into several smaller problems and to combine sub-model results into an overall solution. Both approximate and exact techniques are known for dealing with such multilevel models. The flow of information needed among sub-models may be acyclic, in which case a hierarchical model [STP96] results. If the flow of needed information is non-acyclic, a fixed-point iteration may be necessary [CiTr93]. Other well-known techniques applicable for limiting model sizes are *state truncation* [BVDT88, GCS+86] and *state lumping* [Nico90].

1.2.4 Summary

Figure 1.3 summarizes the different phases and activities of the model-based performance evaluation process. Two main scenarios are considered: In the first one, model-based performance evaluation is applied during the early phases of the system development process to predict the performance or reliability properties of the final product. If the predicted properties do not fulfill the given requirements, the proposed design has to be changed in order to avoid the expected performance problems. In the second scenario, the final product is already available and model-based performance evaluation is applied to derive optimal system configuration parameters, to solve capacity planning problems, or to check whether the existing system would still operate satisfactorily after a modification of its environment. In both scenarios the first activity in the evaluation process is to collect information about the structure and functional behavior of an existing or planned system. The participation of a domain expert in this initial step is very helpful and rather indispensable for complex applications. Usually, the collected information is stated informally and stored in a document using either a textual or a combined textual/graphical form. According to the goal of the evaluation study, the informal description also contains the performance, reliability, availability, dependability, or performability requirements that the system has to fulfill. Based on an informal system description, the formalization process is carried out by the modeler who needs both modeling and application-specific knowledge.

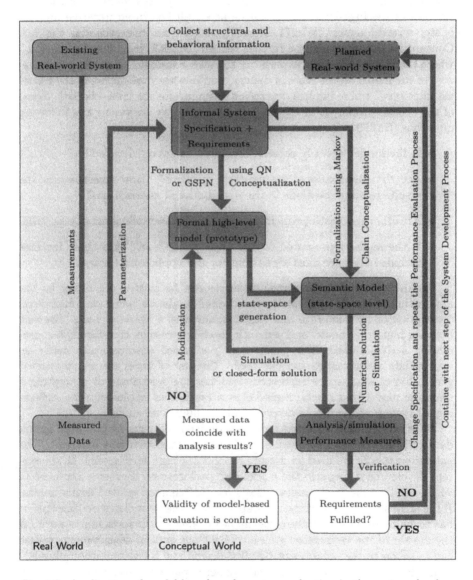

Fig. 1.3 Application of model-based performance evaluation in the system development process and configuration management.

If he selects a Markov chain conceptualization, a formal representation on the state-space level is created. If he chooses instead a queueing network or stochastic Petri net conceptualization, a formal high-level model is derived. The formal model represents the system as well as the interaction with its environment on a conceptual level and, as a result, the model abstracts from all details which are considered to be irrelevant for the evaluation. In the scenario of an existing real-world system, a *conceptual validation* of the correctness of the high-level model can be accomplished in an iterative process of step-wise refinement [Wirt71, Morr87], which is not represented in Fig. 1.3. Conceptual validation can be further refined [NaFi67] into "face validation," where the involved domain experts try to reach consensus on the appropriateness of the model on the basis of dialogs, and into the "validation of the model assumptions," where implicit or explicit assumptions are cross-checked. Some of the crucial properties of the model are checked by answering the following questions [HMT91]:

- Is the model logically correct, complete, or overly detailed?

- Are the distributional assumptions justified? How sensitive are the results to simplifications in the distributional assumptions?

- Are other stochastic properties, such as independence assumptions, valid?

- Is the model represented on the appropriate level? Are those features included that are most significant for the application context?

The deduction of performance measures can be carried out either by the application of closed-form solution methods/simulation based on a high-level description or by numerical analysis/simulation of a Markov chain. The link between high-level model and semantic model represents the state-space generation activity which is in most cases performed automatically inside an appropriate performance analysis tool. The use of tools for an automated generation of state-space representations has the advantage of generating a semantic model that can be regarded as a lower-level implementation reflecting all system properties as described in the high-level model.

After performance measures have been derived, they can be *validated* against data collected through *measurements*, if an existing system is available. The *validation results* are used for *modification* of the high-level model. In the system development scenario the calculated performance measures are used to verify whether the final product meets its performance related requirements. If the answer from the verification step is "NO," the formal system description has to be redesigned and the performance evaluation process starts anew. If the answer of the verification step is "YES," the system development process can move on to the subsequent detailed design and implementation phases.

In the system configuration and capacity planning scenarios there is a strong relation between *measurements* and modeling. Measurement data are to be used for model validation. Furthermore, model *parameterization* relies heavily on input from measurement results. Model parameters are frequently derived from measurements of earlier studies on similar systems also in the system development scenario. Conversely, measurement studies can often be better planned and executed if they are complemented and guided by a model-based evaluation.

1.3 BASICS OF PROBABILITY AND STATISTICS

We begin by giving a brief overview of the more important definitions and results of probability theory. The reader can find additional details in books such as [Alle90, Fell68, Koba78, Triv01]. We assume that the reader is familiar with the basic properties and notations of probability theory.

1.3.1 Random Variables

A random variable is a function that reflects the result of a random experiment. For example, the result of the experiment "toss a single die" can be described by a random variable that can assume the values one through six. The number of requests that arrive at an airline reservation system in one hour or the number of jobs that arrive at a computer system are also examples of a random variable. So is the time interval between the arrivals of two consecutive jobs at a computer system, or the throughput in such a system. The latter two examples can assume continuous values, whereas the first two only assume discrete values. Therefore, we have to distinguish between continuous and discrete random variables.

1.3.1.1 Discrete Random Variables A random variable that can only assume discrete values is called a *discrete random variable*, where the discrete values are often non-negative integers. The random variable is described by the possible values that it can assume and by the probabilities for each of these values. The set of these probabilities is called the *probability mass function* (pmf) of this random variable. Thus, if the possible values of a random variable X are the non-negative integers, then the pmf is given by the probabilities:

$$p_k = P(X = k), \quad \text{for} \quad k = 0, 1, 2 \ldots , \tag{1.1}$$

the probability that the random variable X assumes the value k.

The following is required:

$$P(X = k) \geq 0, \qquad \sum_{\text{all } k} P(X = k) = 1.$$

For example, the following pmf results from the experiment "toss a single die":

$$P(X = k) = \frac{1}{6}, \quad \text{for } k = 1, 2, \ldots, 6.$$

The following are other examples of discrete random variables:

- Bernoulli random variable: Consider a random experiment that has two possible outcomes, such as tossing a coin ($k = 0, 1$). The pmf of the random variable X is given by

$$P(X = 0) = 1 - p \quad \text{and} \quad P(X = 1) = p, \quad \text{with } 0 < p < 1. \quad (1.2)$$

- Binomial random variable: The experiment with two possible outcomes is carried out n times where successive trials are independent. The random variable X is now the number of times the outcome 1 occurs. The pmf of X is given by

$$P(X = k) = \binom{n}{k} p^k (1 - p)^{n-k}, \quad k = 0, 1, \ldots, n. \quad (1.3)$$

- Geometric random variable: The experiment with two possible outcomes is carried out several times, where the random variable X now represents the number of trials it takes for the outcome 1 to occur (the current trial included). The pmf of X is given by

$$P(X = k) = p(1 - p)^{k-1}, \quad k = 1, 2, \ldots. \quad (1.4)$$

- Poisson random variable: The probability of having k events (Poisson pmf) is given by

$$P(X = k) = \frac{(\alpha)^k}{k!} \cdot e^{-\alpha}, \quad k = 0, 1, 2, \ldots; \alpha > 0. \quad (1.5)$$

The Poisson and geometric random variables are very important to our topic; we will encounter them very often. Several important parameters can be derived from a pmf of a discrete random variable:

- Mean value or expected value:

$$\overline{X} = E[X] = \sum_{\text{all } k} k \cdot P(X = k). \quad (1.6)$$

The function of a random variable is another random variable with the expected value of

$$E[f(X)] = \sum_{\text{all } k} f(k) \cdot P(X = k). \quad (1.7)$$

- nth moments:

$$\overline{X^n} = E[X^n] = \sum_{\text{all } k} k^n \cdot P(X = k), \quad (1.8)$$

that is, the nth moment is the expected value of the nth power of X. The first moment of X is simply the mean of X.

- nth central moment:

$$\overline{(X - \overline{X})^n} = E[(X - E[X])^n] = \sum_{\text{all } k}(k - \overline{X})^n \cdot P(X = k). \qquad (1.9)$$

The nth central moment is the expected value of the nth power of the difference between X and its mean. The first central moment is equal to zero.

- The second central moment is called the variance of X:

$$\sigma_X^2 = \text{var}(X) = \overline{(X - \overline{X})^2} = \overline{X^2} - \overline{X}^2, \qquad (1.10)$$

where σ_X is called the standard deviation.

- The coefficient of variation is the normalized standard deviation:

$$c_X = \frac{\sigma_X}{\overline{X}}. \qquad (1.11)$$

- The second moment can be easily obtained from the coefficient of variation:

$$\overline{X^2} = \overline{X}^2 \cdot (1 + c_X^2). \qquad (1.12)$$

Information on the average deviation of a random variable from its expected value is provided by c_X, σ_X, and $\text{var}(X)$. If $c_X = \sigma_X = \text{var}(X) = 0$, then the random variable assumes a fixed value with probability one.

Table 1.1 Properties of several discrete random variables

Random Variables	Parameter	$\overline{X}$	var(X)	c_X^2
Bernoulli	p	p	$p(1-p)$	$\dfrac{1-p}{p}$
Binomial	n, p	np	$np(1-p)$	$\dfrac{1-p}{np}$
Geometric	p	$\dfrac{1}{p}$	$\dfrac{1-p}{p^2}$	$1-p$
Poisson	α	α	α	$\dfrac{1}{\alpha}$

Table 1.1 gives a list of random variables, their mean values, their variances, and the squared coefficients of variation for some important discrete random variables.

1.3.1.2 Continuous Random Variables   A random variable X that can assume all values in the interval $[a, b]$, where $-\infty \le a < b \le +\infty$, is called a *continuous random variable*. It is described by its *distribution function* (also called CDF or cumulative distribution function):

$$F_X(x) = P(X \le x), \tag{1.13}$$

which specifies the probability that the random variable X takes values less than or equal to x, for every x.

From Eq. (1.13) we get for $x < y$:

$$F_X(x) \le F_X(y), \qquad P(x < X \le y) = F_X(y) - F_X(x).$$

The *probability density function* (pdf) $f_X(x)$ can be used instead of the distribution function, provided the latter is differentiable:

$$f_X(x) = \frac{d\,F_X(x)}{d\,x}. \tag{1.14}$$

Some properties of the pdf are

$$f_X(x) \ge 0 \quad \text{for all } x, \qquad \int_{-\infty}^{\infty} f_X(x)\,d\,x = 1,$$

$$P(x_1 \le X \le x_2) = \int_{x_1}^{x_2} f_X(x)\,d\,x, \qquad P(X = x) = \int_{x}^{x} f_X(x)\,d\,x = 0,$$

$$P(X > x_3) = \int_{x_3}^{\infty} f_X(x)\,d\,x.$$

Note that for so-called *defective* random variables [STP96] we have

$$\int_{-\infty}^{\infty} f_X(x)\,d\,x < 1.$$

The density function of a continuous random variable is analogous to the pmf of a discrete random variable. The formulae for the mean value and moments of continuous random variables can be derived from the formulae for discrete random variables by substituting the pmf by the pdf and the summation by an integral:

- Mean value or expected value:

$$\overline{X} = E[X] = \int_{-\infty}^{\infty} x \cdot f_X(x)\,d\,x \tag{1.15}$$

and

$$E[g(X)] = \int_{-\infty}^{\infty} g(x) \cdot f_X(x) \, dx. \tag{1.16}$$

- nth moment:

$$\overline{X^n} = E[X^n] = \int_{-\infty}^{\infty} x^n \cdot f_X(x) \, dx. \tag{1.17}$$

- nth central moment:

$$\overline{(X - \overline{X})^n} = E[(X - E[X])^n] = \int_{-\infty}^{\infty} (x - \overline{X})^n f_X(x) \, dx. \tag{1.18}$$

- Variance:

$$\sigma_X^2 = \text{var}(X) = \overline{(X - \overline{X})^2} = \overline{X^2} - \overline{X}^2, \tag{1.19}$$

with σ_X as the standard deviation.

- Coefficient of variation:

$$c_X = \frac{\sigma_X}{\overline{X}}. \tag{1.20}$$

A very well-known and important continuous distribution function is the *normal distribution*. The CDF of a normally distributed random variable X is given by

$$F_X(x) = \frac{1}{\sqrt{2\pi\sigma_X^2}} \int_{-\infty}^{x} \exp\left(-\frac{(u - \overline{X})^2}{2\sigma_X^2}\right) du, \tag{1.21}$$

and the pdf by

$$f_X(x) = \frac{1}{\sqrt{2\pi\sigma_X^2}} \exp\left(-\frac{(x - \overline{X})^2}{2\sigma_X^2}\right).$$

The standard normal distribution is defined by setting $\overline{X} = 0$ and $\sigma_X = 1$:

$$\text{CDF: } \Phi(x) = \frac{1}{\sqrt{2\pi}} \cdot \int_{-\infty}^{x} \exp\left(-\frac{u^2}{2}\right) du, \tag{1.22}$$

$$\text{pdf: } \phi(x) = \frac{1}{\sqrt{2\pi}} \cdot \exp\left(-\frac{x^2}{2}\right).$$

A plot of the preceding pdf is shown in Fig. 1.4.

For an arbitrary normal distribution we have

$$F_X(x) = \Phi\left(\frac{x - \overline{X}}{\sigma_X}\right) \quad \text{and} \quad f_X(x) = \phi\left(\frac{x - \overline{X}}{\sigma_X}\right),$$

respectively.

Other important continuous random variables are described as follows.

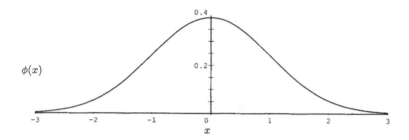

$\phi(x)$

Fig. 1.4 pdf of the standard normal random variable.

(a) Exponential Distribution

The exponential distribution is the most important and also the easiest to use distribution in queueing theory. Interarrival times and service times can often be represented exactly or approximately using the exponential distribution. The CDF of an exponentially distributed random variable X is given by Eq. (1.23):

$$F_X(x) = \begin{cases} 1 - \exp\left(-\dfrac{x}{\overline{X}}\right), & 0 \le x < \infty, \\ 0, & \text{otherwise.} \end{cases}$$

$$\text{with } \overline{X} = \begin{cases} \dfrac{1}{\lambda}, & \text{if } X \text{ represents interarrival times}, \\ \dfrac{1}{\mu}, & \text{if } X \text{ represents service times.} \end{cases} \tag{1.23}$$

Here λ or μ denote the parameter of the random variable. In addition, for an exponentially distributed random variable with parameter λ the following relations hold:

$$\text{pdf:} \quad f_X(x) = \lambda e^{-\lambda x},$$

$$\text{mean:} \quad \overline{X} = \frac{1}{\lambda},$$

$$\text{variance:} \quad \text{var}(X) = \frac{1}{\lambda^2},$$

$$\text{coefficient of variation:} \quad c_X = 1.$$

Thus, the exponential distribution is completely determined by its mean value.

The importance of the exponential distribution is based on the fact that it is the only continuous distribution that possesses the memoryless property:

$$P(X \leq u + t \mid X > u) = 1 - \exp\left(-\frac{t}{\overline{X}}\right) = P(X \leq t). \qquad (1.24)$$

As an example for an application of Eq. (1.24), consider a bus stop with the following schedule: Buses arrive with exponentially distributed interarrival times and identical mean $\overline{X}$. Now if you have already been waiting in vain for u units of time for the bus to come, the probability of a bus arrival within the next t units of time is the same as if you had just shown up at the bus stop, that is, you can forget about the past or about the time already spent waiting.

Another important property of the exponential distribution is its relation to the discrete Poisson random variable. If the interarrival times are exponentially distributed and successive interarrival times are independent with identical mean $\overline{X}$, then the random variable that represents the number of buses that arrive in a fixed interval of time $[0, t)$ has a Poisson distribution with parameter $\alpha = t/\overline{X}$.

Two additional properties of the exponential distribution can be derived from the Poisson property:

1. If we merge n Poisson processes with distributions for the interarrival times $1 - e^{-\lambda_i t}$, $1 \leq i \leq n$, into one single process, then the result is a Poisson process for which the interarrival times have the distribution $1 - e^{-\lambda t}$ with $\lambda = \sum_{i=1}^{n} \lambda_i$ (see Fig. 1.5).

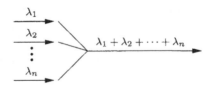

Fig. 1.5 Merging of Poisson processes.

2. If a Poisson process with interarrival time distribution $1 - e^{-\lambda t}$ is split into n processes so that the probability that the arriving job is assigned to the ith process is q_i, $1 \leq i \leq n$, then the ith subprocess has an interarrival time distribution of $1 - e^{-q_i \lambda t}$, i.e., n Poisson processes have been created, as shown in Fig 1.6.

The exponential distribution has many useful properties with analytic tractability, but is not always a good approximation to the observed distribution. Experiments have shown deviations. For example, the coefficient of variation of the service time of a processor is often greater than one, and for a peripheral device it is usually less than one. This observed behavior leads directly to the need to consider the following other distributions:

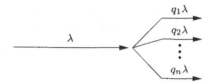

Fig. 1.6 Splitting of a Poisson process.

(b) Hyperexponential Distribution, H_k

This distribution can be used to approximate empirical distributions with a coefficient of variation larger than one. Here k is the number of phases.

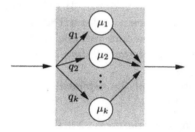

Fig. 1.7 A random variable with H_k distribution.

Figure 1.7 shows a model with hyperexponentially distributed time. The model is obtained by arranging k phases with exponentially distributed times and rates $\mu_1, \mu_2, \ldots, \mu_k$ in parallel. The probability that the time span is given by the jth phase is q_j, where $\sum_{j=1}^{k} q_j = 1$. However, only one phase can be occupied at any time. The resulting CDF is given by

$$F_X(x) = \sum_{j=1}^{k} q_j (1 - e^{-\mu_j x}), \qquad x \geq 0. \tag{1.25}$$

In addition, the following relations hold:

$$\text{pdf:} \quad f_X(x) = \sum_{j=1}^{k} q_j \mu_j\, e^{-\mu_j x}, \quad x > 0,$$

$$\text{mean:} \quad \overline{X} = \sum_{j=1}^{k} \frac{q_j}{\mu_j} = \frac{1}{\mu}, \quad x > 0,$$

$$\text{variance:} \quad \text{var}(X) = 2 \sum_{j=1}^{k} \frac{q_j}{\mu_j^2} - \frac{1}{\mu^2},$$

$$\text{coefficient of variation:} \quad c_X = \sqrt{2\mu^2 \sum_{j=1}^{k} \frac{q_j}{\mu_j^2} - 1} \quad \geq 1.$$

For example, the parameters μ_1, μ_2 of an H_2 distribution can be estimated to approximate an unknown distribution with given mean $\overline{X}$ and sample coefficient of variation c_X as follows:

$$\mu_1 = \frac{1}{\overline{X}} \left[1 - \sqrt{\frac{q_2}{q_1} \frac{c_X^2 - 1}{2}} \right]^{-1}, \tag{1.26}$$

$$\mu_2 = \frac{1}{\overline{X}} \left[1 + \sqrt{\frac{q_1}{q_2} \frac{c_X^2 - 1}{2}} \right]^{-1}. \tag{1.27}$$

The parameters q_1 and q_2 can be assigned any values that satisfy the restrictions $q_1, q_2 \geq 0, q_1 + q_2 = 1$ and $\mu_1, \mu_2 > 0$.

(c) Erlang-k Distribution, E_k

Empirical distributions with a coefficient of variation less than one can be approximated using the Erlang-k distribution. Here k is the number of exponential phases in series. Figure 1.8 shows a model of a time duration that has an E_k distribution. It contains k identical phases connected in series, each with exponentially distributed time. The mean duration of each phase is $\overline{X}/k$, where $\overline{X}$ denotes the mean of the whole time span.

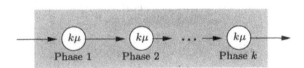

Fig. 1.8 A random variable with E_k distribution.

If the interarrival times of some arrival process like our bus stops are identical exponentially distributed, it follows that the time between the first arrival and the $(k+1)$th arrival is Erlang-k distributed.

The CDF is given by

$$F_X(x) = 1 - e^{-k\mu x} \cdot \sum_{j=0}^{k-1} \frac{(k\mu x)^j}{j!}, \quad x \geq 0, \ k = 1, 2 \ldots . \tag{1.28}$$

Further characteristic measures are

$$\text{pdf:} \quad f_X(x) = \frac{k\mu(k\mu x)^{k-1}}{(k-1)!}\, e^{-k\mu x},\ x > 0,\ k = 1, 2, \ldots,$$

$$\text{mean:} \quad \overline{X} = \frac{1}{\mu},$$

$$\text{variance:} \quad \text{var}(X) = \frac{1}{k\mu^2},$$

$$\text{coefficient of variation:} \quad c_X = \frac{1}{\sqrt{k}} \le 1.$$

If the sample mean $\overline{X}$ and the sample coefficient of variation c_X are given, then the parameters k and μ of the corresponding Erlang distribution are estimated by

$$k = \left\lceil \frac{1}{c_X^2} \right\rceil, \tag{1.29}$$

and

$$\mu = \frac{1}{c_X^2 k \overline{X}}. \tag{1.30}$$

(d) Hypoexponential Distribution

The hypoexponential distribution arises when the individual phases of the E_k distribution are allowed to have assigned different rates. Thus, the Erlang distribution is a special case of the hypoexponential distribution.

For a hypoexponential distributed random variable X with two phases and the rates μ_1 and μ_2 ($\mu_1 \ne \mu_2$), we get the CDF as

$$F_X(x) = 1 - \frac{\mu_2}{\mu_2 - \mu_1}\, e^{-\mu_1 x} + \frac{\mu_1}{\mu_2 - \mu_1}\, e^{-\mu_2 x}, \qquad x \ge 0. \tag{1.31}$$

Furthermore, the following relations hold:

$$\text{pdf:} \quad f_X(x) = \frac{\mu_1 \mu_2}{\mu_1 - \mu_2}(e^{-\mu_2 x} - e^{-\mu_1 x}), \quad x > 0,$$

$$\text{mean:} \quad \overline{X} = \frac{1}{\mu_1} + \frac{1}{\mu_2},$$

$$\text{variance:} \quad \text{var}(X) = \frac{1}{\mu_1^2} + \frac{1}{\mu_2^2},$$

$$\text{coefficient of variation:} \quad c_X = \frac{\sqrt{\mu_1^2 + \mu_2^2}}{\mu_1 + \mu_2} < 1.$$

The values of the parameters μ_1 and μ_2 of the hypoexponential CDF can be estimated given the sample mean $\overline{X}$ and sample coefficient of variation by

$$\mu_1 = \frac{2}{\overline{X}}\left[1 + \sqrt{1 + 2(c_X^2 - 1)}\,\right]^{-1},$$

$$\mu_2 = \frac{2}{\overline{X}}\left[1 - \sqrt{1 + 2(c_X^2 - 1)}\,\right]^{-1},$$

with $0.5 \leq c_X^2 \leq 1$.

For a hypoexponential distribution with k phases and the phase rates μ_1, $\mu_2, \ldots, \mu_k$, we get [Bega93]

$$\text{pdf:} \quad f_X(x) = \sum_{i=1}^{k} a_i \mu_i \, e^{-\mu_i x}, \quad x > 0,$$

$$\text{with } a_i = \prod_{j=1, j \neq i}^{k} \frac{\mu_j}{\mu_j - \mu_i}, \quad 1 \leq i \leq k,$$

$$\text{mean:} \quad \overline{X} = \sum_{i=1}^{k} \frac{1}{\mu_i},$$

$$\text{coefficient of variation:} \quad c_X = \left(1 + 2 \frac{\sum_{i=1}^{k} \left(\mu_i \sum_{j=i+1}^{k} \mu_j \right)}{\sum_{i=1}^{k} \mu_i^2} \right)^{-\frac{1}{2}}.$$

(e) Gamma Distribution

Another generalization of the Erlang-k distribution for arbitrary coefficient of variation is the gamma distribution. The distribution function is given by

$$F_X(x) = \int_0^x \frac{\alpha \mu \cdot (\alpha \mu u)^{\alpha-1}}{\Gamma(\alpha)} \cdot e^{-\alpha \mu u} \, du, \quad x \geq 0, \ \alpha > 0, \tag{1.32}$$

$$\text{with } \Gamma(\alpha) = \int_0^{\infty} u^{\alpha-1} \cdot e^{-u} \, du, \quad \alpha > 0.$$

If $\alpha = k$ is a positive integer, then $\Gamma(k) = (k-1)!$

Thus, the Erlang-k distribution can be considered as a special case of the gamma distribution:

$$\text{pdf:} \quad f_X(x) = \frac{\alpha \mu \cdot (\alpha \mu x)^{\alpha-1}}{\Gamma(\alpha)} e^{-\alpha \mu x}, \quad x > 0, \ \alpha > 0,$$

$$\text{mean:} \quad \overline{X} = \frac{1}{\mu},$$

$$\text{variance:} \quad \text{var}(X) = \frac{1}{\alpha \mu^2},$$

$$\text{coefficient of variation:} \quad c_X = \frac{1}{\sqrt{\alpha}}.$$

It can be seen that the parameters α and μ of the gamma distribution can easily be estimated from c_X and $\overline{X}$:

$$\mu = \frac{1}{\overline{X}} \quad \text{and} \quad \alpha = \frac{1}{c_X^2}.$$

(f) Generalized Erlang Distribution

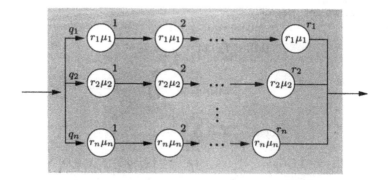

Fig. 1.9 A random variable with generalized Erlang distribution.

Rather complex distributions can be generated by combining hyperexponential and Erlang-k distributions; these are known as generalized Erlang distributions. An example is shown in Fig. 1.9 where n parallel levels are depicted and each level j contains a series of r_j phases connected, each with exponentially distributed time and rate $r_j\mu_j$. Each level j is selected with probability j. The pdf is given by

$$f_X(x) = \sum_{j=1}^{n} q_j \cdot \frac{r_j\mu_j (r_j\mu_j x)^{r_j-1}}{(r_j-1)!} \cdot e^{-r_j\mu_j x}, \qquad x \geq 0. \qquad (1.33)$$

Another type of generalized Erlang distribution can be obtained by lifting the restriction that all the phases within the same level have the same rate, or, alternatively, if there are nonzero exit probabilities assigned such that remaining phases can be skipped. These generalizations lead to very complicated equations that are not further described here.

(g) Cox Distribution, C_k (Branching Erlang Distribution)

In [Cox55] the principle of the combination of exponential distributions is generalized to such an extent that any distribution that possesses a rational Laplace transform can be represented by a sequence of exponential phases with possibly complex probabilities and complex rates. Figure 1.10 shows a model of a Cox distribution with k exponential phases.

The model consists of k phases in series with exponentially distributed times and rates $\mu_1, \mu_2, \ldots, \mu_k$. After phase j, another phase $j+1$ follows with probability a_j and with probability $b_j = 1 - a_j$ the total time span is completed. As described in [SaCh81], there are two cases that must be distinguished when using the sample mean value $\overline{X}$ and the sample coefficient of variation c_X to estimate the parameters of the Cox distribution:

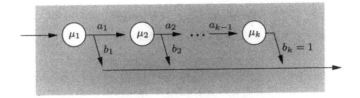

Fig. 1.10 A random variable with C_k distribution.

Case 1: $c_X \leq 1$

To approximate a distribution with $c_X \leq 1$, we suggest using a special Cox distribution with (see Fig. 1.11):

$$\mu_j = \mu \quad j = 1, \ldots, k \,,$$
$$a_j = 1 \quad j = 2, \ldots, k - 1 \,.$$

From the probabilistic definitions, we obtain

$$\text{mean:} \quad \overline{X} = \frac{b_1 + k(1 - b_1)}{\mu} \,,$$

$$\text{variance:} \quad \text{var}\,(X) = \frac{k + b_1(k - 1)\,(b_1(1 - k) + k - 2)}{\mu^2} \,,$$

$$\text{squared coeff. of variation:} \quad c_X^2 = \frac{k + b_1(k - 1)\,(b_1(1 - k) + k - 2)}{[b_1 + k(1 - b_1)]^2} \,.$$

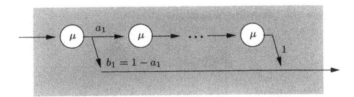

Fig. 1.11 A random variable with C_k distribution and $c_X \leq 1$.

In order to minimize the number of stages, we choose k such that

$$k = \left\lceil \frac{1}{c_X^2} \right\rceil \,. \tag{1.34}$$

Then we obtain the parameters b_1 and μ from the preceding equations:

$$b_1 = \frac{2kc_X^2 + (k - 2) - \sqrt{k^2 + 4 - 4kc_X^2}}{2(c_X^2 + 1)(k - 1)} \,, \tag{1.35}$$

$$\mu = \frac{k - b_1 \cdot (k - 1)}{\overline{X}} \,. \tag{1.36}$$

Case 2: $c_X > 1$

For the case $c_X > 1$, we use a Cox-2 model (see Fig. 1.12).

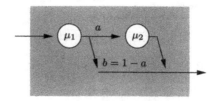

Fig. 1.12 A random variable with Cox-2 distribution.

From classical formulae, we obtain

$$\text{mean:} \quad \overline{X} = \frac{1}{\mu_1} + \frac{a}{\mu_2} ,$$

$$\text{variance:} \quad \text{var}(X) = \frac{\mu_2^2 + a\mu_1^2(2 - a)}{\mu_1^2 \cdot \mu_2^2} ,$$

$$\text{squared coefficient of variation:} \quad c_X^2 = \frac{\mu_2^2 + a\mu_1^2(2 - a)}{(\mu_2 + a\mu_1)^2} .$$

We have two equations for the three parameters μ_1, μ_2, and a, and therefore obtain an infinite number of solutions. To obtain simple equations we choose

$$\mu_1 = \frac{2}{\overline{X}} , \tag{1.37}$$

and obtain from the preceding equations

$$\mu_2 = \frac{1}{\overline{X} c_X^2} \quad \text{and} \quad a = \frac{1}{2c_X^2} . \tag{1.38}$$

(h) Weibull Distribution

The Weibull distribution is at present the most widely used failure distribution. Its CDF is given by

$$F_X(x) = 1 - \exp(-(\lambda x)^\alpha), \quad x \geq 0 . \tag{1.39}$$

From this definition, the following relations can be derived:

$$\text{pdf:} \quad f_X(x) = \alpha\lambda(\lambda x)^{\alpha-1} \exp(-(\lambda x)^\alpha), \quad \lambda > 0 ,$$

$$\text{mean:} \quad \overline{X} = \frac{1}{\lambda}\Gamma\left(1 + \frac{1}{\alpha}\right) ,$$

$$\text{squared coefficient of variation:} \quad c_X^2 = \frac{\Gamma(1 + 2/\alpha)}{\{\Gamma(1 + 1/\alpha)\}^2} - 1 .$$

with the shape parameter $\alpha > 0$, which determines the type of the failure behavior ($\alpha < 0$ means infant mortality and $\alpha > 0$ means wear out) and the so-called scale parameter $\lambda > 0$. $\frac{1}{\lambda}$ is often referred to as the characteristic lifetime of the Weibull distribution, as, independent of the value of α, $F(\frac{1}{\lambda}) = 1 - \exp(-1) \approx 0.632$.

(i) Pareto Distribution

The Pareto distribution acquired importance due to the nature of Internet traffic. Its CDF is given by

$$F_X(x) = \begin{cases} 1 - \left(\dfrac{k}{x}\right)^\alpha, & x \geq k, \\ 0, & x < k. \end{cases} \tag{1.40}$$

Additionally, the following relations hold:

$$\text{pdf:} \quad f_X(x) = \frac{\alpha k^\alpha}{x^{\alpha+1}}, \quad x \geq k, \ \alpha, k > 0,$$

$$\text{mean:} \quad \overline{X} = \begin{cases} \dfrac{\alpha k}{\alpha - 1}, & \alpha > 1, \\ \infty, & \alpha \leq 1. \end{cases}$$

$$\text{variance:} \quad \text{var}(X) = \begin{cases} \dfrac{\alpha k^2}{(\alpha - 1)^2(\alpha - 2)}, & \alpha > 2, \\ \infty, & \alpha \leq 2. \end{cases}$$

$$\text{squared coefficient of variation:} \quad c_X^2 = \frac{1}{\alpha(\alpha - 2)}, \alpha > 2.$$

In the case of the Pareto distribution, α is called the shape parameter and k is referred to as scale parameter. Note that Pareto distribution from Eq. (1.40) distribution can exhibit infinite mean and infinite variance depending on the value of the shape parameter α.

One typical property of the Pareto distribution is that it is heavy-tailed, as also the Weibull distribution with a shape parameter smaller than one. A distribution is said to be heavy-tailed (or long-tailed) if

$$\lim_{x \to \infty} \frac{P(X > x)}{e^{-\varepsilon x}} = \infty, \tag{1.41}$$

for any $\varepsilon > 0$, i.e., if its complementary distribution function, which is defined as $F_X^c(x) = 1 - F_X(x)$, decays more slowly than exponentially.

Heavy-tailed distributions have gained attention recently as being useful for characterizing many types of Internet traffic. This traffic was considered to have a Poisson-like nature in the past. In the meantime it has been shown [LTWW94] that this kind of traffic is self-similar. Informally speaking, self-similarity means that the shape of the traffic is invariant with respect to the time-scale; for example, the histogram of packets per time unit has the same shape if a histogram is built over 1000 seconds or 10 seconds if the time-step for building the classes in the histogram is changed in the same scale. In contrast to this, if a Poisson process is assumed, peaks are flattened out when the time unit is enlarged. This implies, that a long-range dependence may play an important role when Internet traffic is analyzed. However, there are some recent papers [Down01] that question the self-similarity in specific situations.

Heavy-tailed distributions can be used to model self-similar processes and the Pareto distribution is one of the most common heavy-tailed distributions. For details on self-similarity the reader is referred to [LTWW94] and for more information about the effect of heavy-tails to [BBQH03].

(j) Lognormal Distribution

The lognormal distribution is given by

$$F_X(x) = \Phi\left(\frac{\ln(x) - \lambda}{\alpha}\right), \quad x > 0. \tag{1.42}$$

Additionally, the following relations hold:

$$\text{pdf:} \quad f_X(x) = \frac{1}{\alpha x \sqrt{2\pi}} \exp(-\{\ln(x) - \lambda\}^2/2\alpha^2), \quad x > 0,$$

$$\text{mean:} \quad \overline{X} = \exp(\lambda + \alpha^2/2),$$

$$\text{squared coefficient of variation:} \quad c_X^2 = \exp(\alpha^2) - 1,$$

where the shape parameter α is positive and the scale parameter λ may assume any real value. As the Pareto distribution, the lognormal distribution is also a heavy-tailed distribution.

The parameters α and λ can easily be calculated from c_X^2 and $\overline{X}$:

$$\alpha = \sqrt{\ln(c_X^2 + 1)}, \qquad \lambda = \ln \overline{X} - \frac{\alpha^2}{2}. \tag{1.43}$$

The importance of this distribution arises from the fact that the product of n mutually independent random variables has a lognormal distribution in the limit $n \to \infty$. In Table 1.2 the formulae for the expectation $E[X]$, the variance $var(X)$, and the coefficient of variation c_X for some important distribution functions are summarized. Furthermore, in Table 1.3 formulae for estimating the parameters of these distributions are given.

1.3.2   Multiple Random Variables

In some cases, the result of one random experiment determines the values of several random variables, where these values may also affect each other. The *joint probability mass function* of the discrete random variables

Table 1.2 Expectation $E[X]$, variance $\text{var}(X)$, and coefficient of variation c_X of important distributions

Distribution	Parameter	$E[X]$	$\text{var}(X)$	c_X
Exponential	μ	$\dfrac{1}{\mu}$	$\dfrac{1}{\mu^2}$	1
Erlang	μ, k $k=1,2,\dots$	$\dfrac{1}{\mu}$	$\dfrac{1}{k\mu^2}$	$\dfrac{1}{\sqrt{k}} \leq 1$
Gamma	μ, α $(0<\alpha<\infty)$	$\dfrac{1}{\mu}$	$\dfrac{1}{\alpha\mu^2}$	$0 < \dfrac{1}{\sqrt{\alpha}} < \infty$
Hypoexponential	μ_1, μ_2	$\dfrac{1}{\mu_1} + \dfrac{1}{\mu_2}$	$\dfrac{1}{\mu_1^2} + \dfrac{1}{\mu_2^2}$	$\dfrac{\sqrt{\mu_1^2 + \mu_2^2}}{\mu_1 + \mu_2} < 1$
Hyperexponential	k, μ_i, q_i	$\displaystyle\sum_{i=1}^{k} \dfrac{q_i}{\mu_i} = \dfrac{1}{\mu}$	$\displaystyle 2\sum_{i=1}^{k} \dfrac{q_i}{\mu_i^2} - \dfrac{1}{\mu^2}$	$\displaystyle\sqrt{2\mu^2 \sum_{i=1}^{k} \dfrac{q_i}{\mu_i^2} - 1} > 1$

$X_1, X_2, \ldots, X_n$ is given by

$$P(X_1 = x_1, X_2 = x_2, \ldots, X_n = x_n) \qquad (1.44)$$

and represents the probability that $X_1 = x_1$, $X_2 = x_2, \ldots X_n = x_n$. In the continuous case, the *joint distribution function*:

$$F_{\mathbf{X}}(\mathbf{x}) = P(X_1 \leq x_1, X_2 \leq x_2, \ldots, X_n \leq x_n) \qquad (1.45)$$

represents the probability that $X_1 \leq x_1$, $X_2 \leq x_2$, $\ldots X_n \leq x_n$, where $\mathbf{X} = (X_1, \ldots, X_n)$ is the n-dimensional random variable and $\mathbf{x} = (x_1, x_2, \ldots, x_n)$.

A simple example of an experiment with multiple discrete random variables is tossing a pair of dice. The following random variables might be determined:

X_1 number that shows on the first die,

X_2 number that shows on the second die,

$X_3 = X_1 + X_2$,   sum of the numbers of both dice.

1.3.2.1 Independence   The random variables $X_1, X_2, \ldots, X_n$ are called (*statistically*) *independent* if, for the continuous case:

$$\begin{aligned} P(X_1 \leq x_1, X_2 \leq x_2, \ldots, X_n \leq x_n) \\ = P(X_1 \leq x_1) \cdot P(X_2 \leq x_2) \cdot \ldots \cdot P(X_n \leq x_n), \end{aligned} \qquad (1.46)$$

or the discrete case:

$$\begin{aligned} P(X_1 = x_1, X_2 = x_2, \ldots, X_n = x_n) \\ = P(X_1 = x_1) \cdot P(X_2 = x_2) \cdot \ldots \cdot P(X_n = x_n). \end{aligned} \qquad (1.47)$$

Table 1.3 Formulae for estimating the parameters of important distributions

Distribution	Parameter	Calculation of the Parameters
Exponential	μ	$\mu = 1/\overline{X}$
Erlang	μ, k $k = 1, 2, \ldots$	$k = \text{ceil}(1/c_X^2)$ $\mu = 1/(c_X^2 \cdot k\overline{X})$
Gamma	μ, α $0 < \alpha < \infty$	$\alpha = 1/c_X^2$ $\mu = 1/\overline{X}$
Hypoexponential	μ_1, μ_2	$\mu_{1/2} = \dfrac{2}{\overline{X}} \left[1 \pm \sqrt{1 + 2(c_X^2 - 1)} \right]^{-1}$
Hyperexponential (H_2)	μ_1, μ_2, q_1, q_2	$\mu_1 = \dfrac{1}{\overline{X}} \left[1 - \sqrt{\dfrac{q_2}{q_1} \dfrac{c_X^2 - 1}{2}} \right]^{-1}$ $\mu_2 = \dfrac{1}{\overline{X}} \left[1 + \sqrt{\dfrac{q_1}{q_2} \dfrac{c_X^2 - 1}{2}} \right]^{-1}$ $q_1 + q_2 = 1, \mu_2 > 0$
Cox $(c_X \leq 1)$	k, b_i, μ_i	$k = \text{ceil}(1/c_X^2)$ $b_1 = \dfrac{2kc_X^2 + k - 2 - \sqrt{k^2 + 4 - 4kc_X^2}}{2(c_X^2 + 1)(k - 1)}$ $b_2 = b_3 = \ldots = b_{k-1} = 0, \quad b_k = 1$ $\mu_1 = \mu_2 = \ldots = \mu_k = \dfrac{k - b_1 \cdot (k - 1)}{\overline{X}}$
Cox $(c_X > 1)$	k, b, μ_1, μ_2	$k = 2$ $b = c_X^2 \left[1 - \sqrt{1 - \dfrac{2}{1 + c_X^2}} \right]$ $\mu_{1/2} = \dfrac{1}{\overline{X}} \left[1 \pm \sqrt{1 - \dfrac{2}{1 + c_X^2}} \right]$

Otherwise, they are (*statistically*) *dependent*. In the preceding example of tossing a pair of dice, the random variables X_1 and X_2 are independent, whereas X_1 and X_3 are dependent on each other. For example:

$$P(X_1 = i, X_2 = j) = \frac{1}{36}, \qquad i, j = 1, \ldots, 6,$$

since there are 36 different possible results altogether that are all equally probable. Because the following is also valid:

$$P(X_1 = i) \cdot P(X_2 = j) = \frac{1}{6} \cdot \frac{1}{6} = \frac{1}{36},$$

$P(X_1 = i, X_2 = j) = P(X_1 = i) \cdot P(X_2 = j)$ is true, and therefore X_1 and X_2 are independent. On the other hand, if we observe the dependent variables X_1 and X_3, then

$$P(X_1 = 2, X_3 = 4) = P(X_1 = 2, X_2 = 2) = \frac{1}{36},$$

since X_1 and X_2 are independent. Having

$$P(X_1 = 2) = \frac{1}{6}$$
$$\text{and} \quad P(X_3 = 4) = P(X_1 = 1, X_2 = 3) + P(X_1 = 2, X_2 = 2)$$
$$+ P(X_1 = 3, X_2 = 1) = \frac{3}{36} = \frac{1}{12}$$

results in $P(X_1 = 2) \cdot P(X_3 = 4) = \frac{1}{6} \cdot \frac{1}{12} = \frac{1}{72}$, which shows the dependency of the random variables X_1 and X_3:

$$P(X_1 = 2, X_3 = 4) \neq P(X_1 = 2) \cdot P(X_3 = 4).$$

1.3.2.2 Conditional Probability   A *conditional probability*:

$$P(X_1 = x_1 \mid X_2 = x_2, \ldots, X_n = x_n)$$

is the probability for $X_1 = x_1$ under the conditions $X_2 = x_2$, $X_3 = x_3$, etc. Then we get

$$P(X_1 = x_1 \mid X_2 = x_2, \ldots, X_n = x_n)$$
$$= \frac{P(X_1 = x_1, X_2 = x_2, \ldots, X_n = x_n)}{P(X_2 = x_2, \ldots, X_n = x_n)}. \qquad (1.48)$$

For continuous random variables we have

$$P(X_1 \leq x_1 \mid X_2 \leq x_2, \ldots, X_n \leq x_n)$$
$$= \frac{P(X_1 \leq x_1, X_2 \leq x_2, \ldots, X_n \leq x_n)}{P(X_2 \leq x_2, \ldots, X_n \leq x_n)}. \qquad (1.49)$$

We demonstrate this also with the preceding example of tossing a pair of dice. The probability that $X_3 = j$ under the condition that $X_1 = i$ is given by

$$P(X_3 = j \mid X_1 = i) = \frac{P(X_3 = j, X_1 = i)}{P(X_1 = i)} .$$

For example, with $j = 4$ and $i = 2$:

$$P(X_3 = 4 \mid X_1 = 2) = \frac{1/36}{1/6} = \frac{1}{6} .$$

If we now observe both random variables X_1 and X_2, we will see that because of the independence of X_1 and X_2,

$$P(X_1 = j \mid X_2 = i) = \frac{P(X_1 = j, X_2 = i)}{P(X_2 = i)} = \frac{P(X_1 = j) \cdot P(X_2 = i)}{P(X_2 = i)}$$

$$= P(X_1 = j) .$$

1.3.2.3 Important Relations

The following relations concerning multiple random variables will be used frequently in the text:

- The expected value of a sum of random variables is equal to the sum of the expected values of these random variables. If $c_1, \ldots, c_n$ are arbitrary constants and $X_1, X_2, \ldots, X_n$ are (not necessarily independent) random variables, then

$$E\left[\sum_{i=1}^{n} c_i X_i\right] = \sum_{i=1}^{n} c_i E[X_i] . \qquad (1.50)$$

- If the random variables $X_1, X_2, \ldots, X_n$ are stochastically independent, then the expected value of the product of the random variables is equal to the product of the expected values of the random variables:

$$E\left[\prod_{i=1}^{n} X_i\right] = \prod_{i=1}^{n} E[X_i] . \qquad (1.51)$$

- The *covariance* of two random variables X and Y is a way of measuring the dependency between X and Y. It is defined by

$$\text{cov}[X, Y] = E\left[(X - E[X])(Y - E[Y])\right] = \overline{(X - \overline{X})(Y - \overline{Y})}$$

$$= E[X \cdot Y] - E[X] \cdot E[Y] = \overline{XY} - \overline{X} \cdot \overline{Y} . \qquad (1.52)$$

If $X = Y$, then the covariance is equal to the variance:

$$\text{cov}[X, X] = \overline{X^2} - \overline{X}^2 = \text{var}(X) = \sigma_X^2 .$$

If X and Y are statistically independent, then Eq. (1.51) gives us

$$E[X \cdot Y] = E[X] \cdot E[Y] \,,$$

and Eq. (1.52) gives us

$$\mathrm{cov}(X, Y) = 0 \,.$$

In this case, X and Y are said to be uncorrelated.

- The *correlation coefficient* is the covariance normalized to the product of standard deviations:

$$\mathrm{cor}[X, Y] = \frac{\mathrm{cov}[X, Y]}{\sigma_X \cdot \sigma_Y} \,. \tag{1.53}$$

Therefore, if $X = Y$, then

$$\mathrm{cor}[X, X] = \frac{\sigma_X^2}{\sigma_X^2} = 1 \,.$$

- The variance of a sum of random variables can be expressed using the covariance:

$$\mathrm{var}\left[\sum_{i=1}^{n} c_i X_i\right] = \sum_{i=1}^{n} c_i^2 \, \mathrm{var}(X_i) + 2 \cdot \sum_{i=1}^{n-1} \sum_{j=i+1}^{n} c_i c_j \, \mathrm{cov}[X_i, X_j] \,. \tag{1.54}$$

For independent (uncorrelated) random variables we get

$$\mathrm{var}\left[\sum_{i=1}^{n} c_i X_i\right] = \sum_{i=1}^{n} c_i^2 \, \mathrm{var}(X_i) \,. \tag{1.55}$$

1.3.2.4   The Central Limit Theorem   Let $X_1, X_2, \ldots, X_n$ be independent, identically distributed random variables with an expected value of $E[X_i] = \overline{X}$, and a variance of $\mathrm{var}(X_i) = \sigma_X^2$. Then their arithmetic mean is defined by

$$S_n = \frac{1}{n} \sum_{i=1}^{n} X_i \,.$$

We have $E[S_n] = \overline{X}$ and $\mathrm{var}(S_n) = \sigma_X^2/n$. Regardless of the distribution of X_i, the random variable S_n approaches, with increasing n, a normal distribution with a mean of $\overline{X}$ and a variance of $\mathrm{var}(X)/n$:

$$f_{S_n}(x) \xrightarrow[n \to \infty]{} \frac{1}{\sigma_X \sqrt{2\pi/n}} \cdot \exp\left(-\frac{(x - \overline{X})^2}{2\sigma_X^2/n}\right) \,. \tag{1.56}$$

1.3.3   Transforms

Determining the mean values, variance, or moments of random variables is often rather tedious. However, these difficulties can sometimes be avoided by using transforms, since the pmf of a discrete or the density function of a continuous random variable are uniquely determined by their z- or Laplace transform, respectively. Thus, if two random variables have the same transform, they also have the same distribution function and vice versa. In many cases, the parameters of a random variable are, therefore, more easily computed from the transform.

1.3.3.1 z-Transform   Let X be a discrete random variable with the probability mass function $p_k = P(X = k)$ for $k = 0, 1, 2, \ldots$, then the *z-transform* of X is defined by

$$G_X(z) = \sum_{k=0}^{\infty} p_k \cdot z^k \quad \text{with } |z| \leq 1. \tag{1.57}$$

The function $G_X(z)$ is also called the *probability generating function*. Table 1.4 specifies the probability generating functions of some important discrete random variables. If the z-transform $G_X(z)$ of a random variable is given, then the probabilities p_k can be calculated immediately:

$$p_0 = G_X(0), \tag{1.58}$$

$$p_k = \frac{1}{k!} \cdot \frac{\mathrm{d}^k G_X(z)}{\mathrm{d} z^k}\bigg|_{z=0}, \quad k = 1, 2, 3, \ldots. \tag{1.59}$$

And for the moments

$$\overline{X} = \frac{\mathrm{d} G_X(z)}{\mathrm{d} z}\bigg|_{z=1}, \tag{1.60}$$

$$\overline{X^2} = \frac{\mathrm{d}^2 G_X(z)}{\mathrm{d} z^2}\bigg|_{z=1} + \overline{X}, \tag{1.61}$$

$$\overline{X^k} = \frac{\mathrm{d}^k G_X(z)}{\mathrm{d} z^k}\bigg|_{z=1} + \sum_{i=1}^{k-1} a_{k,i} \cdot \overline{X^i}, \quad k = 3, 4, \ldots, \tag{1.62}$$

with

$$3a_{k,1} = (-1)^k (k-1)!, \qquad\qquad k = 3, 4, \ldots,$$

$$a_{k,k-1} = \frac{k(k-1)}{2}, \qquad\qquad k = 3, 4, \ldots,$$

$$a_{k,i} = a_{k-1,i-1} - (k-1)a_{k-1,i}, \quad k = 4, 5, \ldots \text{ and } i = 2, \ldots, k-2.$$

Table 1.4 Generating functions of some discrete random variables

Random Variable	Parameter	$G_X(z)$
Binomial	n, p	$[(1-p) + pz]^n$
Geometric	p	$\dfrac{pz}{1-(1-p)z}$
Poisson	α	$e^{\alpha(z-1)}$

The generating function of a sum of independent random variables is the product of the generating functions of each random variable, so for $X = \sum_{i=1}^{n} X_i$ we get

$$G_X(z) = \prod_{i=1}^{n} G_{X_i}(z). \tag{1.63}$$

1.3.3.2 Laplace Transform The Laplace transform plays a similar role for non-negative continuous random variables as the z-transform does for discrete random variables. If X is a continuous random variable with the density function $f_X(x)$, then the *Laplace transform* (LT) of the density function (pdf) of X is defined by

$$L_X(s) = \int_0^\infty f_X(x) e^{-sx} \, dx, \qquad |e^{-s}| \le 1, \tag{1.64}$$

where s is a complex parameter. $L_X(s)$ is also called the Laplace–Stieltjes transform (LST) of the distribution function (CDF) and can also be written as

$$L_X(s) = \int_0^\infty e^{-sx} \, dF_X(x). \tag{1.65}$$

LST of a random variable X will also be denoted by $X^{\sim}(s)$.

The moments can again be determined by differentiation:

$$\overline{X^k} = (-1)^k \frac{d^k L_X(s)}{d s^k}\bigg|_{s=0}, \qquad k = 1, 2, \dots . \tag{1.66}$$

The Laplace transforms of several pdfs of well-known continuous random variables as well as their moments are given in Table 1.5.

Analogous to the z-transform, the Laplace transform of a sum of independent random variables is the product of the Laplace transform of these

Table 1.5 Laplace transform and moments of several continuous random variables

Random Variable	Parameter	$L_X(s) = X^{\sim}(s)$	$\overline{X^n}$
Exponential	λ	$\dfrac{\lambda}{\lambda + s}$	$\dfrac{n!}{\lambda^n}$
Erlang	k, λ	$\left(\dfrac{\lambda}{\lambda + s}\right)^k$	$\dfrac{k(k+1)\dots(k+n-1)}{\lambda^n}$
Gamma	α, λ	$\left(\dfrac{\lambda}{\lambda + s}\right)^\alpha$	$\dfrac{\alpha(\alpha+1)\dots(\alpha+n-1)}{\lambda^n}$
Hyperexponential	λ	$\displaystyle\sum_{j=1}^{k} q_j \dfrac{\lambda_j}{\lambda_j + s}$	$\displaystyle\sum_{j=1}^{k} q_j \dfrac{n!}{\lambda_j^n}$
Hypoexponential	λ_1, λ_2	$\dfrac{\lambda_1 \lambda_2}{\lambda_1 - \lambda_2}\left(\dfrac{1}{\mu_2 + s} - \dfrac{1}{\mu_1 + s}\right)$	$\dfrac{n!}{\mu_1 - \mu_2}\left(\dfrac{\mu_1}{\mu_2^n} - \dfrac{\mu_2}{\mu_1^n}\right)$

random variables:

$$L_X(s) = \prod_{i=1}^{n} L_{X_i}(s). \tag{1.67}$$

More properties and details of Laplace and z-transforms are listed in [Klei75].

1.3.4   Parameter Estimation

Probability calculations discussed in this book require that the parameters of all the relevant distributions are known in advance. For practical use, these parameters have to be estimated from measured data. In this subsection, we briefly discuss statistical methods of parameter estimation. For further details see [Triv01].

Definition 1.6   Let the set of random variables $X_1, X_2, \dots, X_n$ be given. This set is said to constitute a *random sample* of size n from a population with associated distribution function $F_X(x)$. Assume that the random variables are mutually independent and identically distributed with distribution function (CDF) $F_{X_i}(x) = F_X(x), \quad \forall i, x$.

We are interested in estimating the value of some parameter θ (e.g., the mean or the variance of the distribution or the parameters of the distribution themselves) of the population based on the random samples.

Definition 1.7   A function $\widehat{\Theta} = \widehat{\Theta}(X_1, X_2, \dots, X_n)$ is called an *estimator* of the parameter θ, and an observed value $\hat{\theta} = \hat{\theta}(x_1, x_2, \dots, x_n)$ is known as an *estimate* of θ. An estimator $\widehat{\Theta}$ is considered unbiased, if

$$E[\widehat{\Theta}(X_1, X_2, \dots, X_n)] = \theta.$$

It can be shown that the sample mean $\overline{X}$, defined as

$$\overline{X} = \frac{1}{n} \sum_{i=1}^{n} X_i \,, \tag{1.68}$$

is an unbiased estimator of the population mean μ, and that the sample variance S^2, defined as

$$S^2 = \frac{1}{n-1} \sum_{i=1}^{n} (X_i - \overline{X})^2 \,, \tag{1.69}$$

is an unbiased estimator of the population variance σ^2.

Definition 1.8   An estimator $\widehat{\Theta}$ of a parameter θ is said to be *consistent* if $\widehat{\Theta}$ converges in probability to θ (so-called stochastic convergence); that is

$$\lim_{n \to \infty} P(|\widehat{\Theta} - \theta| \geq \epsilon) = 0 \quad \forall \epsilon > 0 \,.$$

1.3.4.1 Method of Moments   The kth sample moment of a random variable X is defined as

$$M'_k = \sum_{i=1}^{n} \frac{X_i^k}{n} \,, \quad k = 1, 2, \dots \,. \tag{1.70}$$

To estimate one or more parameters of the distribution of X, using the *method of moments,* we equate the first few population moments with their corresponding sample moments to obtain as many equations as the number of unknown parameters. These equations can then be solved to obtain the required estimates. The method yields estimators that are consistent, but they may be biased.

As an example, suppose the time to failure of a computer system has a gamma distributed pdf:

$$f_X(x) = \frac{\lambda^\alpha x^{\alpha-1} e^{-\lambda x}}{\Gamma(\alpha)} \,, \quad k = 1, 2, \dots \,, \tag{1.71}$$

with parameters λ and α. We know that the first two population moments are given by

$$\mu_1 = \frac{\alpha}{\lambda} \quad \text{and} \quad \mu_2 = \frac{\alpha}{\lambda^2} + \mu_1{}^2 \,.$$

Then the estimates $\hat{\lambda}$ and $\hat{\alpha}$ can be obtained by solving

$$\frac{\hat{\alpha}}{\hat{\lambda}} = M_1 \quad \text{and} \quad \frac{\hat{\alpha}}{\hat{\lambda}^2} + \frac{\hat{\alpha}^2}{\hat{\lambda}^2} = M_2 \,.$$

Hence

$$\hat{\alpha} = \frac{M_1{}^2}{M_2 - M_1{}^2} \quad \text{and} \quad \hat{\lambda} = \frac{M_1}{M_2 - M_1{}^2} \,. \tag{1.72}$$

One of the drawbacks of the method of moments is that the solution of the emerging equations may become problematic when there are more than two parameters to be estimated as there may exist no exact solution and therefore optimization methods or least-squares fits would have to be employed. A preferred method of estimation is based on linear regression or maximum likelihood.

1.3.4.2 Linear Regression

Linear regression is one of the most common statistical inference procedures and the reasons for that are its simplicity and computational efficiency. The basic idea of linear regression is to transform the CDF of the distribution to be fitted into a linear form $y = a \cdot x + b$. Then, a least-squares fit through the sample data which defines a straight line is sought for. Finally, the parameters of the distribution itself are computed from the parameters of the straight line by a back-transformation.

As an example consider the estimation of the shape parameter α and the scale parameter λ of a Weibull distribution from a given sample. The CDF is given as

$$F_X(t) = 1 - \exp(-(\lambda t)^\alpha), \quad t \geq 0.$$

Taking the natural logarithm twice we get

$$\underbrace{\ln\left(\ln\left(\frac{1}{1 - F_X(t)}\right)\right)}_{:=y} = \underbrace{\alpha}_{:=a} \cdot \underbrace{\ln(t)}_{:=x} + \underbrace{\alpha \ln(\lambda)}_{:=b}.$$

The x_i used for the least-squares fit are the values in the given sample in ascending order. For the y_i, the empirical CDF needs to be obtained from the measured data. Usually median ranks, which can be easily computed by the approximation

$$y_i \approx \frac{i - 0.3}{n + 0.4} \tag{1.73}$$

are used [Aber94], when the sample size n is small. The i in Eq. (1.73) denotes the index of the failure in the ordered sample of failures. From the least-squares solution the Weibull parameters can then be computed by the back-transformation $\lambda = e^{b/a}$ and $\alpha = a$.

One of the reasons for the popularity of linear regression may be that one does not necessarily need a computer to perform the estimation as there is also a graphical solution, the so-called "probability-paper." In the case of the Weibull distribution this is a coordinate system with logarithmic x-axis and double-logarithmic y-axis, in which the sample points are plotted. By eye-balling a straight line through the sample points one can determine the distribution parameters from the slope and the y-axis intercept.

The only limitation of the linear regression is that a transformation into linear form has to exist which, for example, is not the case for a weighted sum of exponential distributions.

1.3.4.3 Maximum-Likelihood Estimation The principle of this method is to
select as an estimate of θ the value for which the observed sample is most likely
to occur. We define the likelihood function as the product of the marginal
densities:

$$L(\theta) = \prod_{i=1}^{n} f_{X_i}(x_i \mid \theta), \tag{1.74}$$

where $\theta = (\theta_1, \theta_2, \ldots, \theta_k)$ is the vector of parameters to be estimated. As
shown in [ReWa84] the likelihood function is itself a density function. There-
fore, it is a measure for the probability that the observed sample comes from
the distribution the parameters of which have to be estimated. This probability corresponds to the estimate θ that maximizes the likelihood function.

Under some regularity conditions, the maximum-likelihood estimate of θ is
the simultaneous solution of the set of equations:

$$\frac{\partial L(\theta)}{\partial \theta_i} = 0, \quad i = 1, 2, \ldots, k. \tag{1.75}$$

As an example, assume an arrival process where the interarrival time, X,
is exponentially distributed with arrival rate λ. To estimate the arrival rate
λ from a random sample of n interarrival times, we define:

$$L(\lambda) = \prod_{i=1}^{n} \lambda \exp(-\lambda x_i) = \lambda^n \exp(-\lambda \sum_{i=1}^{n} x_i), \tag{1.76}$$

$$\frac{dL}{d\lambda} = n\lambda^{n-1} \exp(-\lambda \sum_{i=1}^{n} x_i) - \left(\sum_{i=1}^{n} x_i \right) \lambda^n \exp(-\lambda \sum_{i=1}^{n} x_i) = 0, \tag{1.77}$$

from which we get the maximum-likelihood estimator of the arrival rate that
is equal to the reciprocal of the sample mean $\overline{X}$.

In the cases where the solution of the system of equations (1.75) is too
complex, a nonlinear optimization algorithm can be used that maximizes the
likelihood function from Eq. (1.74). This optimization algorithm needs to
incorporate any constraints on the parameters to be estimated (for example,
for the exponential distribution there is the constraint $\lambda > 0$). Furthermore,
it is not guaranteed that the likelihood function is unimodal. It may have
several local maxima in addition to a global maximum. Therefore, a global
optimization algorithm such as a genetic algorithm [Gold89] or interval arithmetic [LüLl00] may be in order. From a practical point of view, however, such
approaches may imply very long calculation times. As an alternative it has
become common to use a local optimization algorithm and to run it repeatedly with different initial values. Such an approach is particularly justifiable
if a rough estimate of at least the order of magnitude of the parameter values
is known in advance.

Since the likelihood function is defined as a product of densities, numerical
difficulties may arise. Assume for example a sample of 100 elements with

density values in an order of magnitude of 10^{-5} each. This leads to a likelihood function in an order of magnitude of about 10^{-500} which is hardly insight any computing precision range. This difficulty is alleviated by using the logarithm of the likelihood function and which is given as

$$LN(\theta) = \sum_{i=1}^{n} \ln(f_{X_i}(x_i \mid \theta)) .$$

As the logarithm is a monotone function, the position of the maximum is not changed by this transformation.

1.3.4.4 Confidence Intervals So far we have concentrated on obtaining the point estimates of a desired parameter. However, it is very rare in practice for the point estimate to actually coincide with the exact value of the parameter θ under consideration. In this subsection, we introduce into the technique of computing an *interval estimate*. Such an interval, called the *confidence interval*, is constructed in a way that we have a certain confidence that this interval contains the true value of the unknown parameter θ. In other words, if the estimator $\widehat{\Theta}$ satisfies the condition

$$P(\widehat{\Theta} - \epsilon_1 < \theta < \widehat{\Theta} + \epsilon_2) = \gamma, \tag{1.78}$$

then the interval $A(\theta) = (\widehat{\Theta} - \epsilon_1, \widehat{\Theta} + \epsilon_2)$ is a $100 \times \gamma$ percent confidence interval for the parameter θ. Here γ is called the confidence coefficient (the probability that the confidence interval contains θ).

Consider obtaining confidence intervals when using the sample mean $\overline{X}$ as the estimator for the population mean. If the population distribution is normal with unknown mean μ and a known variance σ^2, then we can show that the sample mean is normally distributed with mean μ and variance σ^2/n, so that the random variable $Z = (\overline{X} - \mu)/(\sigma/\sqrt{n})$ has a standard normal distribution $N(0,1)$. To determine a $100 \times \gamma$ percent confidence interval for the population mean μ, we find a number $z_{\alpha/2}$ (using $N(0,1)$ tables) such that $P(Z > z_{\alpha/2}) = \alpha/2$, where $\alpha = 1 - \gamma$. Then we get

$$P(-z_{\alpha/2} < Z < z_{\alpha/2}) = \gamma. \tag{1.79}$$

We then obtain the $100 \times \gamma$ percent confidence interval as

$$-z_{\alpha/2} < (\overline{X} - \mu)/(\sigma/\sqrt{n}) < z_{\alpha/2} \tag{1.80}$$

or

$$\overline{X} - z_{\alpha/2} \frac{\sigma}{\sqrt{n}} < \mu < \overline{X} + z_{\alpha/2} \frac{\sigma}{\sqrt{n}} . \tag{1.81}$$

The assumption that the population is normally distributed does not always hold. But when the sample size is large, by central limit theorem, the statistic $(\overline{X} - \mu)/(\sigma/\sqrt{n})$ is asymptotically normal (under appropriate conditions).

Also, Eq. (1.81) requires the knowledge of the population variance σ^2. When σ^2 is unknown, we use the sample variance S^2 to get the approximate confidence interval for μ:

$$\overline{X} - z_{\alpha/2}\frac{S}{\sqrt{n}} < \mu < \overline{X} + z_{\alpha/2}\frac{S}{\sqrt{n}}. \tag{1.82}$$

When the sample size is relatively small and the population is normally distributed, then the random variable $T = (\overline{X} - \mu)/(S/\sqrt{n})$ has the Student t distribution with $n-1$ degrees of freedom. We can then obtain the $100(1-\alpha)$ percent confidence interval of μ from

$$\overline{X} - t_{n-1;\alpha/2}\frac{S}{\sqrt{n}} < \mu < \overline{X} + t_{n-1;\alpha/2}\frac{S}{\sqrt{n}}, \tag{1.83}$$

where $P(T > t_{n-1;\alpha/2}) = \alpha/2$.

If a maximum likelihood estimation is used for the point estimation it is convenient to use likelihood ratio confidence bounds as they can directly reuse the result of the point estimation. These are defined as [MeEs98]

$$\frac{L(\theta)}{L(\hat{\theta})} \geq e^{-\frac{\chi^2_{k;\alpha}}{2}}. \tag{1.84}$$

$\hat{\theta}$ is the parameter-vector that was found by the point estimation, $\chi^2_{k;\alpha}$ is the χ^2-quantile with k degrees of freedom, which is just the number of estimated parameters, and the error-probability α.

An easy interpretation of Eq. (1.84) is obtained by transforming the above equation into the form

$$\underbrace{LN(\theta) + \frac{\chi^2_{k;\alpha}}{2} - LN(\hat{\theta})}_{:=g(\theta)} \geq 0, \tag{1.85}$$

by taking the natural logarithm ($g(\theta)$ is referred to as *confidence function* in the remainder of this section). This means that all points are within the confidence interval for which the logarithmic likelihood function takes on values that are at most $\chi^2_{(k;\alpha)}/2$ smaller than the value at the maximum likelihood point. The boundary of the confidence region is therefore given by $g(\theta) = 0$, but unfortunately this is fulfilled at infinitely many points so that a normal search for zeroes does not help as we are searching for the whole zero surface of the function.

For a two-parameter distribution there exists a graphical solution, and the same idea can be used for higher dimensional problems. Figure 1.13 shows the confidence region for a sample of 50 random values for a Weibull distribution. The error probability is 5% and all points, for which the confidence function yields values smaller than zero are set to zero. Figure 1.14 shows a

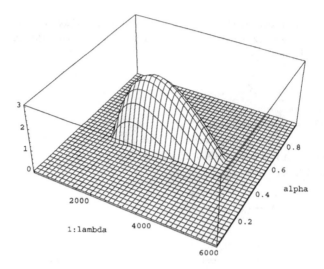

Fig. 1.13   Example for a confidence function.

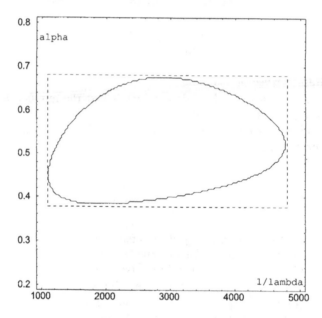

Fig. 1.14   Contour plot of the confidence function.

contour plot of that function where the zero level is contained in a rectangle of minimum size. An algorithm which constructs this minimal surrounding rectangle, starts with the inital values that were found by the point estimation and pushes the boundaries to the outside as long as there are points for which the logarithmic likelihood function yields a positive value. In oder to determine whether a boundary has been pushed far enough, the maximum value on that boundary has to be determined which — in case of a surrounding boundary — must not be geater than zero. For this purpose the same optimization algorithm as the one included in the point estimation can be reused. One just has to add the function which sets the parameter that defines the boundary under investigation to the current value.

1.3.4.5 Real Data All estimation algorithms presented so far have implicitly assumed a so-called *full sample* which means that all events of interest are included in the sample. Furthermore, in all cases there is the underlying assumption of knowing the exact occurrence time of an event x_i. Data that are collected in real-world applications are not always of that kind but may be *grouped* and *censored*.

Grouping (or clustering) refers to the case of knowing the exact event occurrence time. For example, instead of having the exact occurrence times $(x_1, x_2, \ldots, x_n)$ of events, we may only have the information that k_1 events have occurred in the interval between x_1 and x_2, k_2 events in the interval between x_2 and x_3, and so forth.

Censored means that some events are missing in the sample. This is often the case for data that comes from reliability tests where the events in the sample are the times to failure. Imagine, for example, a parallel test-series of 100 parts the reliability of which should be estimated. Every time a part fails the failure time is registered. After a certain time the test is truncated and there may be some parts that did not yet fail. These parts are referred to as so-called suspended elements. The information on these parts, that have survived for example to a time of k hours, needs to be considered in the parameter-estimation algorithm as there is obviously a difference between the statements "There were 2 failures at times x_1 and x_2" and "There were 2 failures at times x_1 and x_2 and there are 98 parts that survived the time x_3."

The method of moments presented in Section 1.3.4.1 can neither handle grouped nor censored data. The linear regression can cope with suspended elements by an adaptation of the calculation of the median ranks using the notion of adjusted ranks [Aber94]. There have been some attempts to enable the regression to treat grouped data directly [Lawl82]. But in practical applications it has become common to distribute the number of events over the interval and thus resulting in the ungrouped case. The most flexible method is the maximum likelihood estimation which can treat suspended elements and grouped data. For example for grouped events the likelihood function can be modified as follows: If there are k_i events within the interval x_i to x_{i+1} the likelihood function becomes

$$L(\theta) = \prod_{i=1}^{n} \left(F_{X_{i+1}}(x_{i+1} \mid \theta) - F_{X_i}(x_i \mid \theta) \right)^{k_i}.$$

For censored data see [Triv01].

Problem 1.1   Show that the sample mean $\overline{X}$, which is defined in (1.68), is an unbiased estimator of the population mean μ.

Problem 1.2   Show that the sample variance S^2, which is defined in (1.69) is an unbiased estimator of the population variance σ^2.

1.3.5   Order Statistics

Let $X_1, X_2 \ldots, X_n$ be mutually independent, identically distributed continuous random variables, each having the distribution function $F(.)$ and density $f(.)$. Let $Y_1, Y_2, \ldots, Y_n$ be random variables obtained by permuting the set $X_1, X_2, \ldots, X_n$ so as to be in increasing order. Thus, for instance

$$Y_1 = \min\{X_1, X_2, \ldots, X_n\}$$

and

$$Y_n = \max\{X_1, X_2, \ldots, X_n\}.$$

The random variable Y_k is generally called the *kth-order statistic*. Because $X_1, X_2, \ldots, X_n$ are continuous random variables, it follows that $Y_1 < Y_2 < \cdots < Y_n$ (as opposed to $Y_1 \leq Y_2 \leq \cdots \leq Y_n$) with a probability of one.

As examples of the use of order statistics, let X_i be the time to failure of the ith component in a system of n independent components. If the system is a series system, then Y_1 will be the system lifetime. Similarly, Y_n will denote the lifetime of a parallel system and Y_{n-m+1} will be the lifetime of an m-out-of-n system.

The distribution function of Y_k is given by

$$F_{Y_k}(y) = \sum_{j=k}^{n} \binom{n}{j} F^j(y)[1 - F(y)]^{n-j}, \quad -\infty < y < \infty. \tag{1.86}$$

In particular, the distribution functions of Y_n and Y_1 can be obtained from Eq. (1.86) as

$$F_{Y_n}(y) = [F(y)]^n, \quad -\infty < y < \infty$$

and

$$F_{Y_1}(y) = 1 - [1 - F(y)]^n, \quad -\infty < y < \infty.$$

1.3.6   Distribution of Sums

Assume that X and Y are continuous random variables with joint probability density function f. Suppose we are interested in the density of a random variable $Z = \Phi(X, Y)$. The distribution of Z may be written as

$$F_Z(z) = P(Z \leq z) = \int\!\!\int_{A_z} f(x, y)\, \mathrm{d}x\, \mathrm{d}y, \tag{1.87}$$

where A_z is a subset of $\mathbb{R}^2$ given by

$$A_z = \{(x,y) \mid \Phi(x,y) \le z\} = \Phi^{-1}((-\infty, z]).$$

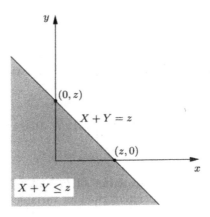

Fig. 1.15   Area of integration for the convolution of X and Y.

A function of special interest is $Z = X + Y$ with

$$A_z = \{(x,y) \mid x + y \le z\},$$

which is the half-plane to the lower left of the line $x+y = z$, shown in Fig. 1.15. Then we get

$$F_Z(z) = \int\!\!\int_{A_z} f(x,y)\,\mathrm{d}x\,\mathrm{d}y = \int_{-\infty}^{\infty} \int_{-\infty}^{z-x} f(x,y)\,\mathrm{d}y\,\mathrm{d}x.$$

Making a variable substitution $y = t - x$, we get

$$F_Z(z) = \int_{-\infty}^{\infty} \int_{-\infty}^{z} f(x,t-x)\,\mathrm{d}t\,\mathrm{d}x = \int_{-\infty}^{z} \int_{-\infty}^{\infty} f(x,t-x)\,\mathrm{d}x\,\mathrm{d}t$$

$$= \int_{-\infty}^{z} f_Z(t)\,\mathrm{d}t.$$

Thus, the density of Z is given by

$$f_Z(z) = \int_{-\infty}^{\infty} f(x, z-x)\,\mathrm{d}x, \quad -\infty < z < \infty. \tag{1.88}$$

Now if X and Y are independent random variables, then $f(x, y) = f_X(x) f_Y(y)$, and Formula (1.88) reduces to

$$f_Z(z) = \int\limits_{-\infty}^{\infty} f_X(x) f_Y(z - x)\, dx, \quad -\infty < z < \infty. \tag{1.89}$$

Further, if both X and Y are nonnegative random variables, then

$$f_Z(z) = \int\limits_{0}^{z} f_X(x) f_Y(z - x)\, dx, \quad 0 < z < \infty. \tag{1.90}$$

This integral is known as the *convolution* of f_X and f_Y. Thus, the pdf of the sum of two non-negative independent random variables is the convolution of their individual pdfs.

If $X_1, X_2, \ldots, X_r$ are mutually independent, identically distributed exponential random variables with parameter λ, then the random variable $X_1 + X_2 + \cdots + X_r$ has an Erlang-r distribution with parameter λ.

Consider an asynchronous transfer mode (ATM) switch with cell interarrival times that are independent from each other and exponentially distributed with parameter λ. Let X_i be the random variable denoting the time between the $(i - 1)$st and ith arrival. Then $Z_r = X_1 + X_2 + \cdots + X_r$ is the time until the rth arrival and has an Erlang-r distribution. Another way to obtain this result is to consider N_t, the number of arrivals in the interval $(0, t]$. As pointed out earlier, N_t has a Poisson distribution with parameter λt. Since the events $[Z_r > t]$ and $[N_t < r]$ are equivalent, we have

$$P(Z_r > t) = P(N_t < r) = \sum_{j=0}^{r-1} e^{-\lambda t} \left[\frac{(\lambda t)^j}{j!} \right],$$

which implies that

$$F_{Z_r}(t) = P(Z_r \leq t) = 1 - \sum_{j=0}^{r-1} \frac{(\lambda t)^j}{j!}\, e^{-\lambda t},$$

which is the Erlang-r distribution function.

If $X \sim EXP(\lambda_1)$, $Y \sim EXP(\lambda_2)$, X and Y are independent, and $\lambda_1 \neq \lambda_2$, then $Z = X + Y$ has a two-stage hypoexponential distribution with two phases and parameters λ_1 and λ_2. To see this, we use

$$f_Z(z) = \int_0^z f_X(x) f_Y(z-x)\,\mathrm{d}x\,, \quad z > 0 \quad \text{(by Eq. (1.90))}$$

$$= \int_0^z \lambda_1 \mathrm{e}^{-\lambda_1 x}\, \lambda_2 \mathrm{e}^{-\lambda_2(z-x)}\,\mathrm{d}x = \lambda_1 \lambda_2 \mathrm{e}^{-\lambda_2 z} \int_0^z \mathrm{e}^{(\lambda_2 - \lambda_1)x}\,\mathrm{d}x$$

$$= \lambda_1 \lambda_2 \mathrm{e}^{-\lambda_2 z} \left[\frac{\mathrm{e}^{-(\lambda_1 - \lambda_2)x}}{-(\lambda_1 - \lambda_2)} \right]_{x=0}^{x=z}$$

$$= \frac{\lambda_1 \lambda_2}{\lambda_1 - \lambda_2} \mathrm{e}^{-\lambda_2 z} + \frac{\lambda_1 \lambda_2}{\lambda_2 - \lambda_1} \mathrm{e}^{-\lambda_1 z} \quad \text{(see Eq. (1.31))}.$$

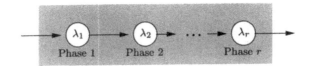

Fig. 1.16 Hypoexponential distribution as a series of exponential stages.

A more general version of this result follows: Let $Z = \sum_{i=1}^{r} X_i$, where $X_1, X_2, \ldots, X_r$ are mutually independent and X_i is exponentially distributed with parameter λ_i ($\lambda_i \neq \lambda_j$ for $i \neq j$). Then Z has a hypoexponential distribution with r phases and the density function:

$$f_Z(z) = \sum_{i=1}^{r} a_i \lambda_i \mathrm{e}^{-\lambda_i z}\,, \quad z > 0\,, \tag{1.91}$$

where

$$a_i = \prod_{\substack{j=1 \\ j \neq i}}^{r} \frac{\lambda_j}{\lambda_j - \lambda_i}\,, \quad 1 \leq i \leq r\,. \tag{1.92}$$

Such a phase-type distribution can be visualized as shown in Fig. 1.16 (see also Section 1.3.1.2).

2

Markov Chains

2.1 MARKOV PROCESSES

Markov processes provide very flexible, powerful, and efficient means for the description and analysis of dynamic (computer) system properties. Performance and dependability measures can be easily derived. Moreover, Markov processes constitute the fundamental theory underlying the concept of queueing systems. In fact, the notation of queueing systems has been viewed sometimes as a high-level *specification* technique for (a sub-class of) Markov processes. Each queueing system can, in principle, be mapped onto an instance of a Markov process and then mathematically evaluated in terms of this process. But besides highlighting the *computational* relation between Markov processes and queueing systems, it is worthwhile pointing out also that fundamental *properties* of queueing systems are commonly *proved* in terms of the underlying Markov processes. This type of use of Markov processes is also possible even when queueing systems exhibit properties such as nonexponential distributions that cannot be represented directly by discrete-state Markov models. Markovizing methods, such as embedding techniques or supplementary variables, can be used in such cases. Here Markov processes serve as a mere *theoretical* framework to prove the correctness of computational methods applied directly to the analysis of queueing systems. For the sake of efficiency, an explicit creation of the Markov process is preferably avoided.

2.1.1 Stochastic and Markov Processes

There exist many textbooks like those from King [King90], Trivedi [Triv01], Allen [Alle90], Gross and Harris [GrHa85], Cinlar [Cinl75], Feller (his two volume classic [Fell68]), and Howard [Howa71] that provide excellent introductions into the basics of stochastic and Markov processes. Besides the theoretical background, many motivating examples are also given in those books. Consequently, we limit discussion here to the essentials of Markov processes and refer to the literature for further details.

Markov processes constitute a special, perhaps the most important, subclass of stochastic processes, while the latter can be considered as a generalization of the concept of random variables. In particular, a stochastic process provides a relation between the elements of a possibly infinite family of random variables. A series of random experiments can thus be taken into consideration and analyzed as a whole.

Definition 2.1   A *stochastic process* is defined as a family of random variables $\{X_t : t \in T\}$ where each random variable X_t is indexed by parameter $t \in T$, which is usually called the *time parameter* if $T \subseteq \mathbb{R}_+ = [0, \infty)$. The set of all possible values of X_t (for each $t \in T$) is known as the state space S of the stochastic process.

If a countable, discrete-parameter set T is encountered, the stochastic process is called a *discrete-parameter* process and T is commonly represented by (a subset of) $\mathbb{N}_0 = \{0, 1, \dots\}$; otherwise we call it a *continuous-parameter* process. The *state space* of the stochastic process may also be continuous or discrete. Generally, we restrict ourselves here to the investigation of discrete state spaces and in that case refer to the stochastic processes as *chains*, but both continuous- and discrete-parameter processes are considered.

Definition 2.2   Continuous-parameter stochastic processes can be probabilistically characterized by the *joint (cumulative) distribution function* (CDF) $F_{\mathbf{X}}(\mathbf{s}; \mathbf{t})$ for a given set of random variables $\{X_{t_1}, X_{t_2}, \dots, X_{t_n}\}$, parameter vector $\mathbf{t} = (t_1, t_2, \dots, t_n) \in \mathbb{R}^n$, and state vector $\mathbf{s} = (s_1, s_2, \dots, s_n) \in \mathbb{R}^n$, where $t_1 < t_2 < \cdots < t_n$:

$$F_{\mathbf{X}}(\mathbf{s}; \mathbf{t}) = P(X_{t_1} \le s_1, X_{t_2} \le s_2, \dots, X_{t_n} \le s_n). \tag{2.1}$$

The *joint probability density function* (pdf),

$$f_{\mathbf{X}}(\mathbf{s}; \mathbf{t}) = \frac{\partial^n F_{\mathbf{X}}(\mathbf{s}; \mathbf{t})}{\partial s_1 \partial s_2 \dots \partial s_n},$$

is defined correspondingly if the partial derivatives exist. If the so-called Markov property is imposed on the *conditional CDF* of a stochastic process, a Markov process results:

Definition 2.3   A stochastic process $\{X_t : t \in T\}$ constitutes a *Markov process* if for all $0 = t_0 < t_1 < \cdots < t_n < t_{n+1}$ and all $s_i \in S$ the conditional CDF of $X_{t_{n+1}}$ depends only on the last previous value X_{t_n} and not on the earlier values $X_{t_0}, X_{t_1}, \dots, X_{t_{n-1}}$:

$$\begin{aligned} P(X_{t_{n+1}} &\le s_{n+1} \mid X_{t_n} = s_n, X_{t_{n-1}} = s_{n-1}, \dots, X_{t_0} = s_0) \\ &= P(X_{t_{n+1}} \le s_{n+1} \mid X_{t_n} = s_n). \end{aligned} \tag{2.2}$$

This most general definition of a Markov process can be adopted to special cases. In particular, we focus here on discrete state spaces and on both discrete- and continuous-parameter Markov processes. As a result, we deal primarily with *continuous-time Markov chains* (CTMC), and with *discrete-time Markov chains* (DTMC), which are introduced in the next section. Finally, it is often sufficient to consider only systems with a time independent, i.e., *time-homogeneous*, pattern of dynamic behavior. Note that time-homogeneous system dynamics is to be discriminated from stationary system behavior, which relates to time independence in a different sense. The former refers to the stationarity of the *conditional* CDF while the latter refers to the stationarity of the CDF *itself*.

Definition 2.4   Letting $t_0 = 0$ without loss of generality, a Markov process is said to be *time-homogeneous* if the conditional CDF of $X_{t_{n+1}}$ does not depend on the observation time, that is, it is invariant with respect to time epoch t_n:

$$P(X_{t_{n+1}} \le s_{n+1} \mid X_{t_n} = s_n) = P(X_{t_{n+1}-t_n} \le s_{n+1} \mid X_0 = s_n). \qquad (2.3)$$

2.1.2 Markov Chains

Equation (2.2) describes the well-known Markov property. Informally this can be interpreted in the sense that the whole history of a Markov chain is summarized in the current state X_{t_n}. Equivalently, given the present, the future is conditionally independent of the past. Note that the Markov property does not prevent the conditional distribution from being dependent on the time variable t_n. Such a dependence is prevented by the definition of homogeneity (see Eq. (2.3)). A unique characteristic is implied, namely, the sojourn time distribution in any state of a homogeneous Markov chain exhibits the memoryless property. An immediate, and somewhat curious, consequence is that the mean sojourn time equals the mean residual and the mean elapsed time in any state and at any time [Triv01].

If not explicitly stated otherwise, we consider Markov processes with discrete state spaces only, that is, Markov chains, in what follows. Note that in this case we are inclined to talk about probability mass functions, pmf, rather than probability density functions, pdf. Refer back to Sections 1.3.1.1 and 1.3.1.2 for details.

2.1.2.1 Discrete-Time Markov Chains   We are now ready to proceed to the formal definition of Markov chains. Discrete-parameter Markov chains are considered first, that is, Markov processes restricted to a discrete, finite, or countably infinite state space, S, and a discrete-parameter space T. For the sake of convenience, we set $T \subseteq \mathbb{N}_0$. The conditional pmf reflecting the Markov property for discrete-time Markov chains, corresponding to Eq. (2.2), is summarized in the following definition:

Definition 2.5   A given stochastic process $\{X_0, X_1, \ldots, X_{n+1}, \ldots\}$ at the consecutive points of observation $0, 1, \ldots, n+1$ constitutes a *DTMC* if the following relation on the *conditional pmf*, that is, the Markov property, holds for all $n \in \mathbb{N}_0$ and all $s_i \in S$:

$$
\begin{aligned}
&P(X_{n+1} = s_{n+1} \mid X_n = s_n, X_{n-1} = s_{n-1}, \ldots, X_0 = s_0) \\
&= P(X_{n+1} = s_{n+1} \mid X_n = s_n).
\end{aligned}
\tag{2.4}
$$

Given an initial state s_0, the DTMC evolves over time, that is, step by step, according to *one-step transition probabilities*. The right-hand side of Eq. (2.4) reveals the conditional pmf of transitions from state s_n at time step n to state s_{n+1} at time step $(n+1)$. Without loss of generality, let $S = \{0, 1, 2, \ldots\}$ and write conveniently the following shorthand notation for the conditional pmf of the process's one-step transition from state i to state j at time n:

$$
p_{ij}^{(1)}(n) = P(X_{n+1} = s_{n+1} = j \mid X_n = s_n = i).
\tag{2.5}
$$

In the homogeneous case, when the conditional pmf is independent of epoch n, Eq. (2.5) reduces to

$$
p_{ij}^{(1)} = p_{ij}^{(1)}(n) = P(X_{n+1} = j \mid X_n = i) = P(X_1 = j \mid X_0 = i), \quad \forall n \in T.
\tag{2.6}
$$

For the sake of convenience, we usually drop the superscript, so that $p_{ij} = p_{ij}^{(1)}$ refers to a one-step transition probability of a homogeneous DTMC.

Starting with state i, the DTMC will go to some state j (including the possibility of $j = i$), so that it follows that $\sum_j p_{ij} = 1$, where $0 \le p_{ij} \le 1$. The one-step transition probabilities p_{ij} are usually summarized in a nonnegative, stochastic[1] transition matrix $\mathbf{P}$:

$$
\mathbf{P} = \mathbf{P}^{(1)} = [p_{ij}] = \begin{pmatrix}
p_{00} & p_{01} & p_{02} & \cdots \\
p_{10} & p_{11} & p_{12} & \cdots \\
p_{20} & p_{21} & p_{22} & \cdots \\
\vdots & \vdots & \vdots & \ddots
\end{pmatrix}.
$$

Graphically, a finite-state DTMC is represented by a *state transition diagram* (also referred to as *state diagram*), a finite directed graph, where state i of the chain is depicted by a vertex, and a one-step transition from state i to state j by an edge marked with one-step transition probability p_{ij}. As an example, consider the one-step transition probability matrix in Eq. (2.7) with state space $S = \{0, 1\}$ and the corresponding graphical representation in Fig. 2.1.

---

[1]The elements in each row of the matrix sum up to 1.

Example 2.1   The one-step transition probability matrix of the two-state DTMC in Fig. 2.1 is given by

$$\mathbf{P}^{(1)} = \begin{pmatrix} \frac{3}{4} & \frac{1}{4} \\ \frac{1}{2} & \frac{1}{2} \end{pmatrix} = \begin{pmatrix} 0.75 & 0.25 \\ 0.5 & 0.5 \end{pmatrix}. \tag{2.7}$$

Conditioned on the current DTMC state, a transition is made from state 0 to state 1 with probability 0.25, and with probability 0.75, the DTMC remains in state 0 at the next time step. Correspondingly, a transition occurs from state 1 to state 0 with probability 0.5, and with probability 0.5 the chain remains in state 1 at the next time step.

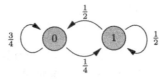

Fig. 2.1   Example of a discrete-time Markov chain referring to Eq. (2.7).

Repeatedly applying one-step transitions generalizes immediately to *n-step transition probabilities*. More precisely, let $p_{ij}^{(n)}(k, l)$ denote the probability that the Markov chain transits from state i at time k to state j at time l in exactly $n = l - k$ steps:

$$p_{ij}^{(n)}(k, l) = P(X_l = j \mid X_k = i), \quad 0 \le k \le l. \tag{2.8}$$

Again, the theorem of total probability applies for any given state i and any given time values k and l such that $\sum_j p_{ij}^{(n)}(k, l) = 1$, where $0 \le p_{ij}^{(n)}(k, l) \le 1$. This fact, together with the Markov property, immediately leads us to a procedure for computing the n-step transition probabilities recursively from the one-step transition probabilities: The transition of the process from state i at time k to state j at time l can be split into sub-transitions from state i at time k to an intermediate state[2] h, say, at time m and from there, *independently of the history* that led to that state, from state h at time m to state j at time l, where $k < m < l$ and $n = l - k$. This condition leads to the well-known system of *Chapman–Kolmogorov equations*:

$$p_{ij}^{(n)}(k, l) = \sum_{h \in S} p_{ih}^{(m-k)}(k, m) p_{hj}^{(l-m)}(m, l), \quad 0 \le k < m < l. \tag{2.9}$$

Note that the conditional independence assumption, i.e., the Markov property, is reflected by the product of terms on the right-hand side of Eq. (2.9).

---

[2]The Markov chain must simply traverse *some* state at any time.

Similar to the one-step case, the n-step transition probabilities can be simplified for homogeneous DTMC such that $p_{ij}^{(n)} = p_{ij}^{(n)}(k,l)$ depend only on the difference $n = l - k$ and not on the actual values of k and l:

$$p_{ij}^{(n)} = P(X_{k+n} = j \mid X_k = i) = P(X_n = j \mid X_0 = i), \quad \forall k \in T. \quad (2.10)$$

Under this condition, the Chapman–Kolmogorov Eq. (2.9) for *homogeneous* DTMC simplifies to

$$p_{ij}^{(n)} = \sum_{h \in S} p_{ih}^{(m)} p_{hj}^{(n-m)}, \quad 0 < m < n. \quad (2.11)$$

Because Eq. (2.11) holds for all $m < n$, let $m = 1$ and get

$$p_{ij}^{(n)} = \sum_{h \in S} p_{ih}^{(1)} p_{hj}^{(n-1)}.$$

With $\mathbf{P}^{(n)}$ as the matrix of n-step transition probabilities $p_{ij}^{(n)}$, Eq. (2.11) can be rewritten in matrix form for the particular case of $m = 1$ as $\mathbf{P}^{(n)} = \mathbf{P}^{(1)}\mathbf{P}^{(n-1)} = \mathbf{P}\mathbf{P}^{(n-1)}$. Applying this procedure recursively results in the following equation[3]:

$$\mathbf{P}^{(n)} = \mathbf{P}\mathbf{P}^{(n-1)} = \mathbf{P}^n. \quad (2.12)$$

The n-step transition probability matrix can be computed by the $(n-1)$-fold multiplication of the one-step transition matrix by itself.

Example 2.2   Referring back to the example in Fig. 2.1 and Eq. (2.7), the four-step transition probability matrix, for instance, can be derived according to Eq. (2.12):

$$\mathbf{P}^{(4)} = \mathbf{P}\mathbf{P}^{(3)} = \mathbf{P}^2\mathbf{P}^{(2)}$$

$$= \begin{pmatrix} 0.75 & 0.25 \\ 0.5 & 0.5 \end{pmatrix}^2 \mathbf{P}^{(2)} = \begin{pmatrix} 0.6875 & 0.3125 \\ 0.625 & 0.375 \end{pmatrix} \mathbf{P}\mathbf{P}^{(1)} \quad (2.13)$$

$$= \begin{pmatrix} 0.67188 & 0.32813 \\ 0.65625 & 0.34375 \end{pmatrix} \mathbf{P}^{(1)} = \begin{pmatrix} 0.66797 & 0.33203 \\ 0.66406 & 0.33594 \end{pmatrix}.$$

Ultimately, we wish to compute the pmf of the random variable X_n, that is, the probabilities $\nu_i(n) = P(X_n = i)$ that the DTMC is in state i, at time step n. These probabilities, called *transient state probabilities* at time n, will then allow us to derive the desired performance measures. Given the n-step transition probability matrix $\mathbf{P}^{(n)}$, the vector of the state probabilities at time n, $\boldsymbol{\nu}(n) = (\nu_0(n), \nu_1(n), \nu_2(n), \dots)$ can be obtained by un-conditioning $\mathbf{P}^{(n)}$ on the *initial probability vector* $\boldsymbol{\nu}(0) = (\nu_0(0), \nu_1(0), \nu_2(0), \dots)$:

$$\boldsymbol{\nu}(n) = \boldsymbol{\nu}(0)\mathbf{P}^{(n)} = \boldsymbol{\nu}(0)\mathbf{P}^n = \boldsymbol{\nu}(n-1)\mathbf{P}. \quad (2.14)$$

---

[3]It is important to keep in mind that $\mathbf{P}^{(n)} = \mathbf{P}^n$ holds only for homogeneous DTMC.

Note that both $\boldsymbol{\nu}(n)$ and $\boldsymbol{\nu}(0)$ are represented as row vectors in Eq. (2.14).

Example 2.3 Assume that the DTMC from Eq. (2.7) under investigation is initiated in state 1, then the initial probability vector $\boldsymbol{\nu}^{(1)}(0)=(0,1)$ is to be applied in the un-conditioning according to Eq. (2.14). With the already computed four-step transition probabilities in Eq. (2.13) the corresponding pmf $\boldsymbol{\nu}^{(1)}(4)$ can be derived:

$$\boldsymbol{\nu}^{(1)}(4) = (0,1) \begin{pmatrix} 0.66797 & 0.33203 \\ 0.66406 & 0.33594 \end{pmatrix} = (0.66406, 0.33594) \,.$$

Example 2.4 Alternatively, with another initial probability vector:

$$\boldsymbol{\nu}^{(2)}(0) = \left(\tfrac{2}{3}, \tfrac{1}{3}\right) = (0.6\bar{6}, 0.3\bar{3}) \,,$$

according to Eq. (2.14), we would get

$$\boldsymbol{\nu}^{(2)}(4) = (0.6\bar{6}, 0.3\bar{3})$$

as the probabilities at time step 4.

Of particular importance are homogeneous DTMC on which a so-called *stationary probability vector* can be imposed in a suitable way:

Definition 2.6 State probabilities $\boldsymbol{\nu} = (\nu_0, \nu_1, \ldots \nu_i, \ldots)$ of a discrete-time Markov chain are said to be *stationary*, if any transitions of the underlying DTMC according to the given one-step transition probabilities $\mathbf{P} = [p_{ij}]$ have no effect on these state probabilities, that is, $\nu_j = \sum_{i \in S} \nu_i p_{ij}$ holds for all states $j \in S$. This relation can also be expressed in matrix form:

$$\boldsymbol{\nu} = \boldsymbol{\nu}\mathbf{P}, \quad \sum_{i \in S} \nu_i = 1 \,. \tag{2.15}$$

Note that according to the preceding definition, more than one stationary pmf can exist for a given, unrestricted, DTMC.

Example 2.5 By substituting the one-step transition matrix from Eq. (2.7) in Eq. (2.15), it can easily be checked that

$$\boldsymbol{\nu}^{(2)} = \left(\tfrac{2}{3}, \tfrac{1}{3}\right) = (0.6\bar{6}, 0.3\bar{3})$$

is a stationary probability vector while $\boldsymbol{\nu}^{(1)} = (0,1)$ is not.

Definition 2.7 For an efficient analysis, we are interested in the *limiting state probabilities* $\tilde{\boldsymbol{\nu}}$ as a particular kind of stationary state probabilities, which are defined by

$$\tilde{\boldsymbol{\nu}} = \lim_{n \to \infty} \boldsymbol{\nu}(n) = \lim_{n \to \infty} \boldsymbol{\nu}(0)\mathbf{P}^{(n)} = \boldsymbol{\nu}(0) \lim_{n \to \infty} \mathbf{P}^{(n)} = \boldsymbol{\nu}(0)\tilde{\mathbf{P}} \,. \tag{2.16}$$

Definition 2.8   As $n \to \infty$, we may require both the n-step transition probability matrix $\mathbf{P}^{(n)}$ and the state probability vector $\boldsymbol{\nu}(n)$ to *converge independently* of the initial probability vector $\boldsymbol{\nu}(0)$ to $\tilde{\mathbf{P}}$ and $\tilde{\boldsymbol{\nu}}$, respectively. Also, we may only be interested in the case where the state probabilities $\tilde{\nu}_i > 0, \forall i \in S$, are *strictly positive* and $\sum_i \tilde{\nu}_i = 1$, that is, $\tilde{\boldsymbol{\nu}}$ constitutes a pmf.

If all these restrictions apply to a given probability vector, it is said to be the *unique steady-state probability vector* of the DTMC.

Example 2.6   Returning to Example 2.1, we note that the n-step transition probabilities converge as $n \to \infty$:

$$\tilde{\mathbf{P}} = \lim_{n \to \infty} \mathbf{P}^{(n)} = \lim_{n \to \infty} \begin{pmatrix} 0.75 & 0.25 \\ 0.5 & 0.5 \end{pmatrix}^n = \begin{pmatrix} 0.6\bar{6} & 0.3\bar{3} \\ 0.6\bar{6} & 0.3\bar{3} \end{pmatrix} .$$

Example 2.7   With this result, the limiting state probability vector $\tilde{\boldsymbol{\nu}}$, which is independent of any initial probability vector $\boldsymbol{\nu}(0)$, can be derived according to Eq. (2.16):

$$\tilde{\boldsymbol{\nu}} = (0.6\bar{6}, 0.3\bar{3}) .$$

Example 2.8   Since all probabilities in the vector $(0.6\bar{6}, 0.3\bar{3})$ are strictly positive, this vector constitutes the unique steady-state probability vector of the DTMC.

Eventually, the limiting state probabilities become *independent of time steps*, such that once the limiting probability vector is reached, further transitions of the DTMC do not change this vector, i.e., it is stationary. Note that such a probability vector does not necessarily exist for all DTMCs.

If Eq. (2.16) holds and $\tilde{\boldsymbol{\nu}}$ is independent of $\boldsymbol{\nu}(0)$, it follows that the limit $\mathbf{P}^{(n)} = [p_{ij}^{(n)}]$ is independent of time n and of index i. All rows of $\tilde{\mathbf{P}}$ would be identical, that is, the rows would match element by element. Furthermore, the jth element $\tilde{p}_{ij}$ of row i equals $\tilde{\nu}_j$ for all $i \in S$:

$$\tilde{\mathbf{P}} = \lim_{n \to \infty} \mathbf{P}^{(n)} = \lim_{n \to \infty} \mathbf{P}^n = \begin{pmatrix} \tilde{\nu}_0 & \tilde{\nu}_1 & \tilde{\nu}_2 & \cdots \\ \tilde{\nu}_0 & \tilde{\nu}_1 & \tilde{\nu}_2 & \cdots \\ \tilde{\nu}_0 & \tilde{\nu}_1 & \tilde{\nu}_2 & \cdots \\ \vdots & \vdots & \vdots & \ddots \end{pmatrix} . \tag{2.17}$$

If the unique steady-state probability vector of a DTMC *exists*, it can be determined by the solution of the system of linear Eqs. (2.15), so that $\tilde{\mathbf{P}}$ need not be determined explicitly. From Eq. (2.14), we have $\boldsymbol{\nu}(n) = \boldsymbol{\nu}(n-1)\mathbf{P}$. If the limit exists, we can take it on both sides of the equation and get

$$\lim_{n \to \infty} \boldsymbol{\nu}(n) = \tilde{\boldsymbol{\nu}} = \lim_{n \to \infty} \boldsymbol{\nu}(n-1)\mathbf{P} = \tilde{\boldsymbol{\nu}}\mathbf{P} . \tag{2.18}$$

In the steady-state case no ambiguity can arise, so that, for the sake of convenience, we may drop the annotation and refer to steady-state probability vector by using the notation $\boldsymbol{\nu}$ instead of $\tilde{\boldsymbol{\nu}}$. Steady-state and stationarity coincide in that case, i.e., there is only a unique stationary probability vector.

The computation of the steady-state probability vector $\boldsymbol{\nu}$ of a DTMC is usually significantly simpler and less expensive than a time-dependent computation of $\boldsymbol{\nu}(n)$. It is therefore the steady-state probability vector of a DTMC that is preferably taken advantage of in modeling endeavors. But a steady-state probability vector does not exist for all DTMCs.[4] Additionally, it is not always appropriate to restrict the analysis to the steady-state case, even if it does exist. Under some circumstances, time-dependent (i.e., transient) analysis would result in more meaningful information with respect to an application. Transient analysis has special relevance if short-term behavior is of more importance than long-term behavior. In modeling terms, "short term" means that the influence of the initial state probability vector $\boldsymbol{\nu}(0)$ on $\boldsymbol{\nu}(n)$ has not yet disappeared by time step n.

Before continuing, some simple example DTMCs are presented to clarify the definitions of this section. The following four one-step transition matrices are examined for the conditions under which stationary, limiting state, and steady-state probabilities exist for them.

Example 2.9 Consider the DTMC shown in Fig. 2.2 with the one-step transition probability matrix (TPM):

$$\mathbf{P} = \begin{pmatrix} 1 & 0 \\ 0 & 1 \end{pmatrix}. \tag{2.19}$$

Fig. 2.2 Example of a discrete-time Markov chain referring to Eq. (2.19).

- For this one-step TPM, an infinite number of stationary probability vectors exists: Any arbitrary probability vector is stationary in this case according to Eq. (2.15).

- The n-step TPM $\mathbf{P}^{(n)}$ converges in the limit to

$$\lim_{n \to \infty} \mathbf{P}^{(n)} = \tilde{\mathbf{P}} = \begin{pmatrix} 1 & 0 \\ 0 & 1 \end{pmatrix}.$$

Furthermore, all n-step TPMs are identical:

$$\mathbf{P} = \tilde{\mathbf{P}} = \mathbf{P}^{(n)}, \quad \forall n \in T.$$

---

[4]The conditions under which DTMCs converge to steady-state is precisely stated in the following section.

The limiting state probabilities $\tilde{\nu}$ do exist and are identical to the initial probability vector $\nu(0)$ in all cases:

$$\tilde{\nu} = \nu(0)\tilde{P} = \nu(0).$$

- A unique steady-state probability vector does not exist for this example.

Example 2.10   Next, consider the DTMC shown in Fig. 2.3 with the TPM:

$$P = \begin{pmatrix} 0 & 1 \\ 1 & 0 \end{pmatrix}. \tag{2.20}$$

Fig. 2.3 Example of a discrete-time Markov chain referring to Eq. (2.20).

- For this one-step TPM, a stationary probability vector, which is unique in this case, does exist according to Eq. (2.15):

$$\nu = (0.5, 0.5).$$

- The n-step transition matrix $P^{(n)}$ does not converge in the limit to any $\tilde{P}$. Therefore, the limiting state probabilities $\tilde{\nu}$ do not exist.

- Consequently, a unique steady-state probability vector does not exist.

Example 2.11   Consider the DTMC shown in Fig. 2.4 with the TPM:

$$P = \begin{pmatrix} 0.5 & 0.5 \\ 0.5 & 0.5 \end{pmatrix}. \tag{2.21}$$

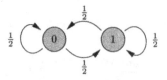

Fig. 2.4 Example of a discrete-time Markov chain referring to Eq. (2.21).

- For this one-step TPM a unique stationary probability vector does exist according to Eq. (2.15):

$$\nu = (0.5, 0.5).$$

Note that this is the same unique stationary probability vector as for the different DTMC in Eq. (2.20).

- The n-step TPM $\mathbf{P}^{(n)}$ converges in the limit to

$$\lim_{n \to \infty} \mathbf{P}^{(n)} = \tilde{\mathbf{P}} = \begin{pmatrix} 0.5 & 0.5 \\ 0.5 & 0.5 \end{pmatrix} .$$

Furthermore, all n-step TPMs are identical:

$$\mathbf{P} = \tilde{\mathbf{P}} = \mathbf{P}^{(n)} , \quad \forall n \in T .$$

The limiting state probabilities $\tilde{\boldsymbol{\nu}}$ do exist, are independent of the initial probability vector, and are unique:

$$\tilde{\boldsymbol{\nu}} = \boldsymbol{\nu}(0)\tilde{\mathbf{P}} = (0.5, 0.5) .$$

- A unique steady-state probability vector does exist. All probabilities are strictly positive and identical to the stationary probabilities, which can be derived from the solution of Eq. (2.15):

$$\boldsymbol{\nu} = \tilde{\boldsymbol{\nu}} = (0.5, 0.5) .$$

Example 2.12 Now consider the DTMC shown in Fig. 2.5 with the TPM:

$$\mathbf{P} = \begin{pmatrix} 0 & 1 \\ 0 & 1 \end{pmatrix} . \tag{2.22}$$

Fig. 2.5 Example of a discrete-time Markov chain referring to Eq. (2.22).

- For this one-step TPM a unique stationary probability vector does exist according to Eq. (2.15):
$$\boldsymbol{\nu} = (0, 1) .$$

- The n-step TPM $\mathbf{P}^{(n)}$ converges in the limit to

$$\tilde{\mathbf{P}} = \begin{pmatrix} 0 & 1 \\ 0 & 1 \end{pmatrix} .$$

Furthermore, all n-step TPMs are identical:

$$\mathbf{P} = \tilde{\mathbf{P}} = \mathbf{P}^{(n)} , \quad \forall n \in T .$$

The limiting state probability vector $\tilde{\boldsymbol{\nu}}$ does exist, is independent of the initial probability vector, and is identical to the unique stationary probability vector $\boldsymbol{\nu}$:

$$\tilde{\boldsymbol{\nu}} = \boldsymbol{\nu} = (0, 1) \ .$$

- A unique steady-state probability vector does not exist for this example. The elements of the unique stationary probability vector are not strictly positive.

We proceed now to identify necessary and sufficient conditions for the existence of a steady-state probability vector of a DTMC. The conditions can be given immediately in terms of properties of the DTMC.

2.1.2.1.1 Classifications of DTMC
DTMCs are categorized based on the classifications of their constituent states.

Definition 2.9   Any state j is said to be *reachable* from any other state i, where $i, j \in S$, if it is possible to transit from state i to state j in a finite number of steps according to the given transition probability matrix. For some integer $n \geq 1$, the following relation must hold for the n-step transition probability:

$$p_{ij}^{(n)} > 0, \quad \exists n, n \geq 1 \ . \tag{2.23}$$

A DTMC is called *irreducible* if all states in the chain can be reached pairwise from each other, i.e., $\forall i, j \in S, \exists n, n \geq 1 : p_{ij}^{(n)} > 0$.

A state $i \in S$ is said to be an *absorbing state*[5] if and only if no other state of the DTMC can be reached from it, i.e., $p_{ii} = 1$.

Note that a DTMC containing at least one absorbing state cannot be irreducible. If countably infinite state models are encountered, we have to discriminate more accurately how states are reachable from each other. The recurrence time and the probability of recurrence must also be taken into account.

Definition 2.10   Let $f_i^{(n)}$, called the *n-step recurrence probability*, denote the conditional probability of the first return to state $i \in S$ in exactly $n \geq 1$ steps after leaving state i. Then, the probability f_i of ever returning to state i is given by

$$f_i = \sum_{n=1}^{\infty} f_i^{(n)} \ . \tag{2.24}$$

Any state $i \in S$ to which the DTMC will return with probability $f_i = 1$, is called a *recurrent state*; otherwise, if $f_i < 1$, i is called a *transient state*.

---

[5] Absorbing states play an important role in the modeling of dependable systems where transient analysis is of primary interest.

Given a recurrent state i, the *mean recurrence time* m_i of state i of a DTMC is given by:

$$m_i = \sum_{n=1}^{\infty} n f_i^{(n)} . \tag{2.25}$$

If the mean recurrence time is finite, that is, $m_i < \infty$, i is called *positive recurrent* or *recurrent non-null*; otherwise, if $m_i = \infty$, state i is said to be *recurrent null*. For any recurrent state $i \in S$, let d_i denote the *period* of state i, then d_i is the greatest common divisor of the set of positive integers n such that $p_{ii}^{(n)} > 0$. A recurrent state i is called *aperiodic* if its period $d_i = 1$, and *periodic* with period d_i if $d_i > 1$.

It has been shown by Feller [Fell68] that the states of an *irreducible* DTMC are all of the same type. Hence, all states are periodic, aperiodic, transient, recurrent null, or recurrent non-null.

Definition 2.11   If one of the states i of an irreducible DTMC is aperiodic then so are all the other states $j \in S$, that is, $d_j = 1, \forall j \in S$, and the DTMC itself is called *aperiodic*; otherwise it is said to be *periodic* with unique period d.
An irreducible, aperiodic, discrete-time Markov chain with all states i being recurrent non-null with finite mean recurrence time m_i is called an *ergodic* Markov chain.

We are now ready to summarize the main results for the classification of discrete-time Markov chains:

- The states of a *finite-state, irreducible* Markov chain are all recurrent non-null.

- Given an *aperiodic* DTMC, the limits $\tilde{\boldsymbol{\nu}} = \lim_{n \to \infty} \boldsymbol{\nu}(n)$ do exist.

- For any *irreducible* and *aperiodic* DTMC, the limit $\tilde{\boldsymbol{\nu}}$ exists and is independent of the initial probability vector $\boldsymbol{\nu}(0)$.

- For an *ergodic* DTMC, the limit $\tilde{\boldsymbol{\nu}} = (\tilde{\nu}_0, \tilde{\nu}_1, \tilde{\nu}_2, \dots)$ exists and comprises the unique steady-state probability vector $\boldsymbol{\nu}$.

- The steady-state probabilities $\nu_i > 0$, $i \in S$, of an *ergodic* Markov chain can be obtained by solving the system of linear Eq. (2.15) or, if the (finite) mean recurrence times m_i are known, by exploiting the relation

$$\nu_i = \frac{1}{m_i} , \quad \forall i \in S . \tag{2.26}$$

If *finite-state* DTMCs are investigated, the solution of Eq. (2.15) can be obtained by applying standard methods for the solution of linear systems. In the *infinite* case, either *generating functions* can be applied or special *structure* of the one-step transition probability matrix $\mathbf{P}^{(1)} = \mathbf{P}$ may be exploited to find the solution in closed form. An example of the latter technique is given in Section 3.1 where we investigate the important class of birth–death processes. The special (tridiagonal) structure of the matrix will allow us to derive closed-form solutions for the state probabilities that are not restricted to any fixed matrix size so that the limiting state probabilities of infinite state DTMCs are captured by the closed-form formulae as well.

2.1.2.1.2 DTMC State Sojourn Times The state *sojourn times* – the time between state changes – play an important role in the characterization of DTMCs. Only homogeneous DTMCs are considered here. We have already pointed out that the transition behavior reflects the memoryless property, that is, it only depends on the current state and neither on the history that led to the state nor on the time already spent in the current state. At every instant of time, the probability of leaving current state i is independently given by $(1 - p_{ii}) = \sum_{i \neq j} p_{ij}$. Applying this repeatedly leads to a description of a random experiment in form of a sequence of Bernoulli trials with probability of success $(1 - p_{ii})$, where "success" denotes the event of leaving current state i. Hence, the *sojourn time* R_i during a single visit to state i is a geometrically distributed random variable[6] with pmf:

$$P(R_i = k) = (1 - p_{ii})p_{ii}^{k-1}, \quad \forall k \in \mathbb{N}^+. \tag{2.27}$$

We can therefore immediately conclude that the expected sojourn time $E[R_i]$, that is, the mean number of time steps the process spends in state i per visit, is

$$E[R_i] = \frac{1}{1 - p_{ii}}. \tag{2.28}$$

Accordingly, the variance $\mathrm{var}[R_i]$ of the sojourn time per visit in state i is given by

$$\mathrm{var}[R_i] = \frac{p_{ii}}{(1 - p_{ii})^2}. \tag{2.29}$$

2.1.2.2 Continuous-Time Markov Chains Continuous- and discrete-time Markov chains provide different yet related modeling paradigms, each of them having their own domain of applications. For the definition of CTMCs we refer back to the definition of general Markov processes in Eq. (2.2) and specialize it to the continuous parameter, discrete state-space case. CTMCs are distinct from DTMCs in the sense that state transitions may occur at arbitrary instants of time and not merely at fixed, discrete time points, as is the case with DTMCs. Therefore, we use a subset of the set of non-negative real numbers $\mathbb{R}_0^+$ to refer to the parameter set T of a CTMC, as opposed to $\mathbb{N}_0$ for DTMCs:

---

[6]Note that the geometric distribution is the discrete-time equivalent of the exponential distribution; i.e., it is the only discrete distribution with the memoryless property.

Definition 2.12   A given stochastic process $\{X_t : t \in T\}$ constitutes a CTMC if for arbitrary $t_i \in \mathbb{R}_0^+$, with $0 = t_0 < t_1 < \cdots < t_n < t_{n+1}, \forall n \in \mathbb{N}$, and $\forall s_i \in S = \mathbb{N}_0$ for the conditional pmf, the following relation holds:

$$
\begin{aligned}
P(X_{t_{n+1}} &= s_{n+1} \mid X_{t_n} = s_n, X_{t_{n-1}} = s_{n-1}, \dots, X_{t_0} = s_0) \\
&= P(X_{t_{n+1}} = s_{n+1} \mid X_{t_n} = s_n).
\end{aligned}
\tag{2.30}
$$

Similar to Eq. (2.4) for DTMCs, Eq. (2.30) expresses the *Markov* property of continuous-time Markov chains. If we further impose homogeneity, then because the exponential distribution is the only continuous-time distribution that provides the memoryless property, the state sojourn times of a CTMC are necessarily exponentially distributed.

Again, the right-hand side of Eq. (2.30) is referred to as the *transition probability* [7] $p_{ij}(u, v)$ of the CTMC to travel from state i to state j during the period of time $[u, v)$, with $u, v \in T$ and $u \le v$:

$$
p_{ij}(u, v) = P(X_v = j \mid X_u = i).
\tag{2.31}
$$

For $u = v$ we define

$$
p_{ij}(u, u) =
\begin{cases}
1, & i = j, \\
0, & \text{otherwise.}
\end{cases}
\tag{2.32}
$$

If the transition probabilities $p_{ij}(u, v)$ depend only on the time difference $t = v - u$ and not on the actual values of u and v, the simplified transition probabilities for *time-homogeneous* CTMC result:

$$
p_{ij}(t) = p_{ij}(0, t) = P(X_{u+t} = j \mid X_u = i) = P(X_t = j \mid X_0 = i), \quad \forall u \in T.
\tag{2.33}
$$

Given the transition probabilities $p_{ij}(u, v)$ and the probabilities $\pi_i(u)$ of the CTMC at time u, the unconditional state probabilities $\pi_j(v), j \in S$ of the process at time v can be derived:

$$
\pi_j(v) = \sum_{i \in S} p_{ij}(u, v) \pi_i(u), \quad \forall u, v \in T \ (u \le v).
\tag{2.34}
$$

With $\mathbf{P}(u, v) = [p_{ij}(u, v)]$ as the matrix of the transition probabilities, for any pair of states $i, j \in S$ and any time interval $[u, v)$, $u, v \in T$, from the parameter domain, and the vector $\boldsymbol{\pi}(u) = (\pi_0(u), \pi_1(u), \pi_2(u), \dots)$ of the state probabilities at any instant of time u, Eq. (2.34) can be given in vector–matrix form:

$$
\boldsymbol{\pi}(v) = \boldsymbol{\pi}(u)\mathbf{P}(u, v), \quad \forall u, v \in T \ (u \le v).
\tag{2.35}
$$

---

[7]Note that, as opposed to the discrete-time case, there is no fixed, discrete number of transition steps considered here.

Note that for all $u \in T$, $\mathbf{P}(u, u) = \mathbf{I}$ is the identity matrix.

In the time-homogeneous case, Eq. (2.34) reduces to

$$\pi_j(t) = \sum_{i \in S} p_{ij}(t)\pi_i(0) = \sum_{i \in S} p_{ij}(0, t)\pi_i(0), \qquad (2.36)$$

or in vector–matrix notation:

$$\boldsymbol{\pi}(t) = \boldsymbol{\pi}(0)\mathbf{P}(t) = \boldsymbol{\pi}(0)\mathbf{P}(0, t). \qquad (2.37)$$

Similar to the discrete-time case (Eq. (2.9)), the *Chapman–Kolmogorov* equation (Eq. (2.38)) for the transition probabilities of a CTMC can be derived from Eq. (2.30) by applying again the theorem of total probability:

$$p_{ij}(u, v) = \sum_{k \in S} p_{ik}(u, w)p_{kj}(w, v), \quad 0 \le u \le w < v. \qquad (2.38)$$

But, *unlike* the discrete-time case, Eq. (2.38) cannot be solved easily and used directly for computing the state probabilities. Rather, it has to be transformed into a system of differential equations which, in turn, leads us to the required results. For this purpose, we define the instantaneous *transition rates* $q_{ij}(t)$ $(i \ne j)$ of the CTMC traveling from state i to state j. These transition rates are related to conditional transition probabilities. Consider the period of time $[t, t + \Delta t)$, where Δt is chosen such that $\sum_{j \in S} q_{ij}(t)\Delta t + o(\Delta t) = 1$.[8] The non-negative, finite, continuous functions $q_{ij}(t)$ can be shown to exist under rather general conditions. For all states i, j, $i \ne j$, we define

$$q_{ij}(t) = \lim_{\Delta t \to 0} \frac{p_{ij}(t, t + \Delta t)}{\Delta t}, \quad i \ne j, \qquad (2.39)$$

$$q_{ii}(t) = \lim_{\Delta t \to 0} \frac{p_{ii}(t, t + \Delta t) - 1}{\Delta t}. \qquad (2.40)$$

If the limits do exist, it is clear from Eqs. (2.39) and (2.40) that, since $\sum_{j \in S} p_{ij}(t, t + \Delta t) = 1$, at any instant of time t:

$$\sum_{j \in S} q_{ij}(t) = 0, \quad \forall i \in S. \qquad (2.41)$$

The quantity $-q_{ii}(t)$ can be interpreted as the total rate at which state i is exited (to any other state) at time t. Accordingly, $q_{ij}(t)$, $(i \ne j)$, denotes the rate at which the CTMC leaves state i in order to transit to state j at time t. As an equivalent interpretation, we can regard $q_{ij}(t)\Delta t + o(\Delta t)$ as the transition probability $p_{ij}(t, t + \Delta t)$ of the Markov chain to transit from

---

[8]The notation $o(\Delta t)$ is defined such that $\lim_{\Delta t \to 0} \frac{o(\Delta t)}{\Delta t} = 0$; that is, we might substitute any function for $o(\Delta t)$ that approaches zero faster than the linear function Δt.

state i to state j in $[t, t+\Delta t)$, where Δt is chosen appropriately. Having these definitions, we return to the Chapman–Kolmogorov equation (Eq. (2.38)). Substituting $v + \Delta t$ for v in (2.38) and subtracting both sides of the original Eq. (2.38) from the result gives us

$$p_{ij}(u, v + \Delta t) - p_{ij}(u, v) = \sum_{k \in S} p_{ik}(u, w)[p_{kj}(w, v + \Delta t) - p_{kj}(w, v)]. \quad (2.42)$$

Dividing both sides of Eq. (2.42) by Δt, taking $\lim_{\Delta t \to 0}$ of the resulting quotient of differences, and letting $w \to v$, we derive a differential equation, the well-known *Kolmogorov forward equation*:

$$\frac{\partial p_{ij}(u, v)}{\partial v} = \sum_{k \in S} p_{ik}(u, v) q_{kj}(v), \quad 0 \le u < v. \quad (2.43)$$

In the homogeneous case, we let $t = v - u$ and get from Eqs. (2.39) and (2.40) *time-independent* transition rates $q_{ij} = q_{ij}(t), \forall i, j \in S$, such that simpler versions of the Kolmogorov forward differential equation for homogeneous CTMCs result:

$$\frac{\mathrm{d}\, p_{ij}(t)}{\mathrm{d}\, t} = \sum_{k \in S} p_{ik}(t) q_{kj} = \sum_{k \in S} p_{ik}(0, t) q_{kj}. \quad (2.44)$$

Instead of the forward equation (2.43), we can equivalently derive and use the *Kolmogorov backward equation* for further computations, both in the homogeneous and nonhomogeneous cases, by letting $w \to u$ in Eq. (2.42) and taking $\lim_{\Delta t \to 0}$ to get

$$\frac{\partial p_{ij}(u, v)}{\partial u} = \sum_{k \in S} p_{kj}(u, v) q_{ik}(u), \quad 0 \le u < v. \quad (2.45)$$

Differentiating Eq. (2.34) on both sides gives Eq. (2.47), using the Kolmogorov (forward) equation (Eq. (2.43)) yields Eq. (2.48), and then, again, applying Eq. (2.34) to Eq. (2.49), we derive the differential equation for the *unconditional state probabilities* $\pi_j(v), \forall j \in S$, at time v in Eq. (2.50):

$$\frac{\mathrm{d}\, \pi_j(v)}{\mathrm{d}\, v} = \frac{\partial \sum_{i \in S} p_{ij}(u, v) \pi_i(u)}{\partial v} \quad (2.46)$$

$$= \sum_{i \in S} \frac{\partial p_{ij}(u, v)}{\partial v} \pi_i(u) \quad (2.47)$$

$$= \sum_{i \in S} \left(\sum_{k \in S} p_{ik}(u, v) q_{kj}(v) \right) \pi_i(u) \quad (2.48)$$

$$= \sum_{k \in S} q_{kj}(v) \sum_{i \in S} p_{ik}(u, v) \pi_i(u) \quad (2.49)$$

$$= \sum_{k \in S} q_{kj}(v) \pi_k(v). \quad (2.50)$$

In the time-homogeneous case, a simpler version of Eq. (2.50) results by assuming $t = v - u$ and using time-independent transition rates q_{ij}. So we get the system of differential Eqs. (2.51):

$$\frac{\mathrm{d}\,\pi_j(t)}{\mathrm{d}\,t} = \sum_{i \in S} q_{ij}\pi_i(t)\,, \quad \forall j \in S\,, \tag{2.51}$$

which is repeatedly used throughout this text. Usually, we prefer vector–matrix form rather than the notation used in Eq. (2.51). Therefore, for the homogeneous case, we define the *infinitesimal generator matrix* $\mathbf{Q}$ of the transition probability matrix $\mathbf{P}(t) = [p_{ij}(0, t)] = [p_{ij}(t)]$ by referring to Eqs. (2.39) and (2.40). The matrix $\mathbf{Q}$

$$\mathbf{Q} = [q_{ij}]\,, \quad \forall i, j \in S\,, \tag{2.52}$$

contains the transition rates q_{ij} from any state i to any other state j, where $i \neq j$, of a given continuous-time Markov chain. The elements q_{ii} on the main diagonal of $\mathbf{Q}$ are defined by $q_{ii} = -\sum_{j, j \neq i} q_{ij}$. With the definition in Eq. (2.52), Eq. (2.51) can be given in vector–matrix form as

$$\dot{\boldsymbol{\pi}}(t) = \frac{\mathrm{d}\,\boldsymbol{\pi}(t)}{\mathrm{d}\,t} = \boldsymbol{\pi}(t)\mathbf{Q}\,. \tag{2.53}$$

For the sake of completeness, we include also the matrix form of the Kolmogorov differential equations in the time-homogeneous case. The Kolmogorov *forward* equation (Eq. (2.44)) can be written as

$$\dot{\mathbf{P}}(t) = \frac{\mathrm{d}\,\mathbf{P}(t)}{\mathrm{d}\,t} = \mathbf{P}(t)\mathbf{Q}\,. \tag{2.54}$$

The Kolmogorov *backward* equation in the homogeneous case results in matrix form as

$$\dot{\mathbf{P}}(t) = \frac{\mathrm{d}\,\mathbf{P}(t)}{\mathrm{d}\,t} = \mathbf{Q}\mathbf{P}(t)\,. \tag{2.55}$$

As in the discrete-time case, often the *steady-state probability vector* of a CTMC is of primary interest. The required properties of the steady-state probability vector, which is also called the *equilibrium probability vector*, are equivalent to the discrete-time case. For all states $i \in S$, the steady-state probabilities π_i are:

1. Independent of time t

2. Independent of the initial state probability vector $\boldsymbol{\pi}(0)$

3. Strictly positive, $\pi_i > 0$

4. Given as the time limits, $\pi_i = \lim_{t \to \infty} \pi_i(t) = \lim_{t \to \infty} p_{ji}(t)$, of the state probabilities $\pi_i(t)$ and of the transition probabilities $p_{ji}(t)$, respectively

If existing for a given CTMC, the steady-state probabilities are independent of time, we immediately get

$$\lim_{t \to \infty} \frac{d\,\boldsymbol{\pi}(t)}{d\,t} = 0\,. \tag{2.56}$$

Under condition (2.56), the differential equation (2.51) for determining the unconditional state probabilities resolves to a much simpler *system of linear equations*:

$$0 = \sum_{i \in S} q_{ij}\pi_i\,, \quad \forall j \in S\,. \tag{2.57}$$

In vector–matrix form, we get accordingly

$$\mathbf{0} = \boldsymbol{\pi}\mathbf{Q}\,. \tag{2.58}$$

Definition 2.13   In analogy to the discrete-time case, a CTMC for which a unique steady-state probability vector exists is called an *ergodic* CTMC.

The strictly positive steady-state probabilities can be gained by the unique solution of Eq. (2.58), when an additional normalization condition is imposed.[9] To express it in vector form, we introduce the *unit vector* $\mathbf{1} = [1, 1, \ldots, 1]^T$ so that the following relation holds:

$$\boldsymbol{\pi}\mathbf{1} = \sum_{i \in S} \pi_i = 1\,. \tag{2.59}$$

Another possibility for determining the steady-state probabilities π_i for all states $i \in S$ of a CTMC, is to take advantage of a well-known relation between the π_i and the *mean recurrence time*[10] $M_i < \infty$, that is, the mean time elapsed between two successive visits of the CTMC to state i:

$$\pi_i = -\frac{1}{M_i q_{ii}}\,, \quad \forall i \in S\,. \tag{2.60}$$

In the time-homogeneous case, we can derive from Eq. (2.41) that for any $j \in S$ $q_{jj} = -\sum_{i, i \neq j} q_{ji}$ and from Eq. (2.57) we get $\sum_{i, i \neq j} q_{ij}\pi_i = -q_{jj}\pi_j$. Putting these together immediately yields the system of *global balance equations*:

$$\sum_{i, i \neq j} q_{ij}\pi_i = \pi_j \sum_{i, i \neq j} q_{ji}\,, \quad \forall j \in S\,. \tag{2.61}$$

---

[9] Note that besides the trivial solution $\pi_i = 0$, $\forall i \in S$, any vector, obtained by multiplying a solution of Eq. (2.58) by an arbitrary real-valued constant, would also yield a solution of Eq. (2.58).

[10] In contrast to DTMCs, where lowercase notation is used to refer to recurrence time, uppercase notation is used for CTMCs.

On the left-hand side of Eq. (2.61), the *total flow* from any other state $i \in S$ *into* state j is captured. On the right-hand side, the *total flow out* of state j into any other state i is summarized. The flows are *balanced* in steady state, i.e., they are in equilibrium.

The conditions under which a CTMC is called ergodic are similar to those for a DTMC. Therefore, we can briefly summarize the criteria for classifying CTMCs and for characterizing their states.

2.1.2.2.1 Classifications of CTMC

Definition 2.14   As for DTMCs, we call a CTMC *irreducible* if every state i is *reachable* from every other state j, where $i, j \in S$; that is, $\forall i, j, i \neq j, \exists t :$ $p_{ji}(t) > 0$. In other words, no proper subset $\hat{S} \subset S$, $\hat{S} \neq S$, of state space S exists, such that $\sum_{j \in \hat{S}} \sum_{i \in S \setminus \hat{S}} q_{ji} = 0$.

Definition 2.15   An irreducible, homogeneous CTMC is called *ergodic* if and only if the unique steady-state probability vector $\boldsymbol{\pi}$ exists.

As opposed to DTMCs, CTMCs cannot be periodic. Therefore, it can be shown that for an irreducible, homogeneous CTMC:

- The limits $\tilde{\pi}_i = \lim_{t \to \infty} \pi_i(t) = \lim_{t \to \infty} p_{ji}(t)$ exist $\forall i, j \in S$ and are independent of the initial probability vector $\boldsymbol{\pi}(0)$. [11]

- The steady-state probability vector $\boldsymbol{\pi}$, if existing, can be uniquely determined by the solution of the linear system of Eq. (2.58) constrained by normalization condition (2.59).

- A unique steady-state, or equilibrium, probability vector $\boldsymbol{\pi}$ exists, if the irreducible, homogeneous CTMC is finite.

- The mean recurrence times M_i are finite for all states $i \in S$, $M_i < \infty$, if the steady-state probability vector exists.

2.1.2.2.2 CTMC State Sojourn Times

We have already mentioned that the distribution of the state *sojourn times* of a homogeneous CTMC must have the memoryless property. Since the exponential distribution is the only continuous distribution with this property, the random variables denoting the sojourn times, or holding times, must be exponentially distributed. Note that the same is true for the random variable referred to as the *residual* state holding time, that is, the time remaining until the next state change occurs.[12] Furthermore, the means of the two random variables are equal to $1/(-q_{ii})$.

Let the random variable R_i denote either the sojourn time or the residual time in state i, then the CDF is given by

$$F_{Ri}(r) = 1 - e^{q_{ii}r}, \quad r \geq 0. \tag{2.62}$$

---

[11] The limits do not necessarily constitute a steady-state probability vector.
[12] The residual time is often referred to as the forward recurrence time.

The mean value of R_i, the mean sojourn time or the mean residual time, is given by

$$E[R_i] = -\frac{1}{q_{ii}}, \qquad (2.63)$$

where q_{ii} is defined in Eq. (2.40).

2.1.2.3 Recapitulation We have introduced Markov chains and indicated their modeling power. The most important feature of homogeneous Markov chains is their unique *memoryless* property that makes them remarkably attractive. Both continuous- and discrete-time Markov chains have been defined and their properties discussed.

The most important algorithms for computation of their state probabilities are discussed in following chapters. Different types of algorithms are related to different categories of Markov chains such as ergodic, absorbing, finite, or infinite chains. Furthermore, the algorithms can be divided into those applicable for computing the steady-state probabilities and those applicable for computing the time-dependent state probabilities. Others provide approximate solutions, often based on an implicit transformation of the state space. Typically, these methods fall into the categories of aggregation/disaggregation techniques. Note that this modeling approximation has to be discriminated from the mathematical properties of the core algorithms, which, in turn, can be numerically exact or approximate, independent of their relation to the underlying model. Typical examples include round-off errors in direct methods such as Gaussian elimination and convergence errors in iterative methods for the solution of linear systems.

2.2 PERFORMANCE MEASURES

We begin by introducing a simple example and then provide an introduction to Markov reward models as a means to obtain performance measures.

2.2.1 A Simple Example

As an example adapted from Heimann, Mittal, and Trivedi [HMT91], consider a multiprocessor system with n processor elements processing a given workload. Each processor is subject to failures with a *mean time to failure* (MTTF), $1/\gamma$. In case of a failure, recovery can successfully be performed with probability c. Typically, recovery takes a brief period of time with mean $1/\beta$. Sometimes, however, the system does not successfully recover from a processor failure and suffers from a more severe impact. In this case, we assume the system needs to be rebooted with longer average duration of $1/\alpha$. Probability c is called the *coverage factor* and is usually close to 1. Unsuccessful recovery is most commonly caused by error propagation when the effect of a failure is

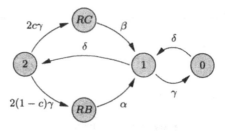

Parameter	Meaning
$1/\gamma$	mean time to failure
$1/\delta$	mean time to repair
$1/\alpha$	mean time to reboot
$1/\beta$	mean time to recover
c	coverage probability

State	Meaning
$i \in \{0,1,2\}$	i working processors
RC	recovery
RB	reboot

Fig. 2.6 A simple model of a multiprocessor system.

not sufficiently shielded from the rest of the system. Failed processors need to be repaired, with the *mean time to repair* (MTTR) of $1/\delta$. Only one processor can be repaired at a time and repair of one processor does not affect the proper working of the remaining processors. If no processor is running correctly, the whole system is out of service until first repair is completed. Neither reboot nor recovery is performed when the last processor fails. If all times are assumed to be independent, exponentially distributed random variables, then a CTMC can be used to model the scenario. In Fig. 2.6 an example is given for the case of $n = 2$ processors and state space $S = \{2, RC, RB, 1, 0\}$.

Since CTMC in Fig. 2.6 is ergodic, the unique steady-state probability vector $\boldsymbol{\pi} = (\pi_2, \pi_{RC}, \pi_{RB}, \pi_1, \pi_0)$ is the solution of Eqs. (2.58) and (2.59). From Fig. 2.6 the infinitesimal generator matrix can be derived:

$$\mathbf{Q} = \begin{pmatrix} -2\gamma & 2c\gamma & 2(1-c)\gamma & 0 & 0 \\ 0 & -\beta & 0 & \beta & 0 \\ 0 & 0 & -\alpha & \alpha & 0 \\ \delta & 0 & 0 & -(\gamma+\delta) & \gamma \\ 0 & 0 & 0 & \delta & -\delta \end{pmatrix}.$$

Clearly, the model in Fig. 2.6 is already an abstract representation of the system. With Eqs. (2.53) and (2.58), transient and steady-state probability vectors could be computed, but it is not yet clear what kind of measures should be calculated because we have said nothing about the application context and the corresponding system requirements. We take this model as a high-level description, which may need further elaboration so that a computational model can be generated from it. We consider four classes of system requirements for the example.

1. *System availability* is the probability of an adequate level of service, or, in other words, the long-term fraction of time of actually delivered service. Usually, short outages can be accepted, but interruptions of longer duration or accumulated outages exceeding a certain threshold may not be tolerable. Accordingly, the model in Fig. 2.6 must be evaluated with respect to requirements from the application context. First, tolerance thresholds must be specified as to what extent total outages can be accepted. Second, the states in the model must be partitioned into two sets: one set comprising the states where the system is considered "up," i.e., the service being actually delivered, and the complementary set comprising the states where the system is classified as "down." In our example, natural candidates for down states are in the set: $\{0, RC, RB\}$. But not all of them are necessarily classified as down states. Since reconfiguration generally takes a very short time, applications may well not be susceptible to such short interruptions. As a consequence, the less significant state RC could even be eliminated in the generation of the computational model. Finally, the measures to obtain from the computational model must be decided. For example, the transient probability of the system being in an up state at a certain time t conditioned on some initial state, the proportion of time being in one of the up states in a given finite time horizon $[0, T)$ conditioned on an initial state, or the long-term proportion of time being in up states are some of the desired measures.

2. *System reliability* is the probability of uninterrupted service exceeding a certain length of time. By definition, no interruption of service at all can be tolerated. But note that it still needs to be exactly specified as to what kind of event is to be considered as an interruption. In the most restrictive application context, reconfiguration (RC) might not be acceptable. In contrast, nothing prevents the assumption to be made that even a reboot (RB) can be tolerated and only the failure of the last component leads to the single down state (0). As a third alternative, reconfiguration may be tolerated but not reboot and not the failure of the last component. The three possible scenarios are captured in Fig. 2.7. The model structures have also been adapted by introducing absorbing down states that reflect the fact that down states are considered as representing catastrophic events.

3. *System performance* takes the capacity of different configurations into account. Typical measures to be calculated are the utilization of the resources or system throughput. Other measures of interest relate to the frequency with which certain incidents occur. With respect to the model in Fig. 2.6, the individual states need to be characterized by their contribution to a successful task completion. The higher the degree of parallelism, the higher the expected accomplishment will be. But it is a nontrivial task to find the right performance indices attributable to each

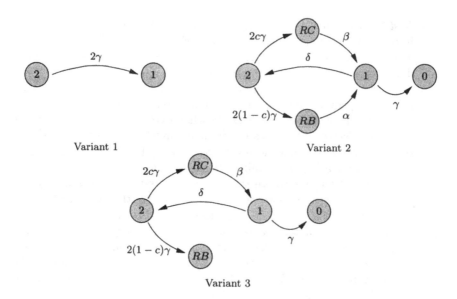

Variant 1   Variant 2

Variant 3

Fig. 2.7   Model variants with absorbing states capturing reliability requirements.

particular state in the computational model. The easiest way would be to assign a capacity proportional to the number of working processors. But because each such state represents a whole configuration, where each system resource and the imposed workload can have an impact on the overall performance, more accurate characterization is needed. One way would be to execute separate modeling studies for every possible configuration and to derive some more detailed measures such as state-dependent effective throughputs or response time percentiles. Another way is to expand the states and to replace some of them with a more detailed representation of the actual configuration and the workload. This approach will lead to a model of enormous size.

4. *Task completion* is reflected in the probability that a user will receive service at the required quality, or in other words, in the proportion of users being satisfied by the received service and its provided quality. Many different kinds of measures could be defined in this category. It could be the proportion of tasks being correctly processed or the probability that computation-time thresholds are not exceeded. With advanced applications such as continuous media (e.g., audio and video streams), measures are investigated relating the degree of user's satisfaction to the quality of the delivered service. Usually, such application-oriented measures are composed of different constituents such as timeliness, delay-variation, loss, and throughput measures.

In this subsection, we have indicated how system requirements affect model formulation. Furthermore, different types of performance measures were motivated and their relation to the model representation outlined. Next, we will show how the performance measures derived from system requirements can be explicitly specified and integrated into the representation of the computational model.

2.2.2 Markov Reward Models

MRMs provide a unifying framework for an integrated specification of model structure and system requirements. We consider the explicit specification of system requirements as an essential part of the computational model. Once the model structure has been defined so that the infinitesimal generator matrix is known, the basic equations can be written depending on the given system requirements and the structure of the matrix. For the sake of completeness, we repeat here the fundamental Eqs. (2.53), (2.58), and (2.59) for the analysis of CTMCs:

$$\frac{d\,\boldsymbol{\pi}(t)}{d\,t} = \boldsymbol{\pi}(t)\mathbf{Q}, \qquad \boldsymbol{\pi}(0) = \boldsymbol{\pi}_0,$$

$$\mathbf{0} = \boldsymbol{\pi}\mathbf{Q}, \qquad \boldsymbol{\pi}\mathbf{1} = 1,$$

where $\boldsymbol{\pi}(t)$ is the transient state probability vector of the CTMC and $\boldsymbol{\pi}$ is the steady-state probability vector (assuming it exists). In addition to the transient state probabilities, sometimes *cumulative* probabilities are of interest. Let

$$\mathbf{L}(t) = \int_0^t \boldsymbol{\pi}(u)\,d\,u, \qquad (2.64)$$

then $\mathbf{L}_i(t)$ denotes the *expected total time* the CTMC spends in state i during the interval $[0, t)$. A more convenient way to calculate the cumulative state probabilities is by solution of differential equation (2.65) [RST89]:

$$\frac{d\,\mathbf{L}(t)}{d\,t} = \mathbf{L}(t)\mathbf{Q} + \boldsymbol{\pi}(0), \quad \mathbf{L}(0) = \mathbf{0}. \qquad (2.65)$$

Closely related to the vector of cumulative state probabilities is the vector describing the time-average behavior of the CTMC [SmTr90]:

$$\mathbf{M}(t) = \frac{1}{t}\mathbf{L}(t). \qquad (2.66)$$

With $\mathbf{I}$ denoting the identity matrix, $\mathbf{M}(t)$ can be seen to satisfy the differential equation (2.67):

$$\frac{d\,\mathbf{M}(t)}{d\,t} = \mathbf{M}(t)\left(\mathbf{Q} - \frac{1}{t}\mathbf{I}\right) + \frac{1}{t}\boldsymbol{\pi}(0), \quad \mathbf{M}(0) = \mathbf{0}. \qquad (2.67)$$

With these definitions, most of the interesting measures can be defined. But the special case of models containing absorbing states, like those in Fig. 2.7, deserves additional attention. Here, it would be interesting to compute measures based on the time a CTMC spends in nonabsorbing states before an absorbing state is ultimately reached. Let the state space $S = A \cup N$ be partitioned into the set A of absorbing and the set N of nonabsorbing (or transient) states. Then, the time spent before absorption can be calculated by taking the limit $\lim_{t \to \infty} \mathbf{L}_N(t)$ restricted to the states of the set N. Note that, in general, unless very special cases for the initial probability vector are considered, the limit does not exist for the states in A. Therefore, to calculate $\mathbf{L}_N(\infty)$, the initially given infinitesimal generator matrix $\mathbf{Q}$ is restricted to those states in N, so that matrix $\mathbf{Q}_N$ of size $|N| \times |N|$ results. Note that $\mathbf{Q}_N$ is *not* an infinitesimal generator matrix. Restricting also the initial probability vector $\boldsymbol{\pi}(0)$ to the nonabsorbing states N results in $\boldsymbol{\pi}_N(0)$ and allows the computation of $\lim_{t \to \infty}$ on both side of differential equation (2.65) so that following linear equation (2.68) results [STP96]:

$$\mathbf{L}_N(\infty)\mathbf{Q}_N = -\boldsymbol{\pi}_N(0). \tag{2.68}$$

To give an example, consider the variant three in Fig. 2.7. Here the state space S is partitioned into the set of absorbing states $A = \{RB, 0\}$ and nonabsorbing states $N = \{2, RC, 1\}$ so that $\mathbf{Q}$ reduces to

$$\mathbf{Q}_N = \begin{pmatrix} -2\gamma & 2c\gamma & 0 \\ 0 & -\beta & \beta \\ \delta & 0 & -(\gamma + \delta) \end{pmatrix}.$$

With $\mathbf{L}_N(\infty)$, the mean time to absorption (MTTA) can then be written as

$$\text{MTTA} = \sum_{i \in N} L_i(\infty). \tag{2.69}$$

MRMs have long been used in Markov decision theory to assign cost and reward structures to states of Markov processes for an optimization [Howa71]. Meyer [Meye80] adopted MRMs to provide a framework for an integrated approach to performance and dependability characteristics. He coined the term *performability* to refer to measures characterizing the ability of fault-tolerant systems, that is, systems that are subject to component failures and that can perform certain tasks in the presence of failures.

With MRMs, rewards can be assigned to states or to transitions between states of a CTMC. In the former case, these rewards are referred to as *reward rates* and in the latter as *impulse rewards*. In this text we consider state-based rewards only.

The reward rates are defined based on the system requirements, be it availability-, reliability-, or task-oriented. Let the reward rate r_i be assigned to state $i \in S$. Then, a reward $r_i\tau_i$ is accrued during a sojourn of time τ_i

in state i. Let $\{X(t), t \geq 0\}$ denote a homogeneous finite-state CTMC with state space S. Then, the random variable

$$Z(t) = r_{X(t)} \tag{2.70}$$

refers to the *instantaneous reward rate of the MRM* at time t. Note the difference between reward rates r_i assigned to individual states i and the overall reward rate $Z(t)$ of the MRM characterizing the stochastic process as a whole. With the instantaneous reward rate of the CTMC defined as in Eq. (2.70), the *accumulated reward* $Y(t)$ in the finite time horizon $[0, t)$ is given by

$$Y(t) = \int_0^t Z(\tau) \, d\tau = \int_0^t r_{X(\tau)} \, d\tau. \tag{2.71}$$

For example, consider the sample paths of $X(t)$, $Z(t)$, and $Y(t)$ processes in Fig. 2.8 adapted from [SmTr90]. A simple three-state MRM is presented, consisting of a CTMC with infinitesimal generator matrix $\mathbf{Q}$, and the reward rate vector $\mathbf{r} = (3, 1, 0)$ assigning reward rates to the states 0, 1, and 2, respectively. Assuming an initial probability vector $\boldsymbol{\pi}(0) = (1, 0, 0)$, the process is initiated in state 0, that is, $X(0) = 0$. Since the sojourn time of the first visit to state 0 is given by t_1, the reward $y_1 = 3t_1$ is accumulated during this period. After transition to state 1, additional reward $y_2 - y_1 = 1(t_2 - t_1)$ is earned, and so forth. While process $X(t)$ is assumed to follow the state transition pattern as shown in Fig. 2.8, the processes $Z(t)$ and $Y(t)$ necessarily show the behavior as indicated, because they are depending on $X(t)$ through the given reward vector $\mathbf{r}$. While $X(t)$ and $Z(t)$ are discrete valued and non-monotonic functions, $Y(t)$, in contrast, is a continuous-valued, monotonically nondecreasing function.

Based on the definitions of $X(t)$, $Z(t)$, and $Y(t)$, which are non-independent random variables, various measures can be defined. The most general measure is referred to as the *performability* [Meye80]:

$$\Psi(y, t) = P[Y(t) \leq y], \tag{2.72}$$

where $\Psi(y, t)$ is the distribution of the accumulated reward over a finite time $[0, t)$. Unfortunately, the performability is difficult to compute for unrestricted models and reward structures. Smaller models can be analyzed via double Laplace transform [SmTr90], while references to other algorithms are summarized in Table 2.17. The same mathematical difficulties arise if the distribution $\Phi(y, t)$ of the *time-average accumulated reward* is to be computed:

$$\Phi(y, t) = P\left[\frac{1}{t}Y(t) \leq y\right]. \tag{2.73}$$

The problem is considerably simplified if the system requirements are limited to expectations and other moments of random variables rather than distribution functions of cumulative rewards. Alternatively, efficient solution

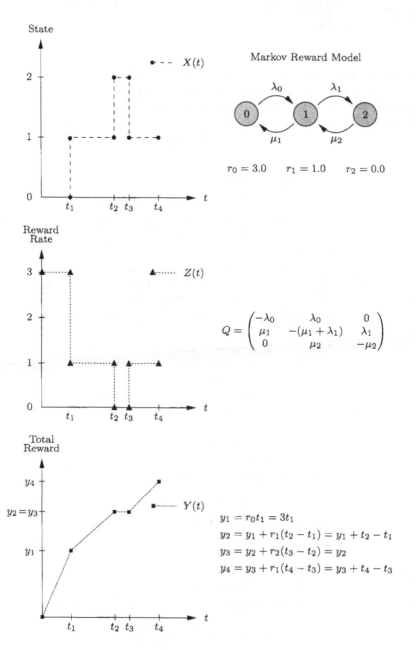

Fig. 2.8 A three-state Markov reward model with sample paths of the $X(t)$, $Z(t)$, and $Y(t)$ processes.

algorithms are available if the rewards can be limited to a binary structure or if the model structure is acyclic.

The *expected instantaneous reward rate* can be computed from the solution of differential equation (2.53):

$$E[Z(t)] = \sum_{i \in S} r_i \pi_i(t) \,. \tag{2.74}$$

If the underlying model is ergodic, the solution of linear equations (2.58) and (2.59) can be used to calculate the *expected reward rate in the limit* as $t \to \infty$:

$$\lim_{t \to \infty} \left(\frac{1}{t} E[Y(t)] \right) = E[Z(\infty)] = E[Z] = \sum_{i \in S} r_i \pi_i \,. \tag{2.75}$$

To compute the first moment of the performability, or the *expected accumulated reward*, advantage can be taken of the solution of differential equation (2.65):

$$E[Y(t)] = \sum_{i \in S} r_i L_i(t) \,. \tag{2.76}$$

For models with absorbing states, the limit as $t \to \infty$ of the expected accumulated reward exists and is called the *expected accumulated reward until absorption*. It follows from the solution of linear equation (2.68) that

$$E[Y(\infty)] = \sum_{i \in N} r_i L_i(\infty) \,. \tag{2.77}$$

Also, higher moments of the random variables can be derived. As an example, the *variance* of the instantaneous reward rate $Z(t)$ is obtained from

$$\mathrm{var}[Z(t)] = \sum_{i \in S} r_i^2 \pi_i(t) - \left(\sum_{i \in S} r_i \pi_i(t) \right)^2 \,. \tag{2.78}$$

Probabilities of $Z(t)$ and Z can also be easily written [TMWH92]:

$$P[Z(t) = x] = \sum_{i \in S, r_i = x} \pi_i(t) \,,$$

$$P[Z = x] = \sum_{i \in S, r_i = x} \pi_i \,.$$

Two different distribution functions can also be easily derived. With the transient state probabilities gained from the solution of Eq. (2.53), the probability that the instantaneous reward rate $Z(t)$ does not exceed a certain level x is given as

$$P[Z(t) \le x] = \sum_{i \in S, r_i \le x,} \pi_i(t) \,. \tag{2.79}$$

The distribution $P[Y(\infty) \leq y]$ of the accumulated reward until absorption can be derived by applying a substitution according to Beaudry [Beau78] and Ciardo et al. [CMST90]. The essential idea is to generate a (computational) model with a *binary* reward structure and infinitesimal generator matrix $\hat{\mathbf{Q}}$ from the given (high-level) CTMC with a general reward structure and infinitesimal generator matrix $\mathbf{Q}$. This transformation is achieved by applying a normalization in the time domain according to the reward rates so that the distribution of the time until absorption of the new CTMC can be used as a substitute measure:

$$P[Y(\infty) \leq y] = 1 - \sum_{i \in N} \hat{\pi}_i(y). \tag{2.80}$$

Returning to the example model in Fig. 2.6, we apply the MRM framework with respect to different system requirements to demonstrate how the model can be instantiated in several ways to yield meaningful computational models. The following case study [HMT91] serves as a guideline as to how to describe models in light of different system requirements, such as availability and reliability or task-oriented and system performance.

2.2.3   A Case Study

We now illustrate the use of MRMs by means of the example in Fig. 2.6.

2.2.3.1   System Availability   Availability measures are based on a *binary* reward structure. Assuming for the model in Fig. 2.6 that one processor is sufficient for the system to be "up" (set of states $U = \{2, 1\}$) and that otherwise it is considered as being down (set of states $D = \{RC, RB, 0\}$, where $S = U \cup D$), a reward rate 1 is attached to the states in U and a reward rate 0 to those in D. The resulting reward function r is summarized in Table 2.1.

The instantaneous availability is given by

$$A(t) = E[Z(t)] = \sum_{i \in S} r_i \pi_i(t) = \sum_{i \in U} \pi_i(t) = \pi_1(t) + \pi_2(t).$$

Unavailability can be calculated with a reverse reward assignment to that for availability. The instantaneous unavailability, therefore, is given by

$$UA(t) = E[Z(t)] = \sum_{i \in S} r_i \pi_i(t) = \sum_{i \in D} \pi_i(t) = \pi_{RC}(t) + \pi_{RB}(t) + \pi_0(t).$$

Steady-state availability, on the other hand, is given by

$$A = E[Z] = \sum_{i \in S} r_i \pi_i = \sum_{i \in U} \pi_i = \pi_2 + \pi_1.$$

Table 2.1 Reward assignment for computation of availability	
State i	Reward Rate r_i
2	1
RC	0
RB	0
1	1
0	0

Table 2.2 Reward assignment for mean uptimes	
State i	Reward Rate r_i
2	t
RC	0
RB	0
1	t
0	0

The interval availability provides a time average value:

$$\bar{A}(t) = \frac{1}{t} E[Y(t)] = \frac{1}{t} \int_0^t A(x)\, \mathrm{d}x = \frac{1}{t} \sum_{i \in U} \int_0^t \pi_i(x)\, \mathrm{d}x = \frac{1}{t}[L_2(t) + L_1(t)].$$

Note that a different classification of the model states would lead to other values of the availability measures. Instantaneous, interval, and steady-state availabilities are the fundamental measures to be applied in this context, but there are related measures that do not rely on the binary reward structure. Other measures related to availability can be considered. To compute the mean transient uptime $T^{(1)}(t)$ and the mean steady-state uptime $T^{(2)}(t)$ in a given time horizon $[0, t)$, reward rates 1 in Table 2.1 are replaced by reward rates t in Table 2.2:

$$T^{(1)}(t) = \frac{1}{t} E[Y(t)] = \frac{1}{t} t \sum_{i \in U} L_i(t) = \sum_{i \in U} L_i(t) = t\bar{A}(t),$$

$$T^{(2)}(t) = E[Z] = tA.$$

Very important measures relate to the *frequency* of certain events of interest. Any event represented in the model can be captured here. For example, we could be interested in the average number of repair calls made in a given period of time $[0, t)$. With assumed repair rate δ and the fact that repair calls are made from states 0 and 1 of the model of Fig. 2.6, the transient average number of repair calls $N^{(1)}(t)$ and steady-state average number of repair calls $N^{(2)}(t)$ can again be determined, with the reward function as defined in Table 2.3:

$$N^{(1)}(t) = \frac{1}{t} E[Y(t)] = \delta(L_0(t) + L_1(t)),$$

$$N^{(2)}(t) = E[Z] = t\delta(\pi_0 + \pi_1).$$

Sometimes, local tolerance levels $\tau_i, i \in S$ are considered such that certain events, such as outages, can be tolerated as long as they do not last longer than a specified threshold τ_i. In our context it would be natural to assume a tolerance with respect to duration of reconfiguration and possibly also with respect to reboot. Based on these assumptions, the reward function as shown

Table 2.3   Reward assignment for number of repair calls in $[0, t)$

State i	Reward Rate r_i
2	0
RC	0
RB	0
1	$t\delta$
0	$t\delta$

Table 2.4   Reward assignment for computation of mean number of outages exceeding certain state-dependent thresholds in finite time $[0, t)$ with the CTMC from Fig 2.6

State i	Reward Rate r_i
2	0
RC	$t\beta\,e^{-\beta\tau_{RC}}$
RB	$t\alpha\,e^{-\alpha\tau_{RB}}$
1	0
0	$t\delta\,e^{-\delta\tau_0}$

in Table 2.4 is defined to compute the mean number of *severe* interruptions $I^{(1)}(t), I^{(2)}(t)$ in finite time $[0, t)$, i.e., interruptions lasting longer than certain thresholds for the transient and the steady-state based cases, respectively:

$$I^{(1)}(t) = \frac{1}{t}E[Y(t)] = \beta\,e^{-\beta\tau_{RC}}\,L_{RC}(t) + \alpha\,e^{-\alpha\tau_{RB}}\,L_{RB}(t) + \delta\,e^{-\delta\tau_0}\,L_0(t),$$

$$I^{(2)}(t) = E[Z] = t\beta\,e^{-\beta\tau_{RC}}\,\pi_{RC} + t\alpha\,e^{-\alpha\tau_{RB}}\,\pi_{RB} + t\delta\,e^{-\delta\tau_0}\,\pi_0.$$

The reward rates in Table 2.4 are specified based on the observation that event duration times are exponentially distributed in CTMCs, such as the one depicted in Fig. 2.6. Therefore, if the parameter were given by λ, the probability of event duration lasting longer than τ_i units of time would be $e^{-\lambda\tau_i}$. This probability is used to weight the average number of events $t\lambda$. But note that the reward rates are by *no* means limited to exponential distributions, any known distributions could be applied here. It is useful to specify events related to timeouts or events that have a (deterministic) deadline attached to them.

2.2.3.2 System Reliability

Again, a *binary* reward function r is defined that assigns reward rates 1 to up states and reward rates 0 to (absorbing) down states, as given in Table 2.5. Recalling that reliability is the likelihood that an unwanted event has not yet occurred since the beginning of the system operation, reliability can be expressed as

$$R(t) = P[T > t] = P[Z(t) = 1] = 1 - P[Z(t) \le 0] = E[Z(t)],$$

where the random variable T characterizes the time to the next occurrence of such an unwanted (failure) event. Referring to the scenarios depicted in Fig. 2.7, three different reliability functions can be identified:

$$R_1(t) = \pi_2(t),$$
$$R_2(t) = \pi_2(t) + \pi_{RC}(t) + \pi_{RB}(t) + \pi_1(t),$$
$$R_3(t) = \pi_2(t) + \pi_{RC}(t) + \pi_1(t).$$

Remember that $R_1(t), R_2(t)$, and $R_3(t)$ are computed on the basis of three different computational models. The function to be used depends on the application requirements.

With a known reliability function $R(t)$, the mean time to the occurrence of an unwanted (failure) event is given by

$$E[T] = \text{MTTF} = \text{MTTA} = \int_0^\infty R(t)\,\mathrm{d}t.$$

MTTF and MTTA are acronyms for *mean time to failure* and *mean time to absorption*, respectively. With the formula of Eq. (2.69), the MTTF (MTTA) can be efficiently computed.

Table 2.5   Reward assignment for computation of reliability with variant-2 CTMC from Fig 2.7

State i	Reward Rate r_i
2	1
RC	1
RB	1
1	1
0	0

Table 2.6   Reward assignment for predicting the number of catastrophic incidents with variant-2 CTMC from Fig 2.7

State i	Reward Rate r_i
2	0
RC	0
RB	0
1	$t\gamma$
0	0

With the reliability $R(t) = E[Z(t)]$ given, the unreliability follows as the complement:

$$UR(t) = 1 - E[Z(t)] = 1 - R(t) = 1 - P[T > t] = P[T \le t].$$

It is an important observation that the unreliability is given as a *distribution function* of a time variable. Note that we have gained this result simply by computing $E[Z(t)]$. The trick was to modify the computational model in the right way such that absorbing states were introduced and the appropriate reward structure was chosen, as the one depicted in Table 2.5. Rather general time distribution functions can be effectively computed in this way. This method allows the calculation of *percentiles* and other powerful performance measures like response-time distribution functions, which are not necessarily of exponential type [MMKT94]. It is also worth mentioning that the unreliability could be calculated based on a reward assignment complementing the one in Table 2.5. In this case, we would get $UR(t) = E[Z(t)] = \pi_0(t)$.

Related to reliability measures, we would often be interested in knowing the expected number of catastrophic events $C(t)$ to occur in a given time interval $[0, t)$. To this end

$$C(t) = \frac{1}{t}E[Y(t)] = \gamma L_1(t)$$

needs to be calculated for some interesting initial up state, $i \in U$ with the reward assignment, as given in Table 2.6.

2.2.3.3 System and Task Performance

In the preceding two subsections, where availability and reliability measures were considered, binary reward functions were of predominant importance. However, other reward functions have already been used to model availability- and reliability-oriented measures such as frequencies of outages or other related incidents of interest. A more general view is presented in this section.

First, we look at a particular task with some assumed task execution time x. Under the given circumstances, we would like to know the probability of successful completion of this task. Two different aspects come into the picture here: the dynamics of the system and the susceptibility of the task (or its user) to these dynamic changes in the configuration. In the case of the example in Fig 2.6, just outages and their impact on the task completion are considered. But this situation can be generalized to different degrees of the task's susceptibility to loss of data or other disturbing events as well.

In the case of a user requiring uninterrupted service for some time x, the probabilities that *no* interruption occurs in the respective states are assigned as reward rates. We allow task requirement to be state dependent as well so that in state 2 the task requires x_2 time units to complete and likewise for state 1. The probabilities can be easily concluded from Fig. 2.6. In state 2, for example, the probability $P[O > x_2]$ that the system will be operational for more than x_2 units of time is given by $e^{-2\gamma x_2}$. Therefore, the task interruption probability is

$$P[O \le x_2] = 1 - P[O > x_2] = 1 - e^{-2\gamma x_2} \,,$$

conditioned on the assumption the system is running in state 2 and on the hypothesis that the first failure causes a task's interruption. These probabilities can be assigned as reward rates as shown in Table 2.7, so that the interruption probability $IP^{(1)}(x_1, x_2)$ of a task of length x_1, x_2, respectively is computed with $E[Z]$, when assuming that probability of a user finding the system initially being up is given by the steady-state probabilities:

$$IP^{(1)}(x_1, x_2) = E[Z] = (1 - e^{-2\gamma x_2})\pi_2 + (1 - e^{-\gamma x_1})\pi_1 \,.$$

Another scenario would be that a running task is constrained to using a single processing element, even if two were available. In this case, a task would only be susceptible to a failure of its own required processor or if an uncovered failure occurs that also has severe impacts on the currently running one. With the reward assignment as shown in Table 2.8 the desired interruption probability $IP^{(2)}$ is obtained:

$$IP^{(2)}(x_1, x_2) = E[Z] = (1 - e^{-(\gamma(1-c)+\gamma)x_2})\pi_2 + (1 - e^{-\gamma x_1})\pi_1 \,.$$

Recall also the reward function as defined in Table 2.4. Based on these reward rates, it is also possible to compute the mean number of *severe* inter-

Table 2.7   Reward assignment for computing a task's interruption probability with duration x for the CTMC from Fig 2.6

State i	Reward Rate r_i
2	$1 - e^{-2\gamma x_2}$
RC	0
RB	0
1	$1 - e^{-\gamma x_1}$
0	0

Table 2.8   Reward assignment for computing a task's interruption probability in case of uncovered error or loss of required processor for the CTMC from Fig 2.6

State i	Reward Rate $r(i)$
2	$1 - e^{-(\gamma(1-c)+\gamma)x_2}$
RC	0
RB	0
1	$1 - e^{-\gamma x_1}$
0	0

ruptions in a finite time duration. These measures can also be applied as success measures for running a task with finite execution time x and with certain interruption tolerance patterns attached to it. In particular, continuous media exhibit such a kind of susceptibility profile. Interruptions or losses can be tolerated up to a certain threshold. Beyond that threshold, a severe degradation in quality is perceived by users.

Referring back to the example in Fig.2.6, it could be of interest to take into account different levels of performance that the modeled system may exhibit in various states. One simple method is to use the processing power (proportional to the number of processing elements actually working). This would lead to the reward function depicted in Table 2.9. With this reward assignment, the expected available computing capacity in steady state $E[Z]$, the expected instantaneous capacity at time t, $E[Z(t)]$, or the cumulative capacity deliverable in finite time $[0, t)$, $E[Y(t)]$, or its time average $\frac{1}{t}E[Y(t)]$ could be computed. These availability measures are known as capacity-oriented availability. But this reward assignment is too coarse to capture performance effects accurately. First, it is overly optimistic with respect to the actual performance and, second, effects due to outages tend to be masked by the coarse reward structure and tend to become insignificant as a consequence. From this point of view, either the reward function or the model structure are ill-posed with respect to the system requirements.

More sophisticated approaches either take into account the user's requirements related to the tasks to be processed or incorporate more information about the investigated system into the model. A hierarchical method is possible, where every working configuration is modeled separately with some (queueing) models so that more interesting parameters can be derived. Trivedi et al. [TMWH92], for example, use a finite queueing system that is subject to losses due to buffer overflow and to response time deadline violation [TSIH90], while de Meer et al. [MTD94] apply infinite queueing systems with limits imposed on the system's response time for processing the tasks, so that losses occur if a deadline is missed. In both cases, closed-form solutions are

Table 2.9   Reward assignment
for computation of capacity oriented availability measures for
the CTMC from Fig 2.6

State i	Reward Rate r_i
2	2
RC	0
RB	0
1	1
0	0

Table 2.10   Reward assignment
for computing the total loss
probability for the CTMC from
Fig 2.6

State i	Reward Rate $r(i)$
2	l_2
RC	1
RB	1
1	l_1
0	1

available to calculate the loss probabilities l_i for all working configurations or
up states $i \in U$ [GrHa85]. Of course, knowledge of the processing power of the
working components and of the traffic characteristics are needed to parameterize the underlying performance models. Many of the details are given in
later chapters when appropriate.

Table 2.11   Reward assignment
for computing the normalized
throughput for the CTMC from
Fig 2.6

State i	Reward Rate $r(i)$
2	$1 - l_2$
RC	0
RB	0
1	$1 - l_1$
0	0

Table 2.12   Reward assignment
for computing of the expected
number of tasks lost in finite time
for the CTMC from Fig 2.6

State i	Reward Rate r_i
2	$l_2 \lambda$
RC	λ
RB	λ
1	$l_1 \lambda$
0	λ

The values l_i are used to characterize the percentage loss of tasks arriving
at the system in state $i \in U$. Consequently, tasks arriving in down states
are lost with a probability of 1.0, that is, $l_i = 1, i \in D$. Furthermore, it is
assumed that (quasi) steady state in the performance model is maintained in
all configuration states, such as those depicted in Fig. 2.6. In other words,
tasks arriving at the system in a certain state are assumed to leave the system
in the same state. This assumption is admissible because the events related
to traffic effects usually occur in orders of magnitude faster than events related to changes in configurations (due to failure-repair events). The resulting
reward assignment is summarized in Table 2.10. Of course, a complementary
reward structure as in Table 2.11 could also be chosen, so that the normalized
throughput would be calculated as has been done by Meyer [Meye80].

The expected total loss probability, TLP, in steady state and the transient expected total loss probability $TLP(t)$ are then given by:

$$TLP(t) = E[Z(t)] = \sum_{i \in U} l_i \pi_i(t) + \sum_{i \in D} 1 \pi_i(t)$$

$$= l_2 \pi_2(t) + l_1 \pi_1(t) + \pi_{RC}(t) + \pi_{RB}(t) + \pi_0(t),$$

$$TLP = E[Z] = \sum_{i \in U} l_i \pi_i + \sum_{i \in D} 1 \pi_i$$

$$= l_2 \pi_2 + l_1 \pi_1 + \pi_{RC} + \pi_{RB} + \pi_0.$$

To compute the expected number of tasks lost in finite time $[0, t)$, two choices are available: the interval-based measure $TL^{(1)}(t)$ or the steady-state-based measure $TL^{(2)}(t)$ can be applied. In the first case, the reward rates in Table 2.10 are multiplied by the tasks' arrival rate λ to yield Table 2.12 and to compute $E[Y(t)]$. In the second case, $E[Z]$ is calculated on the basis of a modified reward assignment. All reward rates from Table 2.12 need to be multiplied by the length of the investigated time horizon $[0, t)$ to get Table 2.13. Of course, the same reward assignment could be used for the interval-based case and $\frac{1}{t} E[Y(t)]$ could be evaluated instead:

$$TL^{(1)}(t) = E[Y(t)] = \int_0^t \left(\sum_{i \in U} l_i \lambda \pi_i(\tau) + \sum_{i \in D} 1 \lambda \pi_i(\tau) \right) d\tau$$

$$= \sum_{i \in U} \lambda l_i L_i(t) + \sum_{i \in D} \lambda L_i(t)$$

$$= \lambda \left(l_2 L_2(t) + l_1 L_1(t) + L_{RC}(t) + L_{RB}(t) + L_0(t) \right),$$

$$TL^{(2)}(t) = E[Z] = t\lambda \left(\sum_{i \in U} l_i \pi_i + \sum_{i \in D} 1 \pi_i \right)$$

$$= t\lambda \left(l_2 \pi_2 + l_1 \pi_1 + \pi_{RC} + \pi_{RB} + \pi_0 \right).$$

Response time distributions and percentiles of more complicated systems can rarely be computed by simple elementary queueing systems, and closed-form solutions are rarely available. Muppala et al. have shown how CTMCs with absorbing states (or their equivalent SRNs) can be exploited to compute these measures numerically in rather general cases [MuTr91, MMKT94]. Once the probabilities that the response time exceeds a certain threshold have been derived with this method, the reward assignments can be made corresponding to Tables 2.10, 2.12, or 2.13 and the appropriate measures can be derived.

The use of reward rates is not restricted to reliability, availability, and performability models. This concept can also be used in pure (failure-free) performance models to conveniently describe performance measures of interest. In many computer performance studies, expected throughput, mean response

Table 2.13   Reward assignment
for computing the expected total
number of tasks lost in finite time
for the CTMC from Fig 2.6

State i	Reward Rate r_i
2	$tl_2\lambda$
RC	$t\lambda$
RB	$t\lambda$
1	$tl_1\lambda$
0	$t\lambda$

Table 2.14   Reward assignment for throughput computation with the CTMC of Fig 2.9

State i	Reward Rate r_i
3	μ
2	μ
1	μ
0	0

time, or utilization are the most important measures. These measures can easily be specified by means of an appropriate reward function. Throughput characterization, for example, can be achieved by assigning the state transition rate corresponding to departure from a queue (service completion) as reward rates to the state where the transition originates. Mean response time of queueing systems can be represented indirectly and computed in two steps. First, by assigning the number of customers present in a state as the reward rate to that state and to compute the measures as required. Second, it can be proven that there exists a linear correspondence between mean response time and mean number of customers present so that Little's theorem [Litt61, King90] can be used for the computation of the mean response times from the mean number of customers. Finally, it is worth mentioning that the utilization is also based on a binary reward assignment. If a particular resource is occupied in a given state, reward rate 1 is assigned, otherwise reward rate 0 indicates the idleness of the resources.

To illustrate these measures, reward definitions for a simple CTMC are considered. Imagine customers arriving at a system with exponentially distributed interarrival times with mean $1/\lambda$. In the system they compete for service from a single server station. Since service is exclusively received by each customer, if more than one customer is in the system at the same time, the others have to wait in line until their turn comes. Service times are independent exponentially distributed with mean $1/\mu$. To keep the example simple, we limit the maximum number of customers in the system to three. The system is described by the MRM shown in Fig. 2.9.

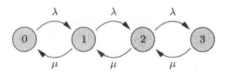

Fig. 2.9   A simple CTMC with arrivals and services.

Table 2.15   Reward assignment for computing the mean number of customers with the CTMC of Fig 2.9

State i	Reward Rate r_i
3	3
2	2
1	1
0	0

Table 2.16   Reward assignment for computing utilization measures with the CTMC of Fig 2.9

State i	Reward Rate r_i
3	1
2	1
1	1
0	0

Every state in $S = \{3, 2, 1, 0\}$ of the MRM represents the number of customers in the system. A state transition occurs if a customer's service is completed by the service station (arc annotated with μ) or if a new customer arrives (arc annotated with λ). Tables 2.14, 2.15, and 2.16 summarize the reward assignments for throughput, mean number of customers, and utilization measure computations, respectively. The formulae for the computation of throughput measures $\lambda^{(1)}(t)$, $\lambda^{(2)}(t)$, $\lambda^{(3)}$; the mean number of customers $\overline{K^{(1)}}(t)$, $\overline{K^{(2)}}(t)$, $\overline{K^{(3)}}$; mean response time measures $\overline{T^{(1)}}(t)$, $\overline{T^{(2)}}(t)$, $\overline{T^{(3)}}$; and utilization measures $\rho^{(1)}(t)$, $\rho^{(2)}(t)$, $\rho^{(3)}$ are also given. Note that response time measures $\overline{T^{(i)}}$ can be calculated with the help of Little's theorem [Litt61]:

$$\overline{T^{(i)}} = \frac{1}{\lambda}\overline{K^{(i)}}. \tag{2.81}$$

$$\lambda^{(1)}(t) = E[Y(t)] = \int_0^t \left(\sum_i r_i \pi_i(\tau) \right) d\tau = \sum_i r_i L_i(t)$$
$$= \mu \left(L_3(t) + L_2(t) + L_1(t) \right),$$

$$\lambda^{(2)}(t) = E[Z(t)] = \sum_i r_i \pi_i(t) = \mu \left(\pi_3(t) + \pi_2(t) + \pi_1(t) \right),$$

$$\lambda^{(3)} = E[Z] = \sum_i r_i \pi_i = \mu \left(\pi_3 + \pi_2 + \pi_1 \right).$$

$$\overline{K^{(1)}}(t) = E[Z(t)] = \sum_i r_i \pi_i(t) = 3\pi_3(t) + 2\pi_2(t) + 1\pi_1(t),$$

$$\overline{T^{(1)}}(t) = \frac{1}{\lambda}\overline{K^{(1)}}(t) = \frac{1}{\lambda}\left(3\pi_3(t) + 2\pi_2(t) + 1\pi_1(t)\right),$$

$$\overline{K^{(2)}}(t) = \frac{1}{t}E[Y(t)] = \frac{1}{t}\int\limits_0^t \left(\sum_i r_i\pi_i\right)d\tau = \frac{1}{t}\sum_i r_iL_i(t)$$

$$= \frac{1}{t}\left(3L_3(t) + 2L_2(t) + 1L_1(t)\right),$$

$$\overline{T^{(2)}}(t) = \frac{1}{\lambda}\overline{K^{(2)}}(t) = \frac{1}{\lambda t}\left(3L_3(t) + 2L_2(t) + 1L_1(t)\right),$$

$$\overline{K^{(3)}} = E[Z] = \sum_i r_i\pi_i = 3\pi_3 + 2\pi_2 + 1\pi_1,$$

$$\overline{T^{(3)}} = \frac{1}{\lambda}\overline{K^{(3)}} = \frac{1}{\lambda}\left(3\pi_3 + 2\pi_2 + 1\pi_1\right).$$

$$\rho^{(1)}(t) = E[Z(t)] = \sum_i r_i\pi_i(t) = \pi_3(t) + \pi_2(t) + \pi_1(t),$$

$$\rho^{(2)}(t) = \frac{1}{t}E[Y(t)] = \frac{1}{t}\int\limits_0^t \left(\sum_i r_i\pi_i(\tau)\right)d\tau = \frac{1}{t}\sum_i r_iL_i(t)$$

$$= \frac{1}{t}\left(L_3(t) + L_2(t) + L_1(t)\right),$$

$$\rho^{(3)} = E[Z] = \sum_i r_i\pi_i = \pi_3 + \pi_2 + \pi_1.$$

More MRM examples with all these types of reward assignments can be found in [MTBH93] and [MuTr92], as well as in [TMWH92], where multiprocessor architectures with different interconnection techniques are studied by defining corresponding reward functions. As another example, queues with breakdown are investigated.

Tables 2.17, 2.18, 2.19, and 2.20 summarize the different reward assignments discussed in this section.

2.3   GENERATION METHODS

We have already pointed out the importance of model generation. There are many reasons for the separation of high-level model description and lower-level computational model. For one thing, it will allow the modeler to focus more on the system being modeled rather than on low-level modeling details. Recall

Table 2.17 Overview of important MRM measures

Measure	Formula	Literature
$E[Z]$	$\sum\limits_{i \in S} r_i \pi_i$	[GaKe79], [Kuba86], [MuTr91], [MTBH93], [TSIH90]
$E[Z(t)]$	$\sum\limits_{i \in S} r_i \pi_i(t)$	[BRT88], [DaBh85], [HMT91], [MuTr91], [NaGa88]
$E[Y(t)]$	$\sum\limits_{i \in S} r_i L_i(t)$	[BRT88], [NaGa88], [MTD94], [MTBH93]
$E[Y(\infty)]$	$\sum\limits_{i \in S} r_i L_i(\infty)$	[TMWH92]
$P[Z(t) = x]$	$\sum\limits_{r_i = x, i \in S} \pi_i(t)$	[Husl81], [LeWi88], [Wu82]
$P[Z(t) \le x]$	$\sum\limits_{r_i \le x, i \in S} \pi_i(t)$	[SoGa89]
$P[Y(\infty) \le y]$	$1 - \sum\limits_{i \in N} \hat{\pi}_i(y)$	[Beau78], [CMST90]
$P[Y(t) \le y]$	$(sI + uR - Q)\tilde{\Psi}^*(u, s) = e$ (for small models)	[STR88], [GoTa87], [CiGr87], [SoGa89], [DoIy87], [Meye82]

that CTMCs and similar other state-space based models tend to become very large when real-world problems are studied. Therefore, it is attractive to be able to specify such large models in a compact way and to avoid the error-prone and tedious creation of models manually. The compact description is based on the identification of regularities or repetitive structures in the system model. Besides reducing the size of the description, such abstractions provide visual and conceptual clarity.

It is worth repeating that regularities must be present in order to take advantage of automatic model generation techniques. If repetitive structures cannot be identified, the "higher-level" representation would hardly be smaller and may even be more cumbersome. Furthermore, since software tools, hardware resources, human labor, and time are necessary to install and run such an automatic generation facility, its application is recommended only beyond a certain model size. Finally, the structure of the high-level description language generally makes it more suited for a specific domain of applications. As an example, effects of resource sharing and queueing are naturally represented by means of queueing systems, while synchronization can be easily specified with some Petri net variant.

Many high-level description languages exist, each with its own merits and limitations. A comprehensive review of specification techniques for MRMs can be found in [HaTr93]. In the remainder of this chapter, we discuss two types of stochastic Petri nets: generalized stochastic Petri nets (GSPNs) and stochastic reward nets (SRNs). We also discuss algorithms to convert a GSPN into a CTMC and an SRN into an MRM.

Table 2.18   Use of reward rates to compute different measures – part I

Requirement	Reward Assignments	Measure
Availability [HMT91], [MTD94], [MCT94], [DaBh85]	$r_i = \begin{cases} 1 & \text{if } i \in U \\ 0 & \text{else} \end{cases}$ [Table 2.1]	$E[Z]$(steady state) $E[Z(t)] = P[Z(t) = 1]$ (instantaneous) $\frac{1}{t}E[Y(t)]$(interval) $E[Y(\infty)]^*$, $P[Y(\infty) \leq y]^*$ $P[\frac{1}{t}Y(t) \leq y]$
Unavailability [HMT91], [MTD94], [MCT94]	$r_i = \begin{cases} 1 & \text{if } i \in D \\ 0 & \text{else} \end{cases}$	$E[Z]$(steady state) $E[Z(t)] = P[Z(t) = 1]$ (instantaneous) $\frac{1}{t}E[Y(t)]$(interval) $P[\frac{1}{t}Y(t) \leq y]$
Mean uptime in $[0, t)$	$r_i = \begin{cases} t & \text{if } i \in U \\ 0 & \text{else} \end{cases}$ [Table 2.2]	$E[Z]$ (approx.) $\frac{1}{t}E[Y(t)]$
Mean uptime in $[0, t)$	$r_i = \begin{cases} 1 & \text{if } i \in U \\ 0 & \text{else} \end{cases}$ [Table 2.1]	$E[Y(t)]$ $E[Y(\infty)]^*$, $P[Y(\infty) \leq y]^*$
Approx. frequency of repair calls in $[0, t)$ with mean duration $\frac{1}{\delta}$ [HMT91]	$r_i = \begin{cases} t\delta & \text{if } i \in S_R \\ 0 & \text{else} \end{cases}$ (S_R: states where repair is needed) [Table 2.3]	$E[Z]$
Frequency of repair calls in $[0, t)$ with mean duration $\frac{1}{\delta}$ [HMT91]	$r_i = \begin{cases} \delta & \text{if } i \in S_R \\ 0 & \text{else} \end{cases}$ (S_R: states where repair is needed) [Table 2.3]	$E[Y(t)]$
Mean percentage of severe interruptions [HMT91]	$r_i = \begin{cases} P[T > \tau_i] & \text{if } i \in U \\ 0 & \text{else} \end{cases}$ ($\tau_i \geq 0$: threshold for state i, T: duration of outage) [Table 2.4]	$E[Y(t)]$

*The measure is valid for models with absorbing states.

N: nonabsorbing states; A: absorbing states; U: up states; D: down states.

Table 2.19   Use of reward rates to compute different measures – part II

Requirement	Reward Assignments	Measure
Reliability [HMT91], [MCT94], [DaBh85]	$r_i = \begin{cases} 1 & \text{if } i \in N \\ 0 & \text{else} \end{cases}$ [Table 2.5]	$E[Z(t)]$, $P[Z(t) = 1]$
Unreliability [STP96]	$r_i = \begin{cases} 1 & \text{if } i \in A \\ 0 & \text{else} \end{cases}$	$E[Z(t)]$, $P[Z(t) = 1]$
System MTTF [HMT91]	$r_i = \begin{cases} 1 & \text{if } i \in N \\ 0 & \text{else} \end{cases}$ [Table 2.5]	$E[Y(\infty)]^*$
Expected number of catastrophic events in $[0, t)$ [HMT91]	$r_i = \begin{cases} \lambda & \text{if } i \in S_C \\ 0 & \text{else} \end{cases}$ (S_C: states where catastrophe may occur, λ: rate of catastrophe) [Table 2.5]	$E[Y(t)]$
Task interruption probability [HMT91], [TSIH90]	$r_i = \begin{cases} P[O_i \leq x_i] & \text{if } i \in U \\ 0 & \text{else} \end{cases}$ (x_i: task execution time in state i, O_i: sojourn time in state i) [Table 2.7, 2.8]	$E[Z]$
Capacity oriented availability [HMT91], [MTD94], [CiGr87]	$r_i = \begin{cases} cap(i) & \text{if } i \in U \\ 0 & \text{else} \end{cases}$ ($cap(i)$: capacity of state i) [Table 2.9]	$E[Z], E[Z(t)], E[Y(t)],$ $\frac{1}{t}E[Y(t)]$ $P[Z(t) = x], P[Z(t) \leq x],$ $P[Y(t) \leq y]$
Total loss probability [MuTr91], [TMWH92], [TSIH90]	$r_i = \begin{cases} 1 & \text{if } i \in D \\ l_i & \text{else} \end{cases}$ (l_i: percentage loss in state i) [Table 2.10]	$E[Z], E[Z(t)], E[Y(t)],$ $\frac{1}{t}E[Y(t)],$ $P[Z(t) = x], P[Z(t) \leq x],$ $P[Y(t) \leq y]$
Normalized throughput [Meye80], [Meye82], [GaKe79], [TMWH92], [STP96]	$r_i = \begin{cases} 1 - l_i & \text{if } i \in U \\ 0 & \text{else} \end{cases}$ [Table 2.11]	$E[Z], E[Z(t)], E[Y(t)],$ $\frac{1}{t}E[Y(t)],$ $P[Z(t) = x], P[Z(t) \leq x],$ $P[Y(t) \leq y]$

*The measure is valid for models with absorbing states.

Table 2.20   Use of reward rates to compute different measures – part III

Requirement	Reward Assignments	Measure
Approx. total loss [MTD94], [MeSe97]	$r_i = \begin{cases} t\lambda l_i & \text{if } i \in U \\ t\lambda & \text{else} \end{cases}$ [Table 2.13]	$E[Z]$
Loss [HMT91]	$r_i = \begin{cases} \delta l_i & \text{if } i \in U \\ \delta & \text{else} \end{cases}$	$E[Z], E[Z(t)], E[Y(t)],$ $\frac{1}{t}E[Y(t)],$ $P[Z(t) = x], P[Z(t) \le x],$ $P[Y(t) \le y]$
Mean number of customers [MTBH93]	$r_i = c(i)$ ($c(i)$: number of customers in state i)	$E[Z], E[Z(t)],$ $P[Z(t) = x]$
Expected throughput [MeSe97], [DoIy87], [MTBH93]	$r_i = \begin{cases} \mu_i & \text{if } i \in S_T \\ 0 & \text{else} \end{cases}$ (μ_i: service rate while in state i)	$E[Z], E[Z(t)], E[Y(t)],$ $\frac{1}{t}E[Y(t)], E[Y(\infty)]^*,$ $P[Z(t) = x], P[Z(t) \le x],$ $P[Y(\infty) \le y]^*, P[Y(t) \le y]$
Utilization [Husl81], [GoTa87], [MTBH93]	$r_i = \begin{cases} 0 & \text{if resource is idle} \\ 1 & \text{else} \end{cases}$	$E[Z(t)], \frac{1}{t}E[Y(t)],$ $E[Z], P[\frac{1}{t}Y(t) \le y]$

*The measure is valid for models with absorbing states

2.3.1   Petri Nets

Petri nets (PNs) were originally introduced by C.A. Petri in 1962 [Petr62]. A PN is a bipartite directed graph, consisting of two types of nodes, namely *places*, P, and *transition*, T. Arcs between the nodes fall in two categories: *input arcs* lead from an *input place* to a transition, and *output arcs* connect a transition to an *output place*. Arcs always have to lead from one type of nodes to the complementary one. A finite number of *tokens* can be distributed among the places and every place may contain an arbitrary (natural) number of tokens. A *marking* $m \in \mathcal{M}$ is defined as a possible assignment of tokens to all places in the PN. Markings are sometimes referred to as *states* of the PN. If P denotes the set of places, then a marking m is a multiset, $m \in \mathcal{M} \subset \mathbb{N}^{|P|}$, describing the number of tokens in each place. To refer to the number of tokens at a particular place $P_i, 1 \le i \le |P|$, in a marking m, the notation $\#(P_i, m)$ is used.

A transition is *enabled* if all of its input places contain at least one token. Note that a transition having no input arc is always enabled. An enable transition can *fire* by removing from each input place one token and by adding to each output place one token. The firing of a transition may transform a PN from one marking into another. With respect to a given *initial* marking, the *reachability set*, $\mathcal{RS}$, is defined as the set of all markings reachable through any possible firing sequences of transitions, starting from the initial marking. The $\mathcal{RS}$ could be infinite if no further restrictions are imposed. Even PNs appearing as being relatively simple can give rise to large or infinite $\mathcal{RS}$s. If more than one transition is simultaneously enabled but cannot fire in parallel, a *conflict* between transitions arises that needs to be resolved through some selection policy, for example, based on a priority scheme or according to a given pmf. Markings in which no transition is enabled are called *absorbing* markings.

For a graphical presentation, places are usually depicted as circles and transitions are depicted as bars. A simple PN with an initial marking

$$m_0 = (2,0,0,0) \ ,$$

and with subsequent marking

$$m_1 = (1,1,1,0) \ ,$$

reachable through firing of transition t_1, is shown in Fig. 2.10.

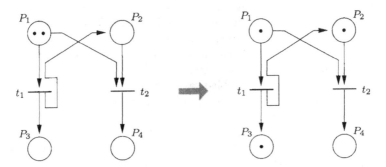

Fig. 2.10   A simple PN before and after the firing of transition t_1.

Transition t_1 is enabled because its single input place P_1 contains two tokens. Since P_2 is empty, transition t_2 is disabled. When firing takes place, one token is removed from input place P_1 and one token is deposited in both output places P_2 and P_3. In the new marking, both transitions t_1 and t_2 are enabled. The apparent conflict between t_1 and t_2 must be solved to decide which of the alternative (absorbing) markings will be reached:

$$m_2 = (0,2,2,0)$$
$$\text{or} \quad m_3 = (0,0,1,1) \ .$$

The reachability set in this example is given by $\{m_0, m_1, m_2, m_3\}$, where the markings are given as defined earlier.

Definition 2.16   A PN is given by a 5-tuple $PN = \{P, T, D^-, D^+, m_0\}$ with:

- A finite set of places $P = \{P_1, P_2, \ldots, P_{|P|}\}$, where each place contains a non-negative number of tokens.

- A finite set of transitions $T = \{t_1, \ldots, t_{|T|}\}$, such that $T \cap P = \emptyset$.

- The set of input arcs $D^- \in \{0, 1\}^{|P \times T|}$ and the set of output arcs $D^+ \in \{0, 1\}^{|P \times T|}$. If $D^-(P_i, t_k) = 1$, then there is an input arc from place P_i to transition t_k, and if $D^+(P_j, t_l) = 1$, then there is an output arc from transition t_l to place P_j.

- The initial marking $m_0 \in \mathcal{M}$.

2.3.2 Generalized Stochastic Petri Nets

GSPNs generalize PNs in such a way that each transition has a *firing time* assigned to it, which may be exponentially distributed or constant zero. Transitions with exponentially distributed firing times are called *timed* transitions, while the others are referred to as *immediate* transitions. The new type of transitions require enabling and firing rules to be adapted accordingly.

The markings $\mathcal{M} = \mathcal{V} \cup \mathcal{T}$ in the reachability set $\mathcal{RS}$ of a GSPN are partitioned into two sets, the *vanishing* markings $\mathcal{V}$ and the *tangible* markings $\mathcal{T}$. Vanishing markings comprise those in which *at least one* immediate transition is enabled. If no immediate transition is enabled, that is, only timed transitions or no transitions are enabled, a tangible marking results. Vanishing markings are not resided in for any finite nonzero time and firings are performed instantaneously in this case. Therefore, immediate transitions always have priority over timed transitions to fire. If several immediate transitions compete for firing, a specified pmf is used to break the tie. If timed transitions compete, a *race model* is applied so that the transition whose firing time elapses first is the next one to fire.

From a given GSPN, an *extended reachability graph* ($\mathcal{ERG}$) is generated containing the markings of the reachability set as nodes and some stochastic information attached to the arcs, thereby relating the markings to each other. Absorbing loops of vanishing markings are a priori excluded from the $\mathcal{ERG}$, because a stochastic discontinuity would result otherwise. The GSPNs we are interested in are bounded; i.e., the underlying reachability set is finite. Marsan, Balbo, and Conte proved that exactly one CTMC corresponds to a given GSPN under condition that only a finite number of transitions can fire in finite time with nonzero probability [ABC84].

Thick bars are used to represent timed transitions graphically, while thin bars are reserved for immediate transitions. An example of a GSPN is shown in Fig. 2.11. In contrast to the PN from Fig. 2.10, an exponentially distributed firing time with rate λ_1 has been assigned to transitions t_1. While in the PN

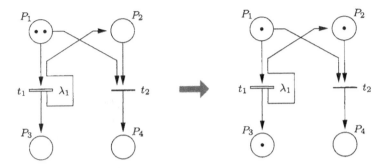

Fig. 2.11   A simple GSPN before and after the firing of transition t_1.

of the right part in Fig. 2.10 both transitions t_1 and t_2 were enabled, only immediate transition t_2 is enabled in GSPN of the right part in Fig. 2.11, because immediate transitions have priority over timed transitions.

2.3.3   Stochastic Reward Nets

Many proposals have been launched to provide extensions to GSPNs (PNs). Some of the most prominent ones are revisited in this section: arc multiplicity, inhibitor arcs, priorities, guards, marking-dependent arc multiplicity, marking-dependent firing rates, and reward rates defined at the net level. With these extensions we obtain the formalism of stochastic reward nets. We note that the first three of these extensions are already present in the GSPN formalism.

Arc Multiplicity.   Frequently more than one token needs to be removed from a place or deposited into a place, which can be easily represented with arc multiplicities. It could be awkward, otherwise, to draw an arc for each token to be moved. Arc multiplicity is a syntactical extension that does not increase the principal modeling power but makes the use of GSPNs more convenient. Arcs with multiplicity are represented by a small number attached to the arc or by a small line cutting through the arc. By default, a multiplicity of 1 is assumed. An example of arcs with multiplicities is depicted in Fig. 2.12.

Since there are initially three tokens in place P_1, transition t_1 is enabled, and after elapsing of the firing time, two tokens are removed from place P_1 (multiplicity 2) and one token is deposited into place P_2. No other transition is enabled in the initial marking. Since in the new marking there are two tokens in P_2 and one token in P_1, only t_2 is enabled and fires immediately, thereby removing two tokens from P_2 (multiplicity 2) and adding one token to P_1. After some exponentially distributed firing time has elapsed, transition t_1 fires for the last time, because an absorbing marking will then be reached. Finally, there will be zero tokens in P_1 and one token in P_2.

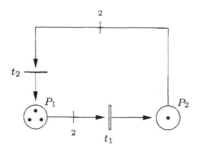

Fig. 2.12 Simple example of multiple arcs.

Inhibitor Arcs. An inhibitor arc from a place to a transition *disables* the transition in any marking where the number of tokens in the place is equal to or greater than the multiplicity of the inhibitor arc. An inhibitor arc exhibits a complementary semantic to the one of a corresponding ordinary arc (with the same multiplicity). If a transition were enabled under the conditions imposed by an ordinary arc it would never be enabled under the conditions imposed by its counterpart, an inhibitor arc, and vice versa. Graphically, an inhibitor arc is indicated by a small circle instead of an arrowhead. The use of an inhibitor arc is demonstrated in Fig. 2.13.

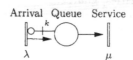

Fig. 2.13 A GSPN with an inhibitor arc having a multiplicity assigned.

The figure entails a queue with finite capacity, which customers enter with a mean interarrival time of $1/\lambda$ as long as the queue is not full; that is, fewer than k customers are already in the system. Arrivals to a full system may not enter and are considered as lost. The arrival transition is disabled in this case. The effective arrival rate, therefore, turns out to be less than λ. Note that we have made use of these kind of scenarios in earlier sections when specifying reward assignments capturing throughput losses or related performance requirements. Customers leave the system after receiving service with exponentially distributed length of mean duration $1/\mu$.

Priorities. Although inhibitor arcs can be used to specify priority relations, it is easier if priority assignments are explicitly introduced in the paradigm. Priorities are specified by integer numbers assigned to transitions. A transition is enabled only if no other transition with a higher priority is enabled. In Fig. 2.14 a simple example of a GSPN with priorities is shown. An idle server is assigned with different priority to each type of waiting customers.

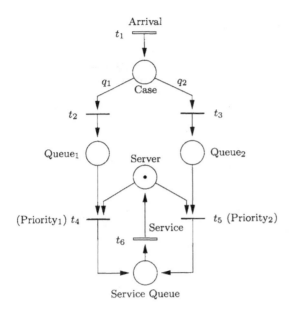

Fig. 2.14   A GSPN with priorities.

Arriving customers belong to priority class i with probability $q_i, \sum_i q_i = 1$. Recall that immediate transitions must always have priority over timed transitions.

Guards.   Guards are general predicates that determine when transitions are to be enabled. This feature provides a powerful means to simplify the graphical representation and to make GSPNs easier to understand, for example, by identifying modular substructures. With these extensions, we have moved out from GSPN class to SRNs [CBC+93]. An example for an SRN with guards is given in Fig. 2.15. A system with two types of resources being subject to failure and repair is represented. The scenario modeled captures the situation of a finite capacity queueing system with finite number of servers (net 1) and additional buffer space (net 2) usable if the active servers are temporarily occupied. The total number of customers in the system is limited by the total of failure-free buffer spots plus the failure-free server units. Hence, arriving customers, which enter with a mean interarrival time $1/\lambda$ (net 3), can only be admitted if fewer customers are already present than given by the current capacity. This condition is described by the guard attached to the "arrival" transition. The guard characterizing the enabling conditions of the "service" transition explicates the fact that service can only be delivered if at least one server is failure free (besides the apparent condition of the presence of at least one customer).

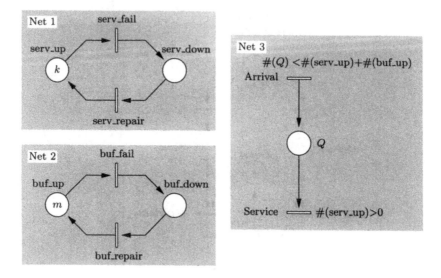

Fig. 2.15   An SRN with guards.

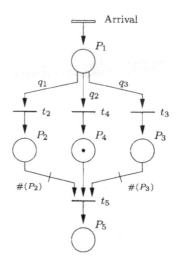

Fig. 2.16   Place flushing modeled with marking-dependent arc multiplicity.

Marking-Dependent Arc Multiplicity.   This feature can be applied when the number of tokens to be transferred depends on the current marking. A useful application is given when places need to be "flushed" at the occurrence of a certain event, such as the appearance of a token at a particular place. Such a case is shown in Fig. 2.16. Customers arrive from outside at place P_1 and are immediately forwarded and deposited at places P_2, P_3, or P_4, with probabilities q_1, q_2, and q_3, respectively. Whenever there is at least one token at place P_4, transition t_5 is enabled, regardless of the number of tokens in the other input places P_2 and P_3. When t_5 fires, one token is removed from place P_4, *all* tokens are removed from place P_2 and P_3, and a single token appears in place P_5.

Marking-Dependent Firing Rates. A common extension of GSPNs is to allow marking-dependent firing rates, where a firing rate can be specified as a function of the number of tokens in any place of the Petri net. Restricted variants do exist, where the firing rate of a transition may depend only on the number of tokens in the input places. As an example, the firing rate of transition service in net 3 of Fig. 2.15 is a function of the number of servers currently up, that is, the number of tokens in place serv_up.

Reward Rate Specification. Traditional output measures obtained from a GSPN are the throughput of a transition and the mean number of tokens in a place. Very often it is useful to have more general information such as:

- The probability that a place[1] P_i is empty while the transition t_j is enabled.

- The probability that two different transitions t_1 and t_2 are simultaneously enabled.

Very general reward rates can be specified on an SRN so as to allow the user to compute such custom measures. Examples of the use of reward rates can be found in [CFMT94, MCT94, MMKT94, MaTr93, MuTr91, MuTr92, WLTV96].

2.3.4 GSPN/SRN Analysis

In this section we introduce the techniques of automated generation of stochastic processes underlying a GSPN/SRN. But it is worthwhile remembering that GSPNs can also be evaluated with discrete-event simulation techniques, a non-state-space based method, and that each technique has its merits.

In general, the analysis of a GSPN/SRN can be decomposed into four subtasks:

1. Generation of the $\mathcal{ERG}$, thereby defining the underlying stochastic process. A semi-Markov process (SMP) results, with zero or exponentially distributed sojourn times. In the case of an SRN, a reward rate for each tangible marking is also generated in this step.

2. Transformation of the SMP into a CTMC by elimination of the vanishing markings with zero sojourn times and the corresponding state transitions.

3. Steady-state, transient, or cumulative transient analyses of the CTMC with methods presented in Chapters 3, 4, and 5.

4. Computation of measures. In the the case of a GSPN, standard measures such as the average number of tokens in each place and the throughput of each timed transition are computed. For the SRN, the specification of reward rates at the net level enables the computation of very general custom measures.

In the $\mathcal{ERG}$, which reflects the properties of the underlying stochastic process, arcs representing the firing of timed transitions are labeled with the corresponding rates, and arcs representing the firing of immediate transitions are marked with probabilities. The $\mathcal{ERG}$ can be transformed into a reduced reachability graph ($\mathcal{RG}$) by eliminating the vanishing markings and the corresponding transitions. Finally, a CTMC will result. In the case of an SRN, a reward vector is also generated and, hence, an MRM is produced.

2.3.4.0.1 $\mathcal{ERG}$ Generation   An $\mathcal{ERG}$ is a directed graph with nodes corresponding to the markings (states) of a GSPN/SRN and weighted arcs representing the probabilities or rates with which marking changes occur. The basic $\mathcal{ERG}$ generation algorithm is summarized in Fig. 2.17.

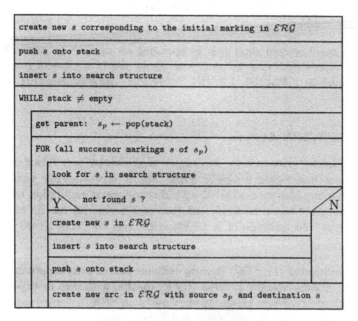

Fig. 2.17   Basic $\mathcal{ERG}$ generation algorithm.

Starting with an initial marking m_0, subsequent markings are generated by systematically following all possible firing patterns. If a new marking is generated, the arcs labeled with the corresponding rates and probabilities are added to the $\mathcal{ERG}$. The algorithm terminates if all possible markings have been generated. Three main data structures are used in this algorithm: a stack (S), the $\mathcal{ERG}$, and a search structure (D).

- **Stack (S):** Newly generated markings are buffered in a stack until they are processed.

- **$\mathcal{ERG}$:** This data structure is used to represent the generated extended reachability graph.

- **Search structure (D):** This data structure contains all earlier generated markings so that a current marking can be checked as to whether it is new or already included in the set D. For efficiency reasons, memory requirements and access time to the data structure should be of concern in an implementation. If the data structure is stored in hash tables, fast access results but memory requirements could be excessive. In contrast, tree structures are more flexible and can be stored more compactly, but access time could be intolerably high for unbalanced trees. Balanced trees such as AVL trees and B-trees are used for this reason [AKH97].

2.3.4.1 Elimination of Vanishing Markings   The elimination of vanishing markings is an important step to be accomplished for generation of a CTMC from a given GSPN/SRN. Two different techniques can be distinguished, "elimination on the fly" and "post-elimination" [CMT91]. To give an example, consider the GSPN in Fig. 2.18a.

The underlying $\mathcal{ERG}$ in Fig. 2.18b contains five markings. Tangible markings are represented by oval boxes, while rectangular boxes are reserved for vanishing markings. The elimination of the vanishing markings $(0,1,0,0,0)$ and $(0,0,1,0,0)$ leads to an $\mathcal{RG}$, shown in Fig. 2.18c. The $\mathcal{RG}$ is a CTMC (redrawn in Fig 2.18d). The elimination of vanishing markings can either be integrated into the generation of the $\mathcal{ERG}$ (elimination on the fly) or performed afterwards (post-elimination). The merits of both methods are discussed in the subsequent subsections.

2.3.4.2 Elimination on the Fly   Consider the snapshot in Fig. 2.18b. Generation of vanishing markings such as $(0,1,0,0,0)$ is avoided by splitting the arc and its attached rate, which would be leading to this vanishing marking, and redirecting the resulting new arcs to subsequent markings. If at least one of these subsequent markings is also of vanishing type, the procedure must be iterated accordingly. Such a situation is indicated in Fig. 2.18b. The splitting is accomplished by weighing the rates in relation to the firing probabilities of the immediate transitions (in our example, t_2 and t_3). As a result, only tangible markings and the arcs with adjusted rates need to be stored. Vanishing markings and the associated firing probabilities are discarded immediately.

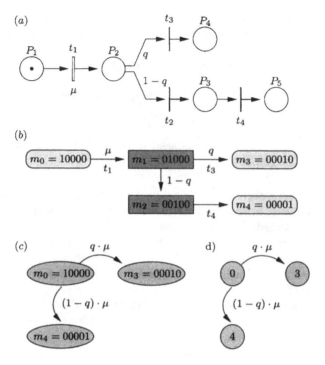

Fig. 2.18 The elimination of vanishing markings demonstrated by (a) a GSPN, (b) the underlying $\mathcal{ERG}$, (c) the resulting $\mathcal{RG}$, and (d) the corresponding CTMC.

The principle of rate splitting and elimination of vanishing markings is illustrated in Fig. 2.19.

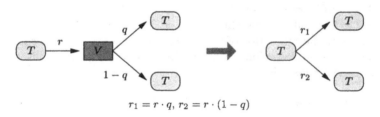

$$r_1 = r \cdot q, \ r_2 = r \cdot (1 - q)$$

Fig. 2.19 On-the-fly elimination of vanishing markings.

Elimination on the fly is efficient with respect to memory requirements, because vanishing markings need not be stored at all. But these savings in space have to be traded with additional cost in time. The same vanishing markings may be hit several times through the $\mathcal{ERG}$-generation process. The elimination step would be repeated although it had been perfectly executed before. This repetition could amount to a significant wast of execution time.

2.3.4.3 Post-Elimination In post-elimination, the complete $\mathcal{ERG}$ is stored during generation. With this technique, it is easier to recognize and resolve (transient) loops consisting of vanishing markings only [CMT90]. Furthermore, no work is duplicated during $\mathcal{ERG}$ generation, and, as an additional advantage, the complete information included in the $\mathcal{ERG}$ is kept for further use. Once the $\mathcal{ERG}$ has been generated, it is transformed into a CTMC by simply applying the same *rate splitting method* as has been described in Section 2.3.4.2.

The success of the post-elimination method largely depends on the use of efficient matrix algorithms. Let

$$\mathbf{P}^{\mathcal{V}} = \left[\mathbf{P}^{\mathcal{VV}} \mid \mathbf{P}^{\mathcal{VT}} \right]$$

denote a matrix, which is split into transition probabilities between vanishing markings only ($\mathbf{P}^{\mathcal{VV}}$) and between vanishing markings and tangible markings ($\mathbf{P}^{\mathcal{VT}}$), where $\mathbf{P}^{\mathcal{V}}$ is of dimension $|\mathcal{V}| \times |\mathcal{M}|$ and the set of markings is partitioned such that $\mathcal{M} = \mathcal{V} \cup \mathcal{T}$. Furthermore, let

$$\mathbf{U}^{\mathcal{T}} = \left[\mathbf{U}^{\mathcal{TV}} \mid \mathbf{U}^{\mathcal{TT}} \right]$$

denote a matrix, which is split into transition rates from tangible markings to vanishing markings ($\mathbf{U}^{\mathcal{TV}}$) and between tangible markings only ($\mathbf{U}^{\mathcal{TT}}$), where $\mathbf{U}^{\mathcal{T}}$ is of dimension $|\mathcal{T}| \times |\mathcal{M}|$. Precisely the same information as contained in the $\mathcal{ERG}$ is provided by $\mathbf{U}^{\mathcal{T}}$ together with $\mathbf{P}^{\mathcal{V}}$.

With representation in matrix form, it is easy to complete the set of transition rates by formally applying the rate-splitting method as indicated earlier. The complete set of transition rates, gained after the rate-splitting method has been applied, is summarized in the rate matrix $\mathbf{U}$, which is of dimension $|\mathcal{T}| \times |\mathcal{T}|$ [CMT91]:

$$\mathbf{U} = \mathbf{U}^{\mathcal{TT}} + \mathbf{U}^{\mathcal{TV}} \left(\mathbf{I} - \mathbf{P}^{\mathcal{VV}} \right)^{-1} \mathbf{P}^{\mathcal{VT}} . \tag{2.82}$$

Rate matrix $\mathbf{U} = [u_{ij}]$, which contains all initially present rates between tangible markings plus the derived rates between tangible markings, obtained by applying the rate-splitting method, needs to be completed to derive the infinitesimal generator matrix $\mathbf{Q} = [q_{ij}]$, the entries of which are given by

$$q_{ij} = \begin{cases} u_{ij} & \text{if } i \neq j , \\ -\sum_{k \in \mathcal{T}, k \neq i} u_{ik} & \text{if } i = j , \end{cases} \tag{2.83}$$

where $\mathcal{T}$ denotes the set of tangible markings. Note that $\mathbf{U}$ may contain nonzero entries on its diagonal, which can be simply ignored as redundant information when creating the infinitesimal generator matrix $\mathbf{Q}$ as shown in Eq. (2.83).

Once the generator matrix has been derived with the described method, numerical solution methods can be applied to yield performance measures as required. Finally, it may be of interest to realize that instead of eliminating the vanishing markings, they can be preserved and the underlying semi-Markov process can be solved numerically [CMT90, MTD94].

```
Algorithm generate_ERG()

initially: S = {}, D = {}, R = {}

create new marking s in ERG: R = {(2,0)}
push s onto the stack: S = {(2,0)}
insert s into the search structure: D = {(2,0)}

WHILE (S != {})
    get parent marking: s_p = pop(S) = {(2,0)},
    R = {(2,0)} S = {}, D = {(2,0)}
    FOR (all successor markings s of s_p)
        s = succ(2,0) = (1,1)
        (not found s in D) = true
            create new marking in ERG: R = {(2,0)(1,1)}
            insert marking into the search structure: D = {(2,0), (1,1)}
            push s onto stack: S = {(1,1)}
        create new arc in ERG: R = {(2,0) --> (1,1)}
    get parent marking: s_p = pop(S) = {(1,1)},
    R = {(2,0) --> (1,1)} S = {}, D = {(2,0), (1,1)}
    FOR (all successor markings s of s_p)
        s = succ(1,1) = (0,2)
        (not found s in D) = true
            create new marking in ERG: R = {(2,0) --> (1,1)(0,2)}
            insert marking into the search structure: D = {(2,0), (1,1), (0,2)}
            push s onto stack: S = {(0,2)}
        create new arc in ERG: R = {(2,0) --> (1,1) --> (0,2)}
        s = succ(1,1) = (2,0)
        (not found s in D) = false
        create new arc in ERG: R = {(2,0) <--> (1,1) --> (0,2)}
    get parent marking: s_p = pop(S) = {(0,2)}
    R = {(2,0) <--> (1,1) --> (0,2)}, S = {}, D = {(2,0), (1,1), (0,2)}
    FOR (all successor markings s of s_p)
        s = succ(0,2) = (1,1)
        (not found s in D) = true
        create new arc in ERG: R = {(2,0) <--> (1,1) <--> (0,2)}

S == {} ==> the generation algorithm stops
```

Fig. 2.20 $\mathcal{ERG}$ generation for Example 2.13.

Example 2.13   A simple example is given to demonstrate the generation of an $\mathcal{ERG}$ by applying the algorithm from Fig. 2.17. Consider the GSPN in Fig. 2.21a that models a simple network of queues. An equivalent queueing network representation is shown in Fig. 2.21b (for a detailed introduction to queueing networks see Chapter 6). The network contains $K = 2$ customers, which are assumed to be initially in node 1. Let (i, j) denote a state of the network, where i represents the number of customers at node 1 and j those at node 2. The CTMC can be directly derived from the $\mathcal{ERG}$ whose generation is shown in Fig. 2.20 since there are no vanishing markings in this $\mathcal{ERG}$.

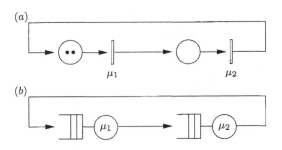

Fig. 2.21 Simple queueing network.

Example 2.14 With the example GSPN from Fig. 2.22a, the intermediate steps to create the rate matrix $\mathbf{U}$ and the infinitesimal generator matrix $\mathbf{Q}$ are demonstrated. Post-elimination is assumed so that the complete $\mathcal{ERG}$ is available, as indicated in Fig. 2.22b. The four matrices $\mathbf{P}^{\mathcal{VV}}$, $\mathbf{P}^{\mathcal{VT}}$, $\mathbf{U}^{\mathcal{TV}}$, and $\mathbf{U}^{\mathcal{TT}}$ are given by

$$\mathbf{P}^{\mathcal{VV}} = \begin{pmatrix} 0 & q_2 \\ 0 & 0 \end{pmatrix}, \qquad \mathbf{P}^{\mathcal{VT}} = \begin{pmatrix} 0 & q_1 & 0 & 0 \\ 0 & 0 & q_4 & q_3 \end{pmatrix},$$

$$\mathbf{U}^{\mathcal{TV}} = \begin{pmatrix} \mu_1 & \mu_2 \\ 0 & 0 \\ 0 & 0 \\ 0 & 0 \end{pmatrix}, \qquad \mathbf{U}^{\mathcal{TT}} = \begin{pmatrix} 0 & 0 & 0 & 0 \\ 0 & 0 & 0 & \mu_3 \\ 0 & 0 & 0 & 0 \\ 0 & 0 & 0 & 0 \end{pmatrix}.$$

Then we get

$$(\mathbf{I} - \mathbf{P}^{\mathcal{VV}})^{-1} = \begin{pmatrix} 1 & q_2 \\ 0 & 1 \end{pmatrix}, \qquad \mathbf{U}^{\mathcal{TV}}(\mathbf{I} - \mathbf{P}^{\mathcal{VV}})^{-1} = \begin{pmatrix} \mu_1 & q_2\mu_1 + \mu_2 \\ 0 & 0 \\ 0 & 0 \\ 0 & 0 \end{pmatrix},$$

$$\mathbf{U}^{\mathcal{TV}}(\mathbf{I} - \mathbf{P}^{\mathcal{VV}})^{-1}\mathbf{P}^{\mathcal{VT}} = \begin{pmatrix} 0 & q_1\mu_1 & q_4(q_2\mu_1 + \mu_2) & q_3(q_2\mu_1 + \mu_2) \\ 0 & 0 & 0 & 0 \\ 0 & 0 & 0 & 0 \\ 0 & 0 & 0 & 0 \end{pmatrix},$$

$$\mathbf{U} = \begin{pmatrix} 0 & q_1\mu_1 & q_4(q_2\mu_1 + \mu_2) & q_3(q_2\mu_1 + \mu_2) \\ 0 & 0 & 0 & \mu_3 \\ 0 & 0 & 0 & 0 \\ 0 & 0 & 0 & 0 \end{pmatrix},$$

$$\mathbf{Q} = \begin{pmatrix} -(q_1\mu_1 + q_4(q_2\mu_1 + \mu_2) & q_1\mu_1 & q_4(q_2\mu_1 + \mu_2) & q_3(q_2\mu_1 + \mu_2) \\ \quad +q_3(q_2\mu_1 + \mu_2)) & & & \\ 0 & -\mu_3 & 0 & \mu_3 \\ 0 & 0 & 0 & 0 \\ 0 & 0 & 0 & 0 \end{pmatrix}.$$

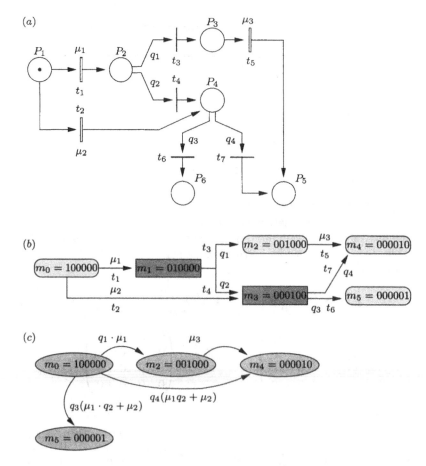

Fig. 2.22 (a) Simple GSPN, (b) the underlying ERG, and (c) the corresponding CTMC.

With Eq. (2.83), the entries of the generator matrix $\mathbf{Q}$ can be determined from $\mathbf{U}$.

2.3.5   A Larger Example

To demonstrate the automated generation of CTMCs from high-level GSPN description, an example of a polling system adapted from [IbTr90] is now presented. (For an introduction to polling systems see [Taka93].)

Consider a two-station single-buffer polling system as modeled by the GSPN shown in Fig. 2.23. The places in the GSPN are interpreted as follows. A token in place P_1 represents the case that station 1 is idle and a token in P_2 correspondingly indicates the idleness of station 2. Tokens in places P_5 (P_6)

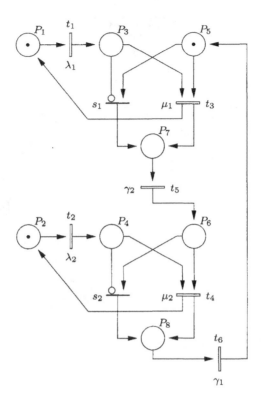

Fig. 2.23 GSPN Model for the two-station single-buffer polling system.

indicate that the server is serving station 1 (station 2). Tokens appearing in Places P_3 (P_4) indicate that station 1 (station 2) has generated a message. If a token appears in place P_7, the server is polling station 2, while a token in place P_8 indicates polling of station 1. In the situation illustrated in Fig. 2.23, both stations are idle and the server has finished polling station 1 (the server is ready to serve station 1). We assume that the time until station i generates a message is exponentially distributed with mean $1/\lambda_i$. Service and polling times at station i are assumed to be exponentially distributed with mean $1/\mu_i$ and $1/\gamma_i$, respectively ($i = 1, 2$). As soon as station 1 (station 2) generates a message, the timed transition t_1 (t_2) fires.

Consider initial marking m_0 shown in Fig. 2.23, where P_5 contains a token and P_3 contains no token. In this situation the immediate transition s_1 is enabled. It fires immediately and a token is deposited in place P_7. This represents the condition that the server, when arriving at station 1 and finding no message there, immediately proceeds to poll station 2. In the case that a token is found in place P_3, the timed transition t_3 is enabled and its firing causes a token to be deposited in place P_1 and another token to be deposited

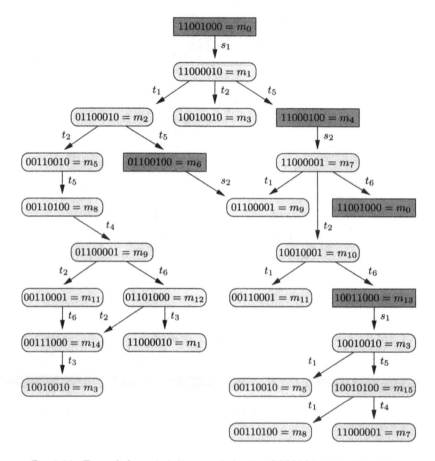

Fig. 2.24 Extended reachability graph for the GSPN Model in Fig. 2.23.

in place P_7. This represents the condition that station 1 returns to the idle state while the server proceeds to poll station 2.

Let l_i denote the number of tokens in place P_i and $m_j = (l_1, \ldots, l_8)$ denote the jth marking of the GSPN. The $\mathcal{ERG}$, obtained from the initial marking m_0, is shown in Fig. 2.24. Labels attached to directed arcs represent those transitions whose firing generates the successor markings. If a previously found marking is obtained again (like m_3), further generations from such a marking are not required. All sixteen possible and unique markings are included in Fig. 2.24. To reduce the clutter, several markings appear more than once in the figure. The $\mathcal{ERG}$ consists of four vanishing markings $\{m_0, m_4, m_6, m_{13}\}$, represented in the graph by rectangles and tangible markings represented by ovals. We can eliminate the vanishing markings by applying the algorithm presented in Section 2.3.4.1.

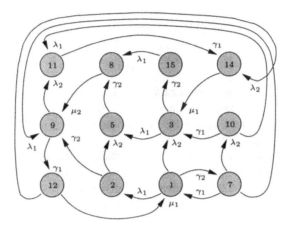

Fig. 2.25   CTMC for the GSPN in Fig. 2.23.

From Fig. 2.24 we can see that all the tangible markings containing a token in place P_1 indicate the condition that station 1 is idle. The set of markings in which station 1 is idle is $S^{(1)} = \{m_1, m_3, m_7, m_{10}, m_{15}\}$. The set of markings $S^{(2)} = \{m_1, m_2, m_7, m_9, m_{12}\}$ gives the condition that station 2 is idle.

In Fig. 2.25, the finite and irreducible CTMC corresponding to the GSPN of Fig. 2.23 is shown, where each state j corresponds to the tangible marking m_j of the $\mathcal{ERG}$ in Fig. 2.24. With the ergodic CTMC in Fig. 2.25, the steady-state probability vector $\boldsymbol{\pi}$ can be derived according to Eq. (2.58). Letting π_k denote the steady-state probability of the CTMC being in state k and letting $p^{(i)}$ denote the steady-state probability that station i, $i = 1, 2$ is idle, we conclude that

$$
\begin{aligned}
p^{(1)} &= \pi_1 + \pi_3 + \pi_7 + \pi_{10} + \pi_{15}, \\
p^{(2)} &= \pi_1 + \pi_2 + \pi_7 + \pi_9 + \pi_{12}.
\end{aligned}
\tag{2.84}
$$

As can be seen in Fig. 2.23, the model is structurally symmetric relative to each station and therefore it can easily be extended to more than two stations.

With Eq. (2.84), the mean number of customers $E[K^{(i)}] = \overline{K^{(i)}}$ at single-buffer station i is determined as

$$
\overline{K^{(i)}} = 1 - p^{(i)}.
$$

Also with Eq. (2.84), the effective arrival rate at single-buffer station i follows as

$$
\lambda_e^{(i)} = p^{(i)} \lambda.
$$

From Little's theorem [Litt61], formulae can be derived for the computation of mean response times of simple queueing models such as the polling system we are investigating in this section. The mean response time $E[T^{(i)}] = \overline{T^{(i)}}$

at station i can be computed as

$$\overline{T^{(i)}} = \frac{\overline{K^{(i)}}}{\lambda_e^{(i)}}.$$ (2.85)

As an extension of the two-station model in Fig. 2.23, let us consider a symmetric polling system with $n = 3, 5, 7, 10$ stations, with $\mu_i = \mu = 1$, $1/\gamma = 1/\gamma_i = 0.005$, and $\lambda_i = \lambda$ for all stations $1 \leq i \leq n$. The offered load is defined by $\rho = \sum_{j=1}^{n} \lambda/\mu = n\lambda/\mu$.

For a numerical computation of steady-state probabilities of CTMCs such as the one depicted in Fig. 2.25, algorithms introduced in Sections 3.4 and 3.5 can be used. In Table 2.21, the mean response times $\overline{T^{(i)}} = \overline{T}$ of single buffer schemes are given for different number of stations. For many other versions of polling system models, see [IbTr90].

Table 2.21   Mean response times $\overline{T}$ at a station

ρ	3 Stations	5 Stations	7 Stations	10 Stations
0.1	1.07702	1.09916	1.11229	1.12651
0.2	1.14326	1.18925	1.21501	1.24026
0.3	1.20813	1.28456	1.32837	1.37051
0.4	1.27122	1.38418	1.45216	1.51879
0.5	1.33222	1.48713	1.58582	1.68618
0.6	1.39096	1.59237	1.72836	1.87310
0.7	1.44732	1.69890	1.87850	2.07914
0.8	1.50126	1.80573	2.03464	2.30293
0.9	1.55279	1.91201	2.19507	2.54220

Problem 2.1   With Definition 2.64, prove differential Eq. (2.65) by integration of Eq. (2.53) on both sides.

Problem 2.2   With Definition 2.66, prove differential Eq. (2.67).

Problem 2.3   Prove linear Eq. (2.68) by observing $\frac{d}{dt}\mathbf{L}_N(\infty) = \mathbf{0}_N$.

Problem 2.4   Classify the measure mean time to absorption (MTTA) as being of transient, steady-state, or stationary type. Refer back to Eqs. (2.65) and (2.68) and give reasons for your decision.

Problem 2.5   Does $lim_{t\to\infty}E[Z(t)]$ exist if the underlying model has at least one absorbing state? If so, what would this measure mean?

Problem 2.6   Specify the reward assignments to model variants one and three in Fig. 2.7 for reliability computations.

Problem 2.7   Compare reliability (unreliability) $R(t)$ $(UR(t))$ functions from Section 2.2.3.2 with the computation formula of the distribution of the accumulated reward until absorption, $P[Y(\infty) \leq y]$ in Eq. (2.80) and comment on it.

2.3.6   Stochastic Petri Net Extensions

In the preceeding sections the most fundamental classes of Petri nets have already been introduced. The intention behind this section is to give an overview of other extensions of Petri nets which can be used as high-level modeling formalisms for performance evaluation.

2.3.6.1 DSPNs Deterministic and stochastic Petri nets (DSPNs) are among common extensions of GSPNs. Besides exponentially distributed and immediate transitions, transitions with deterministic firing times are also allowed. Often the definition of DSPNs already contains some restrictions regarding the race policy or, for example, that in each marking at most one deterministic transition is allowed to be activated at a time. Furthermore, the firing policy of deterministic transitions has to be enabling memory [Germ00]. However, these restrictions are not implied by the concept of native (stochastic) Petri nets but by the solution method that is to be used later on. DSPNs are quite popular as they belong to the class of non-Markovian systems for which specialized solvers exist. Section 2.3.7.2 discusses some solution techniques for the efficient analysis of DSPNs.

2.3.6.2 MRSPNs Markov regenerative stochastic Petri nets (MRSPNs) (sometimes also called *extended DSPNs*) are an extension of SPNs that allow the association of nonexponential distributions with timed transitions. These nonexponentially distributed firing time transitions are referred to as general transitions in the remainder of this section.

With the introduction of generally distributed firing times, the behavior of the marking process depends on the past. Therefore, some memory is needed to control the firing of the generally distributed transitions. The usual way to implement this [TBJ+95] is to associate a memory variable a_g with each general transition that monitors the time for which the transition has been enabled. Additionally, a memory policy is needed for defining how the memory variable of a general transition is to be reset, depending on the enabling and disabling of the corresponding transition. There are three memory policies that have been defined in [ABB+89] and which have become somewhat of a standard:

- Resampling policy: The memory variable is reset to zero at any change of the marking.

- Enabling memory policy: The memory variable is increased as long as the corresponding transition is enabled. When the transition is disabled the memory variable is reset to zero. This policy is also often referred to as prd (preemptive repeat different) in literature [Germ00].

- Age memory policy: The memory variable is increased as long as the corresponding transition is enabled. The memory variable is only reset when the transition fires. This policy is also known as prs (preemptive resume) in literature [Germ00].

At the entrance of a tangible marking, the residual firing time of each enabled transition is computed and the next marking is determined by the minimum residual firing time among the enabled transitions (race policy [ABB+89]). For exponential transitions the value of the age variable can be assumed to be zero because of the memoryless property (the memory is irrelevant for determining the residual firing time). Therefore, the history of the marking process is accounted for through the age variables associated to the general transitions.

According to [TBJ+95], a *regeneration time point* in the marking process is the epoch of entrance in a marking in which all memory variables are equal to zero. This means that at regeneration time points no information about the history is needed.

The sequence of regeneration time points (if existent) is a Markov renewal sequence and the marking process itself is a Markov regenerative process MRGP [CGL94]. A Markov Regenerative Stochastic Petri Net (MRSPN) is now defined as a SPN with general transitions for which an embedded Markov renewal sequence exists [CKT94].

Note that DSPNs form a special sub-class of MRSPNs.

2.3.6.3 FSPNs Fluid Stochastic Petri Nets (FSPNs) represent an expansion of the SPN-formalism where places are allowed to contain fluid tokens and arcs can also represent a fluid flow. There are two main reasons for the introduction of FSPNs:

- There may be physical systems that cannot be modeled by discrete-state SPNs as they contain continuous fluid like quantities and a discrete controlling logic [HKNT98].

- The fluid in the net is used to approximate discrete token movement and can therefore contribute to a smaller reachability graph [TrKu93].

An FSPN therefore has two types of places: Discrete places containing a non-negative number of tokens (as in SPNs) and continuous places containing fluid. Continuous places are usually depicted by two concentric circles.

The transitions in an FSPN can be exponentially distributed or immediate. Transition firings are determined by both discrete and continuous places. Fluid flow is only permitted through the enabled timed (exponential) transitions in the net. Therefore, two different types of arcs exist: Arcs connecting discrete places with transitions, drawn as single arrows, and arcs connecting continuous places with exponential transitions, depicted as double-arrows or dotted arrows. Figure 2.26 shows an example of an FSPN that is given in [TrKu93] and models a machine breakdown system. Work is considered to be a non-negative real quantity and is therefore accumulated in the continuous place r. On the other side there are N machines performing the work, which

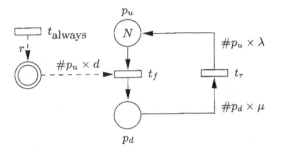

Fig. 2.26 Example of an FSPN.

are originally all functioning, represented by N discrete tokens in the place p_u. If one of these machines fails, a token is moved to p_d and after the repair, modeled by transition t_r, back to p_u.

The state of an FSPN is therefore built up by two parts: The marking of the discrete places of the net, which is given by the number of tokens in the corresponding places, and the fluid level in each of the continuous places.

As well as in the GSPN case, guards are allowed to be associated with transitions, but the enabling of immediate transitions is only allowed to be based on discrete places. A dependence of guard functions for immediate transitions on continuous places would lead to vanishing markings that depend on the current fluid level in continuous places. This type of vanishing markings could not be eliminated in a manner analogous to that of GSPNs.

A further addition compared to GSPNs is that a flow rate function is associated with arcs connecting a continuous place with a timed transition. This rate may be dependent on the current state of the net and defines the rate at which fluid leaves the input place and enters the output place.

[HKNT98] and [TrKu93] give algorithms for the analysis of FSPNs. In most cases, examples in the literature are limited to FSPNs that contain only one continuous place. The reason for this is that the resulting system of equations becomes too hard to solve if there is more than one continuous place. [TrKu93] gives some ideas about the solution of FSPNs with more than one continuous place. Extensions exist to second order FSPN [Wolt99], simulation-based solution of FSPNs [CNT99] and solution of FSPNs with complex boundary conditions [CHT02].

2.3.7 Non-Markovian Models

While Markov chains have been well explored in the preceeding sections, this section is intended to give a brief overview of the ideas behind the most common methods that are available today to solve non-Markovian models.

2.3.7.1   The Method of Supplementary Variables   The method of supplementary variables is one attempt to solve non-Markovian systems analytically. The basic idea behind the method, which is well explained in [Germ00] for non-Markovian stochastic Petri nets, is to lead the non-Markovian case back to the Markovian case. Recall that the analysis of Markovian models is simple because the current state of the system gives a full description about the behavior of the system due to the memoryless-property of the exponential distribution.

Therefore, the idea is to supplement the state of the system with additional information so that the original marking together with the additional information fully describes the current system state and therefore the supplemented state again has memoryless property. For this purpose, supplementary variables, the so-called memory variables, are introduced for nonexponential distributions. These supplementary variables should remember how long the distributions they are connected to have already been enabled or, from a dual viewpoint, they should remember the remaining firing time of a transition. In the first case, the supplementary variables are called *age variables* in the latter case they are called *rest variables* [Germ00]. Both forms of supplementary variables can be used but most often age variables are meant if someone talks about supplementary variables.

With the introduction of the age variables, the expanded state of the system is now memoryless but the construction of the state equations is much more complex. In a stochastic Petri net, the simplest case for the method of supplementary variables would be where in each marking at most one nonexponential transition is enabled and all transitions have race policy enable repeat (prd, enabling memory policy). This leads to a solution method that only needs one supplementary variable. Even in this case the transient state equations lead to a system of partial differential equations but in this special case this system can be transformed into a system of ordinary differential equations and therefore be solved relatively easily [Germ00].

Theoretically, this method has no limitations and seems therefore appropriate for the basis of a universal solution method. However, with each additional supplementary variable that is introduced, the emerging state equations become more and more complex and it seems unlikely that there will exist an algorithm for the construction of the state equations if there are no limitations on the number of concurrently enabled nonexponential transitions or on the race-policies of these transitions. Even if one will implement a specialized solver that allows at most n concurrently enabled transitions in each marking there is the drawback that it is not possible to check at the SPN level if this system can be analyzed by the solver or not. One has to build up the state space of the SPN first, a time- and memory-consuming process, just to check if this SPN belongs to a class that can or cannot be analyzed with the existing solver. Additionally, there is the problem that the method can suffer from long computation times if there are distributions that have infinite support. One advantage however, for an application where the given limitations are satisfied, is that the problem of state-space explosion is not as bad as in the case when phase-type distributions are used for the solution (see Section 2.3.7.3). Furthermore, since the complete extended reachability graph of the SPN is generated first, questions about the existence of particular states can be answered which is not the case in a simulative solution.

2.3.7.2 Specialized Solvers for Particular Classes of Systems

Most specialized solvers concentrate on a class of SPNs that emerges from a restriction of Deterministic and Stochastic Petri nets (DSPNs), i.e., Petri nets in which the transitions are either immediate or fire according to a distribution that is exponentially or deterministically distributed. Other SPN classes have been excluded in the past as they seem to be either not relevant for practical applications or not tractable. The only major extension that has been made was the possibility to use phase-type distributions in addition to deterministic distributions [CiLi98].

In the case of a DSPN there are two major directions for the steady-state solution: One approach based on the method of supplementary variables was proposed in [Germ00]. This approach suffers from the restriction that at most one deterministic transition may be enabled in each state. The second approach is based on the idea of general state-space Markov chains (GSSMCs) as presented by Lindemann et al. in [LiSh96]. This approach allows the analysis of SPNs with concurrently enabled deterministic transitions and is more efficient than the method of supplementary variables. However, it is currently limited to the race with enabling memory execution policy.

In case of the transient solution of DSPNs, there are currently three different directions that are treated in the literature: a solution based on the method of supplementary variables [HeGe97], a method based on Markov regenerative theory [CKT93], and a method based on general state-space Markov chains [LiTh99]. The general ideas of the supplementary variable approach and the method based on Markov regenerative theory are compared in [GLT95]. There are two variants of the method based on Markov regenerative theory, depending on whether the analysis is performed in the time-domain, leading to a discretization of the continuous time-variable, or in the Laplace domain, leading to a numerical inversion of a Laplace–Stieltjes transform. As both of these methods are problematic due the size of the state space and the needed computation time, recent research has mostly concentrated on developing methods that allow an approximation of the true solution with less space and computation time. One example of such an approximation method is given in [CiLi98]. The third direction is the extension of the GSSMC approach that was utilized for the steady-state solution earlier. This approach allows concurrent deterministic transitions but is, as in the steady-state case, currently limited to the race with enabling memory execution policy. As the emerging state equations are a system of multidimensional Fredholm integral equations which can be solved numerically more efficiently than a system of partial differential equations, which emerge in the supplementary variable case, this method is substantially faster than the method of supplementary variables.

There are many specialized solvers for different classes of SPNs, but all of them have two disadvantages in common: First, a huge effort is necessary to solve a very limited class of SPNs and often only a slight variation of the SPN specification may render the solution method unusable. Additionally, these specialized solvers are not very intuitive so that the development of such a solver is a time-consuming and maybe error-prone process. Second, the applicability of most of the solvers can only be decided after the extended reachability graph of the SPN is generated. Recall that time and storage needed for the state-space generation is huge and thus wasted if the solvers are later found to be not applicable.

One advantage of all of these specialized solvers is, however, that they are nearly ideal if applied to their special problem domain. This includes time and storage requirements on the one hand and accuracy of the computed solution on the other.

2.3.7.3 *Solution Using Phase-Type Distributions*   Like the supplementary variable approach, the phase-type distribution (PH) also is a method of Markovizing a non-Markovian model and as such has a wide applicability. But unlike the supplementary variable approach, the PH approach encodes the memory of a nonexponential distribution in a set of discrete phases.

A discrete phase-type distribution (DPH) is defined as the time to absorption in a discrete state discrete-time Markov chain (DTMC) and a continuous phase-type distribution (CPH) is defined as the time to absorption in a discrete-state continuous-time Markov chain (CTMC). As shown by Cox in [Cox55] every continuous distribution can be approximated arbitrarily well by a phase-type distribution. Informally speaking, phase-type distributions are a combination of distributions with memoryless property. In the discrete case the geometric distribution is therefore used and in the continuous case the exponential distribution is used.

Unfortunately, there is no unique representation for phase-type distributions and sometimes a special phase-type distribution could be more appropriate to approximate the behavior of a given nonexponential (or nongeometric) distribution than another one. Canonical forms are shown in [BHST02] for DPHs and for CPHs in [Cuma82, BoCu92]. Figure 2.27 shows one of these canonical forms.

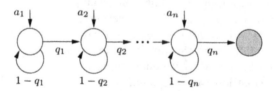

Fig. 2.27   Canonical form of a DPH.

Here, a_i is the probability of starting in the phase i and q_i is the probability of leaving phase i. The constraints are as follows:

$$\sum_{i=1}^{n} a_i = 1.$$
$$0 < q_i < q_{i+1} \leq 1, \ 1 \leq i \leq n - 1.$$

One of the main problems when using this approach is to find a phase-type distribution that approximates a given distribution in an optimal way. Several approximation methods for this purpose have been developed [BHST02, BuHe77, ANO96, FeWh96, MaRe93]. The replacement of the nonexponential distribution by a phase-type distribution can be done on the model level [Cuma85] or on the state-space level.

Because of the known disadvantages of the replacement on the model level the replacement on the state-space level has become common. The detailed algorithms for constructing the phase-expanded reachability graph are given in [BoTe97] for the CPH-case and in [Ciar95] for the DPH-case correspondingly.

The advantage of using DPHs instead of CPHs is that in general less number of phases are needed for the fit of a non-Markovian distribution. However, this advantage comes at the costs of a more complicated algorithm to build up the phase-expanded state space and a more serious state-space explosion problem than in the CPH-case as now simultaneous events within one time step have to be considered.

The main advantage of using phase-type distributions is the mathematical simplicity. The system to solve is lead back algorithmically to a Markovian system and therefore one only has to deal with systems of ODEs or linear equations, a field that has been covered in the past and for which a lot of evaluated solvers have already been developed and are widely available. Therefore, the use of phase-type distributions is very intuitive what in turn leads to an easy extendibility of the method. If one uses, for example, the method of supplementary variables for the analysis of stochastic Petri nets and changes from race policy race enable (prd) to race age (prs) the analysis algorithm has to be completely changed. When using phase-type distributions instead, one just has to change the setting of the phases for the distributions the race policies of which are changed. As a consequence, phase-type distributions seem to be an ideal means for constructing a universal solver for non-Markovian systems.

However, there are some drawbacks that have to be mentioned: First of all, for the use of phase-type distributions a fitting method is needed that can approximate all nonexponential distributions (and in the case of DPHs also exponential distributions) used in the system under investigation by phase-type distributions. Some examples for fitting methods have been given above, but each of these methods is rather limited to a special class of distributions that can be approximated fairly well such that one needs a whole bunch of fitting algorithms to cover a wide range of distributions. A universal fitting method that can approximate CPHs and DPHs arbitrarily well is still lacking today.

The second drawback is that the construction of the phase-expanded reachability graph is time- and memory-consuming. The reason for the high memory consumption is that the state-space explosion problem is increased by the introduction of phase-type distributions. If one had a state space with, for example, 10^7 states, which can be handled with standard state-space storage techniques, and uses five phase-type distributions, each of which has 10 phases, the state space can blow up in the worst case to 10^{12} states. This will require specialized storage techniques that can exploit the structure of the state space to treat these large state spaces. Section 2.3.8 will introduce techniques which are suitable to handle such state spaces.

Moreover, phase-type distributions are not really appropriate for all types of systems. Theoretically, each continuous distribution function can be approximated arbitrarily well by a phase-type distribution [Cox55]. In reality however, the exact approximation of a deterministic distribution by an Erlang distribution for example would need infinitely many phases. As this is impossible one always has to find the balance between the number of phases that are used for the approximation (and which aggravate the state-space explosion problem) and the accuracy that is needed in the analysis. Therefore, it may happen in the phase-type-analysis, that in a situation where two deterministic distributions are enabled concurrently and one fires after one time unit and the other one after 1.1 time units, the one with the larger firing time may fire first, although this is impossible from a theoretical point of view.

2.3.8 Symbolic State Space Storage Techniques

The main limitation when using phase-type distributions is the aggravation of the state-space explosion problem which seems to render this method unusable for real world problems. However, this state-space explosion problem often exists when generating the reachability graph of an SPN, even without phase expansions.

When the reachability graph of an SPN is generated, there are two major structures that have to be stored. All the reachable states have to be stored and indices assigned to these states to be able to subsequently calculate per state performance quantities. Likewise, all transitions between the states have to be stored. In the case of a CTMC or a phase-expanded CTMC, this information constitutes its generator matrix and the dimension of the matrix is the size of the state space. There have been some attempts in the recent past to deal with such huge state spaces and reachability graphs if they are sufficiently structured.

The basic idea behind these techniques is to use binary decision diagrams (BDDs) or extensions of BDDs [HMS99, Park02]. BDDs have become famous for the compact encoding of Boolean functions, i.e., $f : \{0,1\}^n \rightarrow \{0,1\}$. Multi-Terminal Binary Decision Diagrams (MTBDDs) (also called algebraic decision diagrams) are an extension of BDDs that allow as leaves any finite set, for example a subset of $\mathbb{R}$, i.e., $f : \{0,1\}^n \to \mathbb{R}$.

The use of MTBDDs for coding the generator matrix is quite straightforward: The different matrix entries represent the leaves of the MTBDD while a Boolean coding is used for the row and column indices of these entries which define the nonterminal nodes of the MTBDD. Besides the coding of row and column indices into a binary form, an appropriate variable-ordering is also needed.

An example of the coding of the matrix

$$
\begin{array}{c|cccc}
x_1x_2/y_1y_2 & 00 & 01 & 10 & 11 \\
\hline
00 & 0 & \cdot\frac{1}{2} & \frac{1}{3} & \frac{1}{2} \\
01 & \frac{1}{3} & \frac{1}{3} & \frac{1}{2} & 0 \\
10 & 0 & 0 & 0 & \frac{1}{2} \\
11 & 0 & 0 & \frac{1}{3} & \frac{1}{3}
\end{array}
$$

into an MTBDD is given in Fig. 2.28. Row indices are encoded into the variables x_i while column indices are encoded into the variables y_i. A simple binary coding for the row and column indices is used in this example (if the size of the matrix is not a power of two, the matrix has to be padded with an appropriate number of rows and columns of zeroes but these additional entries do not contribute to the size of the MTBDD).

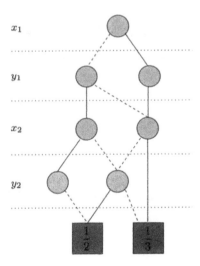

Fig. 2.28 Example for coding a matrix with an MTBDD.

Each nonterminal level of the MTBDD represents one of the row or column variables. Edges encoding the else-branch (zeroes) are drawn as dotted lines while edges encoding the then-branch (ones) are drawn solid. Edges going to the terminal node zero are omitted. The matrix entries together with their coding are also listed in Table 2.22.

Table 2.22   Coding of the matrix into an MTBDD

x_1	y_1	x_2	y_2	Value
0	0	0	1	$\frac{1}{2}$
0	0	1	0	$\frac{1}{3}$
0	0	1	1	$\frac{1}{3}$
0	1	1	0	$\frac{1}{2}$
0	1	0	1	$\frac{1}{2}$
0	1	0	0	$\frac{1}{3}$
1	1	0	1	$\frac{1}{2}$
1	1	1	0	$\frac{1}{3}$
1	1	1	1	$\frac{1}{3}$

In this example, a reduced and ordered MTBDD (ROMTBDD) is implicitly used as it has become common to use ROMTBDDs. An MTBDD is reduced iff

- for each nonterminal vertex v the two children are distinct and each terminal vertex has a distinct value.

- no two nodes on the same level that have the same subgraph.

For further information on this topic, the reader is encouraged to study [HMS99, Park02].

3

Steady-State Solutions of Markov Chains

In this chapter, we restrict ourselves to the computation of the steady-state probability vector[1] of *ergodic* Markov chains. Most of the literature on solution techniques of Markov chains assumes ergodicity of the underlying model. A comprehensive source on algorithms for steady-state solution techniques is the book by Stewart [Stew94].

From Eq. (2.15) and Eq. (2.58), we have $\nu = \nu\mathbf{P}$ and $\mathbf{0} = \pi\mathbf{Q}$, respectively, as points of departure for the study of steady-state solution techniques. Equation (2.15) can be transformed so that

$$\mathbf{0} = \nu(\mathbf{P} - \mathbf{I}). \tag{3.1}$$

Therefore, both for CTMC and DTMC, a linear system of the form

$$\mathbf{0} = \mathbf{x}\mathbf{A} \tag{3.2}$$

needs to be solved. Due to its type of entries representing the parameters of a Markov chain, matrix $\mathbf{A}$ is singular and it can be shown that $\mathbf{A}$ is of rank $n - 1$ for any Markov chain of size $|S| = n$. It follows immediately that the resulting set of equations is not linearly independent and that one of the equations is redundant. To yield a unique, positive solution, we must impose a normalization condition on the solution $\mathbf{x}$ of equation $\mathbf{0} = \mathbf{x}\mathbf{A}$. One way to approach the solution of Eq. (3.2) is to directly incorporate the normalization

---

[1]For the sake of convenience we sometimes use the term "steady-state analysis" as a shorthand notation.

condition

$$\mathbf{x}\mathbf{1} = 1 \tag{3.3}$$

into Eq. (3.2). This can be regarded as substituting one of the columns (say, the last column) of matrix $\mathbf{A}$ by the unit vector $\mathbf{1} = [1, 1, \ldots, 1]^T$. With a slight abuse of notation, we denote the new matrix also by $\mathbf{A}$. The resulting linear system of nonhomogeneous equations is

$$\mathbf{b} = \mathbf{x}\mathbf{A}, \quad \mathbf{b} = [0, 0, \ldots, 0, 1]. \tag{3.4}$$

An alternative to solving Eq. (3.2) is to separately consider normalization Eq. (3.3) as an additional step in numerical computations. We demonstrate both ways when example studies are presented. It is worthwhile pointing out that for any given ergodic CTMC, a DTMC can be constructed that yields an identical steady-state probability vector as the CTMC, and vice versa. Given the generator matrix $\mathbf{Q} = [q_{ij}]$ of a CTMC, we can define

$$\mathbf{P} = \mathbf{Q}/q + \mathbf{I}, \tag{3.5}$$

where q is chosen such that $q > \max_{i,j \in S} |q_{ij}|$. Setting $q = \max_{i,j \in S} |q_{ij}|$ should be avoided in order to assure *aperiodicity* of the resulting DTMC [GLT87]. The resulting matrix $\mathbf{P}$ can be used to determine the steady-state probability vector $\boldsymbol{\pi} = \boldsymbol{\nu}$, by solving $\boldsymbol{\nu} = \boldsymbol{\nu}\mathbf{P}$ and $\boldsymbol{\nu}\mathbf{1} = 1$. This method, used to reduce a CTMC to a DTMC, is called *randomization* or sometimes *uniformization* in the literature. If, on the other hand, a transition probability matrix $\mathbf{P}$ of an ergodic DTMC were given, a generator matrix $\mathbf{Q}$ of a CTMC can be defined by

$$\mathbf{Q} = \mathbf{P} - \mathbf{I}. \tag{3.6}$$

By solving $\mathbf{0} = \boldsymbol{\pi}\mathbf{Q}$ under the condition $\boldsymbol{\pi}\mathbf{1} = 1$, the desired steady-state probability vector $\boldsymbol{\pi} = \boldsymbol{\nu}$ can be obtained.

To determine the steady-state probabilities of finite Markov chains, three different approaches for the solution of a linear system of the form $\mathbf{0} = \mathbf{x}\mathbf{A}$ are commonly used: *direct* or *iterative numerical* methods and techniques that yield *closed-form* results. Both types of numerical methods have merits of their own. Whereas direct methods yield exact results,[2] iterative methods are generally more efficient, both in time and space. Disadvantages of iterative methods are that for some of these methods no guarantee of convergence can be given in general and that determination of suitable error bounds for termination of the iterations is not always easy. Since iterative methods are considerably more efficient in solving Markov chains, they are commonly used for larger models. For smaller models with fewer than a few thousand states, direct methods are reliable and accurate. Though closed-form results are highly desirable, they can be obtained for only a small class of models that have some structure in their matrix.

---

[2]Modulo round-off errors resulting from finite precision arithmetic.

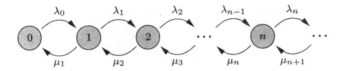

Fig. 3.1 Birth–death process.

Problem 3.1   Show that $\mathbf{P} - \mathbf{I}$ has the properties of a CTMC generator matrix.

Problem 3.2   Show that $\mathbf{Q}/q + \mathbf{I}$ has the properties of a stochastic (DTMC) matrix.

Problem 3.3   Define a CTMC and its generator matrix $\mathbf{Q}$ so that the corresponding DTMC would be periodic if randomization were applied with $q = \max_{i,j \in S} |q_{ij}|$ in Eq. (3.5).

3.1   SOLUTION FOR A BIRTH–DEATH PROCESS

Birth–death processes are Markov chains where transitions are allowed only between neighboring states. We treat the continuous-time case here, but analogous results for the discrete-time case are easily obtained (see [Triv01] for details).

A one-dimensional birth–death process is shown in Fig. 3.1 and its generator matrix is shown as Eq. (3.7):

$$\mathbf{Q} = \begin{pmatrix} -\lambda_0 & \lambda_0 & 0 & 0 & \cdots \\ \mu_1 & -(\lambda_1 + \mu_1) & \lambda_1 & 0 & \cdots \\ 0 & \mu_2 & -(\lambda_2 + \mu_2) & \lambda_2 & \cdots \\ 0 & 0 & \mu_3 & -(\lambda_3 + \mu_3) & \cdots \\ \vdots & \vdots & \vdots & \vdots & \ddots \end{pmatrix}. \tag{3.7}$$

The transition rates $\lambda_k, k \geq 0$ are state-dependent *birth rates* and $\mu_l, l \geq 1$, are referred to as state dependent *death rates*. Assuming ergodicity, the steady-state probabilities of CTMCs of the form depicted in Fig. 3.1 can be uniquely determined from the solution of Eq. (2.58):

$$0 = -\pi_0 \lambda_0 + \pi_1 \mu_1 \,, \tag{3.8}$$

$$0 = -\pi_k (\lambda_k + \mu_k) + \pi_{k-1} \lambda_{k-1} + \pi_{k+1} \mu_{k+1} \,, \quad k \geq 1. \tag{3.9}$$

Solving Eq. (3.8) for π_1, and then using this result for substitution with $k = 1$ in Eq. (3.9) and solving it for π_2 yields

$$\pi_1 = \frac{\lambda_0}{\mu_1} \pi_0 \,, \qquad \pi_2 = \frac{\lambda_0 \lambda_1}{\mu_1 \mu_2} \pi_0 \,. \tag{3.10}$$

Equation (3.10) together with Eq. (3.9) suggest a general solution of the following form:

$$\pi_k = \pi_0 \cdot \prod_{i=0}^{k-1} \frac{\lambda_i}{\mu_{i+1}}, \quad k \geq 1. \tag{3.11}$$

Indeed, Eq. (3.11) provides the unique solution of a one-dimensional birth–death process. Since it is not difficult to prove this hypothesis by induction, we leave this as an exercise to the reader. From the law of total probability, $\sum_i \pi_i = 1$, we get for the probability π_0 of the CTMC being in State 0:

$$\pi_0 = \frac{1}{1 + \sum\limits_{k=1}^{\infty} \prod\limits_{i=0}^{k-1} \frac{\lambda_i}{\mu_{i+1}}} = \frac{1}{\sum\limits_{k=0}^{\infty} \prod\limits_{i=0}^{k-1} \frac{\lambda_i}{\mu_{i+1}}}. \tag{3.12}$$

The condition for convergence of the series in the denominator of Eq. (3.12), which is also the condition for the ergodicity of the birth–death CTMC, is

$$\exists k_0, \quad \forall k > k_0 : \quad \frac{\lambda_k}{\mu_k} < 1. \tag{3.13}$$

Equations (3.11) and (3.12) are used extensively in Chapter 6 to determine the probabilities π_k for many different queueing systems. These probabilities are then used to calculate performance measures such as mean queue length, or mean waiting time for these queueing systems. We deal with multi-dimensional birth–death processes in Section 3.2 and in Chapter 7.

A special case of a birth–death process arises from M/M/1 queueing systems that will be treated in Chapter 7. These systems can be modeled as birth–death processes with state-independent birth and death rates:

$$\lambda_k = \lambda, \quad k \geq 0, \tag{3.14}$$

$$\mu_l = \mu, \quad l \geq 1. \tag{3.15}$$

Equation (3.11) then boils down to the geometric pmf:

$$\pi_k = \pi_0 \left(\frac{\lambda}{\mu}\right)^k, \quad k \geq 0, \tag{3.16}$$

Problem 3.4   Consider a discrete-time birth–death process with birth probability b_i, the death probability d_i, and no state change probability $1 - b_i - d_i$ in state i. Derive expressions for the steady-state probabilities and conditions for convergence [Triv01].

3.2   MATRIX-GEOMETRIC METHOD: QUASI-BIRTH–DEATH PROCESS

In analogy to the scalar (and continuous-time) birth–death process presented in Section 3.1, that had the geometric solution of Eq. (3.11), a more general case is investigated next. This case is suitable for a class of vector state processes that are called *quasi-birth–death processes* (QBDs). It will be shown that the QBDs' infinite infinitesimal generator matrix can be described in a block-tridiagonal form with repetitive elements, where the solution of the steady-state probability vector can be given in matrix-geometric form [Neut81]. The repetitive structure of this form allows the calculation of the state probabilities quite conveniently and efficiently. This approach is known as the *matrix-geometric method* (MGM).

The MGM is found to be useful not only for finding solutions of continuous- and discrete-time QBDs, but also for models of a wide range of systems that include repetitive structures. These systems normally do not meet the requirements needed for a "classical" closed-form solution as presented in Section 3.1. In Chapter 6 the MGM is used for the analysis of GI/M/1-type Markov chains that can be seen as generalizations of QBDs. Furthermore, a slightly more complex but related method, the *matrix-analytic method* (MAM), is introduced in Chapter 6 to analyze M/G/1-type Markov chains.

3.2.1   The Concept

The MGM (and the MAM) can be applied to a multidimensional state space. The method also applies to DTMCs (see Chapter 6), but here we restrict ourselves to two-dimensional CTMCs. The state space S is then given as set of pairs:

$$S = \{(i,j)|0 \leq i, 0 \leq j \leq m\}. \tag{3.17}$$

The parameter i of the first dimension is called the *level* of the state. There may be an infinite number of levels. The parameter j of the second dimension refers to one of $m + 1$ states within a level (also called *inter-level states* or *phases*).

The MGM is based on dividing the levels of the underlying CTMC into two parts[3]:

- The *boundary* (or initial) levels $0 \leq i < l$: This part must be finite but can have a nonregular structure, i.e., the levels may differ from each other.

- The *repeating* (or repetitive) levels $i \geq l$: This part has to have a regular structure but may be infinite.

---

[3]We denote the first repeating level as l.

The elements of the generator matrix $\mathbf{Q}$ are sorted lexicographically and can then be grouped into finite submatrices. Each submatrix represents the rates for a special class of transitions[4]:

The submatrices of the boundary part are denoted by $\mathbf{B}$:

- $\mathbf{B}_{i,i}$: transitions within level $i < l$.

- $\mathbf{B}_{i,i'}$: transitions from level $i < l$ to level $i', 0 \leq i' < \infty, i' \neq i$.

The submatrices of the repeating part are denoted by $\mathbf{A}$:

- $\mathbf{A}_0$: transitions from level i to level $i + 1, \forall i \geq l$.

- $\mathbf{A}_1$: transitions within level $i, \forall i \geq l$.

- $\mathbf{A}_k$: transitions from level i to level $i - (k - 1)$, for $1 < k < i$ and $\forall i \geq l$.

Note that all $\mathbf{A}_0$, $\mathbf{A}_1$, and $\mathbf{A}_k$ are invariant with respect to the absolute value of $i \geq l$. Due to structure of the generator matrix $\mathbf{Q}$, the matrices $\mathbf{B}_{i,i'}$, $\mathbf{A}_0$, and $\mathbf{A}_k$ are non-negative and the matrices $\mathbf{B}_{i,i}$ and $\mathbf{A}_1$ have non-negative off-diagonal elements and negative diagonal elements.

To obtain the steady-state probability vector, the infinite infinitesimal generator matrix $\mathbf{Q}$ as a whole need not be explicitly taken into account. Instead, the finite submatrices of $\mathbf{Q}$ can be used. For this the so-called *rate matrix* $\mathbf{R}$ is constructed depending on the actual form of the repeating part of the generator matrix $\mathbf{Q}$. The main remaining problem is to derive the matrix $\mathbf{R}$ either directly by explicit knowledge of the systems behavior or iteratively (e.g., using successive approximation) and solving the boundary equations of the Markov chain (e.g., by using Gaussian elimination).

3.2.2   Example: The QBD Process

As a simple example of a process that can be analyzed by using the MGM, one-step transitions are restricted to phases within the same level or between phases of adjacent levels. The resulting CTMC is then a QBD. The state transition diagram of a QBD is shown in Fig. 3.2, which is comparable to the CTMC of the birth–death process from Fig. 3.1. This QBD can be interpreted as a birth–death process in a *random environment* with $m + 1$ different states (see, e.g., [LaRa99, Neut79]). The states of the process are ordered lexicographically according to levels and phases:

$$S = \{(0,0), (0,1), \ldots, (0,m), (1,0), (1,1), \ldots, (1,m), (2,0), \ldots\}, \quad (3.18)$$

---

[4]The partitioning described here holds for the MGM. The MAM is treated separately in Chapter 6.

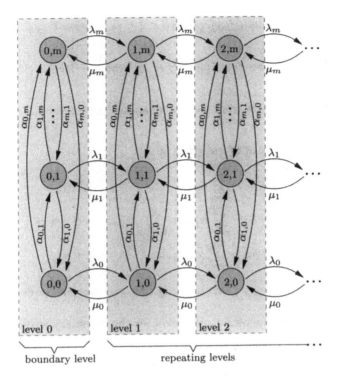

Fig. 3.2 Quasi-birth–death process.

which yields the following structured generator matrix $\mathbf{Q}$ is derived:

$$
\mathbf{Q} =
\left(
\begin{array}{cccc|cccc|ccc}
a_0 & \alpha_{0,1} & \cdots & \alpha_{0,m} & \lambda_0 & 0 & \cdots & 0 & 0 & 0 & \cdots \\
\alpha_{1,0} & a_1 & \cdots & \alpha_{1,m} & 0 & \lambda_1 & \cdots & 0 & 0 & 0 & \cdots \\
\vdots & \vdots & \ddots & \vdots & \vdots & \vdots & \ddots & \vdots & \vdots & \vdots & \ddots \\
\alpha_{m,0} & \alpha_{m,1} & \cdots & a_m & 0 & 0 & \cdots & \lambda_m & 0 & 0 & \cdots \\
\hline
\mu_0 & 0 & \cdots & 0 & b_0 & \alpha_{0,1} & \cdots & \alpha_{0,m} & \lambda_0 & 0 & \cdots \\
0 & \mu_1 & \cdots & 0 & \alpha_{1,0} & b_1 & \cdots & \alpha_{1,m} & 0 & \lambda_1 & \cdots \\
\vdots & \vdots & \ddots & \vdots & \vdots & \vdots & \ddots & \vdots & \vdots & \vdots & \ddots \\
0 & 0 & \cdots & \mu_m & \alpha_{m,0} & \alpha_{m,1} & \cdots & b_m & 0 & 0 & \cdots \\
\hline
0 & 0 & \cdots & 0 & \mu_0 & 0 & \cdots & 0 & b_0 & \alpha_{0,1} & \cdots \\
0 & 0 & \cdots & 0 & 0 & \mu_1 & \cdots & 0 & \alpha_{1,0} & b_1 & \cdots \\
\vdots & \vdots & \ddots & \vdots & \vdots & \vdots & \ddots & \vdots & \vdots & \vdots & \ddots \\
\end{array}
\right).
$$

$$(3.19)$$

The off-diagonal elements of $\mathbf{Q}$ are given by the state transition rates of the QBD:

- $\alpha_{j,j'}$: Transition rate from phase $0 \le j \le m$ to phase $0 \le j' \le m$, $j' \ne j$, within a level.

- λ_j: Transition rate from phase j of an arbitrary level $i \ge 0$ to phase j of level $i + 1$.

- μ_j: Transition rate from phase j of an arbitrary level $i > 0$ to phase j of level $i - 1$.

The diagonal elements of $\mathbf{Q}$ are given by

$$a_j = \begin{cases} -\lambda_0 - \sum_{n=1}^{m} \alpha_{0,n}\,, & j = 0\,, \\ -\lambda_j - \sum_{n=0}^{j-1} \alpha_{j,n} - \sum_{n=j+1}^{m} \alpha_{j,n}\,, & 0 < j < m\,, \\ -\lambda_m - \sum_{n=0}^{m-1} \alpha_{m,n}\,, & j = m\,, \end{cases} \tag{3.20}$$

and $b_j = a_j - \mu_j$, $0 \le j \le m$, ensuring that the elements of each row of $\mathbf{Q}$ sum up to 0.

After defining the following submatrix for the boundary level $i = 0$ (here: $l = 1$):

$$\mathbf{B}_{0,0} = \begin{pmatrix} a_0 & \alpha_{0,1} & \cdots & \alpha_{0,m} \\ \alpha_{1,0} & a_1 & \cdots & \alpha_{1,m} \\ \vdots & \vdots & \ddots & \vdots \\ \alpha_{m,0} & \alpha_{m,1} & \cdots & a_m \end{pmatrix}, \tag{3.21}$$

and the following submatrices for the repeating levels $i \ge 1$:

$$\mathbf{A}_0 = \begin{pmatrix} \lambda_0 & 0 & \cdots & 0 \\ 0 & \lambda_1 & \cdots & 0 \\ \vdots & \vdots & \ddots & \vdots \\ 0 & 0 & \cdots & \lambda_m \end{pmatrix}, \quad \mathbf{A}_1 = \begin{pmatrix} b_0 & \alpha_{0,1} & \cdots & \alpha_{0,m} \\ \alpha_{1,0} & b_1 & \cdots & \alpha_{1,m} \\ \vdots & \vdots & \ddots & \vdots \\ \alpha_{m,0} & \alpha_{m,1} & \cdots & b_m \end{pmatrix},$$

$$\mathbf{A}_2 = \begin{pmatrix} \mu_0 & 0 & \cdots & 0 \\ 0 & \mu_1 & \cdots & 0 \\ \vdots & \vdots & \ddots & \vdots \\ 0 & 0 & \cdots & \mu_m \end{pmatrix}, \tag{3.22}$$

the generator matrix can be written in block form:

$$\mathbf{Q} = \begin{pmatrix} \mathbf{B}_{0,0} & \mathbf{A}_0 & 0 & 0 & \cdots \\ \mathbf{A}_2 & \mathbf{A}_1 & \mathbf{A}_0 & 0 & \cdots \\ 0 & \mathbf{A}_2 & \mathbf{A}_1 & \mathbf{A}_0 & \cdots \\ 0 & 0 & \mathbf{A}_2 & \mathbf{A}_1 & \cdots \\ \vdots & \vdots & \vdots & \vdots & \ddots \end{pmatrix}. \tag{3.23}$$

Matrix $\mathbf{Q}$ in Eq. (3.23) is block-tridiagonal, block analog of the tridiagonal generator matrix of the birth–death process from Eq. (3.7), with the simplification that the birth and death rates of the presented QBD are not level-dependent.

Again, the task is to determine the steady-state probabilities. These are organized lexicographically into the steady-state probability vector $\boldsymbol{\pi}$:

$$\boldsymbol{\pi} = [\pi_{0,0}, \pi_{0,1}, \ldots, \pi_{0,m}, \pi_{1,0}, \pi_{1,1}, \ldots, \pi_{1,m}, \pi_{2,0}, \ldots]. \qquad (3.24)$$

Vector $\boldsymbol{\pi}$ is partitioned according to levels:

$$\boldsymbol{\pi} = [\boldsymbol{\pi}_0, \boldsymbol{\pi}_1, \boldsymbol{\pi}_2, \ldots], \qquad (3.25)$$

where

$$\boldsymbol{\pi}_i = [\pi_{i,0}, \pi_{i,1}, \ldots, \pi_{i,m}], \quad 0 \leq i < \infty. \qquad (3.26)$$

To yield $\boldsymbol{\pi}$, Eq. (2.58), $\mathbf{0} = \boldsymbol{\pi}\mathbf{Q}$, needs to be solved, while obeying the normalization condition of Eq. (2.59), $\boldsymbol{\pi}\mathbf{1} = 1$. Fortunately, the repetitive structure of the QBD can be exploited instead of having to handle the whole infinite generator matrix $\mathbf{Q}$. First, the global balance equations of the repeating levels $i \geq 1$ are considered:

$$\boldsymbol{\pi}_{i-1}\mathbf{A}_0 + \boldsymbol{\pi}_i\mathbf{A}_1 + \boldsymbol{\pi}_{i+1}\mathbf{A}_2 = \mathbf{0}, \quad i > 0. \qquad (3.27)$$

In analogy to Eq. (3.16) it can be shown that the vectors $\boldsymbol{\pi}_i$ can be defined in terms of the vectors $\boldsymbol{\pi}_{i-1}$. In addition, the transitions between the levels of the QBD are independent of level i. Therefore, a constant *rate matrix* $\mathbf{R}$ can be introduced that leads to the matrix-geometric equation:

$$\boldsymbol{\pi}_i = \boldsymbol{\pi}_{i-1}\mathbf{R} = \boldsymbol{\pi}_1\mathbf{R}^{i-1}, \quad i > 0. \qquad (3.28)$$

For any level $i > 0$ the element (j, j') of matrix $\mathbf{R}$ can be interpreted as the expected time spent in state (i, j') divided by the expected time spent in state $(i - 1, j)$ before returning to level $i - 1$, provided that the process observation started from state $(i - 1, j)$. Due to the fact that the elements are independent of the level $i > 0$, i can be set to 1 in this interpretation. To obtain the unknown rate matrix $\mathbf{R}$, $\boldsymbol{\pi}_i$ from Eq. (3.28) is inserted into Eq. (3.27):

$$\begin{aligned} \boldsymbol{\pi}_1\mathbf{R}^{i-2}\mathbf{A}_0 + \boldsymbol{\pi}_1\mathbf{R}^{i-1}\mathbf{A}_1 + \boldsymbol{\pi}_1\mathbf{R}^i\mathbf{A}_2 = \mathbf{0}, \quad i > 0, \\ \boldsymbol{\pi}_1(\mathbf{R}^{i-2}\mathbf{A}_0 + \mathbf{R}^{i-1}\mathbf{A}_1 + \mathbf{R}^i\mathbf{A}_2) = \mathbf{0}, \quad i > 0, \end{aligned} \qquad (3.29)$$

which can be simplified to

$$\mathbf{A}_0 + \mathbf{R}\mathbf{A}_1 + \mathbf{R}^2\mathbf{A}_2 = \mathbf{0}, \qquad (3.30)$$

because Eq. (3.29) holds for $i \geq 1$. We choose $i = 2$. A matrix $\mathbf{R}$ that satisfies Eq. (3.30) also satisfies Eq. (3.29).

Quadratic Eq. (3.30) can be solved for the rate matrix $\mathbf{R}$ that can then be used within Eq. (3.28) to calculate the steady-state probability vectors of the repeating levels. How this is done is described at the end of this section.

To derive $\boldsymbol{\pi}_0$ and $\boldsymbol{\pi}_1$, needed for using Eq. (3.28), the boundary level is considered, where $i = 0$. The boundary part of the equation $\boldsymbol{\pi}\mathbf{Q} = \mathbf{0}$ leads to

$$\boldsymbol{\pi}_0\mathbf{B}_{0,0} + \boldsymbol{\pi}_1\mathbf{A}_2 = \mathbf{0}\,,$$
$$\boldsymbol{\pi}_0\mathbf{A}_0 + \boldsymbol{\pi}_1\mathbf{A}_1 + \boldsymbol{\pi}_2\mathbf{A}_2 = \boldsymbol{\pi}_0\mathbf{A}_0 + \boldsymbol{\pi}_1\left(\mathbf{A}_1 + \mathbf{R}\mathbf{A}_2\right)) = \mathbf{0}\,. \tag{3.31}$$

In analogy to the scalar case in Eq. (3.12), $\boldsymbol{\pi}_0$ and $\boldsymbol{\pi}_1$ need to satisfy the normalization condition $\boldsymbol{\pi}\mathbf{1} = 1$. The normalization condition can be written as

$$\sum_{n=0}^{\infty} \boldsymbol{\pi}_n\mathbf{1} = 1\,. \tag{3.32}$$

Inserting Eq. (3.28) into Eq. (3.32) yields

$$\boldsymbol{\pi}_0\mathbf{1} + \sum_{n=0}^{\infty} \boldsymbol{\pi}_1\mathbf{R}^n\mathbf{1} = \boldsymbol{\pi}_0\mathbf{1} + \boldsymbol{\pi}_0\left(\sum_{n=0}^{\infty} \mathbf{R}^n\right)\mathbf{1} = 1\,. \tag{3.33}$$

For stability and ergodicity[5] reasons, the *drift* of the process to higher levels must be smaller than the drift to lower levels (see [Neut81]), similar to Eq. (3.13) where $\lambda_k < \mu_k$:

$$\boldsymbol{\pi}_A\mathbf{A}_0\mathbf{1} < \boldsymbol{\pi}_A\mathbf{A}_2\mathbf{1}\,, \tag{3.34}$$

where $\boldsymbol{\pi}_A$ is the steady-state probability vector of the generator matrix $\mathbf{A} = \mathbf{A}_0 + \mathbf{A}_1 + \mathbf{A}_2$. This leads to the condition that all eigenvalues of the rate matrix $\mathbf{R}$ are located within the unit circle; i.e., the spectral radius of $\mathbf{R}$ is smaller than one: $sp(\mathbf{R}) < 1$. This allows for the following simplification. The definition of the sum

$$\sum_{n=0}^{\infty} \mathbf{R}^n = \mathbf{I} + \mathbf{R} + \mathbf{R}^2 + \ldots \tag{3.35}$$

is equivalent to

$$\left(\sum_{n=0}^{\infty} \mathbf{R}^n\right)\mathbf{R} = \mathbf{R} + \mathbf{R}^2 + \mathbf{R}^3 + \ldots\,. \tag{3.36}$$

Subtracting Eq. (3.36) from Eq. (3.35) yields

$$\sum_{n=0}^{\infty} \mathbf{R}^n - \left(\sum_{n=0}^{\infty} \mathbf{R}^n\right)\mathbf{R} = \mathbf{I}\,. \tag{3.37}$$

---

[5]Stability conditions of QBDs are covered extensively in [LaRa99].

Factoring out $\sum\limits_{n=0}^{\infty} \mathbf{R}^n$ in Eq. (3.37) results in

$$(\sum_{n=0}^{\infty} \mathbf{R}^n)(\mathbf{I} - \mathbf{R}) = \mathbf{I} \,. \tag{3.38}$$

$(\mathbf{I} - \mathbf{R})$ is invertible because 0 is not one of its eigenvalues. Proof by contradiction: If $(\mathbf{I} - \mathbf{R})$ has an eigenvector v for eigenvalue 0 then $(\mathbf{I} - \mathbf{R})v = 0$. Therefore, $\mathbf{I}v = \mathbf{R}v$ and therefore, $v = \mathbf{R}v$, i.e., v is an eigenvector of $\mathbf{R}$ for eigenvalue 1. This contradicts the precondition $sp(\mathbf{R}) < 1$. The invertibility of $(\mathbf{I} - \mathbf{R})$ enables the following transformation of Eq. (3.38):

$$\sum_{n=0}^{\infty} \mathbf{R}^n = \mathbf{I}(\mathbf{I} - \mathbf{R})^{-1} \,. \tag{3.39}$$

Finally, using Eq. (3.39), Eq. (3.33) can be simplified to

$$\pi_0 \mathbf{1} + \pi_1 (\mathbf{I} - \mathbf{R})^{-1} \mathbf{1} = 1 \,. \tag{3.40}$$

To summarize, the steady-state probability vector of the QBD can be obtained using Eq. (3.30) for deriving $\mathbf{R}$, and Eq. (3.28) together with (3.31) and (3.40) for deriving π. The computation can be carried out in several ways:

Deriving $\mathbf{R}$: $\mathbf{R}$ can be derived directly, iteratively by successive substitution, or by logarithmic reduction.

Deriving $\mathbf{R}$ *Directly:* Some problems lend themselves well to the explicit determination of the matrix $\mathbf{R}$ explicitly, based on the stochastic interpretation of its elements. Examples of these special cases are given in [RaLa86] and [LeWi04].

Deriving $\mathbf{R}$ *Iteratively by Successive Substitution:* This basic method, described in [Neut81], makes use of

$$\mathbf{R}_{(0)} = \mathbf{0} \,, \tag{3.41}$$

and:

$$\mathbf{R}_{(n)} = -(\mathbf{A}_0 + \mathbf{R}_{(n-1)}^2 \mathbf{A}_2)\mathbf{A}_1^{-1} \,. \tag{3.42}$$

Eq. (3.42) is derived from Eq. (3.30) by multiplying $\mathbf{A}_1^{-1}$ on the right. Starting with Eq. (3.41) successive approximations of $\mathbf{R}$ can be obtained by using Eq. (3.42). The iteration is repeated until two consecutive iterates of $\mathbf{R}$ differ by less than a predefined tolerance ϵ:

$$||\mathbf{R}_{(n+1)} - \mathbf{R}_{(n)}|| < \epsilon \,, \tag{3.43}$$

where $||\cdot||$ is an appropriate matrix norm. The sequence $\{\mathbf{R}_{(n)}\}$ is entry-wise nondecreasing which can be proven by induction:

- Basis step:

$$\mathbf{R}_{(1)} = (\mathbf{A}_0 + \mathbf{R}_{(0)}^2 \mathbf{A}_2)(-\mathbf{A}_1^{-1}) = -\mathbf{A}_0 \mathbf{A}_1^{-1} \geq 0 = \mathbf{R}_{(0)}.$$

The matrices $-\mathbf{A}_1^{-1}$ and $\mathbf{A}_2$ are non-negative. For $\mathbf{A}_2$ this is readily seen considering the structure of $\mathbf{Q}$. $\mathbf{A}_1^{-1}$ is non-positive because $\mathbf{A}_1$ is diagonally dominant with negative diagonal and non-negative off-diagonal elements.

- Induction hypothesis:

$$\mathbf{R}_{(n+1)} \geq \mathbf{R}_{(n)}.$$

- Induction proposition:

$$\text{if} \quad \mathbf{R}_{(n+1)} \geq \mathbf{R}_{(n)} \quad \text{then} \quad \mathbf{R}_{(n+2)} \geq \mathbf{R}_{(n+1)} :$$

$$\begin{aligned}
\mathbf{R}_{(n+2)} &= (\mathbf{A}_0 + \mathbf{R}_{(n+1)}^2 \mathbf{A}_2)(-\mathbf{A}_1^{-1}) \\
&\geq (\mathbf{A}_0 + \mathbf{R}_{(n)}^2 \mathbf{A}_2)(-\mathbf{A}_1^{-1}) = \mathbf{R}_{(n+1)}, \\
\mathbf{R}_{(n+2)} &\geq \mathbf{R}_{(n+1)}. \quad q.e.d.
\end{aligned}$$

The monotone convergence of $\{\mathbf{R}_n\}$ towards $\mathbf{R}$ is shown in [Neut81].

Deriving $\mathbf{R}$ by Logarithmic Reduction [LaRa93, LaRa99]:   One of the most efficient methods to compute $\mathbf{R}$ is the logarithmic reduction algorithm presented by Latouche and Ramaswami [LaRa93]. Instead of focusing on the matrix $\mathbf{R}$, this method uses the auxiliary matrix $\mathbf{G}$ that can be derived more easily and can then be used to find $\mathbf{R}$. For each level $i > 0$ the elements (j, j') of matrix $\mathbf{G}$ can be interpreted as the probability that starting from state (i, j) the process returns to state $(i - 1, j')$ in finite time. Similar to the interpretation of the elements of $\mathbf{R}$, we can set $i = 1$. $\mathbf{G}$ can be shown to satisfy

$$\mathbf{A}_2 + \mathbf{A}_1 \mathbf{G} + \mathbf{A}_0 \mathbf{G}^2 = 0. \tag{3.44}$$

$\mathbf{G}$ can again be computed using successive approximations $\mathbf{G}_{(k)}, k = 1, 2, 3, \ldots$, with $\lim_{k \to \infty} \mathbf{G}_{(k)} = \mathbf{G}$. Due to the fact the QBD process under consideration being ergodic and therefore recurrent, the matrix $\mathbf{G}$ is stochastic, i.e., the elements of each row sum up to 1. The approximation algorithm terminates with $\|\mathbf{1} - \mathbf{G}\mathbf{1}\| < \epsilon$. Invariant to the value of $i > 0$ the element (j, j') of $\mathbf{G}_{(k)}$ can be interpreted as the probability that the process enters state $(i - 1, j')$, if it started from state (i, j), without having visited levels greater than level $i + k$. While the successive substitution algorithm is similar to the case where the value of k is increased by 1 in every iteration, the logarithmic reduction algorithm converges faster because k can be doubled in every iteration ($k = 1, 2, 4, 8, \ldots$). The details of the algorithm can be found in [LaRa93]. Having computed the matrix $\mathbf{G}$, the matrix $\mathbf{R}$ can be obtained from

$$\mathbf{R} = -\mathbf{A}_0 (\mathbf{A}_1 + \mathbf{A}_0 \mathbf{G})^{-1}. \tag{3.45}$$

Other algorithms for obtaining **R** were proposed, e.g., the *U*-algorithm in [Lato93], several linear and quadratically convergent algorithms in [LaRa99], and revised and improved forms of the logarithmic reduction algorithm in [Ye01] and [NKW97]. Complexity and performance comparisons of the above and other algorithms for solving QBD processes are provided in [Lato93, LaRa93, Have98, HaOs97, Ost01, TrDo00, Gün89].

Deriving $\pi_i, i \geq 0$: As soon as the rate matrix **R** is computed by the use of one of the methods mentioned above, the steady-state probability vectors π_i can be obtained quite easily by using Eqs. (3.28), (3.31), and (3.40). Due to the limited number and compact size of the boundary equations, i.e., Eq. (3.31) and Eq. (3.40), these equations can in general be solved using standard methods for solving linear systems of equations, e.g., the Gaussian elimination. Starting with Section 3.4 several of these methods will be described and compared. The usage of Eq. (3.28) to obtain the steady-state probability vectors π_i is straightforward. Of course not all π_i can be computed due to their infinite number, but the elements of π_i converge towards 0 for increasing i since $sp(\mathbf{R}) < 1$.

QBDs can be seen as an intersection of M/G/1-type and GI/M/1-type Markov chains. These classes of Markov chains are introduced in Chapter 6. Applications of QBDs and the MGM can be found in [KrNa99] (tele-traffic analysis of ATM networks), [NKW97] (modeling B-ISDN switching fabrics), [LeVe90] (analyzing a single link dynamic model supporting rate-adaptive and constant bit rate service), [DPL02] (analyzing a CDMA system with dynamic priorities), [XF03] (analyzing partial and complete sharing schemes of GSM/GPRSnetworks), [KiLi97] (modeling a downlink/uplink data transmission in a cellular voice network), or [Nels90] (analyzing a parallel processing system). Tool support for analyzing QBDs is provided by MAMSolver [RiSm02a], MGMtool [BRH92], its successor Xmgm [BRH94], SPN2MGM [Have96], MAGIC [Squi91], MAGUS [NeSq94], TELPACK [SOL], and SMAQ [LiHw97, LPA98]. Finite QBDs, i.e., QBDs with a finite number of repeating levels, are covered in [AkSo97, Naou97, NKW97, KrNa99] and QBDs with level-dependent birth-and-death rates are covered in [BrTa97, LaRa99]. QBDs with a infinite number of phases per level are discussed in [DAB00, Rama98].

Example 3.1 We adapt an example given by [LeWi04]. Consider a production line with two phases that are both processed by a single server. The first phase duration is exponentially distributed with rate $\mu_1 = 0.5$ and is identical for all products. The second phase duration is exponentially distributed with rate $\mu_2 = 0.6$, is order-specific, and has higher priority than the first phase. After finishing the first phase at most $m = 3$ semi-finished products can be stored to be finished by the second phase as soon as a new order arrives and the server is free. The order arrival process is Poisson with rate $\lambda = 0.15$.

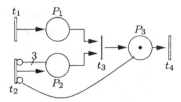

Fig. 3.3 GSPN of two-phase production system.

The production process can be modeled as the GSPN shown in Fig. 3.3. Transition t_1 models the order arrival process, t_2 the first phase, and t_4 the second phase. The immediate transition t_3 is an auxiliary transition for place P_3. This place contains all products that are currently in service of the second phase. Place P_2 has capacity 3 (handled by the inhibitor arc with multiplicity 3) and contains all products that have finished the first phase and are currently waiting for an incoming order. Orders are stored in place P_1. The inhibitor arc from place P_3 to transition t_2 ensures that the server does not produce new semi-finished products while working in phase 2.

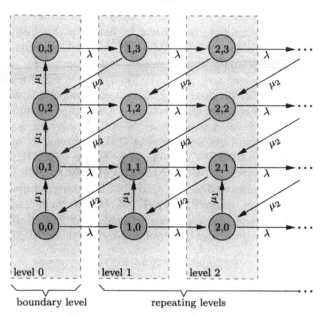

Fig. 3.4 State diagram of two-phase production system.

The CTMC underlying the GSPN is the QBD with state space $S = \{(i,j)|0 \leq i, 0 \leq j \leq m\}$, where the level i is the number of orders in the system (waiting or in service, $i = \#P_1 + \#P_3$) and the phase j is the number of semi-finished products (waiting for an order or in service of the second phase, $j = \#P_2 + \#P_3$). The state diagram of this QBD is shown in Fig. 3.4.

The generator matrix of this QBD follows from Eq. (3.23) and comprises the following submatrices:

$$
\mathbf{B}_{0,0} = \begin{pmatrix} -\lambda - \mu_1 & \mu_1 & 0 & 0 \\ 0 & -\lambda - \mu_1 & \mu_1 & 0 \\ 0 & 0 & -\lambda - \mu_1 & \mu_1 \\ 0 & 0 & 0 & -\lambda \end{pmatrix}
$$
$$
= \begin{pmatrix} -0.65 & 0.5 & 0 & 0 \\ 0 & -0.65 & 0.5 & 0 \\ 0 & 0 & -0.65 & 0.5 \\ 0 & 0 & 0 & -0.15 \end{pmatrix}, \tag{3.46}
$$

$$
\mathbf{A}_0 = \begin{pmatrix} \lambda & 0 & 0 & 0 \\ 0 & \lambda & 0 & 0 \\ 0 & 0 & \lambda & 0 \\ 0 & 0 & 0 & \lambda \end{pmatrix} = \begin{pmatrix} 0.15 & 0 & 0 & 0 \\ 0 & 0.15 & 0 & 0 \\ 0 & 0 & 0.15 & 0 \\ 0 & 0 & 0 & 0.15 \end{pmatrix}, \tag{3.47}
$$

$$
\mathbf{A}_1 = \begin{pmatrix} -\lambda - \mu_1 & \mu_1 & 0 & 0 \\ 0 & -\lambda - \mu_2 & 0 & 0 \\ 0 & 0 & -\lambda - \mu_2 & 0 \\ 0 & 0 & 0 & -\lambda - \mu_2 \end{pmatrix}
$$
$$
= \begin{pmatrix} -0.65 & 0.5 & 0 & 0 \\ 0 & -0.75 & 0 & 0 \\ 0 & 0 & -0.75 & 0 \\ 0 & 0 & 0 & -0.75 \end{pmatrix}, \tag{3.48}
$$

$$
\mathbf{A}_2 = \begin{pmatrix} 0 & 0 & 0 & 0 \\ \mu_2 & 0 & 0 & 0 \\ 0 & \mu_2 & 0 & 0 \\ 0 & 0 & \mu_2 & 0 \end{pmatrix} = \begin{pmatrix} 0 & 0 & 0 & 0 \\ 0.6 & 0 & 0 & 0 \\ 0 & 0.6 & 0 & 0 \\ 0 & 0 & 0.6 & 0 \end{pmatrix}. \tag{3.49}
$$

To see if this QBD is tractable by the matrix-geometric method, the stability of the production system needs to be proven by checking the stability condition of Eq. (3.34). For this, at first the generator matrix $\mathbf{A}$ is obtained:

$$\mathbf{A} = \mathbf{A}_0 + \mathbf{A}_1 + \mathbf{A}_2$$

$$
= \begin{pmatrix} \lambda & 0 & 0 & 0 \\ 0 & \lambda & 0 & 0 \\ 0 & 0 & \lambda & 0 \\ 0 & 0 & 0 & \lambda \end{pmatrix} + \begin{pmatrix} -\lambda - \mu_1 & \mu_1 & 0 & 0 \\ 0 & -\lambda - \mu_2 & 0 & 0 \\ 0 & 0 & -\lambda - \mu_2 & 0 \\ 0 & 0 & 0 & -\lambda - \mu_2 \end{pmatrix}
$$

$$
+ \begin{pmatrix} 0 & 0 & 0 & 0 \\ \mu_2 & 0 & 0 & 0 \\ 0 & \mu_2 & 0 & 0 \\ 0 & 0 & \mu_2 & 0 \end{pmatrix} = \begin{pmatrix} -\mu_1 & \mu_1 & 0 & 0 \\ \mu_2 & -\mu_2 & 0 & 0 \\ 0 & \mu_2 & -\mu_2 & 0 \\ 0 & 0 & \mu_2 & -\mu_2 \end{pmatrix}.
$$

Using $\mathbf{A}$, the steady-state probability vector $\boldsymbol{\pi}_A = [\pi_{A,0}, \pi_{A,1}, \pi_{A,2}, \pi_{A,3}]$ can be obtained:

$$\boldsymbol{\pi}_A \mathbf{A} = \mathbf{0}, \quad \sum_{i=0}^{3} \pi_{A,i} = 1, \quad \boldsymbol{\pi}_A = \left[\frac{\mu_2}{\mu_1 + \mu_2}, \frac{\mu_1}{\mu_1 + \mu_2}, 0, 0\right].$$

So we get

$$\boldsymbol{\pi}_A \mathbf{A}_0 \mathbf{1} = \lambda \frac{\mu_2}{\mu_1 + \mu_2} + \lambda \frac{\mu_1}{\mu_1 + \mu_2} = \lambda, \quad \boldsymbol{\pi}_A \mathbf{A}_2 \mathbf{1} = \mu_2 \frac{\mu_1}{\mu_1 + \mu_2},$$

and finally we check

$$\boldsymbol{\pi}_A \mathbf{A}_0 \mathbf{1} < \boldsymbol{\pi}_A \mathbf{A}_2 \mathbf{1}, \quad \lambda < \frac{\mu_1 \mu_2}{\mu_1 + \mu_2}, \quad 0.15 < 0.27.$$

Therefore, the stability condition holds and the QBD is stable.

To derive the rate matrix $\mathbf{R}$ the successive substitution algorithm is used. As termination condition $\|\mathbf{R}_{(n+1)} - \mathbf{R}_{(n)}\|_\infty < \epsilon = 0.002$ is used where the matrix norm $\| \cdot \|_\infty$ is the row-sum norm. Starting with Eq. (3.41) and following Eq. (3.42) the rate matrix $\mathbf{R}$ is computed iteratively:

$$\mathbf{R}_{(0)} = \mathbf{0},$$

$$\mathbf{R}_{(1)} = -(\mathbf{A}_0 + \mathbf{R}_{(0)}^2 \mathbf{A}_2)\mathbf{A}_1^{-1} = -\mathbf{A}_0 \mathbf{A}_1^{-1}$$

$$= -\begin{pmatrix} 0.15 & 0 & 0 & 0 \\ 0 & 0.15 & 0 & 0 \\ 0 & 0 & 0.15 & 0 \\ 0 & 0 & 0 & 0.15 \end{pmatrix} \begin{pmatrix} -0.65 & 0.5 & 0 & 0 \\ 0 & -0.75 & 0 & 0 \\ 0 & 0 & -0.75 & 0 \\ 0 & 0 & 0 & -0.75 \end{pmatrix}^{-1}$$

$$= \begin{pmatrix} 0.231 & 0.154 & 0 & 0 \\ 0 & 0.200 & 0 & 0 \\ 0 & 0 & 0.200 & 0 \\ 0 & 0 & 0 & 0.200 \end{pmatrix},$$

$$\|\mathbf{R}_{(1)} - \mathbf{R}_{(0)}\|_\infty = 0.385 > \epsilon,$$

and similarly:

$$\mathbf{R}_{(2)} = \begin{pmatrix} 0.292 & 0.195 & 0 & 0 \\ 0.037 & 0.225 & 0 & 0 \\ 0 & 0.032 & 0.200 & 0 \\ 0 & 0 & 0.032 & 0.200 \end{pmatrix},$$

$$\|\mathbf{R}_{(2)} - \mathbf{R}_{(1)}\|_\infty = 0.102 > \epsilon.$$

After ten iterations, the rows converge, so that we have

$$\mathbf{R}_{(10)} = \begin{pmatrix} 0.372 & 0.248 & 0 & 0 \\ 0.074 & 0.249 & 0 & 0 \\ 0.026 & 0.050 & 0.200 & 0 \\ 0.011 & 0.018 & 0.032 & 0.200 \end{pmatrix},$$

$$||\mathbf{R}_{(10)} - \mathbf{R}_{(9)}||_\infty = 0.0023 > \epsilon,$$

$$\mathbf{R}_{(11)} = \begin{pmatrix} 0.373 & 0.249 & 0 & 0 \\ 0.074 & 0.250 & 0 & 0 \\ 0.027 & 0.050 & 0.200 & 0 \\ 0.011 & 0.018 & 0.032 & 0.200 \end{pmatrix}.$$

Due to $||\mathbf{R}_{(11)} - \mathbf{R}_{(10)}||_\infty = 0.0015 < \epsilon$, the iteration is stopped and $\mathbf{R} = \mathbf{R}_{(11)}$ results. Note that the eigenvalues of $\mathbf{R}$ are 0.462, 0.162, and 0.200 (double); all located within the unit circle.

To obtain π_i the boundary Eqs. (3.31) and (3.40) are taken into account:

$$(\pi_0, \pi_1) \begin{pmatrix} \mathbf{B}_{0,0} & (\mathbf{A}_0)^* & \mathbf{1} \\ \mathbf{A}_2 & (\mathbf{A}_1 + \mathbf{R}\mathbf{A}_2)^* & (\mathbf{I} - \mathbf{R})^{-1}\mathbf{1} \end{pmatrix} = (\mathbf{0}, 1), \qquad (3.50)$$

where $(\cdot)^*$ indicates that the last column of the included matrix is removed to avoid linear dependency. The removed column is replaced by the normalization condition. Equation (3.50) is solved for computing π_0 and π_1:

$$(\pi_0, \pi_1) = (\mathbf{0}, 1) \cdot \begin{pmatrix} \mathbf{B}_{0,0} & (\mathbf{A}_0)^* & \mathbf{1} \\ \mathbf{A}_2 & (\mathbf{A}_1 + \mathbf{R}\mathbf{A}_2)^* & (\mathbf{I} - \mathbf{R})^{-1}\mathbf{1} \end{pmatrix}^{-1}$$

$$= (\mathbf{0}, 1)$$

$$\cdot \begin{pmatrix} -0.65 & 0.5 & 0 & 0 & 0.15 & 0 & 0 & 1 \\ 0 & -0.65 & 0.5 & 0 & 0 & 0.15 & 0 & 1 \\ 0 & 0 & -0.65 & 0.5 & 0 & 0 & 0.15 & 1 \\ 0 & 0 & 0 & -0.15 & 0 & 0 & 0 & 1 \\ 0 & 0 & 0 & 0 & -0.501 & 0.5 & 0 & 2.211 \\ 0.6 & 0 & 0 & 0 & 0.150 & -0.75 & 0 & 1.552 \\ 0 & 0.6 & 0 & 0 & 0.030 & 0.120 & -0.75 & 1.420 \\ 0 & 0 & 0.6 & 0 & 0.011 & 0.019 & 0.120 & 1.370 \end{pmatrix}^{-1}$$

$$= (\mathbf{0}, 1)$$

$$
\begin{pmatrix}
-2.138 & -1.250 & 0.699 & 2.329 & -0.770 & -0.649 & 0.427 & 1.798 \\
-0.308 & -1.550 & 0.100 & 0.334 & -0.147 & -0.333 & 0.244 & 1.400 \\
0.123 & 0.150 & -0.499 & -1.663 & 0.102 & 0.133 & 0.060 & 1.001 \\
0.253 & 0.450 & 0.901 & -3.662 & 0.187 & 0.275 & 0.276 & 0.601 \\
-1.825 & -0.700 & 1.792 & 5.972 & -3.035 & -1.977 & 0.762 & 2.524 \\
-1.996 & -1.000 & 1.196 & 3.988 & -1.164 & -2.162 & 0.580 & 2.129 \\
-0.566 & -1.300 & 0.599 & 1.997 & -0.372 & -0.613 & -0.936 & 1.732 \\
0.038 & 0.068 & 0.135 & 0.451 & 0.028 & 0.041 & 0.042 & 0.090
\end{pmatrix}
$$

$$ = (0.038, 0.068, 0.135, 0.451, 0.028, 0.041, 0.042, 0.090) \ . $$

Using the steady-state probability vectors $\pi_0 = (0.038, 0.068, 0.135, 0.451)$ and $\pi_1 = (0.028, 0.041, 0.042, 0.090)$ with Eq. (3.28) all other interesting steady-state probability vectors π_i can be calculated, e.g.,

$$ \pi_2 = \pi_1 \mathbf{R} = (0.016, 0.021, 0.011, 0.018) \ , $$

$$ \pi_3 = \pi_2 \mathbf{R} = (0.008, 0.010, 0.003, 0.004) \ , $$

$$ \pi_6 = \pi_3 \mathbf{R}^3 = (0.001, 0.001, 0.000, 0.000) \ . $$

3.3 HESSENBERG MATRIX: NON-MARKOVIAN QUEUES

Section 3.1 shows that an infinite state CTMC (or DTMC) with a tridiagonal matrix structure can be solved to obtain a closed-form result. In this section we consider two other infinite state DTMCs. However, the structure is more complex so as to preclude the solution by "inspection" that we adopted in Section 3.1. Here we use the method of generating functions (or z-transform) to obtain a solution. The problems we tackle originate from non-Markovian queueing systems where the underlying stochastic process is Markov regenerative [Kulk96]. One popular method for the steady-state analysis of Markov regenerative processes is to apply *embedding technique* so as to produce an *embedded* DTMC from the given Markov regenerative process. An alternative approach is to use the *method of supplementary variables* [Hend72, Germ00]. We follow the embedded DTMC approach.

We are interested in the analysis of a queueing system, where customers arrive according to a Poisson process, so that successive interarrival times are independent, exponentially distributed random variables with parameter λ. The customers experience service at a single server with the only restriction on the service time distribution being that its first two moments are finite. Order of service is first-come-first-served and there is no restriction on the size of the waiting room. We shall see later in Chapter 6 that this is the M/G/1 queueing system. Characterizing the system in such a way that the whole history is summarized in a state, we need to specify the number of customers in the system plus the elapsed service time received by the current customer in service. This description results in a continuous-state stochastic process that is difficult to analyze.

But it is possible to identify time instants where the elapsed time is always known so they need not be explicitly represented. A prominent set of these time instants is given by the departure instants, i.e., when a customer has just completed receiving service and before the turn of the next customer has come. In this case, elapsed time is always zero. As a result, a state description given by the number of customers is sufficient. Furthermore, because the service time distribution is known and arrivals are Poisson, the state transition probabilities can be easily computed. It is not difficult to prove that the stochastic process defined in the indicated way constitutes a DTMC. This DTMC is referred to as *embedded* into the more general continuous-state stochastic process.

Conditions can be identified under which the embedded DTMC is ergodic, i.e., a unique steady-state pmf does exist. In Section 3.3.1 we show how the steady-state probability vector of this DTMC can be computed under given constraints. Fortunately, it can be proven that the steady-state pmf of the embedded DTMC is the same as the limiting probability vector of the original non-Markovian stochastic process we started with. The proof of this fact relies on the so-called PASTA theorem, stating that "Poisson arrivals see time averages" [Wolf82]. The more difficult analysis of a stochastic process of non-Markovian type can thus be reduced to the analysis of a related DTMC, yielding the same steady-state probability vector as of the original process.

3.3.1 Nonexponential Service Times

The service times are given by independent, identically distributed (i.i.d.) random variables and they are independent of the arrival process. Also, the first moment $E[S] = \overline{S}$ and the second moment $E[S^2] = \overline{S^2}$ of the service time S must be finite. Upon completion of service, the customers leave the system.

To define the embedded DTMC $X = \{X_n; n = 0, 1 \ldots\}$, the state space is chosen as the number of customers in the system $\{0, 1, \ldots\}$. As time instants where the DTMC is defined, we select the departure epochs, that is, the points in time when service is just completed and the corresponding customer leaves the system.

Consider the number of customers $X = \{X_n; n = 0, 1, \ldots\}$ left behind by a departing customer, labeled n. Then the state evolution until the next epoch $n + 1$, where the next customer, labeled $n + 1$, completes service, is probabilistically governed by the Poisson arrivals.

Let the random variable Y describe the number of arrivals during a service epoch. The following one-step state transitions are then possible:

$$X_{n+1} = \begin{cases} X_n + Y - 1, & \text{if } X_n > 0, \\ Y, & \text{if } X_n = 0. \end{cases} \tag{3.51}$$

Let $a_k = P[Y = k]$ denote the probability of k arrivals in a service period with given distribution $B(t)$. If we fix the service time at t, then Y is Poisson distributed with parameter λt. To obtain the unconditional probability a_k, we un-condition using the service time distribution:

$$a_k = \int_0^\infty e^{-\lambda t} \frac{(\lambda t)^k}{k!} \, d B(t) . \tag{3.52}$$

Now the transition probabilities are given as

$$P[X_{n+1} = j \mid X_n = i] = \begin{cases} a_{j-i+1} , & i > 0, \ j \geq i - 1, \\ a_j , & i = 0, \ j \geq 0. \end{cases} \tag{3.53}$$

The transition probability matrix $\mathbf{P}$ of the embedded stochastic process is a Hessenberg matrix given by

$$\mathbf{P} = \begin{pmatrix} a_0 & a_1 & a_2 & \cdots \\ a_0 & a_1 & a_2 & \cdots \\ 0 & a_0 & a_1 & \cdots \\ 0 & 0 & a_0 & \cdots \\ \vdots & \vdots & \vdots & \ddots \end{pmatrix} . \tag{3.54}$$

$\mathbf{P}$ can be effectively created, because the arrivals are Poisson and the service time distribution $B(t)$ is known. With $\mathbf{P}$ given as defined in Eq. (3.54), it is not difficult to prove that a DTMC has been defined, i.e., the Markov property of Eq. (2.2) holds. The reader should also verify that this DTMC is aperiodic and irreducible. It is intuitively clear that state 0 is positive recurrent if the server can keep up with arrivals, i.e., the mean service time $\overline{S}$ is smaller than than mean interarrival time $1/\lambda$. Equivalently, the mean number of arrivals $E[N] = \overline{N}$ during mean service period $\overline{S}$ should be less than one [Triv01]:

$$\overline{N} = \lambda \overline{S} < 1 . \tag{3.55}$$

We know from Section 2.1.2.1 that the states of an irreducible DTMC are all of the same type. Therefore, positive recurrence of state 0 implies positive recurrence of the DTMC constructed in the indicated way. A formal proof of the embedded DTMC being positive recurrent if and only if relation (3.55) holds has been given, for example, by Cinlar [Cinl75]. We also know from Section 2.1.2.1 that irreducible, aperiodic, and positive recurrent DTMCs are ergodic. Therefore, the embedded DTMC is ergodic if relation (3.55) holds. Assuming the DTMC to be ergodic, the corresponding infinite set of global balance equations is given as

$$\nu_0 = \nu_0 a_0 + \nu_1 a_0 ,$$

$$\nu_k = \nu_0 a_k + \sum_{i=1}^{k+1} \nu_i a_{k-i+1} , \quad k \geq 1 . \tag{3.56}$$

To compute the steady-state probability vector $\boldsymbol{\nu}$ of the DTMC, we use the method of generating functions wherein we also need to use the LST of the service time random variable. Let the service time distribution be denoted by $B(t)$ and its LST $B^{\sim}(s)$ be given by

$$B^{\sim}(s) = \int_0^\infty e^{-st}\,\mathrm{d}\,B(t). \tag{3.57}$$

Defining generating functions of the state probabilities

$$G(z) = \sum_{k=0}^\infty \nu_k z^k, \tag{3.58}$$

and of the $\{a_j\}$ sequence

$$G_A(z) = \sum_{j=0}^\infty a_j z^j, \tag{3.59}$$

from Eq. (3.56), we have

$$G(z) = \sum_{k=0}^\infty \nu_k z^k = \nu_0 \sum_{k=0}^\infty a_k z^k + \sum_{k=0}^\infty \sum_{i=1}^{k+1} \nu_i a_{k-i+1} z^k\,,$$

or

$$G(z) = \nu_0 G_A(z) + \sum_{i=1}^\infty \sum_{k=i-1}^\infty \nu_i a_{k-i+1} z^k$$

$$= \nu_0 G_A(z) + \sum_{i=1}^\infty \sum_{j=0}^\infty \nu_i a_j z^{j+i-1}$$

$$= \nu_0 G_A(z) + \frac{1}{z}\left(\sum_{i=1}^\infty \nu_i z^i\right)\left(\sum_{j=0}^\infty a_j z^j\right)$$

$$= \nu_0 G_A(z) + \frac{1}{z}(G(z) - \nu_0)G_A(z)\,,$$

or

$$G(z)\frac{z - G_A(z)}{z} = \nu_0 \frac{(z G_A(z) - G_A(z))}{z}\,,$$

or

$$G(z) = \nu_0 \frac{z G_A(z) - G_A(z)}{z - G_A(z)}\,. \tag{3.60}$$

Now, because the arrival process is Poisson, we can obtain the generating function of the $\{a_j\}$ sequence by first conditioning on a fixed service time and then un-conditioning with the service time distribution:

$$G_A(z) = \sum_{j=0}^{\infty} a_j z^j = \sum_{j=0}^{\infty} \int_0^{\infty} P[Y = j \mid \text{service} = t] \, d\,B(t) z^j$$

$$= \sum_{j=0}^{\infty} \int_0^{\infty} e^{-\lambda t} \frac{(\lambda t)^j}{j!} \, d\,B(t) z^j = \int_0^{\infty} \left(\sum_{j=0}^{\infty} \frac{(\lambda t z)^j}{j!} \, e^{-\lambda t} \right) d\,B(t) \quad (3.61)$$

$$= \int_0^{\infty} e^{-\lambda t + \lambda t z} \, d\,B(t) = \int_0^{\infty} e^{-\lambda t (1-z)} \, d\,B(t) = B^{\sim}(\lambda(1-z)).$$

Thus, the generating function for the $\{a_j\}$ sequence is given by the LST of the service time distribution at $\lambda(1-z)$. Hence, we get an expression for the generating function of the DTMC steady-state probabilities:

$$G(z) = \nu_0 \frac{z B^{\sim}(\lambda(1-z)) - B^{\sim}(\lambda(1-z))}{z - B\tilde{3}x(\lambda(1-z))} = \nu_0 \frac{B^{\sim}(\lambda(1-z))(z-1)}{z - B^{\sim}(\lambda(1-z))}. \quad (3.62)$$

The generating function in Eq. (3.62) allows the computation of the infinite steady-state probability vector ν of the DTMC embedded into a continuous-time stochastic process at the departure epochs. These epochs are given by the time instants when customers have just completed their generally distributed service period. The steady-state probabilities are obtained by repeated differentiations of the probability generating function $G(z)$, evaluated at $z = 1$:

$$\nu_i = \frac{1}{i!} \frac{d^i}{d z^i} G(z) \Big|_{z=1}. \quad (3.63)$$

If we set $z = 1$ on the right-hand side of Eq. (3.62) and since $G_A(1) = \sum_{j=0}^{\infty} a_j = 1$, we have $0/0$. Differentiating the numerator and the denominator, as per L'Hospital's rule, we obtain

$$G(1) = 1 = \nu_0 \frac{B^{\sim}(\lambda(1-z)) + z(-1)\lambda B^{\sim\prime}(\lambda(1-z)) + \lambda B^{\sim\prime}(\lambda(1-z))}{1 + \lambda B^{\sim\prime}(\lambda(1-z))}$$

$$= \nu_0 \frac{B^{\sim}(\lambda(1-z)) + \lambda B^{\sim\prime}(\lambda(1-z))[1-z]}{1 + \lambda B^{\sim\prime}(\lambda(1-z))} = \nu_0 \frac{B^{\sim}(0)}{1 + \lambda B^{\sim\prime}(0)}.$$

From Eq. (3.61) it can be easily shown $B^{\sim}(0) = 1$ and $-B^{\sim\prime}(0) = \overline{S}$, so that we have

$$1 = \nu_0 \frac{1}{1 - \lambda \overline{S}}$$

or

$$\nu_0 = 1 - \lambda \overline{S}. \quad (3.64)$$

Taking the derivative of $G(z)$ and setting $z = 1$, we get the expected number of customers in the system, $E[X] = \overline{X}$, in steady state:

$$\overline{X} = \lambda \overline{S} + \frac{\lambda^2 \overline{S^2}}{2(1 - \lambda \overline{S})} . \tag{3.65}$$

Equation (3.65) is known as the *Pollaczek–Khintchine* formula. Remember that this formula has been derived by restricting observation to departure epochs. It is remarkable that this formula holds at random observation points in steady state, if the arrival process is Poisson. The proof is based on the PASTA theorem by Wolff, stating that "Poisson arrivals see time averages" [Wolf82].

We refrain from presenting numerical examples at this point, but refer the reader to Chapter 6 where many examples are given related to the stochastic process introduced in this section.

Problem 3.5   Specialize Eq. (3.65) to

(a) Exponential service time distribution with parameter μ

(b) Deterministic service time with value $1/\mu$

(c) Erlang service time distribution with mean service time $1/\mu$ and k phases

In the notation of Chapter 6, these three cases correspond to M/M/1, M/D/1, and M/E$_k$/1 queueing systems, respectively.

Problem 3.6   Given a mean interarrival time of 1 second and a mean service time of 2 seconds, compute the steady-state probabilities for the three queueing systems M/M/1, M/D/1, and M/E$_k$/1 to be idle. Compare the results for the three systems and comment.

Problem 3.7   Show that the embedded discrete-time stochastic process $X = \{X_n; n = 0, 1, \dots\}$ defined at the departure time instants (with transition probability matrix $\mathbf{P}$ given by Eq. (3.54)), forms a DTMC, i.e., it satisfies the Markov property in Eq. (2.2).

Problem 3.8   Give a graphical representation of the DTMC defined by transition probability matrix $\mathbf{P}$ in Eq. (3.54).

Problem 3.9   Show that the embedded DTMC $X = \{X_n; n = 0, 1, \dots\}$ defined in this section is aperiodic and irreducible.

Problem 3.10   Consider a single-server queueing system with independent, exponentially distributed service times. Furthermore, assume an arrival process with independent, identically distributed interarrival times; a general service time distribution is allowed. Service times and interarrival times are also independent. In the notation of Chapter 6, this is the GI/M/1 queueing system.

1. Select a suitable state space and identify appropriate time instants where a DTMC X^* should be embedded into this non-Markovian continuous-state process.

2. Define a DTMC $X^* = \{X_n^*; n = 0, 1, \dots\}$ to be embedded into the non-Markovian continuous-state process. In particular, define a DTMC X^* by specifying state transitions by taking Eq. (3.51) as a model. Specify the transition probabilities of the DTMC X^* and its transition probability matrix $\mathbf{P}^*$.

3.3.2   Server with Vacations

Server vacations can be modeled as an extension of the approach presented in Section 3.3.1. In particular, the impact of vacations on the investigated performance parameters is of additive nature and can be derived with decomposition techniques. Non-Markovian queues with server vacations have proven to be a very useful class of models. They have been applied in a great variety of contexts [Dosh90], some of which are:

- Analysis of *server breakdowns*, which may occur randomly and preempt a customer (if any) in service. Since breakdown (vacation) has priority over customer service, it is interesting to find out how the overall service capacity is affected by such breakdowns. Such insights can be provided through the analysis of queueing systems with vacations.

- Investigation of *maintenance* strategies of computer, communication, or manufacturing systems. In contrast to breakdowns, which occur randomly, maintenance is usually scheduled at certain fixed intervals in order to optimize system dependability.

- Application of *polling* systems or *cyclic server queues*. Different types of polling systems that have been used include systems with *exhaustive service, limited service, gated service*, or some combinations thereof.

3.3.2.1 Polling Systems   Because polling systems are often counted as one of the most important applications of queueing systems with server vacations, some remarks are in order here. While closed-form expressions are derived later in this section, numerical methods for the analysis of polling systems on the basis of GSPNs is covered in Section 2.3.

The term *polling* comes from the *polling* data link control scheme in which a central computer interrogates each terminal on a multidrop communication line to find out whether it has data to transmit. The addressed terminal transmits data, and the computer examines the next terminal. Here, the server represents the computer, and a queue corresponds to a terminal.

Basic polling models have been applied to analyze the performance of a variety of systems. In the late 1950s, a polling model with a single buffer for each queue was first used in an analysis of a problem in the British cotton industry involving a patrolling machine repairman [MMW57]. In the 1960s, polling models with two queues were investigated for the analysis of vehicle-actuated traffic signal control [Newe69, NeOs69]. There were also some early studies from the viewpoint of queueing theory, apparently independent of traffic analysis [AMM65]. In the 1970s, with the advent of computer communication networks, extensive research was carried out on a polling scheme for data transfer from terminals on multidrop lines to a central computer. Since the early 1980s, the same model has been revived by [Bux81] and others for token passing schemes (e.g., token ring and token bus) in local area networks (LANs). In investigations of asynchronous transfer mode (ATM) for broadband ISDN (integrated services digital network), cyclic scheduling is often proposed. Polling models have been applied for scheduling moving arms in secondary storage devices [CoHo86] and for resource arbitration and load sharing in multiprocessor computers. A great number of applications exist in manufacturing systems, in transportation including moving passengers on circular and on back-and-forth routes, internal mail delivery, and shipyard loading, to mention a few. A major reason for the ubiquity of these applications is that the cyclic allocation of the server (resource) is natural and fair (since no station has to wait arbitrarily long) in many fields of engineering.

The main aim of analyzing polling models is to find the *message waiting time*, defined as the time from the arrival of a randomly chosen message to the beginning of its service. The mean waiting time plus the mean service time is the mean *message response time*, which is the *single most important performance measure* in most computer communication systems. Another interesting characteristic is the *polling cycle time*, which is the time between the server's visit to the same queue in successive cycles. Many variants and related models exist and have been studied. Due to their importance, the following list includes some polling systems of interest:

- Single-service polling systems, in which the server serves only one message and continues to poll the next queue.

- Exhaustive-service polling systems, in which the server serves all the messages at a queue until it is empty before polling the next queue.

- Gated-service polling systems, in which the server serves only those messages that are in the queue at the polling instant before moving to the next queue. In particular, message requests arriving after the server starts serving the queue will wait at the queue until the next time the server visits this queue.

- Mixed exhaustive- and single-service polling systems.

- Symmetric and asymmetric limited-service polling systems, in which the server serves at most $l(i)$ customers in in each service cycle at station i, with $l(i) = l$ for all stations i in the symmetric case.

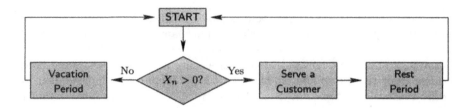

Fig. 3.5 Phases of an M/G/1 queue with vacations.

3.3.2.2 Analysis As in Section 3.3.1, we embed a DTMC into a more general continuous-time stochastic process. Besides nonexponential service times S with its distribution given by $B(t)$ and its LST given by $B^\sim(s)$, we consider *vacations* of duration V with distribution function $C(t)$ and LST $C^\sim(s)$, and *rest* periods of the server of duration R, with distribution $D(t)$ and LST $D^\sim(s)$. Rest periods and vacations are both given by i.i.d. random variables. Finally, the arrivals are again assumed to be Poisson. Our treatment here is based on that given in [King90].

If the queue is inspected and a customer is found to be present in inspection in epoch n, that is $X_n > 0$, the customer is served according to its required service time, followed by a rest period of the server (see Fig. 3.5). Thus, after each service completion the server takes a rest before returning for inspection of the queue. If the queue is found to be empty on inspection, the server takes a vacation before it returns for another inspection. We embed a DTMC into this process at the inspection time instants. With E denoting the arrivals during service period including following rest period and F denoting the arrivals during a vacation, Eq. (3.51) describing possible state transitions from state X_n at embedding epoch n to state X_{n+1} at embedding epoch $n+1$ can be restated:

$$X_{n+1} = \begin{cases} X_n + E - 1, & \text{if } X_n > 0, \\ F, & \text{if } X_n = 0. \end{cases} \tag{3.66}$$

With $e_k = P[E = k]$ and $f_l = P[F = l]$ denoting the probabilities of k or l arrivals during the respective time periods, the transition probabilities of the embedded DTMC can be specified in analogy to Eq. (3.53):

$$P[X_{n+1} = j \mid X_n = i] = \begin{cases} e_{j-i+1}, & i > 0, \ j \geq i-1 \geq 0, \\ f_j, & i = 0, \ j \geq 0. \end{cases} \tag{3.67}$$

The resulting transition probability matrix of the embedded DTMC is given by

$$\mathbf{P} = \begin{pmatrix} f_0 & f_1 & f_2 & \cdots \\ e_0 & e_1 & e_2 & \cdots \\ 0 & e_0 & e_1 & \cdots \\ 0 & 0 & e_0 & \cdots \\ \vdots & \vdots & \vdots & \ddots \end{pmatrix}. \tag{3.68}$$

Note that this transition probability matrix is also in Hessenberg form like the M/G/1 case.

Let $\mathbf{p}$ denote the state probability vector of the embedded DTMC at inspection instants. Then its generating function is given by

$$G(z) = \sum_{k=0}^{\infty} p_k z^k = p_0 \sum_{k=0}^{\infty} f_k z^k + \sum_{k=0}^{\infty} \sum_{i=1}^{k+1} p_i e_{k-i+1} z^k \qquad (3.69)$$

$$= p_0 G_F(z) + \sum_{i=1}^{\infty} \sum_{k=i-1}^{\infty} p_i e_{k-i+1} z^k \qquad (3.70)$$

$$= p_0 \frac{z G_F(z) - G_E(z)}{z - G_E(z)}. \qquad (3.71)$$

In analogy to Eq. (3.61), the generating functions $G_E(z)$ and $G_F(z)$ can be derived:

$$G_E(z) = \sum_{j=0}^{\infty} e_j z^j = \sum_{j=0}^{\infty} \int_0^{\infty} P[E = j \mid \text{service} = t] \, \mathrm{d}[B(t) + D(t)] z^j$$

$$= \sum_{j=0}^{\infty} \int_0^{\infty} \mathrm{e}^{-\lambda t} \frac{(\lambda t)^j}{j!} \, \mathrm{d}[B(t) + D(t)] z^j = \int_0^{\infty} \mathrm{e}^{-\lambda t(1-z)} \, \mathrm{d}[B(t) + D(t)]$$

$$= B^{\sim}(\lambda(1-z)) D^{\sim}(\lambda(1-z)), \qquad (3.72)$$

$$G_F(z) = C^{\sim}(\lambda(1-z)). \qquad (3.73)$$

With generating functions $G(z)$, $G_E(z)$, and $G_F(z)$ defined, and the transition probability matrix as given in Eq. (3.68), an expression for the probability generation function at inspection time instants can be derived for the server with vacation:

$$G(z) = p_0 \frac{z C^{\sim}(\lambda(1-z)) - B^{\sim}(\lambda(1-z)) D^{\sim}(\lambda(1-z))}{z - B^{\sim}(\lambda(1-z)) D^{\sim}(\lambda(1-z))}. \qquad (3.74)$$

Remembering $G(1) = 1$ and by evaluating Eq. (3.74) at $z = 1$, again by differentiating the denominator and the numerator, we have

$$p_0 = \frac{1 - \lambda(\overline{S+R})}{1 - \lambda(\overline{S+R}) + \lambda \overline{V}}. \qquad (3.75)$$

So far, we have derived an expression for the probabilities p_i at *inspection* time instants. But to arrive at the steady-state probabilities, we infer the state probabilities ν_n at *departure* instants. Then an inspection at a time instant immediately preceding the departure must find at least one customer present, $i > 0$, with probability p_i, because a departure occurred. With $a_k = P[Y = k]$ denoting again the probability of k arrivals in a service period, the probabilities ν_n, p_i, and a_k can be related as follows:

$$\nu_n = \sum_{i=1}^{n+1} \frac{p_i a_{n+1-i}}{1 - p_0}.$$

Then, the generating function of the steady-state queue length

$$X(z) = \sum_{j=0}^{\infty} \nu_j z^j$$

is given by

$$
\begin{aligned}
X(z) &= \sum_{n=0}^{\infty} z^n \sum_{i=1}^{n+1} \frac{p_i}{1 - p_0} a_{n+1-i} = \sum_{i=1}^{\infty} \sum_{n=i-1}^{\infty} z^n \frac{p_i}{1 - p_0} a_{n+1-i} \\
&= \frac{1}{1 - p_0} \sum_{i=1}^{\infty} \sum_{j=0}^{\infty} z^{i+j-1} p_i a_j = \frac{1}{z(1 - p_0)} \sum_{i=1}^{\infty} p_i z^i \sum_{j=0}^{\infty} a_j z^j \qquad (3.76) \\
&= \frac{1}{z(1 - p_0)} (G(z) - p_0) G_A(z).
\end{aligned}
$$

Recalling $G_A(z) = B^{\sim}(\lambda(1 - z))$ and the formula for $G(z)$, we have

$$
\begin{aligned}
G(z) &= p_0 \left[\frac{zC^{\sim}(\lambda(1 - z)) - B^{\sim}(\lambda(1 - z)) D^{\sim}(\lambda(1 - z))}{z - B^{\sim}(\lambda(1 - z)) D^{\sim}(\lambda(1 - z))} \right. \\
&\qquad \left. - \frac{z - B^{\sim}(\lambda(1 - z)) D^{\sim}(\lambda(1 - z))}{z - B^{\sim}(\lambda(1 - z)) D^{\sim}(\lambda(1 - z))} \right] \\
&= p_0 \left[\frac{zC^{\sim}(\lambda(1 - z)) - z}{z - B^{\sim}(\lambda(1 - z)) D^{\sim}(\lambda(1 - z))} \right] \qquad (3.77)
\end{aligned}
$$

and

$$
\begin{aligned}
X(z) &= \frac{p_0}{z(1 - p_0)} \frac{z[C^{\sim}(\lambda(1 - z)) - 1] B^{\sim}(\lambda(1 - z))}{z - B^{\sim}(\lambda(1 - z)) D^{\sim}(\lambda(1 - z))} \\
&= \frac{1 - \lambda E[S + R]}{\lambda E[V]} \frac{[C^{\sim}(\lambda(1 - z)) - 1] B^{\sim}(\lambda(1 - z))}{z - B^{\sim}(\lambda(1 - z)) D^{\sim}(\lambda(1 - z))}. \qquad (3.78)
\end{aligned}
$$

Taking the derivative and setting $z = 1$, we have

$$\overline{X} = \lambda \overline{S} + \frac{\lambda^2 \overline{(S + R)^2}}{2(1 - \lambda \overline{S + R})} + \frac{\lambda \overline{V^2}}{2\overline{V}}. \qquad (3.79)$$

If there is no rest period, that is, $R = 0$, and if the server takes no vacation, that is, $V = 0$, then Eq. (3.79) reduces to the Pollaczek–Khintchine formula of Eq. (3.65). Furthermore, the average number of arrivals during a vacation, given by the term

$$\frac{\lambda \overline{V^2}}{2\overline{V}} \, ,$$

is simply added to the average number of customers that would be in the system without vacation. With respect to accumulating arrivals, the rest period can simply be considered as an extension of the service time.

Problem 3.11   Give an interpretation of servers with vacation in terms of polling systems as discussed in Section 3.3.2.1. Give this interpretation for all types of polling systems enumerated in Section 3.3.2.1.

Problem 3.12   How can servers with breakdown and maintenance strategies be modeled by servers with vacation as sketched in Fig. 3.5? What are the differences with polling systems?

Problem 3.13   Give the ergodicity condition for an M/G/1 queue with vacation defined according to Fig. 3.5.

Problem 3.14   Derive the steady-state probabilities ν_0 and ν_1 of an M/G/1 queue with vacation. Assume that the following parameters are given: Rest period is constant at 1 second, vacation is a constant 2 seconds, arrival rate $\lambda = 1$, and service time distribution is Erlang k with $k = 2$ and mean $1/\mu = 0.2$ seconds. Check for ergodicity first.

Problem 3.15   Use the same assumptions as specified in Problem 3.14 and, if steady state exists, compute the mean number of customers in system for the following cases:

1. M/G/1 queue with vacation according to Fig. 3.5.

2. Parameters same as in Problem 3.14 above but with rest period being constant at 0 seconds.

3. Parameters same as in Problem 3.15 part 2 above with vacation being also constant at 0 seconds.

3.4 NUMERICAL SOLUTION: DIRECT METHODS

The closed-form solution methods explored in Sections 3.1 and 3.3 exploited special structures of the Markov chain (or, equivalently, of its parameter matrix). For Markov chains with a more general structure, we need to resort to numerical methods. There are two broad classes of numerical methods to solve the linear systems of equations that we are interested in: direct methods and iterative methods. Direct methods operate and modify the parameter matrix. They use a fixed amount of computation time independent of the parameter values and there is no issue of convergence. But they are subject to fill-in of matrix entries, that is, original zero entries can become nonzeros. This makes the use of sparse storage difficult. Direct methods are also subject to the accumulation of round-off errors.

There are many direct methods for the solution of a system of linear equations. Some of these are restricted to certain regular structures of the parameter matrix that are of less importance for Markov chains, since these structures generally cannot be assumed in the case of a Markov chain. Among the techniques most commonly applied are the well-known Gaussian elimination (GE) algorithm and, a variant thereof, Grassmann's algorithm. The original version of the algorithm, which was published by Grassmann, Taksar, and Heyman, is usually referred to as the GTH algorithm [GTH85] and is based on a renewal argument. We introduce a newer variant by Kumar, Grassmann, and Billington [KGB87] where interpretation gives rise to a simple relation to the GE algorithm. The GE algorithm suffers sometimes from numerical difficulties created by subtractions of nearly equal numbers. It is precisely this property that is avoided by the GTH algorithm and its variant through reformulations relying on regenerative properties of Markov chains. Cancellation errors are nicely circumvented in this way.

3.4.1 Gaussian Elimination

As the point of departure for a discussion of Gaussian elimination, we refer back to Eq. (3.4). The idea of the algorithm is to transform the system of Eq. (3.80), which corresponds in matrix notation to Eq. (3.4), into an equivalent one by applying elementary operations on the parameter matrix that preserve the rank of the matrix:

$$
\begin{aligned}
a_{0,0}x_0 + a_{1,0}x_1 + \ldots + a_{n-1,0}x_{n-1} &= b_0 , \\
a_{0,1}x_0 + a_{1,1}x_1 + \ldots + a_{n-1,1}x_{n-1} &= b_1 ,
\end{aligned}
\tag{3.80}
$$
$$
\vdots
$$
$$
a_{0,n-1}x_0 + a_{1,n-1}x_1 + \ldots + a_{n-1,n-1}x_{n-1} = b_{n-1} .
$$

As a result, an equivalent system of linear equations specified by Eq. (3.81) with a triangular matrix structure is derived, from which the desired solution $\mathbf{x}$, which is identical to the solution of the original system given by Eq. (3.80), can be obtained:

$$
\begin{aligned}
a_{0,0}^{(n-1)} x_0 &= b_0^{(n-1)} , \\
a_{0,1}^{(n-2)} x_0 + a_{1,1}^{(n-2)} x_1 &= b_1^{(n-2)} ,
\end{aligned}
\tag{3.81}
$$
$$
\vdots
$$
$$
a_{0,n-1}^{(0)} x_0 + a_{1,n-1}^{(0)} x_1 + \ldots + a_{n-1,n-1}^{(0)} x_{n-1} = b_{n-1}^{(0)} .
$$

If the system of linear equations has been transformed into a triangular structure, as indicated in Eq. (3.81), the final results can be obtained by means of a straightforward substitution process. Solving the first equation for x_0, substituting the result in the second equation and solving it for x_1, and so on, finally leads to the calculation of x_{n-1}. Hence, the x_i are recursively computed according to Eq. (3.82):

$$x_0 = \frac{b_0^{(n-1)}}{a_{0,0}^{(n-1)}},$$

$$x_j = \frac{b_j^{(n-j)}}{a_{j,j}^{(n-j)}} - \sum_{k=0}^{j-1} \frac{a_{k,j}^{(n-j)}}{a_{j,j}^{(n-j)}} x_k, \quad j = 1, 2, \ldots, n-1. \tag{3.82}$$

To arrive at Eq. (3.81), an elimination procedure first needs to be performed on the original system of Eq. (3.80). Informally, the algorithm can be described as follows: First, the nth equation of Eq. (3.80) is solved for x_{n-1}, and then x_{n-1} is eliminated from all other $n-1$ equations. Next, the $(n-1)$th equation is used to solve for x_{n-2}, and, again, x_{n-2} is eliminated from the remaining $n-2$ equations, and so forth. Finally, Eq. (3.81) results, where $a_{i,j}^{(k)}$ denotes the coefficient of x_i in the $(j+1)$th equation, obtained after the kth elimination step.[6] For the sake of completeness, it is pointed out that $a_{i,j}^{(0)} = a_{i,j}$.

More formally, for the kth elimination step, i.e, the elimination of x_{n-k} from equations $j, j = n - k, n - k - 1, \ldots, 1$, the $(n - k + 1)$th equation is to be multiplied on both sides by

$$-\frac{a_{n-k,j-1}^{(k-1)}}{a_{n-k,n-k}^{(k-1)}}, \tag{3.83}$$

and the result is added to both sides of the jth equation. The computation of the coefficients shown in the system of Eq. (3.84) and Eq. (3.85) for the kth elimination step is

$$a_{ij}^{(k)} = \begin{cases} 0, & \begin{aligned} & j = n-k-1, n-k-2, \ldots, 0, \\ & i = n-1, n-2, \ldots, n-k, \end{aligned} \\[2ex] a_{ij}^{(k-1)} - a_{i,n-k}^{(k-1)} \dfrac{a_{n-k,j}^{(k-1)}}{a_{n-k,n-k}^{(k-1)}}, & \text{otherwise}, \end{cases} \tag{3.84}$$

$$b_j^{(k)} = b_j^{(k-1)} - b_{n-k}^{(k-1)} \frac{a_{n-k,j}^{(k-1)}}{a_{n-k,n-k}^{(k-1)}}, \quad j = n-k-1, n-k, \ldots, 0. \tag{3.85}$$

---

[6]Note that in the system of Eq. (3.81) relation $k = n - j \geq 0$ holds.

In matrix notation we begin with the system of equations:

$$(x_0, x_1, \ldots, x_{n-1}) \begin{pmatrix} a_{0,0} & a_{0,1} & \cdots & a_{0,n-1} \\ a_{1,0} & a_{1,1} & \cdots & a_{1,n-1} \\ \vdots & \vdots & & \vdots \\ a_{n-1,0} & a_{n-1,1} & \cdots & a_{n-1,n-1} \end{pmatrix} = (b_0, b_1, \ldots, b_{n-1}) \ .$$

After the elimination procedure, a modified system of equations results, which is equivalent to the original one. The resulting parameter matrix is in upper triangular form, where the parameters of the matrix are defined according to Eq. (3.84) and the vector representing the right-hand side of the equations according to Eq. (3.85):

$$(x_0, x_1, \ldots, x_{n-1}) \begin{pmatrix} a_{0,0}^{(n-1)} & a_{0,1}^{(n-2)} & a_{0,2}^{(n-3)} & \cdots & a_{0,n-1} \\ 0 & a_{1,1}^{(n-2)} & a_{1,2}^{(n-3)} & \cdots & a_{1,n-1} \\ 0 & 0 & a_{2,2}^{(n-3)} & \cdots & a_{2,n-1} \\ \vdots & \vdots & & & \vdots \\ 0 & 0 & \cdots & 0 & a_{n-1,n-1} \end{pmatrix}$$

$$= (x_0, x_1, \ldots, x_{n-1}) \begin{pmatrix} u_{0,0} & u_{0,1} & u_{0,2} & \cdots & u_{0,n-1} \\ 0 & u_{1,1} & u_{1,2} & \cdots & u_{1,n-1} \\ 0 & 0 & u_{2,2} & \cdots & u_{2,n-1} \\ \vdots & \vdots & & & \vdots \\ 0 & 0 & \cdots & 0 & u_{n-1,n-1} \end{pmatrix} \qquad (3.86)$$

$$= \mathbf{x}\mathbf{U}$$

$$= \left(b_0^{(n-1)}, b_1^{(n-2)}, \ldots, b_{n-1} \right) \ .$$

The Gaussian elimination procedure takes advantage of elementary matrix operations that preserve the rank of the matrix. Such elementary operations correspond to interchanging of equations, multiplication of equations by a real-valued constant, and addition of a multiple of an equation to another equation. In matrix terms, the essential part of Gaussian elimination is provided by the factorization of the parameter matrix $\mathbf{A}$ into the components of an *upper triangular matrix* $\mathbf{U}$ and a *lower triangular matrix* $\mathbf{L}$. The elements of matrix $\mathbf{U}$ are resulted from the elimination procedure while the entries of $\mathbf{L}$ are the terms from Eq. (3.83) by which the columns of the original matrix $\mathbf{A}$ were multiplied during the elimination process:

$$
\mathbf{A} = \begin{pmatrix}
a_{0,0} & a_{0,1} & \cdots & a_{0,n-1} \\
a_{1,0} & a_{1,1} & \cdots & a_{1,n-1} \\
\vdots & \vdots & & \vdots \\
a_{n-1,0} & a_{n-1,1} & \cdots & a_{n-1,n-1}
\end{pmatrix}
$$

$$
= \begin{pmatrix}
a_{0,0}^{(n-1)} & a_{0,1}^{(n-2)} & a_{0,2}^{(n-3)} & \cdots & a_{0,n-1} \\
0 & a_{1,1}^{(n-2)} & a_{1,2}^{(n-3)} & \cdots & a_{1,n-1} \\
0 & 0 & a_{2,2}^{(n-3)} & \cdots & a_{3n} \\
\vdots & \vdots & & & \vdots \\
0 & 0 & \cdots & 0 & a_{n-1,n-1}
\end{pmatrix}
\begin{pmatrix}
1 & 0 & \cdots & & & 0 \\
\dfrac{a_{1,0}^{(n-2)}}{a_{1,1}^{(n-2)}} & 1 & 0 & & \cdots & 0 \\
\vdots & & 1 & 0 & & \vdots \\
\dfrac{a_{n-2,0}^{(1)}}{a_{n-2,n-2}^{(1)}} & \cdots & \dfrac{a_{n-2,n-3}^{(1)}}{a_{n-2,n-2}^{(1)}} & 1 & 0 \\
\dfrac{a_{n-1,0}}{a_{n-1,n-1}} & & \cdots & \dfrac{a_{n-1,n-2}}{a_{n-1,n-1}} & 1
\end{pmatrix}
$$

$$
= \mathbf{UL}. \tag{3.87}
$$

As a result of the factorization of the parameter matrix $\mathbf{A}$, the computation of the result vector $\mathbf{x}$ can split into two simpler steps:

$$
\mathbf{b} = \mathbf{xA} = \mathbf{xUL} = \mathbf{yL}.
$$

The solutions of both equations, first of

$$
\mathbf{yL} = \mathbf{b} \tag{3.88}
$$

for the vector of unknowns $\mathbf{y}$ and then finally of

$$
\mathbf{xU} = \mathbf{y} \tag{3.89}
$$

$$
= \left(b_0^{(n-1)}, b_1^{(n-2)}, \ldots, b_{n-1} \right) \tag{3.90}
$$

for the vector of unknowns $\mathbf{x}$, is required.

Note that the intermediate result $\mathbf{y}$ from Eq. (3.88), being necessary to compute the final results in Eq. (3.89), is identical to Eq. (3.86) and can readily be calculated with the formulae presented in Eq. (3.85). Since only the coefficients of matrix $\mathbf{U}$ are used in this computation, it is not necessary to compute and to represent explicitly the lower triangular matrix $\mathbf{L}$. It is finally worth mentioning that pivoting is not necessary due to the structure of the underlying generator matrix, which is weakly diagonal dominant, since $|q_{i,i}| \geq q_{i,j}, \forall i, j$. This property is inherited by the parameter matrices.

Now the *Gaussian elimination algorithm* can be summarized as follows:

STEP 1 Construct the parameter matrix $\mathbf{A}$ and the right-side vector $\mathbf{b}$ according to Eq. (3.4) as discussed in Chapter 3.

STEP 2 Carry out elimination steps or, equivalently, apply the standard algorithm to split the parameter matrix **A** into upper triangular matrix **U** and lower triangular matrix **L** such that Eq. (3.87) holds. Note that the parameters of **U** can be computed with the recursive formulae in Eq. (3.84) and the computation of **L** can be deliberately avoided.

STEP 3 Compute the intermediate results **y** according to Eq. (3.88) or, equivalently, compute the intermediate results with the result from Eq. (3.90) according to Eq. (3.85).

STEP 4 Perform the substitution to yield the final result **x** according to Eq. (3.89) by recursively applying the formulae shown in Eq. (3.82).

Example 3.2   Consider the CTMC depicted in Fig. 3.6. This simple finite birth–death process is ergodic for any finite λ and μ so that their unique steady-state probabilities can be computed. Since closed-form formulae have

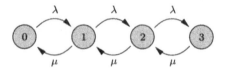

Fig. 3.6 A simple finite birth–death process.

been derived for this case, we can easily compute the steady-state probabilities π_i as summarized in Table 3.1 with $\lambda = 1$ and $\mu = 2$ and with

$$\pi_0 = \frac{1}{\sum\limits_{i=0}^{3} \left(\frac{\lambda}{\mu}\right)^i}, \quad \pi_k = \pi_0 \left(\frac{\lambda}{\mu}\right)^k, \quad k = 1, 2, 3.$$

Table 3.1 Steady-state probabilities computed using closed-form expressions

π_0	π_1	π_2	π_3
$\frac{8}{15}$	$\frac{4}{15}$	$\frac{2}{15}$	$\frac{1}{15}$

Alternatively, the state probabilities can be computed applying the Gaussian elimination method introduced in this section. The results from Table 3.1 can then be used to verify the correctness of the results.

STEP 1 First, the generator matrix $\mathbf{Q}$ is derived from Fig. 3.6:

$$\mathbf{Q} = \begin{pmatrix} -1 & 1 & 0 & 0 \\ 2 & -3 & 1 & 0 \\ 0 & 2 & -3 & 1 \\ 0 & 0 & 2 & -2 \end{pmatrix}.$$

In order to include the normalization condition and to derive the parameter matrix $\mathbf{A}$ of the linear system to be solved, the last column of $\mathbf{Q}$ is replaced by the unit vector:

$$\mathbf{A} = \begin{pmatrix} -1 & 1 & 0 & 1 \\ 2 & -3 & 1 & 1 \\ 0 & 2 & -3 & 1 \\ 0 & 0 & 2 & 1 \end{pmatrix}.$$

The resulting system of linear equations is fully specified according to Eq. (3.4), when vector $\mathbf{b} = (0, 0, 0, 1)$ is given.

STEP 2 With Eq. (3.84) and Eq. (3.85) in mind, matrix $\mathbf{A}$ is transformed into upper-triangular matrix $\mathbf{U}$ via intermediate steps $\mathbf{A}^{(1)}, \mathbf{A}^{(2)}, \mathbf{A}^{(3)} = \mathbf{U}$. In parallel, vector $\mathbf{b}$ is transformed via $\mathbf{b}^{(1)}$ and $\mathbf{b}^{(2)}$ into $\mathbf{b}^{(3)}$, resulting in the following sequence of matrices:

$$\mathbf{A}^{(1)} = \begin{pmatrix} -1 & 1 & -2 & 1 \\ 2 & -3 & -1 & 1 \\ 0 & 2 & -5 & 1 \\ 0 & 0 & 0 & 1 \end{pmatrix}, \qquad \mathbf{b}^{(1)} = (0, 0, -2, 1),$$

$$\mathbf{A}^{(2)} = \begin{pmatrix} -1 & \frac{1}{5} & -2 & 1 \\ 2 & -3\frac{2}{5} & -1 & 1 \\ 0 & 0 & -5 & 1 \\ 0 & 0 & 0 & 1 \end{pmatrix}, \qquad \mathbf{b}^{(2)} = \left(0, -\frac{4}{5}, -2, 1\right),$$

$$\mathbf{A}^{(3)} = \begin{pmatrix} -\frac{15}{17} & \frac{1}{5} & -2 & 1 \\ 0 & -3\frac{2}{5} & -1 & 1 \\ 0 & 0 & -5 & 1 \\ 0 & 0 & 0 & 1 \end{pmatrix}, \qquad \mathbf{b}^{(3)} = \left(-\frac{8}{17}, -\frac{4}{5}, -2, 1\right).$$

STEP 3 For the sake of completeness we also present the lower triangular matrix $\mathbf{L}$ containing the factors from Eq. (3.83) used in the elimination steps:

$$\mathbf{L} = \begin{pmatrix} 1 & 0 & 0 & 0 \\ -\frac{10}{17} & 1 & 0 & 0 \\ 0 & -\frac{2}{5} & 1 & 0 \\ 0 & 0 & 2 & 1 \end{pmatrix}.$$

It can be easily verified that

$$\mathbf{y} = \mathbf{b}^{(3)} = \left(-\frac{8}{17}, -\frac{4}{5}, -2, 1\right)$$

is the intermediate solution vector $\mathbf{y}$ of Eq. (3.88) with given lower triangular matrix $\mathbf{L}$.

STEP 4 The last step is the substitution according to the recursive Eq. (3.82) to yield the final results $\mathbf{x}$ as a solution of Eqs. (3.89) and (3.90):

$$x_0 = \frac{b_0^{(3)}}{a_{0,0}^{(3)}} = \frac{y_0}{u_{0,0}} = \frac{-\frac{8}{17}}{-\frac{15}{17}} = \frac{8}{15},$$

$$x_1 = \frac{b_1^{(2)}}{a_{1,1}^{(2)}} - \frac{a_{0,1}^{(2)}}{a_{1,1}^{(2)}}x_0 = \frac{y_1}{u_{1,1}} - \frac{u_{0,1}}{u_{1,1}}x_0$$

$$= \frac{-\frac{4}{5}}{-\frac{17}{5}} - \frac{\frac{1}{5}}{-\frac{17}{5}}\frac{8}{15} = \frac{4}{15},$$

$$x_2 = \frac{b_2^{(1)}}{a_{2,2}^{(1)}} - \frac{a_{0,2}^{(1)}}{a_{2,2}^{(1)}}x_0 - \frac{a_{1,2}^{(1)}}{a_{2,2}^{(1)}}x_1 = \frac{y_2}{u_{2,2}} - \frac{u_{0,2}}{u_{2,2}}x_0 - \frac{u_{1,2}}{u_{2,2}}x_1$$

$$= \frac{-2}{-5} - \frac{-2}{-5}\frac{8}{15} - \frac{-1}{-5}\frac{4}{15} = \frac{2}{15},$$

$$x_3 = \frac{b_3}{a_{3,3}} - \frac{a_{0,3}}{a_{3,3}}x_0 - \frac{a_{1,3}}{a_{3,3}}x_1 - \frac{a_{2,3}}{a_{3,3}}x_2$$

$$= \frac{y_3}{u_{3,3}} - \frac{u_{0,3}}{u_{3,3}}x_0 - \frac{u_{1,3}}{u_{3,3}}x_1 - \frac{u_{2,3}}{u_{3,3}}x_2$$

$$= 1 - x_0 - x_1 - x_2 = 1 - \frac{8+4+2}{15} = \frac{1}{15}.$$

The computational complexity of the Gaussian elimination algorithm can be characterized by $O(n^3/3)$ multiplications or divisions and a storage requirement of $O(n^2)$, where n is the number of states, and hence the number of equations.

Note that cancellation and rounding errors possibly induced by Gaussian elimination can adversely affect the results. This difficulty is specially relevant true if small parameters have to be dealt with, as is often the case with the analysis of large Markov chains, where relatively small state probabilities may result. For some analyses though, such as in dependability evaluation studies, we may be particularly interested in the probabilities of states that are relatively rarely entered, e.g., system down states or unsafe states. In this case, the accuracy of the smaller numbers can be of predominant interest.

3.4.2   The Grassmann Algorithm

Grassmann's algorithm constitutes a numerically stable variant of the Gaussian elimination procedure. The algorithm completely avoids subtractions and it is therefore less sensitive to rounding and cancellation errors caused by the subtraction of nearly equal numbers. Grassmann's algorithm was originally introduced for the analysis of ergodic, *discrete-time* Markov chains $X = \{X_n; n = 0, 1, \ldots\}$ and was based on arguments from the theory of regenerative processes [GTH85]. A modification of the GTH algorithm has been suggested by Marsan, Meo, and de Souza e Silva [MSA96]. We follow the variant presented by Kumar et al., which allows a straightforward interpretation in terms of continuous-time Markov chains [KGB87].

We know from Eq. (2.41) that the following relation holds:

$$-q_{i,i} = \sum_{j, j \neq i} q_{i,j}. \tag{3.91}$$

Furthermore, Eq. (2.57) can be suitably rearranged:

$$-\pi_i q_{i,i} = \sum_{j=0, j \neq i}^{n-1} \pi_j q_{j,i}. \tag{3.92}$$

Letting $i = n - 1$ and dividing Eq. (3.92) on both sides by $q_{n-1,n-1}$ yields

$$-\pi_{n-1} = \sum_{j=0}^{n-2} \pi_j \frac{q_{j,n-1}}{q_{n-1,n-1}}.$$

This result can be used to eliminate π_{n-1} on the right-hand side of Eq. (3.92):

$$
\begin{aligned}
-\pi_i q_{i,i} &= \sum_{j=0, j \neq i}^{n-2} \pi_j q_{j,i} - \sum_{j=0}^{n-2} \pi_j \frac{q_{j,n-1} q_{n-1,i}}{q_{n-1,n-1}} \\
&= \sum_{j=0, j \neq i}^{n-2} \pi_j \left(q_{j,i} - \frac{q_{j,n-1} q_{n-1,i}}{q_{n-1,n-1}} \right) - \pi_i \frac{q_{i,n-1} q_{n-1,i}}{q_{n-1,n-1}}.
\end{aligned}
\tag{3.93}
$$

Adding the last term of Eq. (3.93) on both sides of that equation results in an equation that can be interpreted similarly as Eq. (3.92):

$$-\pi_i \left(q_{i,i} - \frac{q_{i,n-1} q_{n-1,i}}{q_{n-1,n-1}} \right) = \sum_{j=0, j \neq i}^{n-2} \pi_j \left(q_{j,i} - \frac{q_{j,n-1} q_{n-1,i}}{q_{n-1,n-1}} \right), \quad 0 \leq i \leq n - 2. \tag{3.94}$$

With

$$\bar{q}_{j,i} = q_{j,i} - \frac{q_{j,n-1} q_{n-1,i}}{q_{n-1,n-1}} = q_{j,i} + \frac{q_{j,n-1} q_{n-1,i}}{\sum\limits_{l=0}^{n-2} q_{n-1,l}},$$

the transition rates of a new Markov chain, having one state less than the original one, are defined. Note that this elimination step, i.e., the computation of $\bar{q}_{j,i}$, is achieved merely by *adding non-negative* quantities to originally non-negative values $q_{j,i}, j \neq i$. Only the diagonal elements $q_{i,i}$ and $\bar{q}_{i,i}$ are negative. It should be noted that the computation of the rates $\bar{q}_{i,i}$ on the diagonal of the reduced Markov chain can be completely avoided due to the property of Markov chains reflected in Eq. (3.91). In order to assure the $[\bar{q}_{j,i}]$ properly define a CTMC, it remains to be proven that the following equation:

$$\sum_{i=0}^{n-2} \bar{q}_{j,i} = 0,$$

holds for all j:

$$\sum_{i=0}^{n-2} \bar{q}_{j,i} = \sum_{i=0}^{n-2} q_{j,i} + \sum_{i=0}^{n-2} \frac{q_{j,n-1} q_{n-1,i}}{\sum_{l=0}^{n-2} q_{n-1,l}} = \sum_{i=0}^{n-2} q_{j,i} + q_{j,n-1} \sum_{i=0}^{n-2} \frac{q_{n-1,i}}{\sum_{l=0}^{n-2} q_{n-1,l}}$$

$$= \sum_{i=0}^{n-2} q_{j,i} + q_{j,n-1} = \sum_{i=0}^{n-1} q_{j,i} = 0.$$

The new transition rates $\bar{q}_{j,i}$ can be interpreted as being composed of rate $q_{j,i}$, describing direct transitions from state j to state i, and of rate $q_{j,n-1}$ describing transitions from state j via state $n-1$ to state i with conditional branching probability:

$$\frac{q_{n-1,i}}{\sum_{l=0}^{n-2} q_{n-1,l}}.$$

This interpretation implies that no time is spent in (the otherwise eliminated) state $n-1$. The elimination procedure is iteratively applied to the generator matrix with entries $q_{j,i}^{(k)}$ of stepwise reduced state spaces until an upper triangular matrix results, where $q_{j,i}^{(k)}$ denotes the matrix entries after having applied elimination step $k, 1 \leq k \leq n-1$. Finally, each element $q_{i,i}^{(n-1)}$ on the main diagonal is equal to -1.

As usual, the elimination is followed by a substitution process to express the relations of the state probabilities to each other. Since the normalization condition has not been included initially, it must still be applied to yield the final state probability vector. Grassmann's algorithm has been presented in terms of a CTMC generator matrix and, therefore, the parameter matrix must initially be properly defined:

STEP 1

$$\mathbf{A} = \begin{cases} \mathbf{Q}, & \text{for a CTMC} \\ \mathbf{P} - \mathbf{I}, & \text{for a DTMC} \end{cases}$$

STEP 2 For $l = n - 1, n - 2, \ldots, 1$ do

$$a_{j,i}^{(n-l)} = \begin{cases} \dfrac{a_{j,i}^{(n-l-1)}}{\sum_{m=0}^{l-1} a_{l,m}^{(n-l-1)}}, & j < l, i = l, \\[4ex] a_{j,i}^{(n-l-1)} + \dfrac{a_{j,l}^{(n-l-1)} a_{l,i}^{(n-l-1)}}{\sum_{m=0}^{l-1} a_{l,m}^{(n-l-1)}}, & j \neq i, 1 \leq j, i \leq l - 1, \\[4ex] -1, & j = i = l, \\[1ex] 0, & j = l, i < l.^7 \end{cases}$$

STEP 3 For $l = 1, 2, \ldots, n - 1$ do

$$x_l = \sum_{i=0}^{l-1} x_i a_{il}^{(n-l)}.$$

STEP 4 For $i = 0, 1, \ldots, n - 1$ do

$$\left. \begin{array}{c} \pi_i \\ \nu_i \end{array} \right\} = \frac{x_i}{\sum_{j=0}^{n-1} x_j}.$$

In matrix notation, the parameter matrix $\mathbf{A}$ is decomposed into factors of an upper triangular matrix $\mathbf{U}$ and a lower triangular matrix $\mathbf{L}$ such that the following equations hold:

$$\mathbf{0} = \mathbf{xA} = \mathbf{xUL}.$$

Of course, any (nontrivial) solution of $\mathbf{0} = \mathbf{xU}$ is also a solution of the original equation $\mathbf{0} = \mathbf{xA}$. Therefore, there is no need to represent $\mathbf{L}$ explicitly.

Example 3.3   Consider the CTMC presented in Fig. 3.7. It represents a finite queueing system with capacity 3, where customers arrive according to independent, exponentially distributed interarrival times with rate λ as long as buffer space is available. Service is received from a single server in two consecutive phases, both phases having mean duration $1/(2\mu)$. The pair of integers (k, l) assigned to a state i represents the number of customers $k(i)$ in the system and the phase $l(i)$ of the customer currently in service. To construct the generator matrix $\mathbf{Q}$, the states are ordered as indicated in Table 3.2.

---

[7]Setting the elements on the diagonal equal to -1 and below the diagonal equal to 0 is only included here for the sake of completeness. These assignments can be skipped in an actual implementation for efficiency reasons.

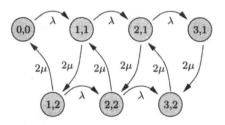

Fig. 3.7 A CTMC for a system with Erlang-2 service time distribution.

Table 3.2 A possible state ordering for Fig. 3.7

(k, l)	$(0,0)$	$(1,1)$	$(2,1)$	$(3,1)$	$(1,2)$	$(2,2)$	$(3,2)$
#	0	1	2	3	4	5	6

Grassmann's algorithm is applied with $\lambda = \mu = 1$:

STEP 1

$$\mathbf{A} = \begin{pmatrix} -1 & 1 & 0 & 0 & 0 & 0 & 0 \\ 0 & -3 & 1 & 0 & 2 & 0 & 0 \\ 0 & 0 & -3 & 1 & 0 & 2 & 0 \\ 0 & 0 & 0 & -2 & 0 & 0 & 2 \\ 2 & 0 & 0 & 0 & -3 & 1 & 0 \\ 0 & 2 & 0 & 0 & 0 & -3 & 1 \\ 0 & 0 & 2 & 0 & 0 & 0 & -2 \end{pmatrix}.$$

STEP 2

$$\mathbf{A}^{(1)} = \begin{pmatrix} -1 & 1 & 0 & 0 & 0 & 0 & 0 \\ 0 & -3 & 1 & 0 & 2 & 0 & 0 \\ 0 & 0 & -3 & 1 & 0 & 2 & 0 \\ 0 & 0 & 2 & -2 & 0 & 0 & 1 \\ 2 & 0 & 0 & 0 & -3 & 1 & 0 \\ 0 & 2 & 1 & 0 & 0 & -3 & \frac{1}{2} \\ 0 & 0 & 0 & 0 & 0 & 0 & -1 \end{pmatrix},$$

$$\mathbf{A}^{(2)} = \begin{pmatrix} -1 & 1 & 0 & 0 & 0 & 0 & 0 \\ 0 & -3 & 1 & 0 & 2 & 0 & 0 \\ 0 & \frac{4}{3} & -3 & 1 & 0 & \frac{2}{3} & 0 \\ 0 & 0 & 2 & -2 & 0 & 0 & 1 \\ 2 & \frac{2}{3} & \frac{1}{3} & 0 & -3 & \frac{1}{3} & 0 \\ 0 & 0 & 0 & 0 & 0 & -1 & \frac{1}{2} \\ 0 & 0 & 0 & 0 & 0 & 0 & -1 \end{pmatrix},$$

$$\mathbf{A}^{(3)} = \begin{pmatrix} -1 & 1 & 0 & 0 & 0 & 0 & 0 \\ \frac{4}{3} & -3 & \frac{11}{9} & 0 & \frac{2}{3} & 0 & 0 \\ 0 & \frac{4}{3} & -3 & 1 & 0 & \frac{2}{3} & 0 \\ 0 & 0 & 2 & -2 & 0 & 0 & 1 \\ 0 & 0 & 0 & 0 & -1 & \frac{1}{3} & 0 \\ 0 & 0 & 0 & 0 & 0 & -1 & \frac{1}{2} \\ 0 & 0 & 0 & 0 & 0 & 0 & -1 \end{pmatrix},$$

$$\mathbf{A}^{(4)} = \begin{pmatrix} -1 & 1 & 0 & 0 & 0 & 0 & 0 \\ \frac{4}{3} & -3 & \frac{11}{9} & 0 & \frac{2}{3} & 0 & 0 \\ 0 & \frac{4}{3} & -3 & \frac{1}{2} & 0 & \frac{2}{3} & 0 \\ 0 & 0 & 0 & -1 & 0 & 0 & 1 \\ 0 & 0 & 0 & 0 & -1 & \frac{1}{3} & 0 \\ 0 & 0 & 0 & 0 & 0 & -1 & \frac{1}{2} \\ 0 & 0 & 0 & 0 & 0 & 0 & -1 \end{pmatrix},$$

$$\mathbf{A}^{(5)} = \begin{pmatrix} -1 & 1 & 0 & 0 & 0 & 0 & 0 \\ \frac{4}{3} & -3 & \frac{11}{12} & 0 & \frac{2}{3} & 0 & 0 \\ 0 & 0 & -1 & \frac{1}{2} & 0 & \frac{2}{3} & 0 \\ 0 & 0 & 0 & -1 & 0 & 0 & 1 \\ 0 & 0 & 0 & 0 & -1 & \frac{1}{3} & 0 \\ 0 & 0 & 0 & 0 & 0 & -1 & \frac{1}{2} \\ 0 & 0 & 0 & 0 & 0 & 0 & -1 \end{pmatrix},$$

$$\mathbf{A}^{(6)} = \begin{pmatrix} 0 & \frac{3}{4} & 0 & 0 & 0 & 0 & 0 \\ 0 & -1 & \frac{11}{12} & 0 & \frac{2}{3} & 0 & 0 \\ 0 & 0 & -1 & \frac{1}{2} & 0 & \frac{2}{3} & 0 \\ 0 & 0 & 0 & -1 & 0 & 0 & 1 \\ 0 & 0 & 0 & 0 & -1 & \frac{1}{3} & 0 \\ 0 & 0 & 0 & 0 & 0 & -1 & \frac{1}{2} \\ 0 & 0 & 0 & 0 & 0 & 0 & -1 \end{pmatrix}.$$

STEP 3 Since $\mathbf{A}^{(6)}$ is in upper triangular form the solution of $\mathbf{x}\mathbf{A}^{(6)} = \mathbf{0}$ can be easily obtained:

$$x_1 = \frac{3}{4}x_0 \quad x_2 = \frac{11}{12}x_1 \quad\quad x_3 = \frac{1}{2}x_2$$
$$x_4 = \frac{2}{3}x_1 \quad x_5 = \frac{1}{3}x_4 + \frac{2}{3}x_2 \quad x_6 = \frac{1}{2}x_5 + x_3 .$$

Because only six equations with seven unknowns were derived, one equation is deliberately added. For convenience, we let $x_0 = \frac{4}{3}$, thereby yielding the intermediate solution:

$$\mathbf{x} = \left(\frac{4}{3}, 1, \frac{11}{12}, \frac{11}{24}, \frac{2}{3}, \frac{5}{6}, \frac{21}{24} \right) .$$

STEP 4 With

$$\frac{1}{\sum\limits_{i=0}^{n-1} x_i} = \frac{12}{73},$$

and with

$$\pi_i = \frac{x_i}{\sum\limits_{i=0}^{n-1} x_i},$$

the state probability vector π results:

$$\pi = (0.2192, 0.1644, 0.1507, 0.0753, 0.1096, 0.1370, 0.1438).$$

Measures of interest could be the average number of customers in the system $E[N] = \overline{N}$

$$\overline{N} = \sum_{i=0}^{n-1} k(i)\pi_i = \pi_1 + \pi_4 + 2(\pi_2 + \pi_5) + 3(\pi_3 + \pi_6) = 1.507,$$

the effective throughput

$$\lambda_e = \lambda(\pi_0 + \pi_1 + \pi_2 + \pi_4 + \pi_5) = 0.7809,$$

or the mean response time $E[T] = \overline{T}$

$$\overline{T} = \frac{\overline{N}}{\lambda_e} = 1.9298.$$

Comparing these results to the measures obtained from an analysis of the birth–death process for this queueing system assuming exponentially (rather than Erlang) distributed service times with same parameters $\lambda = \mu = 1$ shows that response time has decreased and the effective throughput has increased simply by reducing the variance of the service process:

$$\pi_k^{(\text{EXP})} = \frac{1}{4} = 0.25, \quad k = 0, 1, 2, 3,$$

$$\overline{N}^{(\text{EXP})} = (1 + 2 + 3) \cdot \frac{1}{4} = \frac{3}{2} = 1.5,$$

$$\lambda_e^{(\text{EXP})} = 3 \cdot \frac{1}{4} = 0.75,$$

$$\overline{T}^{(\text{EXP})} = \frac{3}{2} \cdot \frac{4}{3} = 2.$$

The computational complexity of Grassmann's algorithm is, similar to Gaussian elimination, $O(n^3/3)$ multiplications or divisions, and the storage requirement is $O(n^2)$. Sparse storage techniques can be applied under some circumstances, when regular matrix structures, like those given with diagonal matrices, are preserved during computation. Although cancellation errors are being avoided with Grassmann's algorithm, rounding errors can still occur, propagate, and accumulate during the computation. Therefore, applicability of the algorithm is also limited to medium-sized (around 500 states) Markov models.

3.5 NUMERICAL SOLUTION: ITERATIVE METHODS

The main advantage of iterative methods over direct methods is that they preserve the sparsity the parameter matrix. Efficient sparse storage schemes and efficient sparsity-preserving algorithms can thus be used. Further advantages of iterative methods are based on the property of successive convergence to the desired solution. A good initial estimate can speed up the computation considerably. The evaluation can be terminated if the iterations are sufficiently close to the exact value; i.e., a prespecified tolerance level is not exceeded. Finally, because the parameter matrix is not altered in the iteration process, iterative methods do not suffer from the accumulation of round-off errors. The main disadvantage of iterative methods is that convergence is not always guaranteed and depending on the method, the rate of convergence is highly sensitive to the values of entries in the parameter matrix.

3.5.1 Convergence of Iterative Methods

Convergence is a very important issue for iterative methods that must be dealt with consciously. There are some heuristics that can be applied for choosing appropriate techniques for decisions on convergence, but no general algorithm for the selection of such a technique exists. In what follows we mention a few of the many issues to be considered with respect to convergence. A complete picture is beyond the scope of this text.

A tolerance level ϵ must be specified to provide a measure of how close the current iteration vector $\mathbf{x}^{(k)}$ is to the desired solution vector $\mathbf{x}$. Because the desired solution vector is not known, an *estimate* of the error must be used to determine convergence. Some distance measures are commonly used to evaluate the current iteration vector $\mathbf{x}^{(k)}$ in relation to some earlier iteration vectors $\mathbf{x}^{(l)}, l < k$. If the current iteration vector is "close enough" to earlier ones with respect to ϵ, then this condition is taken as an *indicator* of convergence to the final result. If ϵ were too small, convergence could become very slow or not take place at all. If ϵ were too large, accuracy requirements could be violated or, worse, convergence could be wrongly assumed. Some appropriate norm functions have to be applied in order to compare different iteration vectors. Many such norm functions exist, all having a different impact on speed and pattern of convergence. Size and type of the parameter matrix should be taken into consideration for the right choice of such a norm function. Concerning the right choice of ϵ and the norm function, it should be further noted that the components x_i of the solution vector can differ by many orders of magnitude from each other in their values. An appropriate definition must take these differences into account and relate them to the accuracy requirements derived from the modeling context.

3.5.2   Power Method

Referring to Eq. (2.15) immediately leads to a first, reliable, though sometimes slowly converging, iterative method for the computation of the steady-state probability vector of finite *ergodic* Markov chains. For the sake of completeness we may mention here that for the power method to converge the transition probability, matrix $\mathbf{P}$ only needs to be *aperiodic*; irreducibility is not necessary. The power method mimics the transient behavior of the underlying DTMC until some stationary, not necessarily steady-state, convergence is reached. Therefore, it can also be exploited as a method for computing the transient state probability vector $\boldsymbol{\nu}(n)$ of a DTMC.

Equation $\boldsymbol{\nu} = \boldsymbol{\nu}\mathbf{P}$ suggests starting with an initial guess of some probability vector $\boldsymbol{\nu}^{(0)}$ and repeatedly multiplying it by the transition probability matrix $\mathbf{P}$ until convergence to $\boldsymbol{\nu}$ is assured, with $\lim_{i \to \infty} \boldsymbol{\nu}^{(i)} = \boldsymbol{\nu}$. Due to the assumed ergodicity, or at least aperiodicity, of the underlying Markov chain, this procedure is guaranteed to converge to the desired fixed point of the unique steady-state probability vector. A single iteration step is as follows:

$$\boldsymbol{\nu}^{(i+1)} = \boldsymbol{\nu}^{(i)}\mathbf{P}, \quad i \geq 0. \tag{3.95}$$

Hence, the iteration vector at step i is related to the initial probability vector via the $\mathbf{P}^i$, the ith power of the transition probability matrix $\mathbf{P}$:

$$\boldsymbol{\nu}^{(i)} = \boldsymbol{\nu}^{(0)}\mathbf{P}^i, \quad i \geq 0. \tag{3.96}$$

The power method is justified by the theory of eigenvalues and eigenvectors. The vector $\boldsymbol{\nu}$ can be interpreted as the left eigenvector $\mathbf{y}_1$ corresponding to the unit eigenvalue $\lambda_1 = 1$, where λ_1 is the dominant eigenvalue, which is the largest of all n eigenvalues, $|\lambda_1| \geq |\lambda_2| \geq \cdots \geq |\lambda_n|$. Since the eigenvectors are mutually orthogonal, $\boldsymbol{\nu}^{(0)}$ can be expressed as a linear combination of all left eigenvectors of $\mathbf{P}$ so that the following relation holds:

$$\boldsymbol{\nu}^{(0)} = c_1\mathbf{y}_1 + c_2\mathbf{y}_2 + \cdots + c_n\mathbf{y}_n, \quad c_j \in \mathbb{R}, \ 1 \leq j \leq n. \tag{3.97}$$

Accordingly, after the first iteration step, the estimate $\boldsymbol{\nu}^{(1)} = \boldsymbol{\nu}^{(0)}\mathbf{P}$ can again be expressed in terms of the eigenvectors $\mathbf{y}_i$ and eigenvalues λ_i:

$$\boldsymbol{\nu}^{(1)} = c_1\lambda_1\mathbf{y}_1 + c_2\lambda_2\mathbf{y}_2 + \cdots + c_n\lambda_n\mathbf{y}_n, \quad c_j \in \mathbb{R}, \ 1 \leq j \leq n. \tag{3.98}$$

Repeatedly applying this procedure yields

$$\boldsymbol{\nu}^{(i)} = c_1\lambda_1^i\mathbf{y}_1 + c_2\lambda_2^i\mathbf{y}_2 + \cdots + c_n\lambda_n^i\mathbf{y}_n, \quad c_j \in \mathbb{R}, \ 1 \leq j \leq n. \tag{3.99}$$

Since all eigenvalues $|\lambda_j| < 1$ for $j \geq 2$, the ith power of the eigenvalues λ_j, $j \geq 2$, approaches 0 in the limit, $\lim_{i \to \infty} |\lambda_j|^i = 0$. With respect to the unit eigenvalue $|\lambda_1| = 1$ we obtain $|\lambda_1|^i = 1$, for all $i \geq 1$. Therefore, relation (3.100) holds and proves our assumption of the correct convergence of the power method:

$$\lim_{i \to \infty} \boldsymbol{\nu}^{(i)} = \lim_{i \to \infty} \{c_1 \lambda_1^i \mathbf{y}_1 + c_2 \lambda_2^i \mathbf{y}_2 + \cdots + c_n \lambda_n^i \mathbf{y}_n\}$$
$$= \lim_{i \to \infty} c_1 \lambda_1^i \mathbf{y}_1 \qquad (3.100)$$
$$= c_1 \mathbf{y}_1 .$$

Only a renormalization remains to be performed to yield the final result of the steady-state probability vector $\boldsymbol{\nu}$.

An immediate consequence of the nature of the power method is that speed of convergence depends on the relative sizes of the eigenvalues. The closer the non-dominant eigenvalues are to 1, the slower the convergence, as can be concluded from Eq. (3.100). The second largest eigenvalue, $|\lambda_2|$, is the dominating factor in this process, i.e., if it is close to 1, convergence will be slow.

The algorithm of the power method is outlined as follows:

STEP 1 Select q appropriately:

$$\mathbf{A} = \begin{cases} \mathbf{P}, \\ \mathbf{Q}/q + \mathbf{I}; \end{cases}$$
$$\boldsymbol{\nu}^{(0)} = \left(\nu_0^{(0)}, \nu_1^{(0)}, \ldots, \nu_{n-1}^{(0)}\right) .$$

Select convergence criterion ϵ, and let $n = 0$. Define some vector norm function $f\left(\|\boldsymbol{\nu}^{(n)}, \boldsymbol{\nu}^{(l)}\|\right)$, $n \geq l$.
Set *convergence* = false.

STEP 2 Repeat until *convergence*:

STEP 2.1 $\boldsymbol{\nu}^{(n+1)} = \boldsymbol{\nu}^{(n)} \mathbf{A}$;

STEP 2.2 IF $f\left(\|\boldsymbol{\nu}^{(n+1)}, \boldsymbol{\nu}^{(l+1)}\|\right) < \epsilon$, $l \leq n$
THEN *convergence* = true;

STEP 2.3 $n = n + 1, l = l + 1$.

STEP 3 $\left.\begin{matrix} \boldsymbol{\pi} \\ \boldsymbol{\nu} \end{matrix}\right\} \approx \boldsymbol{\nu}^{(n)} .$

Example 3.4 Assume three customers circulating indefinitely between two service stations, alternately receiving service from both stations having exponentially distributed service times with parameters μ_1 and μ_2, respectively. Let (i, j) denote a possible state of the system, where $i + j = 3$ and i, j refer

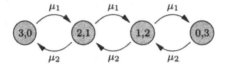

Fig. 3.8 A birth–death model of a two station cyclic queue with three customers.

to the number of customers at the first and the second stations, respectively. The CTMC shown in Fig. 3.8 models this two station cyclic queueing network. With $\mu_1 = 1$ and $\mu_2 = 2$ the following generator matrix $\mathbf{Q}$ results:

$$\mathbf{Q} = \begin{pmatrix} -1 & 1 & 0 & 0 \\ 2 & -3 & 1 & 0 \\ 0 & 2 & -3 & 1 \\ 0 & 0 & 2 & -2 \end{pmatrix}.$$

Together with an arbitrarily given initial probability vector

$$\boldsymbol{\pi}(0) = (\pi_{3,0}(0), \pi_{2,1}(0), \pi_{1,2}(0), \pi_{0,3}(0))$$
$$= (0.25, 0.25, 0.25, 0.25),$$

and, given $q = 3.0000003$, a transition probability matrix

$$\mathbf{P} = \begin{pmatrix} 0.6666667 & 0.3333333 & 0 & 0 \\ 0.6666666 & 0.0000001 & 0.3333333 & 0 \\ 0 & 0.6666666 & 0.0000001 & 0.3333333 \\ 0 & 0 & 0.6666666 & 0.3333334 \end{pmatrix},$$

the power method can be applied, recalling $\boldsymbol{\nu}^{(0)} = \boldsymbol{\pi}(0)$.

Since we are dealing here with a small ergodic model, Gaussian elimination or Grassmann's algorithm could also be applied to yield the following exact steady-state probability vector:

$$\boldsymbol{\pi} = (0.53\overline{3}, 0.26\overline{6}, 0.13\overline{3}, 0.6\overline{6}).$$

Using this result, the iteration vector can be compared to the exact probability vector $\boldsymbol{\pi}$ at intermediate steps for the purpose of demonstrating convergence as shown in Table 3.3. The results are gained by choosing $\epsilon = 10^{-7}$ and by applying the following test of convergence, presuming $n \geq 1$:

$$\frac{\|\boldsymbol{\nu}^{(n)} - \boldsymbol{\nu}^{n-1}\|_2}{\|\boldsymbol{\nu}^{n-1}\|_2} < \epsilon.$$

The number of iteration steps necessary to yield convergence to the exact results depends strongly on the uniformization factor q. Some examples demonstrating this dependency are summarized in Table 3.4.

Table 3.3 Intermediate steps and convergence of the power method

$\nu^{(0)}$	0.25	0.25	0.25	0.25
$\nu^{(1)}$	0.333333325	0.25	0.25	0.166666675
$\nu^{(2)}$	0.3888888778	0.2777777722	0.1944444556	0.1388888944
$\nu^{(3)}$	0.4444444278	0.2592592648	0.1851851880	0.1111111194
$\nu^{(4)}$	0.4691357926	0.2716049333	0.1604938370	0.09876543704
$\nu^{(5)}$	0.4938271481	0.2633744897	0.1563786029	0.08641975926
$\nu^{(10)}$	0.5276973339	0.2671002038	0.1357177959	0.06948466636
$\nu^{(20)}$	0.5332355964	0.2666741848	0.1333746836	0.06671553511
$\nu^{(30)}$	0.5333316384	0.2666667970	0.1333340504	0.06666751412
$\nu^{(38)}$	0.5333332672	0.2666666718	0.1333333613	0.06666669973

Table 3.4 Convergence of the power
method as a function uniformization
factor q

q	Iterations
3.000003	54
3.003	55
3.03	55
3.3	61
6	115
30	566

3.5.3 Jacobi's Method

The system of linear equations of interest to us is

$$\mathbf{b} = \mathbf{x}\mathbf{A}.\tag{3.101}$$

The normalization condition may or may not be incorporated in Eq. (3.101). The parameters of both DTMC and CTMC are given by the entries of the matrix $\mathbf{A} = [a_{ij}]$. The solution vector $\mathbf{x}$ will contain the unconditional state probabilities, or, if not yet normalized, a real-valued multiple thereof. If the normalization is incorporated, we have $\mathbf{b} = [0, 0, \ldots, 0, 1]$, and $\mathbf{b} = \mathbf{0}$ otherwise. Consider the jth equation from the system of equations (3.101):

$$b_j = \sum_{i \in S} a_{ij} x_i.\tag{3.102}$$

Solving Eq. (3.102) for x_j immediately results in

$$x_j = \frac{b_j - \sum_{i, i \neq j} a_{ij} x_i}{a_{jj}}.\tag{3.103}$$

Any given approximate solution $\hat{\mathbf{x}} = [\hat{x}_0, \hat{x}_1, \ldots, \hat{x}_{n-1}]$ can be inserted for the variables $x_i, i \neq j$, on the right-hand side of Eq. (3.103). From these intermediate values, better estimates of the x_j on the left-hand side of the equation may be obtained. Applying this procedure repeatedly and in parallel for all n equations results in our first iterative method. The values $x^{(k)}$ of the kth iteration step are computed from the values obtained from the $(k-1)$st step for each equation independently:

$$x_j^{(k)} = \frac{b_j - \sum\limits_{i, i \neq j} a_{ij} x_i^{(k-1)}}{a_{jj}}, \quad \forall j \in S. \qquad (3.104)$$

The iteration may be started with an arbitrary initial vector $\mathbf{x}^{(0)}$. The closer the initial vector is to the solution, the faster the algorithm will generally converge. Note that the equations can be evaluated in parallel, a fact that can be used as a means for computational speed-up. The method is therefore called the *method of simultaneous displacement* or, simply, the *Jacobi method*. Although the method is strikingly simple, it suffers from poor convergence and hence is rarely applied in its raw form.

Splitting the matrix $\mathbf{A} = \mathbf{D} - \mathbf{L} - \mathbf{U}$ into its constituents of the *diagonal* matrix $\mathbf{D}$, the *strictly lower-triangular* matrix $-\mathbf{L}$, and the *strictly upper-triangular* matrix $-\mathbf{U}$ provides a way to present the main computation step of the Jacobi method in matrix notation:

$$\mathbf{x}^{(k)} = \left(\mathbf{b} + \mathbf{x}^{(k-1)} (\mathbf{U} + \mathbf{L})\right) \mathbf{D}^{-1}. \qquad (3.105)$$

Before the algorithm is presented, we may recall that we can deliberately incorporate the normalization condition into the parameter matrix $\mathbf{A}$. Note that repeated normalization could have an adverse effect on convergence if large models are investigated, since relatively small numbers could result. Therefore, we leave it open whether to repeat normalization in each iteration step or not.

STEP 1 Define parameter matrix $\mathbf{A}$ and $\mathbf{b}$ properly from generator matrix $\mathbf{Q}$ or transition probability matrix $\mathbf{P}$.

- Choose initial vector $\mathbf{x}^{(0)}$.
- Choose convergence criterion ϵ.
- Choose some norm function $f\left(\|\mathbf{x}^{(k)}, \mathbf{x}^{(l)}\|\right)$, $k \geq l$.
- Split parameter matrix $\mathbf{A} = \mathbf{D} - \mathbf{L} - \mathbf{U}$.
- *convergence* = false, and $k = l = 1$.

STEP 2 Repeat until *convergence*:

STEP 2.1 $\mathbf{x}^{(k)} = \left(\mathbf{b} + \mathbf{x}^{(k-1)} (\mathbf{U} + \mathbf{L})\right) \mathbf{D}^{-1}$;

STEP 2.2 • IF $f\left(\|\mathbf{x}^{(k)} - \mathbf{x}^{(k-l)}\|\right) < \epsilon$
  – THEN *convergence* = true

   − ELSE $k = k + 1$ and $l \in \{1, \ldots, k\}$.

$$\left.\begin{array}{c} \pi \\ \nu \end{array}\right\} = \frac{\mathbf{x}^{(k)}}{\sum_{j=0}^{n-1} x_j^{(k)}} \, .$$

Example 3.5 Consider two customers circulating among three service stations according to the following pattern. When a customer has received service of mean duration $1/\mu_1$ at the first station, it then queues with probability p_{12} at station two for service of mean duration $1/\mu_2$, or with p_{13} at station three with a mean service duration of $1/\mu_3$. After completion of service at stations two or three, customers return with probability one back to station one.

This scenario is captured by a CTMC with the state diagram in Fig. 3.9, where the notation (k, l, m) indicates that there are k customers in station one, l customers in station two, and m customers in station three. We number the states of the CTMC of Fig. 3.9 as per Table 3.5.

Table 3.5 A possible state ordering according to Fig. 3.9

(k, l, m)	$(2, 0, 0)$	$(1, 1, 0)$	$(1, 0, 1)$	$(0, 2, 0)$	$(0, 1, 1)$	$(0, 0, 2)$
#	1	2	3	4	5	6

With $\mu_1 = 1$, $\mu_2 = 2$, $\mu_3 = 3$, $p_{12} = 0.4$, and $p_{13} = 0.6$, the generator matrix $\mathbf{Q}$ can be obtained:

$$\mathbf{Q} = \begin{pmatrix} -1 & 0.4 & 0.6 & 0 & 0 & 0 \\ 2 & -3 & 0 & 0.4 & 0.6 & 0 \\ 3 & 0 & -4 & 0 & 0.4 & 0.6 \\ 0 & 2 & 0 & -2 & 0 & 0 \\ 0 & 3 & 2 & 0 & -5 & 0 \\ 0 & 0 & 3 & 0 & 0 & -3 \end{pmatrix} \, .$$

To provide a controlled experiment with respect to the iteration process, we first compute the exact state probabilities in Table 3.6 with Grassmann's algorithm introduced in Section 3.4.2. With Table 3.6 we know the exact state probabilities $\mathbf{x} = \boldsymbol{\pi}$ and can use them for comparison with the iteration vector $\mathbf{x}^{(k)}$.

With the following constraints concerning equation vector $\mathbf{b}$, initial probability vector $\mathbf{x}^{(0)}$, normalization after each iteration step, convergence criterion ϵ, and the norm for test of convergence, Jacobi's method is applied:

$$\mathbf{b} = (0, 0, 0, 0, 0, 0) \, , \quad \mathbf{x}^{(0)} = \left(\frac{1}{6}, \frac{1}{6}, \frac{1}{6}, \frac{1}{6}, \frac{1}{6}, \frac{1}{6}\right) ,$$

$$\mathbf{x}^{(k)} = \frac{\mathbf{x}^{(k)}}{\|\mathbf{x}^{(k)}\|_1} \, , \quad \epsilon = 10^{-7} \, , \quad \frac{\|\mathbf{x}^{(k)} - \mathbf{x}^{(k-1)}\|_2}{\|\mathbf{x}^{(k-1)}\|_2} < \epsilon \, .$$

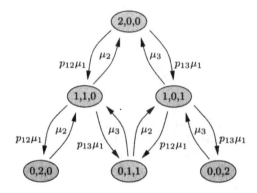

Fig. 3.9 A birth–death model of two customers traversing three stations.

Table 3.6 State probabilities for Fig. 3.9 computed using Grassmann's algorithm	
π_0	0.6578947368
π_1	0.1315789474
π_2	0.1315789474
π_3	0.02631578947
π_4	0.02631578947
π_5	0.02631578947

Table 3.7 State probabilities for Fig. 3.9 computed using Jacobi's method	
π_0	0.6578947398
π_1	0.1315789172
π_2	0.1315789478
π_3	0.02631578619
π_4	0.02631578846
π_5	0.02631582059

The results of using Jacobi algorithm, needing 324 iterations are given in Table 3.7.

The method of Jacobi is of less practical importance due to its slow pattern of convergence. But techniques have been derived to speed up its convergence, resulting in well-known algorithms such as Gauss–Seidel iteration or the successive over-relaxation (SOR) method.

3.5.4 Gauss–Seidel Method

To improve convergence, a given method often needs to be changed only slightly. Instead of deploying full parallelism by evaluating Eq. (3.104) for each state $j \in S$ independently, we can serialize the procedure and take advantage of the already updated new estimates in each step. Assuming the computations to be arranged in the order $0, 1, \ldots, n-1$, where $|S| = n$, it immediately follows, that for the calculation of the estimates $x_j^{(k)}$, all j previously computed estimates $x_i^{(k)}$, $i < j$, can be used in the computation. Taking advantage of the more up-to-date information, we can significantly speed up the convergence.

The resulting method is called the *Gauss–Seidel* iteration:

$$x_j^{(k)} = \frac{b_j - \left(\sum_{i=0}^{j-1} a_{ij} x_i^{(k)} + \sum_{i=j+1}^{n-1} a_{ij} x_i^{(k-1)} \right)}{a_{jj}}, \quad \forall j \in S. \tag{3.106}$$

Note that the order in which the estimates $x_j^{(k)}$ are calculated in each iteration step can have a decisive impact on the speed of convergence. In particular, it is generally a matter of fact that matrices of Markov chains are sparse. The interdependencies between the equations are therefore limited to a certain degree, and parallel evaluation might still be possible, even if the most up-to-date information is incorporated in each computation step. The equations can deliberately be arranged so that the interdependencies become more or less effective for the convergence process. Equivalently, the possible degree of parallelism can be maximized or minimized by a thoughtful choice of that order. Apparently, a trade-off exists between the pattern of convergence and possible speedup due to parallelism. In matrix notation, the Gauss–Seidel iteration step is written as

$$\mathbf{x}^{(k)} = \left(\mathbf{b} + \mathbf{x}^{(k-1)} \mathbf{L} \right) (\mathbf{D} - \mathbf{U})^{-1}, \quad k \geq 1. \tag{3.107}$$

Equivalently, we could rewrite Eq. (3.107), thereby reflecting the Gauss–Seidel step from Eq. (3.106) more obviously:

$$\mathbf{x}^{(k)} = \left(\mathbf{b} + \mathbf{x}^{(k)} \mathbf{U} + \mathbf{x}^{(k-1)} \mathbf{L} \right) \mathbf{D}^{-1}, \quad k \geq 1. \tag{3.108}$$

3.5.5 The Method of Successive Over-Relaxation

Though Gauss–Seidel is widely applied, further variants of the iteration scheme do exist. A promising approach for an increased rate of convergence is to apply extrapolations via the so-called method of *successive over-relaxation*, SOR for short. The SOR method provides means to weaken or to enhance the impact of a Gauss–Seidel iteration step. A new estimate is calculated by weighting (using a relaxation parameter ω) a previous estimate with the newly computed Gauss–Seidel estimate. The iteration step is presented in matrix notation in Eq. (3.109), from which it can be concluded that the SOR method coincides with Gauss–Seidel if ω is set equal to *one*:

$$\mathbf{x}^{(k)} = \left(\omega \mathbf{b} + \mathbf{x}^{(k-1)} \left(\omega \mathbf{L} + (1 - \omega) \mathbf{D} \right) \right) (\mathbf{D} - \omega \mathbf{U})^{-1}, \quad k \geq 1. \tag{3.109}$$

Equation (3.109) can be easily verified. Remembering the splitting of matrix $\mathbf{A} = \mathbf{D} - \mathbf{L} - \mathbf{U}$ and introducing a scalar ω, we can immediately derive the equation $\omega \mathbf{b} = \mathbf{x} \omega (\mathbf{D} - \mathbf{L} - \mathbf{U})$ from Eq. (3.101) as a starting point. Adding $\mathbf{x} \mathbf{D}$ to both sides of the equation yields

$$\omega \mathbf{b} + \mathbf{x} \mathbf{D} = \mathbf{x} \omega (\mathbf{D} - \mathbf{L} - \mathbf{U}) + \mathbf{x} \mathbf{D}. \tag{3.110}$$

By applying simple arithmetic manipulations we get

$$\omega\mathbf{b} + \mathbf{x}\left(\omega\mathbf{L} + (1 - \omega)\,\mathbf{D}\right) = \mathbf{x}\left(\mathbf{D} - \omega\mathbf{U}\right), \qquad (3.111)$$

and, finally, by multiplying both sides of Eq. (3.111) with $(\mathbf{D} - \omega\mathbf{U})^{-1}$, we get Eq. (3.109). To complete the description of the SOR algorithm, we need to focus on the impact of the numerical properties of the iterative method on the convergence process. First of all, a sensible choice of the relaxation parameter ω is crucial for an accelerated convergence. Then we need to carefully specify appropriate criteria to test convergence in each iteration step. It is wise to choose these criteria as a function of the problem domain, i.e., in a model-specific manner. Finally, overflows and underflows have to be dealt with, especially if the investigated models contain states with very small probabilities associated with them. This situation commonly occurs if very large or so-called stiff models are investigated, such that some states have a very small probability mass.

3.5.5.1 The Relaxation Parameter

The relaxation parameter ω was introduced to gain an increased speed of convergence. It was mentioned that the incorporation of the normalization condition $\mathbf{x}\mathbf{1} = 1$ in each iteration step can have a negative impact on the rate of convergence. From this point of view, normalization should, for practical reasons, be postponed until convergence has been assured. Henceforth, we continue discussion of the iterative method on the basis of a homogeneous linear system $\mathbf{0} = \mathbf{x}\mathbf{A}$ according to Eq. (3.101). With this variant in mind, the solution of Eq. (3.109) can be identified as the largest left eigenvector $\mathbf{y}_1$ of the matrix $(\omega\mathbf{L} + (1 - \omega)\,\mathbf{D})(\mathbf{D} - \omega\mathbf{U})^{-1}$. For the corresponding eigenvalue, we have $|\lambda_1| = 1$. (Note that the normalized eigenvector equals the desired solution vector $\mathbf{x}$.) It is known from [StGo85] that the speed of convergence is a function of the second largest eigenvalue λ_2, with $|\lambda_1| \geq |\lambda_2|$, of the iteration matrix, which can be influenced by an accurate choice of ω. To guarantee convergence, all eigenvalues should be real valued, the second largest eigenvalue $|\lambda_2|$ should be *smaller than one* in magnitude because the smaller $|\lambda_2|$ is, the faster the convergence would be. Conditioned on the assumption that $|\lambda_2| < 1$, which is not always true, the following relation holds for the number of iteration steps i, the second largest eigenvalue $|\lambda_2|$, and the required accuracy ϵ:

$$i \geq \frac{\log(\epsilon)}{\log(|\lambda_2|)}. \qquad (3.112)$$

If all the eigenvalues were known, an *optimal* ω_0 that minimizes $|\lambda_2|$ of the iteration matrix could be derived. But this computation is generally more expensive than the gain due to the implied acceleration of the convergence. Therefore, heuristics are often applied and instead of an exact value ω_0, an approximation $\omega_0^{\approx}$ is derived. Sometimes advantage is taken of certain regularities of the parameter matrix for a simplified estimation $\omega_0^{\approx}$ of

ω_0. Equivalently, the parameter matrix can sometimes be rearranged in a problem-dependent way so that a more favorable structure results.

For the SOR iteration method to converge, it has been shown in [HaYo81] that the relaxation parameter must obey the relation $0 < \omega < 2$. It is usually a promising approach to choose ω adaptively by successive re-estimations of $\omega_0^{\approx}$. In particular, we follow [StGo85] for such a technique. With the subsequently introduced approach, ω should be chosen such that $\omega \leq \omega_0^{\approx}$, otherwise convergence is not guaranteed because of possible oscillation effects. Furthermore, we assume $\omega_0^{\approx} \geq 1$, i.e., underrelaxation is not considered.

The computation would be initiated with $\omega^{(1)} = 1$, which is fixed for some, say 10, steps of iterations. Using the intermediate estimate $|\lambda_2^{\approx(k)}|$ of the second largest eigenvalue, which is also called the subdominant eigenvalue, in the iteration process an intermediate estimate $\omega_0^{\approx(k)}$ of the optimal relaxation parameter can be derived by evaluating Eq. (3.113):

$$\omega_0^{\approx(k)} = \frac{2}{1 + \sqrt{1 - \left(|\lambda_2^{\approx(k)}|\right)^2}} . \tag{3.113}$$

The estimated subdominant eigenvalue $|\lambda_2^{\approx(k)}|$ is approximated as a function of the recent relaxation parameter $\omega^{(k-1)}$ and the relative change in values $\delta^{(k-1)}$ of the successive iterates $\mathbf{x}^{(k-3)}, \mathbf{x}^{(k-2)}, \mathbf{x}^{(k-1)}$ as shown in Eq. (3.114):

$$\left|\lambda_2^{\approx(k)}\right| = \frac{\delta^{(k-1)} + \omega^{(k-1)} - 1}{\omega^{(k-1)}\sqrt{\delta^{(k-1)}}} . \tag{3.114}$$

Finally, the relative change in values $\delta^{(k-1)}$ is calculated from the intermediate results as presented in Eq. (3.115):

$$\delta^{(k-1)} = \frac{\|\mathbf{x}^{(k-1)} - \mathbf{x}^{(k-2)}\|_\infty}{\|\mathbf{x}^{(k-2)} - \mathbf{x}^{(k-3)}\|_\infty} . \tag{3.115}$$

Note that the maximum vector norm $\|\mathbf{x}\|_\infty$ is defined as $\|\mathbf{x}\|_\infty = \max_{1 \leq i \leq n} |x_i|$, where $\mathbf{x} \in \mathbb{R}^n$. We hasten to add that the preceding heuristic for the estimation of the relaxation parameter has to be applied carefully. The setting in which it had originally been introduced was restricted to a certain regular model structure. If it is applied otherwise, we must be aware of the heuristic nature of the approach. The choice of the relaxation parameter ω has a decisive impact on the speed of convergence. Due to the lack of a universally applicable formula for an accurate computation of the parameter, its nontrivial estimation is the most critical task for a successful application of the SOR method. In addition, we would like to point out that ω does not have to be re-estimated in each iteration step k but, rather, periodically with some period of, say, 10 iteration steps.

3.5.5.2 The Test of Convergence

To terminate an iterative computation process, appropriate criteria need to be provided by means of which it can be assured that the iteration process converges to the desired solution $\mathbf{x}$. Generally, some vector norm $\|\mathbf{x} - \mathbf{x}^{(k)}\|$ would be used to estimate the error that should be smaller than some threshold ϵ. Since $\mathbf{x}$ is not known, it must either be estimated as shown in [StGo85], or some different heuristics have to be applied. Intuitively, convergence is reached if successive iterates do not differ substantially. But what "substantial difference" means can only be answered in the context of the model under consideration. Convergence could be relatively slow, that is, the vectors $\mathbf{x}^{(k-1)}, \mathbf{x}^{(k)}$ would not be much apart, even though still being far from the final solution $\mathbf{x}$. Furthermore, if the underlying model is large, the elements $x_i^{(k)}$ of $\mathbf{x}^{(k)}$ could easily become smaller than ϵ and convergence could be mistakenly assumed.

To overcome these two problems, it is often a good choice to use the criterion from Eq. (3.116) of the relative change in values of the most recent iterate $\mathbf{x}^{(k)}$ and the vector obtained $m > 1$ steps earlier, $\mathbf{x}^{(k-m)}$, namely

$$\frac{\|\mathbf{x}^{(k)} - \mathbf{x}^{(k-m)}\|_\infty}{\|\mathbf{x}^{(k)}\|_\infty} < \epsilon. \tag{3.116}$$

According to [StGo85], m should be chosen as a function of the number of necessary iterations, $5 < m < 50$. If SOR converges quickly, the much simpler and computationally more efficient criterion as given in Eq. (3.117) might be applicable:

$$\|\mathbf{x}^{(k)} - \mathbf{x}^{(k-1)}\| < \epsilon. \tag{3.117}$$

Commonly applied vector norms are

$$\|\mathbf{x}\|_1 = \sum_{i=1}^n |x_i|, \tag{3.118}$$

$$\|\mathbf{x}\|_2 = \sqrt{\sum_{i=1}^n |x_i|^2}. \tag{3.119}$$

3.5.5.3 Overflow and Underflow

If the normalization condition is incorporated in the transition matrix, as discussed in the context of Eq. (3.101), overflow and underflow are less likely to occur, but convergence is much slower and computational overhead is higher. Therefore, some heuristics are applied to circumvent these numerical problems. To avoid underflow, elements $x_i^{(k)}$ shrinking below a lower threshold ϵ_l are simply set equal to zero. Conversely, if there exist elements $x_i^{(k)}$ that exceed an upper threshold ϵ_u, a normalization is adaptively performed and the values are scaled back to a more convenient range.

3.5.5.4 The Algorithm

The algorithm presented in Fig. 3.10 is in a skeletal form. As we have discussed at length, it needs to be carefully adapted for the peculiarities of specific cases to which it is applied. Since for $\omega = 1$

SOR coincides with Gauss–Seidel, we do not separately present the Gauss–Seidel algorithm. Furthermore, for the sake of simplicity, we do not consider scaling and rescaling of the parameter matrix, which is a technique that is commonly performed in order to avoid divisions by diagonal elements a_{jj} (cf. Eq. (3.106)). It should be noted that a sum indexed from 0 to -1 is interpreted to be zero, $\sum_0^{-1} \cdots = 0$.

3.6 COMPARISON OF NUMERICAL SOLUTION METHODS

The solution algorithms discussed in earlier sections are herein compared by the means of the model introduced in Fig. 2.6 (Section 2.2). Patterns of convergence and numerical accuracy are compared. According to Eq. (2.58), the following system of steady-state equations results:

$$0 = -2\gamma\pi_2 + \delta\pi_1 \,,$$
$$0 = -\beta\pi_{RC} + 2c\gamma\pi_2 \,,$$
$$0 = -\alpha\pi_{RB} + 2(1-c)\gamma\pi_2 \,,$$
$$0 = -(\gamma + \delta)\pi_1 + \alpha\pi_{RB} + \beta\pi_{RC} + \delta\pi_0 \,,$$
$$0 = -\delta\pi_0 + \gamma\pi_1 \,.$$

With the normalization condition

$$1 = \pi_2 + \pi_{RC} + \pi_{RB} + \pi_1 + \pi_0 \,,$$

and the parameters from Table 3.8, three computational experiments were performed using the Gaussian elimination algorithm. The results are summa-

Table 3.8   Experimental parameters for the model with two processor elements

Parameter	Experiment 1	Experiment 2	Experiment 3
$1/\gamma$	24 hr 54 sec	1 yr 12 hr	100 yr 262 hr 48 min
$1/\delta$	4 hr 10 min	4 hr 10 min	4 hr 10 min
$1/\alpha$	10 min	10 min	10 min
$1/\beta$	30 sec	30 sec	30 sec
c	0.99	0.99	0.99

rized in Table 3.9. From the data of the first experiment it is concluded that, for example, with probability $\pi_2 + \pi_1 \approx .71 + .25 = .96$, at least one processor is up in steady state. This quantity could provide an important measure for the availability of a modeled system. Other measures of interest can be similarly derived. By comparing the results of different experiments in Table 3.9 it can be seen how the state probabilities depend on the MTTF, being varied from approximately 1 day to a little less than 100 years and 11 days. Although the probability mass is captured to a larger extent by π_2, the other

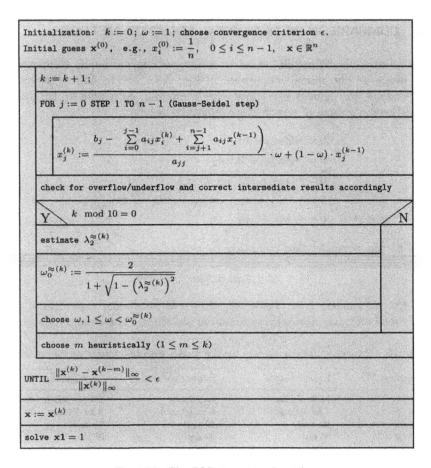

Fig. 3.10 The SOR iteration algorithm.

Table 3.9   Steady-state probabilities of the model with two processor elements

Probability	Experiment 1	Experiment 2	Experiment 3
π_2	0.7100758490525	0.9990481945467	0.9999904674119
π_{RB}	0.0000986153339	0.0000003796383	0.0000000038040
π_{RC}	0.0004881459029	0.0000018792097	0.0000000188296
π_1	0.2465383347910	0.0009490957848	0.0000095099093
π_0	0.0427990549197	0.0000004508205	0.0000000000452

probabilities are of no less interest to designers of dependable systems because the effectiveness of redundancy and recovery techniques is measured by the values of these small probabilities. Note, in the case of the third experiment in particular, that the numbers differ by many orders of magnitude. This property could impose challenges on the numerical algorithms used to analyze the underlying CTMC since an appropriate accuracy has to be assured. For instance, the results in Table 3.9 are exact values up to 13 digits. While such an accuracy is most likely not very significant for the probability π_2, it seems to be a minimum requirement for the computation of π_0, since the 10 leading digits of π_0 are zeroes in the third experiment. The more the values of the model parameters diverge from each other and the larger the investigated model is, the more this property tends to manifest itself. Note that for the analysis of larger models, iterative methods are usually preferred. The convergence criteria have to be chosen very carefully in cases where the resulting values differ by orders of magnitude. On the one hand, the convergence criteria should be stringent if the significant measures result in very small values. One the other hand, computational complexity increases substantially with a more stringent convergence criteria. Even worse, the probability of convergence might significantly decrease. More details are presented in the following sections.

3.6.1   Case Studies

In the following, we refer back to the model in Fig. 2.6 as a starting point. The model can easily be extended to the case of n processors, while the basic structure is preserved. Furthermore, the same parameters as in Table 3.8 are used in three corresponding experiments. The computations are performed with *Gaussian elimination*, the *power* method, the *Gauss–Seidel* algorithm, and different variants of the *SOR* method. The intention is to study the accuracy and pattern of convergence of the iterative methods in a comparative manner. Concerning the SOR method, for example, no sufficiently efficient algorithm is known for a computation of the optimum ω. With the help of these case studies some useful insights can be obtained.

There are many possibilities to specify convergence criteria, as has been discussed earlier. In our example, convergence is assumed if both a *maximum norm* variant

$$\|\mathbf{x}\|_\infty = \max_i |x_i| = \max_i |x_i^{(n)} - x_i^{(n-1)}|, \qquad (3.120)$$

of the difference between the most recent iterates $\mathbf{x}^{(n)}$ and $\mathbf{x}^{(n-1)}$ and the *square root norm*

$$\|\mathbf{r}\|_2 = \sqrt{\sum_{i=0}^{n-1}(r_i)^2} \qquad (3.121)$$

of the *residues*

$$r_i = \sum_{j=0}^{i-1} q_{j,i}x_j^{(n)} + \sum_{j=i}^{n-1} q_{j,i}x_j^{(n-1)} - b_i \qquad (3.122)$$

are less than the specified ϵ:

$$\|\mathbf{x}\|_\infty < \epsilon \wedge \|\mathbf{r}\|_2 < \epsilon.$$

Note that the pattern of convergence of iterative methods can depend on the ordering of the states as well, and thus on the iteration matrix. This dependence is particularly important for the SOR-type methods where in each iteration step the most up-to-date information is used in the computation of the components of the intermediate solution vector. Therefore, we provide in our comparative study the results of two different state ordering strategies. The resulting number of iterations are summarized in Table 3.10 for the cases of $\epsilon = 10^{-10}$, 10^{-15}, 10^{-16}, 10^{-17}, and 10^{-20} as some examples.

Comparing corresponding experiments, it can be seen from Table 3.10 how the number of iteration steps of the power method increases as the convergence criterion is tightened from $\epsilon = 10^{-10}$ to $\epsilon = 10^{-20}$. Furthermore, the power method tends to converge faster, the more the model parameters differ quantitatively from each other. In these examples, convergence is relatively slow in experiment one, medium in experiment two, and fast in the third experiment for fixed ϵ. A similar pattern of convergence using the Gauss–Seidel method and the SOR variants can be observed. The relative best results are achieved by SOR with $\omega = 0.8$. Gauss–Seidel and SOR ($\omega = 1.2$) do not converge if $\epsilon \leq 10^{-17}$.

Another view is provided in Fig. 3.11 where the number of iterations is depicted as a function of ω that is systematically varied, $0 < \omega < 2$, and the criterion of convergence $\epsilon = 10^{-15}$ is chosen. Entries on the graph indicate that convergence was observed within $20,000$ iterations for a particular ω. The results confirm that convergence is relatively slow in experiment one. Furthermore, in experiment one, SOR converges only if $\omega < 1$ while the number of iterations are quite sensitive to variations in ω. Fewer iteration steps are needed in experiments two and three. Relatively fewer iterations, 279, are observed in experiment one with $\omega = 0.95$, in experiment two, 25, with $\omega = 1.0$; and in experiment three, 21, with $\omega = 1.01$. It is interesting to note that in experiments one and three, even if ω is chosen only slightly larger

Table 3.10   Number of iterations required to analyze the model with ten processors

Methods	$\epsilon = 1 \cdot e^{-10}$			$\epsilon = 1 \cdot e^{-15}$		
	Exp. 1	Exp. 2	Exp. 3	Exp. 1	Exp. 2	Exp. 3
Power	32119	16803	16703	61013	24305	24083
Gauss–Seidel	653	23	17	*	25	17
SOR ($\omega = 1.2$)	*	49	–	*	55	–
SOR ($\omega = 0.8$)	239	41	41	367	47	41

Methods	$\epsilon = 1 \cdot e^{-16}$			$\epsilon = 1 \cdot e^{-17}$		
	Exp. 1	Exp. 2	Exp. 3	Exp. 1	Exp. 2	Exp. 3
Power	66547	26487	25309	70221	26883	26625
Gauss–Seidel	*	25	17	*	*	*
SOR ($\omega = 1.2$)	*	57	*	*	*	*
SOR ($\omega = 0.8$)	369	47	43	*	47	43

Methods	$\epsilon = 1 \cdot e^{-20}$		
	Exp. 1	Exp. 2	Exp. 3
Power	71573	30039	30703
Gauss–Seidel	*	*	*
SOR ($\omega = 1.2$)	*	*	*
SOR ($\omega = 0.8$)	*	*	47

Note: An asterisk (*) indicates no convergence in 20,000 iterations and a dash (–) indicates convergence with negative results.

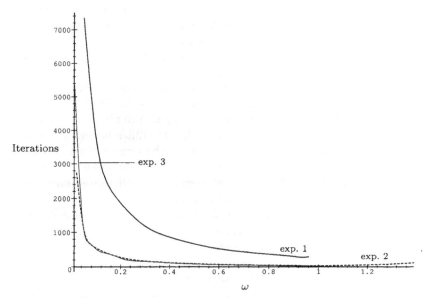

Fig. 3.11   The number of SOR iterations as a function of ω.

than 0.95 and 1.01, respectively, no convergence results. Similar outcomes are observed for different criteria of convergence ϵ. Over-relaxation, that is $\omega > 1$, can only be successfully applied in experiment two.

Table 3.11   Accuracy (A) in the number of digits and number of iterations (I) needed by different SOR variants with $\epsilon = 10^{-15}$

	Exp. 2 (d)		Exp. 2 (r)		Exp. 3 (d)		Exp. 3 (r)	
ω	A	I	A	I	A	I	A	I
0.1	*	*	20	539	*	*	20	507
0.2	21	279	20	273	22	265	20	247
0.5	22	97	20	91	23	91	21	83
0.7	23	59	22	57	23	55	21	51
0.8	23	47	23	47	24	43	23	41
1.0	33	27	23	25	49	21	30	17
1.1	28	41	23	39	45	45	–	–
1.2	23	57	23	55	*	*	–	–
1.3	23	81	23	79	*	*	–	–

Note: An asterisk (*) indicates no convergence in 20,000 iterations and a dash (–) indicates convergence with negative results.

Using different norms affects pattern of convergence and accuracy of the numerical results even for the same criterion of convergence ϵ. In Fig. 3.12 the number of iterations is depicted as a function of ω in experiment three, comparing the residue norm (r) to the difference norm (d) with $\epsilon = 10^{-15}$. It can be seen that applying the difference norm consistently results in slightly more iterations for a given ω. A similar pattern was observed in all the experiments. There are ranges of ω where the application of the difference norm *does not* lead to convergence while the application of the residue norm *does* lead to convergence, and vice versa.

The accuracy in terms of the number of correct digits of the numerical values resulting from the application of different norms in the SOR algorithm is indicated in Table 3.11. The numbers are compared to the results obtained using Gaussian elimination. It can be seen that the difference norm, both in experiments two and three, consistently results in the same or higher accuracy as the residue norm for a given ω. A trade-off is observed between the number of necessary iterations and the achieved accuracy. Both factors have to be taken into account when designing calculations.

In experiment four, the data from Table 3.12 are applied in a series of computations. The number of iterations versus ω is depicted in Fig. 3.13. The sensitivity of the convergence behavior to ω is impressive in this study. While the minimum number of iterations are observed with $\omega \approx 0.8$, the number of iterations grows quickly as ω approaches one.

We do not claim to be complete in the discussion of factors that have an impact on the speed of convergence of numerical techniques. We only presented *some* issues of practical relevance that have to be taken into account

Table 3.12   Experimental parameters for
the model with ten processors

Parameter	Experiment 4
$1/\gamma$	9512 yr 8243 hr 24 min
$1/\delta$	50 min
$1/\alpha$	5 min
$1/\beta$	30 sec
c	0.99
ϵ	10^{-20}
norm	residue

while considering convergence. The results presented are conditioned on the
fact that other convergence dominating issues such as state ordering were not
taken into account. The methods could have been tuned further, of course,
resulting in a different pattern of convergence.

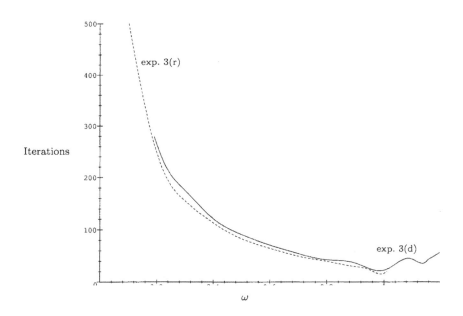

Fig. 3.12   Comparison of difference (d) and residue (r) norm in experiment three.

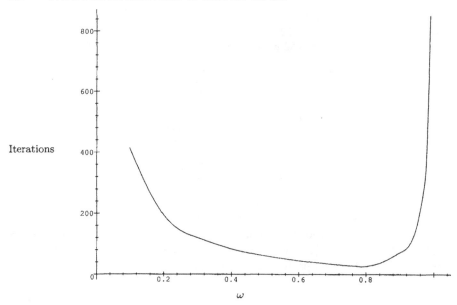

Fig. 3.13 Number of iterations applying the square root norm $\|\mathbf{r}\|_2 < 10^{-20}$ of the residue in experiment four.

4

Steady-State Aggregation/Disaggregation Methods

In this chapter we consider two main methods: Courtois' decomposition method and Takahashi's iterative aggregation/disaggregation method.

4.1 COURTOIS' APPROXIMATE METHOD

In this section we introduce an efficient method for the steady-state analysis of Markov chains. Whereas direct and iterative techniques can be used for the *exact* analysis of Markov chains as previously discussed, the method of Courtois [Cour75, Cour77] is mainly applied to *approximate* computations $\nu^{\approx}$ of the desired state probability vector ν. Courtois' approach is based on decomposability properties of the models under consideration. Initially, substructures are identified that can separately be analyzed. Then, an aggregation procedure is performed that uses independently computed subresults as constituent parts for composing the final results. The applicability of the method needs to be verified in each case. If the Markov chain has tightly coupled subsets of states, where the states within each subset are tightly coupled to each other and weakly coupled to states outside the subset, it provides a strong intuitive indication of the applicability of the approach. Such a subset of states might then be aggregated to form a *macro-state* as a basis for further analysis. The macro-state probabilities, together with the conditional micro-state probabilities from within the subsets, can be composed to yield the micro-state probabilities of the initial model. Details of the approach are clarified through the following example.

4.1.1   Decomposition

Since the method of Courtois is usually expressed in terms of a DTMC, whereas we are emphasizing the use of a CTMC in our discussion of methodologies, we would like to take advantage of this example and bridge the gap by choosing a CTMC as a starting point for our analysis. With the following model in mind, we can explain the CTMC depicted in Fig. 4.1. We assume a system in which two customers are circulating among three stations according to some stochastic regularities. Each arbitrary pattern of distribution of the customers among the stations is represented by a state. In general, if there are N such stations over which K customers are arbitrarily distributed, then from simple combinatorial reasoning we know that $\binom{N+K-1}{N-1} = \binom{N+K-1}{K}$ such combinations exist. Hence, in our example with $N = 3$ and $K = 2$ we have $\binom{4}{2} = 6$ states.

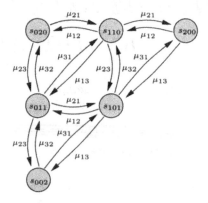

Fig. 4.1   CTMC subject to decomposition.

In state 020, for example, two customers are in station two, while stations one and three are both empty. After a time period of exponentially distributed length, a customer travels from station two to station one or station three. The transition behavior is governed by the transition rates μ_{21} or μ_{23} to states 110 or 011, respectively. The transition behavior between the other states can be explained similarly.

The analysis of such a simple model could be easily carried out by using one of the standard direct or iterative methods. Indeed, we use an exact method to validate the accuracy of the decomposition/aggregation approach for our example. We use the simple example to illustrate Courtois' method. To this end, the model needs to be explored further as to whether it is *nearly completely decomposable*, that is, whether we can find state subsets that represent tightly coupled structures.

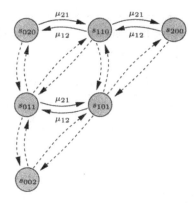

Fig. 4.2   Decomposition of the CTMC with regard to station three.

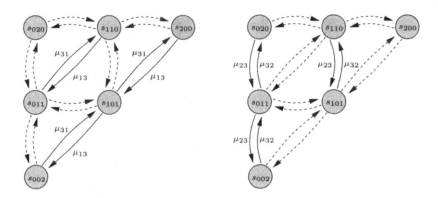

Fig. 4.3   Decompositions of the CTMC with regard to stations two and one.

The application may suggest a state set partitioning along the lines of the customers' circulation pattern among the visited stations. This would be a promising approach if the customers are preferably staying within the bounds of a subset of two stations and only relatively rarely transfer to the third station, i.e., the most isolated one. All such possibilities are depicted in Fig. 4.2 and 4.3. In Fig. 4.2, station three is assumed to be the most isolated one, i.e., the one with the least interactions with the others. Solid arcs emphasize the tightly coupled states, whereas dotted arcs are used to represent the "loose coupling." When customers are in the first station, they are much more likely to transit to the second station, then return to the first station, before visiting the third station. Hence, we would come up with three subsets $\{020, 110, 200\}$, $\{011, 101\}$, and $\{002\}$ in each of which the number of customers in the third station is fixed at 0, 1, and 2, respectively. Alternatively, in Fig. 4.3, we have shown the scenario where stations two and three are isolated.

Table 4.1   Transition rates

$\mu_{12} = 4.50$	$\mu_{21} = 2.40$	$\mu_{31} = 0.20$
$\mu_{13} = 0.30$	$\mu_{23} = 0.15$	$\mu_{32} = 0.20$

Now we proceed to discuss Courtois' method on the basis of the set of parameters in Table 4.1. It suggests a decomposition according to Fig. 4.2. Clearly, the parameter values indicate strong interactions between stations one and two, whereas the third station seems to interact somewhat less with the others. This level of interaction should be reflected at the CTMC level representation. The corresponding infinitesimal generator matrix $\mathbf{Q}$ is given as

$$\mathbf{Q} = \begin{pmatrix} -4.80 & 4.50 & 0 & 0.30 & 0 & 0 \\ 2.40 & -7.35 & 4.50 & 0.15 & 0.30 & 0 \\ 0 & 2.40 & -2.55 & 0 & 0.15 & 0 \\ 0.20 & 0.20 & 0 & -5.20 & 4.50 & 0.30 \\ 0 & 0.20 & 0.20 & 2.40 & -2.95 & 0.15 \\ 0 & 0 & 0 & 0.20 & 0.20 & -0.40 \end{pmatrix},$$

and is depicted symbolically in Table 4.2.

Table 4.2   Generator matrix $\mathbf{Q}$ of the CTMC structured for decomposition

	200	110	020	101	011	002
200	$-\Sigma$	μ_{12}	0	μ_{13}	0	0
110	μ_{21}	$-\Sigma$	μ_{12}	μ_{23}	μ_{13}	0
020	0	μ_{21}	$-\Sigma$	0	μ_{23}	0
101	μ_{31}	μ_{32}	0	$-\Sigma$	μ_{12}	μ_{13}
011	0	μ_{31}	μ_{32}	μ_{21}	$-\Sigma$	μ_{23}
002	0	0	0	μ_{31}	μ_{32}	$-\Sigma$

It is known from Eq. (3.5) that we can transform any CTMC to a DTMC by defining $\mathbf{P} = \mathbf{Q}/q + \mathbf{I}$, with $q > \max_{i,j \in S} |q_{ij}|$. Next we solve $\boldsymbol{\nu} = \boldsymbol{\nu}\mathbf{P}$ instead of $\boldsymbol{\pi}\mathbf{Q} = \mathbf{0}$ and we can assert that $\boldsymbol{\nu} = \boldsymbol{\pi}$. Of course, we have to fulfill the normalization condition $\boldsymbol{\nu}\mathbf{1} = 1$. For our example, the transformation results in the transition probability matrix as shown in Table 4.3, where q is appropriately chosen.

Table 4.3 Transition-probability matrix $\mathbf{P}$ of the DTMC structured for decomposition

	200	110	020	101	011	002
200	$1-\frac{\Sigma}{q}$	$\frac{\mu_{12}}{q}$	0	$\frac{\mu_{13}}{q}$	0	0
110	$\frac{\mu_{21}}{q}$	$1-\frac{\Sigma}{q}$	$\frac{\mu_{12}}{q}$	$\frac{\mu_{23}}{q}$	$\frac{\mu_{13}}{q}$	0
020	0	$\frac{\mu_{21}}{q}$	$1-\frac{\Sigma}{q}$	0	$\frac{\mu_{23}}{q}$	0
101	$\frac{\mu_{31}}{q}$	$\frac{\mu_{32}}{q}$	0	$1-\frac{\Sigma}{q}$	$\frac{\mu_{12}}{q}$	$\frac{\mu_{13}}{q}$
s011	0	$\frac{\mu_{31}}{q}$	$\frac{\mu_{32}}{q}$	$\frac{\mu_{21}}{q}$	$1-\frac{\Sigma}{q}$	$\frac{\mu_{23}}{q}$
002	0	0	0	$\frac{\mu_{31}}{q}$	$\frac{\mu_{32}}{q}$	$1-\frac{\Sigma}{q}$

Given the condition $q > \max_{i,j} |q_{ij}| = 7.35$, we conveniently fix $q = 10$. Substituting the parameter values from Table 4.1 in Table 4.3 yields the transition probability matrix $\mathbf{P}$:

$$\mathbf{P} = \begin{pmatrix} 0.52 & 0.45 & 0 & 0.03 & 0 & 0 \\ 0.24 & 0.265 & 0.45 & 0.015 & 0.03 & 0 \\ 0 & 0.24 & 0.745 & 0 & 0.015 & 0 \\ 0.02 & 0.02 & 0 & 0.48 & 0.45 & 0.03 \\ 0 & 0.02 & 0.02 & 0.24 & 0.705 & 0.015 \\ 0 & 0 & 0 & 0.02 & 0.02 & 0.96 \end{pmatrix}. \tag{4.1}$$

As indicated, $\mathbf{P}$ is partitioned into $|M \times M|$ number of submatrices $\mathbf{P}_{IJ}$, with $M = 3$ and $0 \le I, J \le 2$ in our example. The submatrix with macro row index I and column index J is denoted by $\mathbf{P}_{IJ}$, while the elements of each submatrix can be addressed via double subscription p_{IJij}. The M diagonal submatrices $\mathbf{P}_{II}$ represent for $I = 0, 1, 2$ the interesting cases of zero, one, and two customers, respectively, staying in the third station. For example, the second diagonal submatrix $\mathbf{P}_{11}$ is

$$\mathbf{P}_{11} = \begin{pmatrix} 0.48 & 0.45 \\ 0.24 & 0.705 \end{pmatrix}. \tag{4.2}$$

The elements of the submatrix can be addressed by the schema just introduced: $p_{1110} = 0.24$. Since the indices $I, 0 \le I \le M - 1$ are used to uniquely refer to subsets as elements of a partition of the state space S, we conveniently denote the corresponding subset of states by S_I; for example, $S_1 = \{101, 011\}$.

Of course, each diagonal submatrix could possibly be further partitioned into substructures according to this schema. The depth to which such multilayer decomposition/aggregation technique is carried out depends on the number of stations N and on the degree of coupling between the stations. We further elaborate on criteria for the decomposability in the following.

A convenient way to structure the transition probability matrix in our example is to number the states using radix K notation:

$$\sum_{i=1}^{N} k_i K^i, \quad \text{where} \quad \sum_{i=1}^{N} k_i = K.$$

where k_i denotes the number of customers in station i and N is the index of the least coupled station, i.e., the one to be isolated. For instance, in our example a value of 4 would be assigned to state 200 and a value of 12 to state 011. Note that the states in Tables 4.2 and 4.3 have been numbered according to this rule.

The transition probability matrix in block form is given as

$$\mathbf{P} = \begin{pmatrix} \mathbf{P}_{00} & \mathbf{P}_{01} & \cdots & & \mathbf{P}_{0(M-1)} \\ \mathbf{P}_{10} & \mathbf{P}_{11} & \cdots & & \mathbf{P}_{1(M-1)} \\ \vdots & & & \vdots & \\ \mathbf{P}_{I0} & & \cdots & \mathbf{P}_{IJ} & \cdots & \mathbf{P}_{I(M-1)} \\ \vdots & & & \vdots & \\ \mathbf{P}_{(M-2)0} & & \cdots & & \mathbf{P}_{(M-2)(M-1)} \\ \mathbf{P}_{(M-1)0} & & \cdots & & \mathbf{P}_{(M-1)(M-1)} \end{pmatrix}. \quad (4.3)$$

Generally, the partitioning of matrix $\mathbf{P}$ in submatrices $\mathbf{P}_{IJ}$ is done as a function of the number of customers in the least coupled station N. Given that K customers are in the system, there will be $M = K + 1$ diagonal submatrices. Each diagonal submatrix $\mathbf{P}_{II}, 0 \leq I \leq K$, is then considered to be a basic building block for defining transition probability matrices $\mathbf{P}_{II}^*$ of a reduced system of $N-1$ stations and $K-I$ customers. For example, $\mathbf{P}_{11}^*$ would be used to describe a system with $K - 1$ customers circulating among $N - 1$ stations, while one customer stays in station N all the time. The transitions from station N to all other stations would be temporarily ignored.[1] In terms of our example, we would only consider transitions between states 101 and 011.

To derive stochastic submatrices, the transition probability matrix $\mathbf{P}$ is decomposed into two matrices $\mathbf{A}$ and $\mathbf{B}$ that add up to the original one. Matrix $\mathbf{A}$ comprises the diagonal submatrices $\mathbf{P}_{II}$, and matrix $\mathbf{B}$ the complementary off-diagonal submatrices $\mathbf{P}_{IJ}, I \neq J$:

$$\mathbf{P} = \mathbf{A} + \mathbf{B}. \quad (4.4)$$

---

[1]Note that $\mathbf{P}_{11}$ is not a stochastic matrix; it has to be modified to obtain a stochastic matrix $\mathbf{P}_{11}^*$.

In our example, Eq. (4.4) has the following form:

$$\mathbf{P} = \begin{pmatrix} 0.52 & 0.45 & 0 & & & \\ 0.24 & 0.265 & 0.45 & & \mathbf{0} & \\ 0 & 0.24 & 0.745 & & & \\ & & & 0.48 & 0.45 & \\ & \mathbf{0} & & 0.24 & 0.705 & \\ & & & & & 0.96 \end{pmatrix} + \begin{pmatrix} & & & 0.03 & 0 & 0 \\ & \mathbf{0} & & 0.015 & 0.03 & 0 \\ & & & 0 & 0.015 & 0 \\ 0.02 & 0.02 & 0 & & & 0.03 \\ 0 & 0.02 & 0.02 & \mathbf{0} & & 0.015 \\ 0 & 0 & 0 & 0.02 & 0.02 & 0 \end{pmatrix}.$$

$$\mathbf{A} \qquad\qquad\qquad\qquad\qquad \mathbf{B}$$

Since we wish to transform each diagonal submatrix $\mathbf{P}_{II}$ into a stochastic matrix $\mathbf{P}^*_{II}$, we need to define a matrix $\mathbf{X}$, such that Eq. (4.5) holds and matrices $\mathbf{P}^*, \mathbf{P}^*_{00}, \ldots \mathbf{P}^*_{(M-1)(M-1)}$ are all stochastic:

$$\mathbf{P}^* = (\mathbf{A} + \mathbf{X}) = \begin{pmatrix} \mathbf{P}^*_{00} & & & & \\ & \ddots & & \mathbf{0} & \\ & & \mathbf{P}^*_{II} & & \\ & \mathbf{0} & & \ddots & \\ & & & & \mathbf{P}^*_{(M-1)(M-1)} \end{pmatrix}. \qquad (4.5)$$

There are multiple ways to define matrix $\mathbf{X}$. Usually, accuracy and computational complexity of the method depend on the way $\mathbf{X}$ is defined. It is most important, though, to ensure that the matrices $\mathbf{P}^*_{II}, 0 \le I \le M-1$, are all ergodic; i.e., they are aperiodic and irreducible. Incorporating matrix $\mathbf{X}$ in matrix $\mathbf{P}$, we get

$$\begin{aligned} \mathbf{P} &= (\mathbf{A} + \mathbf{X}) + (\mathbf{B} - \mathbf{X}) \\ &= \mathbf{P}^* + \mathbf{C}. \end{aligned} \qquad (4.6)$$

In our example, we suggest the use of following matrix $\mathbf{X}$ in order to create a stochastic matrix $\mathbf{P}^*$:

$$\mathbf{X} = \begin{pmatrix} 0 & 0 & 0.03 & 0 & 0 & 0 \\ 0 & 0.045 & 0 & 0 & 0 & 0 \\ 0.015 & 0 & 0 & 0 & 0 & 0 \\ 0 & 0 & 0 & 0 & 0.07 & 0 \\ 0 & 0 & 0 & 0.055 & 0 & 0 \\ 0 & 0 & 0 & 0 & 0 & 0.04 \end{pmatrix}.$$

This gives us

$$\mathbf{P} = \begin{pmatrix} 0.52 & 0.45 & 0.03 & & & \\ 0.24 & 0.31 & 0.45 & & \mathbf{0} & \\ 0.015 & 0.24 & 0.745 & & & \\ & & & 0.48 & 0.52 & \\ & \mathbf{0} & & 0.295 & 0.705 & \\ & & & & & 1 \end{pmatrix} + \begin{pmatrix} 0 & 0 & -0.03 & 0.03 & 0 & 0 \\ 0 & -0.045 & 0 & 0.015 & 0.03 & 0 \\ -0.015 & 0 & 0 & 0 & 0.015 & 0 \\ 0.02 & 0.02 & 0 & 0 & -0.07 & 0.03 \\ 0 & 0.02 & 0.02 & -0.055 & 0 & 0.015 \\ 0 & 0 & 0 & 0.02 & 0.02 & -0.04 \end{pmatrix}.$$

$$\mathbf{P}^* \qquad\qquad\qquad\qquad\qquad \mathbf{C}$$

4.1.2 Applicability

The applicability of Courtois' method needs to be checked in each case. In general, partitioning of the state space and subsequent aggregation of the resulting subsets into macro-states can be exactly performed if the DTMC under consideration has a lumpable transition probability matrix. In such a case, the application of Courtois' method will not be an approximation. A transition probability matrix $\mathbf{P} = [p_{ij}]$ is *lumpable* with respect to a partition of S in subsets S_I, $0 \leq I \leq M - 1$, if for each submatrix $\mathbf{P}_{IJ}, \forall I, J, 0 \leq I \neq J \leq M - 1$ real-valued numbers $0 \leq r_{IJ} \leq 1$ exist such that Eq. (4.7) holds [KeSn78]:

$$\sum_{j \in S_J} p_{IJ_{ij}} = r_{IJ}, \qquad \forall i \in S_I. \tag{4.7}$$

Note that the diagonal submatrices $\mathbf{P}_{II}$ need not be completely decoupled from the rest of the system, but rather the matrix has to exhibit regularities imposed by the lumpability condition in order to allow an exact aggregation. In fact, if the $\mathbf{P}_{II}$ are completely decoupled from the rest of the system, i.e., the $r_{IJ} = 0$, $I \neq J$, then this can be regarded as a special case of lumpability of $\mathbf{P}$. More details on and an application example of state lumping techniques are given in Section 4.2, particularly in Section 4.2.2.

From our example in Eq. (4.1), $\mathbf{P}$ is not lumpable with respect to the chosen partition. Hence Courtois' method will be an approximation. A measure of accuracy can be derived according to Courtois from Eqs. (4.4) and (4.6). The degree ϵ of coupling between macro-states can be computed from matrix $\mathbf{B} = [b_{ij}]$ in Eq. (4.4). If ϵ is sufficiently small it can be shown that the error induced by Courtois' method is bounded by $O(\epsilon)$.

Let ϵ be defined as follows:

$$\epsilon = \max_{i \in S} \left\{ \sum_{j \in S} b_{ij} \right\}. \tag{4.8}$$

Equation (4.6) can be rewritten as

$$\mathbf{P} = \mathbf{P}^* + \epsilon \tilde{\mathbf{C}}. \tag{4.9}$$

In our example, ϵ can be computed to be

$$\epsilon = \max \left\{ \begin{array}{c} 0.03 \\ 0.015 + 0.03 \\ 0.015 \\ 0.02 + 0.02 + 0.03 \\ 0.02 + 0.02 + 0.015 \\ 0.02 + 0.02 \end{array} \right\} = 0.07.$$

Equation (4.9) thus results in

$$
\mathbf{P} = \underbrace{\begin{pmatrix} 0.52 & 0.45 & 0.03 & & & \\ 0.24 & 0.31 & 0.45 & & 0 & \\ 0.015 & 0.24 & 0.745 & & & \\ & & & 0.48 & 0.52 & \\ & 0 & & 0.295 & 0.705 & \\ & & & & & 1 \end{pmatrix}}_{\mathbf{P^*}} + \underbrace{0.07}_{\epsilon} \underbrace{\begin{pmatrix} 0 & 0 & -0.429 & 0.429 & 0 & 0 \\ 0 & -0.643 & 0 & 0.214 & 0.429 & 0 \\ -0.214 & 0 & 0 & 0 & 0.214 & 0 \\ 0.286 & 0.286 & 0 & 0 & -1 & 0.429 \\ 0 & 0.286 & 0.286 & -0.786 & 0 & 0.214 \\ 0 & 0 & 0 & 0.286 & 0.286 & -0.572 \end{pmatrix}}_{\tilde{\mathbf{C}}} .
$$

To prove that $\mathbf{P}$ is nearly completely decomposable, it is sufficient to show that relation (4.10) holds between ϵ and the maximum of the second largest eigenvalues $\lambda_I^*(2)$ of $\mathbf{P}_{II}^*$ for all $I, 0 \le I \le M - 1$ [Cour77]:

$$
\epsilon < \frac{1 - \max_I |\lambda_I^*(2)|}{2}. \tag{4.10}
$$

For each $\mathbf{P}_{II}^*$, the eigenvalues $\lambda_I^*(k)$ can be arranged according to their decreasing absolute values:

$$
|\lambda_I^*(1)| > |\lambda_I^*(2)| > \cdots > |\lambda_I^*(n(I))|. \tag{4.11}
$$

Since $\mathbf{P}_{II}^*$, $0 \le I \le 2$, are all stochastic, we can immediately conclude $|\lambda_I^*(1)| = 1$ for all I. In our example, the eigenvalues of the three diagonal submatrices

$$
\mathbf{P}_{00}^* = \begin{pmatrix} 0.52 & 0.45 & 0.03 \\ 0.24 & 0.31 & 0.45 \\ 0.015 & 0.24 & 0.745 \end{pmatrix}, \quad \mathbf{P}_{11}^* = \begin{pmatrix} 0.48 & 0.52 \\ 0.295 & 0.705 \end{pmatrix}, \quad \mathbf{P}_{22}^* = (1)
$$

need to be computed. The eigenvalues of $\mathbf{P}_{00}^*$ are the roots of the following equation:

$$
\begin{aligned}
\det(\mathbf{P}_{00}^* - \lambda_0^*\mathbf{I}) &= \begin{vmatrix} (0.52 - \lambda_1^*) & 0.45 & 0.03 \\ 0.24 & (0.31 - \lambda_1^*) & 0.45 \\ 0.015 & 0.24 & (0.745 - \lambda_1^*) \end{vmatrix} \\
&= (\lambda_1^* - 1)(\lambda_1^* - 0.595)(\lambda_1^* + 0.02) \\
&= 0,
\end{aligned}
$$

where $\mathbf{I}$ denotes the identity matrix. The resulting eigenvalues of $\mathbf{P}_{00}^*$ are arranged according to their absolute value as

$$
|\lambda_0^*(1)| = 1, \quad |\lambda_0^*(2)| = 0.595, \quad |\lambda_0^*(3)| = 0.02.
$$

The eigenvalues of $\mathbf{P}_{11}^*$ are determined by the solution of

$$
\begin{aligned}
\det(\mathbf{P}_{11}^* - \lambda_1^*\mathbf{I}) &= \begin{vmatrix} (0.48 - \lambda_1^*) & 0.52 \\ 0.295 & (0.705 - \lambda_1^*) \end{vmatrix} \\
&= (\lambda_1^* - 1)(\lambda_1^* - 0.185) \\
&= 0,
\end{aligned}
$$

which, in turn, gives

$$|\lambda_1^*(1)| = 1 \quad \text{and} \quad |\lambda_1^*(2)| = 0.185\,.$$

The eigenvalue of $\mathbf{P}_{22}^*$ immediately evaluates to

$$|\lambda_2^*(1)| = 1\,.$$

With the second largest eigenvalues computed, we can examine whether decomposability condition (4.10) holds:

$$\max_I |\lambda_I^*(2)| = \max \left\{ \begin{array}{c} 0.595 \\ 0.185 \end{array} \right\} = 0.595$$

$$0.07 = \epsilon < \frac{1 - \max_I |\lambda_I^*(2)|}{2}$$

$$0.07 = \epsilon < \frac{1 - 0.595}{2} = 0.2025\,.$$

Because $\epsilon < 0.2025$, condition (4.10) holds and the transition probability matrix $\mathbf{P}$ is nearly completely decomposable with respect to the chosen decomposition strategy depicted in Fig. 4.2.

4.1.3 Analysis of the Substructures

As a first step toward the computation of the approximate state probability vector $\boldsymbol{\nu}^{\approx} \approx \boldsymbol{\nu}$, we analyze each submatrix $\mathbf{P}_{II}^*$ separately and compute the conditional state probability vector $\boldsymbol{\nu}_I^*$, $0 \leq I \leq M - 1$:

$$\boldsymbol{\nu}_I^*(\mathbf{P}_{II}^* - \mathbf{I}) = \mathbf{0}, \qquad \boldsymbol{\nu}_I^*\mathbf{1} = 1. \tag{4.12}$$

Thus, $\boldsymbol{\nu}_I^*$ is the left eigenvector of $\mathbf{P}_{II}^*$ corresponding to the eigenvalue $\lambda_I^*(1) = 1$. Substituting the parameters from $\mathbf{P}_{00}^*$ of our example, the solution of

$$(\nu_{00}^*, \nu_{01}^*, \nu_{02}^*) \cdot \begin{pmatrix} (0.52 - 1) & 0.45 & 0.03 \\ 0.24 & (0.31 - 1) & 0.45 \\ 0.015 & 0.24 & (0.745 - 1) \end{pmatrix} = \mathbf{0}\,, \quad \nu_0^*\mathbf{1} = 1$$

yields the conditional steady-state probabilities of the micro-states aggregated to the corresponding macro-state 0:

$$\nu_{00}^* = 0.164\,, \quad \nu_{01}^* = 0.295\,, \quad \nu_{02}^* = 0.54\,.$$

Similarly, $\mathbf{P}_{11}^*$ is used in

$$(\nu_{10}^*, \nu_{11}^*) \cdot \begin{pmatrix} (0.48 - 1) & 0.52 \\ 0.295 & (0.705 - 1) \end{pmatrix} = \mathbf{0}\,, \quad \nu_1^*\mathbf{1} = 1$$

to obtain the conditional state probabilities of the micro-states aggregated to the corresponding macro-state 1:

$$\nu_{10}^* = 0.362, \quad \nu_{11}^* = 0.638.$$

Finally

$$\nu_2^* = \nu_{20}^* = 1.$$

Nearly completely decomposable systems can be characterized with respect to "long-term" and "short-term" behavior:

- From a "short-term" perspective, systems described by their probability transition matrix $\mathbf{P}$ can be decomposed into M independent subsystems, each of whose dynamics is governed by a stochastic process approximately described by the matrices $\mathbf{P}_{II}^*, 0 \leq I \leq M - 1$. As an outcome of the analyses of M independent subsystems the conditional micro-state probability vector $\boldsymbol{\nu}_I^*$ results.

- In the "long run," the impact of the interdependencies between the subsystems cannot be neglected. The interdependencies between subsystems, or macro-states, I and J, for all I, J, are described by the transition probabilities Γ_{IJ} that can be – approximately – derived from the transition probability matrix $\mathbf{P}$ and the conditional state probability vector $\boldsymbol{\nu}_I^*$. Solving the transition matrix $\boldsymbol{\Gamma} = [\Gamma_{IJ}]$ for the macro-states yields the macro-state probability vector $\boldsymbol{\gamma}$. The macro-state probability vector $\boldsymbol{\gamma}$ can, in turn, be used for unconditioning of $\boldsymbol{\nu}_I^*, 0 \leq I \leq M-1$, to yield the final result, the approximate state probability vector $\boldsymbol{\nu}^{\approx} \approx \boldsymbol{\nu}$.

4.1.4 Aggregation and Unconditioning

Having obtained the steady-state probability vector for each subset, we are now ready for the next step in Courtois' method. The transition probability matrix over the macro-states, $\boldsymbol{\Gamma} = [\Gamma_{IJ}]$ is approximately computed as

$$\Gamma_{IJ} = \sum_{i \in S_I} \left(\nu_{Ii}^* \sum_{j \in S_J} p_{IJij} \right). \tag{4.13}$$

Comparing Eq. (4.13) with the lumpability condition (4.7), it is clear that the Γ_{IJ} can be exactly determined if the model under consideration is lumpable, that is

$$\Gamma_{IJ} = r_{IJ} \tag{4.14}$$

holds, independent of $\boldsymbol{\nu}_I^*$.

In our example, the macro-state transition probabilities Γ_{IJ} are derived according to Eq. (4.13):

$$
\begin{aligned}
\Gamma_{00} &= \nu_{00}^* \cdot (p_{0000} + p_{0001} + p_{0002}) + \nu_{01}^* \cdot (p_{0010} + p_{0011} + p_{0012}) \\
&\quad + \nu_{02}^* \cdot (p_{0020} + p_{0021} + p_{0022}) = 0.9737\,, \\
\Gamma_{01} &= \nu_{00}^* \cdot (p_{0100} + p_{0101}) + \nu_{01}^* \cdot (p_{0110} + p_{0111}) \\
&\quad + \nu_{02}^* \cdot (p_{0120} + p_{0121}) = 0.0263\,, \\
\Gamma_{02} &= \nu_{00}^* \cdot (p_{0200}) + \nu_{01}^* \cdot (p_{0210}) + \nu_{13}^* \cdot (p_{0220}) = 1 - \Gamma_{00} - \Gamma_{01} = 0\,, \\
\Gamma_{10} &= \nu_{10}^* \cdot (p_{1000} + p_{1001} + p_{1002}) + \nu_{11}^* \cdot (p_{1010} + p_{1011} + p_{1012}) \approx 0.04\,, \\
\Gamma_{11} &= \nu_{10}^* \cdot (p_{1100} + p_{1101}) + \nu_{11}^* \cdot (p_{1110} + p_{1111}) = 0.9396\,, \\
\Gamma_{12} &= 1 - \Gamma_{10} - \Gamma_{11} = 0.0204\,, \\
\Gamma_{20} &= \nu_{20}^* \cdot (p_{2000} + p_{2001} + p_{2002}) = 0\,, \\
\Gamma_{21} &= \nu_{20}^* \cdot (p_{2100} + p_{2101}) = 0.04\,, \\
\Gamma_{33} &= 1 - \Gamma_{20} - \Gamma_{21} = 0.96\,.
\end{aligned}
$$

Accordingly, the macro-state transition probability matrix $\boldsymbol{\Gamma}$ is

$$
\boldsymbol{\Gamma} = \begin{pmatrix} 0.9737 & 0.0263 & 0 \\ 0.04 & 0.9396 & 0.0204 \\ 0 & 0.04 & 0.96 \end{pmatrix}.
$$

The next step is to compute the macro steady-state probability vector $\boldsymbol{\gamma}$ using:

$$
\boldsymbol{\gamma}\boldsymbol{\Gamma} = \boldsymbol{\gamma}, \qquad \boldsymbol{\gamma}\mathbf{1} = 1. \tag{4.15}
$$

In our example, Eq. (4.15) results in

$$
(\gamma_0, \gamma_1, \gamma_2) \begin{pmatrix} (0.9737 - 1) & 0.0263 & 0 \\ 0.04 & (0.9396 - 1) & 0.0204 \\ 0 & 0.04 & (0.96 - 1) \end{pmatrix} = \mathbf{0}, \quad \boldsymbol{\gamma}\mathbf{1} = 1,
$$

from which the macro steady-state probabilities are calculated as

$$
\gamma_0 = 0.502\,, \quad \gamma_1 = 0.330\,, \quad \gamma_2 = 0.168\,.
$$

The final step is to obtain the approximate steady-state probabilities for the original model. For convenience, the elements of the state probability vector $\boldsymbol{\nu}^{\approx} \approx \boldsymbol{\nu}$ are partitioned along the lines of the decomposition strategy. Hence, we refer to the elements of $\boldsymbol{\nu}^{\approx} = [\nu_{I\,i}^{\approx}]$ in compliance with the usual notation in this context. The unconditioning in Eq. (4.16) of the $\boldsymbol{\nu}_I^*$ is to be performed for all macro-states I:

$$
\nu_{I\,i}^{\approx} = \gamma_I \nu_{I\,i}^*, \qquad 0 \leq I \leq M - 1 \text{ and } \forall i \in S_I \tag{4.16}
$$

The use of Eq. (4.16) in our example gives the results summarized in Table 4.4. For the sake of completeness, the exact state probabilities are also shown for comparison.

Table 4.4   State probabilities computed using Courtois' method and exact ones

State	Probability	Formula	Approx.	Exact	% Error
200	$\nu^{\approx}_{0\,0}$	$\gamma_0\nu^*_{0\,0}$	0.0828	0.0787	+ 5.2
110	$\nu^{\approx}_{0\,1}$	$\gamma_0\nu^*_{0\,1}$	0.1480	0.1478	+ 0.1
020	$\nu^{\approx}_{0\,2}$	$\gamma_0\nu^*_{0\,2}$	0.2710	0.2777	- 2.4
101	$\nu^{\approx}_{1\,0}$	$\gamma_1\nu^*_{1\,0}$	0.1194	0.1144	+ 4.4
011	$\nu^{\approx}_{1\,1}$	$\gamma_1\nu^*_{1\,1}$	0.2108	0.2150	- 2.0
002	$\nu^{\approx}_{2}$	$\gamma_2\nu^*_{2}$	0.1680	0.1664	+ 0.1

For models with large state spaces, Courtois' method can be very efficient if the underlying model is nearly completely decomposable. If the transition probability matrix obeys certain regularity conditions the results can be exact. The error induced by the method can, in principle, be bounded [CoSe84]. But since the computational complexity of this operation is considerable, a formal error bounding is often omitted. The efficiency of Courtois' method is due to the fact that instead of solving one linear system of equations of the size of the state space S, several much smaller linear systems are solved independently, one system for each subset S_I of the partitioned state space S, and one for the aggregated chain.

We conclude this section by summarizing the entire algorithm. For the sake of simplicity, only one level of decomposition is considered. Of course, the method can be iteratively applied on each diagonal submatrix $\mathbf{P}^*_{II}$.

4.1.5   The Algorithm

STEP 1   Create the state space and organize it appropriately according to a pattern of decomposition.

STEP 2   Build the transition probability matrix $\mathbf{P}$ (by use of randomization technique $\mathbf{P} = \mathbf{Q}/q + \mathbf{I}$ if the starting point is a CTMC), and partition $\mathbf{P}$ into $M \times M$ number of submatrices $\mathbf{P}_{IJ}, 0 \leq I, J, \leq M - 1$, appropriately.

STEP 3   Verify the nearly complete decomposability of $\mathbf{P}$ according to relation (4.10) with the chosen value of ϵ.

STEP 4   Decompose $\mathbf{P}$ such that $\mathbf{P} = \mathbf{P}^* + \epsilon\tilde{\mathbf{C}}$ according to Eq. (4.9). Matrix $\mathbf{P}^*$ contains only stochastic diagonal submatrices $\mathbf{P}^*_{II}$, and ϵ is a measure of the accuracy of Courtois' method. It is defined as the maximum sum of the entries of the nondiagonal submatrices $\mathbf{P}_{IJ}$, $I \neq J$, of $\mathbf{P}$.

STEP 5   For each I, $0 \leq I \leq M - 1$, solve equation $\nu^*_I\mathbf{P}^*_{II} = \nu^*_I$ with $\nu^*_I\mathbf{1} = 1$ to obtain the conditional state probability vectors ν^*_I.

STEP 6   Compute the coupling between the decomposed macro-states:

 STEP 6.1   Generate the transition probability matrix $\boldsymbol{\Gamma} = [\Gamma_{IJ}]$ according to Eq. (4.13).

 STEP 6.2   Solve Eq. (4.15) to obtain the macro steady-state probability vector $\boldsymbol{\gamma}$.

 STEP 7   Compute the approximate steady-state probability vector $\boldsymbol{\nu}^{\approx}$ of the micro-states by unconditioning of the conditional state probability vectors $\boldsymbol{\nu}_I^*, 0 \leq I \leq M - 1$, according to Eq. (4.16).

 STEP 8   From $\boldsymbol{\nu}^{\approx}$ compute steady-state performance and dependability measures along the lines of a specified reward structure.

Problem 4.1   Verify that the transition probability matrix from Eq. (4.1) is not lumpable according to the chosen partition.

Problem 4.2   Recall Eq. (4.7) and modify the parameters in Table 4.1 of the example in Fig. 4.1 such that the resulting model is lumpable and nearly completely decomposable. Derive the generator matrix according to Table 4.2 and the transition probability matrix according to Table 4.3 for the resulting model.

Problem 4.3   Discuss the relationship between lumpability and nearly complete decomposability. Under what condition does Courtois' method yield the exact state probabilities if the model is both lumpable and nearly completely decomposable? As a hint, compare Eq. (4.13) with lumpability condition (4.7) and relate it to Eq. (4.16).

Problem 4.4   Modify Courtois' method such that it becomes directly applicable for an analysis of CTMCs. Apply the new algorithm directly, that is, without using uniformization, to the original model in Fig. 4.1 and solve for the steady-state probability vector.

4.2   TAKAHASHI'S ITERATIVE METHOD

Although closely related to and under some circumstances even coinciding with Courtois' approach, Takahashi's method [Taka75] differs substantially both with respect to the methodology used and the applicability conditions. While Courtois' method is non-iterative and is applied for the approximate computation of the steady-state probability vector $\boldsymbol{\nu}^{\approx}$ for a given ergodic DTMC (or $\boldsymbol{\pi}^{\approx}$ in the case of a CTMC), Takahashi's iterative method allows a computation of the *exact* state probability vector. (Note that numerical errors are still encountered even in such "exact" methods. The method is exact in that there are no modeling approximations.) To allow a straightforward comparison of the two methods, we prefer a continued discussion in terms of DTMCs. Recall that any given ergodic CTMC can easily be transformed into an ergodic DTMC.

Much like Courtois' method, Takahashi's approach is to partition the state space S into M disjoint subsets of states $S_I \subset S, 0 \leq I \leq M - 1$ such that each subset S_I is *aggregated* into a *macro* state I. The criteria, however, used to cluster states differ in the two approaches. The calculation of transition probabilities Γ_{IJ} among the macro-states I and J is performed on the basis of conditional micro-state probability vector ν_I^* and the originally given transition probabilities p_{ij}, or $p_{IJ_{ij}}$ according to Eq. (4.13). With Courtois' method, the conditional probability vectors ν_I^*, given partition element I, can be separately computed for each subset of micro-states $S_I, 0 \leq I \leq M - 1$, if the original model is nearly completely decomposable. By contrast, in Takahashi's method, the partitioning of the state space is performed on the basis of *approximate lumpability*.[2]

Note that if the state space is approximately lumpable with respect to a partition, it does not necessarily imply that the subsets are nearly decoupled as needed in Courtois' approach. As an immediate consequence, the conditional micro-state probabilities cannot be calculated independently for each subset of states. Instead, the complementary subset of states and the interactions with this complement is taken into account when approximating conditional micro-state probabilities of a partition element. The whole complementary set is aggregated into a single state representing all external interactions from the states within the particular subset. Since the micro-state probabilities as well as the macro-state probabilities are not known in the beginning, initial estimates are needed. The macro- and micro-state probabilities are *iteratively* calculated with Takahashi's method. Aggregation and disaggregation phases are repeated alternately until some convergence criterion is satisfied. By adding an extra computational step, Schweitzer [Schw84] shows that geometric convergence can be guaranteed. Usually, the extra computation takes the form of a power or a Gauss–Seidel iteration step.

4.2.1 The Fundamental Equations

Our discussion of Takahashi's method is based on that given by Schweitzer [Schw84]. Further details can also be found in [Stew94]. Let a given state space S be partitioned into M subsets:

$$S = \bigcup_{I=0}^{M-1} S_I \quad \text{and} \quad S_I \cap S_J = \emptyset, \quad \forall I \neq J, \quad 0 \leq I, J \leq M - 1. \quad (4.17)$$

---

[2]While lumpability has been introduced in Eq. (4.7), approximate lumpability is defined later in Section 4.2.2.

Formally, the macro-state probability vector γ could be calculated from the micro-state probability vector ν for each component I of the vector:

$$\gamma_I = \sum_{i \in S_I} \nu_i, \quad 0 \le I \le M - 1. \tag{4.18}$$

Since ν is not known in advance, Eq. (4.18) cannot be directly used. The macro-state probabilities can be derived, however, if the macro-state transition probability matrix $\mathbf{\Gamma} = [\Gamma_{IJ}]$ were known via the solution of Eq. (4.19):

$$\gamma = \gamma\mathbf{\Gamma}, \quad \gamma\mathbf{1} = 1. \tag{4.19}$$

The implied aggregation step is based on iterative approximations of the probabilities $\nu_{I\,i}^{*}$ of states $i \in S_I$ given an estimate $\nu^{\approx}$ of the micro-state probability vector:

$$\nu_{I\,i}^{*} \approx \frac{\nu_i^{\approx}}{\sum_{k \in S_I} \nu_k^{\approx}}, \forall i \in S_I. \tag{4.20}$$

Therefore, from Eq. (4.20), and Eq. (4.13), the macro-state transition probabilities Γ_{IJ} can be approximated as

$$\Gamma_{IJ} \approx \sum_{i \in S_I} \sum_{j \in S_J} \frac{\nu_i^{\approx} p_{ij}}{\sum_{k \in S_I} \nu_k^{\approx}} = \sum_{i \in S_I} \frac{\nu_i^{\approx}}{\sum_{k \in S_I} \nu_k^{\approx}} \sum_{j \in S_J} p_{ij}$$
$$\approx \sum_{i \in S_I} \left(\nu_{I\,i}^{*} \sum_{j \in S_J} p_{IJij} \right). \tag{4.21}$$

For the disaggregation step, the probabilities Γ_{Ii} of transitions from macro-state I to micro-state i are pairwise needed for all $0 \le I \le M - 1$ and for all $i \in S$. The transition probabilities Γ_{Ii} can be derived from Eq. (4.21) and are given by Eq. (4.22):

$$\Gamma_{Ij} = \sum_{i \in S_I} \frac{\nu_i^{\approx} p_{ij}}{\sum_{k \in S_I} \nu_k^{\approx}}. \tag{4.22}$$

To complete the disaggregation step, the micro-state probabilities ν_i are expressed in terms of macro-state probabilities γ_I and transition probabilities Γ_{Ii} for each subset of states S_I separately. To accomplish this task, $|S_I|$ linear equations need to be solved for each subset S_I:

$$\nu_i = \sum_{j \in S_I} \nu_j p_{ji} + \sum_{K=0, K \ne I}^{M-1} \gamma_K \Gamma_{Ki}, \quad \forall i \in S_I. \tag{4.23}$$

The second term in Eq. (4.23) represents the aggregation of the complementary subset of macro-states into a single external "super" state and the interactions in terms of transition probabilities with this state.

Macro-state probability vector Eq. (4.19), together with transition probability aggregation steps as given in Eq. (4.21), constitutes one major part of Takahashi's method, the *aggregation* step. Micro-state probability vector Eq. (4.23), together with transition probability disaggregations, as indicated in Eq. (4.22), forms the other major part of Takahashi's method, the *disaggregation* step.

4.2.2   Applicability

Given partition

$$S = \bigcup_{J=0}^{M-1} S_J \,,$$

a subset S_I is *lumpable* if and only if real-valued constants r_{Ij} exist such that condition (4.24) holds $\forall j \in S - S_I$:

$$p_{ij} = r_{Ij} \quad \forall i \in S_I \,. \tag{4.24}$$

Condition (4.24) is necessary and sufficient for the lumpability of S_I with respect to $S - S_I$. The lumpability condition may be restated by referring to Eq. (4.7) if pairwise lumpability is required such that real-valued constants r_{IJ} exist and Condition (4.25) holds $\forall I, J, 0 \leq I \neq J \leq M - 1$:

$$\sum_{j \in S_J} p_{ij} = r_{IJ} = \frac{1}{|S_I|} \sum_{i \in S_I} \sum_{j \in S_J} p_{ij}, \quad \forall i \in S_I \,. \tag{4.25}$$

If the state space can be partitioned such that the lumpability condition holds for all pairs of subsets $S_I, S_J \subset S$, then Takahashi's algorithm terminates in one iteration with the exact state probability vector ν. Unfortunately, lumpability implies very strict structural limitations on the underlying Markov model. More often, however, Markov chains exhibit *approximate lumpability* such that there exists a sufficiently small lumpability error ϵ:

$$\epsilon = \max_{0 \leq I \leq M-1} \sum_{J=0, J \neq I}^{M-1} \sum_{i \in S_I} \left| \sum_{j \in S_J} p_{ij} - \frac{1}{|S_I|} \sum_{i \in S_I} \sum_{j \in S_J} p_{ij} \right| . \tag{4.26}$$

Lumpability error ϵ is a measure for the speed of convergence of Takahashi's method. The problem remains, however, to find a good partitioning among the states of a given DTMC. Some hints concerning convergence and state-space partitioning can be found in Stewart's book [Stew94]. The underlying Markov models of some so-called product-form queueing networks, which are discussed in Chapter 7, exhibit the exact lumpability property. Furthermore, the lumpability equations can be exploited theoretically to investigate accuracy of approximate product-form queueing network algorithms.

In general, the efficiency of Takahashi's method can benefit from an exploitable structure of the underlying transition probability matrix. This fact, however, imposes the burden on the user of finding a good clustering strategy, and no generally applicable rule is known for a good clustering strategy.

Finally, it is worth mentioning that approximate lumpability is different from weak lumpability, which can be applied for an aggregation/disaggregation-based transient analysis of CTMCs. A comprehensive study on lumpability and weak lumpability is presented by Nicola [Nico90].

4.2.3   The Algorithm

Since convergence of Takahashi's method is a nontrivial problem, some care is needed here. We follow an approach that is often applied to enforce geometric convergence, namely, to incorporate an intermediate power, Gauss–Seidel, or SOR step between successive aggregations and disaggregations. If the decrease in residual error is not sufficient from iteration to iteration, such a step is suggested to be included in the computation. Usually, the corresponding condition is stated in terms of a real-valued lower improvement bound c, $0 < c < 1$ on the error ratio from iteration step $n - 1$ to iteration step n:

$$\frac{\text{residual error}\left(\boldsymbol{\nu}^{(n)}\right)}{\text{residual error}\left(\boldsymbol{\nu}^{(n-1)}\right)} \leq c. \tag{4.27}$$

The complete algorithm is presented as Fig. 4.4.

4.2.4   Application

We demonstrate Takahashi's method with the same example used in Section 4.1 to illustrate Courtois' method to allow an easy comparison of the numerical behavior of the two algorithms. Recall that we considered a network in which two customers were circulating among three stations according to some stochastic regularities. The resulting state transition diagram of the set S of six states was depicted in Fig. 4.1. Again, we will apply the decomposition strategy indicated in Fig. 4.2 and partition S into three disjoint subsets: $S_0 = \{020, 110, 200\}$, $S_1 = \{011, 101\}$, and $S_2 = \{002\}$. Arranging the states in the same order as defined in Table 4.3 and using the same numbers in our example results in the following transition probability matrix:

$$\mathbf{P} = \left(\begin{array}{ccc|cc|c} 0.52 & 0.45 & 0 & 0.03 & 0 & 0 \\ 0.24 & 0.265 & 0.45 & 0.015 & 0.03 & 0 \\ 0 & 0.24 & 0.745 & 0 & 0.015 & 0 \\ \hline 0.02 & 0.02 & 0 & 0.48 & 0.45 & 0.03 \\ 0 & 0.02 & 0.02 & 0.24 & 0.705 & 0.015 \\ \hline 0 & 0 & 0 & 0.02 & 0.02 & 0.96 \end{array}\right). \tag{4.28}$$

It has been shown that the transition probability matrix $\mathbf{P}$ in Eq. (4.28) is nearly completely decomposable so that Courtois' method could be applied. We now investigate the lumpability error induced for the same partition with respect to approximate lumpability as defined in Eq. (4.26). To calculate the

Create the state space S and organize it appropriately according to a pattern of decomposition along the lines of approximate lumpability into transition probability matrix $\mathbf{P} = [P_{IJ_{ij}}]$.

Initialization:
a) $n := 0$;
b) estimate $\boldsymbol{\nu}^{(0)}$;
c) choose ϵ and $0 < c < 1$;
d) choose some vector-norm function $f(\| \ldots \|)$;

$f\left(\| \boldsymbol{\nu}^{(n)} - \boldsymbol{\nu}^{(n)} \mathbf{P} \| \right) \geq \epsilon$

Geometric Convergence:

$n \geq 1$ Y   N

residualerror $\left(\boldsymbol{\nu}^{(n)} \right) > c * $ residualerror $\left(\boldsymbol{\nu}^{(n-1)} \right)$ Y   N

$\boldsymbol{\nu}' = \boldsymbol{\nu}^{(n)}$;

$\boldsymbol{\nu}^{(n)} = \boldsymbol{\nu}' \mathbf{P}$;

$n := n + 1$;

Aggregation for all $0 \leq I, J \leq M - 1$:
a) create $\boldsymbol{\Gamma}^{(n)} = \left[\Gamma_{IJ}^{(n)} \right] = \left[\sum_{i \in S_I} \sum_{j \in S_J} \frac{\nu_i^{(n-1)} p_{ij}}{\sum_{k \in S_I} \nu_k^{(n-1)}} \right]$

according to Eq. (4.21);
b) solve $\boldsymbol{\gamma}^{(n)} = \boldsymbol{\gamma}^{(n)} \boldsymbol{\Gamma}^{(n)}$ according to Eq. (4.19);

Disaggregation for all $0 \leq I \leq M - 1$:
a) calculate $\Gamma_{Ij}^{(n)} = \sum_{i \in S_I} \frac{\nu_i^{(n-1)} p_{ij}}{\sum_{k \in S_I} \nu_k^{(n-1)}}, \quad \forall j \in S$

according to Eq. (4.22);
b) calculate $\boldsymbol{\nu}_I^{(n)} = [\nu_{Ii}]$ with Eq. (4.23) by solving the system of equations:

$$\nu_i^{(n)} = \sum_{j \in S_I} \nu_j^{(n)} p_{ji} + \sum_{K=0, K \neq I}^{M-1} \gamma_K^{(n)} \Gamma_{Ki}^{(n)}, \quad \forall i \in S_I;$$

Normalization:
With $\boldsymbol{\nu}^{(n)} = \left[\nu_I^{(n)} \right]$ from the previous step solve: $\boldsymbol{\nu}^{(n)} \mathbf{1} = 1$.

Fig. 4.4 Takahashi's algorithm.

lumpability error, we first let $I = 0$, $J = 1$ and denote with ϵ_{IJ} the inner sum:

$$\epsilon_{01} = \sum_{i \in S_0} \left| \sum_{j \in S_1} p_{ij} - \frac{1}{|S_0|} \sum_{i \in S_0} \sum_{j \in S_1} p_{ij} \right|$$

$$= \left| 0.03 - \frac{0.03 + 0.015 + 0.03 + 0.015}{3} \right|$$

$$+ \left| 0.015 + 0.03 - \frac{0.03 + 0.015 + 0.03 + 0.015}{3} \right|$$

$$+ \left| 0.015 - \frac{0.03 + 0.015 + 0.03 + 0.015}{3} \right|$$

$$= |0.03 - 0.03| + |0.045 - 0.03| + |0.015 - 0.03|$$

$$= 0 + 0.015 + 0.015 = 0.03 \,.$$

Similarly, we get

$$\epsilon_{02} = 0 \,.$$

Therefore, by letting $I = 0$, $J \neq I$, the total lumpability error is given by:

$$\epsilon_0 = \epsilon_{01} + \epsilon_{02} = 0.03 + 0 = 0.03 \,.$$

For the other elements of the partition, we get

$$\epsilon_1 = 0.015, \qquad \epsilon_2 = 0 \,.$$

With

$$\epsilon = \max_{0 \leq I \leq M-1} \epsilon_I = \max\{0.03, 0.015, 0\} = 0.03 \,,$$

measures can be derived for the speed of convergence if Takahashi's method is applied with respect to the given partition of $S = \bigcup_{I=0}^{2} S_I$.

4.2.4.1 Aggregation

With the initial probability vector

$$\boldsymbol{\nu}^{(0)} = \boldsymbol{\nu}(0) = \left(\frac{1}{6}, \frac{1}{6}, \frac{1}{6}, \frac{1}{6}, \frac{1}{6}, \frac{1}{6} \right) ,$$

the first aggregation step, i.e., the computation of

$$\boldsymbol{\Gamma}^{(1)} = \left[\Gamma_{IJ}^{(1)} \right]$$

can be accomplished according to Eq. (4.21). As an example, the calculation of $\Gamma_{00}^{(1)}$ is shown:

$$\Gamma_{00}^{(1)} = \frac{1/6}{1/2} \cdot \sum_{i \in S_0} \sum_{j \in S_0} p_{ij}$$

$$= \frac{1}{3} \cdot (0.52 + 0.45 + 0.24 + 0.265 + 0.45 + 0.24 + 0.745)$$

$$= \frac{1}{3} \cdot 2.91 = 0.97.$$

Thus, in the first aggregation step, we get the following macro-state transition probability matrix:

$$\mathbf{\Gamma}^{(1)} = \begin{pmatrix} 0.97 & 0.03 & 0 \\ 0.04 & 0.9375 & 0.0225 \\ 0 & 0.04 & 0.96 \end{pmatrix} . \tag{4.29}$$

Then the macro-state probability vector $\boldsymbol{\gamma}^{(1)}$ is given by the solution of the following system of linear equations:

$$\left(\gamma_0^{(1)}, \gamma_1^{(1)}, \gamma_2^{(1)} \right) \begin{pmatrix} -0.03 & 0.03 & 0 \\ 0.04 & -0.0625 & 0.0225 \\ 0 & 0.04 & -0.04 \end{pmatrix} = 0, \quad \boldsymbol{\gamma}^{(1)}\mathbf{1} = 1. \tag{4.30}$$

From Eq. (4.30) we obtain the macro-state probabilities up to the fourth decimal digit:

$$\boldsymbol{\gamma}^{(1)} = (0.4604, 0.3453, 0.1943) . \tag{4.31}$$

4.2.4.2 Disaggregation
In the first part of the disaggregation step, transition probabilities $\Gamma_{Ij}^{(1)}$, are computed:

$$\Gamma_{10}^{(1)} = \frac{1}{2}0.02 + \frac{1}{2}0 = 0.01, \quad \Gamma_{20}^{(1)} = 0.$$

From there we compute

$$
\begin{aligned}
\nu_0^{(1)} &= \sum_{j=0,1,2} \nu_j^{(1)} p_{j0} + \gamma_1^{(1)}\Gamma_{10}^{(1)} + \gamma_2^{(1)}\Gamma_{20}^{(1)} \\
&= (\nu_0^{(1)}\cdot 0.52 + \nu_1^{(1)}\cdot 0.24 + \nu_2^{(1)}\cdot 0) + (0.3453\cdot 0.01 + 0.1943\cdot 0) \\
&= (\nu_0^{(1)}\cdot 0.52 + \nu_1^{(1)}\cdot 0.24 + \nu_2^{(1)}\cdot 0) + 0.003453 .
\end{aligned}
$$

The remaining transition probabilities from macro-states to micro-states are summarized as follows:

$$
\begin{aligned}
\Gamma_{11}^{(1)} &= 0.02, & \Gamma_{21}^{(1)} &= 0, \\
\Gamma_{12}^{(1)} &= 0.01, & \Gamma_{22}^{(1)} &= 0, \\
\Gamma_{03}^{(1)} &= 0.015, & \Gamma_{04}^{(1)} &= 0.015, \\
\Gamma_{24}^{(1)} &= 0.02, & \Gamma_{05}^{(1)} &= 0, \\
\Gamma_{15}^{(1)} &= 0.0225.
\end{aligned}
$$

To complete disaggregation in the first iteration step, the following three sets of equations, for $\{\nu_0^{(1)}, \nu_1^{(1)}, \nu_2^{(1)}\}$, $\{\nu_3^{(1)}, \nu_4^{(1)}\}$, and $\{\nu_5^{(1)}\}$ need to be solved separately for a calculation of micro-state probabilities:

$$\nu_0^{(1)} = (\nu_0^{(1)} \cdot 0.52 + \nu_1^{(1)} \cdot 0.24) + 0.003453 \,,$$
$$\nu_1^{(1)} = (\nu_0^{(1)} \cdot 0.45 + \nu_1^{(1)} \cdot 0.265 + \nu_2^{(1)} \cdot 0.24) + 0.006906 \,,$$
$$\nu_2^{(1)} = (\nu_0^{(1)} \cdot 0 + \nu_1^{(1)} \cdot 0.45 + \nu_2^{(1)} \cdot 0.745) + 0.003453 \,,$$
$$\nu_3^{(1)} = (\nu_3^{(1)} \cdot 0.48 + \nu_4^{(1)} \cdot 0.24) + 0.010792 \,,$$
$$\nu_4^{(1)} = (\nu_3^{(1)} \cdot 0.45 + \nu_4^{(1)} \cdot 0.705) + 0.010792 \,,$$
$$\nu_5^{(1)} = (\nu_5^{(1)} \cdot 0.96) + 0.00776925 \,.$$

With the solutions of these systems of equations and the normalization condition

$$\boldsymbol{\nu}^{(1)} \mathbf{1} = 1 \,,$$

the probability vector $\boldsymbol{\nu}^{(1)}$ at the end of the first iteration with a precision up to the fourth digit is

$$\boldsymbol{\nu}^{(1)} = (0.0812, 0.1486, 0.2753, 0.1221, 0.1864, 0.1864) \,. \qquad (4.32)$$

In Table 4.5, the micro-state probabilities obtained after one iteration of Takahashi's method are compared to the results gained if Courtois' method is applied. The percentage error in the results obtained with Courtois' approach are also included in the table. It can be seen that even after the first iteration step, the results are already relatively close to the exact values. After four iterations, the results gained with Takahashi's method resemble the exact values up to the fourth decimal digit.

4.2.5   Final Remarks

Takahashi's method was presented in terms of DTMCs. However, this choice implies no limitation since it is a well-known fact that ergodic DTMCs and CTMCs are equivalent with respect to their steady-state probability computations, when applying the transformations given in Eq. (3.6) and Eq. (3.5). But Takahashi's method can be more conveniently applied directly on the generator matrix $\mathbf{Q} = [q_{ij}]$ of a CTMC. As usual, we refer to the steady-state probability vector of the given micro-state CTMC through $\boldsymbol{\pi}$ or $\boldsymbol{\pi}^{\approx}$, respectively. In the continuous-time case, the resulting equations correspond directly to their discrete-time counterparts. Instead of using Eq. (4.19), the macro-state probability vector $\boldsymbol{\sigma}$ is now calculated based on infinitesimal generator matrix $\boldsymbol{\Sigma}$ of the continuous-time macro-state process:

$$0 = \boldsymbol{\sigma}\boldsymbol{\Sigma} \,, \quad \boldsymbol{\sigma}\mathbf{1} = 1 \,. \qquad (4.33)$$

Table 4.5   State probabilities: Takahashi's, Courtois', and exact methods

State	$\nu^{(0)}$	$\nu^{(1)}$	$\nu^{(2)}$	$\nu^{(3)}$	$\nu^{(4)}$	Exact	% Error[1]	% Error[4]
200	0.16$\overline{6}$	0.0785	0.0787	0.0787	0.0787	0.0787	- 0.3	0.0
110	0.16$\overline{6}$	0.1436	0.1476	0.1478	0.1478	0.1478	- 2.8	0.0
020	0.16$\overline{6}$	0.2660	0.2770	0.2777	0.2777	0.2777	- 4.2	0.0
101	0.16$\overline{6}$	0.1179	0.1146	0.1144	0.1144	0.1144	+ 3.1	0.0
011	0.16$\overline{6}$	0.2138	0.2150	0.2150	0.2150	0.2150	- 0.6	0.0
002	0.16$\overline{6}$	0.1801	0.1672	0.1665	0.1664	0.1664	+ 8.2	0.0

State	Courtois	Exact	% Error
200	0.0828	0.0787	+ 5.2
110	0.1480	0.1478	+ 0.1
020	0.2710	0.2777	- 2.4
101	0.1194	0.1144	+ 4.4
011	0.2108	0.2150	- 2.0
002	0.1680	0.1664	+ 0.1

The matrix entries Σ_{IJ} of Σ are calculated with an interpretation corresponding to the one applied to Eq. (4.21):

$$\Sigma_{IJ} = \sum_{i \in S_I} \sum_{j \in S_J} \frac{\pi_i^{\approx} q_{ij}}{\sum_{k \in S_I} \pi_k^{\approx}}, \quad I \neq J. \tag{4.34}$$

The disaggregation steps are analogous to Eq. (4.22) and Eq. (4.23) $\forall I, 0 \leq I \leq M - 1$:

$$\Sigma_{Ij} = \sum_{i \in S_I} \frac{\pi_i^{\approx} q_{ij}}{\sum_{k \in S_I} \pi_k^{\approx}}, \quad j \in S_J, J \neq I, \tag{4.35}$$

$$\pi_i \sum_{j \in S, j \neq i} q_{ij} = \sum_{j \in S_I, j \neq i} \pi_j q_{ji} + \sum_{J=0, J \neq I}^{M-1} \sigma_J \Sigma_{Ji}, \quad \forall i \in S_I. \tag{4.36}$$

Problem 4.5   Compare Courtois' method to Takahashi's method for numerical accuracy on the example studies performed in Sections 4.1 and 4.2.4. In particular, explain the observed numerical differences in the corresponding macro-state transition probability matrices Γ, in the corresponding macro-state probability vectors γ, and in the corresponding micro-state probability vectors ν.

Problem 4.6   Apply decomposition with regard to station two, as indicated in Fig. 4.3, to the DTMC underlying the discussions in Sections 4.1 and 4.2.4. Build up and organize a transition probability matrix that properly reflects the chosen strategy of decomposition. Calculate the approximate lumpability error ϵ according to Eq. (4.26), compare it to the results obtained in Section 4.2.4, and interpret the differences.

Problem 4.7   Apply Takahashi's algorithm on the basis of the decomposition strategy with regard to station two, as discussed in the previous problem. Compare the resulting macro-state transition probability matrices, the macro-state probability vectors, and the micro-state probability vectors obtained in the iteration step to those obtained under the original decomposition strategy in Section 4.2.4. Comment on and interpret the results.

Problem 4.8   Use the modified algorithm of Takahashi and apply it for a numerical analysis of the CTMC specified by its generator matrix $\mathbf{Q}$ in Table 4.2 and the parameters in Table 4.1. Apply the same partition as indicated in Table 4.2.

5

Transient Solution of Markov Chains

Transient solution is more meaningful than steady-state solution when the system under investigation needs to be evaluated with respect to its short-term behavior. Using steady-state measures instead of transient measures could lead to substantial errors in this case. Furthermore, applying transient analysis is the only choice if nonergodic models are investigated. Transient analysis of Markov chains has been attracting increasing attention and is of particular importance in dependability modeling.

Unlike steady-state analysis, CTMCs and DTMCs have to be treated differently while performing transient analysis. Surprisingly, not many algorithms exist for the transient analysis of DTMCs. Therefore, we primarily focus on methods for computing the transient state probability vector $\pi(t)$ for CTMCs as defined in Eq. (2.53). Furthermore, additional attention is given to the computation of quantities related to transient probabilities such as cumulative measures.

Recall from Eq. (2.53) that for the computation of transient state probability vector $\pi(t)$, the following linear differential equation has to be solved, given infinitesimal generator matrix $\mathbf{Q}$ and initial probability vector $\pi(0)$:

$$\frac{\mathrm{d}\,\pi(t)}{\mathrm{d}\,t} = \pi(t)\mathbf{Q}\,, \quad \pi(0) = (\pi_0(0), \pi_1(0), \dots)\,. \tag{5.1}$$

Measures that can be immediately derived from transient state probabilities are often referred to as *instantaneous* measures. However, sometimes measures based on cumulative accomplishments during a given period of time

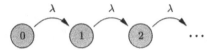

Fig. 5.1 A pure birth process.

$[0, t)$ could be more relevant. Let

$$\mathbf{L}(t) = \int_0^t \boldsymbol{\pi}(u)\,\mathrm{d}\,u$$

denote the vector of the total expected times spent in the states of the CTMC during the indicated period of time. By integrating Eq. (2.53) on both sides, we obtain a new differential equation for $\mathbf{L}(t)$:

$$\frac{\mathrm{d}\,\mathbf{L}(t)}{\mathrm{d}\,t} = \mathbf{L}(t)\mathbf{Q} + \boldsymbol{\pi}(0)\,, \quad \mathbf{L}(0) = \mathbf{0}\,. \tag{5.2}$$

Cumulative measures can be directly computed from the transient solution of Eq. (5.2).

5.1 TRANSIENT ANALYSIS USING EXACT METHODS

We introduce transient analysis with a simple example of a pure birth process.

5.1.1 A Pure Birth Process

As for birth–death processes in steady-state case, we derive a transient closed-form solution in this special case. Consider the infinite state CTMC depicted in Fig. 5.1 representing a pure birth process with constant birth rate λ. The number of births, $N(t)$, at time t is defined to be the state of the system. The only transitions possible are from state k to state $k + 1$ with rate λ. Note that this is a nonirreducible Markov chain for any finite value of λ, so the steady-state solution does not exist.

From Fig. 5.1, the infinitesimal generator matrix $\mathbf{Q}$ is given by:

$$\mathbf{Q} = \begin{pmatrix} -\lambda & \lambda & 0 & 0 & \ldots \\ 0 & -\lambda & \lambda & 0 & \ldots \\ 0 & 0 & -\lambda & \lambda & \ldots \\ \vdots & \vdots & \vdots & \vdots & \ddots \end{pmatrix}\,.$$

Due to the special structure of the generator matrix, it is possible to obtain a closed-form transient solution of this process. With generator matrix $\mathbf{Q}$ and

Eq. (2.53), we derive a system of linear differential equations for our example:

$$\frac{d}{dt}\pi_0(t) = -\lambda\pi_0(t),$$ (5.3)

$$\frac{d}{dt}\pi_k(t) = -\lambda\pi_k(t) + \lambda\pi_{k-1}(t), \quad k \geq 1.$$ (5.4)

With the initial state probabilities:

$$\pi_k(0) = \begin{cases} 1 & k = 0, \\ 0 & k \geq 1, \end{cases}$$ (5.5)

the solution of the differential equations can be obtained. From differentiation and integration theory the unique solution of Eq. (5.3) is:

$$\pi_0(t) = e^{-\lambda t}.$$ (5.6)

By letting $k = 1$ in Eq. (5.4) and subsequently substituting the solution for $\pi_0(t)$ into the differential equation, we get:

$$\frac{d}{dt}\pi_1(t) = -\lambda\pi_1(t) + \lambda\pi_0(t)$$
$$= -\lambda\pi_1(t) + \lambda e^{-\lambda t}.$$

Applying again elementary differentiation and integration rules to this differential equation results in another simple unique solution:

$$\pi_1(t) = \lambda t\, e^{-\lambda t}.$$ (5.7)

If this process is repeated, we get a closed-form solution for each transient state probability $\pi_k(t)$:

$$\pi_k(t) = \frac{(\lambda t)^k}{k!} e^{-\lambda t}, \quad k \geq 0.$$ (5.8)

The proof of Eq. (5.8) is straightforward using the principle of induction and we leave it as an exercise to the reader. We recognize Eq. (5.8) as the *Poisson pmf*. Poisson probabilities are plotted as functions of t in Fig. 5.2 for several values of λ. Thus, the random variable $N(t)$ at time t is Poisson distributed with parameter λt, while the stochastic process $\{N(t)|t \geq 0\}$ is the Poisson process with rate λ. So the transient analysis of a very simple birth process provided as a by-product one of the most important results of Markov chain and queueing theory; i.e., the stochastic process under consideration is the Poisson process. One important measure of the Poisson process is its mean value function $m(t)$, defined as the expected number of births in the

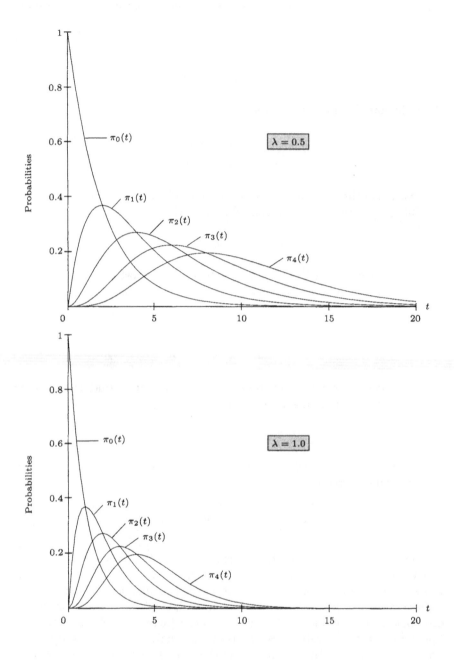

Fig. 5.2   Selected Poisson probabilities with parameters $\lambda = 0.5$ and 1.0.

interval $[0, t)$. This measure is computed as:

$$m(t) = \sum_{i=0}^{\infty} i\pi_i(t) = \sum_{i=0}^{\infty} i\frac{(\lambda t)^i}{i!} e^{-\lambda t}$$

$$= e^{-\lambda t} \sum_{i=1}^{\infty} \frac{(\lambda t)^{i-1}}{(i-1)!}(\lambda t) = (\lambda t) e^{-\lambda t} \sum_{i=0}^{\infty} \frac{(\lambda t)^i}{i!} \qquad (5.9)$$

$$= (\lambda t) e^{-\lambda t} e^{\lambda t} = \lambda t.$$

Eq. (5.8) together with Eq. (5.9) characterize the Poisson pmf and its mean λt.

Problem 5.1   Using the principle of mathematical induction, show that the Poisson pmf as given in Eq. (5.8) is the solution to Eq. (5.4).

Problem 5.2   Derive the solution of Eq. (5.2) of the pure birth process of Fig. 5.1, assuming $\pi_0(0) = 1$ and $\pi_k(0) = 0, \quad \forall k \geq 1$.

5.1.2   A Two-State CTMC

While in the previous section transient analysis based on closed-form expressions was possible due to very simple and regular model structure, in this section we carry out closed-form transient analysis of a CTMC with two states. Our treatment closely follows that in [STP96]. Figure 5.3 shows a homogeneous, continuous-time Markov chain model for this system with state space $S = \{0, 1\}$. A typical application of this model is to a system subject to failure and repair.

Fig. 5.3   A simple two-state CTMC.

The generator matrix of the CTMC of Fig. 5.3 is

$$\mathbf{Q} = \begin{bmatrix} -\mu & \mu \\ \lambda & -\lambda \end{bmatrix}.$$

From the system of Eqs. (2.51), we get

$$\frac{d}{dt}\pi_0(t) = -\mu\pi_0(t) + \lambda\pi_1(t),$$

$$\frac{d}{dt}\pi_1(t) = \mu\pi_0(t) - \lambda\pi_1(t). \qquad (5.10)$$

Applying the law of total probability, we conclude

$$\pi_0(t) = 1 - \pi_1(t)\,,$$

and rewrite Eq. (5.10) as

$$\frac{\mathrm{d}}{\mathrm{d}\,t}\pi_1(t) = \mu(1 - \pi_1(t)) - \lambda\pi_1(t)$$

or:

$$\frac{\mathrm{d}}{\mathrm{d}\,t}\pi_1(t) + (\mu + \lambda)\pi_1(t) = \mu\,. \tag{5.11}$$

A standard method of analysis for linear differential Eq. (5.11) is to use the method of integrating factors [RaBe74]. Both sides of the (rearranged) equation are multiplied by the integrating factor

$$\mathrm{e}^{\int(\mu+\lambda)\,\mathrm{d}\,t} = \mathrm{e}^{(\mu+\lambda)t}\,,$$

to get

$$\mu\,\mathrm{e}^{(\mu+\lambda)t} = \mathrm{e}^{(\mu+\lambda)t}\frac{\mathrm{d}}{\mathrm{d}\,t}\pi_1(t) + (\mu + \lambda)\,\mathrm{e}^{(\mu+\lambda)t}\,\pi_1(t) \tag{5.12}$$

$$= \frac{\mathrm{d}}{\mathrm{d}\,t}(\mathrm{e}^{(\mu+\lambda)t}\,\pi_1(t))\,. \tag{5.13}$$

By inspection of Eq. (5.12) it can be seen that the sum expression on the right-hand side of the equation equals the derivative of the product of the subterms so that Eq. (5.13) results. Integrating Eq. (5.13) on both sides gives us as an intermediate step:

$$\frac{\mu}{\mu + \lambda}\,\mathrm{e}^{(\mu+\lambda)t} + c = \mathrm{e}^{(\mu+\lambda)t}\,\pi_1(t)\,. \tag{5.14}$$

Multiplying Eq. (5.14) with

$$\mathrm{e}^{-(\mu+\lambda)t}$$

provides the desired result:

$$\pi_1(t) = \frac{\mu}{\mu + \lambda} + c\,\mathrm{e}^{-(\mu+\lambda)t}\,. \tag{5.15}$$

Integration constant c reflects the dependency of the transient state probabilities on the initial probability vector $\boldsymbol{\pi}(0)$. Assuming, for example,

$$\boldsymbol{\pi}^{(1)}(0) = (0, 1)\,,$$

results in

$$1 = \pi_1^{(1)}(0) = \frac{\mu}{\mu + \lambda} + c\,\mathrm{e}^{-(\mu+\lambda)0}$$

$$= \frac{\mu}{\mu + \lambda} + c\,.$$

Thus, unconditioning with an initial pmf $\pi^{(1)}(0)$ yields the corresponding integration constant:

$$c = 1 - \frac{\mu}{\mu + \lambda} = \frac{\lambda}{\mu + \lambda}.$$

The final expressions of the transient state probabilities in this case results in

$$\pi_1^{(1)}(t) = \frac{\mu}{\mu + \lambda} + \frac{\lambda}{\mu + \lambda} \left(e^{-(\mu+\lambda)t} \right), \tag{5.16}$$

$$\pi_0^{(1)}(t) = 1 - \pi_1^{(1)}(t)$$

$$= \frac{\mu + \lambda}{\mu + \lambda} - \frac{\mu + \lambda \left(e^{-(\mu+\lambda)t} \right)}{\mu + \lambda}$$

$$= \frac{\lambda - \lambda \left(e^{-(\mu+\lambda)t} \right)}{\mu + \lambda}$$

$$= \frac{\lambda}{\mu + \lambda} - \frac{\lambda}{\mu + \lambda} \left(e^{-(\mu+\lambda)t} \right). \tag{5.17}$$

With a closed-form expression of the transient state probabilities given, performance measure can be easily derived.

Problem 5.3   First, derive expressions for the steady-state probabilities of the CTMC shown in Fig. 5.3 by referring to Eq. (2.58). Then, take the limits as $t \rightarrow \infty$ of the transient pmf $\pi^{(1)}(t)$ and compare the results to the steady-state case π.

Problem 5.4   Derive the transient state probabilities of the CTMC shown in Fig. 5.3 assuming initial pmf $\pi^{(2)}(0) = (0.5, 0.5)$. Take the limit as $t \rightarrow \infty$ of the resulting transient pmf $\pi^{(2)}(t)$ and compare the results to those gained with $\pi^{(1)}(t)$ when taking the limit.

Problem 5.5   Assume the system modeled by the CTMC in Fig. 5.3 to be "up" in state 1 and to be "down" in state 0.

(a) Define the reward assignments for a computation of availability measures based on Table 2.1.

(b) Derive formulas for the instantaneous availability $A(t)$, the interval unavailability $\overline{UA}(t)$, and the steady-state availability A for this model, based both on $\pi^{(1)}(t)$ and $\pi^{(2)}(t)$. Refer to Section 2.2.3.1 for details on how these availability measures are defined.

(c) Let $\mu = 1$, $\lambda = 0.2$ and compute all measures that have been specified in this problem. Evaluate the transient measures at time instants $t \in \{1, 2, 3, 5, 10\}$.

5.1.3   Solution Using Laplace Transforms

It is well-known that *Laplace transforms* can be used for the solution of linear differential equations such as those in Eq. (5.1) or (5.2). Unfortunately, the applicability of this method is limited to problem domains with small state spaces or to those that imply a regular matrix structure. These restrictions are due to the difficulties immanent in computations of roots of a polynomial and in the computation of the inverse Laplace transform. We will not go into detail with respect to the applicability of the Laplace transform method for the computation of transient state probabilities but refer the reader to [RaTr95, STP96, Triv01], where further information is given.

With the given initial probability vector $\pi(0)$, the Laplace transform of Eq. (2.53) yields the following equation in terms of transform variable s:

$$s\pi(s) - \pi(0) = \pi(s)\mathbf{Q}. \qquad (5.18)$$

Equation (5.18) can be solved for the transformed probability vector $\pi(s)$:

$$\pi(s) = \pi(0)(s\mathbf{I} - \mathbf{Q})^{-1}. \qquad (5.19)$$

Note that the closed-form solution obtainable from Eq. (5.19) is completely symbolic with respect to both the system parameters and the time variable t. However, the actual use of this expression requires the computation of all the roots of the characteristic polynomial (equivalently, all the eigenvalues of $\mathbf{Q}$) and subsequent symbolic inversion of Laplace transform [Triv01]. Thus, fully symbolic solution is restricted to very small CTMCs or a CTMC with a highly structured matrix. Seminumerical solution, where all model parameters are required to be in numerical form while the time variable t is symbolic, is possible for models of moderate sizes (about 500 states) [RaTr95, STP96]. For larger models, numerical solution is the only viable approach.

5.1.4   Numerical Solution Using Uniformization

We first introduce uniformization method for the computation of instantaneous state probabilities. Next we discuss a method to deal with stiff Markov chains. Finally, we consider the computation of cumulative state probabilities.

5.1.4.1   The Instantaneous Case   Transient uniformization, originally also referred to as Jensen's method [Jens53] for a numerical solution of Eq. (2.53), is commonly considered as the method of choice under most circumstances. This method has been adapted by Muppala et al. [MMT94, MuTr92] in such a way that the important sub-class of so-called *stiff* Markov chains can be analyzed more efficiently and accurately. Further effort has been devoted to extend the original algorithm so that *cumulative* measures can be efficiently computed [MMT96, ReTr89, SoGa89].

Transient uniformization takes advantage of Eq. (3.5) to embed a DTMC into a given CTMC as we have seen earlier. Hence, the transient state probability vector $\boldsymbol{\pi}(t)$ of the CTMC is expressed in terms of a power series of the one-step transition probability matrix $\mathbf{P} = \mathbf{Q}/q + \mathbf{I}$ of the embedded DTMC. Since the transient state probability vector is computed at time t, an unconditioning needs to be performed with respect to the number of jumps the DTMC makes during the interval $[0, t)$. As a result of the transformation, state transitions occur from all states with a *uniform* rate q.

The one-step transition probabilities of the DTMC are defined as follows:

$$
p_{ji} = \begin{cases} \dfrac{q_{ji}}{q}, & j \neq i, \\[2mm] 1 - \displaystyle\sum_{k \neq j} \dfrac{q_{jk}}{q}, & j = i. \end{cases} \tag{5.20}
$$

The number of transitions of the DTMC during the interval $[0, t)$ is governed by a Poisson random variable with parameter qt. Thus, the desired CTMC probability vector at time t can be written in terms of the DTMC state probability vector $\boldsymbol{\nu}(i)$ at step i:

$$
\boldsymbol{\pi}(t) = \sum_{i=0}^{\infty} \boldsymbol{\nu}(i) e^{-qt} \frac{(qt)^i}{i!}, \quad \boldsymbol{\nu}(0) = \boldsymbol{\pi}(0). \tag{5.21}
$$

The DTMC state probability vector can be iteratively computed:

$$
\boldsymbol{\nu}(i) = \boldsymbol{\nu}(i-1)\mathbf{P}. \tag{5.22}
$$

Recall that in order to avoid periodicity of the embedded DTMC, parameter q has to be chosen so that $q > \max_{i,j} |q_{i,j}|$. The infinite series in Eq. (5.21) needs to be truncated for the purposes of numerical computation. Fortunately, a right truncation point r can be determined with a numerically bounded error ϵ_r for an approximation $\tilde{\boldsymbol{\pi}}(t)$ with respect to some vector norm, such as the maximum norm:

$$
\begin{aligned}
\|\boldsymbol{\pi}(t) - \tilde{\boldsymbol{\pi}}(t)\|_{\infty} &= \left\| \sum_{i=0}^{\infty} \boldsymbol{\nu}(i) e^{-qt} \frac{(qt)^i}{i!} - \sum_{i=0}^{r} \boldsymbol{\nu}(i) e^{-qt} \frac{(qt)^i}{i!} \right\|_{\infty} \\
&= \left\| \sum_{i=r+1}^{\infty} \boldsymbol{\nu}(i) e^{-qt} \frac{(qt)^i}{i!} \right\|_{\infty} \\
&\leq \sum_{i=r+1}^{\infty} e^{-qt} \frac{(qt)^i}{i!} \\
&= 1 - \sum_{i=0}^{r} e^{-qt} \frac{(qt)^i}{i!} \\
&\leq \epsilon_r.
\end{aligned} \tag{5.23}
$$

For large values of qt, r tends to increase and the lower terms contribute increasingly less to the sum. To avoid numerical inaccuracies stemming from computations with small numbers, a left truncation point l can be introduced. To this end, an overall error tolerance $\epsilon = \epsilon_l + \epsilon_r$ is partitioned to cover both left and right truncation errors. Applying again a vector norm, a left truncation point l can be determined similarly to the right truncation point r:

$$\sum_{i=0}^{l-1} e^{-qt} \frac{(qt)^i}{i!} \leq \epsilon_l . \tag{5.24}$$

With appropriately defined truncation points l and r as in Eqs. (5.23) and (5.24), the original uniformization Eq. (5.21) can be approximated for an implementation:

$$\boldsymbol{\pi}(t) \approx \sum_{i=l}^{r} \boldsymbol{\nu}(i) e^{-qt} \frac{(qt)^i}{i!} , \quad \boldsymbol{\nu}(0) = \boldsymbol{\pi}(0). \tag{5.25}$$

In particular, $O(\sqrt{qt})$ terms are needed between the left and the right truncation points and the transient DTMC state probability vector $\boldsymbol{\nu}(l)$ must be computed at the left truncation point l. The latter operation according to Eq. (5.22) requires $O(qt)$ computation time. Thus, the overall complexity of uniformization is $O(\eta qt)$ [ReTr88]. (Here η denotes the number of nonzero entries in $\mathbf{Q}$.) To avoid underflow, the method of Fox and Glynn can be applied in the computation of l and r [FoGl88]. Matrix squaring can be exploited for a reduction of computational complexity [ReTr88, AbMa93]. But because matrix fill-ins might result, this approach is often limited to models of medium size (around 500 states).

5.1.4.2 Stiffness Tolerant Uniformization

As noted, uniformization is plagued by the stiffness index qt [ReTr88]. Muppala and Trivedi observed that the most time consuming part of uniformization is the iteration to compute $\boldsymbol{\nu}(i)$ in Eq. (5.22). But this iteration is the main step in the power method for computing the steady-state probabilities of a DTMC as discussed in Chapter 3. Now if and when the values of $\boldsymbol{\nu}(i)$ converge to a stationary value $\tilde{\boldsymbol{\nu}}$, the iteration Eq. (5.22) can be terminated resulting in considerable savings in computation time [MuTr92]. We have also seen in Chapter 3 that speed of convergence to a stationary probability vector is governed by the second largest eigenvalue of the DTMC, but not by qt. However, since the probability vectors $\boldsymbol{\nu}(i)$ are computed iteratively according to the power method, convergence still remains to be effectively determined. Because an a priori determination of time of convergence is not feasible, three cases have to be differentiated for a computation of the transient state probability vector $\boldsymbol{\pi}(t)$ [MuTr92, MMT94]:

1. Convergence occurs beyond the right truncation point. In this case, computation of a stationary probability vector is not effective and the transient state probability vector $\boldsymbol{\pi}(t)$ is calculated according to Eq. (5.25) without modification.

2. Convergence occurs between left and right truncation points at time c, $l < c \le r$. Here advantage can be taken of the fact that $\boldsymbol{\nu}(i) = \boldsymbol{\nu}(c) = \tilde{\boldsymbol{\nu}}$, $\forall i > c$. By letting the right truncation point $r \to \infty$, we can rewrite Eq. (5.25) as

$$
\begin{aligned}
\boldsymbol{\pi}(t) \approx \sum_{i=l}^{\infty} \boldsymbol{\nu}(i) &= \mathrm{e}^{-qt} \frac{(qt)^i}{i!} \\
&= \sum_{i=l}^{c} \boldsymbol{\nu}(i)\, \mathrm{e}^{-qt} \frac{(qt)^i}{i!} + \boldsymbol{\nu}(c) \sum_{i=c+1}^{\infty} \mathrm{e}^{-qt} \frac{(qt)^i}{i!} \\
&= \sum_{i=l}^{c} \boldsymbol{\nu}(i)\, \mathrm{e}^{-qt} \frac{(qt)^i}{i!} + \boldsymbol{\nu}(c)\Big(1 - \sum_{i=0}^{c} \mathrm{e}^{-qt} \frac{(qt)^i}{i!}\Big) .
\end{aligned}
\tag{5.26}
$$

3. Convergence occurs before the left truncation point, that is, $c < l$. In this case, we get

$$
\boldsymbol{\pi}(t) = \boldsymbol{\nu}(c) = \tilde{\boldsymbol{\nu}} .
\tag{5.27}
$$

Recall that stationary pmfs do not necessarily exhibit the properties of steady-state probability vectors in our terminology. In particular, the CTMCs need not be ergodic for the approach of stiffness tolerant uniformization to work. The only required property is that the DTMC must be aperiodic for a stationary pmf to exist (see Section 2.1.2 for more details). Issues of steady-state or stationary pmf detection by means of applying different norms are discussed in Chapters 2 and 4. More details can be found in [CBC+93, StGo85, Stew94]. Error bounds and complexity estimations of the stiffness tolerant uniformization algorithm discussed in this section are given in [MMT94].

5.1.4.3 The Cumulative Case Uniformization can also be extended to the cumulative case in Eq. (5.2) [ReTr89]. By integration of Eq. (5.21) with respect to time parameter t, we get:

$$
\mathbf{L}(t) = \frac{1}{q} \sum_{i=0}^{\infty} \boldsymbol{\nu}(i) \sum_{j=i+1}^{\infty} \mathrm{e}^{-qt} \frac{(qt)^j}{j!} = \frac{1}{q} \sum_{i=0}^{\infty} \boldsymbol{\nu}(i) \left(1 - \sum_{j=0}^{i} \mathrm{e}^{-qt} \frac{(qt)^j}{j!}\right) .
\tag{5.28}
$$

The infinite series needs to be truncated again, so that Eq. (5.28) reduces to:

$$
\mathbf{L}(t) \approx \frac{1}{q} \sum_{i=0}^{r} \boldsymbol{\nu}(i) \left(1 - \sum_{j=0}^{i} \mathrm{e}^{-qt} \frac{(qt)^j}{j!}\right) .
\tag{5.29}
$$

Again, techniques for detecting stationarity of pmfs can be applied for making uniformization less susceptible to stiffness and more efficient [MMT96]. Two cases need to be differentiated with respect to the relation of convergence time step c to truncation point r:

1. Convergence occurs beyond the truncation point, that is, $c > r$. As a consequence, Eq. (5.29) remains unaffected and must be evaluated as is.

2. If convergence occurs before truncation point, Eq. (5.29) is adjusted to

$$
\begin{aligned}
\mathbf{L}(t) &= \frac{1}{q} \sum_{i=0}^{\infty} \boldsymbol{\nu}(i) \mathbf{L}'(t) \\
&= \frac{1}{q} \sum_{i=0}^{c} \boldsymbol{\nu}(i) \mathbf{L}'(t) + \frac{1}{q} \boldsymbol{\nu}(c) \sum_{i=c+1}^{\infty} \mathbf{L}'(t) \\
&= \frac{1}{q} \sum_{i=0}^{c} \boldsymbol{\nu}(i) \mathbf{L}'(t) + \frac{1}{q} \boldsymbol{\nu}(c) \left(\sum_{i=0}^{\infty} \mathbf{L}'(t) - \sum_{i=0}^{c} \mathbf{L}'(t) \right) \\
&= \frac{1}{q} \sum_{i=0}^{c} \boldsymbol{\nu}(i) \mathbf{L}'(t) + \frac{1}{q} \boldsymbol{\nu}(c) \left(qt - \sum_{i=0}^{c} \mathbf{L}'(t) \right) \\
&= \frac{1}{q} \sum_{i=0}^{c} \boldsymbol{\nu}(i) \left(1 - \sum_{j=0}^{i} e^{-qt} \frac{(qt)^j}{j!} \right) \\
&\quad + \frac{1}{q} \boldsymbol{\nu}(c) \left(qt - \sum_{i=0}^{c} \left(1 - \sum_{j=0}^{i} e^{-qt} \frac{(qt)^j}{j!} \right) \right),
\end{aligned}
\tag{5.30}
$$

with

$$
\mathbf{L}'(t) = \sum_{j=i+1}^{\infty} e^{-qt} \frac{(qt)^j}{j!} = \left(1 - \sum_{j=0}^{i} e^{-qt} \frac{(qt)^j}{j!} \right).
$$

In case truncation needs to be performed, a time-dependent error-bound estimate $\epsilon^{(r)}(t)$ related to Eq. (5.29) can be given as a function of the truncation point r. Assuming an error tolerance allowance, the corresponding number of required terms r can be determined:

$$
\begin{aligned}
\epsilon^{(r)}(t) &\leq \frac{1}{q} \sum_{i=r+1}^{\infty} \sum_{j=i+1}^{\infty} e^{-qt} \frac{(qt)^j}{j!} \\
&\leq \frac{1}{q} \sum_{i=r+1}^{\infty} (i - (r+1)) e^{-qt} \frac{(qt)^i}{i!} \\
&\leq \frac{1}{q} \sum_{i=r+1}^{\infty} i e^{-qt} \frac{(qt)^i}{i!} - \frac{1}{q} \sum_{i=r+1}^{\infty} (r+1) e^{-qt} \frac{(qt)^i}{i!} \\
&\leq t \sum_{i=r}^{\infty} e^{-qt} \frac{(qt)^i}{i!} - \frac{r+1}{q} \sum_{i=r+1}^{\infty} e^{-qt} \frac{(qt)^i}{i!}.
\end{aligned}
\tag{5.31}
$$

5.1.5 Other Numerical Methods

We briefly discuss numerical methods based on ordinary differential equation approach followed by methods exploiting weak lumpability.

5.1.5.1 Ordinary Differential Equations Standard techniques for the solution of ordinary differential equations (ODE) can be utilized for the numerical solution of the Kolmogorov differential equations of a CTMC. Such ODE solution methods discretize the solution interval into a finite number of time intervals $\{t_1, t_2, \ldots, t_i, \ldots, t_n\}$. The difference between successive time points, called the step size h, can vary from step to step. There are two basic types of ODE solution methods: *explicit* and *implicit*.

In an explicit method, the solution $\pi(t_i)$ is approximated based on values $\pi(t_j)$ for $j < i$. The computational complexity of explicit methods such as Runge–Kutta is $O(\eta q t)$ [ReTr88]. Although explicit ODE-based methods generally provide good results for non-stiff models, they are inadequate if stiff models need to be studied. Note that stiff CTMCs are commonly encountered in dependability modeling.

In an implicit method, $\pi(t_i)$ is approximated based on values $\pi(t_j)$ for $j \leq i$. Examples of implicit methods are TR-BDF2 [ReTr88] and implicit Runge–Kutta [MMT94]. At each time step a linear system solution is required in an implicit method. The increased overhead is compensated for by better stability properties and lower computational complexity on stiff models [MMT94, ReTr88].

For non-stiff models, Uniformization is the method of choice, while for moderately stiff models, uniformization with steady-state detection is recommended [MMT94]. For extremely stiff models, TR-BDF2 works well if the accuracy required is low (eight decimal digits). For high accuracy on extremely stiff models, implicit Runge–Kutta is recommended [MMT94].

5.1.5.2 Weak Lumpability An alternative method for transient analysis based on weak lumpability has been introduced by Nicola [Nico90]. Nicola's method is the transient counterpart of Takahashi's steady-state method, which was introduced in Section 4.2. Recall that state lumping is an exact approach for an analysis of a CTMC with reduced state space and, therefore, with reduced computational requirements. State lumping can be a very efficient method for a computation of transient state probabilities, if the lumpability conditions apply. Note that state lumping can also be orthogonally combined with other computational methods of choice, such as stiff uniformization, to yield an overall highly accurate and efficient method.

For the sake of completeness, we present the basic definitions of weak lumpability without going into further detail here. Given conditional probability vector $\boldsymbol{\nu}_J^*$ and an initial probability vector $\boldsymbol{\nu}(0)$, with

$$\nu_{Jj}^* = \frac{\nu_j(0)}{\sum\limits_{k \in S_J} \nu_k(0)} , \tag{5.32}$$

then a subset $S_J \subset S$ is *weakly lumpable* with respect to $\boldsymbol{\nu}(0)$ if for all $i \in S - S_J$ real-valued constants $0 \leq r_{iJ}, \leq 1$ exist such that conditions (5.33) and (5.34) hold:

$$\frac{p_{ij}}{\nu_{Jj}^*} = r_{iJ}, \quad \forall j \in S_J , \tag{5.33}$$

$$\sum\limits_{k \in S_J} p_{kj} \frac{\nu_{Jk}^*}{\nu_{Jj}^*} = r_{JJ}, \quad \forall j \in S_J . \tag{5.34}$$

A CTMC is weakly lumpable for a given initial pmf, while (strong) lumpability, as it was introduced in Section 4.2, implied a lumpability of a CTMC irrespective of the initial pmf [Abde82, KeSn78]. Nicola also extended (strong and weak) lumpability of CTMCs to (strong and weak) lumpability of MRMs. In particular, he identified measure-dependent conditions for lumping an MRM (see Section 2.2 for details on relevant measures). Lumping of an MRM requires conditions on lumping of the underlying CTMC plus further conditions on the reward rates attached to the states.

5.2 AGGREGATION OF STIFF MARKOV CHAINS

Bobbio and Trivedi [BoTr86] have introduced an approximate method for the computation of transient state probabilities $\boldsymbol{\pi}^{\approx}(t)$ of finite CTMCs that can be regarded as an extension of Courtois' method, which was presented in Section 4.1. Since transient analysis does not require ergodicity of the investigated models, the aggregation and disaggregation have to be applied differently with respect to different subsets of states. Therefore, in contrast to Courtois' method, aggregation/disaggregation also needs to be performed for the *initial state probability vector*, and for *transient states*. The aggregation/disaggregation of *recurrent* subsets of states is performed analogous to Courtois' method. It is worth emphasizing that the method introduced in this section not only provides a technique for an efficient, approximate computation of transient state probabilities of possibly large CTMCs, but is also often applicable where other transient numerical techniques fail. This application is especially important in cases where the transition rates of the infinitesimal generator matrix differ by several orders of magnitude.

5.2.1 Outline and Basic Definitions

Bobbio and Trivedi motivate their approach by observing that transient analysis is often performed for models that exhibit the so-called *stiffness* property. These models usually include fast states and slow states such that the transition rates differ by several orders of magnitude. A state is called *fast* if at least one rate leaving that state is classified to be high. If all its outgoing rates are classified to be low, the state is called *slow*. Subsets of fast recurrent states are aggregated to macro-states with a similar argument as had been used by Courtois when taking advantage of near complete decomposability. Any slow state is by definition a macro-state. Subsets of transient fast states are eliminated by adjusting the transition rates out of the resulting macro-states.

It is convenient to characterize a transition rate by comparison to the ratio $1/T$, where the transient computation is to be performed for the period of time $[0, T)$. Rates are high if they are larger than some threshold τ, with $\tau >> 1/T$. Low rates, on the other hand, are of the order of magnitude of $1/T$. The actual computation of transient state probabilities is only performed for the slow or aggregated macro-states. It is assumed that the fast states within an aggregate reach steady state at a time scale much shorter than that describing the dynamics among the macro-states.

A subset $S_I \subset S$ of fast states is aggregated into a macro-state I if and only if the following conditions are satisfied:

1. Between any pairs $i, j \in S_I$ of states within the subset a directed path from i to j, and vice versa, exists such that the path is defined along fast transitions only.

2. No other state outside of S_I can be reached from within S_I via a transition with a fast rate.

An immediate consequence of this definition is that transitions from the states of a fast recurrent subset S_I to any state in a disjoint subset of states $S_J, J \neq I$, have only low rates attached. These subsets of states can therefore be regarded as separable from the remaining states in the sense of Courtois' near complete decomposability property. In terms of matrix notation, the infinitesimal generator matrix $\mathbf{Q}$ can be reorganized in a nearly block diagonally dominant form. The transition rates inside each diagonal block dominate the transition rates between blocks.

Assume that the state space S has been partitioned into $F + 1$ subsets $S_I, 0 \leq I \leq F$. The set of slow states initially present in the model is contained in S_0. $S_1, S_2, \ldots, S_{F-1}$ are the subsets of fast recurrent states, and S_F is the possibly empty set of fast transient states. The cardinality of any set of states $S_J, 0 \leq J \leq F$, is denoted by $n_J = |S_J|$. Without loss of generality, the generator matrix $\mathbf{Q}$ can be restructured in a block form reflecting the previously described state partition:

$$
\mathbf{Q} = \begin{pmatrix}
\mathbf{A}_{00} & \mathbf{A}_{01} & \cdots & & \mathbf{A}_{0F} \\
\mathbf{A}_{10} & \mathbf{Q}_1 & \cdots & & \mathbf{A}_{1F} \\
\vdots & & \ddots & & \vdots \\
\mathbf{A}_{(F-1)0} & \cdots & & \mathbf{Q}_{F-1} & \mathbf{A}_{(F-1)F} \\
\mathbf{A}_{F0} & \cdots & & \mathbf{A}_{F(F-1)} & \mathbf{A}_{FF}
\end{pmatrix} .
\tag{5.35}
$$

Let $\mathbf{Q}_I = \mathbf{Q}_{II}$, $1 \le I \le F - 1$, denote submatrices containing transition rates between grouped states within fast recurrent subset S_I only. Submatrices $\mathbf{A}_{IJ}$, $I \ne J$, contain entries of transitions rates from subset S_I to S_J. Finally, $\mathbf{A}_{00}$ and $\mathbf{A}_{FF}$ collect intra-slow states and intra-fast transient states transition rates, respectively.

Matrices $\mathbf{Q}_I$, $1 \le I \le F - 1$, contain at least one fast entry in each row. Furthermore, there must be at least one fast entry in one of the matrices $\mathbf{A}_{FI}$, $0 \le I \le F - 1$ if $S_F \ne \emptyset$. Matrix $\mathbf{A}_{FF}$ may contain zero or more fast entries. Finally, all other matrices, i.e., $\mathbf{A}_{00}, \mathbf{A}_{IJ}$, $I \ne J, 0 \le I \le F - 1, 0 \le J \le F$, contain only slow entries. By definition, only slow transitions are possible among these subsets S_I and S_J. An approximation to $\boldsymbol{\pi}(t)$ is derived, where $\boldsymbol{\pi}(t)$ is the solution to the differential equation (5.36):

$$
\frac{\mathrm{d}}{\mathrm{d}t} \boldsymbol{\pi}(t) = \boldsymbol{\pi}(t)\mathbf{Q}, \quad \boldsymbol{\pi}(0) = \left(\pi_{0_0}(0), \pi_{0_1}(0), \ldots, \pi_{F_{n_F - 1}}(0) \right) .
\tag{5.36}
$$

The reorganized matrix $\mathbf{Q}$ forms the basis to create the macro-state generator matrix $\boldsymbol{\Sigma}$. Three steps have to be carried out for this purpose: first, aggregation of the fast recurrent subsets into macro-states and the corresponding adaptation of the transition rates among macro-states and remaining fast transient states, resulting in intermediate generator matrix $\tilde{\boldsymbol{\Sigma}}$. In a second step, the fast transient states are eliminated and the transition rates between the remaining slow states are adjusted, yielding the final generator matrix $\boldsymbol{\Sigma}$. Finally, the initial state probability vector $\boldsymbol{\pi}(0)$ is condensed into $\boldsymbol{\sigma}(0)$ as per the aggregation pattern. Transient solution of differential Eq. (5.37) describing the long-term interactions between macro-states is carried out:

$$
\frac{\mathrm{d}}{\mathrm{d}t} \boldsymbol{\sigma}(t) = \boldsymbol{\sigma}(t)\boldsymbol{\Sigma},
\tag{5.37}
$$

$$
\boldsymbol{\sigma}(0) = \left(\sigma_{0_0}(0), \ldots, \sigma_{0_{n_0 - 1}}(0), \sigma_1(0), \ldots, \sigma_{F-1}(0) \right) .
$$

Once the macro-state probability vector $\boldsymbol{\sigma}(t)$ has been computed, disaggregations can be performed to yield an approximation $\boldsymbol{\pi}^{\approx}(t)$ of the complete probability vector $\boldsymbol{\pi}(t)$.

5.2.2   Aggregation of Fast Recurrent Subsets

Each subset of fast recurrent states is analyzed in isolation from the rest of the system by cutting off all slow transitions leading out of the aggregate. Collecting all states from such a subset S_I and arranging them together with the corresponding entries of the originally given infinitesimal generator matrix $\mathbf{Q}$ into a submatrix $\mathbf{Q}_I$ gives rise to the possibility of computing the conditional steady-state probability vectors $\boldsymbol{\pi}_I^*$, $1 \leq I \leq F - 1$. Since $\mathbf{Q}_I$ does not, in general, satisfy the properties of an infinitesimal generator matrix, it needs to be modified to matrix $\mathbf{Q}_I^*$:

$$\mathbf{Q}_I^* = \mathbf{Q}_I + \mathbf{D}_I. \tag{5.38}$$

Matrix $\mathbf{D}_I$ is a diagonal matrix whose entries are the sum of all cut-off transition rates for the corresponding state. Note that these rates are, by definition, orders of magnitude smaller than the entries on the diagonal of $\mathbf{Q}_I$. The inverse of this quantity is used as a measure of coupling between the subset S_I and the rest of the system.

Since $\mathbf{Q}_I^*$ is the infinitesimal generator matrix of an ergodic CTMC, the solution of

$$\boldsymbol{\pi}_I^* \mathbf{Q}_I^* = \mathbf{0}, \quad \boldsymbol{\pi}_I^* \mathbf{1} = 1 \tag{5.39}$$

yields the desired conditional steady-state probability vector $\boldsymbol{\pi}_I^*$ for the subset of fast states S_I. For each such subset S_I, a macro-state I is defined. The set of aggregated states $\{I, 1 \leq I \leq F-1\}$, constructed in this way, together with the initially given slow states S_0 form the set of macro-states $M = S_0 \cup \{I, 1 \leq I \leq F-1\}$ for which a transient analysis is performed in order to account for long-term effects among the set of all macro-states.

To create the intermediate generator matrix $\tilde{\boldsymbol{\Sigma}}$ of size $(|M|+n_F)$ x $(|M|+n_F)$, unconditionings and aggregations of the transition rates have to be performed. The dimensions $[n_I$ x $n_J]$ of submatrices $\tilde{\boldsymbol{\Sigma}}_{IJ}$ are added to the corresponding equations in what follows. Let us define matrices of appropriate size:

$$\mathbf{E} = \begin{pmatrix} 1 & 1 & \cdots & 1 \\ 1 & 1 & \cdots & 1 \\ \vdots & \vdots & \ddots & \vdots \\ 1 & 1 & \cdots & 1 \end{pmatrix}, \quad [n_0 \text{ x } n_I],$$

containing 1s only as entries. The following cases are distinguished:

- Transitions remain unchanged among slow states in S_0:

$$\tilde{\boldsymbol{\Sigma}}_{00} = \mathbf{A}_{00}, \quad [n_0 \text{ x } n_0]. \tag{5.40}$$

- Transitions involving fast transient states from S_F:

  - Transitions between slow states and fast transient states, and vice versa, remain unchanged:

$$\tilde{\boldsymbol{\Sigma}}_{F0} = \mathbf{A}_{F0}, \quad [n_F \text{ x } n_0], \tag{5.41}$$

$$\tilde{\boldsymbol{\Sigma}}_{0F} = \mathbf{A}_{0F}, \quad [n_0 \text{ x } n_F]. \tag{5.42}$$

- Transitions from aggregated macro-states I, $1 \leq I \leq F - 1$, to fast transient states in S_F need to be weighted and added with the help of the conditional micro-state probability vector $\boldsymbol{\pi}_I^*$:

$$\tilde{\boldsymbol{\Sigma}}_{IF} = \boldsymbol{\pi}_I^* \mathbf{A}_{IF}, \quad [1 \times n_F]. \tag{5.43}$$

- Transition rates from fast transient states to aggregated macro-states need to be aggregated into a single transition rate for all I, $1 \leq I \leq F - 1$:

$$\tilde{\boldsymbol{\Sigma}}_{FI} = \mathbf{A}_{FI} \mathbf{1}, \quad [n_F \times 1]. \tag{5.44}$$

- Transitions from aggregated macro-states I, $1 \leq I \leq F - 1$, to slow states in S_0 need to be weighted and added with the help of the conditional micro-state probability vector $\boldsymbol{\pi}_I^*$:

$$\tilde{\boldsymbol{\Sigma}}_{I0} = \boldsymbol{\pi}_I^* \mathbf{A}_{I0}, \quad [1 \times n_0]. \tag{5.45}$$

- Transitions from slow states in S_0 to aggregated macro-states I, $1 \leq I \leq F - 1$ need to be aggregated into a single transition rate originating from each slow state in S_0:

$$\tilde{\boldsymbol{\Sigma}}_{0I} = \mathbf{A}_{0I} \mathbf{E}, \quad [n_0 \times 1]. \tag{5.46}$$

- Transition rates from aggregated macro-state I to aggregated macro-state J, $I \neq J$ and $1 \leq I, J \leq F - 1$ have to be weighted and added with the help of the conditional micro-state probability vector $\boldsymbol{\pi}_I^*$ for each $j \in S_J$ separately. The resulting values are aggregated to yield the transition rates between macro-states I and J:

$$\tilde{\boldsymbol{\Sigma}}_{IJ} = \boldsymbol{\pi}_I^* \mathbf{A}_{IJ} \mathbf{1}, \quad [1 \times 1]. \tag{5.47}$$

- To complete the infinitesimal generator matrix $\tilde{\boldsymbol{\Sigma}}$, the diagonal elements need to be computed for all I, $1 \leq I \leq F - 1$, as usual:

$$\tilde{\boldsymbol{\Sigma}}_{II} = - \sum_{J, I \neq J} \tilde{\boldsymbol{\Sigma}}_{IJ}, \quad [1 \times 1]$$

$$= - \left(\tilde{\boldsymbol{\Sigma}}_{I0} \mathbf{1} + \sum_{J, J \geq 1, J \neq I} \tilde{\boldsymbol{\Sigma}}_{IJ} + \tilde{\boldsymbol{\Sigma}}_{IF} \mathbf{1} \right)$$

$$= - \left(\boldsymbol{\pi}_I^* \mathbf{A}_{I0} \mathbf{1} + \sum_{J, J \geq 1, J \neq I} \boldsymbol{\pi}_I^* \mathbf{A}_{IJ} \mathbf{1} + \boldsymbol{\pi}_I^* \mathbf{A}_{IF} \mathbf{1} \right). \tag{5.48}$$

Collecting the results from Eq. (5.40) through Eq. (5.48) yields the intermediate infinitesimal generator matrix $\tilde{\boldsymbol{\Sigma}}$. The matrix $\tilde{\boldsymbol{\Sigma}}$ is structured in such a way that it reflects the partitioning into macro-states and fast transient states:

$$\tilde{\mathbf{\Sigma}} = \begin{pmatrix} \tilde{\mathbf{\Sigma}}_{MM} & \tilde{\mathbf{\Sigma}}_{MF} \\ \tilde{\mathbf{\Sigma}}_{FM} & \tilde{\mathbf{\Sigma}}_{FF} \end{pmatrix} . \tag{5.49}$$

If no fast transient states are included, then $\mathbf{\Sigma} = \tilde{\mathbf{\Sigma}}$ already constitutes the final generator matrix. In any case, a transient analysis can be carried out based on a model with reduced state space. Although the intermediate model may still be stiff because fast and slow states are still simultaneously included, the computational savings can be considerable due to smaller state space. But more aggregation is possible, as is highlighted in the following section. With $\tilde{\mathbf{\Sigma}}$ given, a differential equation can be derived for the computation of the transient state probability vector $\tilde{\boldsymbol{\sigma}}(t)$, conditioned on initial probability vector $\tilde{\boldsymbol{\sigma}}(0)$:

$$\frac{\mathrm{d}}{\mathrm{d}t}\tilde{\boldsymbol{\sigma}}(t) = \tilde{\boldsymbol{\sigma}}(t)\tilde{\mathbf{\Sigma}}, \tag{5.50}$$

$$\tilde{\boldsymbol{\sigma}}(0) = \left(\tilde{\sigma}_{0_0}(0), \ldots, \tilde{\sigma}_{0_{n_0-1}}(0), \tilde{\sigma}_1(0), \ldots, \tilde{\sigma}_{F-1}(0), \tilde{\sigma}_{F_0}(0), \ldots, \tilde{\sigma}_{F_{n_F-1}}(0) \right) .$$

5.2.3 Aggregation of Fast Transient Subsets

Recall that a fast transient subset of states is connected to states in other subsets by at least one fast transition. Thus, if such a set is not nearly completely decomposable from the rest of the system, it is tightly coupled to at least one other non-empty subset of states.

Now consider an intermediate state space consisting of a non-empty fast transient subset of states $S_F \subset S$, and the set of macro-states $M = S_0 \cup \{I, 1 \leq I \leq F-1\}$ resulting from aggregation of fast recurrent subsets and collecting them together with the slow states. Furthermore, assume the intermediate transition rates among the states in $M \cup S_F$ to be already calculated as shown in the previous section.

From Eq. (5.50) and Eq. (5.49), the following two equations can be derived:

$$\frac{\mathrm{d}}{\mathrm{d}t}\tilde{\boldsymbol{\sigma}}_M(t) = \tilde{\boldsymbol{\sigma}}_M(t)\tilde{\mathbf{\Sigma}}_{MM} + \tilde{\boldsymbol{\sigma}}_F(t)\tilde{\mathbf{\Sigma}}_{FM}, \tag{5.51}$$

$$\frac{\mathrm{d}}{\mathrm{d}t}\tilde{\boldsymbol{\sigma}}_F(t) = \tilde{\boldsymbol{\sigma}}_M(t)\tilde{\mathbf{\Sigma}}_{MF} + \tilde{\boldsymbol{\sigma}}_F(t)\tilde{\mathbf{\Sigma}}_{FF}. \tag{5.52}$$

It is reasonable to assume that for the fast transient states, the derivative on the left-hand side of Eq. (5.52) approaches zero (i.e., $d/dt\tilde{\boldsymbol{\sigma}}_F(t) = 0$), with respect to the time scale of slow states. Hence from Eq. (5.52) we obtain an approximation of $\tilde{\boldsymbol{\sigma}}_F$:

$$\tilde{\boldsymbol{\sigma}}_F^{\approx} = -\tilde{\boldsymbol{\sigma}}_M^{\approx}\tilde{\mathbf{\Sigma}}_{MF}\tilde{\mathbf{\Sigma}}_{FF}^{-1}. \tag{5.53}$$

Using the result from Eq. (5.53) in Eq. (5.51) provides us with the following approximation:

$$\frac{\mathrm{d}}{\mathrm{d}t}\tilde{\boldsymbol{\sigma}}_{M}^{\approx}(t) = \tilde{\boldsymbol{\sigma}}_{M}^{\approx}(t)\tilde{\boldsymbol{\Sigma}}_{MM} - \tilde{\boldsymbol{\sigma}}_{M}^{\approx}(t)\tilde{\boldsymbol{\Sigma}}_{MF}\tilde{\boldsymbol{\Sigma}}_{FF}^{-1}\tilde{\boldsymbol{\Sigma}}_{FM} \tag{5.54}$$

$$= \tilde{\boldsymbol{\sigma}}_{M}^{\approx}(t)\left(\tilde{\boldsymbol{\Sigma}}_{MM} - \tilde{\boldsymbol{\Sigma}}_{MF}\tilde{\boldsymbol{\Sigma}}_{FF}^{-1}\tilde{\boldsymbol{\Sigma}}_{FM}\right) \tag{5.55}$$

$$= \tilde{\boldsymbol{\sigma}}_{M}^{\approx}(t)\tilde{\boldsymbol{\Sigma}}_{MM}^{\approx}. \tag{5.56}$$

Hence, $\boldsymbol{\Sigma} \approx \tilde{\boldsymbol{\Sigma}}_{MM}^{\approx}$ is taken as the infinitesimal generator matrix of the final aggregated Markov chain. Bobbio and Trivedi [BoTr86] have shown that:

$$\mathbf{P}_{FM} = -\tilde{\boldsymbol{\Sigma}}_{FF}^{-1}\tilde{\boldsymbol{\Sigma}}_{FM} \tag{5.57}$$

is the *asymptotic exit probability* matrix from the subset of fast transient states S_F to macro-states M. The matrix entries of $\mathbf{P}_{FM} = [p_{iI}(0,\infty)]$, $i \in S_F, I \in M$ are the conditional transition probabilities that the stochastic process, once initiated in state i at time $t = 0$ will ultimately exit S_F and hit macro-state $I \in M$ as time $t \to \infty$. These probabilities are used to adjust the transition rates among macro-states during elimination of fast transient states. The interpretation is that the fast transient states form a probabilistic switch in the limit, and the time spent in these sets of states can be neglected in the long run. Furthermore, related to this assumption, the exit stationary probability will be reached in a period of time much smaller than the time scale characterizing the dynamics among the macro-states. The error induced by this approximation is inversely proportional to the stiffness ratio.

For the sake of completeness, it is worth mentioning that the entries in matrix $\tilde{\boldsymbol{\Sigma}}_{MF}$ from Eq. (5.55) denote exit rates, leading from macro-states in M via the probabilistic switch, represented by matrix $\mathbf{P}_{FM}$ Eq. (5.57), back to macro-states. Hence, the rate matrix $\tilde{\boldsymbol{\Sigma}}_{MF}\mathbf{P}_{FM}$ is simply added to $\tilde{\boldsymbol{\Sigma}}_{MM}$ in Eq. (5.55) to obtain the rate adjustment.

5.2.4   Aggregation of Initial State Probabilities

Since the transient analysis is performed only on the set of macro-states M, the initial probability vector $\boldsymbol{\pi}(0)$ must be adjusted accordingly. This adjustment is achieved in two steps. First, the probability mass $\boldsymbol{\pi}_I(0)$ assigned to micro-states $i \in S_I$, which are condensed into a single macro-state I, is accumulated for all I, $1 \leq I \leq F - 1$:

$$\tilde{\sigma}_I(0) = \boldsymbol{\pi}_I(0)\mathbf{1}. \tag{5.58}$$

Second, if $S_F \neq \emptyset$, then the probability mass $\boldsymbol{\pi}_F(0)$ assigned to the fast transient states S_F is spread probabilistically across the macro-states M according to rule implied by the probabilistic switch $\mathbf{P}_{FM}$ given in Eq. (5.57):

$$\tilde{\boldsymbol{\sigma}}_M^{\approx}(0) = \tilde{\boldsymbol{\sigma}}(0) + \boldsymbol{\pi}_F(0)\mathbf{P}_{FM}. \tag{5.59}$$

Depending on S_F, the initial probability vector $\boldsymbol{\sigma}(0)$ is given by $\tilde{\boldsymbol{\sigma}}_M^{\approx}(0)$ according to Eq. (5.58) or by $\tilde{\boldsymbol{\sigma}}(0)$ according to Eq. (5.59).

The aggregation part has now completely been described. With this technique, the transient analysis of large stiff Markov chains can be reduced to the analysis of smaller non-stiff chains. Transient analysis is accomplished only for the macro-states in the model, be it either the initially present slow states or the macro-states resulting from an aggregation of fast recurrent subsets of states. The accuracy of the method is inversely proportional to the stiffness ratio of the initial model.

With the assumption that steady state is reached much quicker for the fast states than for the slow states, further approximations become possible. Applying certain disaggregation steps, approximations of the transient fast state probabilities can also be derived, thus resulting in a complete state probability vector.

5.2.5 Disaggregations

5.2.5.1 Fast Transient States With the approximate macro-state probability vector $\tilde{\boldsymbol{\sigma}}_M^{\approx}(t)$, obtained as the solution of Eq. (5.56), the approximate fast transient state probability vector $\tilde{\boldsymbol{\sigma}}_F^{\approx}(t)$ can be derived easily with a transient interpretation of Eq. (5.53):

$$\tilde{\boldsymbol{\sigma}}_F^{\approx}(t) = -\tilde{\boldsymbol{\sigma}}_M^{\approx}(t)\tilde{\boldsymbol{\Sigma}}_{MF}\tilde{\boldsymbol{\Sigma}}_{FF}^{-1} . \tag{5.60}$$

With:

$$c = \tilde{\boldsymbol{\sigma}}_M^{\approx}(t)\mathbf{1} + \tilde{\boldsymbol{\sigma}}_F^{\approx}(t)\mathbf{1} , \tag{5.61}$$

normalization can be taken into account for computation of a probability vector so that the first disaggregation step is accomplished, leading to the intermediate transient state probability vector $\tilde{\boldsymbol{\sigma}}^{\approx(c)}(t)$ as an approximate solution of Eq. (5.50):

$$\tilde{\boldsymbol{\sigma}}(t) \approx \tilde{\boldsymbol{\sigma}}^{\approx(c)}(t) = \frac{1}{c}\left(\tilde{\boldsymbol{\sigma}}_M^{\approx}(t), \tilde{\boldsymbol{\sigma}}_F^{\approx}(t)\right) . \tag{5.62}$$

5.2.5.2 Fast Recurrent States In a second disaggregation step, the transient micro-state probability vector $\boldsymbol{\pi}^{\approx}(t)$ is also calculated as an approximation of $\boldsymbol{\pi}(t)$, which is formally defined as the solution of Eq. (5.36), incorporating the original infinitesimal generator matrix $\mathbf{Q}$ from (5.35). For a computation of $\boldsymbol{\pi}^{\approx}(t) = (\pi_I^{\approx}(t))$, we need to refer to the conditional micro-state probability vectors $\boldsymbol{\pi}_I^* = (\pi_{I_0}^*, \ldots, \pi_{I_{n_I-1}}^*), 1 \leq I \leq F - 1$, from Eq. (5.39). The transient macro-state probabilities $\sigma_I(t), 1 \leq I \leq F - 1$, are used to uncondition the conditional steady-state probability vectors $\boldsymbol{\pi}_I^*, 1 \leq I \leq F - 1$:

$$\boldsymbol{\pi}_I(t) \approx \boldsymbol{\pi}_I^{\approx}(t) = (\sigma_I(t)\boldsymbol{\pi}_I^*) . \tag{5.63}$$

If $S_F \neq \emptyset$, then unconditioning has to be performed on the basis of $\tilde{\sigma}(t) \approx \tilde{\sigma}^{\approx(c)}(t)$ according to Eq. (5.62):

$$\pi_I(t) \approx \pi_I^{\approx}(t) = \left(\tilde{\sigma}_I^{\approx(c)}(t) \pi_I^* \right) . \tag{5.64}$$

Collecting all the intermediate results from Eq. (5.62) and Eq. (5.64) yields the final approximate transient probability vector in its most general form:

$$
\begin{aligned}
\pi(t) \approx \pi^{\approx}(t) = \Big(& \tilde{\sigma}_{0_0}^{\approx(c)}(t), \ldots, \tilde{\sigma}_{0_{n_0-1}}^{\approx(c)}(t), \\
& \tilde{\sigma}_1^{\approx(c)}(t) \pi_{1_0}^*, \ldots, \tilde{\sigma}_{F-1}^{\approx(c)}(t) \pi_{F-1_{n_{F-1}-1}}^*, \\
& \tilde{\sigma}_{F_0}^{\approx(c)}(t), \ldots, \tilde{\sigma}_{F_{n_F-1}}^{\approx(c)}(t) \Big) \\
= \Big(& \tilde{\sigma}_0^{\approx(c)}(t), \pi_1^{\approx}(t), \ldots, \pi_{F-1}^{\approx}(t), \tilde{\sigma}_F^{\approx(c)}(t) \Big) .
\end{aligned} \tag{5.65}
$$

If $S_F = \emptyset$ then Eq. (5.65) simplifies to the following:

$$
\begin{aligned}
\pi(t) \approx \pi^{\approx}(t) = \Big(& \tilde{\sigma}_{0_0}(t), \ldots, \tilde{\sigma}_{0_{n_0-1}}(t), \\
& \tilde{\sigma}_1(t) \pi_{1_0}^*, \ldots, \tilde{\sigma}_{F-1}(t) \pi_{F-1_{n_{F-1}-1}}^* \Big) \\
= \Big(& \tilde{\sigma}_0(t), \pi_1^{\approx}(t), \ldots, \pi_{F-1}^{\approx}(t) \Big) .
\end{aligned} \tag{5.66}
$$

5.2.6 The Algorithm

STEP 1 Initialization:

- Specify $\tau >> 1/T$, the transition rate threshold, as a function of the time horizon T for which the transient analysis is to be carried out.
- Partition the state space S into slow states S_0 and fast states $S - S_0$ with respect to τ.
- Partition the fast states $S - S_0$ further into fast recurrent subsets S_I, $1 \leq I \leq F - 1$ and possibly one subset of fast transient states S_F.
- Arrange the infinitesimal generator matrix $\mathbf{Q}$ according to Eq. (5.35).

STEP 2 Steady-state analysis of fast recurrent subsets:

- From submatrices $\mathbf{Q}_I$ (Eq. (5.35)), obtain infinitesimal generator matrices $\mathbf{Q}_I^*$ according to Eq. (5.38).
- Compute the conditional micro-state probability vectors π_I^* according to Eq. (5.39).

STEP 3 Construct the intermediate generator matrix $\tilde{\Sigma}$ as in Eq. (5.49):

- Use $\mathbf{Q}$ from Eq. (5.35) and π_I^*, $1 \leq I \leq F - 1$ from Eq. (5.39) and apply operations starting from Eq. (5.40) through Eq. (5.48).

STEP 4   Aggregation of fast transient states:

- IF $S_F \neq \emptyset$
  - THEN perform aggregation of fast transient states and construct $\tilde{\boldsymbol{\Sigma}}_{MM}^{\approx}$:
    * Compute the probabilistic switch $\mathbf{P}_{FM}$ according to Eq. (5.57) and $\tilde{\boldsymbol{\Sigma}}_{MM}^{\approx}$ according to Eq. (5.55) and Eq. (5.56) from $\tilde{\boldsymbol{\Sigma}}$.
    * $\boldsymbol{\Sigma} = \tilde{\boldsymbol{\Sigma}}_{MM}^{\approx}$.
  - ELSE $\boldsymbol{\Sigma} = \tilde{\boldsymbol{\Sigma}}$.

STEP 5   Aggregation of the initial state probability vector $\boldsymbol{\sigma}(0)$:

- Accumulate initial probabilities $\pi_{I_i}(0)$ of micro-states $i \in S_I$ into $\tilde{\sigma}_I(0)$, $1 \leq I \leq F - 1$ according to Eq. (5.58).

- IF $S_F \neq \emptyset$
  - THEN calculate approximate initial macro-state probability vector $\boldsymbol{\sigma}(0) = \tilde{\boldsymbol{\sigma}}_M^{\approx}(0)$ according to Eq. (5.59).

  - ELSE $\boldsymbol{\sigma}(0) = \tilde{\boldsymbol{\sigma}}(0)$ according to Eq. (5.58).

STEP 6   Computation of the transient macro-state probability vector $\boldsymbol{\sigma}(t)$:

- Solve Eq. (5.37).

STEP 7   Disaggregations:

- IF $S_F \neq \emptyset$
  - THEN
    * Compute intermediate transient state probability vector $\tilde{\boldsymbol{\sigma}}(t)$ according to Eq. (5.62).
    * Compute the approximate micro-state probability subvectors $\boldsymbol{\pi}_I^{\approx}(t)$ by unconditioning of $\boldsymbol{\pi}_I^*$ according to Eq. (5.64) for all $1 \leq I \leq F - 1$.
  - ELSE compute $\boldsymbol{\pi}_I^{\approx}(t)$ by unconditioning of $\boldsymbol{\pi}_I^*$ according to Eq. (5.63) for all $1 \leq I \leq F - 1$.

STEP 8   Final result:

- Compose the approximate transient probability vector $\boldsymbol{\pi}(t) \approx \boldsymbol{\pi}^{\approx}(t)$ according to Eq. (5.65) if $S_F \neq \emptyset$ or Eq. (5.66) if $S_F = \emptyset$.

5.2.7   An Example: Server Breakdown and Repair

Assume a queueing system with a single server station, where a maximum of m customers can wait for service. Customers arrive at the station with arrival rate λ. Service time is exponentially distributed with mean $1/\mu$. Furthermore, the single server is subject to failure and repair, both times to failure and repair times being exponentially distributed with parameters γ and δ, respectively. It is reasonable to assume that λ and μ differ by orders of magnitude relative to γ and δ. Usually, repair of a failed unit takes much longer than traffic-related events in computer systems. This condition is even more relevant for failure events that are relatively infrequent. While traffic-related events typically take place in the order of microseconds, repair durations are in the order of minutes, hours, or days, and failure events in the order of months, years, or multiple thereof. Thus, the transition rates λ and μ can be classified as being fast, and γ and δ as slow. Note that we are interested in computing performance measures related to traffic events so that for the rate threshold, the relation $\tau < \delta < \gamma$ generally holds and the application of the aggregation/disaggregation approach is admissible.

The described scenario is captured by a CTMC with a state space $S = \{(l, k), 0 \leq l \leq m, k \in \{0, 1\}\}$, where l denotes the number of customers in the queue and k the number of nonfailed servers, as depicted in Fig. 5.4a. The state space S is partitioned according to the classification scheme applied to the rates, into a set of slow states S_0, a set of fast recurrent states S_1, and a set of fast transient states $S_F = S_2$:

$$
\begin{aligned}
S_0 &= \{(m, 0)\}\,, \\
S_1 &= \{(0, 1), (1, 1), \cdots, (m, 1)\}\,, \\
S_2 &= \{(0, 0), (1, 0), \cdots, (m - 1, 0)\}\,.
\end{aligned}
$$

For the case of $m = 2$, the following state ordering is used in compliance with the partitioning:

$$
\begin{array}{cccccc}
0 & 1_0 & 1_1 & 1_2 & 2_0 & 2_1 \\
(2,0) & (0,1) & (1,1) & (2,1) & (0,0) & (1,0)
\end{array}\;.
$$

To complete initialization, the infinitesimal generator matrix $\mathbf{Q}$, $m = 2$, is restructured accordingly to reflect the partitioning:

$$
\mathbf{Q} =
\begin{pmatrix}
-\delta & 0 & 0 & \delta & 0 & 0 \\
0 & -(\lambda + \gamma) & \lambda & 0 & \gamma & 0 \\
0 & \mu & -(\lambda + \gamma + \mu) & \lambda & 0 & \gamma \\
\gamma & 0 & \mu & -(\mu + \gamma) & 0 & 0 \\
0 & \delta & 0 & 0 & -(\lambda + \delta) & \lambda \\
\lambda & 0 & \delta & 0 & 0 & -(\lambda + \delta)
\end{pmatrix}\,.
$$

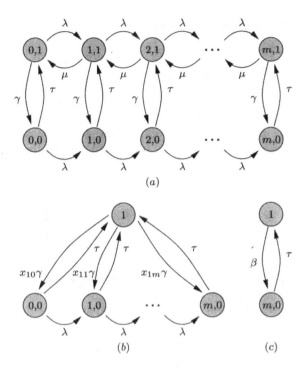

Fig. 5.4 (a) CTMC model of a finite capacity single server queueing system subject to server breakdown and repair. (b) The model of Fig. 5.4a after the aggregation of the fast recurrent subset into macro-state 1. (c) Final aggregated macro-state chain for the model of Fig. 5.4a after elimination of the fast transient subset.

Next, the steady-state analysis of the fast recurrent subset S_1 has to be performed, and the infinitesimal generator matrix $\mathbf{Q}_1^*$ must be created:

$$\mathbf{Q}_1^* = \mathbf{Q}_1 + \mathbf{D}_1$$

$$= \begin{pmatrix} -(\lambda+\gamma) & \lambda & 0 \\ \mu & -(\lambda+\gamma+\mu) & \lambda \\ 0 & \mu & -(\mu+\gamma) \end{pmatrix} + \begin{pmatrix} \gamma & 0 & 0 \\ 0 & \gamma & 0 \\ 0 & 0 & \gamma \end{pmatrix}$$

$$= \begin{pmatrix} -\lambda & \lambda & 0 \\ \mu & -(\lambda+\mu) & \lambda \\ 0 & \mu & -\mu \end{pmatrix}.$$

Closed-form solutions of the steady-state probabilities are well-known from literature (see, e.g., Gross and Harris [GrHa85]) for the case of the finite birth–death process as described by $\mathbf{Q}_1^*$:

$$\pi_{1_i}^* = \left(\frac{\lambda}{\mu}\right)^i \frac{1 - \frac{\lambda}{\mu}}{1 - \left(\frac{\lambda}{\mu}\right)^3}, \quad i = 0, 1, \ldots, 2.$$

To aggregate the set of fast recurrent states S_1 into macro-state 1, $\boldsymbol{\pi}_1^*$ is used to derive the intermediate infinitesimal generator matrix $\tilde{\boldsymbol{\Sigma}}$ for the reduced state space $\{(2,0), 1, (0,0), (1,0)\}$. By applying Eq. (5.40) through Eq. (5.48), we obtain:

$$\tilde{\boldsymbol{\Sigma}} = \begin{pmatrix} -\delta & \delta & 0 & 0 \\ \pi_{1_2}^*\gamma & -\gamma & \pi_{1_0}^*\gamma & \pi_{1_1}^*\gamma \\ 0 & \delta & -(\lambda+\delta) & \lambda \\ \lambda & \delta & 0 & -(\lambda+\delta) \end{pmatrix}.$$

Matrix $\tilde{\boldsymbol{\Sigma}}$ represents the CTMC as depicted in Fig. 5.4b for the case of $m = 2$. For an aggregation of the fast transient states, first $\tilde{\boldsymbol{\Sigma}}_{FF}^{-1}$ and the probabilistic switch $\mathbf{P}_{FM}$ are calculated from $\tilde{\boldsymbol{\Sigma}}$ according to Eq. (5.57):

$$\tilde{\boldsymbol{\Sigma}}_{FF}^{-1} = \begin{pmatrix} -\dfrac{1}{\lambda+\delta} & -\dfrac{\lambda}{(\lambda+\delta)^2} \\ 0 & -\dfrac{1}{\lambda+\delta} \end{pmatrix}, \quad \mathbf{P}_{FM} = \begin{pmatrix} \dfrac{\lambda\delta}{(\lambda+\delta)^2} & \dfrac{\delta(\delta+2\lambda)}{(\lambda+\delta)^2} \\ \dfrac{\lambda}{\lambda+\delta} & \dfrac{\delta}{\lambda+\delta} \end{pmatrix}.$$

With $\tilde{\boldsymbol{\Sigma}}$ and $\mathbf{P}_{FM}$ given, $\boldsymbol{\Sigma} = \tilde{\boldsymbol{\Sigma}}_{MM}^{\approx}$ can be easily derived using some algebraic manipulation according to Eq. (5.55) and Eq. (5.56):

$$\boldsymbol{\Sigma} = \begin{pmatrix} -\delta & \delta \\ \gamma \sum\limits_{i=0}^{2} \pi_{1_i}^* \left(\dfrac{\lambda}{\lambda+\delta}\right)^{2-i} & -\gamma \sum\limits_{i=0}^{2} \pi_{1_i}^* \left(\dfrac{\lambda}{\lambda+\delta}\right)^{2-i} \end{pmatrix},$$

where $\lambda/(\lambda+\delta)$ denotes the probability that a new customer arrives while repair is being performed, and $\gamma\left(\lambda/(\lambda+\delta)\right)^{2-i}$ is the probability that a failure occurs when there are i customers present and that $m - i$ customers arrive during repair.

An initial probability vector $\boldsymbol{\pi}(0) = (0, 1, 0, \ldots, 0)$ is assumed, i.e., $\pi_{1_0} = 1$. This condition implies $\boldsymbol{\sigma}(0) = (0, 1)$ in Eq. (5.58) and Eq. (5.59). From Eqs. (5.16) and (5.17) we know symbolic expressions for the transient state probabilities of a two-state CTMC, given an initial pmf $\boldsymbol{\pi}(0)$. We can therefore apply this result by substituting the rates from $\boldsymbol{\Sigma}$ into these expressions. By letting

$$\alpha = \gamma \sum_{i=0}^{2} \pi_{1_i}^* \left(\frac{\lambda}{\lambda+\delta}\right)^{2-i},$$

we get the following macro-state probabilities:

$$\sigma_0(t) = 1 - \sigma_1(t) \,,$$

$$\sigma_1(t) = \frac{\delta}{\delta + \alpha} + \frac{\alpha}{\delta + \alpha} \left(e^{-(\delta + \alpha)t} \right) \,.$$

Finally, disaggregations must be performed. Given $\tilde{\boldsymbol{\sigma}}_M^{\approx}(t) = \boldsymbol{\sigma}(t)$, Eq. (5.60) needs to be evaluated for a computation of the fast transient probability vector:

$$\tilde{\boldsymbol{\sigma}}_F^{\approx}(t) = -\tilde{\boldsymbol{\sigma}}_M^{\approx}(t) \tilde{\boldsymbol{\Sigma}}_{MF} \tilde{\boldsymbol{\Sigma}}_{FF}^{-1}$$

$$= -\left(1 - \sigma_1(t), \frac{\delta}{\delta + \alpha} + \frac{\alpha \cdot e^{-(\delta + \alpha)t}}{\delta + \alpha} \right) \cdot \begin{pmatrix} 0 & 0 \\ -\frac{\pi_{1_0}^* \gamma}{\lambda + \delta} & -\gamma \left(\frac{\pi_{1_0}^* \lambda}{(\lambda + \delta)^2} + \frac{\pi_{1_1}^*}{\lambda + \delta} \right) \end{pmatrix}$$

$$= \left(\sigma_1(t) \frac{\pi_{1_0}^* \gamma}{\lambda + \delta}, \sigma_1(t) \frac{\gamma}{\lambda + \delta} \left(\frac{\pi_{1_0}^* \lambda}{\lambda + \delta} + \pi_{1_1}^* \right) \right) \,.$$

Next, normalization is accomplished with help of c according to Eq. (5.61):

$$c = 1 + \sigma_1(t) \frac{\gamma}{\lambda + \delta} \left(\pi_{1_0}^* \left(1 + \frac{\lambda}{\lambda + \delta} \right) + \pi_{1_1}^* \right) \,.$$

The intermediate transient state probability vector $\tilde{\boldsymbol{\sigma}}(t)$ from Eq. (5.62) follows immediately:

$$\tilde{\boldsymbol{\sigma}}(t) = \frac{1}{c} \left(1 - \sigma_1(t), \sigma_1(t), \sigma_1(t) \frac{\pi_{1_0}^* \gamma}{\lambda + \delta}, \sigma_1(t) \frac{\gamma}{\lambda + \delta} \left(\frac{\pi_{1_0}^* \lambda}{\lambda + \delta} + \pi_{1_1}^* \right) \right) \,.$$

The final step consists of unconditioning the micro-state probability vector $\boldsymbol{\pi}_1^*$:

$$\boldsymbol{\pi}_1^{\approx}(t) = \frac{\sigma_1(t)}{c} \left(\pi_{1_0}^*, \pi_{1_1}^*, \pi_{1_2}^* \right) \,.$$

Collecting all the results provides an approximation $\boldsymbol{\pi}^{\approx}(t)$ to the transient probability vector, which is represented as the transpose:

$$\boldsymbol{\pi}(t) \approx \boldsymbol{\pi}^{\approx}(t) = \begin{pmatrix} \frac{1}{c}(1 - \sigma_1(t)) \\[2mm] \frac{\sigma_1(t)}{c} \pi_{1_0}^* \\[2mm] \frac{\sigma_1(t)}{c} \pi_{1_1}^* \\[2mm] \frac{\sigma_1(t)}{c} \pi_{1_2}^* \\[2mm] \frac{1}{c} \sigma_1(t) \frac{\pi_{1_0}^* \gamma}{\lambda + \delta} \\[2mm] \frac{1}{c} \sigma_1(t) \frac{\gamma}{\lambda + \delta} \left(\frac{\pi_{1_0}^* \lambda}{\lambda + \delta} + \pi_{1_1}^* \right) \end{pmatrix}^T$$

From the transient state probabilities, all the performance measures of interest can be calculated.

Table 5.1   Parameter set for Table 5.2

δ	γ	λ	μ
0.01	0.0001	0.5	1

The exact transient state probability vector $\pi(t)$ is compared to the approximate state probability vector $\pi^{\approx}(t)$ at time instants $t = 10, 100$ and 1000 in Table 5.2, with the parameter set from Table 5.1. (Note that the initial probability vectors are chosen as has been specified earlier.) The results of the approximate method are generally very close to the exact values for most states. Usually, the accuracy depends on the length of the investigated time horizon for a transient analysis. Short-term results are particularly dependent on the initial probability vector so that some more significant differences can arise as in the case of state $(2,0)$ in our example, where the absolute state probabilities are rather small, and hence the percentage difference is high. Furthermore, whether the error is positive or negative may change over time.

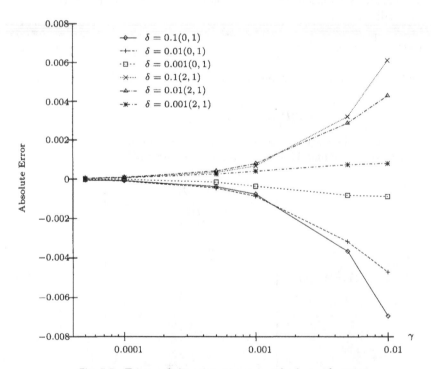

Fig. 5.5   Errors of the approximate method as a function γ.

Table 5.2   State probabilities as a function of time: exact vs. approximate

No.	State	Exact $\pi(t)$			
		$t = 0$	$t = 10$	$t = 100$	$t = 1000$
0	(2,0)	0	0.00067567	0.00601869	0.00962544
1_0	(0,1)	1	0.57099292	0.56777268	0.56567586
1_1	(1,1)	0	0.28540995	0.28392083	0.28289339
1_2	(2,1)	0	0.14264597	0.14201127	0.14152880
2_0	(0,0)	0	0.00011195	0.00011134	0.00011092
2_1	(1,0)	0	0.00016354	0.00016484	0.00016421

No.	State	Approx $\pi^{\approx}(t)$			
		$t = 0$	$t = 10$	$t = 100$	$t = 1000$
0	(2,0)	0	0.00092448	0.00611894	0.00962543
1_0	(0,1)	0.571429	0.57074200	0.56777400	0.56577100
1_1	(1,1)	0.285714	0.28537100	0.28388700	0.28288600
1_2	(2,1)	0.142857	0.14268500	0.14194400	0.14144300
2_0	(0,0)	0	0.00011191	0.00011133	0.00011094
2_1	(1,0)	0	0.00016567	0.00016481	0.00016423

No.	State	Absolute error			Error %		
		$t = 10$	$t = 100$	$t = 1000$	$t = 10$	$t = 100$	$t = 1000$
0	(2,0)	-2.488e-04	-1.003e-04	1e-08	-36.8242	-1.66564	0.00010
1_0	(0,1)	2.509e-04	-1.32e-06	-9.514e-05	0.04394	-0.00023	-0.01682
1_1	(1,1)	3.895e-05	3.383e-05	7.39e-06	0.01365	0.01192	0.00261
1_2	(2,1)	-3.903e-05	6.727e-05	8.58e-05	-0.02736	0.04737	0.06062
2_0	(0,0)	4e-08	1e-08	-2e-08	0.03573	0.00898	-0.01803
2_1	(1,0)	-2.13e-06	3e-08	-2e-08	-1.30243	0.01820	-0.01218

The impact of different degrees of stiffness on the accuracy of the results are indicated in Table 5.3 and in Fig. 5.5. In particular, the empirical results depicted in Fig. 5.5 support the assumption of the approximation being better for stiffer models. Almost no error is observed if $\gamma \leq 10^{-4}$, while the error substantially increases for the less stiff models as γ gets larger. The results depend on the three different parameter sets given in Table 5.4. A closer look at Fig. 5.5 reveals a nonmonotonic ordering of the degree of accuracy. Let $\gamma = 0.01$, then the worst result is obtained for the largest δ, that is, $\delta = 0.1$ in state $(2, 1)$, as anticipated. But this behavior does not occur for $\gamma < 0.001$, where the approximate probability is closer to the exact value for $\delta = 0.1$ than for $\delta = 0.01$. Similar patterns can be observed in other cases, as well.

Problem 5.6   Verify the construction of the intermediate infinitesimal generator matrix $\tilde{\Sigma}$ in the example study by applying Eq. (5.40) through Eq. (5.48) for each submatrix of $\mathbf{Q}$ in the indicated way.

Table 5.3   State probabilities as a function of stiffness: exact vs. approximate

		Exact $\pi(1000)$		
No.	State	(1)	(2)	(3)
0	(2,0)	0.06038036	0.08838416	0.07090560
1_0	(0,1)	0.53671266	0.51861447	0.51254489
1_1	(1,1)	0.26839084	0.25981567	0.26054365
1_2	(2,1)	0.13424710	0.13066102	0.13600236
2_0	(0,0)	0.00010714	0.00101689	0.00854241
2_1	(1,0)	0.00016050	0.00150639	0.01146107

		Approx $\pi^{\approx}(1000)$		
No.	State	(1)	(2)	(3)
0	(2,0)	0.06046620	0.08838110	0.07070710
1_0	(0,1)	0.53672400	0.51948100	0.51948100
1_1	(1,1)	0.26836200	0.25974100	0.25974000
1_2	(2,1)	0.13418100	0.12987000	0.12987000
2_0	(0,0)	0.00010713	0.00101859	0.00865801
2_1	(1,0)	0.00016048	0.00150791	0.01154400

		Absolute Error			Error %		
No.	State	(1)	(2)	(3)	(1)	(2)	(3)
0	(2,0)	-8.584e-05	3.06e-06	1.985e-04	-0.14217	0.00346	0.27995
1_0	(0,1)	-1.134e-05	-8.665e-04	-6.936e-03	-0.00211	-0.16709	-1.35327
1_1	(1,1)	2.884e-05	7.467e-05	8.037e-04	0.01075	0.02874	0.30845
1_2	(2,1)	6.61e-05	7.91e-04	6.132e-03	0.04924	0.60540	4.50901
2_0	(0,0)	1e-08	-1.7e-06	-1.156e-04	0.00933	-0.16718	-1.35325
2_1	(1,0)	2e-08	-1.52e-06	-8.293e-05	0.01246	-0.10090	-0.72358

Table 5.4   Parameter set for Table 5.3

Experiment	δ	γ	λ	μ
(1)	0.001	0.0001		
(2)	0.01	0.001	0.5	1
(3)	0.1	0.01		

Problem 5.7   Verify the correctness of the probabilistic switch $\mathbf{P}_{FM}$ in the example study by relating it to Eq. (5.57).

Problem 5.8   Modify the example in Fig. 5.4a in such a way that no repair is possible, that is, let the repair rate be $\delta = 0$. Let $X^{(s)}(t)$ denote the system's service rate at time t, and let $Y^{(s)}(t) = \int_0^\tau X^{(s)}(\tau)\,d\tau$ be the number of customers serviced in $[0,t)$. Derive all the intermediate probabilities, the approximate transient state probability vector $\pi^{\approx}(t)$, the expected instantaneous reward rate $E[X^{(s)}(t)]$, the cumulative expected reward $E[Y^{(s)}(t)]$ in finite time, and the limit $E[Y^{(s)}(\infty)]$ for the modified model. Perform the analyses of the resulting model with the same parameters that have been used in the example presented this section.

Problem 5.9   Consider the steady-state case as $t \to \infty$ for the example in Fig. 5.4a. Derive the approximate steady-state probability vector $\pi^\approx$ and calculate the expected number of customers in the system in steady state. Then, calculate the probability that the incoming customers find the system full, in which case they must be rejected. Use this probability to calculate the *effective* arrival rate λ_e, defined as λ minus the rejected portion of arrivals. Use Little's law to calculate the average response time.

Problem 5.10   Apply the aggregation technique for the transient analysis of stiff Markov chains of this section to the example in Section 4.1 and derive the approximate transient probability vector. Assume $\pi(0) = \left(\frac{1}{6}, \frac{1}{6}, \frac{1}{6}, \frac{1}{6}, \frac{1}{6}, \frac{1}{6}\right)$, evaluate the approximate transient state probability vector at time instances $t \in \{1, 2, 3, 5, 10\}$, and compare the results to those obtained by applying Courtois' method and to the exact results.

6

Single Station Queueing Systems

A single station queueing system, as shown in Fig. 6.1, consists of a queueing buffer of finite or infinite size and one or more identical servers. Such an elementary queueing system is also referred to as a service station or, simply, as a node. A server can only serve one customer at a time and hence, it is

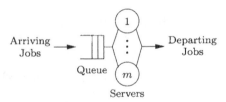

Arriving Jobs → **Queue** → **Servers** (1 ... m) → **Departing Jobs**

Fig. 6.1 Service station with m servers (a multiple server station).

either in a "busy" or an "idle" state. If all servers are busy upon the arrival of a customer, the newly arriving customer is buffered, assuming that buffer space is available, and waits for its turn. When the customer currently in service departs, one of the waiting customers is selected for service according to a *queueing (or scheduling) discipline*. An elementary queueing system is further described by an arrival process, which can be characterized by its sequence of interarrival time random variables $\{A_1, A_2, \cdots\}$. It is common to assume that the sequence of interarrival times is independent and identically distributed, leading to an arrival process that is known as a renewal process [Triv01]. Special classes of arrival processes are introduced in Section 6.8. The distribution function of interarrival times can be continuous or discrete.

We deal only with the former case in this book. For information related to discrete interarrival time distributions, the reader may consult [Dadu96].

The average interarrival time is denoted by $E[A] = \overline{T}_A$ and its reciprocal by the average arrival rate λ:

$$\lambda = \frac{1}{\overline{T}_A}. \tag{6.1}$$

The most common interarrival time distribution is the exponential, in which case the arrival process is Poisson. The sequence $\{B_1, B_2, \ldots\}$ of service times of successive jobs also needs to be specified. We assume that this sequence is also a set of independent random variables with a common distribution function. The mean service time $E[B]$ is denoted by $\overline{T}_B$ and its reciprocal by the service rate μ:

$$\mu = \frac{1}{\overline{T}_B}. \tag{6.2}$$

6.1   NOTATION

6.1.1   Kendall's Notation

The following notation, known as Kendall's notation, is widely used to describe elementary queueing systems:

$$A/B/m \text{ - queueing discipline,}$$

where A indicates the distribution of the interarrival times, B denotes the distribution of the service times, and m is the number of servers ($m \geq 1$). The following symbols are normally used for A and B:

M Exponential distribution (memoryless property)
E_k Erlang distribution with k phases
H_k Hyperexponential distribution with k phases
C_k Cox distribution with k phases
D Deterministic distribution, i.e., the interarrival time
        or service time is constant
G General distribution
GI General distribution with independent interarrival times

Due to proliferation of high-speed networks, there is considerable interest in traffic arrival processes where successive arrivals are correlated. Such non-GI arrival processes include the Markov modulated Poisson process (MMPP) (see Section 6.8.2), Markovian arrival process (see Section 6.8.3), or batch Markovian arrival process (BMAP) (see Section 6.8.4).

The queueing discipline or service strategy determines which job is selected from the queue for processing when a server becomes available. Some commonly used queueing disciplines are:

FCFS (First-Come-First-Served): If no queueing discipline is given in the Kendall notation, then the default is assumed to be the FCFS discipline. The jobs are served in the order of their arrival.

LCFS (Last-Come-First-Served): The job that arrived last is served next.

SIRO (Service-In-Random-Order): The job to be served next is selected at random.

RR (Round Robin): If the servicing of a job is not completed at the end of a time slice of specified length, the job is preempted and returns to the queue, which is served according to FCFS. This action is repeated until the job service is completed.

PS (Processor Sharing): This strategy corresponds to round robin with infinitesimally small time slices. It is as if all jobs are served simultaneously and the service time is increased correspondingly.

IS (Infinite Server): There is an ample number of servers so that no queue ever forms.

Static Priorities: The selection depends on priorities that are permanently assigned to the job. Within a class of jobs with the same priority, FCFS is used to select the next job to be processed.

Dynamic Priorities: The selection depends on dynamic priorities that alter with the passing of time.

Preemption: If priority or LCFS discipline is used, then the job currently being processed is interrupted and preempted if there is a job in the queue with a higher priority.

As an example of Kendall's notation, the expression

$$M/G/1\text{-LCFS preemptive resume (PR)}$$

describes an elementary queueing system with exponentially distributed interarrival times, arbitrarily distributed service times, and a single server. The queueing discipline is LCFS where a newly arriving job interrupts the job currently being processed and replaces it in the server. The servicing of the job that was interrupted is resumed only after all jobs that arrived after it have completed service.

Kendall's notation can be extended in various ways. An additional parameter is often introduced to represent the number of places in the queue (if the queue is finite) and we get the extended notation

$$A/B/m/K\text{-queueing discipline,}$$

where K is the capacity of the station (queue + server). This means that if the number of jobs at server and queue is K, a newly arriving job is lost.

6.1.2 Performance Measures

The different types of queueing systems are analyzed mathematically to determine performance measures from the description of the system. Because a queueing model represents a dynamic system, the values of the performance measures vary with time. Normally, however, we are content with the results in the steady state. The system is said to be in steady state when all transient behavior has ended, the system has settled down, and the values of the performance measures are independent of time. The system is then said to be in statistical equilibrium; i.e., the rate at which jobs enter the system is equal to the rate at which jobs leave the system. Such a system is also called a *stable system*. Transient solutions of simple queueing systems are available in closed form, but for more general cases, we need to resort to Markov chain techniques as described in Chapter 5. Recall that the generation and the solution of large Markov chains can be automated via stochastic reward nets [MuTr92].

The most important performance measures are:

Probability of the Number of Jobs in the System π_k: It is often possible to describe the behavior of a queueing system by means of the probability vector of the number of jobs in the system π_k. The mean values of most of the other interesting performance measures can be deduced from π_k:

$$\pi_k = P[\text{there are } k \text{ jobs in the system}].$$

Utilization ρ: If the queueing system consists of a single server, then the utilization ρ is the fraction of the time in which the server is busy, i.e., occupied. In case there is no limit on the number of jobs in the single server queue, the server utilization is given by

$$\rho = \frac{\text{mean service time}}{\text{mean interarrival time}} = \frac{\text{arrival rate}}{\text{service rate}} = \frac{\lambda}{\mu}. \tag{6.3}$$

The utilization of a service station with multiple servers is the mean fraction of active servers. Since $m\mu$ is the overall service rate, we have:

$$\rho = \frac{\lambda}{m\mu}, \tag{6.4}$$

and ρ can be used to formulate the condition for stationary behavior mentioned previously. The condition for stability is

$$\rho < 1; \tag{6.5}$$

i.e., on average the number of jobs that arrive in a unit of time must be less than the number of jobs that can be processed. All the results given in Chapters 6 – 10 apply only to stable systems.

Throughput λ: The throughput of an elementary queueing system is defined as the mean number of jobs whose processing is completed in a single unit of time, i.e., the departure rate. Since the departure rate is equal to the arrival rate λ for a queueing system in statistical equilibrium, the throughput is given by

$$\lambda = m \cdot \rho \cdot \mu \qquad (6.6)$$

in accordance with Eq. (6.4). We note that in the case of finite buffer queueing system, throughput can be different from the external arrival rate.

Response Time T: The response time, also known as the sojourn time, is the total time that a job spends in the queueing system.

Waiting Time W: The waiting time is the time that a job spends in a queue waiting to be serviced. Therefore, we have

Response time = waiting time + service time.

Since W and T are usually random numbers, their mean is calculated. Then

$$\overline{T} = \overline{W} + \frac{1}{\mu} . \qquad (6.7)$$

The distribution functions of the waiting time, $F_W(x)$, and the response time, $F_T(x)$, are also sometimes required.

Queue Length Q: The queue length, Q, is the number of jobs in the queue.

Number of Jobs in the System K: The number of jobs in the queueing system is represented by K. Then

$$\overline{K} = \sum_{k=1}^{\infty} k \cdot \pi_k . \qquad (6.8)$$

The mean number of jobs in the queueing system $\overline{K}$ and the mean queue length $\overline{Q}$ can be calculated using one of the most important theorems of queueing theory, *Little's theorem*:

$$\overline{K} = \lambda \overline{T}, \qquad (6.9)$$

$$\text{and} \quad \overline{Q} = \lambda \overline{W} . \qquad (6.10)$$

Little's theorem is valid for all queueing disciplines and arbitrary GI/G/m queue. The proof is given in [Litt61].

6.2 MARKOVIAN QUEUES

6.2.1 The M/M/1 Queue

Recall that in this case, the arrival process is Poisson, the service times are exponentially distributed, and there is a single server. The system can be modeled as a birth–death process with birth rate (arrival rate) λ and a constant death rate (service rate) μ. We assume that $\lambda < \mu$ so the underlying CTMC is ergodic and hence the queueing system is stable. Then using Eq. (3.12), we obtain the steady-state probability of the system being empty:

$$\pi_0 = \frac{1}{1 + \sum_{k=1}^{\infty} \prod_{i=0}^{k-1} \frac{\lambda}{\mu}} = \frac{1}{1 + \sum_{k=1}^{\infty} \left(\frac{\lambda}{\mu}\right)^k},$$

which can be simplified to

$$\pi_0 = \frac{1}{1 + \frac{\lambda/\mu}{1 - \lambda/\mu}} = 1 - \frac{\lambda}{\mu}.$$

From Eq. (3.11), for the steady-state probability that there are k jobs in the system we get:

$$\pi_k = \pi_0 \left(\frac{\lambda}{\mu}\right)^k, \quad k \geq 0,$$

$$\pi_k = \left(1 - \frac{\lambda}{\mu}\right) \cdot \left(\frac{\lambda}{\mu}\right)^k,$$

or with the utilization $\rho = \lambda/\mu$ we obtain:

$$\pi_0 = 1 - \rho \tag{6.11}$$

and

$$\pi_k = (1 - \rho)\rho^k, \tag{6.12}$$

the probability mass function (pmf) of the modified geometric random variable. In Fig. 6.2, we plot this pmf for $\rho = 1/2$. The mean number of jobs is obtained using Eqs. (6.12) and (6.8):

$$\overline{K} = \frac{\rho}{1 - \rho}. \tag{6.13}$$

In Fig. 6.3, the mean number of jobs is plotted as a function of the utilization ρ. This is the typical behavior of all queueing systems.

From Eq. (1.10) we obtain the variance of the number of jobs in the system:

$$\sigma_K^2 = \frac{\rho}{(1 - \rho)^2} \tag{6.14}$$

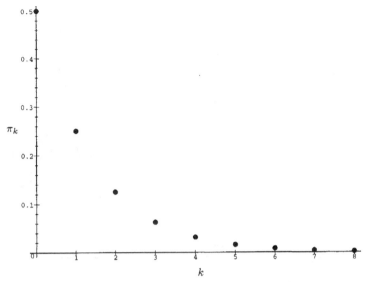

Fig. 6.2   The solution for π_k in an M/M/1 queue.

and the coefficient of variation:

$$c_K = \frac{\sigma_K}{K} = \frac{1}{\sqrt{\rho}}.$$

With Little's theorem (Eqs. (6.9) and (6.10)) we get the following for the mean response time:

$$\overline{T} = \frac{1/\mu}{1-\rho},$$ (6.15)

with Eq. (6.7) for the mean waiting time:

$$\overline{W} = \frac{\rho/\mu}{1-\rho},$$ (6.16)

and with Little's theorem again for the mean queue length:

$$\overline{Q} = \frac{\rho^2}{1-\rho}.$$ (6.17)

The same formulae are valid for M/G/1-PS and M/G/1-LCFS preemptive resume (see [Lave83]). For M/M/1-FCFS we can get a relation for the response time distribution if we consider the response time as the sum of $k+1$ independent exponentially distributed random variables [Triv01]:

$$X + X_1 + X_2 + \cdots + X_k,$$

where X is the service time of the tagged job, X_1 is the remaining service time of the job in service when the tagged job arrives, and $X_2, \ldots, X_k$ are the

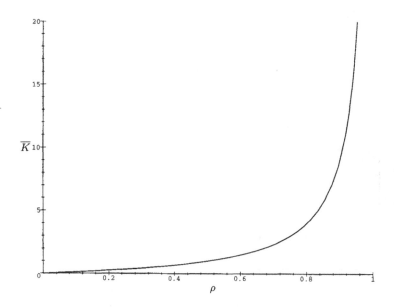

Fig. 6.3 The mean number of jobs $\overline{K}$ in an M/M/1 queue.

service times of the jobs in the queue; each of these is exponentially distributed with parameter μ. Note that X_1 is also exponentially distributed with rate μ due to the memoryless property of the exponential distribution. Noting that the Laplace–Stieltjes transform (LST) of the exponentially distributed service time (see Table 1.5) is

$$L_X(s) = \frac{\mu}{\mu + s},$$

we get for the conditional LST of the response time:

$$L_{T|K}(s|k) = \left(\frac{\mu}{\mu + s}\right)^{k+1}.$$

Unconditioning using the steady-state probability Eq. (6.12), the LST of the response time is

$$L_T(s) = \sum_{k=0}^{\infty} \left(\frac{\mu}{\mu + s}\right)^{k+1} \cdot (1 - \rho)\rho^k$$

$$L_T(s) = \frac{\mu(1 - \rho)}{s + \mu(1 - \rho)}.$$

$$(6.18)$$

Thus, the response time T is exponentially distributed with the parameter $\mu(1 - \rho)$:

$$F_T(x) = 1 - e^{-\mu(1-\rho)x}$$

$$(6.19)$$

and the variance:

$$\text{var}(T) = \frac{1}{\mu^2(1-\rho)^2}. \tag{6.20}$$

Similarly we get the distribution of the waiting time:

$$F_W(x) = \begin{cases} 1 - \rho, & x = 0, \\[2mm] 1 - \rho \cdot e^{-\mu(1-\rho)x}, & x > 0. \end{cases} \tag{6.21}$$

Thus, $F_W(0) = P(W = 0) = 1 - \rho$ is the mass at origin, corresponding to the probability that an arriving customer does not have to wait in the queue.

6.2.2 The M/M/∞ Queue

In an M/M/∞ queueing system we have a Poisson arrival process with arrival rate λ and an infinite number of servers with service rate μ each. If there are k jobs in the system, then the overall service rate is $k\mu$ because each arriving job immediately gets a server and does not have to wait. Once again, the underlying CTMC is a birth–death process. From Eq. (3.11) we obtain the steady-state probability of k jobs in the system:

$$\pi_k = \pi_0 \prod_{i=0}^{k-1} \frac{\lambda}{(i+1)\mu} = \pi_0 \left(\frac{\lambda}{\mu}\right)^k \frac{1}{k!};$$

with Eq. (3.12), we obtain the steady-state probability of no jobs in the system:

$$\pi_0 = \frac{1}{1 + \sum_{k=1}^{\infty} \left(\frac{\lambda}{\mu}\right)^k \cdot \frac{1}{k!}} = e^{-\frac{\lambda}{\mu}}, \tag{6.22}$$

and finally:

$$\pi_k = \frac{\left(\frac{\lambda}{\mu}\right)^k}{k!} \cdot e^{-\frac{\lambda}{\mu}}. \tag{6.23}$$

This is the Poisson pmf, and the expected number of jobs in the system is

$$\overline{K} = \frac{\lambda}{\mu}. \tag{6.24}$$

With Little's theorem the mean response time as expected is:

$$\overline{T} = \frac{1}{\mu}. \tag{6.25}$$

6.2.3   The M/M/m Queue

An M/M/m queueing system with arrival rate λ and service rate μ for each server can also be modeled as a birth–death process with

$$\lambda_k = \lambda, \quad k \geq 0,$$

$$\mu_k = \begin{cases} k\mu, & 0 \leq k \leq m, \\ m\mu, & m \leq k. \end{cases}$$

The condition for the queueing system to be stable (underlying CTMC to be ergodic) is $\lambda < m\mu$. The steady-state probabilities are given by (from Eq. (3.11))

$$\pi_k = \begin{cases} \pi_0 \displaystyle\prod_{i=0}^{k-1} \frac{\lambda}{(i+1)\mu} = \pi_0 \left(\frac{\lambda}{\mu}\right)^k \cdot \frac{1}{k!}, & 0 \leq k \leq m, \\ \pi_0 \displaystyle\prod_{i=0}^{m-1} \frac{\lambda}{(i+1)\mu} \cdot \prod_{i=m}^{k-1} \frac{\lambda}{m\mu}, & k \geq m. \end{cases}$$

With an individual server utilization, $\rho = \lambda/(m\mu)$, we obtain

$$\pi_k = \begin{cases} \pi_0 \dfrac{(m\rho)^k}{k!}, & 0 \leq k \leq m, \\ \pi_0 \dfrac{\rho^k m^m}{m!}, & k \geq m, \end{cases} \tag{6.26}$$

and from Eq. (3.12) we obtain:

$$\pi_0 = \left[\sum_{k=0}^{m-1} \frac{(m\rho)^k}{k!} + \frac{(m\rho)^m}{m!} \frac{1}{1-\rho} \right]^{-1}. \tag{6.27}$$

The steady-state probability that an arriving customer has to wait in the queue is given by

$$P_m = P(K \geq m) = \sum_{k=m}^{\infty} \pi_k$$

$$= \frac{(m\rho)^m}{m!(1-\rho)} \cdot \pi_0 \tag{6.28}$$

Using Eqs. (6.26) and (6.8), for the mean number of jobs in the system we obtain:

$$\overline{K} = m\rho + \frac{\rho}{1-\rho} \cdot P_m, \tag{6.29}$$

and for the mean queue length we obtain:

$$\overline{Q} = \frac{\rho}{1 - \rho} \cdot P_m.\qquad(6.30)$$

From this the mean response time $\overline{T}$ and mean waiting time $\overline{W}$ by Little's theorem (Eqs. (6.9) and (6.10)) can be easily derived. A formula for the distribution of the waiting time is given in [GrHa85]:

$$F_W(x) = \begin{cases} 1 - P_m, & x = 0, \\ 1 - P_m \cdot e^{-m\mu(1-\rho)x}, & x > 0. \end{cases}\qquad(6.31)$$

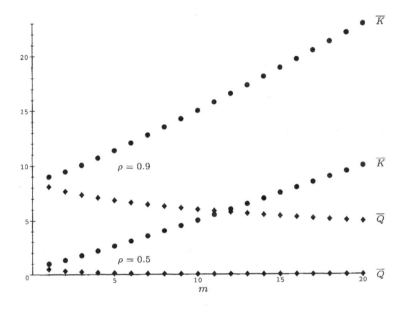

Fig. 6.4 Mean queue length $\overline{Q}$ and mean number of jobs in the system $\overline{K}$ as functions of the number of servers m.

Figure 6.4 shows the interesting and important property of an M/M/m queue that the mean number of jobs in the system $\overline{K}$ increases with the number of servers m if the server utilization is constant but the mean queue length $\overline{Q}$ decreases.

6.2.4   The M/M/1/K Finite Capacity Queue

In an M/M/1/K queueing system, the maximum number of jobs in the system is K, which implies a maximum queue length of $K - 1$. An arriving job enters the queue if it finds fewer than K jobs in the system and is lost otherwise.

This behavior can be modeled by a birth–death process with

$$\lambda_k = \begin{cases} \lambda, & 0 \le k < K, \\ 0, & k \ge K, \end{cases}$$

$$\mu_k = \mu \quad k = 1, \ldots, K.$$

Using Eqs. (3.12) and (3.11), and $a = \lambda/\mu$, we obtain the steady-state probability of k jobs in the system:

$$\pi_k = \begin{cases} \dfrac{(1-a)a^k}{1 - a^{K+1}}, & 0 \le k \le K, \\ 0, & k > K. \end{cases} \tag{6.32}$$

Since there are no more than K customers in the system, the system is stable for all values of λ and μ. That is, we need not assume that $\lambda < \mu$ for the system to be stable. In other words, since the CTMC is irreducible and finite, it is ergodic. If $\lambda = \mu$, then $a = 1$ and

$$\pi_0 = \frac{1}{K+1} = \pi_k, \quad k = 1, 2, \ldots, K. \tag{6.33}$$

The mean number of jobs in the system is given by (using Eq. (6.8))

$$\overline{K} = \begin{cases} \dfrac{a}{1-a} - \dfrac{K+1}{1 - a^{K+1}} \cdot a^{K+1}, & a \ne 1, \\ \dfrac{K}{2}, & a = 1. \end{cases} \tag{6.34}$$

Note that the utilization $\rho = 1 - \pi_0 \ne \lambda/\mu$ in this case; it is for this reason that we labeled the quantity $a = \lambda/\mu$ in the preceding equations. The throughput in this case is not equal to λ, but is equal to $\lambda(1 - \pi_K)$. Similar results have been derived for an M/M/m/K system (see [Klei75], [Alle90], [GrHa85], or [Triv01]).

6.2.5   Machine Repairman Model

Another special case of the birth–death process is obtained when the birth rate λ_j is of the form $(M - j)\lambda$, $j = 0, 1, \ldots, M$, and the death rate, $\mu_j = \mu$. This structure can be useful in the modeling of interactive computer systems where a an individual terminal user issues a request at the rate λ whenever it is in the "thinking state." If j out of the total of M terminals are currently waiting for a response to a pending request, the effective request rate is then $(M - j)\lambda$. The request completion rate is denoted by μ. Similarly, when M machines share a repair facility, the failure rate of each machine is λ and the repair rate is μ (see Fig. 6.5). Since the underlying CTMC is irreducible and finite, it is ergodic implying that the queueing system is always stable. The

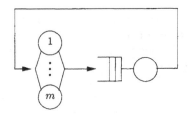

Fig. 6.5 Machine repairman model.

expressions for steady-state probabilities are obtained using Eqs. (3.12) and (3.11) as

$$\pi_k = \pi_0 \prod_{i=0}^{k-1} \frac{\lambda(M-i)}{\mu}, \quad 0 \le k \le M,$$

or

$$\pi_k = \pi_0 \left(\frac{\lambda}{\mu}\right)^k \frac{M!}{(M-k)!}. \tag{6.35}$$

Hence

$$\pi_0 = \frac{1}{\sum_{k=0}^{M} \left(\frac{\lambda}{\mu}\right)^k \frac{M!}{(M-k)!}}. \tag{6.36}$$

The utilization of the computer (or the repairman) is given by $\rho = (1 - \pi_0)$ (see Eq. (6.11)) and the throughput by $\mu(1 - \pi_0)$. With the average thinking time $1/\lambda$, for the mean response time of the computer we get:

$$\overline{T} = \frac{M}{\mu(1 - \pi_0)} - \frac{1}{\lambda} \tag{6.37}$$

and, using Little's theorem, the average number of jobs in the computer:

$$\overline{K} = M - \frac{\mu(1 - \pi_0)}{\lambda}. \tag{6.38}$$

6.2.6 Closed Tandem Network

Consider the closed tandem network in Fig. 6.6 with a total number of K

Fig. 6.6 Closed tandem network.

customers. This network is known as the cyclic queueing network. Assume that node 1 has service rate μ_1 and that node 2 has service rate μ_2. The state space in this case is $\{(k_1, k_2) \mid k_1 \geq 0, \; k_2 \geq 0, \; k_1 + k_2 = K\}$, where k_i ($i = 1, 2$) is the number of jobs at node i. Such multidimensional state spaces, we modify the earlier notation for steady-state probabilities. The joint probability of the state (k_1, k_2) will be denoted by $\pi(k_1, k_2)$ while the marginal pmf at node i will be denoted by $\pi_i(k_i)$. Although the underlying CTMC appears to have a two-dimensional state space, in reality it is one-dimensional birth–death process because of the condition $k_1 + k_2 = K$. Thus, the marginal pmf of node 2 can be derived from Eqs. (3.12) and (3.11):

$$\pi_2(k_2) = \pi_2(0) \cdot \left(\frac{\mu_1}{\mu_2}\right)^{k_2}, \quad k_2 \leq K; \quad \pi_2(k_2) = 0, \; k_2 > K.$$

$$\pi_2(0) = \frac{1}{1 + \sum_{i=1}^{K} \left(\frac{\mu_1}{\mu_2}\right)^i} = \frac{1}{\sum_{i=0}^{K} \left(\frac{\mu_1}{\mu_2}\right)^i},$$

and with $u = \mu_1/\mu_2$:

$$\pi_2(k_2) = \pi_2(0) \cdot u^{k_2}, \tag{6.39}$$

$$\pi_2(0) = \begin{cases} \dfrac{1 - u}{1 - u^{K+1}}, & u \neq 1, \\[3mm] \dfrac{1}{K + 1}, & u = 1. \end{cases} \tag{6.40}$$

Similarly, for node 1 we obtain:

$$\pi_1(k_1) = \pi_1(0) \cdot \frac{1}{u^{k_1}}, \tag{6.41}$$

$$\pi_1(0) = \begin{cases} \dfrac{1 - \frac{1}{u}}{1 - \left(\frac{1}{u}\right)^{K+1}}, & u \neq 1, \\[3mm] \dfrac{1}{K + 1}, & u = 1. \end{cases} \tag{6.42}$$

The utilizations are given by

$$\rho_1 = 1 - \pi_1(0), \qquad \rho_2 = 1 - \pi_2(0), \tag{6.43}$$

and the throughput

$$\lambda = \lambda_1 = \lambda_2 = \rho_1 \mu_1 = \rho_2 \mu_2, \tag{6.44}$$

and the mean number of customers in both nodes is given by:

$$\overline{K}_2 = \pi_2(0) \cdot \sum_{k_2=0}^{K} k_2 \cdot u^{k_2}$$

$$= \pi_2(0) \cdot u \frac{\partial}{\partial u} \cdot \sum_{k_2=0}^{K} u^{k_2}$$

$$= \pi_2(0) \cdot u \frac{\partial}{\partial u} \cdot \frac{1 - u^{k_2+1}}{1 - u}$$

$$= \frac{u}{1 - u} - \frac{(K+1) \cdot u^{K+1}}{1 - u^{K+1}}, \quad u \neq 1, \tag{6.45}$$

$$\overline{K}_1 = K - \overline{K}_2. \tag{6.46}$$

6.3   NON-MARKOVIAN QUEUES

6.3.1   The M/G/1 Queue

The mean waiting time $\overline{W}$ of an arriving job in an M/G/1-FCFS system has two components:

1. The mean remaining service time $\overline{W}_0$ of the job in service (if any),

2. The sum of the mean service times of the jobs in the queue ahead of the tagged job.

We can sum these components to

$$\overline{W} = \overline{W}_0 + \overline{Q} \cdot \overline{T}_B, \tag{6.47}$$

where the mean remaining service time $\overline{W}_0$ is given by

$$\overline{W}_0 = P(\text{server is busy}) \cdot \overline{R} + P(\text{server is idle}) \cdot 0, \tag{6.48}$$

with the mean remaining service time $\overline{R}$ of a busy server (the remaining service time of an idle server is obviously zero). When a job arrives, the job in service needs $\overline{R}$ time units on the average to be finished. This quantity is also called the mean residual life and is given by [Klei75]

$$\overline{R} = \frac{\overline{T_B^2}}{2\overline{T}_B} = \frac{\overline{T}_B}{2}(1 + c_B^2). \tag{6.49}$$

For an M/M/1 queue ($c_B^2 = 1$), we obtain

$$\overline{R}_{\text{M/M/1}} = \overline{T}_B = \frac{1}{\mu},$$

which is related to the memoryless property of the exponential distribution. The probability P(server is busy) is the utilization ρ by definition. From Eq. (6.47) and Little's theorem $\overline{Q} = \lambda \cdot \overline{W}$, we obtain

$$\overline{W} = \frac{\overline{W}_0}{1 - \rho} \qquad (6.50)$$

and with Eq. (6.49) we finally obtain

$$\overline{Q} = \frac{\rho^2}{(1 - \rho)} \cdot \frac{(1 + c_B^2)}{2} , \qquad (6.51)$$

the well-known Pollaczek–Khintchine formula (see Eq. (3.65)) for the mean queue length.

For exponentially distributed service times we have $c_B^2 = 1$ and for deterministic service times we have $c_B^2 = 0$. Thus:

$$\overline{Q}_{M/M/1} = \frac{\rho^2}{1 - \rho} \qquad \text{(see Eq. (6.17))},$$

$$\overline{Q}_{M/D/1} = \frac{\rho^2}{2(1 - \rho)} .$$

The mean queue length is divided by two if we have deterministic service time instead of exponentially distributed service times. Using the theory of embedded DTMC (see Section 3.3.1), the Pollaczek–Khintchine transform equation can be derived (Eq. (3.62)):

$$G(z) = B^\sim(\lambda - \lambda z)\frac{(1 - \rho)(1 - z)}{B^\sim(\lambda - \lambda z) - z} , \qquad (6.52)$$

where $G(z)$ is the z-transform of the steady-state probabilities of the number of jobs in the system π_k and $B^\sim(s)$ is the LST of the service time.

As an example we consider the M/M/1 queue with $B^\sim(s) = \mu/(s + \mu)$ (see Table 1.5). From Eq. (6.52) it follows that

$$G(z) = \frac{1 - \rho}{1 - \rho z}$$

or:

$$G(z) = \sum_{k=0}^{\infty}(1 - \rho)\rho^k \cdot z^k .$$

Hence we find the steady-state probability of k jobs in the system:

$$\pi_k = (1 - \rho)\rho^k .$$

With Eqs. (6.52) and (1.60), for the mean number in the system [Klei75] (see also Eq. (3.65)) we obtain:

$$\overline{K} = \rho + \frac{\rho^2}{1-\rho} \cdot \frac{1+c_B^2}{2} \tag{6.53}$$

or, using Eq. (6.7) and Little's theorem we get:

$$\overline{K} = \rho + \overline{Q}. \tag{6.54}$$

Hence

$$\overline{Q} = \frac{\rho^2}{1-\rho} \cdot \frac{1+c_B^2}{2},$$

which is the same as Eq. (6.51).

Figure 6.7a shows that for an M/G/1 queue the mean number of jobs $\overline{K}$ increases linearly with increasing squared coefficient of variation, and that the rate of increase is very sensitive to the utilization. From Fig. 6.7b we see how dramatically the mean number of jobs $\overline{K}$ increases if the utilization approaches 1 and that this is especially the case if the squared coefficient of variation c_B^2 is large.

6.3.1.1 Heavy-Tailed Distribution

While the M/G/1-queue with heavy-tailed service time distribution is still a current topic of research, some general results are available concerning the tail asymptotics of the distribution of the steady-state waiting time W. Let $B(x)$ be a heavy-tailed distribution for the service time X, i.e.,

$$1 - B(x) = P(X > x) \sim x^{-\nu} \quad x \to \infty.$$

Additionally assume that $B(x)$ is regularly varying, i.e., $P(X > x) = x^{-\nu}L(x)$ with $L(x)$ slowly varying ($\lim_{t\to\infty} \frac{L(tx)}{L(t)} = 1, \forall x > 0$). Then the following relation for the waiting time holds [BBQH03]:

$$P(W > x) \sim Cx^{1-\nu} \quad x \to \infty$$

An example of a regularly varying heavy-tailed distribution is the Pareto while the Weibull and the lognormal are not. An approximation containing the remaining service time X_r is known [BBQZ03]:

$$P(W > x) \sim \frac{\rho}{1-\rho}P(X_r > x) \quad x \to \infty.$$

Boxma [BBQZ03] states that the waiting time distribution in the M/G/1-FCFS queue is "one degree heavier" than the tail of the service time distribution, in the regularly varying case. The fact that the response time of a customer in the M/G/1-FCFS queue equals the sum of its waiting time and its lighter-tailed service requirement, the two quantities being independent, implies that the tail behavior of the response-time distribution is also "one degree heavier" than the service time distribution tail.

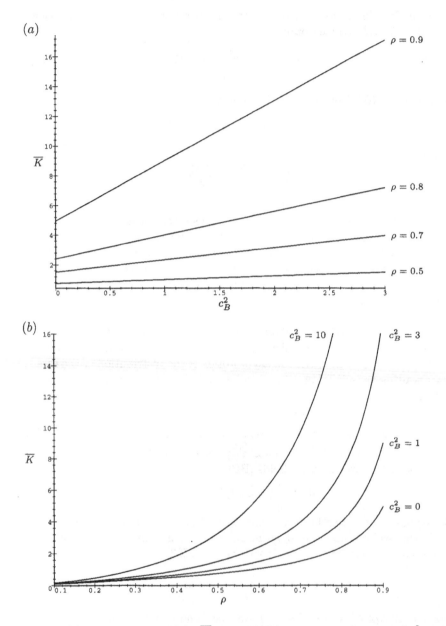

Fig. 6.7 (a) Mean number of jobs $\overline{K}$ in an M/G/1 queue as a function of c_B^2 with parameter ρ. (b) Mean number of jobs $\overline{K}$ in an M/G/1 queue as a function of ρ with parameter c_B^2.

6.3.1.2 Matrix-Analytic Solution of Models of M/G/1-Type As already seen in Section 3.3.1, *embedding techniques* can be applied for the steady-state analysis of processes with nonexponential service times. The transition probability matrix **P** of the embedded DTMC, as shown in Eq. (3.54), is in upper Hessenberg form. A more generalized version of the DTMC's transition probability matrix is of the form

$$\mathbf{P} = \begin{pmatrix} \mathbf{B}_0 & \mathbf{B}_1 & \mathbf{B}_2 & \mathbf{B}_3 & \cdots \\ \mathbf{A}_0 & \mathbf{A}_1 & \mathbf{A}_2 & \mathbf{A}_3 & \cdots \\ 0 & \mathbf{A}_0 & \mathbf{A}_1 & \mathbf{A}_2 & \cdots \\ 0 & 0 & \mathbf{A}_0 & \mathbf{A}_1 & \cdots \\ \vdots & \vdots & \vdots & \vdots & \ddots \end{pmatrix}, \tag{6.55}$$

where the submatrices $\mathbf{A}_k$ and $\mathbf{B}_i$, related but not identical to the submatrices of the matrix-geometric method (MGM) introduced in Section 3.2.1, describe the following transitions:

- $\mathbf{B}_i$: transitions from level 0 to level i, $0 \leq i$.

- $\mathbf{A}_k$: transitions from level i, $\forall i \geq l$, to level $i + (k-1)$, for $0 \leq k < i$.

Similar to the MGM, the states of the system are organized in levels (denoted by the index $0 \leq i < \infty$) and phases. The first repeating level is indexed by $i = l$. Levels with a value of $i < l$ are called boundary levels and levels with $i \geq l$ are repeating levels. In this Section, the case $l = 1$ will be looked at.

For many stochastic processes an embedded DTMC can be identified whose transition probability matrix resembles Eq. (6.55). Some examples are given in [Neut89] and [RiSm02b]. These stochastic processes are often referred to as *Markov chains of M/G/1-type*. They can be seen as a generalization of quasi-birth–death processes, where the transitions are still skip-free to the left but not skip-free to the right, i.e., transitions can occur between phases of the same or adjacent levels or from any phase to any higher level. Unfortunately, the MGM cannot be used for analyzing these processes because a matrix-geometric relation comparable to Eq. (3.28) does not exist for processes of M/G/1-type. Instead, the slightly more complex *matrix-analytic method* (MAM) is used. This method makes use of the auxiliary matrix **G**, which was already introduced in brief in the context of the logarithmic reduction algorithm of the MGM (see Section 3.2). The matrix **G** is derivable from the matrices $\mathbf{A}_k$ by using the matrix equation [Neut89]:

$$\mathbf{G} = \sum_{k=0}^{\infty} \mathbf{A}_k \mathbf{G}^k. \tag{6.56}$$

As soon as the matrix **G** is known, it can then be used to calculate the steady-state probability vector $\boldsymbol{\nu} = [\nu_0, \nu_1, \nu_2, \ldots]$ of the embedded DTMC. The steady-state probability vector $\boldsymbol{\nu}$ is partitioned into its steady-state probability sub-vectors ν_i that contain the steady-state probabilities of all phases within the corresponding level i. The calculation of $\boldsymbol{\nu}$ can be done, e.g., by applying *Ramaswami's formula* [Rama88] iteratively:

$$\boldsymbol{\nu}_i = \left(\boldsymbol{\nu}_0 \bar{\mathbf{B}}_i + \sum_{n=1}^{i-1} \boldsymbol{\nu}_n \bar{\mathbf{A}}_{i+1-n} \right) (\mathbf{I} - \bar{\mathbf{A}}_1)^{-1}, \quad i \geq 1, \qquad (6.57)$$

where

$$\bar{\mathbf{A}}_k = \sum_{n=k}^{\infty} \mathbf{A}_n \mathbf{G}^{n-k}, \quad k \geq 0,$$

and

$$\bar{\mathbf{B}}_k = \sum_{n=k}^{\infty} \mathbf{B}_n \mathbf{G}^{n-k}, \quad k \geq 0.$$

Following [Neut89], the still missing steady-state probability vector $\boldsymbol{\nu}_0$ of the boundary level $i = 0$ can be calculated using

$$\boldsymbol{\nu}_0 = \left(\boldsymbol{\kappa} \tilde{\boldsymbol{\kappa}}^T \right)^{-1} \boldsymbol{\kappa}, \qquad (6.58)$$

where $\boldsymbol{\kappa}$ is the steady-state probability vector of the stochastic matrix $\mathbf{K}$:

$$\mathbf{K} = \mathbf{B}_0 + \hat{\mathbf{B}} \mathbf{X}, \qquad (6.59)$$

solving $\boldsymbol{\kappa} = \boldsymbol{\kappa} \mathbf{K}, \boldsymbol{\kappa} \mathbf{1} = 1$. The auxiliary matrix $\mathbf{X}$ is defined by

$$\mathbf{X} = \left(\mathbf{I} - \hat{\mathbf{A}} \right)^{-1} \mathbf{A}_0, \qquad (6.60)$$

where the matrix $\hat{\mathbf{A}}$ is given by the sum

$$\hat{\mathbf{A}} = \bar{\mathbf{A}}_1 = \sum_{k=1}^{\infty} \mathbf{A}_k \mathbf{G}^{k-1}. \qquad (6.61)$$

The finite vector $\tilde{\boldsymbol{\kappa}}$ needed for solving Eq. (6.58) can be calculated by

$$\tilde{\boldsymbol{\kappa}} = \boldsymbol{\psi}_2 + \boldsymbol{\psi}_1 \left(\hat{\mathbf{B}} \left(\mathbf{I} - \hat{\mathbf{A}} \right)^{-1} \right)^T, \qquad (6.62)$$

where the auxiliary vectors $\boldsymbol{\psi}_1$ and $\boldsymbol{\psi}_2$ are obtained using

$$\boldsymbol{\psi}_1 = \left(\left(\mathbf{I} - \mathbf{A}_0 - \hat{\mathbf{A}} \right) \left(\mathbf{I} - \mathbf{A} + \left(1 - \boldsymbol{\beta}^T \right) \mathbf{g} \right)^{-1} \mathbf{1} + \left(1 - \alpha \boldsymbol{\beta}^T \right)^{-1} \mathbf{A}_0 \mathbf{1} \right)^T, \qquad (6.63)$$

and

$$\boldsymbol{\psi}_2 = \left(1 + \left(\sum_{k=1}^{\infty} \mathbf{B}_k - \hat{\mathbf{B}} \right) \left(\mathbf{I} - \mathbf{A} + \left(1 - \boldsymbol{\beta}^T \right) \mathbf{g} \right)^{-1} \mathbf{1} \right.$$
$$\left. + \left(1 - \alpha \boldsymbol{\beta}^T \right)^{-1} \sum_{k=1}^{\infty} (k-1) \mathbf{B}_k \mathbf{1} \right)^T. \qquad (6.64)$$

The stochastic matrix $\mathbf{A}$ and auxiliary matrix $\hat{\mathbf{B}}$ needed for Eqs. (6.63), (6.63), and (6.64) are given by

$$\mathbf{A} = \sum_{k=0}^{\infty} \mathbf{A}_k \qquad (6.65)$$

and

$$\hat{\mathbf{B}} = \bar{\mathbf{B}}_1 = \sum_{k=1}^{\infty} \mathbf{B}_k \mathbf{G}^{k-1} . \qquad (6.66)$$

α is the steady-state probability vector of the stochastic matrix $\mathbf{A}$, which is given by Eq. (6.65), solving $\alpha = \alpha \mathbf{A}$, $\alpha \mathbf{1} = 1$. $\mathbf{g}$ is the steady-state probability vector of the stochastic matrix $\mathbf{G}$, which is given by Eq. (6.56), solving $\mathbf{g} = \mathbf{g}\mathbf{G}$, $\mathbf{g}\mathbf{1} = 1$.

Finally, the vector β is given by

$$\beta = \left(\sum_{k=1}^{\infty} k\mathbf{A}_k \mathbf{1} \right)^T . \qquad (6.67)$$

To ensure a stable system and positive recurrence of the process under consideration, the stability condition needs to be fulfilled:

$$\alpha \beta^T < 1 . \qquad (6.68)$$

Once ν_0 is found, the main task remaining is to solve Eq. (6.56) and Eq. (6.57). Several algorithms have been proposed to perform these tasks. The interested reader is referred to [Mein98] and the references therein.

We refrain from presenting numerical examples at this point, but refer the reader to Section 6.8.4 where an extensive example is given related to the batch Markovian arrival process (BMAP) introduced in that section.

6.3.2   The GI/M/1 Queue

We state the results for GI/M/1 queues, using the parameter σ given by

$$\sigma = A^{\sim}(\mu - \mu\sigma) , \qquad (6.69)$$

where $A^{\sim}(s)$ is the LST of the interarrival time [Tane95]. For example, the mean number of jobs in the system is:

$$\overline{K} = \frac{\rho}{1 - \sigma} , \qquad (6.70)$$

the variance of the number of jobs in the system is:

$$\sigma_K^2 = \frac{\rho(1 + \sigma - \rho)}{(1 - \sigma)^2} , \qquad (6.71)$$

the mean response time is:

$$T = \frac{1}{\mu} \cdot \frac{1}{1-\sigma}, \tag{6.72}$$

the mean queue length is:

$$\overline{Q} = \frac{\rho \cdot \sigma}{1-\sigma}, \tag{6.73}$$

the variance of the queue length is:

$$\sigma_Q^2 = \frac{\rho\sigma(1+\sigma(1-\rho))}{(1-\sigma)^2}, \tag{6.74}$$

the mean waiting time is:

$$\overline{W} = \frac{1}{\mu} \cdot \frac{\sigma}{1-\sigma}, \tag{6.75}$$

the steady-state probability of the number of jobs in the system is:

$$\begin{aligned} \pi_k &= \rho(1-\sigma)\sigma^{k-1}, \quad k > 0, \\ \pi_0 &= 1-\rho, \end{aligned} \tag{6.76}$$

and the waiting time distribution is:

$$F_W(x) = \begin{cases} 1-\sigma, & x = 0, \\ 1-\sigma \cdot e^{-\mu(1-\sigma)x}, & x > 0. \end{cases} \tag{6.77}$$

In the case of an M/M/1 queue, we have $A\tilde{}(s) = \lambda/(s+\lambda)$ (see Table 1.5). Together with Eq. (6.69) we obtain

$$\sigma = \frac{\lambda}{\mu} = \rho,$$

and with the Eqs. (6.70) to (6.77) we obtain the well-known M/M/1 formulae.

A more interesting example is the $E_2/M/1$ queue. From Table 1.5 we have

$$A\tilde{}(s) = \left(\frac{\lambda}{s+\lambda}\right)^2,$$

and with Eq. (6.69) we have

$$\sigma = \rho + \frac{1}{2} - \sqrt{\rho - \frac{1}{4}}.$$

Using Eqs. (6.70)–(6.77), we obtain explicit formulae for $E_2/M/1$ performance measures.

The behavior of an M/G/1 and of a GI/M/1 queue is very similar, especially if $c_X^2 \leq 1$. This is shown in Fig. 6.8 where we compare the mean

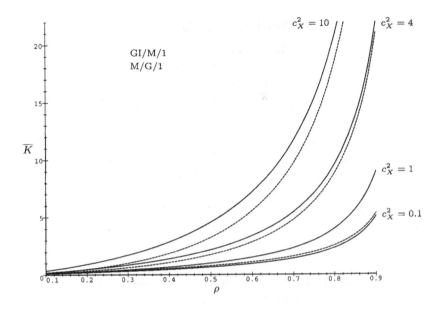

Fig. 6.8 Mean number of jobs $\overline{K}$ in an M/G/1 and a GI/M/1 queue.

number of jobs $\overline{K}$ for M/G/1 and GI/M/1 queues having the same coefficient of variation c_X^2. Note that c_X^2 in the case of GI/M/1 denotes the coefficient of variation of the interarrival times, while in the M/G/1 case it denotes the coefficient of variation of the service times. Note also that in the M/G/1 case, $\overline{K}$ depends only on the first two moments of the service time distribution, while in the GI/M/1 case, the dependence is more extensive. In Fig. 6.8, for the GI/M/1 queue, the gamma distribution is used. For increasing values of c_X^2 the deviation increases.

6.3.2.1 Matrix-Geometric Solution of Models of GI/M/1-Type Similar to Markov chains of the M/G/1-type introduced in Section 6.3.1.2, there exists another class of Markov chains that is called Markov chains of the GI/M/1-type. These processes can be handled by embedding a DTMC that is not skip-free to the left, but to the right; i.e., transitions can occur between phases of the same or adjacent levels or from any phase to any lower level. Analogous to the lower Hessenberg form transition probability matrix of DTMCs embedded at the arrival instants of GI/M/1 queues, the transition probability matrix **P** of this DTMC is in lower block Hessenberg form:

$$
\mathbf{P} = \begin{pmatrix}
\mathbf{B}_{1,0} & \mathbf{A}_0 & \mathbf{0} & \mathbf{0} & \cdots \\
\mathbf{B}_{2,0} & \mathbf{A}_1 & \mathbf{A}_0 & \mathbf{0} & \cdots \\
\mathbf{B}_{3,0} & \mathbf{A}_2 & \mathbf{A}_1 & \mathbf{A}_0 & \cdots \\
\mathbf{B}_{4,0} & \mathbf{A}_3 & \mathbf{A}_2 & \mathbf{A}_1 & \cdots \\
\vdots & \vdots & \vdots & \vdots & \ddots
\end{pmatrix}.
\tag{6.78}
$$

The mapping of the transitions to the elements of the submatrices $\mathbf{A}$ and $\mathbf{B}$, which was given in Section 3.2 in the context of continuous-time QBDs, still holds for Markov chains of the GI/M/1 type. Several instances of this class are presented in [Neut81]. For DTMCs Eq. (2.15) needs to be fulfilled. Applying the global balance equations (cf. Eq. (3.27)) of the repeating levels $i \geq 1$ yields

$$
\sum_{k=0}^{\infty} \boldsymbol{\nu}_{i-1+k} \mathbf{A}_k = \boldsymbol{\nu}_k, \quad i > 0.
\tag{6.79}
$$

For Markov chains of GI/M/1-type a matrix-geometric solution exists, in analogy to the one introduced for QBDs in Eq. (3.28):

$$
\boldsymbol{\nu}_{i+1} = \boldsymbol{\nu}_i \mathbf{R} = \boldsymbol{\nu}_0 \mathbf{R}^i, \quad i \geq 0.
\tag{6.80}
$$

Inserting Eq. (6.80) into Eq. (6.79) leads to[1]

$$
\mathbf{R} = \sum_{k=0}^{\infty} \mathbf{R}^k \mathbf{A}_k.
\tag{6.81}
$$

Equation (6.81) allows us to obtain the rate matrix $\mathbf{R}$, which – compared to the continuous-time QBD case – has a slightly different interpretation in the context of DTMCs: The entry $\mathbf{R}_{j,j'}$ of $\mathbf{R}$ can be interpreted as the expected number of visits to state (i, j') divided by the expected number of visits to state $(i - 1, j)$ before returning to level $i - 1$, provided that the process observation started from state $(i - 1, j)$.

To derive $\boldsymbol{\nu}_0$ the boundary level needs to be taken into account again. $\boldsymbol{\nu}_0$ satisfies

$$
\boldsymbol{\nu}_0 = \boldsymbol{\nu}_0 \sum_{k=0}^{\infty} \mathbf{R}^k \mathbf{B}_k,
\tag{6.82}
$$

with the normalization

$$
\boldsymbol{\nu}_0 (\mathbf{I} - \mathbf{R})^{-1} \mathbf{1} = 1.
\tag{6.83}
$$

To solve Eqs. (6.81), (6.82), and (6.83), several algorithms can be used that were introduced in the context of QBDs in Section 3.2.2.

We refrain from presenting numerical examples at this point, but refer the interested reader to Section 3.2 where a detailed example is given in the context of the matrix-geometric method for QBDs.

---

[1]Same approach as for obtaining Eq. (3.30).

6.3.3   The GI/M/m Queue

Exact results for GI/M/m queueing systems are also available. See [Alle90], [Klei75], or [GrHa85]. The mean waiting time is given by

$$\overline{W} = \frac{J}{m\mu(1 - \sigma)^2},$$ (6.84)

where

$$\sigma = A\tilde{\ }(m\mu - m\mu\sigma)$$ (6.85)

and

$$J = \left(\frac{1}{1 - \sigma} + \sum_{k=0}^{m-2} R_k \right)^{-1}.$$

For R_k, see [Klei75] (page 408). We introduce only a heavy traffic approximation

$$\overline{W} \approx \frac{\sigma_A^2 + \sigma_B^2/m^2}{2(1 - \rho)} \cdot \lambda,$$ (6.86)

which is an upper bound [Klei75] and can be used for GI/G/1 and GI/G/m system as well (see subsequent sections). Here λ is the reciprocal of the mean interarrival time, σ_A^2 is the variance of interarrival time, and σ_B^2 is the variance of service time.

6.3.4 . The GI/G/1 Queue

In the GI/G/1 case only approximation formulae and bounds exist. We can use M/G/1 and GI/M/1 results as upper or lower bounds, depending on the value of the coefficient of variation (see Table 6.1).

Table 6.1 Upper bounds (UB) and lower bounds (LB) for the GI/G/1 mean waiting time

c_A^2	c_B^2	M/G/1	GI/M/1
> 1	> 1	LB	LB
> 1	< 1	LB	UB
< 1	> 1	UB	LB
< 1	< 1	UB	UB

Another upper bound is given by Eq. (6.86) with $m = 1$ [Klei75]

$$\overline{W} < \frac{\sigma_A^2 + \sigma_B^2}{2(1 - \rho)} \cdot \lambda.$$ (6.87)

A modification of this upper bound is [Marc78]:

$$\overline{W} < \frac{1 + c_B^2}{(1/\rho^2) + c_B^2} \cdot \frac{\sigma_A^2 + \sigma_B^2}{2(1 - \rho)} \cdot \lambda.$$ (6.88)

This formula is exact for M/G/1 and is a good approximation for GI/M/1 and GI/G/1 queues if ρ is not too small and c_A^2 or c_B^2 are not too big.

A lower bound is also known [Marc78]:

$$\overline{W} > \frac{\rho^2 \cdot c_B^2 + \rho(\rho - 2)}{2\lambda(1 - \rho)}, \qquad (6.89)$$

but more complex and better lower bounds are given in [Klei76].

Many approximation formulae for the mean waiting time are mentioned in the literature. Four of them that are either very simple and straightforward or good approximations are introduced here. First, the well-known Allen–Cunneen approximation formula for GI/G/m queue is [Alle90]

$$\overline{W} \approx \frac{\rho/\mu}{1 - \rho} \cdot \frac{c_A^2 + c_B^2}{2}. \qquad (6.90)$$

This formula is exact for M/G/1 (Pollaczek–Khintchine formula) and a fair approximation elsewhere and is the basis for many other better approximations. A very good approximation is the Krämer–Langenbach-Belz formula, a direct extension of Eq. (6.90) via a correction factor

$$\overline{W} \approx \frac{\rho/\mu}{1 - \rho} \cdot \frac{c_A^2 + c_B^2}{2} \cdot G_{\text{KLB}}, \qquad (6.91)$$

with the correction factor

$$G_{\text{KLB}} = \begin{cases} \exp\left(-\frac{2}{3} \cdot \frac{1 - \rho}{\rho} \cdot \frac{(1 - c_A^2)^2}{c_A^2 + c_B^2}\right), & 0 \le c_A \le 1, \\[2ex] \exp\left(-(1 - \rho)\frac{c_A^2 - 1}{c_A^2 + 4c_B^2}\right), & c_A > 1. \end{cases} \qquad (6.92)$$

Another extension of the Allen–Cunneen formula is the approximation of Kulbatzki [Kulb89]:

$$\overline{W} \approx \frac{\rho/\mu}{1 - \rho} \cdot \frac{c_A^{f(c_A, c_B, \rho)} + c_B^2}{2}, \qquad (6.93)$$

with

$$f(c_A, c_B, \rho) = \begin{cases} 1, & c_A \in \{0, 1\}, \\[1ex] \left[\rho(14.1c_A - 5.9) + (-13.7c_A + 4.1)\right]c_B^2 \\ \quad + \left[\rho(-59.7c_A + 21.1) + (54.9c_A - 16.3)\right]c_B \\ \quad + \left[\rho(c_A - 4.5) + (-1.5c_A + 6.55)\right], & 0 \le c_A \le 1, \\[1ex] -0.75\rho + 2.775, & c_A > 1, \end{cases}$$
$$\qquad (6.94)$$

It is interesting to note that Eq. (6.94) was obtained using simulation experiments. A good approximation for the case $c_A^2 < 1$ is the Kimura approximation [Kimu85]:

$$\overline{W} \approx \frac{c_A^2 + c_B^2}{2} \overline{W}_{\text{M/M/m}} \left((1 - c_A^2) \exp\left(\frac{2(1 - \rho)}{3\rho} \right) + c_A^2 \right)^{-1}. \qquad (6.95)$$

6.3.5 The M/G/m Queue

We obtain the Martin's approximation formula for M/G/m queues [Mart72] by an extension of the Eq. (6.47) for the mean waiting time for an M/G/1 queue:

$$\overline{W} = \overline{W}_0 + \frac{\overline{Q}}{m} \cdot \overline{T}. \qquad (6.96)$$

Because of the m servers, an arriving customer has to wait, on the average, only for the service of $\overline{Q}/m$ customers. The remaining service time in this case is

$$\overline{W}_0 = P_m \cdot \overline{R}. \qquad (6.97)$$

With Eq. (6.49) we obtain as an approximation

$$\overline{R} \approx \overline{T} \frac{(1 + c_B^2)}{2m}, \qquad (6.98)$$

and with Eqs. (6.96) and (6.97) finally:

$$\overline{W} \approx \frac{P_m/\mu}{1 - \rho} \cdot \frac{(1 + c_B^2)}{2m}. \qquad (6.99)$$

This is a special case of the Allen–Cunneen formula for GI/G/m queues (see next section) and is exact for M/M/m and M/G/1 queues.

For the waiting probability P_m we can use Eq. (6.28) for M/M/m queues or a simple yet good approximation from [Bolc83]:

$$P_m \approx \begin{cases} \dfrac{\rho^m + \rho}{2}, & \rho > 0.7, \\[2mm] \rho^{\frac{m+1}{2}}, & \rho < 0.7. \end{cases} \qquad (6.100)$$

As an example, we compare the exact waiting probability with this approximation for $m = 5$ in Table 6.2.

From Fig. 6.9 we see that the deviation for the mean number of jobs $\overline{K}$ in the system is very small if we use the approximation for P_m instead of the exact value.

Table 6.2 Exact (Eq. (6.28)) and approximate (Eq. (6.100)) values of the probability of waiting

ρ	0.2	0.4	0.6	0.7	0.8	0.9	0.95	0.99
$P_{m_{ex}}$	0	0.06	0.23	0.38	0.55	0.76	0.88	0.97
$P_{m_{app}}$	0	0.06	0.21	0.34	0.56	0.75	0.86	0.97

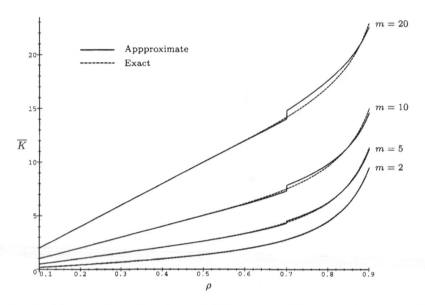

Fig. 6.9 Mean number of jobs in an M/M/m queue with exact and approximative formulae for P_m.

A good approximation for the mean waiting time in M/G/m queues is due to Cosmetatos [Cosm76]:

$$\overline{W}_{\mathrm{M/G/m}} \approx c_B^2 \overline{W}_{\mathrm{M/M/m}} + (1 - c_B^2)\overline{W}_{\mathrm{M/D/m}}. \tag{6.101}$$

In Eq. (6.101), we can use Eq. (6.30) for $\overline{W}_{\mathrm{M/M/m}}$ and use

$$\overline{W}_{\mathrm{M/D/m}} = \frac{1}{2} \cdot \frac{1}{nc_{Dm}} \cdot \overline{W}_{\mathrm{M/M/m}}$$

where

$$nc_{Dm} = \left(1 + (1 - \rho)(m - 1)\frac{\sqrt{4 + 5m} - 2}{16\rho m}\right)^{-1}. \tag{6.102}$$

For $\overline{W}_{\mathrm{M/D/m}}$ we can also use the Crommelin approximation formula [Crom34]:

$$\overline{W}_{\mathrm{M/D/m}} \approx \frac{1}{\mu} \sum_{\nu=1}^{\infty} \left(e^{-\nu\rho m} \left(\frac{(\nu\rho m)^{\nu m}}{(\nu m)!} - (1 - 1/\rho) \sum_{i=1}^{\nu m} \frac{(\nu\rho m)^i}{i!} \right) + \left(1 - \frac{1}{\rho} \right) \right).$$

$$(6.103)$$

Boxma, Cohen, and Huffels [BCH79] also use the preceding formulae for $\overline{W}_{\mathrm{M/D/m}}$ as a basis for their approximation:

$$\overline{W}_{\mathrm{M/G/m}} \approx \frac{1}{2}(1 + c_B^2) \frac{2\overline{W}_{\mathrm{M/D/m}}\overline{W}_{\mathrm{M/M/m}}}{2a\overline{W}_{\mathrm{M/D/m}} + (1 - a)\overline{W}_{\mathrm{M/m/m}}}, \qquad (6.104)$$

where

$$a = \begin{cases} 1, & m = 1, \\[2mm] \dfrac{1}{m-1} \left(\dfrac{(c_B^2 + 1)}{\gamma_1} - m + 1 \right), & m > 1, \end{cases}$$

and

$$\gamma_1 \approx \frac{1 - c_B^2}{m + 1} + \frac{c_B^2}{m}.$$

Tijms [Tijm86] uses γ_1 from the BCH-formula (Eq. (6.104)) in his approximation:

$$\overline{W}_{\mathrm{M/G/m}} \approx \left((1 - \rho)\gamma_1 m + \frac{\rho}{2}(c_B^2 + 1) \right) \overline{W}_{\mathrm{M/M/m}}. \qquad (6.105)$$

In Fig. 6.10 we compare five approximations introduced above with the results obtained from discrete-event simulation (DES). From the figure we see that for $\rho = 0.7$ all approximations are good for $c_B^2 < 2$, and that for higher values of c_B^2 the approximation due to Cosmetatos is very good and the others are fair.

6.3.6 The GI/G/m Queue

For GI/G/m queues only bounds and approximation formulae are available. These are extensions of M/G/m or GI/G/1 formulae. We begin with the well-known upper bound due to Kingman [King70]:

$$\overline{W} \leq \frac{\sigma_A^2 + \sigma_B^2/m + (m-1)/(m^2 \cdot \mu^2)}{2(1 - \rho)} \cdot \lambda \qquad (6.106)$$

and the lower bound of Brumelle [Brum71] and Marchal [Marc74]:

$$\overline{W} \geq \frac{\rho^2 c_B^2 - \rho(2 - \rho)}{2\lambda(1 - \rho)} - \frac{m-1}{m} \cdot \frac{c_B^2 + 1}{2\mu}. \qquad (6.107)$$

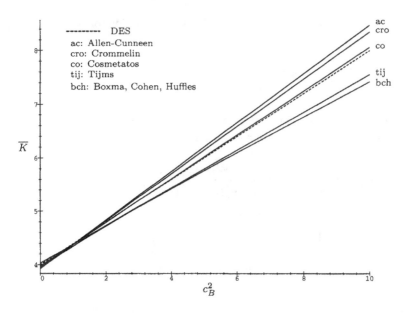

Fig. 6.10   Comparison of different M/G/5 approximations with DES ($\rho = 0.7$).

As a heavy traffic approximation we have Eq. (6.86), which we already introduced for GI/M/m queues:

$$\overline{W} \approx \frac{\sigma_A^2 + \sigma_B^2/m^2}{2(1 - \rho)} \cdot \lambda, \tag{6.108}$$

and we have the Kingman–Köllerström approximation [Köll74] for the waiting time distribution:

$$F_W(x) \approx 1 - \exp\left(-\frac{2(1 - \rho)}{\sigma_A^2 + \sigma_B^2/m^2} \cdot \frac{1}{\lambda}x\right). \tag{6.109}$$

The most known approximation formula for GI/G/m queues is the Allen–Cunneen (A–C) formula [Alle90]. We already introduced it for the special case GI/G/1 (Eq. (6.90)). Note that the A–C formula is an extension of Martin's formula (Eq. (6.99)) where we replace the 1 in the term $(1 + c_B^2)$ by c_A^2 to consider approximately the influence of the distribution of interarrival times:

$$\overline{W} \approx \frac{P_m/\mu}{1 - \rho} \cdot \frac{c_A^2 + c_B^2}{2m}. \tag{6.110}$$

For the probability of waiting we can either use Eq. (6.28) or the good approximation provided by Eq. (6.100). As in the GI/G/1 case, the Allen–Cunneen approximation was improved by Krämer–Langenbach-Belz [KrLa76] using a correction factor:

$$\overline{W} \approx \frac{P_m/\mu}{1-\rho} \cdot \frac{c_A^2 + c_B^2}{2m} \cdot G_{\text{KLB}}, \tag{6.111}$$

$$G_{\text{KLB}} = \begin{cases} \exp\left(-\frac{2}{3}\frac{1-\rho}{P_m}\frac{(1-c_A^2)^2}{c_A^2 + c_B^2}\right), & 0 \le c_A \le 1, \\ \exp\left(-(1-\rho)\frac{c_A^2 - 1}{c_A^2 + 4c_B^2}\right), & c_A > 1, \end{cases} \tag{6.112}$$

and by Kulbatzki [Kulb89] using the exponent $f(c_A, c_B, \rho)$ in place of 2 for c_A in Eq. (6.110):

$$\overline{W} \approx \frac{P_m/\mu}{1-\rho} \cdot \frac{c_A^{f(c_A,c_B,\rho)} + c_B^2}{2m}. \tag{6.113}$$

For the definition of $f(c_A, c_B, \rho)$, see Eq. (6.94). The Kulbatzki formula was further improved by Jaekel [Jaek91]. We start with the Kulbatzki GI/G/1-formula and use a heuristic correction factor to consider the number of servers m:

$$\overline{W} \approx \frac{\rho/\mu}{1-\rho} \cdot \frac{c_A^{f(c_A,c_B,\rho)} + c_B^2}{2} \cdot \rho^{\sqrt{0.5(m-1)}}. \tag{6.114}$$

This formula is applicable even if the values of m and the coefficients of variation are large.

In order to extend the Cosmetatos approximation [Cosm76] from M/G/m to GI/G/m queues, $\overline{W}_{\text{M/M/m}}$ and $\overline{W}_{\text{M/D/m}}$ need to be replaced by $\overline{W}_{\text{GI/M/m}}$ and $\overline{W}_{\text{GI/D/m}}$, respectively:

$$\overline{W}_{\text{GI/G/m}} \approx c_B^2 \overline{W}_{\text{GI/M/m}} + (1 - c_B^2)\overline{W}_{\text{GI/D/m}}, \tag{6.115}$$

where $\overline{W}_{\text{GI/M/m}}$ is given by Eq. (6.84) or by the approximation

$$\overline{W}_{\text{GI/M/m}} \approx \begin{cases} \frac{1}{2}(c_A^2 + 1)\exp\left(-\frac{2}{3}\cdot\frac{1-\rho}{P_m}\cdot\frac{(1-c_A^2)^2}{1+c_A^2}\right)\overline{W}_{\text{M/M/m}} \\ \qquad\qquad\qquad\qquad\qquad \text{for } 0 \le c_A \le 1, \\ \frac{1}{2}\left(c_A^{f(c_A,c_B,\rho)} + 1\right)\overline{W}_{\text{M/M/m}} \\ \qquad\qquad\qquad\qquad\qquad \text{for } c_A > 1, \end{cases} \tag{6.116}$$

and $\overline{W}_{\text{GI/D/m}}$ is given by

$$\overline{W}_{\text{GI/D/m}} = \frac{1}{2} \cdot \frac{1}{nc_{Dm}} \cdot \overline{W}_{\text{GI/M/m}}, \tag{6.117}$$

with nc_{Dm} from Eq. (6.102) or

$$\overline{W}_{\text{GI/D/m}} \approx c_A^{h(\rho,m)f(c_A,0,\rho)} \cdot \overline{W}_{\text{M/D/m}}$$

$$h(\rho,m) = 4\sqrt{(m-1)/(m+4)} \cdot (1-\rho) + 1 \tag{6.118}$$

with $\overline{W}_{M/D/m}$ from Eq. (6.102) or Eq. (6.103). A good approximation for the case $c_A^2 < 1$ is given by Kimura [Kimu85]:

$$\overline{W}_{GI/G/m} = \frac{c_A^2 + c_B^2}{2} \overline{W}_{M/M/m}$$

$$\cdot \left(\frac{1 - c_A^2}{1 - 4c(m,\rho)} \exp\left(\frac{2(1-\rho)}{3\rho} \right) + \frac{1 - c_B^2}{1 + c(m,\rho)} + c_A^2 + c_B^2 - 1 \right)^{-1},$$

$$c(m,\rho) = (1-\rho)(m-1)\frac{\sqrt{4+5m}-2}{16\rho m}.$$

(6.119)

Finally, the Boxma, Cohen, and Huffels formula [BCH79] can be extended to GI/G/m queues:

$$\overline{W}_{GI/G/m} = \frac{1}{2}(1 + c_B^2)\frac{2\overline{W}_{GI/D/m}\overline{W}_{GI/M/m}}{2a\overline{W}_{GI/D/m} + (1-a)\overline{W}_{GI/M/m}} \qquad (6.120)$$

as well as the Tijms formula [Tijm86]:

$$\overline{W}_{GI/G/m} = \left((1-\rho)\gamma_1 m + \frac{\rho}{2}(c_B^2 + 1) \right) \overline{W}_{GI/M/m}, \qquad (6.121)$$

with a and γ_1 from Eq. (6.104).

In Figs. 6.11 and 6.12 the mean queue lengths as functions of the coefficients of variation c_A^2, c_B^2, the utilization ρ, and the number of servers m are shown in a very compact manner using the approximate formula of Allen–Cunneen. Fig. 6.13 shows that the approximations of Allen–Cunneen, Krämer/Langenbach–Belz, and Kulbatzki are close together, especially for higher values of the coefficients of variation c_X^2 with a large number of servers m.

6.4   PRIORITY QUEUES

In a priority queueing system we assume that an arriving customer belongs to a priority class r ($r = 1, 2, \ldots, R$). The next customer to be served is the customer with the highest priority number r. Inside a priority class the queueing discipline is FCFS.

6.4.1   Queue without Preemption

Here we consider the case where a customer already in service is not preempted by an arriving customer with higher priority. The mean waiting time $\overline{W}_r$ of an arriving customer of priority class r has three components:

1. The mean remaining service time $\overline{W}_0$ of the job in service (if any).

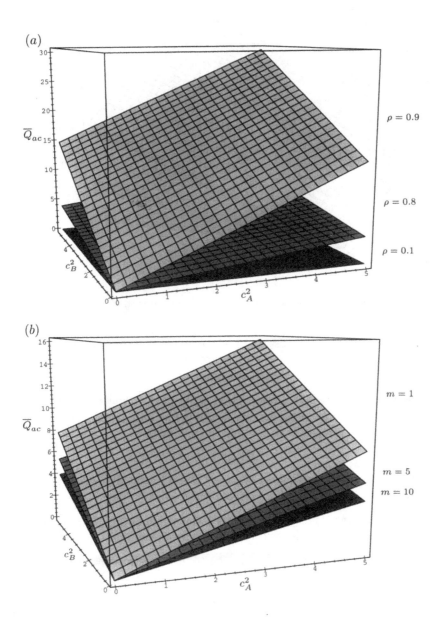

Fig. 6.11 (a) Mean queue length $\overline{Q}$ of a GI/G/10 queueing system. (b) Mean queue length $\overline{Q}$ of a GI/G/m queue with $\rho = 0.8$.

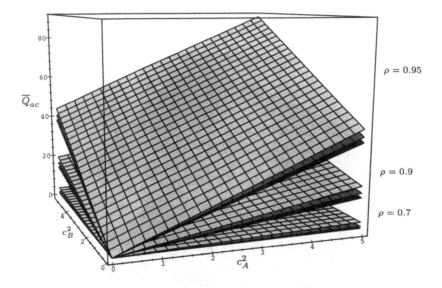

Fig. 6.12   Mean queue length of a GI/G/m queue with $m = 1, 5, 10$.

2. Mean service time of customers in the queue that are served before the tagged customer. These are the customers in the queue of the same and higher priority as the tagged customer.

3. Mean service time of customers that arrive at the queue while the tagged customer is in the queue and are served before him. These are customers with higher priority than the tagged customers.

Define:

$\overline{N}_{ir}$: Mean number of customers of class i found in the queue by the tagged (priority r) customer and receiving service before him,

$\overline{M}_{ir}$: Mean number of customers of class i who arrive during the waiting time of the tagged customer and receive service before him.

Then the mean waiting time of class r customers can be written as the sum of three components:

$$\overline{W}_r = \overline{W}_0 + \sum_{i=1}^{R} \overline{N}_{ir} \cdot \frac{1}{\mu_i} + \sum_{i=1}^{R} \overline{M}_{ir} \cdot \frac{1}{\mu_i}. \qquad (6.122)$$

For multiple server systems $(m > 1)$:

$$\overline{W}_r = \overline{W}_0 + \sum_{i=1}^{R} \frac{\overline{N}_{ir}}{m \cdot \mu_i} \frac{1}{\mu_i} + \sum_{i=1}^{R} \frac{\overline{M}_{ir}}{m} \cdot \frac{1}{\mu_i}, \qquad (6.123)$$

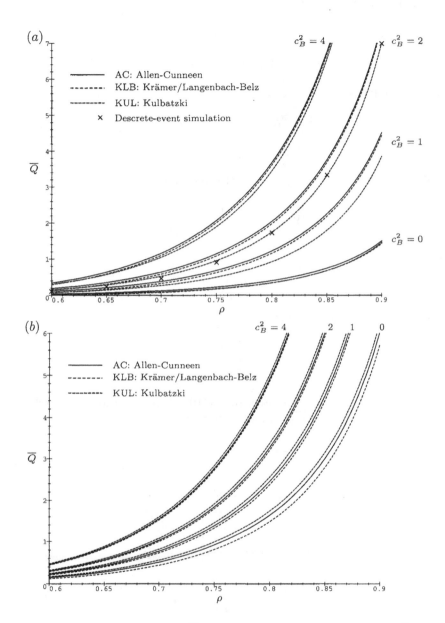

Fig. 6.13 (*a*) Mean queue length $\overline{Q}$ for a GI/G/10 queueing system with $c_A^2 = 0.5$. (*b*) Mean queue length $\overline{Q}$ for a GI/G/10 queueing system with $c_A^2 = 2.0$.

where $\overline{N}_{ir}$ and $\overline{M}_{ir}$ are given by

$$\overline{N}_{ir} = 0 \quad i < r \,,$$
$$\overline{M}_{ir} = 0 \quad i \leq r \,,$$

(6.124)

and, with Little's theorem;

$$\overline{N}_{ir} = \lambda_i \overline{W}_i \quad i \geq r \,,$$
$$\overline{M}_{ir} = \lambda_i \overline{W}_r \quad i > r \,.$$

(6.125)

We can solve Eqs. (6.122) and (6.123) to obtain

$$\overline{W}_r = \frac{\overline{W}_0}{(1 - \sigma_r)(1 - \sigma_{r+1})} \,,$$

(6.126)

where

$$\sigma_r = \sum_{i=r}^{R} \rho_i \,.$$

(6.127)

The overall mean waiting time is

$$\overline{W} = \sum_{i=1}^{R} \frac{\lambda_i}{\lambda} \cdot \overline{W}_i.$$

(6.128)

Values for $\overline{W}_0$ are given by Eqs. (6.48), (6.49), (6.97), and (6.98) and are the weighted sum of the $\overline{W}_{0i}$ of all the classes:

$$\overline{W}_{0,\mathrm{M/M/1}} = \sum_{i=1}^{R} \rho_i \frac{1}{\mu_i} \,,$$

(6.129)

$$\overline{W}_{0,\mathrm{M/G/1}} = \sum_{i=1}^{R} \rho_i \cdot \frac{1 + c_{B_i}^2}{2\mu_i} \,,$$

(6.130)

$$\overline{W}_{0,\mathrm{GI/G/1,AC}} \approx \sum_{i=1}^{R} \rho_i \cdot \frac{c_{A_i}^2 + c_{B_i}^2}{2\mu_i} \,,$$

(6.131)

$$\overline{W}_{0,\mathrm{GI/G/1,KLB}} \approx \sum_{i=1}^{R} \rho_i \cdot \frac{c_{A_i}^2 + c_{B_i}^2}{2\mu_i} \cdot G_{\mathrm{KLB}} \,,$$

(6.132)

$$\overline{W}_{0,\mathrm{GI/G/1,KUL}} \approx \sum_{i=1}^{R} \rho_i \cdot \frac{c_{A_i}^{f(c_{A_i}, c_{B_i}, \rho_i)} + c_{B_i}^2}{2\mu_i} \,,$$

(6.133)

$$\overline{W}_{0,\mathrm{M/M/m}} = \frac{P_m}{m\rho} \sum_{i=1}^{R} \rho_i \cdot \frac{1}{\mu_i} \,,$$

(6.134)

$$\overline{W}_{0,\text{M/G/m}} \approx \frac{P_m}{2m\rho} \cdot \sum_{i=1}^{R} \rho_i \cdot \frac{1 + c_{B_i}^2}{\mu_i} \,, \tag{6.135}$$

$$\overline{W}_{0,\text{GI/G/m,AC}} \approx \frac{P_m}{2m\rho} \cdot \sum_{i=1}^{R} \rho_i \cdot \frac{c_{A_i}^2 + c_{B_i}^2}{\mu_i} \,, \tag{6.136}$$

$$\overline{W}_{0,\text{GI/G/m,KLB}} \approx \frac{P_m}{2m\rho} \sum_{i=1}^{R} \rho_i \cdot \frac{c_{A_i}^2 + c_{B_i}^2}{\mu_i} G_{\text{KLB}} \,, \tag{6.137}$$

$$\overline{W}_{0,\text{GI/G/m,KUL}} \approx \frac{P_m}{2m\rho} \sum_{i=1}^{R} \rho_i \cdot \frac{c_{A_i}^{f(c_{A_i}, c_{B_i}, \rho_i)} + c_{B_i}^2}{\mu_i} \,. \tag{6.138}$$

For $f(c_{A_i}, c_{B_i}, \rho_i)$, see Eq. (6.94); for $G_{\text{KLB,GI/G/1}}$, see Eq. (6.92); and for $G_{\text{KLB,GI/G/m}}$, see Eq. (6.112).

Also, the GI/G/m-FCFS Approximation of Cosmetatos can be extended to priority queues:

$$\overline{W}_{\text{GI/G/m}_r} \approx c_{B_r}^2 \overline{W}_{\text{GI/M/m}_r} + (1 - c_{B_r}^2)\overline{W}_{\text{GI/D/m}_r} \,, \tag{6.139}$$

with

$$\overline{W}_{\text{GI/M/m}_r} = \frac{\overline{W}_{0,\text{GI/M/m}}}{(1 - \sigma_r)(1 - \sigma_{r+1})} \,,$$

$$\overline{W}_{\text{GI/D/m}_r} = \frac{\overline{W}_{0,\text{GI/D/m}}}{(1 - \sigma_r)(1 - \sigma_{r+1})} \,,$$

$$\overline{W}_{0,\text{GI/M/m}} \approx \frac{P_m}{2m\rho} \sum_{i=1}^{R} \rho_i \frac{c_{A_i}^{f(c_{A_i}, 1, \rho_i)} + 1}{\mu_i} \,,$$

$$\overline{W}_{0,\text{GI/D/m}} \approx \frac{P_m}{2m\rho} \sum_{i=1}^{R} \rho_i \frac{c_{A_i}^{f(c_{A_i}, 0, \rho_i)}}{\mu_i} \,.$$

The M/G/m Cosmetatos approximation can similarly be extended. All GI/G/m approximations yield good results for the M/G/m–priority queues as well.

Figure 6.14 shows the mean queue length $\overline{Q}_r$ for different priority classes of a priority queueing system without preemption and $R = 3$ priority classes together with the mean queue length for the same system without priorities. It can be seen that the higher-priority jobs have a much lower queue length, and the lower-priority jobs have a much higher mean queue length than the FCFS system.

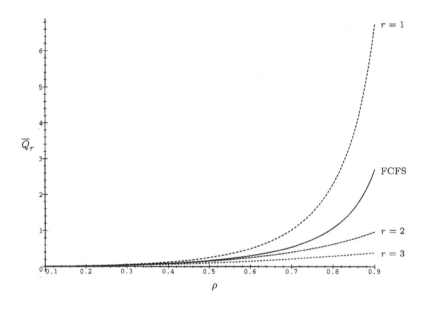

Fig. 6.14 Mean queue lengths $\overline{Q}_r$ for an M/M/1 priority queue without preemption.

6.4.2 Conservation Laws

In priority queueing systems, the mean waiting time of jobs is dependent on their priority class. It is relatively short for jobs with high priority and considerably longer for jobs with low priority because there exists a fundamental conservation law ([Klei65, Schr70, HeSo84]):

$$\sum_{i=1}^{R} \rho_i \overline{W}_i = \frac{\rho \overline{W}_0}{1 - \rho} = \rho \cdot \overline{W}_{\text{FCFS}} \qquad (6.140)$$

or

$$\frac{1}{\rho} \sum_{i=1}^{R} \rho_i \overline{W}_i = \frac{\overline{W}_0}{1 - \rho} = \overline{W}_{\text{FCFS}} . \qquad (6.141)$$

The conservation law to apply the following restrictions must be satisfied:

1. No service facility is idle as long as there are jobs in the queue, i.e., scheduling is work-conserving [ReKo75].

2. No job leaves the system before its service is completed.

3. The distributions of the interarrival times and the service times are arbitrary with the restriction that the first moments of both the distributions and the second moment of the service time distribution exist.

4. The service times of the jobs are independent of the queueing discipline.

5. Preemption is allowed only when all jobs have the same exponential service time distribution and preemption is of the type preemptive resume.

6. For GI/G/m queues all classes have the same service times. This restriction is not necessary for GI/G/1 queues [HeSo84].

If for GI/G/m queues the service times of the classes differ, then the conservation law is an approximation [Jaek91].

6.4.3 Queue with Preemption

Here we consider the case that an arriving customer can preempt a customer of lower priority from service. The preempted customer will be continued later at the point where he was preempted (preemptive resume). Thus, a customer of class r is not influenced by customers of the classes $1, 2, \ldots, r - 1$. Hence we need to consider only a system with the classes $r, r+1, \ldots, R$ to calculate the mean waiting time $\overline{W}_r$ of class r.

In the conservation law we can replace $\rho = \sum\limits_{i=1}^{R} \rho_i$ by $\sigma_r = \sum\limits_{i=r}^{R} \rho_i$ and $\overline{W}_{\mathrm{FCFS}}$ by

$$\overline{W}^r = \frac{\overline{W}_0^r}{1 - \sigma_r}, \tag{6.142}$$

where $\overline{W}_0^r$ is the mean remaining service time for such a system and we obtain it by substitution of ρ by σ_r in the formulae. For example:

$$\overline{W}_{0_{\mathrm{AC}}}^r = \frac{P_m^r}{2m\sigma_r} \cdot \sum\limits_{i=r}^{R} \rho_i \frac{c_A^2 + c_B^2}{\mu_i}. \tag{6.143}$$

We sum only from r and R rather than from 1 to R. For P_m^r, we replace ρ by σ_r in the exact or approximative formula for P_m.

To obtain the mean waiting time of a customer of class r we apply the conservation law twice:

$$\sigma_r \cdot \overline{W}^r = \sum\limits_{i=r}^{R} \rho_i \overline{W}_i, \qquad \sigma_{r+1} \cdot \overline{W}^{r+1} = \sum\limits_{i=r+1}^{R} \rho_i \cdot \overline{W}_i. \tag{6.144}$$

By substitution we obtain

$$\overline{W}_r = \frac{1}{\rho_r} \left(\sigma_r \overline{W}^r - \sigma_{r+1} \overline{W}^{r+1} \right). \tag{6.145}$$

Equation (6.145) is exact for M/M/1 and M/G/1 queues. It is also exact for M/M/m queues if the mean service time of all classes are the same. For other systems, Eq. (6.145) is a good approximation. From Fig. 6.15 it can be seen that even for heavy utilization the mean queue length for the highest priority class is negligible in the case of preemption, at the expense of poorer service for lower-priority classes.

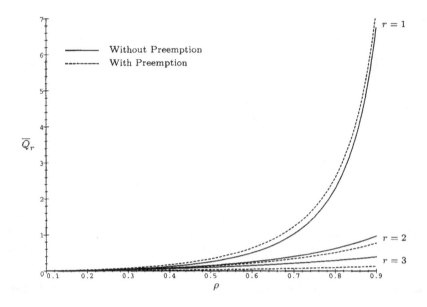

Fig. 6.15   Mean queue lengths for an M/M/1 priority system with and without pre-emption.

6.4.4   Queue with Time-Dependent Priorities

Customers with low priorities have long waiting times especially when the utilization of the system is very high. Sometimes it is advantageous for a customer priority to increase with the time, so that he does not have to wait too long. This possibility can be considered if we do not have a fixed priority function:

$$q_r(t) = \text{Priority of class } r \text{ at time } t\,.$$

Such systems are more flexible but need more expense for the administration.
A popular and easy to handle priority function is the following:

$$q_r(t) = (t - t_0) \cdot b_r \tag{6.146}$$

(see Fig. 6.16a), with $0 \le b_1 \le b_2 \le \ldots \le b_r$. The customer enters the system at time t_0 and then increases the priority at the rate b_r. Customers of the same priority class have the same rate of increase b_r but different values of the arrival time t_0.

We only consider systems without preemption and provide the following recursive formula [BoBr84] for the mean waiting time of priority class-r customer:

$$\overline{W}_r = \frac{\dfrac{\overline{W}_0}{1-\rho} - \sum_{i=1}^{r-1} \rho_i \overline{W}_i \left(1 - \dfrac{b_i}{b_r}\right)}{1 - \sum_{i=r+1}^{R} \rho_i \left(1 - \dfrac{b_r}{b_i}\right)} \qquad (6.147)$$

with the same mean remaining service times $\overline{W}_0$ as for the priority systems with fixed priorities. If these are exact, then Eq. (6.147) is also exact; otherwise it is an approximation formula.

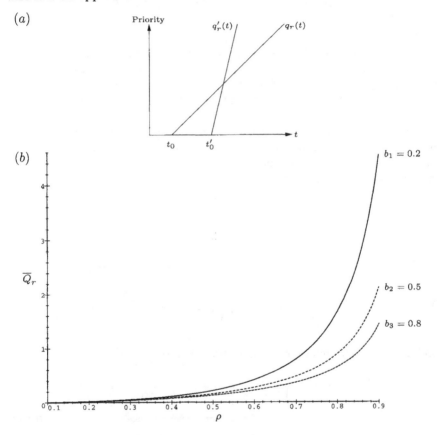

Fig. 6.16   (a) Priority function with slope b_r. (b) Mean queue length $\overline{Q}_r$ for an M/M/1 priority system with time-dependent priorities having slope b_r.

Another important priority function is

$$q_r(t) = r_r + t - t_0, \qquad (6.148)$$

with $0 \le r_1 \le r_2 \le \ldots \le r_r$. Each priority class has a different starting priority r_r and a constant slope (see Fig. 6.17a). In [BoBr84] a heavy traffic approximation (as ρ approaches 1) is given:

$$\overline{W}_r \approx \frac{\overline{W}_0}{1-\rho} - P_m \cdot \sum_{i=1}^{R} \rho_i (r_r - r_i), \tag{6.149}$$

as well as a more accurate recursive formula:

$$\overline{W}_r \approx \frac{\overline{W}_0}{1-\rho} - \sum_{i=1}^{r-1} \rho_i \overline{W}_i \left(1 - \exp\left(\frac{P_m(r_i - r_r)}{\overline{W}_i} \right) \right). \tag{6.150}$$

Note that for $m = 1$, we have $P_m = \rho$.

From Figs. 6.16b and 6.17b it can be seen that systems with time-dependent priorities are more flexible than systems with static priorities. If the values of b_r or r_r, respectively, are close together, then the systems behave like FCFS systems, and if they are very different, then they behave more like systems with static priorities.

In many real-time systems, a job has to be serviced within an *upper time limit*. To achieve such a behavior, it is advantageous to use a priority function that increases from 0 to ∞ between the arrival time and the upper time limit u_r. A convenient priority function is

$$q_r(t) = \begin{cases} (t - t_0)/(u_r - t + t_0) & t_0 < t \le u_r + t_0, \\ \infty & u_r + t_0 \le t. \end{cases} \tag{6.151}$$

Again we get a heavy traffic approximation (as ρ approaches 1):

$$\overline{W}_r \approx \left(\frac{\overline{W}_0}{1-\rho} - P_m \sum_{i=1}^{r-1} \rho_i (u_i - u_r) \right) \left(1 - (1 - P_m) \sum_{i=r+1}^{R} \rho_i \left(1 - \frac{u_i}{u_r} \right) \right)^{-1} \tag{6.152}$$

and a recursive formula for more accurate results:

$$\overline{W}_r \approx \left(\frac{\overline{W}_0}{1-\rho} - \sum_{i=1}^{r-1} \rho_i \cdot \overline{W}_i \left(1 - \frac{u_r}{u_i} \right) \left(1 - P_m \exp\left(-\frac{\rho u_i}{\overline{W}_i} \right) \right) \right)$$
$$\cdot \left(1 - \sum_{i=r+1}^{R} \rho_i \left(1 - \frac{u_i}{u_r} \right) \left(1 - P_m \cdot \exp\left(-\frac{\rho u_r}{\overline{W}_r} \right) \right) \right)^{-1}. \tag{6.153}$$

For more details and other priority functions see [EsBo03, BoBr84] and [Jaek91].

In Fig. 6.18b, the mean waiting times for a system with static priorities are compared to those of a priority system with upper time limits. The figure also contains the upper time limits. For the two higher priorities, the static priority system is better, but the upper time limit system is better over all priorities.

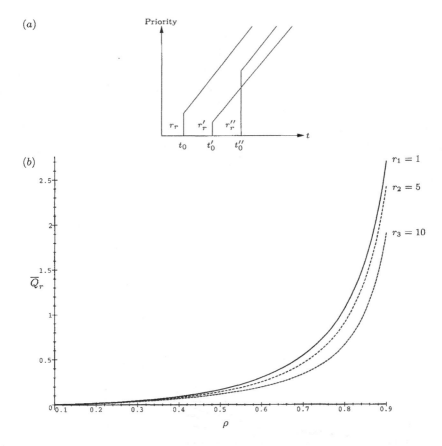

Fig. 6.17   (a) Priority function with starting priority r_r. (b) Mean queue lengths $\overline{Q}_r$ for an M/M/1 priority system with time-dependent priorities having starting priorities r_r.

6.5   ASYMMETRIC QUEUES

The calculation of the performance measures is fairly simple if there is only one server or several servers with identical service times arranged in parallel, and the interarrival time and service time are exponentially distributed (M/M/m queueing systems). However, heterogeneous multiple servers often occur in practice and here the servers have different service times. This situation occurs, for example, when machines of different age or manufacturer are running in parallel. It is therefore useful to be able to calculate the performance measures of such systems. One method for solving this problem is given in [BoSc91] in which the formulae derived provide results whose relative deviation from values obtained by simulation are quite acceptable. The het-

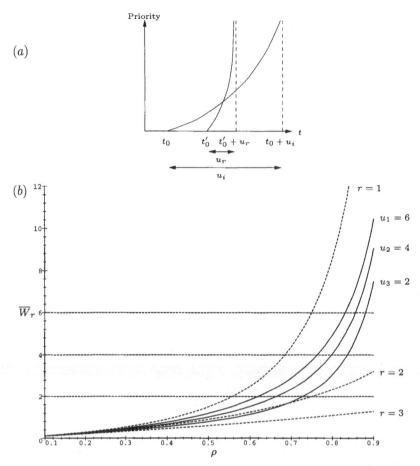

Fig. 6.18 (a) Priority function with upper time limit u_r. (b) Mean waiting time $\overline{W}_r$ for an M/M/1 priority system with static priorities and with upper time limits $u_r = 2, 4, 6$.

erogeneous multiple server is treated in more detail by [BFH88] and [Baer85] who achieve exact results. Also see [Triv01] and [GeTr83] for exact results.

We will assume throughout this section that the arrival process is Poisson with rate λ and that service times are exponentially distributed with rate μ_i on the ith server.

6.5.1   Approximate Analysis

In homogeneous (symmetric) queueing systems, the response time depends only on whether an arriving job finds some free server. A heterogeneous (asymmetric) multiple server system differs since which of the m servers processes the job is now important. The analysis in [BoSc91] is based on the following consideration: Let p_k $(k = 1, \ldots, m)$ be the probability that a job is assigned to the kth server. Because it can be assumed that a faster server processes more jobs than a slower server, it follows that the ratios of the probabilities p_k to each other are the same as the ratios of the service rates to each other. Here no distinction is made between different strategies for selecting a free server. This approximation is justified by the fact that in a system with heavy traffic there is usually one free server and therefore a server selection strategy is not needed. Then p_k $(1 \leq k \leq m)$ is approximately given by

$$p_k \approx \frac{\mu_k}{\sum\limits_{i=1}^{m} \mu_i} . \qquad (6.154)$$

In addition, p_k can be used to determine the utilization ρ_k of the kth server:

$$\rho_k \approx \frac{\lambda}{\mu_k} \cdot p_k = \frac{\lambda}{\mu_k} \cdot \frac{\mu_k}{\sum\limits_{i=1}^{m} \mu_i} = \frac{\lambda}{\sum\limits_{i=1}^{m} \mu_i} . \qquad (6.155)$$

Thus, the utilization of all the servers is the same and the overall utilization is given by

$$\rho = \rho_k . \qquad (6.156)$$

In [BoSc91] it is shown that the known formulae for symmetric M/M/m and GI/G/m queues can be applied to asymmetric M/M/m and GI/G/m queues if Eqs. (6.155) and (6.156) are used to calculate the utilization ρ. Good approximations to the performance measures are obtained provided the values of the service rates do not differ too much, i.e, $\mu_{\min}/\mu_{\max} \leq 10$ (see Table 6.3).

Table 6.3 Approximative and simulative results for (a) M/M/5 and (b) M/E$_2$/5 queueing systems

	λ	μ_k	$\rho(A)$	$\rho(S_1)$	$\rho(S_2)$	$\overline{Q}(A)$	$\overline{Q}(S_1)$	$\overline{Q}(S_2)$
(a)	73	10,15,20,20,25	0.811	0.799	0.819	2.472	2.367	2.495
	73	16,17,18,19,20	0.811	0.807	0.812	2.472	2.476	2.500
	73	5,10,18,22,35	0.811	0.808	0.844	2.472	2.443	2.626
	81.11	4,8,16,32,40	0.811	0.826	0.858	2.472	2.549	2.713
	81.11	8,9,20,31,32	0.811	0.807	0.839	2.472	2.456	2.606
	81.11	8,14,20,26,32	0.811	0.801	0.830	2.472	2.442	2.569
	81.11	10,15,20,25,30	0.811	0.799	0.824	2.472	2.425	2.523
(b)	81.11	10,15,20,25,30	0.811	0.800	0.825	1.854	1.854	1.962

Notes: A = approximation, S_1 = discrete-event simulation with random selection of a free server, and S_2 = discrete-event simulation with selection of the fastest free server.

6.5.2   Exact Analysis

This section describes the methods used for calculating the characteristics of asymmetric queueing systems described in [BFH88] and [Baer85]. These systems are treated as a CTMC, and the performance measures are calculated by solving the underlying CTMCs. In this way exact results are obtained for elementary asymmetric M/M/m queueing systems. This method is used as follows: First, the steady-state probabilities are calculated for systems without queues, known as loss systems. Then the formulae given in [BFH88] and [Baer85] can be used to extend the results to asymmetric, non-lossy systems.

Two strategies for selecting a free server are investigated:

1. The job is assigned randomly to a free server as in [BFH88].

2. The fastest free server is selected as in [Baer85].

6.5.2.1   Analysis of M/M/m Loss Systems   A loss system contains no queues. The name indicates that jobs that cannot find a free server are rejected and disappear from the system. A CTMC is constructed and solved to obtain a closed-form formula for its steady-state probabilities. It should be noted that asymmetric queueing systems differ from symmetric queueing systems in that it is not sufficient when producing the CTMC to describe the states by the number of servers occupied. Due to the different service rates, the state classification depends on which of the m servers are currently occupied. Thus, each state is characterized by a set $g = (k_1, k_2, \ldots, k_i)$, where k_l belongs to g if the lth server is currently occupied. Here $i = |g|$ is the number of servers currently occupied ($1 \leq i \leq m$) and l is an index ($1 \leq l \leq m$).

Random Selection of a Free Server   An M/M/3 queue is used as an example. The state transition diagram of the loss system is shown in Fig. 6.19. The steady-state probabilities are given by [BFH88]:

$$\pi_g = \frac{(m - |g|)!}{m!} \cdot \prod_{k \in g} \frac{\lambda}{\mu_k} \cdot \pi_\emptyset, \qquad (6.157)$$

for all $g \subseteq G$ with $g \neq \emptyset$ and $G = \{1, 2, \ldots, m\}$. In order to determine π_0, we use the normalization condition:

$$\sum_{g \subseteq G} \pi_g = 1 \qquad \text{with} \qquad G = \{1, 2, \ldots, m\}. \qquad (6.158)$$

The probability of loss $\pi^{(L)}$ is the probability that all servers are occupied:

$$\pi_m^{(L)} = \pi_{\{1,2,\ldots,m\}} = \pi_G. \qquad (6.159)$$

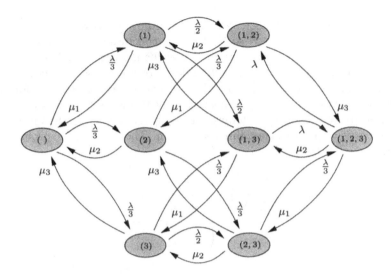

Fig. 6.19 CTMC for an asymmetric M/M/3 loss system with random selection of a free server.

The utilization ρ_k of the kth server is equal to the sum of the state probabilities of all states in which the kth server is occupied:

$$\rho_k^L = \sum_{g: \, k \in g} \pi_g \, . \tag{6.160}$$

Choice of the Fastest Free Server. Here it is always assumed that the individual servers are sorted in descending order of the service rates; i.e., the fastest server has the lowest index. Let $\boldsymbol{\mu}^m = (\mu_1, \mu_2, \cdots, \mu_m)$ and let $(\boldsymbol{\mu}^{k-1}, \mu_m) = (\mu_1, \mu_2, \cdots, \mu_{k-1}, \mu_m)$. The probability of loss $\pi_m^{(L)} = \pi_{\{1,\dots,m\}}(\boldsymbol{\mu}^m)$ is given by [Baer85]:

$$\pi_m^{(L)} = \pi_{\{1,\dots,m\}}(\boldsymbol{\mu}^m) = B_m(\boldsymbol{\mu}^m)$$

$$= B_{m-1}(\boldsymbol{\mu}^{m-1}) \cdot \left[1 + \frac{\mu_m}{\lambda} \cdot \prod_{k=1}^{m-1} \frac{B_k(\boldsymbol{\mu}^{k-1}, \mu_m)}{B_k(\boldsymbol{\mu}^{k-1}, \mu_k + \mu_m)} \right]^{-1}, \tag{6.161}$$

with

$$B_1(\boldsymbol{\mu}^1) = \pi_{\{1\}}(\mu_1) = \frac{\lambda}{\lambda + \mu_1} \, .$$

The utilization ρ_k of the individual servers is given by

$$\rho_k = \frac{\lambda}{\mu_k} [B_{k-1}(\boldsymbol{\mu}^{k-1}) - B_k(\boldsymbol{\mu}^k)] \qquad \text{with} \qquad B_0(\boldsymbol{\mu}^0) = 1 \, . \tag{6.162}$$

6.5.2.2 Extending to Nonlossy System According to [BFH88], the next step is to use the results obtained so far to draw conclusions about the behavior of the corresponding nonlossy system. Here the method used to select a free server of the corresponding loss system does not matter. It is only necessary that the values appropriate to the selected strategy are used in the formulae finally obtained. In what follows, a superscript W means that the variable is a characteristic of a nonlossy system; a superscript L means that the variable is a characteristic of the corresponding loss system. The steady-state probabilities are given by [BFH88]

$$\pi_i^{(W)} = \left(\sum_{k=1}^{m} \frac{\mu_k}{\lambda} \right)^{m-i} \cdot \pi_m^{(W)} \qquad \text{for} \qquad i > m, \tag{6.163}$$

with

$$\pi_m^{(W)} = \frac{\pi_m^{(L)}}{N}, \tag{6.164}$$

where

$$N = 1 + \frac{\pi_m^{(L)} \cdot c}{1 - c} \qquad \text{with} \qquad c = \frac{\lambda}{\sum\limits_{k=1}^{m} \mu_k}. \tag{6.165}$$

It is now possible to calculate all the state probabilities of the asymmetric nonlossy system and all interesting performance measures. It should be noted that c is not the same as the utilization ρ. The method for calculating ρ is described in the following.

The probability P_m that an arriving job is queued is equal to the sum of the probabilities of all states in which there are m or more jobs in the system:

$$P_m = \sum_{i=m}^{\infty} \pi_i^{(W)} = \frac{1}{1-c} \cdot \pi_m^{(W)} = \frac{1}{1-c} \cdot \frac{\pi_m^{(L)}}{N}. \tag{6.166}$$

Analogous to the symmetric case, the mean queue length $\overline{Q}$ is given by

$$\overline{Q} = \sum_{i=m+1}^{\infty} (i - m) \cdot \pi_i^{(W)} = \frac{P_m \cdot c}{1 - c}. \tag{6.167}$$

The mean waiting time can be calculated using Little's theorem:

$$\overline{W} = \frac{\overline{Q}}{\lambda}. \tag{6.168}$$

To calculate the utilization ρ, it is first necessary to calculate the specific utilization ρ_k for each server:

$$\rho_k^{(W)} = \frac{\rho_k^{(L)}}{N} + \sum_{i=m+1}^{\infty} \pi_i^{(W)} = \frac{\rho_k^{(L)} + \pi_m^{(L)} \cdot c/(1-c)}{N}. \tag{6.169}$$

Then ρ is given by:

$$\rho = \frac{m}{\sum\limits_{k=1}^{m} \rho_k^{-1}} . \tag{6.170}$$

Exact Analysis of an Asymmetric M/M/2 Queue

In [Triv01] an exact solution is given for an asymmetric M/M/2 queue with the strategy of selecting the fastest free server. Figure 6.20 shows the CTMC state transition diagram for this system. The state of the system is defined to

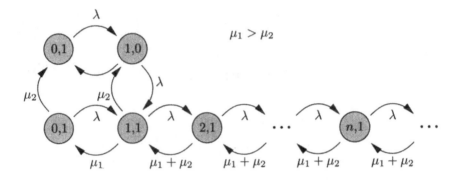

Fig. 6.20 CTMC for the M/M/2 asymmetric system.

be the tuple (k_1, k_2), where $k_1 \geq 0$ denotes the number of jobs in the queue including any at the faster server, and $k_2 \in \{0, 1\}$ denotes the number of jobs at the slower server.

If we use Eq. (2.58) for the underlying CTMC, we can calculate the steady-state probabilities easily with

$$c = \frac{\lambda}{\mu_1 + \mu_2} \tag{6.171}$$

$$\pi(k, 1) = c\pi(k - 1, 1) = c^{k-1}\pi(1, 1), \qquad \text{for} \quad k > 1. \tag{6.172}$$

Furthermore,

$$\pi(0, 1) = \frac{c}{1 + 2c} \frac{\lambda}{\mu_2} \pi(0, 0) \,,$$

$$\pi(1, 0) = \frac{1 + c}{1 + 2c} \frac{\lambda}{\mu_1} \pi(0, 0) \,, \tag{6.173}$$

$$\pi(1, 1) = \frac{c}{1 + 2c} \frac{\lambda(\lambda + \mu_2)}{\mu_1 \mu_2} \pi(0, 0) \,.$$

Finally, using the normalization condition, we get

$$\pi(0,0) = [1 + \frac{\lambda(\lambda + \mu_2)}{\mu_1\mu_2(1 + 2c)(1 - c)}]^{-1}. \tag{6.174}$$

And with these, we obtain the utilizations of the two servers:

$$\begin{aligned} \rho_1 &= 1 - \pi(0,0) - \pi(0,1), \\ \rho_2 &= 1 - \pi(0,0) - \pi(1,0), \end{aligned} \tag{6.175}$$

and the mean number of jobs in system:

$$\overline{K} = \frac{1}{A(1 - c)^2}, \tag{6.176}$$

where

$$A = \left[\frac{\mu_1\mu_2(1 + 2c)}{\lambda(\lambda + \mu_2)} + \frac{1}{1 - c} \right].$$

Example 6.1   To show how to use the introduced formulae for asymmetric M/M/m queues, we apply them to a simple M/M/2 example with $\lambda = 0.2$, $\mu_1 = 0.5$, and $\mu_2 = 0.25$.

Approximation [BoSc91]   With Eq. (6.156) for the utilization,

$$\rho = \rho_1 = \rho_2 = \frac{\lambda}{\mu_1 + \mu_2} = \frac{0.2}{0.5 + 0.25} = \underline{0.267};$$

with Eq. (6.29) for symmetric M/M/2 queues for mean number in system,

$$\overline{K} = 2\rho + \frac{2\rho^3}{1 - \rho^2} = \underline{0.575};$$

and with Little's theorem (Eq. (6.9)),

$$\overline{T} = \underline{2.875}.$$

Exact [Triv01]   With Eq. (6.171),

$$c = \underline{0.267};$$

with Eqs. (6.173), (6.174), and (6.175) for the utilizations of the servers,

$$\pi(0,0) = \left[1 + \frac{0.2(0.2 + 0.25)}{0.25 \cdot 0.5(1 + 2c)(1 - c)}\right]^{-1}$$

$$= \underline{0.6096}\,,$$

$$\pi(0,1) = \frac{0.267}{1 + 0.534} \cdot \frac{0.2}{0.25} \cdot 0.6056 = \underline{0.085}\,,$$

$$\pi(1,0) = \frac{1.267}{1.534} \cdot \frac{0.2}{0.5} \cdot 0.6056 = \underline{0.2014}\,,$$

$$\rho_1 = 1 - 0.6096 + 0.085 = \underline{0.305}\,,$$
$$\rho_2 = 1 - 0.6096 + 0.2014 = \underline{0.189}\,;$$

and with Eq. (6.176),

$$A = \frac{0.5 \cdot 0.25(1 + 0.534)}{0.2(0.2 + 0.25)} + \frac{1}{0.733} = \underline{3.495}\,,$$

$$\overline{K} = \frac{1}{3.495 + 0.733^2} = \underline{0.533}\,,$$

$$\overline{T} = \frac{\overline{K}}{\lambda} = \underline{2.663}\,.$$

Exact [BFH88] First we consider the asymmetric M/M/2 loss system with random selection of a free server: With Eq. (6.157) for the state probabilities,

$$\pi_1 = \frac{(2 - 1)!}{2!} \cdot \frac{\lambda}{\mu_1} \cdot \pi_0 = \frac{1}{2} \cdot \frac{0.2}{0.5} \cdot \pi_0 = 0.2\pi_0\,,$$

$$\pi_2 = \frac{(2 - 1)!}{2!} \cdot \frac{\lambda}{\mu_2} \cdot \pi_0 = \frac{1}{2} \cdot \frac{0.2}{0.25} \cdot \pi_0 = 0.4\pi_0\,,$$

$$\pi_{1,2} = \frac{(2 - 2)!}{2!} \cdot \frac{\lambda}{\mu_1} \cdot \frac{\lambda}{\mu_2} \cdot \pi_0 = 0.16\pi_0\,.$$

With the normalization condition,

$$(1 + 0.2 + 0.4 + 0.16)\pi_0 = 1\,, \qquad \pi_0 = 0.568\,.$$

Now the probability of loss can be determined (Eq. (6.159)):

$$\pi_2^{(L)} = \pi_{1,2} = 0.16 \cdot 0.568 = \underline{0.091}\,;$$

and we can also determine the utilizations of the servers (Eq. (6.160)):

$$\rho_1^L = \pi_1 + \pi_{1,2} = (0.2 + 1.6) \cdot 0.568 = \underline{0.204}\,,$$

$$\rho_2^L = \pi_2 + \pi_{1,2} = (0.4 + 1.6) \cdot 0.568 = \underline{0.319}.$$

This result means that the slower server has a higher utilization as expected. Now we can use these results of the loss system to obtain performance measures for the non-lossy system. For c and N (Eq. (6.165)) we get:

$$c = \frac{0.2}{0.5 + 0.25} = 0.267\,, \qquad N = 1 + \frac{0.091 \cdot 0.267}{0.733} = \underline{1.033}\,,$$

and for the probability of waiting (Eq. (6.166)) we obtain:

$$P_2 = \frac{1}{0.733} \cdot \frac{0.091}{1.033} = \underline{0.120}\,.$$

The mean queue length and the mean waiting time (Eqs. (6.167), (6.168)) are

$$\overline{Q} = \frac{0.120 \cdot 0.267}{0.733} = \underline{0.0437}\,, \qquad \overline{W} = \underline{0.219}\,.$$

Utilization of the servers (Eq. (6.169)):

$$\rho_1^{(W)} = \frac{0.204 + 0.091 \cdot 0.267/0.733}{1.033} = \underline{0.230}\,,$$

$$\rho_2^{(W)} = \frac{0.319 + 0.091 \cdot 0.267/0.733}{1.033} = \underline{0.341}\,.$$

Utilization of the system (Eq. (6.170)):

$$\rho = \frac{2}{\frac{1}{0.230} + \frac{1}{0.341}} = \underline{0.275}\,.$$

Mean number of jobs in system:

$$\overline{K} = \rho_1 + \rho_2 + \overline{Q} = 0.230 + 0.341 + 0.0437 = \underline{0.6147}\,.$$

Mean response time:

$$\overline{T} = \frac{\overline{K}}{\lambda} = \underline{3.074}\,.$$

Exact [Baer85] The last case we consider is the asymmetric M/M/2 queue with selection of the fastest free server using the formulae in [Baer85]. First we again consider the loss system and determine the loss probability (Eq. (6.161)):

$$\pi_2^{(L)} = B_2(\boldsymbol{\mu}^2) = B_1(\boldsymbol{\mu}^1) \left[1 + \frac{\mu_2}{\lambda} \cdot \frac{B_1(\mu_2)}{B_1(\mu_1 + \mu_2)}\right]^{-1}$$

$$= \frac{\lambda}{\lambda + \mu_1} \cdot \left[1 + \frac{\mu_2}{\lambda} \cdot \frac{\frac{\lambda}{\lambda+\mu_2}}{\frac{\lambda}{\lambda+\mu_1+\mu_2}}\right]^{-1}$$

$$= \frac{\lambda}{\lambda + \mu_1} \cdot \left[1 + \frac{0.25}{0.2} \cdot \frac{0.2 + 0.25 + 0.5}{0.2 + 0.25}\right]^{-1}$$

$$= \underline{0.0785},$$

and the utilizations of the servers (Eq. (6.162)):

$$\rho_1^L = \frac{\lambda}{\mu_1}[B_0(\boldsymbol{\mu}^0) - B_1(\boldsymbol{\mu}^1)] = \frac{0.2}{0.5}\left[1 - \frac{0.2}{0.2 + 0.5}\right] = \underline{0.286},$$

$$\rho_2^L = \frac{0.2}{0.25}\left[B_1(\boldsymbol{\mu}^1) - B_2(\boldsymbol{\mu}^2)\right] = \frac{0.2}{0.25}\left[\frac{0.2}{0.2 + 0.5} - 0.0785\right] = \underline{0.166}.$$

Again we use the results of the loss system to obtain performance measures for the queueing system with selection of the fastest server. From Eq. (6.165)

$$c = \underline{0.267}, \qquad N = 1 + \frac{0.0785 \cdot 0.267}{0.733} = \underline{1.0286}.$$

Probability of waiting (Eq. (6.166)):

$$P_2 = \frac{1}{0.733} \cdot \frac{0.0785}{1.0280} = \underline{0.104}.$$

Mean queue length (Eq. (6.167)):

$$\overline{Q} = \frac{0.104 \cdot 0.267}{0.733} = \underline{0.0379}.$$

Mean waiting time:

$$\overline{W} = \frac{\overline{Q}}{\lambda} = \underline{0.190}.$$

Utilizations of the servers (Eq. (6.169)):

$$\rho_1^{(W)} = \frac{0.286 + 0.0785 \cdot 0.267/0.733}{1.0286} = \underline{0.306},$$

$$\rho_2^{(W)} = \frac{0.166 + 0.0785 \cdot 0.267/0.733}{1.0286} = \underline{0.189}.$$

Utilization of the system (Eq. (6.170)):

$$\rho = \frac{2}{\frac{1}{0.306} + \frac{1}{0.189}} = \underline{0.234}.$$

Mean number of jobs in the system:

$$\overline{K} = \rho_1 + \rho_2 + \overline{Q} = 0.306 + 0.189 + 0.0379 = \underline{0.533}\,.$$

Mean system time:

$$\overline{T} = \frac{\overline{K}}{\lambda} = \underline{2.665}\,.$$

In Table 6.4, results of this example are summarized. In Table 6.5, the

Table 6.4   Results for several performance measures with different solution techniques and strategies for a asymmetric M/M/2 queue ($\lambda = 0.2$, $\mu_1 = 0.5$, $\mu_2 = 0.25$)

Strategy Reference	Appr. [BoSc91]	FFS [Triv01]	FFS [Baer85]	Random [BFH88]
ρ	0.267	(0.267)	0.234	0.275
ρ_1	0.267	0.305	0.306	0.230
ρ_2	0.267	0.189	0.189	0.341
$\overline{K}$	0.575	0.533	0.533	0.615
$\overline{T}$	2.875	2.663	2.665	3.074

Notes: FFS = selection of the fastest free server and Random = random selection of the free server.

mean response time $\overline{T}$ for an asymmetric M/M/2 queue is compared with the mean response times $\overline{T}_1$ (for an M/M/1 queue with service rate μ_1) and $\overline{T}_2$ (for an M/M/1 queue with service rate μ_2) [Triv01]. From Table 6.5 we

Table 6.5   Mean response times $\overline{T}, \overline{T}_1$, and $\overline{T}_2$ for an asymmetric M/M/2 queue with strategy FFS and two M/M/1 queues ($\lambda = 0.2$, $\mu_2 = 0.25$, $\mu_1 = \alpha \cdot \mu_2$)

α	1	2	3	4	5
$\overline{T}_1$	20	3.33	1.818	1.25	0.55
$\overline{T}_2$	20	20	20	20	20
$\overline{T}$	4.762	2.662	1.875	1.459	1.20

see that sometimes it is better to disconnect the slower server if we wish to minimize the mean response time. This is the case only for low utilization of the servers and if the service rates differ considerably. This issue is further explored in [GeTr83].

6.6   QUEUES WITH BATCH SERVICE AND BATCH ARRIVALS

6.6.1   Batch Service

An interesting and important way to service customers in queueing systems is to serve them in batches. There are many applications of this kind of service, especially in manufacturing systems. Two different policies are in use. In the case of a full batch policy (FB) with batch size b, the service of the batch is started only after b customers of the batch have arrived. In the case of the minimum batch policy (MB), a minimum number a ($< b$) of customers is sufficient to start the service of the batch. If there are more then b waiting customers, only b customers are collected into a batch and serviced together.

The extended Kendall notation for batch service systems is

$$GI/G^{[a,b]}/m, \text{ Multiserver system with MB policy}$$
$$GI/G^{[b,b]}/m, \text{ Multiserver system with FB policy}$$

A special case of the MB policy is the GREEDY policy where the service begins when $a = 1$ customer is in the queue. To obtain a formula for the mean waiting time and the mean queue length for the $GI/G^{[b,b]}/m$ queueing system, we use the already introduced formulae for $GI/G/m$ queueing systems and calculate the mean queue length $\hat{Q}$ for the batches. From this result we calculate the mean queue length $\overline{Q}_{\text{batch}}$ for the individual customers. If the arrival rate of the individual customers is λ, then the arrival rate of the batches with batch size b is

$$\hat{\lambda} = \frac{\lambda}{b}, \tag{6.177}$$

and the coefficient of variation for the batch interarrival time is

$$\hat{c}_A^2 = \frac{c_A^2}{b}, \tag{6.178}$$

using Eq. (1.55) with the coefficient of variation of the individual customers being c_A. Using the Allen–Cunneen formula (Eq.(6.110)) for the mean waiting time $\overline{W}$ in a $GI/G/m$ FCFS system and Little's law, we obtain for the mean queue length $\hat{Q}$ of the batches, in a $GI/G^{[b,b]}/m$ queueing system:

$$\hat{Q} \approx \frac{\rho P_m}{1 - \rho} \cdot \frac{\hat{c}_A^2 + c_B^2}{2}, \tag{6.179}$$

or with the Kramer–Langenbach-Belz formula (Eq.(6.111)), which produces more accurate results:

$$\hat{Q} \approx \frac{\rho P_m}{1 - \rho} \cdot \frac{\hat{c}_A^2 + c_B^2}{2} \cdot G_{KLB}. \tag{6.180}$$

The mean queue length of the individual customers is then approximately given by [HaSp95, Kirs91]:

$$\overline{Q}_{\text{batch}} \approx b \cdot \hat{Q} + \frac{b-1}{2} \,. \tag{6.181}$$

The first term in this formula is due to the customers in the $\hat{Q}$ batches of the queue, and the second term is the mean number of customers who are still not combined into a batch. The accuracy of Eq. (6.181) depends mainly on the accuracy of the approximation formula for $\hat{Q}$.

A solution for $GI/G^{[a,b]}/m$ queueing systems is given by a heuristic extension of Eq. (6.181) [Kirs91]:

$$\overline{Q}_{\text{batch}}^{[a,b]} \approx b \cdot \hat{Q} + P_m \cdot \frac{b-1}{2} + (1 - P_m) \cdot \frac{a-1}{2} \,. \tag{6.182}$$

Equation (6.182) reduces to Eq. (6.181) if $a = b$. For the probability of waiting, P_m, we use the corresponding value for the $M/M/m$ queueing system (Eq.(6.28)).

In the special case of an $M/M^{[b,b]}/1$ queue, we obtain an $E_b/M/1$ queueing system for the calculation of $\hat{Q}$ since in this case the interarrival time of the batch is the sum of b independent exponentially distributed interarrival times. In the case of an $M/M^{[b,b]}/m$ queue, we obtain an $E_b/M/m$ queueing system for which exact formulae (see Eqs. (6.73) and (6.84), respectively) are known.

Figure 6.21a shows the variation of the mean queue length $\overline{Q}_{\text{batch}}$ for individual jobs with the batch size b in a full batch system. Figure 6.21b shows that for increasing values of ρ, the mean queue length of a minimum batch system approaches that of a full batch system with the same batch size b.

Figure 6.22 shows that the mean queue length $\overline{Q}_{\text{batch}}$ of an individual job in a minimum batch system depends more on the batch size b than on the minimum batch size a, as also can be seen from Eq. (6.182). Moreover, we realize that a has a noticeable influence only on $\overline{Q}_{\text{batch}}^{[a,b]}$ for small values of ρ.

6.6.2   Batch Arrivals

Analysis of systems with batch service will naturally lead to queueing systems with batch arrivals. Whenever customers are served in batches, they will arrive as a batch at the following station. Let X be the i.i.d. random variable describing the size of the arriving batches.

The extended Kendall notation for systems with batch arrivals is

$GI^X/G/m$, Multiserver system with batch arrivals but single service,
$GI^X/G^{[b,b]}/m$, Multiserver system with batch arrivals and FB service policy.

We focus on the generic system with batch arrivals and FB service policy.

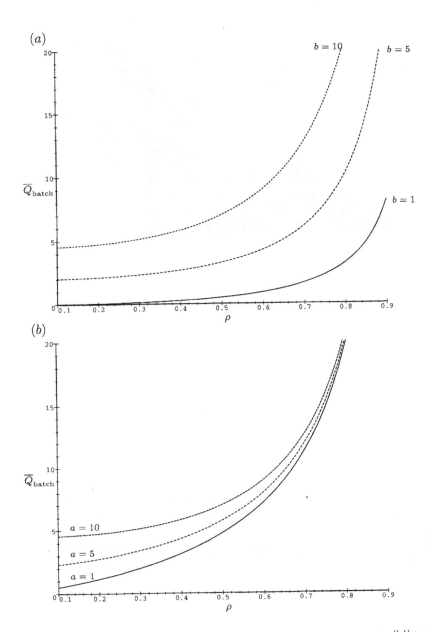

Fig. 6.21 (a) Mean queue length $\overline{Q}_{\text{batch}}$ for a full batch system M/M$^{[b,b]}$/1-FCFS
and (b) $\overline{Q}_{\text{batch}}^{[a,10]}$ for a minimum batch system.

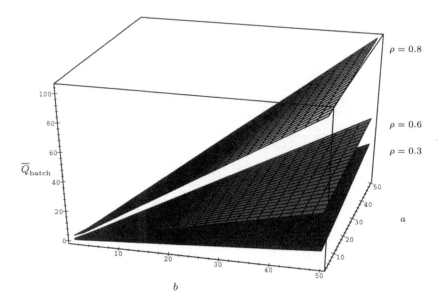

Q_{batch}

Fig. 6.22 Mean queue length $\overline{Q}_{\text{batch}}^{[a,b]}$ of a minimum batch system.

By transforming the stream of arriving batches of different sizes to a stream of batches of the service batch size b, we can derive the performance measures. Using renewal theory the coefficient of variation $c_{A^b}^2$ of the interarrival time of batches with batch size b can be derived (see [Zisg99] for further details):

$$c_{A^b}^2 = \frac{\overline{X}}{b}(c_X^2 + c_{Ax}^2), \qquad (6.183)$$

with the mean value of the size of the arriving batches $\overline{X}$, the coefficient of variation of the size of the arriving batches c_X^2 and the coefficient of variation of the interarrival time of the arriving batches c_{Ax}^2. The mean queue length of the individual customers can now be calculated by replacing the coefficient of variation of the interarrival time of the batches $\hat{c}_A^2$ by $c_{A^b}^2$ (Eq. (6.183)) in Eq. (6.179):

$$\overline{Q}_{\text{batch}} \approx b \cdot \frac{\rho P_m}{1-\rho} \cdot \frac{c_{A^b}^2 + c_B^2}{2} + \frac{b-1}{2}, \qquad (6.184)$$

or in Eq. (6.180):

$$\overline{Q}_{\text{batch}} \approx b \cdot \frac{\rho P_m}{1-\rho} \cdot \frac{c_{A^b}^2 + c_B^2}{2} \cdot G_{KLB} + \frac{b-1}{2}. \qquad (6.185)$$

The accuracy of Eq. (6.184) and specially of Eq. (6.185) is very high for $\overline{X} \leq b$ (see, for example, Table 6.6). In the case of $\overline{X} > b$, Eq. (6.185) gives a lower bound for the mean number of customers in the queue.

Table 6.6   Mean queue length $\overline{Q}_{\text{batch}}$ for a $GI^X/G^{[5,5]}/4$ system with uniformly distributed batch sizes of the arriving batches $\overline{X} = 1$ and $\overline{X} = 3$ (from [Hans03])

ρ	c_X	c_{AX}	Simulation	Approximation (Eq. (6.184))
0.5	1.2	1.2	12.703	12.920
0.5	1.2	0.0	12.183	12.294
0.5	1.1	0.7	12.331	12.467
0.5	0.5	0.5	12.036	12.196
0.7	1.2	1.2	20.643	21.291
0.7	1.2	0.0	17.381	17.690
0.7	1.1	0.7	18.325	18.686
0.7	0.5	0.5	16.622	17.125
0.9	1.2	1.2	56.584	57.505
0.9	1.2	0.0	31.475	31.982
0.9	1.1	0.7	38.422	39.036
0.9	0.5	0.5	26.988	27.976

For $M^X/M^Y/1$ systems with batch arrivals and batch service, FB policy and arbitrarily distributed batch sizes see [CMP99]. A more detailed discussion of batch systems can be found in [Medh03].

6.7   RETRIAL QUEUES

Retrial queues (or queues with repeated calls, returning customers, repeated orders etc.) exhibit the following characteristic behavior [BEA99, FaTe97, Medh03]: A customer arriving when all servers are busy leaves the service area but after some random time repeats the request. This feature plays a special role in some computer and communication systems, in telephone systems, in supermarkets or at airports when aircrafts are waiting for landing permission.

In general a retrial queue may be described as follows. There are a number of servers at which requests arrive. These requests are called primary requests. If an arriving primary request finds a free server, it immediately occupies this server and leaves the system after completion of service. If all servers are busy, then the request is rejected and forms a source of repeated requests (retrials). The sources remain in a so called orbit. Every such source sends repeated requests until it finds a free server, in which case it is served and the source is eliminated. The basic model of a retrial queue is shown in Fig. 6.23:

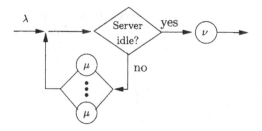

Fig. 6.23   The basic retrial queue.

6.7.1   M/M/1 Retrial Queue

In this section we consider a simple M/M/1 retrial queue with exponentially distributed repeating or delay times (with rate ν). Let C and N be the number of requests in the server and in the orbit, respectively; then:

$$p_{in} = P[C = i, N = n] \quad \text{with} \quad i = 0, 1; \quad n \geq 0. \tag{6.186}$$

For an M/M/1 retrial queue the state diagram is given in Fig. 6.24. After solving the steady-state balance equation, we obtain ([Medh03]):

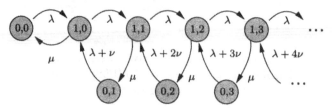

Fig. 6.24   CTMC for the M/M/1 retrial queue.

$$p_{0,n} = \frac{\rho^n}{n!\nu^n} \cdot p_{0,0} \prod_{i=0}^{n-1}(\lambda + i\nu) \quad n \geq 1, \tag{6.187}$$

$$p_{1,n} = \frac{\rho^{n+1}}{n!\nu^n} \cdot p_{0,0} \prod_{i=1}^{n}(\lambda + i\nu) \quad n \geq 0, \tag{6.188}$$

$$p_{0,0} = (1 - \rho)^{\frac{\lambda}{\nu}+1}, \tag{6.189}$$

with $\rho = \lambda/\mu$.

From these equations we obtain for the mean number of requests in the system:

$$\overline{K} = \frac{\rho}{1 - \rho} \cdot \left(1 + \frac{\lambda}{\nu}\right), \tag{6.190}$$

Table 6.7   Mean queue length $\overline{Q}$ in a retrial system with different values of the retrial rate ν

ν	0.0	0.2	0.5	1.0	1.5	2.0	5.0	10.0	∞
$\overline{Q}$	∞	19.2	9.60	6.40	5.30	4.80	3.82	3.52	3.20

$$\text{and} \quad var(K) = \frac{\rho}{(1 - \rho)^2} \cdot \left(1 + \frac{\lambda}{\nu}\right), \tag{6.191}$$

and for the mean queue length we obtain:

$$\overline{Q} = \frac{\rho^2}{1 - \rho} \cdot \left(1 + \frac{\mu}{\nu}\right) \tag{6.192}$$

and:

$$var(Q) = \frac{\rho^2}{(1 - \rho)^2} \cdot \left(1 + \rho - \rho^2 + \frac{\mu}{\nu}\right). \tag{6.193}$$

Using Little's theorem we obtain for the mean response time $\overline{T}$ and the mean waiting time $\overline{W}$:

$$\overline{T} = \frac{1}{\mu} \cdot \frac{1}{1 - \rho} \cdot \left(1 + \frac{\lambda}{\nu}\right), \tag{6.194}$$

$$\overline{W} = \frac{1}{\mu} \cdot \frac{\rho}{1 - \rho} \cdot \left(1 + \frac{\mu}{\nu}\right). \tag{6.195}$$

If the retrial rate $\nu = \infty$, we have an ordinary M/M/1 FCFS system (see Section 6.2.1).

Example 6.2   We consider an M/M/1 retrial queue with arrival rate $\lambda = 0.8$ and service rate $\mu = 1.0$ and compute the mean queue length $\overline{Q}$ (mean number of requests in the orbit) for different values of the retrial rate ν. The results are listed in Table 6.7.

As expected the mean queue length decreases with increasing retrial rate ν. If the retrial rate tends to ∞ the mean queue length tends to the mean queue length of the corresponding M/M/1 FCFS system.

6.7.2   M/G/1 Retrial Queue

Now we consider a single server retrial system with generally distributed service times. The coefficient of variation of the service time is c_B. Then

Table 6.8 Mean queue length $\overline{Q}$ in a retrial system with different
values of the coefficient of variation of the service time c_B

c_B	0.0	0.2	0.5	0.8	1.0	2.0	5.0	10.0	∞
$\overline{Q}$	4.80	4.86	5.20	5.82	6.40	11.2	44.8	160	∞

the mean queue length (mean number of requests in the orbit) is given by
([Medh03])

$$\overline{Q} = \frac{\rho^2}{1 - \rho} \cdot \left(\frac{1 + c_B^2}{2} + \frac{\mu}{\nu} \right) , \qquad (6.196)$$

and the mean number of requests in the system is given by

$$\overline{K} = \overline{Q} + \rho . \qquad (6.197)$$

From these equations we get the mean response time $\overline{T}$ and the mean
waiting time $\overline{W}$ (mean time in orbit) by using again Little's theorem. It
can be seen easily from Eq. (6.196) that we have an ordinary M/G/1 FCFS
system for $\nu = \infty$ (see Eq. (6.51)) and an M/M/1 retrial system for $c_B^2 = 1$
(see Eq. (6.192)).

Example 6.3   We consider an M/G/1 retrial system with arrival rate $\lambda =$
0.8, service rate $\mu = 1.0$ and retrial rate $\nu = 1.0$ and compute the mean queue
length $\overline{Q}$ (mean number of requests in the orbit) for different values of the
coefficient of variation of the service time c_B. The results are listed in Table
6.8.

As expected, the mean queue length increases with increasing coefficient
of variation of the service time c_B. If c_B tends to 0, the mean queue length
tends to the mean queue length of the corresponding M/D/1 retrial system;
and if it tends to 1, the mean queue length tends to the mean queue length
of the corresponding M/M/1 retrial system.

For other retrial systems such as the multiserver retrial systems or retrial
systems with finite orbit size, see [BEA99, FaTe97, Medh03]. Retrial queues
with finite source of customers has also been studied extensively. Some recent
results can be found in [FaTe97, ABS04, RKS05, Sztr05] An exhaustive bibliography on retrial systems is given in [Arta99a] and [Arta99b].

6.8   SPECIAL CLASSES OF POINT ARRIVAL PROCESSES

In this section several classes of (continuous-time) point processes are described.
These processes are commonly used for modeling of arrival processes.

6.8.1   Point, Renewal, and Markov Renewal Processes

Point processes [KöSc92, Gard86] are generalizations of the well-known Poisson processes, which were introduced in Section 1.3.1.2. Point processes are a class of discrete parameter stochastic processes $\{X_n : n \in \mathbb{N}_0\}$ (see Section 2.1.1) in which each (preferably non-negative) random variable $X_n \in T \subseteq \mathbb{R}_+$ is interpreted as the point in time when the nth event occurs. The inter-event times of a point process are then given by $T_n = X_{n+1} - X_n$.

In the special case where $\{T_n : n \in \mathbb{N}_0\}$ are independent and identically distributed, the processes are known as a *renewal processes* [LMN90, Triv01, Medh03]. With the Kendall notation, *GI* (*G*eneral distribution with *I*ndependent inter-event times) is used to indicate such processes. In the field of renewal processes, events occurring at times X_n are also referred to as *renewals*. Renewal processes are Poisson if the distribution of each T_n is exponential.

In this section, arrival events are of special interest. Therefore, the terms *arrivals* and *interarrival times* are used instead of *events* and *inter-event times*. Often we are not interested in the arrival instants but only in the number of arrivals $N(t)$ within a certain time interval $[0, t), t \in T$. The resulting processes $\{N(t) : t \in T\}$ are called *counting processes*.

Unfortunately, renewal processes are not suitable to describe all arrival processes appropriately. For example, today's Internet traffic is said to comprise complex *long-range dependent*, *fractal*, *self-similar* or *bursty* behavior that could be described better by using nonexponential *heavy-tailed* distributions or by processes that are more general than renewal processes. For a description and applicability of these properties we refer the interested reader to [LTWW94, PaFl95, WTE96, Ost01, HeLu03, Vesi04].

One class of processes that is more general than the class of renewal processes is the class of *Markov renewal processes* [Cinl75, King90, LMN90, Kulk96, Germ00, Vesi04]. *Markov renewal processes* are bivariate processes $\{(Y_n, T_n) : n \in \mathbb{N}_0\}$ where, besides the time until next renewal, T_n, the state Y_n at the time of renewal is also kept track of. The sequence $\{Y_n : n \in \mathbb{N}_0\}$ forms a DTMC. The special case that each T_n is exponentially distributed, with a parameter possibly dependent on state Y_n, gives rise to several useful and well-known arrival processes: the *Markov modulated Poisson process* (MMPP) and the *Markovian arrival process* (MAP). The generalization of the MAP that includes batch arrivals is known as *batch Markovian arrival process* (BMAP).

6.8.2   MMPP

Markov modulated Poisson processes (MMPPs) [FiMe93, HeLu03], that are a subclass of the *doubly stochastic Poisson* (or *Cox*) *processes* described in [Gran76], can be used to model time-varying arrival rates and important correlations between interarrival times. Despite these abilities MMPPs are still tractable by analytical methods.

The current arrival rate $\lambda_i, 0 \leq i \leq m$, of an MMPP is defined by the current state i of an underlying CTMC with $m + 1$ states. The counting process of an MMPP is given by the bivariate process $\{(J(t), N(t)) : t \in T\}$, where $N(t)$ again is the number of arrivals within a certain time interval $[0, t)$, $t \in T$, and $0 \leq J(t) \leq m$ is the state of the underlying CTMC. If we consider $\mathbf{Q}_{\mathrm{MMPP}} = [q_{ij}], 0 \leq i, j \leq m$, being the infinitesimal generator matrix of the underlying CTMC, then the rates of the transitions between the states of the CTMC are given by the nondiagonal elements of $\mathbf{Q}_{\mathrm{MMPP}}$. Assuming the underlying CTMC is homogeneous then its steady-state probability vector π_{MMPP} follows from Eq. (2.58) and Eq. (2.59).

If the arrival rates λ_i, $0 \leq i \leq m$, are collected in an arrival rate vector $\boldsymbol{\lambda} = (\lambda_0, \lambda_1, \ldots, \lambda_m)$ the mean steady-state arrival rate generated by the MMPP is

$$\bar{\lambda} = \pi_{\mathrm{MMPP}} \boldsymbol{\lambda}^T . \tag{6.198}$$

Special cases of the MMPP are the *switched Poisson process* (SPP), which is a two-state MMPP [HoSe83, LMN90], and the *interrupted Poisson process* (IPP) [LMN90], which is an SPP with one of the arrival rates being zero.

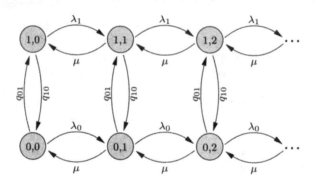

Fig. 6.25   State transition diagram of an MMPP/M/1 queue with two-state MMPP.

The MMPP/M/1 Queue   Consider a queueing system with a two-state MMPP arrival process with arrival rates λ_0, λ_1 and constant service rate μ. The corresponding state transition diagram is shown in Fig. 6.25. A possible GSPN specifying such a state transition diagram is shown in Fig. 6.26. All transitions in this GSPN have exponentially distributed firing times and are denoted with their respective firing rate. Alternative and more compact representations could be constructed by using stochastic reward nets or by the introduction of inhibitor arcs.

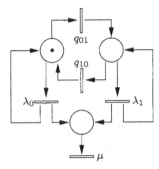

Fig. 6.26 Generalized stochastic Petri net of an MMPP/M/1 queue with two-state MMPP.

To calculate the steady-state probability that the queueing system is empty, we reuse Eq. (6.76) from the GI/M/1 queue, with $\rho = \frac{\bar{\lambda}}{\mu}$:

$$\pi_0 = 1 - \frac{\bar{\lambda}}{\mu}. \tag{6.199}$$

Unfortunately, the steady-state probabilities of the MMPP/M/1 queue, where there are one or more jobs in the system, cannot be derived that easily. A possible way for calculation arises from the fact that the state transition diagram of Fig. (6.25) is skip-free to the left and right and therefore constitutes a QBD process. QBDs can be evaluated using the matrix-geometric method as shown in Section 3.2.

The generator matrix of the underlying two-state CTMC of the MMPP is

$$\mathbf{Q}_{\text{MMPP}} = \begin{pmatrix} -q_{01} & q_{01} \\ q_{10} & -q_{10} \end{pmatrix}. \tag{6.200}$$

The steady-state probabilities $\pi_{\text{MMPP},0}, \pi_{\text{MMPP},1}$ of the underlying CTMC are then given by (cf. [FiMe93])

$$\pi_{\text{MMPP},0} = \frac{q_{10}}{q_{01} + q_{10}}, \tag{6.201}$$

$$\pi_{\text{MMPP},1} = \frac{q_{01}}{q_{01} + q_{10}}. \tag{6.202}$$

According to Eq. (6.198), Eq. (6.201), and Eq. (6.202), $\bar{\lambda}$ can be calculated using:

$$\bar{\lambda} = \boldsymbol{\pi}_{\text{MMPP}} \boldsymbol{\lambda}^T = \pi_{\text{MMPP},0} \lambda_0 + \pi_{\text{MMPP},1} \lambda_1 = \frac{q_{10}\lambda_0 + q_{01}\lambda_1}{q_{01} + q_{10}}. \tag{6.203}$$

Finally, utilizing Eq. (6.203) and Eq. (6.199) the steady-state probability that the queueing system is empty can be calculated:

$$\pi_0 = 1 - \frac{q_{10}\lambda_0 + q_{01}\lambda_1}{(q_{01} + q_{10})\mu}, \tag{6.204}$$

A procedure to solve an MMPP/G/1 queueing system is introduced in [FiMe93]. The MMPP is used to model bursty traffic on an ISDN channel in Section 13.2.4.2.

6.8.3   MAP

A *Markovian arrival process* (MAP) [LMN90] is a generalization of the MMPP. Instead of switching between different arrival rates that depend on the states of the underlying CTMC, an arrival is triggered by specific transitions between states. For example, if the underlying CTMC with $m + 1$ states is in state i, $0 \leq i \leq m$, it leaves this state with rate λ_i. With probability p_{ij} this transition ends in state j, $0 \leq j \leq m$ ($i = j$ may hold), and triggers an arrival. With probability p'_{ij}, $i \neq j$, this transition ends in state j, $0 \leq j \leq m$, without triggering an arrival. All outgoing transition probabilities of a single state sum up to one, i.e.,

$$\sum_{j=0}^{m} p_{ij} + \sum_{\substack{j=0 \\ j \neq i}}^{m} p'_{ij} = 1, \quad 0 \leq i \leq m. \tag{6.205}$$

If we define the variables

$$D_{ij} = \lambda_i p_{ij}, \quad 0 \leq i \leq m, 0 \leq j \leq m,$$
$$C_{ij} = \lambda_i p'_{ij}, \quad 0 \leq i \leq m, 0 \leq j \leq m, i \neq j,$$
$$C_{ii} = -\lambda_i, \quad 0 \leq i \leq m,$$

and the matrices

$$\mathbf{D}_0 = [C_{ij}], \quad 0 \leq i \leq m, 0 \leq j \leq m,$$
$$\mathbf{D}_1 = [D_{ij}], \quad 0 \leq i \leq m, 0 \leq j \leq m,$$

then the infinitesimal generator matrix $\mathbf{Q}_{\text{MAP}}$ of the underlying CTMC is given by

$$\mathbf{Q}_{\text{MAP}} = \mathbf{D}_0 + \mathbf{D}_1. \tag{6.206}$$

The steady-state probability vector of the underlying CTMC are calculated using the well-known equations

$$\boldsymbol{\pi} \mathbf{Q}_{\text{MAP}} = \mathbf{0}, \tag{6.207}$$
$$\boldsymbol{\pi} \mathbf{1} = 1. \tag{6.208}$$

The mean steady-state arrival rate $\bar{\lambda}$ generated by the MAP is then given by

$$\bar{\lambda} = \boldsymbol{\pi} \mathbf{D}_1 \mathbf{1}. \tag{6.209}$$

The superposition of independent MAPs yields a MAP again [LMN90].

The MMPP as MAP If $\mathbf{Q}_{\mathrm{MMPP}}$ is the infinitesimal generator matrix of the underlying CTMC of an MMPP and the arrival rates are organized in a diagonal arrival rate matrix given by

$$
\mathbf{A} = \begin{pmatrix} \lambda_0 & 0 & \cdots & 0 \\ 0 & \lambda_1 & \cdots & 0 \\ \vdots & \vdots & \ddots & \vdots \\ 0 & 0 & \cdots & \lambda_m \end{pmatrix}, \tag{6.210}
$$

then the MMPP can be seen as a special case of MAPs where $\mathbf{D}_0 = \mathbf{Q}_{\mathrm{MMPP}} - \mathbf{A}$ and $\mathbf{D}_1 = \mathbf{A}$ [LMN90, Luca91].

The Poisson Process as MAP Setting $\mathbf{D}_0 = [-\lambda]$, $\mathbf{D}_1 = [\lambda]$ and $\boldsymbol{\pi} = (1)$ and following Eq. (6.209) it can be seen that the MAP is equivalent to a Poisson process with rate $\bar{\lambda} = (1)\mathbf{D}_1\mathbf{1} = \lambda$ [LMN90].

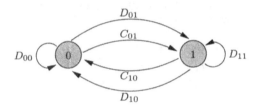

Fig. 6.27 Two-state CTMC underlying a MAP.

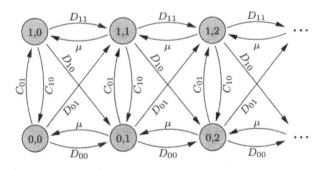

Fig. 6.28 State transition diagram of a MAP/M/1 queue with two-state MAP.

The MAP/M/1 Queue Consider the two-state CTMC underlying a MAP shown in Fig. 6.27, with $D_{00} = \lambda_0 p_{00}$, $D_{01} = \lambda_0 p_{01}$, $C_{01} = \lambda_0 p'_{01}$, $D_{01} = \lambda_1 p_{01}$, $D_{11} = \lambda_1 p_{11}$, $C_{10} = \lambda_1 p'_{10}$. Remember, the transitions denoted by D_{ij} result in an arrival. The state transition diagram of a MAP/M/1 queue comprising this arrival process and a service process with constant rate μ

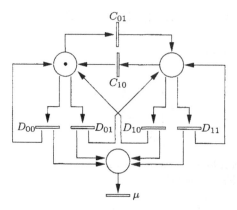

Fig. 6.29 Petri net of a MAP/M/1 queue with two-state MAP.

can be drawn as shown in Fig. 6.28. An equivalent Petri net model of this queueing system is shown in Fig. 6.29. The infinite generator matrix of the underlying CTMC can be given by

$$
\begin{aligned}
\mathbf{Q}_{\text{MAP}} &= \mathbf{D}_0 + \mathbf{D}_1 \\
&= \begin{pmatrix} C_{00} & C_{01} \\ C_{10} & C_{11} \end{pmatrix} + \begin{pmatrix} D_{00} & D_{01} \\ D_{10} & D_{11} \end{pmatrix} \\
&= \begin{pmatrix} -\lambda_0 & \lambda_0 p'_{01} \\ \lambda_1 p'_{10} & -\lambda_1 \end{pmatrix} + \begin{pmatrix} \lambda_0 p_{00} & \lambda_0 p_{01} \\ \lambda_1 p_{10} & \lambda_1 p_{11} \end{pmatrix} \\
&= \begin{pmatrix} \lambda_0(p_{00} - 1) & \lambda_0(p_{01} + p'_{01}) \\ \lambda_1(p_{10} + p'_{10}) & \lambda_1(p_{11} - 1) \end{pmatrix} .
\end{aligned} \tag{6.211}
$$

Because we have:

$$
p_{00} + p_{01} + p'_{01} = 1, \tag{6.212}
$$

$$
p_{11} + p_{10} + p'_{10} = 1, \tag{6.213}
$$

the rows of $\mathbf{Q}_{\text{MAP}}$ sum up to 0. Furthermore, assuming irreducibility of $\mathbf{Q}_{\text{MAP}}$, p_{00} and p_{11} are smaller than 1, i.e., $\mathbf{Q}_{\text{MAP}}$ has negative diagonal elements. Both properties are necessary for generator matrices. To obtain the steady-state probability π_0 that there are is no job in the queueing system, we first have to derive the steady-state probabilities $\pi_{\text{MAP},0}$ and $\pi_{\text{MAP},1}$ of the underlying CTMC by using Eqs. (6.207) and (6.208):

$$
\pi_{\text{MAP}} \mathbf{Q}_{\text{MAP}} = \mathbf{0}, \tag{6.214}
$$

$$
\pi_{\text{MAP}} \mathbf{1} = 1, \tag{6.215}
$$

which are equivalent to the equations

$$
\pi_{\text{MAP},0}(\lambda_0(p_{00} - 1)) + \pi_{\text{MAP},1}(\lambda_1(p_{10} + p'_{10})) = 0, \tag{6.216}
$$

$$\pi_{\text{MAP},0}(\lambda_0(p_{01} + p'_{01})) + \pi_{\text{MAP},1}(\lambda_1(p_{10} - 1)) = 0, \tag{6.217}$$

$$\pi_{\text{MAP},0} = 1 - \pi_{\text{MAP},1}, \tag{6.218}$$

$$\pi_{\text{MAP},1} = 1 - \pi_{\text{MAP},0}. \tag{6.219}$$

Transformation of Eqs. (6.216) to (6.217) under consideration of Eqs. (6.212) and (6.213) yields

$$\pi_{\text{MAP},0} = \frac{\lambda_1(p_{10} + p'_{10})}{\lambda_0(p_{01} + p'_{01}) + \lambda_1(p_{10} + p'_{10})}, \tag{6.220}$$

and symmetrically:

$$\pi_{\text{MAP},1} = \frac{\lambda_0(p_{01} + p'_{01})}{\lambda_0(p_{01} + p'_{01}) + \lambda_1(p_{10} + p'_{10})}. \tag{6.221}$$

Using Eqs. (6.209), (6.220), and (6.221), we get the mean arrival rate $\bar{\lambda}$ generated by the two-state MAP:

$$
\begin{aligned}
\bar{\lambda} &= \boldsymbol{\pi}_{\text{MAP}} \mathbf{D}_1 \mathbf{1} \\
&= (\pi_{\text{MAP},0}, \pi_{\text{MAP},1}) \begin{pmatrix} \lambda_0 p_{00} & \lambda_0 p_{01} \\ \lambda_1 p_{10} & \lambda_1 p_{11} \end{pmatrix} (1,1)^T \\
&= \pi_{\text{MAP},0}\lambda_0 p_{00} + \pi_{\text{MAP},1}\lambda_1 p_{10} + \pi_{\text{MAP},0}\lambda_0 p_{01} + \pi_{\text{MAP},1}\lambda_1 p_{11} \\
&= \frac{\lambda_0\lambda_1((p_{01} + p_{00})(p_{10} + p'_{10}) + (p_{10} + p_{11})(p_{01}p'_{01}))}{\lambda_0(p_{01} + p'_{01}) + \lambda_1(p_{10} + p'_{10})}.
\end{aligned}
\tag{6.222}
$$

Putting $\bar{\lambda}$ of Eq. (6.222) into Eq. (6.199), we get the steady-state probability that there is no job in the MAP/M/1 queue:

$$\pi_0 = 1 - \frac{\lambda_0\lambda_1((p_{01} + p_{00})(p_{10} + p'_{10}) + (p_{10} + p_{11})(p_{01}p'_{01}))}{\lambda_0\mu(p_{01} + p'_{01}) + \lambda_1\mu(p_{10} + p'_{10})}. \tag{6.223}$$

The state transition diagram of the MAP/M/1 queue shown in Fig. 6.28 is skip-free to the left and right. The underlying stochastic processes can therefore be seen as a QBD that can be solved using the matrix-geometric method as introduced in Section 3.2.

6.8.4 BMAP

The *batch Markovian arrival process* (BMAP) [Luca91, Luca93] is equivalent to the *versatile Markovian point process*. The versatile Markovian point process, which has a more complex notation than the BMAP, was introduced by Neuts [Neut79] and is therefore often referred to as *N-process* [Rama80]. In contrast to the MAP, which was introduced in Section 6.8.3, the BMAP includes batch arrivals and can therefore be seen as a generalization of the MAP. To handle batch arrivals, a notation slightly different to the one needed for MAPs is introduced to describe BMAPs. The probability for a transition from state i, $0 \leq i \leq m$, to state j, $0 \leq j \leq m$, without an arrival (which was p'_{ij} in the MAP case) is denoted by $p_i(0, j)$, $i \neq j$. A transition resulting in a single arrival (p_{ij} in the MAP case) has now assigned the probability $p_i(1, j)$. Batch arrivals of size $k \geq 0$ can be generated by transitions occurring with probability $p_i(k, j)$. Considering a single state i, the sum of the probabilities of all outgoing transitions has to be equal to 1, similar to Eq. (6.205):

$$\sum_{k=1}^{\infty} \sum_{j=0}^{m} p_i(k, j) + \sum_{\substack{j=0 \\ j \neq i}}^{m} p_i(0, j) = 1, \quad 0 \leq i \leq m. \qquad (6.224)$$

The infinite generator matrix $\mathbf{Q}_{\text{BMAP}}$ of the CTMC underlying the BMAP can then be given by

$$\mathbf{Q}_{\text{BMAP}} = \begin{pmatrix} \mathbf{D}_0 & \mathbf{D}_1 & \mathbf{D}_2 & \mathbf{0} & \cdots \\ \mathbf{0} & \mathbf{D}_0 & \mathbf{D}_1 & \mathbf{D}_2 & \cdots \\ \mathbf{0} & \mathbf{0} & \mathbf{D}_0 & \mathbf{D}_1 & \cdots \\ \vdots & \vdots & \vdots & \vdots & \ddots \end{pmatrix}, \qquad (6.225)$$

where the matrices $\mathbf{D}_k$ are given by

$$\mathbf{D}_0 = [C_{ij}], \quad 0 \leq i \leq m, 0 \leq j \leq m, \qquad (6.226)$$
$$\mathbf{D}_k = [D_{k,ij}], \quad 0 \leq i \leq m, 0 \leq j \leq m, 0 \leq k. \qquad (6.227)$$

The elements C_{ij} and $D_{k,ij}$ are defined by

$$D_{k,ij} = \lambda_i p_i(k, j), \quad 0 \leq i \leq m, \ 0 \leq j \leq m, \ 0 < k,$$
$$C_{ij} = \lambda_i p_i(0, j), \quad 0 \leq i \leq m, \ 0 \leq j \leq m, \ i \neq j,$$
$$C_{ii} = -\lambda_i, \quad 0 \leq i \leq m.$$

Knowing the matrices $\mathbf{D}_k$ the infinitesimal generator $\mathbf{D}$ can be defined as

$$\mathbf{D} = \sum_{k=0}^{\infty} \mathbf{D}_k. \qquad (6.228)$$

The steady-state probability vector $\boldsymbol{\pi}$ of the CTMC with generator $\mathbf{D}$ can be calculated as usual:

$$\boldsymbol{\pi} \mathbf{D} = \mathbf{0}, \qquad (6.229)$$
$$\boldsymbol{\pi} \mathbf{1} = 1. \qquad (6.230)$$

The mean steady-state arrival rate $\bar{\lambda}$ generated by the BMAP is finally given by [Luca93]

$$\bar{\lambda} = \boldsymbol{\pi} \sum_{k=1}^{\infty} k \mathbf{D}_k \mathbf{1}. \qquad (6.231)$$

The superposition of a finite number of independent BMAPs yields a BMAP again [Luca93].

Many queueing systems with BMAP arrivals, e.g., BMAP/G/1 queues, are of M/G/1-type and can therefore be analyzed using matrix-analytic methods which were introduced in Section 6.3.1.2. See, for instance, Example 6.4.

The MAP as BMAP A MAP is a BMAP where only single arrivals can occur, so for $k > 1$ all $p_i(k, j) = 0$ for all $0 \le i, j \le m$, and therefore $\mathbf{D}_k = \mathbf{0}$, $k > 1$ [Luca91].

Example 6.4 As a numerical example a BMAP/M/1 queue is evaluated. BMAP/M/1 queues are of M/G/1-type and can thus be evaluated using the matrix-analytic method (see Section 6.3.1.2). To keep the calculations traceable, the example is kept quite simple. Its purpose is not to provide the most elegant method for solving BMAP/M/1 queues but to illustrate the concepts of the matrix-analytic method.

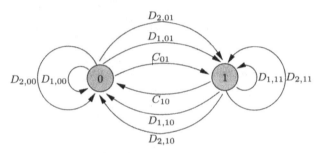

Fig. 6.30 Two-state CTMC underlying a BMAP.

In Fig. 6.30 the CTMC underlying the two-state BMAP arrival process is investigated in this example is shown. The considered BMAP produces batch arrivals with a maximum batch size of $k = 2$.

For this example the following transition rates are defined:

$$C_{01} = \frac{1}{2}, \quad C_{10} = \frac{1}{3},$$

$$D_{1,01} = \frac{1}{2}, \quad D_{1,10} = \frac{1}{3},$$

$$D_{2,01} = \frac{1}{2}, \quad D_{2,10} = 1,$$

$$D_{1,00} = \frac{1}{4}, \quad D_{1,11} = 1,$$

$$D_{2,00} = \frac{1}{4}, \quad D_{2,11} = \frac{1}{3}.$$

Consider $\mu = 11$ to be the rate of the exponentially distributed service time of the BMAP/M/1 queue. Then the resulting state transition diagram of the

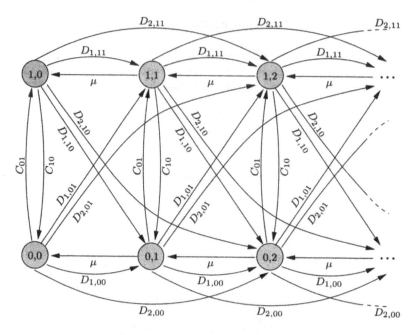

Fig. 6.31 State transition diagram of the BMAP/M/1 queue.

BMAP/M/1 queue is derived as shown in Fig. 6.30. The generator of this BMAP/M/1 queue is given by

$$\mathbf{Q} = \begin{pmatrix} \mathbf{B}_0 & \mathbf{B}_1 & \mathbf{B}_2 & 0 & 0 & \cdots \\ \mathbf{A}_0 & \mathbf{A}_1 & \mathbf{A}_2 & \mathbf{A}_3 & 0 & \cdots \\ 0 & \mathbf{A}_0 & \mathbf{A}_1 & \mathbf{A}_2 & \mathbf{A}_3 & \cdots \\ 0 & 0 & \mathbf{A}_0 & \mathbf{A}_1 & \mathbf{A}_2 & \cdots \\ 0 & 0 & 0 & \mathbf{A}_0 & \mathbf{A}_1 & \cdots \\ \vdots & \vdots & \vdots & \vdots & \vdots & \ddots \end{pmatrix}. \tag{6.232}$$

With

$$C_{00} = -(C_{01} + D_{1,01} + D_{2,01} + D_{1,00} + D_{2,00}) = -2 \tag{6.233}$$

and

$$C_{11} = -(C_{10} + D_{1,10} + D_{2,10} + D_{1,11} + D_{2,11}) = -3 \tag{6.234}$$

the submatrices of $\mathbf{Q}$ can be calculated by

$$\mathbf{B}_0 = \mathbf{D}_0 = \begin{pmatrix} C_{00} & C_{01} \\ C_{10} & C_{11} \end{pmatrix} = \begin{pmatrix} -2 & \frac{1}{2} \\ \frac{1}{3} & -3 \end{pmatrix}, \tag{6.235}$$

$$\mathbf{A}_0 = \begin{pmatrix} \mu & 0 \\ 0 & \mu \end{pmatrix} = \begin{pmatrix} 11 & 0 \\ 0 & 11 \end{pmatrix},$$

$$\mathbf{A}_1 = \begin{pmatrix} C_{00} - \mu & C_{01} \\ C_{10} & C_{11} - \mu \end{pmatrix} = \begin{pmatrix} -13 & \frac{1}{2} \\ \frac{1}{3} & -14 \end{pmatrix}, \tag{6.236}$$

$$\mathbf{B}_1 = \mathbf{A}_2 = \mathbf{D}_1 = \begin{pmatrix} D_{1,00} & D_{1,01} \\ D_{1,10} & D_{1,11} \end{pmatrix} = \begin{pmatrix} \frac{1}{4} & \frac{1}{2} \\ \frac{1}{3} & \frac{1}{3} \end{pmatrix}, \tag{6.237}$$

$$\mathbf{B}_2 = \mathbf{A}_3 = \mathbf{D}_2 = \begin{pmatrix} D_{2,00} & D_{2,01} \\ D_{2,10} & D_{2,11} \end{pmatrix} = \begin{pmatrix} \frac{1}{4} & \frac{1}{2} \\ \frac{1}{3} & \frac{1}{3} \end{pmatrix}. \tag{6.238}$$

The matrices $\mathbf{D}_k$ in Eqs. (6.235), (6.237), and (6.238) are defined according to Eq. (6.226) and Eq. (6.227).

There are two possibilities to evaluate this model. On the one hand, we can use the mean steady-state arrival rate $\bar{\lambda}_{\text{BMAP}}$ generated by the BMAP as given by Eq. (6.231). The second, more complicated approach is to use Neuts' matrix-analytic method as introduced in Section 6.3.1.2.

Utilizing the Mean Arrival Rate The performance measure of interest often is the number of jobs in the queueing system. Many other performance measures can be further derived using Little's law. Therefore, the state probabilities of the BMAP are of no particular interest. Thus, the BMAP can be sufficiently described by the mean steady-state arrival rate $\bar{\lambda}_{\text{BMAP}}$ it generates.

Knowing the infinitesimal generator $\mathbf{D}$ via Eq. (6.228),

$$\mathbf{D} = \sum_{k=0}^{\infty} \mathbf{D}_k = \mathbf{D}_0 + \mathbf{D}_1 + \mathbf{D}_2$$

$$= \begin{pmatrix} -2 & \frac{1}{2} \\ \frac{1}{3} & -3 \end{pmatrix} + \begin{pmatrix} \frac{1}{4} & \frac{1}{2} \\ \frac{1}{3} & 1 \end{pmatrix} + \begin{pmatrix} \frac{1}{4} & \frac{1}{2} \\ 1 & \frac{1}{3} \end{pmatrix} = \begin{pmatrix} -\frac{3}{2} & \frac{3}{2} \\ \frac{5}{3} & -\frac{5}{3} \end{pmatrix}, \tag{6.239}$$

the values $\pi_{\text{BMAP},1}$ and $\pi_{\text{BMAP},2}$ can be obtained by solving (6.229) and (6.230):

$$\pi_{\text{BMAP}} \mathbf{D} = (\pi_{\text{BMAP},1}, \pi_{\text{BMAP},2}) \begin{pmatrix} -\frac{3}{2} & \frac{3}{2} \\ \frac{5}{3} & -\frac{5}{3} \end{pmatrix} = \mathbf{0}. \tag{6.240}$$

As expected, $det(\mathbf{D}) = 0$ and therefore the system of equations given in Eq. (6.240) is linearly dependent and produces

$$\pi_{\text{BMAP},2} = \frac{9}{10} \pi_{\text{BMAP},1}. \tag{6.241}$$

Considering the normalization condition

$$\pi_{\text{BMAP}} \mathbf{1} = \pi_{\text{BMAP},1} + \pi_{\text{BMAP},2} = 1 \tag{6.242}$$

we get

$$\pi_{\text{BMAP},1} = \frac{10}{19} \tag{6.243}$$

and

$$\pi_{\text{BMAP},2} = \frac{9}{19}. \tag{6.244}$$

Knowing the steady-state probabilities, λ_{BMAP} can be calculated according to Eq. (6.231):

$$
\begin{aligned}
\bar{\lambda}_{BMAP} &= \pi_{BMAP} \sum_{k=1}^{\infty} k \mathbf{D}_k \mathbf{1} \\
&= (\pi_{BMAP,1}, \pi_{BMAP,2}) (\mathbf{D}_1 + 2\mathbf{D}_2)(1,1)^T \\
&= \left(\frac{10}{19}, \frac{9}{19}\right)\left(\begin{pmatrix} \frac{1}{4} & \frac{1}{2} \\ \frac{1}{3} & \frac{1}{3} \end{pmatrix} + 2\begin{pmatrix} \frac{1}{4} & \frac{1}{2} \\ \frac{1}{3} & \frac{1}{3} \end{pmatrix}\right)(1,1)^T \\
&= \frac{117}{38} \approx 3.078 \,.
\end{aligned}
\tag{6.245}
$$

Using the mean arrival rate $\bar{\lambda}_{BMAP}$ produced by the BMAP, the utilization of the BMAP/M/1 – FCFS queue can be calculated using the utilization law given by Eq. (6.3):

$$
\rho = \frac{\bar{\lambda}_{BMAP}}{\mu} = \frac{117}{38 \cdot 11} = \frac{117}{418} \approx 0.2799 \,.
\tag{6.246}
$$

With $\rho < 1$ the stability condition of Eq. (6.5) is fulfilled. Inserting ρ into Eq. (6.11), the steady-state probability π_0 for 0 jobs in the system can be derived:

$$
\pi_0 = 1 - \rho = 1 - 0.2799 = 0.7201 \,.
\tag{6.247}
$$

Unfortunately, all other steady-state probabilities with one or more jobs in the system cannot be derived that easily. In contrast to the MMPP/M/1 queue and the MAP/M/1 queue, the state transition diagram of the BMAP/M/1 queue shown in Fig. (6.31) is not skip-free to the right. Therefore, it cannot be classified as a QBD and cannot be evaluated using the matrix-geometric method. Instead, the matrix-analytic method can be used, which was introduced in Section 6.3.1.2.

Using the Matrix-Analytic Method   To evaluate all steady-state probabilities of the number of jobs in the BMAP/M/1 system, the matrix-analytic method can be used. The theory of the matrix-analytic method has already been covered in more detail in Section 6.3.1.2. Its full potential is not exhausted by the example at hand. Nevertheless, the example is illustrative and the results can be compared to the ones obtained above.

To keep the notation comparable to Neuts' book [Neut89], we use the randomization method introduced in Eq. (3.5) to switch from the CTMC given by Fig. (6.30) to an equivalent DTMC. The DTMC is defined by the probability matrix $\mathbf{P}$ and comprises the same steady-state probabilities as the matrix $\mathbf{Q}$ of the BMAP/M/1 queue that is given by Eq. (6.232). $\mathbf{P}$ can be calculated from $\mathbf{Q}$ via

$$
\mathbf{P} = \frac{1}{q}\mathbf{Q} + \mathbf{I} \,,
\tag{6.248}
$$

where q is set to $15 > max|\mathbf{Q}_{ij}| = |-14|$.

Therefore, the randomization yields

$$
\mathbf{P} = \begin{pmatrix}
\tilde{\mathbf{B}}_0 & \tilde{\mathbf{B}}_1 & \tilde{\mathbf{B}}_2 & 0 & 0 & \cdots \\
\tilde{\mathbf{A}}_0 & \tilde{\mathbf{A}}_1 & \tilde{\mathbf{A}}_2 & \tilde{\mathbf{A}}_3 & 0 & \cdots \\
0 & \tilde{\mathbf{A}}_0 & \tilde{\mathbf{A}}_1 & \tilde{\mathbf{A}}_2 & \tilde{\mathbf{A}}_3 & \cdots \\
0 & 0 & \tilde{\mathbf{A}}_0 & \tilde{\mathbf{A}}_1 & \tilde{\mathbf{A}}_2 & \cdots \\
0 & 0 & 0 & \tilde{\mathbf{A}}_0 & \tilde{\mathbf{A}}_1 & \cdots \\
\vdots & \vdots & \vdots & \vdots & \vdots & \ddots
\end{pmatrix},
\tag{6.249}
$$

comprising the submatrices

$$
\tilde{\mathbf{B}}_0 = \frac{1}{q}\mathbf{B}_0 + \mathbf{I} = \begin{pmatrix} \frac{13}{15} & \frac{1}{30} \\ \frac{1}{45} & \frac{4}{5} \end{pmatrix} = \begin{pmatrix} 0.8667 & 0.0333 \\ 0.0222 & 0.8000 \end{pmatrix},
\tag{6.250}
$$

$$
\tilde{\mathbf{A}}_0 = \frac{1}{q}\mathbf{A}_0 = \begin{pmatrix} \frac{11}{15} & 0 \\ 0 & \frac{11}{15} \end{pmatrix} = \begin{pmatrix} 0.7333 & 0.0000 \\ 0.0000 & 0.7333 \end{pmatrix},
\tag{6.251}
$$

$$
\tilde{\mathbf{A}}_1 = \frac{1}{q}\mathbf{A}_1 + \mathbf{I} = \begin{pmatrix} \frac{2}{15} & \frac{1}{30} \\ \frac{1}{45} & \frac{1}{15} \end{pmatrix} = \begin{pmatrix} 0.1333 & 0.0333 \\ 0.0222 & 0.0667 \end{pmatrix},
\tag{6.252}
$$

$$
\tilde{\mathbf{B}}_1 = \tilde{\mathbf{A}}_2 = \frac{1}{q}\mathbf{B}_1 = \begin{pmatrix} \frac{1}{60} & \frac{1}{30} \\ \frac{1}{45} & \frac{1}{15} \end{pmatrix} = \begin{pmatrix} 0.0167 & 0.0333 \\ 0.0222 & 0.0667 \end{pmatrix},
\tag{6.253}
$$

$$
\tilde{\mathbf{B}}_2 = \tilde{\mathbf{A}}_3 = \frac{1}{q}\mathbf{B}_2 = \begin{pmatrix} \frac{1}{60} & \frac{1}{30} \\ \frac{1}{15} & \frac{1}{45} \end{pmatrix} = \begin{pmatrix} 0.0167 & 0.0333 \\ 0.0667 & 0.0222 \end{pmatrix}.
\tag{6.254}
$$

Knowing these submatrices, the matrix $\mathbf{A}$ can be calculated according to Eq. (6.65):

$$
\mathbf{A} = \sum_{k=0}^{\infty} \tilde{\mathbf{A}}_k = \tilde{\mathbf{A}}_0 + \tilde{\mathbf{A}}_1 + \tilde{\mathbf{A}}_2 + \tilde{\mathbf{A}}_3 = \begin{pmatrix} 0.9000 & 0.1000 \\ 0.1111 & 0.8889 \end{pmatrix}.
\tag{6.255}
$$

The steady-state probability vector $\boldsymbol{\alpha} = (\alpha_1 \ \alpha_2)$ can be derived via $\boldsymbol{\alpha}\mathbf{A} = \boldsymbol{\alpha}$, which is a linearly dependent system of equations because $det(\mathbf{A} - I) \approx 0$. Together with the normalization condition $\alpha_1 + \alpha_2 = 1$, we get

$$
\boldsymbol{\alpha} = (0.5263, 0.4737).
\tag{6.256}
$$

To check the stability condition $\boldsymbol{\alpha}\boldsymbol{\beta}^T < 1$ from Eq. (6.68), the vector $\boldsymbol{\beta}$ has to be calculated using Eq. (6.67):

$$
\boldsymbol{\beta} = \left(\sum_{k=1}^{\infty} k\tilde{\mathbf{A}}_k \mathbf{1} \right)^T = \left(\tilde{\mathbf{A}}_1 \mathbf{1} + 2\tilde{\mathbf{A}}_2 \mathbf{1} + 3\tilde{\mathbf{A}}_3 \mathbf{1} \right)^T = (0.4167, 0.5333).
\tag{6.257}
$$

Due to $\boldsymbol{\alpha}\boldsymbol{\beta}^T = 0.4719 < 1$ the system under consideration is stable and can be evaluated using the matrix-analytic method.

To calculate the matrix $\mathbf{G}$, we will use the following iterative method suggested by Neuts [Neut89]:

$$\mathbf{G}_{(0)} = \mathbf{0}, \tag{6.258}$$

$$\mathbf{G}_{(n)} = \sum_{k=0}^{\infty} \tilde{\mathbf{A}}_k \mathbf{G}_{(n-1)}^k. \tag{6.259}$$

After 13 iterations the prior chosen stopping criterion

$$\|\mathbf{G}_{(13)} - \mathbf{G}_{(12)}\|_{\infty} = 3.3 \cdot 10^{-5} < \epsilon = 5 \cdot 10^{-5} \tag{6.260}$$

is fulfilled and the matrix $\mathbf{G}$ converges to

$$\mathbf{G} = \mathbf{G}_{(13)} = \begin{pmatrix} 0.8998 & 0.1002 \\ 0.1168 & 0.8832 \end{pmatrix}. \tag{6.261}$$

Matrix $\mathbf{G}$ is stochastic and $det(\mathbf{G} - \mathbf{I}) \approx 0$. With $\mathbf{gG} = \mathbf{g}$ and $\mathbf{g1} = 1$ the vector $\mathbf{g} = (0.5383, 0.4617)$ can be derived.

Knowing the matrix $\mathbf{G}$, the following steps are needed to derive the steady-state probability vector $\boldsymbol{\nu}_0$ of the boundary level $i = 0$ by Eq. (6.58): After calculating the matrix $\hat{\mathbf{A}}$ using Eq. (6.61),

$$\hat{\mathbf{A}} = \tilde{\mathbf{A}}_1 + \tilde{\mathbf{A}}_2 \mathbf{G} + \tilde{\mathbf{A}}_3 \mathbf{G}^2 = \begin{pmatrix} 0.1729 & 0.0938 \\ 0.1094 & 0.1573 \end{pmatrix}, \tag{6.262}$$

all elements are known for getting the vector $\boldsymbol{\psi}_1$ using Eq. (6.63):

$$\boldsymbol{\psi}_1 = \left(\left(\mathbf{I} - \tilde{\mathbf{A}}_0 - \hat{\mathbf{A}} \right) \left(\mathbf{I} - \mathbf{A} + \left(1 - \boldsymbol{\beta}^T \right) \mathbf{g} \right)^{-1} \mathbf{1} + \left(1 - \alpha \boldsymbol{\beta}^T \right)^{-1} \tilde{\mathbf{A}}_0 \mathbf{1} \right)^T$$
$$= (1.2905, 1.5032). \tag{6.263}$$

Using Eq. (6.66), the auxiliary matrix $\hat{\mathbf{B}}$ can be obtained:

$$\hat{\mathbf{B}} = \tilde{\mathbf{B}}_1 + \tilde{\mathbf{B}}_2 \mathbf{G} = \begin{pmatrix} 0.0356 & 0.0644 \\ 0.0848 & 0.0930 \end{pmatrix}, \tag{6.264}$$

which is then used to calculate the vector $\boldsymbol{\psi}_2$ via Eq. (6.64):

$$\boldsymbol{\psi}_2 = \left(1 + \left(\tilde{\mathbf{A}}_2 + \tilde{\mathbf{A}}_3 - \hat{\mathbf{B}} \right) \left(\mathbf{I} - \mathbf{A} + \left(1 - \boldsymbol{\beta}^T \right) \mathbf{g} \right)^{-1} \mathbf{1} \right.$$
$$\left. + \left(1 - \alpha \boldsymbol{\beta}^T \right)^{-1} \tilde{\mathbf{A}}_3 \mathbf{1} \right)^T = (1.0970, 1.1641). \tag{6.265}$$

The knowledge of the auxiliary vectors $\boldsymbol{\psi}_1$ and $\boldsymbol{\psi}_2$ from Eq. (6.263) and Eq. (6.265) allows us to derive the vector $\tilde{\boldsymbol{\kappa}}$ via Eq. (6.62):

$$\tilde{\boldsymbol{\kappa}} = \boldsymbol{\psi}_2 + \boldsymbol{\psi}_1 \left(\left(\tilde{\mathbf{A}}_2 + \tilde{\mathbf{A}}_3 \mathbf{G} \right) \left(\mathbf{I} - \hat{\mathbf{A}} \right)^{-1} \right)^T = (1.2905, 1.5032). \tag{6.266}$$

The last step before being able to calculate the steady-state probability vector ν_0 using Eq. (6.58) is to derive the steady-state probability vector κ of the stochastic matrix $\mathbf{K}$. For this, at first the auxiliary matrix $\mathbf{X}$ needs to be obtained using Eq. (6.60):

$$\mathbf{X} = (\mathbf{I} - \hat{\mathbf{A}})^{-1}\tilde{\mathbf{A}}_0 = \begin{pmatrix} 0.8998 & 0.1002 \\ 0.1168 & 0.8832 \end{pmatrix}. \tag{6.267}$$

Then the matrix $\mathbf{K}$ can be derived via Eq. (6.59):

$$\mathbf{K} = \tilde{\mathbf{B}}_0 + \hat{\mathbf{B}}\mathbf{X} = \begin{pmatrix} 0.9062 & 0.0938 \\ 0.1094 & 0.8906 \end{pmatrix}. \tag{6.268}$$

Matrix $\mathbf{K}$ is stochastic and $det(\mathbf{K}-\mathbf{I}) \approx 0$. With $\kappa\mathbf{K} = \kappa$ and $\kappa\mathbf{1} = 1$ the vector $\kappa = (0.5383, 0.4617)$ can be derived. Finally, by knowing $\tilde{\kappa}_1$ by Eq. (6.266) and κ the steady-state probability vector ν_0 containing the steady-state probabilities of both phases of level $i = 0$ can be calculated using Eq. (6.58):

$$\nu_0 = \left(\kappa\tilde{\kappa}^T\right)^{-1}\kappa = (0.3877, 0.3324). \tag{6.269}$$

If we are only interested in the number of jobs in the BMAP/M/1 queue and not in the states of the BMAP, we can calculate the overall steady-state probability for level $i = 0$ by

$$\pi_0 = \nu_0\mathbf{1} = 0.3877 + 0.3324 = 0.7201, \tag{6.270}$$

which is equivalent to the steady-state probability π_0 derived in Eq. (6.247) using the mean arrival rate $\bar{\lambda}$.

The steady-state probability vector of level $i = 1$ can be calculated using Eq. (6.57):

$$\nu_1 = \nu_0\bar{\mathbf{B}}_1\left(\mathbf{I} - \bar{\mathbf{A}}_1\right)^{-1} = (0.0604, 0.0730), \tag{6.271}$$

which leads to $\pi_1 = \nu_1\mathbf{1} = 0.0604 + 0.0730 = 0.1335$.

The correctness of ν_0 and ν_1 can also be checked by the following equation, which can be derived directly from $\nu\mathbf{P} = \nu$:

$$\nu_0\tilde{\mathbf{B}}_0 + \nu_1\tilde{\mathbf{A}}_0 = (0.3877, 0.3324) = \nu_0. \tag{6.272}$$

The steady-state probability vectors of all other levels can be calculated equivalently to Eq. (6.271); e.g.,

$$\begin{aligned}
\nu_2 &= \left(\nu_0\bar{\mathbf{B}}_2 + \sum_{k=1}^{1}\nu_k\bar{\mathbf{A}}_{3-k}\right)\left(\mathbf{I} - \bar{\mathbf{A}}_1\right)^{-1} \\
&= \left(\nu_0\bar{\mathbf{B}}_2 + \nu_1\left(\tilde{\mathbf{A}}_2 + \tilde{\mathbf{A}}_3\mathbf{G}\right)\right)\left(\mathbf{I} - \bar{\mathbf{A}}_1\right)^{-1} \\
&= (0.0503, 0.0434)
\end{aligned} \tag{6.273}$$

leads to $\pi_2 = \nu_2 1 = 0.0503 + 0.0434 = 0.0927$.

Problem 6.1   Consider two M/M/1 queues, with arrival rate $\lambda_1 = 0.5/\sec$ and service rate $\mu_1 = 1/\sec$ each, and an M/M/2 queue with arrival rate $\lambda_2 = 2\lambda_1 = 1/\sec$ and service rate $\mu_2 = \mu_1 = 1/\sec$ for each server. Compare utilization ρ_i, mean number of jobs in the system $\overline{K}_i$, mean waiting time $\overline{W}_i$, and mean response time $\overline{T}_i$ for both cases $(i = 1, 2)$. Discuss the results.

Problem 6.2   Consider an M/M/1/K system and determine the through-put and the mean number of jobs $\overline{K}$ as a function of the maximum number of jobs in the system K for the cases $\lambda = \mu/2$, $\lambda = \mu$, and $\lambda = 2\mu$ with $\mu = 1/\sec$. Discuss the results.

Problem 6.3   Compare the mean number of jobs $\overline{K}$ of an M/D/1, an M/M/1, and an M/G/1 queue $(c_B^2 = 2)$ with $\lambda = 0.8/\sec$ and $\mu = 1/\sec$. Give a comprehensible explanation why the values are different although the arrival rate λ and the service rate μ are always the same.

Problem 6.4   Compare the mean number of jobs $\overline{K}$ of an M/G/1 and a GI/M/1 queue with $\lambda = 0.9/\sec$, $\mu = 1/\sec$, $c_A^2 = 0.5$, and $c_B^2 = 0.5$, respectively. Discuss the results.

Problem 6.5   Consider a GI/G/1 queue with $\rho = 0.8$, $c_A^2 = 0.5$, and $c_B^2 = 2$. Compute the mean queue length $\overline{Q}$ using the Allen–Cunneen, the Krämer–Langenbach-Belz, and the Kimura approximations. Also compute upper and lower bounds for the mean queue length (see Section 6.3.4).

Problem 6.6   Consider an M/M/1 priority system with 2 priority classes, and $\mu_1 = \mu_2 = 1/\sec$ and $\lambda_1 = \lambda_2 = 0.4/\sec$. Compute the mean waiting time $\overline{W}$ for the two classes for a system with and without preemption. Discuss the results and compare them to the corresponding M/M/1-FCFS system.

Problem 6.7   Consider an asymmetric M/M/2 queue with $\lambda = 1/\sec$, $\mu_1 = 1/\sec$, and $\mu_2 = 1.5/\sec$. Compute the utilizations ρ_1 and ρ_2 of the two servers and the mean number of jobs $\overline{K}$ using the approximation method and the exact methods for the strategy random selection of the free server and the strategy selection of the fastest free server (see Section 6.5). Discuss the results.

Problem 6.8   Consider an M/M$^{[5,5]}$/1 batch system and an M/M$^{[2,5]}$/1 batch system with $\rho = 0.5$. Compute the mean queue length $\overline{Q}$ for both cases and discuss the results.

Problem 6.9   Write down the steady-state balance equation for the M/M/1 retrial queue using Fig. 6.24 and show that the Eqs. (6.187), (6.188), and (6.189) satisfy the balance equations.

Problem 6.10   Construct a GSPN model for the M/M/1 retrial queue.

Problem 6.11 Consider an M/M/1 retrial queue with arrival rate $\lambda = 1.0$ and service rate $\mu = 1.5$ and compute the mean response time $\overline{T}$ for different values of the retrial rate ν. Discuss the results.

Problem 6.12 Consider an M/G/1 retrial queue with arrival rate $\lambda = 1.0$, service rate $\mu = 1.5$ and retrial rate $\nu = 1.0$. Compute the mean waiting time $\overline{W}$ (mean time in orbit) for different values of the coefficient of variation of the service time c_B. Discuss the results.

Problem 6.13 Construct an SRN model for the MMPP/M/1 queue.

Problem 6.14 Construct an SRN model for the MAP/M/1 queue.

7

Queueing Networks

Queueing networks consisting of several service stations are more suitable for representing the structure of many systems with a large number of resources than models consisting of a single service station. In a queueing network at least two service stations are connected to each other. A station, i.e., a *node*, in the network represents a resource in the real system. Jobs in principle can be transferred between any two nodes of the network; in particular, a job can be directly returned to the node it has just left.

A queueing network is called *open* when jobs can enter the network from outside and jobs can also leave the network. Jobs can arrive from outside the network at every node and depart from the network from any node. A queueing network is said to be *closed* when jobs can neither enter nor leave the network. The number of jobs in a closed network is constant. A network in which a new job enters whenever a job leaves the system can be considered as a closed one.

In Fig. 7.1 an open queueing network model of a simple computer system is shown. An example of a closed queueing network model is shown in Fig. 7.2. This is the central-server model, a particular closed network that has been proposed by [Buze71] for the investigation of the behavior of multiprogramming system with a fixed degree of multiprogramming. The node with service rate μ_1 is the central-server representing the central processing unit (CPU). The other nodes model the peripheral devices: disk drives, printers, magnetic tape units, etc. The number of jobs in this closed model is equal to the degree of multiprogramming. A closed tandem queueing network with two nodes is shown in Fig. 7.3. A very frequently occurring queueing network is the machine repairman model, shown in Fig. 7.4.

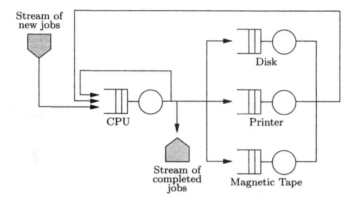

Fig. 7.1   Computer system shown as an open queueing network.

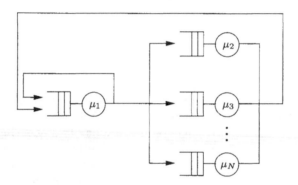

Fig. 7.2   Central-server model.

Fig. 7.3   Closed tandem network.

There are many systems that can be represented by the machine repairman model, for example, a simple terminal system where the M machines represent the M terminals and the repairman represents the computer. Another example is a system in which M machines operate independently of each other and are repaired by a single repairman if they fail. The M machines are modeled by a delay server or an infinite server node. A job does not have to wait; it is immediately accepted by one of the servers. When a machine fails, it sends a repair request to the repair facility. Depending on the service discipline, this request may have to wait until other requests have been dealt with. Since there are M machines, there can be at most M repair requests.

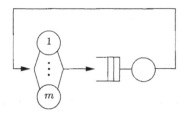

Fig. 7.4 Machine repairman model.

This is a closed two-node queueing network with M jobs that are either being processed, waiting, or are in the working machines.

7.1 DEFINITIONS AND NOTATION

We will consider both single class and multiclass networks.

7.1.1 Single Class Networks

The following symbols are used in the description of queueing networks:

N	Number of nodes
K	The constant number of jobs in a closed network
$(k_1, k_2,\ldots,k_N)$	The state of the network
k_i	The number of jobs at the ith node; for closed networks $\sum_{i=1}^{N} k_i = K$
m_i	The number of parallel servers at the ith node ($m_i \geq 1$)
μ_i	Service rate of the jobs at the ith node
$1/\mu_i$	The mean service time of the jobs at the ith node
p_{ij}	Routing probability, the probability that a job is transferred to the jth node after service completion at the ith node. (In open networks, the node with index 0 represents the external world to the network.)
p_{0j}	The probability that a job entering the network from outside first enters the jth node
p_{i0}	The probability that a job leaves the network just after completing service at node i ($p_{i0} = 1 - \sum_{j=1}^{N} p_{ij}$)
λ_{0i}	The arrival rate of jobs from outside to the ith node

λ The overall arrival rate from outside to an open network

$$(\lambda = \sum_{i=1}^{N} \lambda_{0i})$$

λ_i the overall arrival rate of jobs at the ith node

The *arrival rate* λ_i for node $i = 1, \dots, N$ of an open network is calculated by adding the arrival rate from outside and the arrival rates from all the other nodes. Note that in statistical equilibrium the rate of departure from a node is equal to the rate of arrival, and the overall arrival rate at node i can be written as:

$$\lambda_i = \lambda_{0i} + \sum_{j=1}^{N} \lambda_j p_{ji}, \quad \text{for } i = 1, \dots, N \tag{7.1}$$

for an open network. These are known as traffic equations. For closed networks these equations reduce to

$$\lambda_i = \sum_{j=1}^{N} \lambda_j p_{ji}, \quad \text{for } i = 1, \dots, N, \tag{7.2}$$

since no jobs enter the network from outside.

Another important network parameter is the *mean number of visits* (e_i) of a job to the ith node, also known as the *visit ratio* or *relative arrival rate*:

$$e_i = \frac{\lambda_i}{\lambda}, \quad \text{for } i = 1, \dots, N, \tag{7.3}$$

where λ is the overall throughput of the network (see Eqs. (7.24) and (7.25)). The visit ratios can also be calculated directly from the routing probabilities using Eqs. (7.3) and (7.1), or (7.3) and (7.2). For open networks, since $\lambda_{0i} = \lambda \cdot p_{0i}$, we have

$$e_i = p_{0i} + \sum_{j=1}^{N} e_j p_{ji}, \quad \text{for } i = 1, \dots, N, \tag{7.4}$$

and for closed networks we have

$$e_i = \sum_{j=1}^{N} e_j p_{ji}, \quad \text{for } i = 1, \dots, N. \tag{7.5}$$

Since there are only $(N - 1)$ independent equations for the visit ratios in closed networks, the e_i can only be determined up to a multiplicative constant. Usually we assume that $e_1 = 1$, although other possibilities are used as well. Using e_i, we can also compute the *relative utilization* x_i, which is given by

$$x_i = \frac{e_i}{\mu_i}. \tag{7.6}$$

It is easy to show that [ChLa74] the ratio of server utilizations is given by

$$\frac{x_i}{x_j} = \frac{\rho_i}{\rho_j} . \tag{7.7}$$

7.1.2 Multiclass Networks

The model type discussed in the previous section can be extended by including multiple job classes in the network. The job classes can differ in their service times and in their routing probabilities. It is also possible that a job changes its class when it moves from one node to another. If no jobs of a particular class enter or leave the network, i.e., the number of jobs of this class is constant, then the job class is said to be *closed*. A job class that is not closed is said to be *open*. If a queueing network contains both open and closed classes, then it is said to be a *mixed* network. Figure 7.5 shows a mixed network.

The following additional symbols are needed to describe queueing networks that contain multiple job classes, namely:

R The number of job classes in a network

k_{ir} The number of jobs of the rth class at the ith node; for a closed network:

$$\sum_{i=1}^{N} \sum_{r=1}^{R} k_{ir} = K \tag{7.8}$$

K_r The number of jobs of the rth class in the network; not necessarily constant, even in a closed network:

$$\sum_{i=1}^{N} k_{ir} = K_r \tag{7.9}$$

$\mathbf{K}$ The number of jobs in the various classes, known as the population vector ($\mathbf{K} = (K_1, \ldots, K_R)$)

$\mathbf{S}_i$ The state of the ith node ($\mathbf{S}_i = (k_{i1}, \ldots, k_{iR})$):

$$\sum_{i=1}^{N} \mathbf{S}_i = \mathbf{K} \tag{7.10}$$

$\mathbf{S}$ The overall state of the network with multiple classes ($\mathbf{S} = (\mathbf{S}_1, \ldots, \mathbf{S}_N)$)

μ_{ir} The service rate of the ith node for jobs of the rth class

$p_{ir,js}$ The probability that a job of the rth class at the ith node is transferred to the sth class and the jth node (routing probability)

$p_{0,js}$ The probability in an open network that a job from outside the network enters the jth node as a job of the sth class

$p_{ir,0}$ The probability in an open network that a job of the rth class leaves the network after having been serviced at the ith node, so

$$p_{ir,0} = 1 - \sum_{j=1}^{N} \sum_{s=1}^{R} p_{ir,js} \tag{7.11}$$

λ The overall arrival rate from outside to an open network

$\lambda_{0,ir}$ The arrival rate from outside to node i for class r jobs ($\lambda_{0,ir} = \lambda \cdot p_{0,ir}$)

λ_{ir} The arrival rate of jobs of the rth class at the ith node:

$$\lambda_{ir} = \lambda \cdot p_{0,ir} + \sum_{j=1}^{N} \sum_{s=1}^{R} \lambda_{js} \cdot p_{js,ir} \, ; \tag{7.12}$$

for closed networks, $p_{0,ir} = 0$ ($1 < i < N$, $1 < r < R$) and we obtain

$$\lambda_{ir} = \sum_{j=1}^{N} \sum_{s=1}^{R} \lambda_{js} \cdot p_{js,ir} \, . \tag{7.13}$$

The mean number of visits e_{ir} of a job of the rth class at the ith node of an open network can be determined from the routing probabilities similarly to Eq. (7.4):

$$e_{ir} = p_{0,ir} + \sum_{j=1}^{N} \sum_{s=1}^{R} e_{js} p_{js,ir} \, , \quad \text{for } i = 1, \ldots, N \, , \ r = 1, \ldots, R \, . \tag{7.14}$$

For closed networks, the corresponding equation is

$$e_{ir} = \sum_{j=1}^{N} \sum_{s=1}^{R} e_{js} p_{js,ir} \, , \qquad \text{for } i = 1, \ldots, N \, , \ r = 1, \ldots, R \, . \tag{7.15}$$

Usually we assume that $e_{1r} = 1$, for $r = 1, \ldots, R$, although other settings are also possible.

7.2   PERFORMANCE MEASURES

7.2.1   Single Class Networks

Analytic methods to calculate state probabilities and other performance measures of queueing networks are described in the following sections. The determination of the *steady-state probabilities* $\pi(k_1, \ldots, k_N)$ of all possible states of the network can be regarded as the central problem of queueing theory. The mean values of all other important performance measures of the network can be calculated from these. There are, however, simpler methods for calculating these characteristics directly without using these probabilities.

Note that we use a slightly different notation compared with that used in Chapters 2–5. In Chapters 2-5, $\pi_i(t)$ denoted the transient probability of the CTMC being in state i at time t and π_i as the steady-state probability in state i. Since we now deal with multidimensional state spaces, $\pi(k_1, k_2, \ldots, k_N)$ will denote the steady-state probability of state $(k_1, k_2, \ldots, k_N)$.

The most important performance measures for queueing networks are:

Marginal Probabilities $\pi_i(k)$: For *closed* queueing networks, the *marginal probabilities* $\pi_i(k)$ that the ith node contains exactly $k_i = k$ jobs are calculated as follows:

$$\pi_i(k) = \sum_{\substack{\sum\limits_{j=1}^{N} k_j = K \\ \&\ k_i = k}} \pi(k_1, \ldots, k_N). \tag{7.16}$$

Thus, $\pi_i(k)$ is the sum of the probabilities of all possible states $(k_1, \ldots, k_N)$, $0 \leq k_i \leq K$ that satisfy the condition $\sum\limits_{j=1}^{N} k_j = K$ where a fixed number of jobs, k, is specified for the ith node. The normalization condition that the sum of the probabilities of all possible states $(k_1, \ldots, k_N)$ that satisfy the condition $\sum\limits_{j=1}^{N} k_j = K$ with $(0 \leq k_j \leq K)$ must be 1, that is,

$$\sum_{\sum\limits_{j=1}^{N} k_j = K} \pi(k_1, \ldots, k_N) = 1. \tag{7.17}$$

Correspondingly, for *open* networks we have

$$\pi_i(k) = \sum_{k_i = k} \pi(k_1, \ldots, k_N),$$

with the normalization condition

$$\sum \pi(k_1, \ldots, k_N) = 1.$$

Now we can use the marginal probabilities to obtain other interesting performance measures for open and closed networks. For closed networks we have to take into consideration that $\pi_i(k) = 0 \quad \forall k > K$.

Utilization ρ_i: The *utilization* ρ_i of a single server node with index i is given by

$$\rho_i = \sum_{k=1}^{\infty} \pi_i(k), \tag{7.18}$$

where ρ_i is the probability that the ith node is busy, that is,

$$\rho_i = 1 - \pi_i(0) \,. \tag{7.19}$$

For nodes with multiple servers we have

$$\rho_i = \frac{1}{m_i} \sum_{k=0}^{\infty} \min\,(m_i, k)\pi_i(k) = 1 - \sum_{k=0}^{m_i-1} \frac{m_i - k}{m_i} \cdot \pi_i(k) \,, \tag{7.20}$$

and if the service rate is independent of the load we get (see Eq. (6.4))

$$\rho_i = \frac{\lambda_i}{m_i \mu_i} \,. \tag{7.21}$$

Throughput λ_i: The *throughput* λ_i of an individual node with index i represents in general the rate at which jobs leave the node:

$$\lambda_i = \sum_{k=1}^{\infty} \pi_i(k)\mu_i(k) \,, \tag{7.22}$$

where the service rate $\mu_i(k)$ is, in general, dependent on the load, i.e., on the number of jobs at the node. For example, a node with multiple servers $(m_i > 1)$ can be regarded as a single server whose service rate depends on the load $\mu_i(k) = \min(k, m_i) \cdot \mu_i$, where μ_i is the service rate of an individual server. It is also true for load-independent service rates that (see Eqs. (6.4), and (6.5))

$$\lambda_i = m_i \cdot \rho_i \cdot \mu_i \,. \tag{7.23}$$

We note that for a node in equilibrium, arrival rate and throughput are equal. Also note that when we consider nodes with finite buffers, arriving customers can be lost when the buffer is full. In this case, node throughput will be less than the arrival rate to the node.

Overall Throughput λ: The *overall throughput* λ of an open network is defined as the rate at which jobs leave the network. For a network in equilibrium, this departure rate is equal to the rate at which jobs enter the network, that is

$$\lambda = \sum_{i=1}^{N} \lambda_{0i} \,. \tag{7.24}$$

The overall throughput of a closed network is defined as the throughput of a particular node with index i for which $e_i = 1$. Then the overall throughput of jobs in closed networks is

$$\lambda = \frac{\lambda_i}{e_i} \,. \tag{7.25}$$

Mean Number of Jobs $\overline{K}_i$**:** The *mean number of jobs* at the ith node is given by

$$\overline{K}_i = \sum_{k=1}^{\infty} k \cdot \pi_i(k). \tag{7.26}$$

From Little's theorem (see Eq. (6.9)), it follows that

$$\overline{K}_i = \lambda_i \cdot \overline{T}_i, \tag{7.27}$$

where $\overline{T}_i$ denotes the mean response time.

Mean Queue Length $\overline{Q}_i$**:** The *mean queue length* at the ith node is determined by

$$\overline{Q}_i = \sum_{k=m_i}^{\infty} (k - m_i) \cdot \pi_i(k), \tag{7.28}$$

or, using Little's theorem,

$$\overline{Q}_i = \lambda_i \overline{W}_i, \tag{7.29}$$

where $\overline{W}_i$ is the *mean waiting time*.

Mean Response Time $\overline{T}_i$**:** The *mean response time* of jobs at the ith node can be calculated using Little's theorem (see Eq. (6.9)) for a given mean number of jobs $\overline{K}_i$:

$$\overline{T}_i = \frac{\overline{K}_i}{\lambda_i}. \tag{7.30}$$

Mean Waiting Time $\overline{W}_i$**:** If the service rates are independent of the load, then the *mean waiting time* at node i is

$$\overline{W}_i = \overline{T}_i - \frac{1}{\mu_i}. \tag{7.31}$$

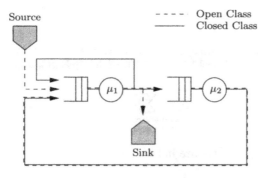

Fig. 7.5   A mixed network.

7.2.2 Multiclass Networks

The extension of queueing networks to include multiple job classes leads to a corresponding extension of the performance measures. The *state probability* of a network with multiple job classes is represented by $\pi(\mathbf{S}_1, \ldots, \mathbf{S}_N)$. The normalization condition that the sum of the probabilities of all possible states $(\mathbf{S}_1, \ldots, \mathbf{S}_N)$ is 1 must also be satisfied here.

Marginal Probability $\pi_i(\mathbf{k})$: For closed networks the *marginal probability*, i.e., the probability that the ith node is in the state $\mathbf{S}_i = \mathbf{k}$, is given by

$$\pi_i(\mathbf{k}) = \sum_{\substack{\sum_{j=1}^{N} \mathbf{S}_j = \mathbf{K} \\ \&\ \mathbf{S}_i = \mathbf{k}}} \pi(\mathbf{S}_1, \ldots, \mathbf{S}_N), \tag{7.32}$$

and for open networks we have

$$\pi_i(\mathbf{k}) = \sum_{s_i = k} \pi(\mathbf{S}_1, \ldots, \mathbf{S}_N). \tag{7.33}$$

The following formulae for computing the performance measures can be applied to open and closed networks.

Utilization ρ_{ir}: The *utilization* of the ith node with respect to jobs of the rth class is

$$\rho_{ir} = \frac{1}{m_i} \sum_{\substack{\text{all states } \mathbf{k} \\ \text{with } k_r > 0}} \pi_i(\mathbf{k}) \frac{k_{ir}}{k_i} \min(m_i, k_i), \quad k_i = \sum_{r=1}^{R} k_{ir}, \tag{7.34}$$

and if the service rates are independent on the load, we have

$$\rho_{ir} = \frac{\lambda_{ir}}{m_i \mu_{ir}}. \tag{7.35}$$

Throughput λ_{ir}: The *throughput* λ_{ir} is the rate at which jobs of the rth class are serviced and leave the ith node [BrBa80]:

$$\lambda_{ir} = \sum_{\substack{\text{all states } \mathbf{k} \\ \text{with } k_r > 0}} \pi_i(\mathbf{k}) \frac{k_{ir}}{k_i} \mu_i(k_i); \tag{7.36}$$

or if the service rates are independent on the load, we have:

$$\lambda_{ir} = m_i \cdot \rho_{ir} \cdot \mu_{ir}. \tag{7.37}$$

Overall Throughput λ_r: The *overall throughput* of jobs of the rth class in closed networks with multiclasses is

$$\lambda_r = \frac{\lambda_{ir}}{e_{ir}}, \qquad (7.38)$$

and for open networks we obtain

$$\lambda_r = \sum_{i=1}^{N} \lambda_{0,ir}. \qquad (7.39)$$

Mean Number of Jobs $\overline{K}_{ir}$: The *mean number of jobs* of the rth class at the ith node is

$$\overline{K}_{ir} = \sum_{\substack{\text{all states } \mathbf{k} \\ \text{with } k_r > 0}} k_r \cdot \pi_i(\mathbf{k}). \qquad (7.40)$$

Little's theorem can also be used here:

$$\overline{K}_{ir} = \lambda_{ir} \cdot \overline{T}_{ir}. \qquad (7.41)$$

Mean Queue Length $\overline{Q}_{ir}$: The *mean queue length* of class r jobs at the ith node can be calculated using Little's theorem as

$$\overline{Q}_{ir} = \lambda_{ir} \overline{W}_{ir}. \qquad (7.42)$$

Mean Response Time $\overline{T}_{ir}$: The *mean response time* of jobs of the rth class at the ith node can also be determined using Little's theorem (see Eq. (7.41)):

$$\overline{T}_{ir} = \frac{\overline{K}_{ir}}{\lambda_{ir}}. \qquad (7.43)$$

Mean Waiting Time $\overline{W}_{ir}$: If the service rates are load-independent, then the *mean waiting time* is given by

$$\overline{W}_{ir} = \overline{T}_{ir} - \frac{1}{\mu_{ir}}. \qquad (7.44)$$

7.3 PRODUCT-FORM QUEUEING NETWORKS

In the rest of this chapter, we will discuss queueing networks that have a special structure such that their solutions can be obtained without generating their underlying state space. Such networks are known as product-form or separable networks.

7.3.1 Global Balance

The behavior of many queueing system models can be described using CTMCs. A CTMC is characterized by the transition rates between the states of the corresponding model (for more details on CTMC see Chapters $2-5$). If the CTMC is ergodic, then a unique steady-state probability vector independent of the initial probability vector exists. The system of equations to determine the steady-state probability vector π is given by $\pi Q = 0$ (see Eq. (2.58)), where Q is the infinitesimal generator matrix of the CTMC. This equation says that for each state of a queueing network in equilibrium, the flux out of a state is equal to the flux into that state. This conservation of flow in steady state can be written as

$$\sum_{j \in S} \pi_j q_{ji} = \pi_i \sum_{j \in S} q_{ij}, \quad \forall i \in S, \tag{7.45}$$

where q_{ij} is the transition rate from state i to state j. After subtracting $\pi_i \cdot q_{ii}$ from both sides of Eq. (7.45) and noting that $q_{ii} = -\sum_{j \neq i} q_{ij}$, we obtain the global balance equation (see Eq. (2.61)):

$$\forall i \in S : \quad \sum_{j \neq i} \pi_j q_{ji} - \pi_i \sum_{j \neq i} q_{ij} = 0, \tag{7.46}$$

which corresponds to the matrix equation $\pi Q = 0$ (Eq. (2.58)).

In the following we use two simple examples to show how to write the global balance equations and use them to obtain performance measures.

Example 7.1   Consider the closed queueing network given in Fig. 7.6. The network consists of two nodes ($N = 2$) and three jobs ($K = 3$). The service times are exponentially distributed with mean values $1/\mu_1 = 5$ sec and $1/\mu_2 = 2.5$ sec, respectively. The service discipline at each node is FCFS. The state space of the CTMC consists of the following four states:

$$\{(3,0), (2,1), (1,2), (0,3)\}.$$

Fig. 7.6 A closed network.

The notation (k_1, k_2) says that there are k_1 jobs at node 1 and k_2 jobs at node 2, and $\pi(k_1, k_2)$ denotes the probability for that state in equilibrium. Consider state $(1, 2)$. A transition from state $(1, 2)$ to state $(0, 3)$ takes place if the job at node 1 completes service (with corresponding rate μ_1). Therefore,

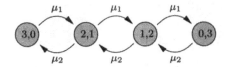

Fig. 7.7   State transition diagram for Example 7.1.

μ_1 is the transition rate from state $(1,2)$ to state $(0,3)$ and, similarly, μ_2 is the transition rate from state $(1,2)$ to state $(2,1)$. The flux into a state of the model is just given by all arcs into the corresponding state, and the flux out of that state is determined from the set of all outgoing arcs from the state. The corresponding state transition diagram is shown in Fig. 7.7. The global balance equations for this example are

$$\pi(3,0)\mu_1 = \pi(2,1)\mu_2\,,$$
$$\pi(2,1)(\mu_1 + \mu_2) = \pi(3,0)\mu_1 + \pi(1,2)\mu_2\,,$$
$$\pi(1,2)(\mu_1 + \mu_2) = \pi(2,1)\mu_1 + \pi(0,3)\mu_2\,,$$
$$\pi(0,3)\mu_2 = \pi(1,2)\mu_1\,.$$

Rewriting this system of equations in the form $\boldsymbol{\pi}\mathbf{Q} = \mathbf{0}$, we have

$$\mathbf{Q} = \begin{pmatrix} -\mu_1 & \mu_1 & 0 & 0 \\ \mu_2 & -(\mu_1 + \mu_2) & \mu_1 & 0 \\ 0 & \mu_2 & -(\mu_1 + \mu_2) & \mu_1 \\ 0 & 0 & \mu_2 & -\mu_2 \end{pmatrix}$$

and the steady-state probability vector $\boldsymbol{\pi} = (\pi(3,0), \pi(2,1), \pi(1,2), \pi(0,3))$. Once the steady-state probabilities are known, all other performance measures and marginal probabilities $\pi_i(k)$ can be computed. If we use $\mu_1 = 0.2$ and $\mu_2 = 0.4$, then the generator matrix $\mathbf{Q}$ has the following values:

$$\mathbf{Q} = \begin{pmatrix} -0.2 & 0.2 & 0 & 0 \\ 0.4 & -0.6 & 0.2 & 0 \\ 0 & 0.4 & -0.6 & 0.2 \\ 0 & 0 & 0.4 & -0.4 \end{pmatrix}\,.$$

Using one of the steady-state solution methods introduced in Chapter 3, the steady-state probabilities are computed to be

$$\pi(3,0) = \underline{0.5333}\,, \quad \pi(2,1) = \underline{0.2667}\,, \quad \pi(1,2) = \underline{0.1333}\,, \quad \pi(0,3) = \underline{0.0667}$$

and are used to determine all other performance measures of the network, as follows:

- Marginal probabilities (see Eq. (7.16)):

$$\pi_1(0) = \pi_2(3) = \pi(0,3) = \underline{0.0667}\,, \quad \pi_1(1) = \pi_2(2) = \pi(1,2) = \underline{0.133}\,,$$

$$\pi_1(2) = \pi_2(1) = \pi(2,1) = \underline{0.2667}, \quad \pi_1(3) = \pi_2(0) = \pi(3,0) = \underline{0.5333}.$$

- Utilizations (see Eq. (7.20)):

$$\rho_1 = 1 - \pi_1(0) = \underline{0.9333}, \quad \rho_2 = 1 - \pi_2(0) = \underline{0.4667}.$$

- Throughput (see Eq. (7.23)):

$$\lambda = \lambda_1 = \lambda_2 = \rho_1\mu_1 = \rho_2\mu_2 = \underline{0.1867}.$$

- Mean number of jobs (see Eq. (7.26)):

$$\overline{K}_1 = \sum_{k=1}^{3} k\pi_1(k) = \underline{2.2667}, \quad \overline{K}_2 = \sum_{k=1}^{3} k\pi_2(k) = \underline{0.7333}.$$

- Mean response time of the jobs (see Eq. (7.30)):

$$\overline{T}_1 = \frac{\overline{K}_1}{\lambda_1} = \underline{12.1429}, \quad \overline{T}_2 = \frac{\overline{K}_2}{\lambda_2} = \underline{3.9286}.$$

Example 7.2   Now we modify the closed network of Example 7.1 so that the service time at node 1 is Erlang-2 distributed, while the service time at node 2 remains exponentially distributed. The modified network is shown in Fig. 7.8. Service rates of the different phases at node 1 are given by $\mu_{11} = \mu_{12} = 0.4 \text{ sec}^{-1}$, and the service rate of node 2 is $\mu_2 = 0.4$. There are $K = 2$ jobs in the system. A state $(k_1, l; k_2)$, $l = 0, 1, 2$ of the network is now not

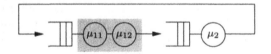

Fig. 7.8 Network with Erlang-2 distributed server.

only given by the number k_i of jobs at the nodes, but also by the phase l in which a job is being served at node 1. This state definition leads to exactly five states in the CTMC underlying the network.

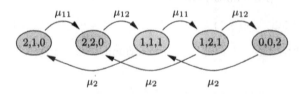

Fig. 7.9 State transition diagram for Example 7.2.

The state transition diagram of the CTMC is shown in Fig. 7.9. The following global balance equations can be derived:

$$\pi(2,1;0)\mu_{11} = \pi(1,1;1)\mu_2 \,,$$
$$\pi(2,2;0)\mu_{12} = \pi(2,1;0)\mu_{11} + \pi(1,2;1)\mu_2 \,,$$
$$\pi(1,1;1)(\mu_{11} + \mu_2) = \pi(2,2;0)\mu_{12} + \pi(0,0;2)\mu_2 \,,$$
$$\pi(1,2;1)(\mu_{12} + \mu_2) = \pi(1,1;1)\mu_{11} \,,$$
$$\pi(0,0;2)\mu_2 = \pi(1,2;1)\mu_{12} \,.$$

The generator matrix is

$$
\mathbf{Q} = \begin{pmatrix}
-\mu_{11} & \mu_{11} & 0 & 0 & 0 \\
0 & -\mu_{12} & \mu_{12} & 0 & 0 \\
\mu_2 & 0 & -(\mu_{11} + \mu_2) & \mu_{11} & 0 \\
0 & \mu_2 & 0 & -(\mu_{12} + \mu_2) & \mu_{12} \\
0 & 0 & \mu_2 & 0 & -\mu_2
\end{pmatrix}.
$$

After inserting the values $\mu_{11} = \mu_{12} = \mu_2 = 0.4$, $\mathbf{Q}$ becomes

$$
\mathbf{Q} = \begin{pmatrix}
-0.4 & 0.4 & 0 & 0 & 0 \\
0 & -0.4 & 0.4 & 0 & 0 \\
0.4 & 0 & -0.8 & 0.4 & 0 \\
0 & 0.4 & 0 & -0.8 & 0.4 \\
0 & 0 & 0.4 & 0 & -0.4
\end{pmatrix}.
$$

Solving the system of equations $\boldsymbol{\pi}\mathbf{Q} = \mathbf{0}$, we get

$$\pi(2,1;0) = \underline{0.2219}\,, \quad \pi(2,2;0) = \underline{0.3336}\,,$$
$$\pi(1,1;1) = \underline{0.2219}\,, \quad \pi(1,2;1) = \underline{0.1102}\,,$$
$$\pi(0,0;2) = \underline{0.1125}\,.$$

These state probabilities are used to determine the marginal probabilities:

$$\pi_1(0) = \pi_2(2) = \pi(0,0;2) = \underline{0.1125}\,,$$
$$\pi_1(1) = \pi_2(1) = \pi(1,1;1) + \pi(1,2;1) = \underline{0.3321}\,,$$
$$\pi_1(2) = \pi_2(0) = \pi(2,1;0) + \pi(2,2;0) = \underline{0.5555}\,.$$

The computation of other performance measures is done in the same way as in Example 7.1.

7.3.2 Local Balance

Numerical techniques based on the solution of the global balance equations (see Eq. (2.61)) can in principle always be used, but for large networks this technique is very expensive because the number of equations can be extremely large. For such large networks, we therefore look for alternative solution techniques.

In this chapter we show that efficient and exact solution algorithms exist for a large class of queueing networks. These algorithms avoid the generation and solution of global balance equations. If all nodes of the network fulfill certain assumptions concerning the distributions of the interarrival and service times and the queueing discipline, then it is possible to derive local balance equations, which describe the system behavior in an unambiguous way. These local balance equations allow an essential simplification with respect to the global balance equations because each equation can be split into a number of single equations, each one related to each individual node.

Queueing networks that have an unambiguous solution of the local balance equations are called *product-form networks*. The steady-state solution to such networks' state probabilities consist of multiplicative factors, each factor relating to a single node. Before introducing the different solution methods for product-form networks, we explain the local balance concept in more detail. This concept is the theoretical basis for the applicability of analysis methods.

Consider global balance equations (Eq. (2.61)) for a CTMC:

$$\forall i \in S: \quad \sum_{j \in S} \pi_j q_{ji} = \pi_i \sum_{j \in S} q_{ij}$$

or

$$\boldsymbol{\pi} \cdot \mathbf{Q} = \mathbf{0} \, ,$$

with the normalization condition

$$\sum_{i \in S} \pi_i = 1 \, .$$

Chandy [Chan72] noticed that under certain conditions the global balance equations can be split into simpler equations, known as *local balance equations*.

Local balance property for a node means: The departure rate from a state of the queueing network due to the departure of a job from node i equals the arrival rate to this state due to an arrival of a job to this node.

This can also be extended to queueing networks with several job classes in the following way:

The departure rate from a state of the queueing network due to the departure of a *class r-job* from node i equals the arrival rate to this state due to an arrival of a *class r-job* to this node.

In the case of nonexponentially distributed service times, arrivals and departures to phases, instead of nodes, have to be considered.

Example 7.3   Consider a closed queueing network (see Fig. 7.10) consisting of $N = 3$ nodes with exponentially distributed service times and the

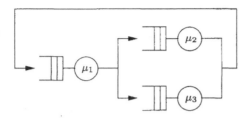

Fig. 7.10 A closed queueing network.

following service rates: $\mu_1 = 4$ sec^{-1}, $\mu_2 = 1$ sec^{-1}, and $\mu_3 = 2$ sec^{-1}. There are $K = 2$ jobs in the network, and the routing probabilities are given as: $p_{12} = 0.4$, $p_{13} = 0.6$, and $p_{21} = p_{31} = 1$. The following states are possible in the network:

$$(2,0,0), \quad (0,2,0), \quad (0,0,2), \quad (1,1,0), \quad (1,0,1), \quad (0,1,1).$$

The state transition diagram of the underlying CTMC is shown in Fig. 7.11.

We set the overall flux into a state equal to the overall flux out of the state for each state to get global balance equations:

(1) $\qquad \pi(2,0,0)(\mu_1 p_{12} + \mu_1 p_{13}) = \pi(1,0,1)\mu_3 p_{31} + \pi(1,1,0)\mu_2 p_{21}\,,$

(2) $\qquad\qquad\qquad \pi(0,2,0)\mu_2 p_{21} = \pi(1,1,0)\mu_1 p_{12}\,,$

(3) $\qquad\qquad\qquad \pi(0,0,2)\mu_3 p_{31} = \pi(1,0,1)\mu_1 p_{13}\,,$

(4) $\quad \pi(1,1,0)(\mu_2 p_{21} + \mu_1 p_{13} + \mu_1 p_{12}) = \pi(0,2,0)\mu_2 p_{21} + \pi(2,0,0)\mu_1 p_{12}$
$$\qquad\qquad\qquad\qquad\qquad\qquad + \pi(0,1,1)\mu_3 p_{31}\,,$$

(5) $\quad \pi(1,0,1)(\mu_3 p_{31} + \mu_1 p_{12} + \mu_1 p_{13}) = \pi(0,0,2)\mu_3 p_{31} + \pi(0,1,1)\mu_2 p_{21}$
$$\qquad\qquad\qquad\qquad\qquad\qquad + \pi(2,0,0)\mu_1 p_{13}\,,$$

(6) $\qquad \pi(0,1,1)(\mu_3 p_{31} + \mu_2 p_{21}) = \pi(1,1,0)\mu_1 p_{13} + \pi(1,0,1)\mu_1 p_{12}\,.$

To determine the local balance equations we start, for example, with the state $(1,1,0)$. The departure rate out of this state, because of the departure of a job from node 2, is given by $\pi(1,1,0)\cdot \mu_2 \cdot p_{21}$. This rate is equated to the arrival rate into this state $(1,1,0)$ due to the arrival of a job at node 2; $\pi(2,0,0)\cdot \mu_1 \cdot p_{12}$. Therefore, we get the local balance equation:

(4') $\quad \pi(1,1,0)\cdot \mu_2 \cdot p_{21} = \pi(2,0,0)\cdot \mu_1 \cdot p_{12}\,.$

Correspondingly, the departure rate of a serviced job at node 1 from state $(1,1,0)$ equals the arrival rate of a job, arriving at node 1 into state $(1,1,0)$:

(4'') $\quad \pi(1,1,0)\cdot \mu_1 \cdot (p_{13} + p_{12}) = \pi(0,1,1)\cdot \mu_3 \cdot p_{31} + \pi(0,2,0)\cdot \mu_2 \cdot p_{21}\,.$

By adding these two local balance equations, (4') and (4''), we get the global balance Eq. (4). Furthermore, we see that the global balance Eqs. (1),

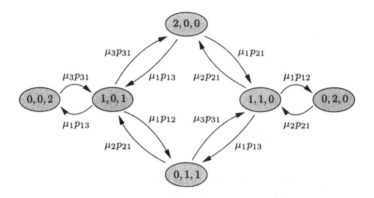

Fig. 7.11 The CTMC for Example 7.3.

(2), and (3) are also local balance equations at the same time. The rest of the local balance equations are given by

(5') $\pi(1,0,1)\mu_1(p_{12}+p_{13}) = \pi(0,1,1)\mu_2 p_{21} + \pi(0,0,2)\mu_3 p_{31}$,

(5'') $\pi(1,0,1)\mu_3 p_{31} = \pi(2,0,0)\mu_1 p_{13}$,

(6') $\pi(0,1,1)\mu_2 p_{21} = \pi(1,0,1)\mu_1 p_{12}$,

(6'') $\pi(0,1,1)\mu_3 p_{31} = \pi(1,1,0)\mu_1 p_{13}$,

with $(5') + (5'') = (5)$ and $(6') + (6'') = (6)$, respectively.

Noting that $p_{12} + p_{13} = 1$ and $p_{21} = p_{31} = 1$, the following relations can be derived from these local balance equations:

$$\pi(1,0,1) = \pi(2,0,0)\frac{\mu_1}{\mu_3}p_{13}, \qquad \pi(1,1,0) = \pi(2,0,0)\frac{\mu_1}{\mu_2}p_{12},$$

$$\pi(0,0,2) = \pi(2,0,0)\left(\frac{\mu_1}{\mu_3}p_{13}\right)^2, \quad \pi(0,2,0) = \pi(2,0,0)\left(\frac{\mu_1}{\mu_2}p_{12}\right)^2,$$

$$\pi(0,1,1) = \pi(2,0,0)\frac{\mu_1^2}{\mu_2\mu_3}p_{12}p_{13}.$$

Imposing the normalization condition, that the sum of probabilities of all states in the network is 1, for $\pi(2,0,0)$ we get the following expression:

$$\pi(2,0,0) = \left[1 + \mu_1\left(\frac{p_{13}}{\mu_3} + \frac{p_{12}}{\mu_2} + \frac{\mu_1 p_{13}^2}{\mu_3^2} + \frac{\mu_1 p_{12}^2}{\mu_2^2} + \frac{\mu_1 p_{12}p_{13}}{\mu_2\mu_3}\right)\right]^{-1}.$$

After inserting the values, we get the following results:

$$\pi(2,0,0) = \underline{0.103}, \quad \pi(0,0,2) = \underline{0.148}, \quad \pi(1,0,1) = \underline{0.123},$$
$$\pi(0,2,0) = \underline{0.263}, \quad \pi(1,1,0) = \underline{0.165}, \quad \pi(0,1,1) = \underline{0.198}.$$

As can be seen in this example, the structure of the local balance equations is much simpler than the global balance equations. However, not every network has a solution of the local balance equations, but there always exists a solution of the global balance equations. Therefore, local balance can be considered as a sufficient (but not necessary) condition for the global balance. Furthermore, if there exists a solution for the local balance equations, the model is then said to have the *local balance property*. Then this solution is also the unique solution of the system of global balance equations.

The computational effort for the numerical solution of the local balance equations of a queueing network is still very high but can be reduced considerably with the help of a characteristic property of local-balanced queueing networks: For the determination of the state probabilities it is not necessary to solve the local balance equations for the whole network. Instead, the state probabilities of the queueing network in these cases can be determined very easily from the state probabilities of individual single nodes of the network. If each node in the network has the local balance property, then the following two very important implications are true:

- The overall network also has the local balance property as proven in [CHT77].

- There exists a product-form solution for the network, that is,

$$\pi(\mathbf{S}_1, \mathbf{S}_2, \ldots, \mathbf{S}_N) = \frac{1}{G} \left[\pi(\mathbf{S}_1) \cdot \pi(\mathbf{S}_2) \cdot \ldots \cdot \pi(\mathbf{S}_N) \right], \qquad (7.47)$$

  in the sense that the expression for the state probability of the network is given by the product of marginal state probabilities of each individual node. The proof of this fact can be found in [Munt73]. The normalization constant G is chosen in such a way that the sum of probabilities over all states in the network equals 1.

Equation (7.47) says that in networks having the local-balance property, the nodes behave as if they were single queueing systems. This characteristic means that the nodes of the network can be examined in isolation from the rest of the network. Networks of the described type belong to the class of so-called *separable networks* or *product-form networks*. Now we need to examine for which types of elementary queueing systems a solution of the local balance equation exists. If the network consists of only these types of nodes then we know, because of the preceding conclusion, that the whole network has the local-balance property and the network has a product-form solution. The local-balance equations for a single node can be presented in a simplified form as follows [SaCh81]:

$$\pi(\mathbf{S}) \cdot \mu_r(\mathbf{S}) = \pi(\mathbf{S} - \mathbf{1}_r) \cdot \lambda_r. \qquad (7.48)$$

In this equation, $\mu_r(\mathbf{S})$ is the rate with which class-r jobs in state $\mathbf{S}$ are serviced at the node, λ_r is the rate at which class-r jobs arrive at the node, and $(\mathbf{S} - \mathbf{1}_r)$ describes the state of the node after a single class-r job leaves it.

It can be shown that for the following types of queueing systems the local balance property holds [Chan72]:

Type 1: **M/M/m–FCFS**. The service rates for different job classes must be equal. Examples of Type 1 nodes are input/output (I/O) devices or disks.

Type 2: **M/G/1–PS**. The CPU of a computer system can very often be modeled as a Type 2 node.

Type 3: **M/G/∞ (infinite server)**. Terminals can be modeled as Type 3 nodes.

Type 4: **M/G/1–LCFS PR**. There is no practical example for the application of Type 4 nodes in computer systems.

For Type 2, Type 3, and Type 4 nodes, different job classes can have *different general service time distributions*, provided that these have rational Laplace transform. In practice, this requirement is not an essential limitation as any distribution can be approximated as accurately as necessary using a Cox distribution.

In the next section we consider product-form solutions of separable networks in more detail.

Problem 7.1   Consider the closed queueing network given in Fig. 7.12. The network consists of two nodes ($N = 2$) and three jobs ($K = 3$). The

Fig. 7.12   A simple queueing network example for Problem 7.1.

service times are exponentially distributed with mean values $1/\mu_1 = 5\,\text{sec}$ and $1/\mu_2 = 2.5\,\text{sec}$. The service discipline at each node is FCFS.

(a) Determine the local balance equations.

(b) From the local balance equations, derive the global balance equations.

(c) Determine the steady-state probabilities using the local balance equations.

7.3.3   Product-Form

The term *product-form* was introduced by [Jack63] and [GoNe67a], who considered open and closed queueing networks with exponentially distributed interarrival and service times. The queueing discipline at all stations was assumed to be FCFS. As the most important result for the queueing theory, it is shown that for these networks the solution for the steady-state probabilities can be expressed as a product of factors describing the state of each node. This solution is called *product-form solution*. In [BCMP75] these results were extended to open, closed, and mixed networks with several job classes, non-exponentially distributed service times and different queueing disciplines. In this section we consider these results in more detail and give algorithms to compute performance measures of product-form queueing networks.

A necessary and sufficient condition for the existence of product-form solutions is given in the previous section but repeated here in a slightly different way:

Local Balance Property: Steady-state probabilities can be obtained by solving steady-state (global) balance equations. These equations balance the rate at which the CTMC leaves that state with the rate at which the CTMC enters it. The problem is that the number of equations increases exponentially in the number of states. Therefore, a new set of balance equations, the so-called *local balance equations*, is defined. With these, the rate at which jobs enter a *single* node of the network is equated to the rate at which they leave it. Thus, local balance is concerned with a local situation and reduces the computational effort.

Moreover, there exist two other characteristics that apply to a queueing network with product-form solution:

M ⇒ M-Property (Markov Implies Markov): A service station has the M ⇒ M-property if and only if the station transforms a Poisson arrival process into a Poisson departure process. In [Munt73] it is shown that a queueing network has a product-form solution if all nodes of the network have the M ⇒ M-property.

Station-Balance Property: A service discipline is said to have station-balance property if the service rates at which the jobs in a position of the queue are served are proportional to the probability that a job enters this position. In other words, the queue of a node is partitioned into positions and the rate at which a job enters this position is equal to the rate with which the job leaves this position. In [CHT77] it is shown that networks that have the station-balance property have a product-form solution. The opposite does not hold.

The relation between station balance (SB), local balance (LB), product-form property (PF), and Markov implies Markov property ($M \Rightarrow M$) is shown in Fig. 7.13.

7.3.4 Jackson Networks

The breakthrough in the analysis of queueing networks was achieved by the works of Jackson [Jack57, Jack63]. He examined open queueing networks and

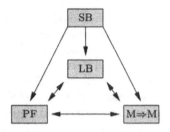

Fig. 7.13 Relation between SB, LB, PF, and $M \Rightarrow M$.

found product-form solutions. The networks examined fulfill the following assumptions:

- There is only one job class in the network.

- The overall number of jobs in the network is unlimited.

- Each of the N nodes in the network can have Poisson arrivals from outside. A job can leave the network from any node.

- All service times are exponentially distributed.

- The service discipline at all nodes is FCFS.

- The ith node consists of $m_i \geq 1$ identical service stations with the service rates μ_i, $i = 1, \ldots, N$. The arrival rates λ_{0i}, as well as the service rates, can depend on the number k_i of jobs at the node. In this case we have *load-dependent service rates* and *load-dependent arrival rates*.

Note: A service station with more than one server and a constant service rate μ_i is equivalent to a service station with exactly one server and load-dependent service rates:

$$\mu_i(k) = \begin{cases} k_i \cdot \mu_i, & k_i \leq m_i, \\ m_i \cdot \mu_i, & k_i \geq m_i. \end{cases} \qquad (7.49)$$

Jackson's Theorem: If in an open network ergodicity ($\lambda_i < \mu_i \cdot m_i$) holds for all nodes $i = 1, \ldots, N$ (the arrival rates λ_i can be computed using Eq. (7.1)), then the steady-state probability of the network can be expressed as the product of the state probabilities of the individual nodes, that is,

$$\pi(k_1, k_2, \ldots, k_N) = \pi_1(k_1) \cdot \pi_2(k_2) \cdot \ldots \cdot \pi_N(k_N). \qquad (7.50)$$

The nodes of the network can be considered as independent M/M/m queues with arrival rate λ_i and service rate μ_i. To prove this theorem, [Jack63] has shown that Eq. (7.50) fulfills the global balance equations. Thus, the marginal probabilities $\pi_i(k_i)$ can be computed with the well-known formulae for M/M/m systems (see Eqs. (6.26), (6.27)):

$$
\pi_i(k_i) = \begin{cases} \pi_i(0)\dfrac{(m_i\rho_i)^{k_i}}{k_i!}\,, & k_i \leq m_i\,, \\[2mm] \pi_i(0)\dfrac{m_i^{m_i}\rho_i^{k_i}}{m_i!}\,, & k_i > m_i\,, \end{cases} \tag{7.51}
$$

where $\pi_i(0)$ is given by the condition $\sum\limits_{k_i=0}^{\infty} \pi_i(k_i) = 1$:

$$
\pi_i(0) = \left(\sum_{k_i=0}^{m_i-1} \frac{(m_i\rho_i)^{k_i}}{k_i!} + \frac{(m_i\rho_i)^{m_i}}{m_i!(1-\rho_i)} \right)^{-1}\,, \quad \rho_i = \frac{\lambda_i}{m_i\mu_i} < 1\,. \tag{7.52}
$$

Proof: We verify that Eq. (7.50) fulfills the following global balance equations:

$$
\left(\sum_{i=1}^{N} \lambda_{0i} + \sum_{i=1}^{N} \alpha_i(k_i)\mu_i \right) \pi(k_1,\ldots,k_N) =
$$
$$
= \sum_{i=1}^{N} \lambda_{0i}\gamma(k_i)\pi(k_1,\ldots,k_i-1,\ldots,k_N)
$$
$$
+ \sum_{i=1}^{N} \alpha_i(k_i+1)\mu_i \left(1 - \sum_{j=1}^{N} p_{ij} \right) \pi(k_1,\ldots,k_i+1,\ldots,k_N) \tag{7.53}
$$
$$
+ \sum_{i=1}^{N}\sum_{j=1}^{N} \alpha_j(k_j+1)\mu_j p_{ji}\pi(k_1,\ldots,k_j+1,\ldots,k_i-1,\ldots,k_N).
$$

The indicator function $\gamma(k_i)$ is given by

$$
\gamma(k_i) = \begin{cases} 0, & k_i = 0, \\ 1, & k_i > 0. \end{cases} \tag{7.54}
$$

The function

$$
\alpha_i(k_i) = \begin{cases} k_i, & k_i \leq m_i\,, \\ m_i, & k_i \geq m_i \end{cases} \tag{7.55}
$$

gives the load-dependent service rate multiplier.

For the proof we use the following relations:

$$\frac{\pi(k_1,\ldots,k_i+1,\ldots,k_N)}{\pi(k_1,\ldots,k_i,\ldots,\ldots,k_N)} = \frac{\pi_1(k_1)\cdots\pi_i(k_i+1)\cdots\pi_N(k_N)}{\pi_1(k_1)\cdots\pi_i(k_i)\cdots\pi_N(k_N)}$$

$$= \frac{\lambda_i}{\mu_i\alpha_i(k_i+1)},$$

$$\frac{\pi(k_1,\cdots,k_i-1,\cdots,k_N)}{\pi(k_1,\cdots,k_i,\cdots,k_N)} = \frac{\mu_i\alpha_i(k_i)}{\lambda_i}, \qquad (7.56)$$

$$\frac{\pi(k_1,\ldots,k_j+1,\ldots,k_i-1,\ldots,k_N)}{\pi(k_1,\ldots,k_j,\ldots,k_i,\ldots,k_N)} = \frac{\lambda_j\mu_i\alpha_i(k_i)}{\lambda_i\mu_j\alpha_j(k_j+1)}.$$

If we divide Eq. (7.53) by $\pi(k_1,\ldots,k_N)$ and insert Eq. (7.56), then, by using the relation $\gamma(k_i)\cdot\alpha_i(k_i)\equiv\alpha_i(k_i)$, we get

$$\sum_{i=1}^{N}\lambda_{0i} + \sum_{i=1}^{N}\alpha_i(k_i)\mu_i = \sum_{i=1}^{N}\left(1-\sum_{j=1}^{N}p_{ij}\right)\lambda_i$$

$$+ \sum_{i=1}^{N}\frac{\lambda_{0i}\mu_i\alpha_i(k_i)}{\lambda_i} + \sum_{i=1}^{N}\sum_{j=1}^{N}\frac{\mu_i\alpha_i(k_i)}{\lambda_i}p_{ji}\lambda_j. \qquad (7.57)$$

The first term can be rewritten as

$$\sum_{i=1}^{N}\left(1-\sum_{j=1}^{N}p_{ij}\right)\lambda_i = \sum_{i=1}^{N}\lambda_{0i},$$

and the last one as

$$\sum_{i=1}^{N}\sum_{j=1}^{N}\frac{\mu_i\alpha_i(k_i)}{\lambda_i}p_{ji}\lambda_j = \sum_{i=1}^{N}\mu_i\alpha_i(k_i) - \sum_{i=1}^{N}\frac{\lambda_{0i}\mu_i\alpha_i(k_i)}{\lambda_i}. \qquad (7.58)$$

Substituting these results on the right side of Eq. (7.57), we get the same result as on the left-hand side. $q.e.d.$

The algorithm based on Jackson's theorem for computing the steady-state probabilities can now be described in the following three steps:

STEP 1 For all nodes, $i = 1,\ldots,N$, compute the arrival rates λ_i of the open network by solving the traffic equations, Eq. (7.1).

STEP 2 Consider each node i as an M/M/m queueing system. Check the ergodicity, Eq. (6.5), and compute the state probabilities and performance measures of each node using the formulae given in Section 6.2.3.

STEP 3 Using Eq. (7.50), compute the steady-state probabilities of the overall network.

We note that for a Jackson network with feedback input, processes to the nodes will, in general, not be Poisson and yet the nodes behave like independent M/M/m nodes. Herein lies the importance of Jackson's theorem [BeMe78]. We illustrate the procedure with the following example.

Example 7.4   Consider the queueing network given in Fig. 7.14, which consists of $N = 4$ single server FCFS nodes. The service times of the jobs at

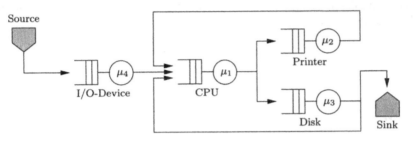

Source

I/O-Device   CPU   Printer   Disk   Sink

Fig. 7.14 Open queueing network model of a computer system.

each node are exponentially distributed with respective means:

$$\frac{1}{\mu_1} = 0.04\,\text{sec}\,, \quad \frac{1}{\mu_2} = 0.03\,\text{sec}\,, \quad \frac{1}{\mu_3} = 0.06\,\text{sec}\,, \quad \frac{1}{\mu_4} = 0.05\,\text{sec}\,.$$

The interarrival time is also exponentially distributed with the parameter:

$$\lambda = \lambda_{04} = 4\ \text{jobs/sec}\,.$$

Furthermore, the routing probabilities are given as follows:

$$p_{12} = p_{13} = 0.5, \quad p_{41} = p_{21} = 1, \quad p_{31} = 0.6, \quad p_{30} = 0.4.$$

Assume that we wish to compute the steady-state probability of state $(k_1, k_2, k_3, k_4) = (3, 2, 4, 1)$ with the help of the Jackson's method. For this we follow the three steps given previously:

STEP 1 Compute the arrival rates from the traffic equations, Eq. (7.1):

$$\lambda_1 = \lambda_2 p_{21} + \lambda_3 p_{31} + \lambda_4 p_{41} = \underline{20}, \quad \lambda_2 = \lambda_1 p_{12} = \underline{10},$$
$$\lambda_3 = \lambda_1 p_{13} = \underline{10}, \quad \lambda_4 = \lambda_{04} = \underline{4}.$$

STEP 2 Compute the state probabilities and important performance measures for each node. For the *utilization* of a single server we use Eq. (6.3):

$$\rho_1 = \frac{\lambda_1}{\mu_1} = \underline{0.8}\,, \quad \rho_2 = \frac{\lambda_2}{\mu_2} = \underline{0.3}\,, \quad \rho_3 = \frac{\lambda_3}{\mu_3} = \underline{0.6}\,, \quad \rho_4 = \frac{\lambda_4}{\mu_4} = \underline{0.2}\,.$$

Thus, ergodicity ($\rho_i < 1$) is fulfilled for all nodes. The *mean number of jobs* at the nodes is given by Eq. (6.13)

$$\overline{K}_1 = \frac{\rho_1}{1 - \rho_1} = \underline{4}, \quad \overline{K}_2 = \underline{0.429}, \quad \overline{K}_3 = \underline{1.5}, \quad \overline{K}_4 = \underline{0.25}.$$

Mean response times, from Eq. (6.15):

$$\overline{T}_1 = \frac{1/\mu_1}{1 - \rho_1} = \underline{0.2}, \quad \overline{T}_2 = \underline{0.043}, \quad \overline{T}_3 = \underline{0.15}, \quad \overline{T}_4 = \underline{0.0625}.$$

The *mean overall response time* of a job is given by using Little's theorem in the following way:

$$\overline{T} = \frac{\overline{K}}{\lambda} = \frac{1}{\lambda} \sum_{1=1}^{4} \overline{K}_i = \underline{1.545}.$$

Mean waiting times, from Eq. (6.16):

$$\overline{W}_1 = \frac{\rho_1/\mu_1}{1 - \rho_1} = \underline{0.16}, \quad \overline{W}_2 = \underline{0.013}, \quad \overline{W}_3 = \underline{0.09}, \quad \overline{W}_4 = \underline{0.0125}.$$

Mean queue lengths, from Eq. (6.17):

$$\overline{Q}_1 = \frac{\rho_1^2}{1 - \rho_1} = \underline{3.2}, \quad \overline{Q}_2 = \underline{0.129}, \quad \overline{Q}_3 = \underline{0.9}, \quad \overline{Q}_4 = \underline{0.05}.$$

The necessary marginal probabilities can be computed using Eq. (6.12):

$$\pi_1(3) = (1 - \rho_1)\rho_1^3 = \underline{0.1024}, \quad \pi_2(2) = (1 - \rho_2)\rho_2^2 = \underline{0.063},$$
$$\pi_3(4) = (1 - \rho_3)\rho_3^4 = \underline{0.0518}, \quad \pi_4(1) = (1 - \rho_4)\rho_4 = \underline{0.16}.$$

STEP 3 Computation of the state probability $\pi(3, 2, 4, 1)$ using Eq. (7.50):

$$\pi(3, 2, 4, 1) = \pi_1(3) \cdot \pi_2(2) \cdot \pi_3(4) \cdot \pi_4(1) = \underline{0.0000534}.$$

7.3.5   Gordon–Newell Networks

Gordon and Newell [GoNe67a] considered closed queueing networks for which they made the same assumptions as in open queueing networks, except that no job can enter or leave the system ($\lambda_{0i} = \lambda_{i0} = 0$). This restriction means that the number K of jobs in the system is always constant:

$$K = \sum_{i=1}^{N} k_i.$$

Thus, the number of possible states is finite, and it is given by the binomial coefficient:

$$\binom{N + K - 1}{N - 1},$$

which describes the number of ways of distributing K jobs on N nodes. The theorem of Gordon and Newell says that the probability for each network state in equilibrium is given by the following product-form expression:

$$\pi(k_1,\ldots,k_N) = \frac{1}{G(K)} \prod_{i=1}^{N} F_i(k_i). \qquad (7.59)$$

Here $G(K)$ is the so-called *normalization constant*. It is given by the condition that the sum of all network state probabilities equals 1:

$$G(K) = \sum_{\substack{N \\ \sum\limits_{i=1} k_i = K}} \prod_{i=1}^{N} F_i(k_i). \qquad (7.60)$$

The $F_i(k_i)$ are functions that correspond to the state probabilities $\pi_i(k_i)$ of the ith node and are given by:

$$F_i(k_i) = \left(\frac{e_i}{\mu_i}\right)^{k_i} \cdot \frac{1}{\beta_i(k_i)}, \qquad (7.61)$$

where the visit ratios e_i are computed using Eq. (7.5). The function $\beta_i(k_i)$ is given by

$$\beta_i(k_i) = \begin{cases} k_i!, & k_i \le m_i, \\ m_i! \cdot m_i^{k_i - m_i}, & k_i \ge m_i, \\ 1, & m_i = 1. \end{cases} \qquad (7.62)$$

For various applications a more general form of the function $F_i(k_i)$ is advantageous. In this generalized function, the service rates depend on the number of jobs at the node. For this function we have

$$F_i(k_i) = \frac{e_i^{k_i}}{A_i(k_i)}, \qquad (7.63)$$

with

$$A_i(k_i) = \begin{cases} \prod_{j=1}^{k_i} \mu_i(j), & k_i > 0, \\ 1, & k_i = 0. \end{cases} \qquad (7.64)$$

With relation (7.49) it can easily be seen that the case of constant service rates, Eq. (7.61), is a special case of Eq. (7.63).

Proof: Gordon and Newell have shown in [GoNe67a] that Eq. (7.59) fulfills the global balance equations (see Eq. (7.53)) that have now the following form:

$$\left(\sum_{i=1}^{N}\gamma(k_i)\alpha_i(k_i)\mu_i\right)\pi(k_1,\ldots,k_N) =$$

$$\sum_{j=1}^{N}\sum_{i=1}^{N}\gamma(k_i)\alpha_j(k_j+1)\mu_j p_{ji}\pi(k_1,\ldots,k_i-1,\ldots,k_j+1,\ldots,k_N),$$

(7.65)

where the left-hand side describes the departure rate out of state $(k_1,\ldots,k_N)$ and the right-hand side describes the arrival rate from successor states into this state. The function $\gamma(k_i)$ and $\alpha_i(k_i)$ are given by Eqs. (7.54) and (7.55).

We define now a variable transformation as follows:

$$\pi(k_1,\ldots,k_N) = \frac{Q(k_1,\ldots,k_N)}{\prod\limits_{i=1}^{N}\beta_i(k_i)}, \qquad (7.66)$$

and

$$\pi(k_1,\ldots,k_i-1,\ldots,k_j+1,\ldots,k_N) =$$

$$\frac{\frac{\alpha_i(k_i)}{\alpha_j(k_j+1)}Q(k_1,\ldots,k_i-1,\ldots,k_j+1,\ldots,k_N)}{\prod\limits_{i=1}^{N}\beta_i(k_i)}. \qquad (7.67)$$

If we substitute these equations into Eq. (7.65), we get

$$\frac{\sum\limits_{i=1}^{N}\gamma(k_i)\alpha_i(k_i)\mu_i Q(k_1,\ldots,k_N)}{\prod\limits_{i=1}^{N}\beta_i(k_i)} =$$

$$\frac{\sum\limits_{j=1}^{N}\sum\limits_{i=1}^{N}\gamma(k_i)\alpha_j(k_j+1)\mu_j p_{ji}\frac{\alpha_i(k_i)}{\alpha_j(k_j+1)}Q(k_1,\ldots,k_i-1,\ldots,k_j+1,\ldots,k_N)}{\prod\limits_{i=1}^{N}\beta_i(k_i)}.$$

This equation can be simplified using the relation $\gamma(k_i)\alpha_i(k_i) \equiv \alpha_i(k_i)$:

$$\sum_{i=1}^{N}\alpha_i(k_i)\mu_i Q(k_1,\ldots,k_N) =$$

$$\sum_{j=1}^{N}\sum_{i=1}^{N}\alpha_i(k_i)\mu_j p_{ji}Q(k_1,\ldots,k_i-1,\ldots,k_j+1,\ldots,k_N),$$

(7.68)

and $Q(k_1, \ldots, k_N)$ can be written in the form

$$Q(k_1, \ldots, k_N) = \left\{ \prod_{i=1}^{N} x_i^{k_i} \right\} \cdot c, \tag{7.69}$$

with the relative utilization $x_i = e_i / \mu_i$ (Eq. (7.6)).

Here c is a constant. Substituting this into (7.68), we have

$$\sum_{i=1}^{N} \alpha_i(k_i) \mu_i (x_1^{k_1} \cdots x_N^{k_N}) \cdot c$$

$$= \sum_{j=1}^{N} \sum_{i=1}^{N} \alpha_i(k_i) \mu_j p_{ji} (x_1^{k_1} \cdots x_i^{k_i - 1} \cdots x_j^{k_j + 1} \cdots x_N^{k_N}) \cdot c$$

$$= \sum_{j=1}^{N} \sum_{i=1}^{N} \alpha_i(k_i) \mu_j p_{ji} (x_1^{k_1} \cdots x_N^{k_N}) \frac{x_j}{x_i} \cdot c,$$

$$\sum_{i=1}^{N} \alpha_i(k_i) \mu_i = \sum_{j=1}^{N} \sum_{i=1}^{N} \alpha_i(k_i) \mu_j p_{ji} \frac{x_j}{x_i}.$$

This expression can be rewritten as

$$\sum_{i=1}^{N} \alpha_i(k_i) \left(\mu_i - \sum_{j=1}^{N} \mu_j p_{ji} \frac{x_j}{x_i} \right) = 0.$$

Since at least one $\alpha_i(k_i)$ will be nonzero, it follows that the factor in the square brackets must be zero. Thus, we are led to consider the system of linear algebraic equations for x_i:

$$\mu_i x_i = \sum_{j=1}^{N} \mu_j x_j p_{ji},$$

where $x_i = e_i / \mu_i$ (Eq. (7.6)) and

$$e_i = \sum_{j=1}^{N} e_j p_{ji}.$$

This is the traffic equation for closed networks (Eq. (7.5)). That means that Eq. (7.69) is correct and with Eqs. (7.66) and (7.61) we obtain Eq. (7.59).$q.e.d.$

Thus, the Gordon–Newell theorem yields a product-form solution. In the general form it says that the state probabilities $\pi(k_1, k_2, \cdots, k_N)$ are given as the product of the functions $F_i(k_i)$, $i = 1 \ldots, N$, defined for single nodes. It is interesting to note that if we substitute in Eq. (7.59) $F_i(k_i)$ by $L_i \cdot F_i(k_i)$, $1 \leq i \leq N$, then this has no influence on the solution of the state probabilities $\pi(k_1, k_2, \ldots, k_N)$ as long as L_i is a positive real number. Furthermore, the use of $\lambda \cdot e_i$, $i = 1, \ldots, N$, with an arbitrary constant $\lambda > 0$, has no influence on the results because the visit ratios e_i are relative values [ShBu77]. (Also see Problems 2 and 3 on page 444 of [Triv01].)

The Gordon–Newell method for computing the state probabilities can be summarized in the following four steps:

STEP 1 Compute the visit ratios e_i for all nodes $i = 1, \ldots, N$ of the closed network using Eq. (7.5).

STEP 2 For all $i = 1, \ldots, N$, compute the functions $F_i(k_i)$ using Eq. (7.61) or Eq. (7.63) (in the case of load-dependent service rates).

STEP 3 Compute the normalization constant $G(K)$ using Eq. (7.60).

STEP 4 Compute the state probabilities of the network using Eq. (7.59). From the marginal probabilities, which can be determined from the state probabilities using Eq. (7.16), all other required performance measures can be determined.

An example of the application of the Gordon–Newell theorem follows:

Example 7.5   Consider the closed queueing network shown in Fig. (7.15) with $N = 3$ nodes and $K = 3$ jobs. The queueing discipline at all nodes is FCFS. The routing probabilities are given as follows:

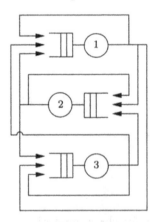

Fig. 7.15   Closed queueing network.

$$p_{11} = 0.6, \quad p_{21} = 0.2, \quad p_{31} = 0.4,$$
$$p_{12} = 0.3, \quad p_{22} = 0.3, \quad p_{32} = 0.1,$$

$$p_{13} = 0.1, \quad p_{23} = 0.5, \quad p_{33} = 0.5.$$

The service time at each node is exponentially distributed with the rates:

$$\mu_1 = 0.8\,\mathrm{sec}^{-1}, \quad \mu_2 = 0.6\,\mathrm{sec}^{-1}, \quad \mu_3 = 0.4\,\mathrm{sec}^{-1}. \tag{7.70}$$

This network consists of

$$\binom{N+K-1}{N-1} = \underline{10}$$

states, namely,

$$(3,0,0), \quad (2,1,0), \quad (2,0,1), \quad (1,2,0), \quad (1,1,1),$$
$$(1,0,2), \quad (0,3,0), \quad (0,2,1), \quad (0,1,2), \quad (0,0,3).$$

We wish to compute the state probabilities using the Gordon–Newell theorem. For this we proceed in the following four steps:

STEP 1 Determine the visit ratios at each node using Eq. (7.5):

$$e_1 = e_1 p_{11} + e_2 p_{21} + e_3 p_{31} = \underline{1},$$
$$e_2 = e_1 p_{12} + e_2 p_{22} + e_3 p_{32} = \underline{0.533},$$
$$e_3 = e_1 p_{13} + e_2 p_{23} + e_3 p_{33} = \underline{0.733}.$$

STEP 2 Determine the functions $F_i(k_i)$ for $i = 1, 2, 3$ using Eq. (7.61):

$$F_1(0) = (e_1/\mu_1)^0 = \underline{1}, \qquad F_1(1) = (e_1/\mu_1)^1 = \underline{1.25},$$
$$F_1(2) = (e_1/\mu_1)^2 = \underline{1.5625}, \quad F_1(3) = (e_1/\mu_1)^3 = \underline{1.953},$$

and correspondingly:

$$F_2(0) = \underline{1}, \quad F_2(1) = \underline{0.889}, \quad F_2(2) = \underline{0.790}, \quad F_2(3) = \underline{0.702},$$
$$F_3(0) = \underline{1}, \quad F_3(1) = \underline{1.833}, \quad F_3(2) = \underline{3.361}, \quad F_3(3) = \underline{6.162}.$$

STEP 3 Determine the normalization constant using Eq. (7.60):

$$\begin{aligned}
G(3) = \ & F_1(3)F_2(0)F_3(0) + F_1(2)F_2(1)F_3(0) + F_1(2)F_2(0)F_3(1) \\
& + F_1(1)F_2(2)F_3(0) + F_1(1)F_2(1)F_3(1) + F_1(1)F_2(0)F_3(2) \\
& + F_1(0)F_2(3)F_3(0) + F_1(0)F_2(2)F_3(1) + F_1(0)F_2(1)F_3(2) \\
& + F_1(0)F_2(0)F_3(3) = \underline{24.733}.
\end{aligned}$$

STEP 4 Determine the state probabilities using the Gordon–Newell theorem, Eq. (7.59):

$$\pi(3,0,0) = \frac{1}{G(3)} F_1(3) \cdot F_2(0) \cdot F_3(0) = \underline{0.079},$$

$$\pi(2,1,0) = \frac{1}{G(3)} F_1(2) \cdot F_2(1) \cdot F_3(0) = \underline{0.056} \,.$$

In the same way we compute

$$\pi(2,0,1) = \underline{0.116}\,, \quad \pi(1,2,0) = \underline{0.040}\,, \quad \pi(1,1,1) = \underline{0.082}\,, \quad \pi(1,0,2) = \underline{0.170}\,,$$
$$\pi(0,3,0) = \underline{0.028}\,, \quad \pi(0,2,1) = \underline{0.058}\,, \quad \pi(0,1,2) = \underline{0.121}\,, \quad \pi(0,0,3) = \underline{0.249}\,.$$

Using Eq. (7.16), all marginal probabilities can now be determined:

$$\pi_1(0) = \pi(0,3,0) + \pi(0,2,1) + \pi(0,1,2) + \pi(0,0,3) = \underline{0.457}\,,$$
$$\pi_1(1) = \pi(1,2,0) + \pi(1,1,1) + \pi(1,0,2) = \underline{0.292}\,,$$
$$\pi_1(2) = \pi(2,1,0) + \pi(2,0,1) = \underline{0.172}\,,$$
$$\pi_1(3) = \pi(3,0,0) = \underline{0.079}\,,$$

$$\pi_2(0) = \pi(2,0,1) + \pi(1,0,2) + \pi(0,0,3) + \pi(3,0,0) = \underline{0.614}\,,$$
$$\pi_2(1) = \pi(2,1,0) + \pi(1,1,1) + \pi(0,1,2) = \underline{0.259}\,,$$
$$\pi_2(2) = \pi(1,2,0) + \pi(0,2,1) = \underline{0.098}\,,$$
$$\pi_2(3) = \pi(0,3,0) = \underline{0.028}\,,$$

$$\pi_3(0) = \pi(3,0,0) + \pi(2,1,0) + \pi(1,2,0) + \pi(0,3,0) = \underline{0.203}\,,$$
$$\pi_3(1) = \pi(2,0,1) + \pi(1,1,1) + \pi(0,2,1) = \underline{0.257}\,,$$
$$\pi_3(2) = \pi(1,0,2) + \pi(0,1,2) = \underline{0.291}\,,$$
$$\pi_3(3) = \pi(0,0,3) = \underline{0.249}\,.$$

For the other performance measures, we get

- Utilization at each node, Eq. (7.19):

$$\rho_1 = 1 - \pi_1(0) = \underline{0.543}\,, \quad \rho_2 = \underline{0.386}\,, \quad \rho_3 = \underline{0.797}\,.$$

- Mean number of jobs at each node, Eq. (7.26):

$$\overline{K}_1 = \sum_{k=1}^{3} k \cdot \pi_1(k) = \underline{0.873}\,, \quad \overline{K}_2 = \underline{0.541}\,, \quad \overline{K}_3 = \underline{1.585}\,.$$

- Throughputs at each node, Eq. (7.23):

$$\lambda_1 = m_1 \rho_1 \mu_1 = \underline{0.435}\,, \quad \lambda_2 = \underline{0.232}\,, \quad \lambda_3 = \underline{0.319}\,.$$

- Mean response time at each node, Eq. (7.43):

$$\overline{T}_1 = \frac{\overline{K}_1}{\lambda_1} = \underline{2.009}\,, \quad \overline{T}_2 = \underline{2.337}\,, \quad \overline{T}_3 = \underline{4.976}\,.$$

7.3.6 BCMP Networks

The results of Jackson and Gordon–Newell were extended by Baskett, Chandy, Muntz, and Palacios in their classic article [BCMP75], to queueing networks with several job classes, different queueing strategies, and generally distributed service times. The considered networks can be open, closed, or mixed.

Following [BrBa80], an allowed state in a queueing model without class switching is characterized by four conditions:

1. The number of jobs in each class at each node is always non-negative, i.e.,

$$k_{ir} \geq 0, \quad 1 \leq r \leq R, \quad 1 \leq i \leq N.$$

2. For all jobs the following condition must hold:

$$k_{ir} > 0 \quad \text{if there exists a way for class-}r \text{ jobs to}$$
$$\text{node } i \text{ with a nonzero probability.}$$

3. For a closed network, the number of jobs in the network is given by

$$K = \sum_{r=1}^{R} \sum_{i=1}^{N} k_{ir}.$$

4. The sum of class-r jobs in the network is constant at any time, i.e.,

$$K_r = \sum_{i=1}^{N} k_{ir} = \text{const.} \quad 1 \leq r \leq R.$$

If class switching is allowed, then conditions $1-3$ are fulfilled, but condition 4 can not be satisfied because the number of jobs within a class is no longer constant but depends on the time when the system is looked at and can have the values $k \in \{0, \ldots, K\}$. In order to avoid this situation, the concept of chains is introduced.

7.3.6.1 The Concept of Chains Consider the routing matrix $\mathbf{P} = [p_{ir,js}]$, $i, j = 1, \ldots, N$, and $r, s = 1, \ldots, R$ of a closed queueing network. The routing matrix $\mathbf{P}$ defines a finite-state DTMC whose states are pairs of form (i, r) (node number i, class index r). We call (i, r) a *node-class pair*. This state space can be partitioned into disjoint sets Γ_i, $i = 1, \ldots, U$:

$$\Gamma = \Gamma_1 + \Gamma_2 + \ldots + \Gamma_U,$$

where Γ_i is a closed communicating class of recurrent states of the DTMC.

With a possible relabeling of nodes, we get the routing matrix of Fig. 7.16, where submatrices P_i contains the transition probabilities in set Γ_i and each

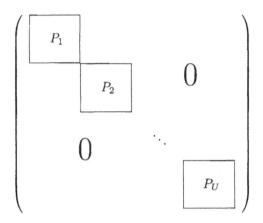

Fig. 7.16 Modified routing matrix.

P_i is disjoint and assumed to be ergodic. Let chain C_i denote the set of job classes in Γ_i. Because of the disjointness of the chains C_i, it is impossible for a job to switch from one chain to another. If a job starts in a chain, it will never leave this chain. With the help of this partitioning technique, the number of jobs in each chain is always constant in closed networks. Therefore, Condition 4 for queueing networks is fulfilled for the chains. Muntz [Munt73] proved that a closed queueing network with U chains is equivalent to a closed queueing network with U job classes. If there is no class switching allowed, then the number of chains equals the number of classes. But if class switching is allowed, the number of chains is smaller than the number of classes. This means that the dimension of the population vector is reduced. An extension of the chains concept to open networks is possible.

The procedure to find the chains from a given solution matrix is the following:

STEP 1 Construct R sets that satisfy the following condition:

$$E_r := \left\{ s : \begin{array}{l} \text{node-class pair } (j,s) \text{ can be reached from} \\ \text{pair } (i,r) \text{ in a finite number of steps} \end{array} \right\},$$

$$1 \leq r \leq R, \quad 1 \leq s \leq R.$$

STEP 2 Eliminate all subsets and identical sets so that we have

$$U \leq R \quad \text{sets to obtain the chains} \quad C_1, \ldots, C_U.$$

STEP 3 Compute the number of jobs in each chain:

$$K_q^* = \sum_{r \in C_q} K_r, \quad 1 \leq q \leq U.$$

The set of all possible states in a closed multiclass network is given by the following binomial coefficient:

$$\prod_{q=1}^{U} \binom{N \cdot |C_q| + K_q^* - 1}{N \cdot |C_q| - 1},$$

where $|C_q|$ is the number of elements in C_q, $1 \le q \le U$.

For open networks, the visit ratios in a chain are given by

$$2e_{ir} = p_{0,ir} + \sum_{\substack{s \in C_q \\ j=1,\dots,N}} e_{js} p_{js,ir} \qquad \text{for} \quad \begin{array}{l} r \in C_q, \\ i = 1, \dots N, \\ 1 \le q \le U, \end{array} \qquad (7.71)$$

and for closed networks we have

$$e_{ir} = \sum_{\substack{s \in C_q \\ j=1,\dots,N}} e_{js} p_{js,ir} \qquad \text{for} \quad \begin{array}{l} r \in C_q, \\ i = 1, \dots N, \\ 1 \le q \le U. \end{array} \qquad (7.72)$$

Example 7.6   Let the following routing matrix **P** be given:

	(1,1)	(1,2)	(1,3)	(2,1)	(2,2)	(2,3)	(3,1)	(3,2)	(3,3)
(1,1)	0	0.4	0	0	0.3	0	0	0.3	0
(1,2)	0.3	0	0	0	0	0	0	0.7	0
(1,3)	0	0	0	0	0	0.3	0	0	0.7
(2,1)	0	0	0	0	1	0	0	0	0
(2,2)	0.3	0	0	0	0	0	0.7	0	0
(2,3)	0	0	0.3	0	0	0	0	0	0.7
(3,1)	1.0	0	0	0	0	0	0	0	0
(3,2)	0	0	0	1.0	0	0	0	0	0
(3,3)	0	0	0.4	0	0	0.6	0	0	0

$\mathbf{P} =$

By using the chaining technique, we get

$$E_1 = \{1,2\}, \quad E_2 = \{1,2\}, \quad E_3 = \{3\},$$

meaning that $R = 3$ classes are reduced to $U = 2$ chains. By eliminating the subsets and identical sets, we get two chains:

$$C_1 = \{1,2\}, \quad C_2 = \{3\}.$$

Then the reorganized routing matrix $\mathbf{P}'$ has the form

$$
\mathbf{P'} =
\begin{array}{c}
\\
(1,1) \\
(1,2) \\
(2,1) \\
(2,2) \\
(3,1) \\
(3,2) \\
(1,3) \\
(2,3) \\
(3,3)
\end{array}
\begin{array}{ccccccccc}
(1,1) & (1,2) & (2,1) & (2,2) & (3,1) & (3,2) & (1,3) & (2,3) & (3,3) \\
\left(\begin{array}{ccccccccc}
0 & 0.4 & 0 & 0.3 & 0 & 0.3 & 0 & 0 & 0 \\
0.3 & 0 & 0 & 0 & 0 & 0.7 & 0 & 0 & 0 \\
0 & 0 & 0 & 1.0 & 0 & 0 & 0 & 0 & 0 \\
0.3 & 0 & 0 & 0 & 0.7 & 0 & 0 & 0 & 0 \\
1.0 & 0 & 0 & 0 & 0 & 0 & 0 & 0 & 0 \\
0 & 0 & 1.0 & 0 & 0 & 0 & 0 & 0 & 0 \\
0 & 0 & 0 & 0 & 0 & 0 & 0 & 0.3 & 0.7 \\
0 & 0 & 0 & 0 & 0 & 0 & 0.3 & 0 & 0.7 \\
0 & 0 & 0 & 0 & 0 & 0 & 0.4 & 0.6 & 0
\end{array} \right)
\end{array}.
$$

If a job starts in one of the chains C_1 or C_2, then it cannot leave them.

Because of the switch from classes to chains, it is necessary to transfer class measures into chain measures [BrBa80] (chain measures are marked with a *):

- Number of visits: In a chain a job can reach different node-class pairs (i,j) (i: node index, j: class index) where

$$
e_{iq}^* = \frac{\sum\limits_{r \epsilon C_q} e_{ir}}{\sum\limits_{r \epsilon C_q} e_{1r}} . \tag{7.73}
$$

- The number of jobs per class is constant in a chain, but within the chain jobs can change their class. Therefore, if a job of chain q visits node i, it is impossible to know to which class it belongs because we consider a chain to be a single entity. If we make the transformation

$$
\text{class} \quad \longrightarrow \quad \text{chain},
$$

the information about the class of a job is lost. Because different job classes have different service times, this missing information is exchanged by the scale factor α. For the service time in a chain we get

$$
s_{iq}^* = \frac{1}{\mu_{iq}^*} = \sum_{r \epsilon C_q} s_{ir} \cdot \alpha_{ir}, \tag{7.74}
$$

$$
\alpha_{ir} = \frac{e_{ir}}{\sum\limits_{s \in C_q} e_{is}} . \tag{7.75}
$$

In Chapter 8 we introduce several algorithms to calculate the performance measures of single and multiclass queueing networks without class switching (such as mean value analysis or convolution). Using the concept of chains, it is possible to also use these algorithms for queueing networks with class switching. To do so, we proceed in the following steps:

STEP 1 Calculate the number of visits e_{ir} in the original network.

STEP 2 Determine the chains $C_1 \ldots C_U$, and calculate the number of jobs K_q^* in each chain.

STEP 3 Compute the number of visits e_{iq}^* for each chain, Eq. (7.73).

STEP 4 Determine the scale factors α_{ir}, Eq. (7.75).

STEP 5 Calculate the service times s_{iq}^* for each chain, Eq. (7.74).

STEP 6 Derive the performance measures per chain [BrBa80] with one of the algorithms introduced later (mean value analysis, convolution, etc.).

STEP 7 Calculate the performance measures per class from the performance measures per chain:

$$\overline{T}_{ir}(\mathbf{K}^*) = s_{ir} \cdot \left(1 + \overline{K}_i(\mathbf{K}^* - \mathbf{1}_q)\right) , \quad r \in C_q$$
$$\lambda_{ir}(\mathbf{K}^*) = \alpha_{ir} \cdot \lambda_{iq}^* ,$$
$$\rho_{ir}(\mathbf{K}^*) = s_{ir} \cdot \lambda_{ir}(\mathbf{K}^*) ,$$

where

$$\mathbf{K}^* = (K_1^*, \ldots, K_U^*) :$$ Population vector containing the number of jobs in each chain,

$$\mathbf{K}^* - \mathbf{1}_q :$$ $\mathbf{K}^*$ with one job less in chain q,

$$\overline{K}_i(\mathbf{K}^* - \mathbf{1}_q) :$$ Mean number of jobs at node i if the number of jobs in the chains is given by $(\mathbf{K}^* - \mathbf{1}_q)$.

All other performance measures can easily be obtained using the formulae of Section 7.2.

In the following section we introduce the BCMP theorem, which is the basis of all analysis techniques to come. If class switching is not allowed in the network, then the BCMP theorem can be applied directly. In the case of class switching, it needs to be applied in combination with the concept of chains.

7.3.6.2 BCMP Theorem The theorems of Jackson and of Gordon and Newell have been extended by [BCMP75] to networks with several job classes and different service strategies and interarrival/service time distributions, and also to mixed networks that contain open and closed classes. The networks considered by BCMP must fulfill the following assumptions:

- Queueing discipline: The following disciplines are allowed at network nodes: FCFS, PS, LCFS-PR, IS (infinite server).

- Distribution of the service times: The service times of an FCFS node must be exponentially distributed and class-independent (i.e., $\mu_{i1} = \mu_{i2} = \ldots = \mu_{iR} = \mu_i$), while PS, LCFS-PR and IS nodes can have any kind of service time distribution with a rational Laplace transform. For the latter three queueing disciplines, the mean service time for different job classes can be different.

- Load-dependent service rates: The service rate of an FCFS node is only allowed to depend on the number of jobs at this node, whereas in a PS, LCFS-PR and IS node the service rate for a particular job class can also depend on the number of jobs of that class at the node but not on the number of jobs in another class.

- Arrival processes: In open networks two kinds of arrival processes can be distinguished from each other.

  Case 1: The arrival process is Poisson where all jobs arrive at the network from one source with an overall arrival rate λ, where λ can depend on the number of jobs in the network. The arriving jobs are distributed over the nodes in the network in accordance to the probability $p_{0,ir}$ where:

  $$\sum_{i=1}^{N} \sum_{r=1}^{R} p_{0,ir} = 1.$$

  Case 2: The arrival process consists of U independent Poisson arrival streams where the U job sources are assigned to the U chains. The arrival rate λ_u from the uth source can be load dependent. A job arrives at the ith node with probability $p_{0,ir}$ so that

  $$\sum_{\substack{r \in C_u \\ i=1,\ldots,N}} p_{0,ir} = 1, \quad \text{for all } u = 1, \ldots, U.$$

These assumptions lead to the four product-form node types and the local balance conditions for BCMP networks (see Section 7.3.2), that is,

Type 1: $-/M/m - FCFS$	Type 2: $-/G/1 - PS$
Type 3: $-/G/\infty$ (IS)	Type 4: $-/G/1 - LCFS$ PR

Note that we use $-/M/m$ notation since we know that, in general, the arrival process to a node in a BCMP network will not be Poisson.

The BCMP Theorem says that networks with the characteristics just described have product-form solution:

For open, closed, and mixed queueing networks whose nodes consist only of the four described node types, the steady-state probabilities have the following product-form:

$$\pi(\mathbf{S}_1, \ldots, \mathbf{S}_N) = \frac{1}{G(\mathbf{K})} d(\mathbf{S}) \prod_{i=1}^{N} f_i(\mathbf{S}_i), \qquad (7.76)$$

where,

$G(\mathbf{K})$ = the normalization constant,

$d(\mathbf{S})$ = a function of the number of jobs in the network, $\mathbf{S} = (\mathbf{S}_1, \ldots, \mathbf{S}_N)$,

and

$$d(\mathbf{S}) = \begin{cases} \displaystyle\prod_{i=0}^{K(\mathbf{S})-1} \lambda(i), & \text{open network with arrival process 1,} \\[2em] \displaystyle\prod_{u=1}^{U} \prod_{i=0}^{K_u(\mathbf{S})-1} \lambda_u(i), & \text{open network with arrival process 2,} \\[2em] 1, & \text{closed networks.} \end{cases}$$

$f_i(\mathbf{S}_i)$ = a function which depends on the type and state of each node

and

$$f_i(\mathbf{S}_i) = \begin{cases} \left(\dfrac{1}{\mu_i}\right)^{k_i} \displaystyle\prod_{j=1}^{k_i} e_{is_{ij}}, & \text{Type 1,} \\[2em] k_i! \displaystyle\prod_{r=1}^{R} \prod_{l=1}^{u_{ir}} \frac{1}{k_{irl}!} \left(\frac{e_{ir} A_{irl}}{\mu_{irl}}\right)^{k_{irl}}, & \text{Type 2,} \\[2em] \displaystyle\prod_{r=1}^{R} \prod_{l=1}^{u_{ir}} \frac{1}{k_{irl}!} \left(\frac{e_{ir} A_{irl}}{\mu_{irl}}\right)^{k_{irl}}, & \text{Type 3,} \\[2em] \displaystyle\prod_{j=1}^{k_i} \frac{e_{ir_j} A_{ir_j m_j}}{\mu_{ir_j m_j}}, & \text{Type 4.} \end{cases} \qquad (7.77)$$

Variables in for Eq. (7.77) have the following meanings:

s_{ij}: Class of the job that is at the jth position in the FCFS queue.

μ_{irl}: Mean service rate in the lth phase ($l = 1, \ldots, u_{ir}$) in a Cox distribution (see Chapter 1).

u_{ir}: Maximum number of exponential phases.

A_{irl}: $\displaystyle\prod_{j=0}^{l-1} a_{irj}$, probability that a class-r job at the ith node reaches the lth service phase ($A_{ir1} = 1$ because of $a_{ir0} = 1$).

a_{irj}: Probability that a class-r job at the ith node moves to the $(j+1)$th phase.

k_{irl}: Number of class-r jobs in the lth phase of node i.

For the load-dependent case, $f_i(\mathbf{S}_i)$ is of the form

$$f_i(\mathbf{S}_i) = \left(\frac{1}{\mu_i}\right)^{k_i} \prod_{j=1}^{k_i} e_{is_{ij}}.$$

Proof: The proof of this theorem is very complex and therefore only the basic idea is given here (for the complete proof see [Munt72]). In order to find a solution for the steady-state probabilities $\pi(\mathbf{S})$, the following global balance equations have to be solved:

$$\pi(\mathbf{S}) \begin{bmatrix} \text{state transition rate} \\ \text{from state } \mathbf{S} \end{bmatrix} = \sum_{\tilde{\mathbf{S}}} \pi(\tilde{\mathbf{S}}) \begin{bmatrix} \text{state transition rate from} \\ \text{state } \tilde{\mathbf{S}} \text{ to state } \mathbf{S} \end{bmatrix},$$

$$(7.78)$$

with the normalization condition

$$\sum_{\mathbf{S}} \pi(\mathbf{S}) = 1. \tag{7.79}$$

Now we insert Eq. (7.76) into Eq. (7.78) to verify that Eq. (7.78) can be written as a system of balance equations that [Chan72] calls *local balance equations*. All local balance equations can be transformed into a system of $N-1$ independent equations. To get unambiguity for the solution the normalization condition, Eq. (7.79), has to be used.

Now we wish to give two simplified versions of the BCMP theorem for open and closed networks.

BCMP Version 1: For a *closed* queueing network fulfilling the assumptions of the BCMP theorem, the steady-state state probabilities have the form

$$\pi(\mathbf{S}_1, \ldots, \mathbf{S}_N) = \frac{1}{G(\mathbf{K})} \prod_{i=1}^{N} F_i(\mathbf{S}_i), \tag{7.80}$$

where the normalization constant is defined as

$$G(\mathbf{K}) = \sum_{\sum_{i=1}^{N} \mathbf{S}_i = \mathbf{K}} \prod_{i=1}^{N} F_i(\mathbf{S}_i), \tag{7.81}$$

and the function $F_i(\mathbf{S}_i)$ is given by

$$
F_i(\mathbf{S}_i) = \begin{cases}
k_i! \dfrac{1}{\beta_i(k_i)} \cdot \left(\dfrac{1}{\mu_i}\right)^{k_i} \cdot \displaystyle\prod_{r=1}^{R} \dfrac{1}{k_{ir}!} e_{ir}^{k_{ir}}, & \text{Type 1,} \\[3ex]
k_i! \displaystyle\prod_{r=1}^{R} \dfrac{1}{k_{ir}!} \cdot \left(\dfrac{e_{ir}}{\mu_{ir}}\right)^{k_{ir}}, & \text{Type 2,4,} \\[3ex]
\displaystyle\prod_{r=1}^{R} \dfrac{1}{k_{ir}!} \cdot \left(\dfrac{e_{ir}}{\mu_{ir}}\right)^{k_{ir}}, & \text{Type 3.}
\end{cases} \tag{7.82}
$$

The quantity $k_i = \sum\limits_{r=1}^{R} k_{ir}$ gives the overall number of jobs of all classes at node i. The visit ratios e_{ir} can be determined with Eq. (7.72), while the function $\beta_i(k_i)$ is given in Eq. (7.62).

For FCFS nodes ($m_i = 1$) with load-dependent service rates $\mu_i(j)$ we get

$$
F_i(\mathbf{S}_i) = \frac{k_i!}{\displaystyle\prod_{j=1}^{k_i} \mu_i(j)} \prod_{r=1}^{R} \frac{1}{k_{ir}!} e_{ir}^{k_{ir}}. \tag{7.83}
$$

BCMP Version 2: For an *open* queueing network fulfilling the assumptions of the BCMP theorem and load-independent arrival and service rates, we have

$$
\pi(k_1, \ldots, k_N) = \prod_{i=1}^{N} \pi_i(k_i), \tag{7.84}
$$

where

$$
\pi_i(k_i) = \begin{cases}
(1 - \rho_i)\rho_i^{k_i}, & \text{Type 1,2,4 } (m_i = 1), \\[2ex]
e^{-\rho_i} \dfrac{\rho_i^{k_i}}{k_i!}, & \text{Type 3,}
\end{cases} \tag{7.85}
$$

and

$$
k_i = \sum_{i=i}^{R} k_{ir},
$$

$$
\rho_i = \sum_{r=1}^{R} \rho_{ir},
$$

with

$$
\rho_{ir} = \begin{cases}
\lambda_r \dfrac{e_{ir}}{\mu_i}, & \text{Type 1 } (m_i = 1), \\[2ex]
\lambda_r \dfrac{e_{ir}}{\mu_{ir}}, & \text{Type 2,3,4.}
\end{cases} \tag{7.86}
$$

Furthermore, we have

$$\overline{K}_{ir} = \frac{\rho_{ir}}{1 - \rho_i} . \tag{7.87}$$

For Type 1 nodes with more than one service unit ($m_i > 1$), Eq. (6.26) can be used to compute the probabilities $\pi_i(k_i)$. For the existence of the steady-state probabilities $\pi_i(k_i)$, ergodicity condition ($\rho_i < 1$) has to be fulfilled for all i, $i = 1, \ldots, N$.

The algorithm to determine the performance measures using the BCMP theorem can now be given in the following five steps:

STEP 1 Compute the visit ratios e_{ir} for all $i = 1, \ldots, N$ and $r = 1, \ldots, R$ using Eq. (7.71).

STEP 2 Compute the utilization of each node using Eq. (7.86).

STEP 3 Compute the other performance measures with the equations given in Section 7.2.

STEP 4 Compute the marginal probabilities of the network using Eq. (7.85).

STEP 5 Compute the state probabilities using Eq. (7.84).

Example 7.7   Consider the open network given in Fig. 7.17 with $N = 3$ nodes and $R = 2$ job classes. The first node is of Type 2 and the second and third nodes are of Type 4. The service times are exponentially distributed with the rates

$$\mu_{11} = 8 \sec^{-1}, \quad \mu_{21} = 12 \sec^{-1}, \quad \mu_{31} = 16 \sec^{-1},$$
$$\mu_{12} = 24 \sec^{-1}, \quad \mu_{22} = 32 \sec^{-1}, \quad \mu_{32} = 36 \sec^{-1}.$$

The interarrival times are also exponentially distributed with the rates

$$\lambda_1 = \lambda_2 = 1 \text{ job/sec}.$$

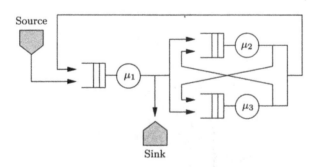

Fig. 7.17 Open queueing network.

The routing probabilities are given by

$$p_{0,11} = 1, \quad p_{21,11} = 0.6, \quad p_{0,12} = 1, \quad p_{22,12} = 0.7,$$
$$p_{11,21} = 0.4, \quad p_{21,31} = 0.4, \quad p_{12,22} = 0.3, \quad p_{22,32} = 0.3,$$
$$p_{11,31} = 0.3, \quad p_{31,11} = 0.5, \quad p_{12,32} = 0.6, \quad p_{32,12} = 0.4,$$
$$p_{11,0} = 0.3, \quad p_{31,21} = 0.5, \quad p_{12,0} = 0.1, \quad p_{32,22} = 0.6,$$

which means that class switching is not allowed in the network. We wish to compute the probability for the state $(k_1, k_2, k_3) = (3, 2, 1)$ by using the BCMP theorem, Eq. (7.84).

STEP 1 Compute the visit ratios e_{ir} for all $i = 1, \ldots, N$ and $r = 1, \ldots, R$ using Eq. (7.71).

$$e_{11} = p_{0,11} + e_{11}p_{11,11} + e_{21}p_{21,11} + e_{31}p_{31,11} = \underline{3.333},$$
$$e_{21} = p_{0,21} + e_{11}p_{11,21} + e_{21}p_{21,21} + e_{31}p_{31,21} = \underline{2.292},$$
$$e_{31} = p_{0,31} + e_{11}p_{11,31} + e_{21}p_{21,31} + e_{31}p_{31,31} = \underline{1.917}.$$

In the same way we get

$$e_{12} = \underline{10}, \quad e_{22} = \underline{8.049}, \quad e_{32} = \underline{8.415}.$$

STEP 2 Compute the utilization of each node using Eq. (7.86):

$$\rho_1 = \lambda_1 \frac{e_{11}}{\mu_{11}} + \lambda_2 \frac{e_{12}}{\mu_{12}} = \rho_{11} + \rho_{12} = \underline{0.833},$$

$$\rho_2 = \lambda_1 \frac{e_{21}}{\mu_{21}} + \lambda_2 \frac{e_{22}}{\mu_{22}} = \rho_{21} + \rho_{22} = \underline{0.442},$$

$$\rho_3 = \lambda_1 \frac{e_{31}}{\mu_{31}} + \lambda_2 \frac{e_{32}}{\mu_{32}} = \rho_{31} + \rho_{32} = \underline{0.354}.$$

STEP 3 Compute the other performance measures of the network. In our case we use Eq. (7.87) to determine the mean number of jobs at each node:

$$\overline{K}_{11} = \frac{\rho_{11}}{1 - \rho_1} = \underline{2.5}, \quad \overline{K}_{21} = \frac{\rho_{21}}{1 - \rho_2} = \underline{0.342}, \quad \overline{K}_{31} = \frac{\rho_{31}}{1 - \rho_3} = \underline{0.186},$$

$$\overline{K}_{12} = \underline{2.5}, \quad \overline{K}_{22} = \underline{0.5}, \quad \overline{K}_{32} = \underline{0.362}.$$

STEP 4 Determine the marginal probabilities using Eq. (7.85):

$$\pi_1(3) = (1 - \rho_1)\rho_1^3 = \underline{0.0965}, \quad \pi_2(2) = (1 - \rho_2)\rho_2^2 = \underline{0.1093},$$
$$\pi_3(1) = (1 - \rho_3)\rho_3 = \underline{0.2287}.$$

STEP 5 Compute the state probabilities for the network using Eq. (7.84):

$$\pi(3, 2, 1) = \pi_1(3) \cdot \pi_2(2) \cdot \pi_3(1) = \underline{0.00241}\,.$$

An example of the BCMP theorem for closed networks, Eq. (7.80), is not given in this chapter. As shown in Example 7.7, the direct use of the BCMP theorem will require that all states of the network have to be considered in order to compute the normalization constant. This is a very complex procedure and only suitable for small networks because for bigger networks the number of possible states in the network becomes prohibitively large. In Chapter 8 we provide efficient algorithms to analyze closed product-form queueing networks. In these algorithms the states of the queueing network are not explicitly involved in the computation and therefore these algorithms provide much shorter computation time.

Several researchers have extended the class of product-form networks to nodes of the following types:

- In [Spir79] the SIRO (service in random order) is examined and it is shown that -/M/1-SIRO nodes fulfill the local balance property and therefore have a product-form solution.

- In [Noet79] the LBPS (last batch processor sharing) strategy is introduced and it is show that -/M/1-LBPS node types have product-form solutions. In this strategy the processor is assigned between the two last batch jobs. If the last batch consists only of one job, then we get the LCFS-PR strategy and if the batch consists of all jobs, we get PS.

- In [ChMa83] the WEIRDP-strategy (**weird, parameterized strategy**) is considered, where the first job in the queue is assigned $100 \cdot p \%$ of the processor and the rest of the jobs are assigned $100 \cdot (1 - p) \%$ of the processor. It is shown that -/M/1-WEIRDP nodes have product-form solution.

- In [CHT77] it is shown that -/G/1-PS, -/G/∞-IS, and -/G/1-LCFS-PR nodes with arbitrary differentiable service time distribution also have product-form solution.

Furthermore, in [Tows80] and [Krze87], the class of product-form networks is extended to networks where the probability to enter a particular node depends on the number of jobs at that node. In this case, we have so-called *load-dependent routing probabilities.*

Problem 7.2   Consider an open queueing network with $N = 3$ nodes, FCFS queueing discipline, and exponentially distributed service times with the mean values

$$\frac{1}{\mu_1} = 0.08\,\text{sec}, \quad \frac{1}{\mu_2} = 0.06\,\text{sec}, \quad \frac{1}{\mu_3} = 0.04\,\text{sec}\,.$$

Only at the first node do jobs arrive from outside with exponentially distributed interarrival times and the rate $\lambda_{01} = 4$ jobs/sec. Node 1 is a multiple server node with $m_1 = 2$ server, nodes 2 and 3 are single server nodes. The routing probabilities are given as follows:

$$p_{11} = 0.2, \quad p_{21} = 1, \quad p_{31} = 0.5,$$
$$p_{12} = 0.4, \quad p_{30} = 0.5,$$
$$p_{13} = 0.4.$$

(a) Draw the queueing network.

(b) Determine the steady-state probability for the state $\mathbf{k} = (4, 3, 2)$.

(c) Determine all performance measures.

Note: Use the Jackson's theorem.

Problem 7.3   Determine the CPU utilization and other performance measures of a central server model with $N = 3$ nodes and $K = 4$ jobs. Each node has only one server and the queueing discipline at each node is FCFS. The exponentially distributed service times have the respective means

$$\frac{1}{\mu_1} = 2 \text{ msec}, \quad \frac{1}{\mu_2} = 5 \text{ msec}, \quad \frac{1}{\mu_3} = 5 \text{ msec},$$

and the routing probabilities are

$$p_{11} = 0.3, \quad p_{12} = 0.5, \quad p_{13} = 0.2, \quad p_{21} = p_{31} = 1.$$

Problem 7.4   Consider a closed queueing network with $K = 3$ nodes, $N = 3$ jobs, and the service discipline FCFS. The first node has $m_1 = 2$ servers and the other two nodes have one server each. The routing probabilities are given by

$$p_{11} = 0.6, \quad p_{21} = 0.5, \quad p_{31} = 0.4,$$
$$p_{12} = 0.3, \quad p_{22} = 0.0, \quad p_{32} = 0.6,$$
$$p_{12} = 0.1, \quad p_{23} = 0.5, \quad p_{33} = 0.0.$$

The service times are exponentially distributed with the rates

$$\mu_1 = 0.4 \sec^{-1}, \quad \mu_2 = 0.6 \sec^{-1}, \quad \mu_3 = 0.3 \sec^{-1}.$$

(a) Draw the queueing network.

(b) What is the steady-state probability that there are two jobs at node 2?

(c) Determine all performance measures.

(d) Solve this problem also with the local and global balance equations respectively and compare the solution complexity for the two different methods.

Note: Use the Gordon–Newell theorem for parts (a), (b), and (c).

Problem 7.5   Consider a closed queueing network with $N = 3$ nodes, $K = 3$ jobs, and exponentially distributed service times with the rates

$$\mu_1 = 0.72 \sec^{-1}, \quad \mu_2 = 0.64 \sec^{-1}, \quad \mu_3 = 1 \sec^{-1},$$

and the routing probabilities

$$p_{31} = 0.4, \quad p_{32} = 0.6, \quad p_{13} = p_{23} = 1.$$

Determine all performance measures.

Problem 7.6   Consider the following routing matrix **P**:

P	(1,1)	(1,2)	(1,3)	(1,4)	(2,1)	(2,2)	(2,3)	(2,4)
(1,1)	0.5	0	0	0	0.25	0	0.25	0
(1,2)	0	0.5	0	0.5	0	0	0	0
(1,3)	0.5	0	0	0	0	0	0.5	0
(1,4)	0	0	0	0.5	0	0.25	0	0.25
(2,1)	0.5	0	0	0	0.5	0	0	0
(2,2)	0	0.5	0	0	0	0.5	0	0
(2,3)	0	0	1	0	0	0	0	0
(2,4)	0	0	0	1	0	0	0	0

At the beginning, the jobs are distributed as follows over the four different classes:

$$K_1' = K_2' = K_3' = K_4' = 5.$$

(a) Determine the disjoint chains.

(b) Determine the number of states in the CTMC underlying the network.

(c) Determine the visit ratios.

Problem 7.7   Consider an open network with $N = 2$ nodes and $R = 2$ job classes. Node 1 is of Type 2 and node 2 of Type 4. The service rates are

$$\mu_{11} = 4 \sec^{-1}, \quad \mu_{12} = 5 \sec^{-1}, \quad \mu_{21} = 6 \sec^{-1}, \quad \mu_{22} = 2 \sec^{-1},$$

and the arrival rates per class are

$$\lambda_1 = \lambda_2 = 1 \sec^{-1}.$$

The routing probabilities are given as

$$p_{0,11} = p_{0,12} = 1, \quad p_{21,11} = p_{22,12} = 1,$$

$$p_{11,21} = 0.6, \quad p_{12,22} = 0.4 \quad p_{11,0} = 0.4, \quad p_{12,0} = 0.6.$$

With the BCMP theorem, Version 2, determine

(a) All visit ratios.

(b) The utilizations of node 1 and node 2.

(c) The steady-state probability for state (2,1).

8

Algorithms for Product-Form Networks

Although product-form solutions can be expressed very easily as formulae, the computation of state probabilities in a closed queueing network is very time consuming if a straightforward computation of the normalization constant using Eq. (7.3.5) is carried out. As seen in Example 7.7, considerable computation is needed to analyze even a single class network with a small number of jobs, primarily because the formula makes a pass through all the states of the underlying CTMC. Therefore, we need to develop efficient algorithms to reduce the computation time [Buze71].

Many efficient algorithms for calculating performance measures of closed product-form queueing networks have been developed. The most important ones are the *convolution algorithm* and the *mean value analysis* (MVA) [ReLa80]. The convolution algorithm is an efficient iterative technique for calculating the normalization constant, which is needed when all performance measures are computed using a set of simple equations. In contrast, the mean value analysis is an iterative technique where the mean values of the performance measures can be computed directly without computing the normalization constant. The last method for analyzing product-form networks presented in detail is the so-called *flow-equivalent server method* [CHW75b, ABP85]. This method is well-suited when we are especially interested in computing the performance of a single station or a part of the network.

There are several other algorithms that we do not cover in detail due to space limitations. The *RECAL algorithm* [CoGe86] (recursion by chain algorithm) is well-suited for networks with many job classes. Based on the mean value analysis, [ChSa80] and [SaCh81] developed an algorithm called *LBANC* (local balance algorithm for normalizing constants). This algorithm iterative-

369

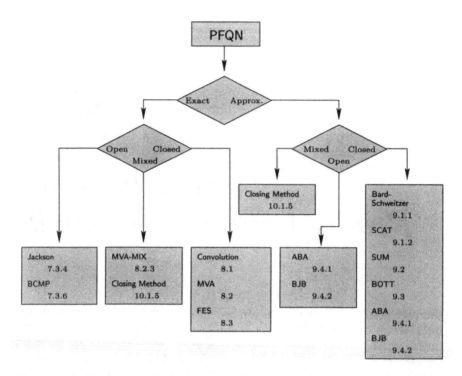

Fig. 8.1 Sections with PFQN solution algorithms in Chapters 7, 8, and 9.

ly computes the normalization constant and the performance measures. It is very well suited for networks with a small number of nodes but a large number of jobs. The *CCNC algorithm* (coalesce computation of normalizing constants)[ChSa80] is especially used when storage space is limited. That the convolution algorithm, the mean value analysis, and LBANC can be derived from each other has been shown by [Lam81]. The *DAC algorithm* (distribution analysis by chain)[SoLa89] is used to directly compute the state probabilities of a product-form queueing network. A generalization of the MVA to higher moments is introduced in [Stre86]. Other algorithms for determining the normalization constant are the algorithms of [Koba78] and [Koba79] (which are based on the enumeration of polynomials of Polya), and the *partial fraction method* of [Moor72].

For very large networks, execution time of the preceding algorithms is not acceptable. Many approximation algorithms have been developed for this purpose. Exact and approximate algorithms are discussed in this chapter and in Chapter 9 and are summarized in Fig. 8.1.

8.1   THE CONVOLUTION ALGORITHM

The convolution algorithm was one of the first efficient algorithms for analyzing closed product-form queueing networks and is still in use today. The name of this technique reflects the method of determining the normalization constant $G(K)$ from the functions $F_i(k_i)$, which is similar to the convolution of two probability mass functions. Once the normalization constant is computed, the system performance measures of interest can be easily derived [Buze71]. We recall that the convolution operator $\otimes$ is defined in the following way: Let A, B, C be vectors of length $K+1$. Then the convolution $C = A \otimes B$ is defined as:

$$C(k) = \sum_{j=0}^{k} A(j) \cdot B(k-j), \quad k = 0, \ldots, K. \tag{8.1}$$

First we consider the convolution algorithm for single class closed queueing networks. Then this technique is extended to multiclass closed queueing networks.

8.1.1   Single Class Closed Networks

According to the BCMP theorem, Eq. (7.80), the state probabilities of a closed single class product-form queueing network with N nodes can be expressed in the following way:

$$\pi(k_1, \ldots, k_N) = \frac{1}{G(K)} \prod_{i=1}^{N} F_i(k_i),$$

where the functions $F_i(k_i)$ are defined in Eq. (7.61), and

$$G(K) = \sum_{\sum_{i=1}^{N} k_i = K} \prod_{i=1}^{N} F_i(k_i)$$

is the normalization constant of the network.

The computation of $G(K)$ is carried out by iterating over the number of nodes in the network and over the number of possible jobs at each node. For this purpose the following auxiliary functions $G_n(k)$, $n = 1, \ldots, N$ and $k = 0, \ldots, K$ are defined as follows:

$$G_n(k) = \sum_{\sum_{i=1}^{n} k_i = k} \prod_{i=1}^{n} F_i(k_i). \tag{8.2}$$

The desired normalization constant is then

$$G(K) = G_N(K). \tag{8.3}$$

From Eq. (8.2) for $n > 1$, it follows that

$$
G_n(k) = \sum_{j=0}^{k} \left(\sum_{\substack{\sum_{i=1}^{n} k_i = k \\ \& k_n = j}} \prod_{i=1}^{n} F_i(k_i) \right) = \sum_{j=0}^{k} F_n(j) \left(\sum_{\substack{\sum_{i=1}^{n} k_i = k_j \\ \& k_n = 0}} \prod_{i=1}^{n-1} F_i(k_i) \right)
$$

$$
= \sum_{j=0}^{k} F_n(j) \cdot G_{n-1}(k-j). \tag{8.4}
$$

For $n = 1$ we have

$$
G_1(k) = F_1(k), \quad k = 1, \ldots, K. \tag{8.5}
$$

The initial condition is of the form

$$
G_n(0) = 1, \quad n = 1, \ldots, N. \tag{8.6}
$$

The convolution method for computing the normalization constant $G(K)$ is fully determined by Eqs. (8.4) to (8.6). The $(K+1)$-dimensional vectors

$$
G_n = \begin{pmatrix} G_n(0) \\ \vdots \\ G_n(K) \end{pmatrix}, \quad n = 1, \ldots, N,
$$

are therefore determined by the convolution

$$
G_n = F_n \otimes G_{n-1}
$$

where

$$
F_n = \begin{pmatrix} F_n(0) \\ \vdots \\ F_n(K) \end{pmatrix}.
$$

The computation of $G_n(k)$ is easily visualized as in Fig. 8.2. In this figure we show the values of the functions needed in the calculation of $G_N(K)$. The real objective of the algorithm is to determine the last value in the last column because this value is the normalization constant $G(K)$ that we are looking for. The values $G_N(k)$ for $k = 1, \ldots, K - 1$ in the last column are also useful in determining the system performance measures. Buzen [Buze71, Buze73] has developed algorithms that are based on the normalization constant $G(K)$ and the functions $F_i(k_i)$ to compute all important performance measures of the queueing network without needing the underlying CTMC state probabilities.

Computation of the performance measures using the normalization constant is now illustrated.

	1	$\cdots$	$n-1$	n	$\cdots$	N
0	1	$\cdots$	$G_{n-1}(0)(\cdot F_n(k))$	1		$1 = G(0)$
1	$F_1(1)$	$\cdots$	$G_{n-1}(1)(\cdot F_n(k-1))$	$G_n(1)$		
$\vdots$	$\vdots$		$\vdots$	$\vdots$		
$k-1$	$F_1(k-1)$	$\cdots$	$G_{n-1}(k-1)(\cdot F_n(1))$	$\sum$		
k	$F_1(k)$	$\cdots$	$G_{n-1}(k)(\cdot F_n(0))$	$G_n(k)$	$\cdots$	$G_N(k) = G(k)$
$\vdots$	$\vdots$			$\vdots$		
K	$F_1(K)$			$G_n(K)$		$G_N(K) = G(K)$

Fig. 8.2 Computation of $G_n(k)$.

(a) The *marginal probability* that there are exactly $k_i = k$ jobs at the ith node is given by Eq. (7.16):

$$\pi_i(k) = \sum_{\substack{\sum_{j=1}^{N} k_j = K \\ \& k_i = k}} \pi(k_1, \ldots, k_N).$$

If we substitute Eq. (7.59), we get

$$\pi_i(k) = \sum_{\substack{\sum_{j=1}^{N} k_j = K \\ \& k_i = k}} \frac{1}{G(K)} \prod_{j=1}^{N} F_j(k_j) = \frac{F_i(k)}{G(K)} \sum_{\substack{\sum_{j=1}^{N} k_j = K \ \& \ j \neq i \\ \& k_i = k}} \prod_{\substack{j=1 \\ j \neq i}}^{N} F_j(k_j)$$

$$= \frac{F_i(k)}{G(K)} \cdot G_N^{(i)}(K - k). \tag{8.7}$$

Then $G_N^{(i)}(k)$ can be interpreted as the normalization constant of the network with k jobs and node i removed from the network.

$$G_N^{(i)}(k) = \sum_{\substack{\sum_{j=1}^{N} k_j = K \ \& \ j \neq i \\ \& k_i = K - k}} \prod_{\substack{j=1 \\ j \neq i}}^{N} F_j(k_j). \tag{8.8}$$

By definition we have

$$G_{N-1}(k) = G_{N-1}^{(N)}(k) = G_N^{(N)}(k) \quad \text{for } k = 0, \ldots, K.$$

Because of Eq. (8.7), for node N we have [Buze71, Buze73]

$$\pi_N(k) = \frac{F_N(k)}{G(K)} \cdot G_{N-1}(K - k). \tag{8.9}$$

In [BBS77], a simple algorithm for computing the $G_N^{(i)}(k)$ for $k = 0, \ldots, K$ is presented. Since the sum of all marginal probabilities is 1, it follows from Eq. (8.7):

$$\sum_{j=0}^{K} \pi_i(j) = \sum_{j=0}^{K} \frac{F_i(j)}{G(K)} G_N^{(i)}(K - j) = 1,$$

and therefore we have

$$G(K) = \sum_{j=0}^{K} F_i(j) \cdot G_N^{(i)}(K - j), \quad j = 1, \ldots, N. \qquad (8.10)$$

With the help of these equations we can derive an iterative formula for computing the $G_N^{(i)}(k)$, for $k = 0, \ldots, K$:

$$G_N^{(i)}(k) = G(k) - \sum_{j=1}^{k} F_i(j) \cdot G_N^{(i)}(k - j), \qquad (8.11)$$

with the initial condition

$$G_N^{(i)}(0) = G(0) = 1, \quad i = 1, \ldots, N. \qquad (8.12)$$

In the case of $m_i = 1$ the preceding formulae can be considerably simplified.

Because of Eq. (8.11) we have

$$
\begin{aligned}
G_N^{(i)}(k) &= G(k) - \sum_{j=0}^{k-1} F_i(j+1) G_N^{(i)}(k - 1 - j) \\
&= G(k) - \frac{e_i}{\mu_i} \sum_{j=0}^{k-1} F_i(j) G_N^{(i)}(k - 1 - j) \\
&= G(k) - \frac{e_i}{\mu_i} G(k - 1).
\end{aligned}
$$

After inserting this result in Eq. (8.7), we get for the marginal probabilities

$$\pi_i(k) = \left(\frac{e_i}{\mu_i}\right)^k \cdot \frac{1}{G(K)} \cdot \left(G(K - k) - \frac{e_i}{\mu_i} \cdot G(K - k - 1)\right), \qquad (8.13)$$

where $G(k) = 0$ for $k < 0$.

(b) The *throughput* of node i in the load-dependent or load-independent case is given by the formula

$$\lambda(K) = \frac{G(K - 1)}{G(K)} \quad \text{and} \quad \lambda_i(K) = e_i \cdot \frac{G(K - 1)}{G(K)}. \qquad (8.14)$$

Proof: By definition, the throughput is given by Eq. (7.22):

$$\lambda_i(k) = \sum_{k=1}^{K} \pi_i(k)\mu_i(k) = \sum_{k=1}^{K} \frac{F_i(k)}{G(K)} G_N^{(i)}(K-k)\mu_i(k)$$

$$= \sum_{k=1}^{K} \frac{e_i}{\mu_i(k)} \frac{F_i(k-1)}{G(K)} G_N^{(i)}(K-k)\mu_i(k)$$

$$= \frac{e_i}{G(K)} \sum_{k=1}^{K} F_i(k-1) G_N^{(i)}(K-k)$$

$$= \frac{e_i}{G(K)} \sum_{k=0}^{K-1} F_i(k) G_N^{(i)}(K-1-k) = e_i \frac{G(K-1)}{G(K)} . \quad q.e.d.$$

(c) The *utilization* of a node in the load-independent case can be determined by inserting Eq. (8.14) in the well-known relation $\rho_i = \lambda_i/(m_i\mu_i)$:

$$\rho_i = \frac{e_i}{m_i\mu_i} \cdot \frac{G(K-1)}{G(K)} . \tag{8.15}$$

(d) The *mean number of jobs* for a single server node can be computed from following simplified equation [Buze71]:

$$\overline{K}_i = \sum_{k=1}^{K} \left(\frac{e_i}{\mu_i}\right)^k \cdot \frac{G(K-k)}{G(K)} . \tag{8.16}$$

(e) The *mean response time* of jobs at node i can be determined with the help of Little's theorem. For a single server node we have

$$\overline{T}_i = \frac{\overline{K}_i}{\lambda_i} = \sum_{k=1}^{K} \left(\frac{e_i}{\mu_i}\right)^k \cdot \frac{G(K-k)}{e_i \cdot G(K-1)} . \tag{8.17}$$

The use of the convolution method for determining the normalization constant is now demonstrated by a simple example. We also show how to compute other performance measures from these results.

Example 8.1   Consider the following closed queueing network (Fig. 8.3) with $N = 3$ nodes and $K = 3$ jobs. The first node has $m_1 = 2$ and the second node has $m_2 = 3$ identical service stations. For the third node we have $m_3 = 1$. The service time at each node is exponentially distributed with respective rates

$$\mu_1 = 0.8\,\text{sec}^{-1}, \quad \mu_2 = 0.6\,\text{sec}^{-1}, \quad \mu_3 = 0.4\,\text{sec}^{-1} .$$

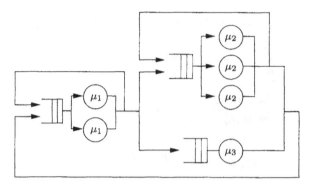

Fig. 8.3   A closed queueing network.

The visit ratios are given as follows:

$$e_1 = 1, \quad e_2 = 0.667, \quad e_3 = 0.2.$$

From Eq. (7.61) for the functions $F_i(k_i)$, $i = 1, 2, 3$, we have:

$$
\begin{array}{lll}
F_1(0) = 1, & F_2(0) = 1, & F_3(0) = 1, \\
F_1(1) = 1.25, & F_2(1) = 1.111, & F_3(1) = 0.5, \\
F_1(2) = 0.781, & F_2(2) = 0.617, & F_3(2) = 0.25, \\
F_1(3) = 0.488, & F_2(3) = 0.229, & F_3(3) = 0.125.
\end{array}
$$

In Table 8.1 the intermediate results as well as the final result of computing $G(K)$ are shown.

Table 8.1   Computation of the normalization constant

	1	2	$N = 3$
0	1	1	1
1	1.25	2.361	2.861
2	0.781	2.787	4.218
$K = 3$	0.488	2.356	4.465

Thus, the normalization constant $G(K) = 4.465$. The marginal probabilities for the single server node 3 can be computed using Eq. (8.13):

$$\pi_3(0) = \left(\frac{e_3}{\mu_3}\right)^0 \cdot \frac{1}{G(3)} \cdot \left(G(3) - \frac{e_3}{\mu_3}G(2)\right) = 0.528,$$

$$\pi_3(1) = \left(\frac{e_3}{\mu_3}\right)^1 \cdot \frac{1}{G(3)} \cdot \left(G(2) - \frac{e_3}{\mu_3}G(1)\right) = 0.312,$$

$$\pi_3(2) = \left(\frac{e_3}{\mu_3}\right)^2 \cdot \frac{1}{G(3)} \cdot \left(G(1) - \frac{e_3}{\mu_3}G(0)\right) = \underline{0.132},$$

$$\pi_3(3) = \left(\frac{e_3}{\mu_3}\right)^3 \cdot \frac{1}{G(3)} \cdot \left(G(0) - \frac{e_3}{\mu_3} \cdot 0\right) = \underline{0.028}.$$

For the computation of the marginal probabilities of nodes 1 and 2, we need the values $G_N^{(i)}(k)$ for $k = 0, 1, 2, 3$ and $i = 1, 2$. With Eq. (8.11) we get

$$G_N^{(1)}(0) = \underline{1},$$
$$G_N^{(1)}(1) = G(1) - F_1(1)G_N^{(1)}(0) = \underline{1.611},$$
$$G_N^{(1)}(2) = G(2) - (F_1(1)G_N^{(1)}(1) + F_1(2)G_N^{(1)}(0)) = \underline{1.423},$$
$$G_N^{(1)}(3) = G(3) - (F_1(1)G_N^{(1)}(2) + F_1(2)G_N^{(1)}(1) + F_1(3)G_N^{(1)}(0)) = \underline{0.940}.$$

In the same way we compute

$$G_N^{(2)}(0) = \underline{1}, \quad G_N^{(2)}(1) = \underline{1.656}, \quad G_N^{(2)}(2) = \underline{1.75}, \quad G_N^{(2)}(3) = \underline{1.316}.$$

With Eq. (8.7) the marginal probabilities are

$$\pi_1(0) = \frac{F_1(0)}{G(3)}G_N^{(1)}(3) = \underline{0.211}, \quad \pi_1(1) = \frac{F_1(1)}{G(3)}G_N^{(1)}(2) = \underline{0.398},$$

$$\pi_1(2) = \frac{F_1(2)}{G(3)}G_N^{(1)}(1) = \underline{0.282}, \quad \pi_1(3) = \frac{F_1(3)}{G(3)}G_N^{(1)}(0) = \underline{0.109},$$

and

$$\pi_2(0) = \underline{0.295}, \quad \pi_2(1) = \underline{0.412}, \quad \pi_2(2) = \underline{0.242}, \quad \pi_2(3) = \underline{0.051}.$$

The throughputs can be computed using Eq. (8.14):

$$\lambda_1 = e_1\frac{G(2)}{G(3)} = \underline{0.945}, \quad \lambda_2 = e_2\frac{G(2)}{G(3)} = \underline{0.630}, \quad \lambda_3 = e_3\frac{G(2)}{G(3)} = \underline{0.189}.$$

The utilizations are given by Eq. (7.21):

$$\rho_1 = \frac{\lambda_1}{m_1\mu_1} = \underline{0.590}, \quad \rho_2 = \frac{\lambda_2}{m_2\mu_2} = \underline{0.350}, \quad \rho_3 = \frac{\lambda_3}{\mu_3} = \underline{0.473}.$$

The mean number of jobs at the multiserver nodes is given by Eq. (7.26):

$$\overline{K}_1 = \pi_1(1) + 2\pi_1(2) + 3\pi_1(3) = \underline{1.290},$$
$$\overline{K}_2 = \pi_2(1) + 2\pi_2(2) + 3\pi_2(3) = \underline{1.050},$$

where we use Eq. (8.16) for the single server node 3:

$$\overline{K}_3 = \left(\frac{e_3}{\mu_3}\right)\frac{G(2)}{G(3)} + \left(\frac{e_3}{\mu_3}\right)^2\frac{G(1)}{G(3)} + \left(\frac{e_3}{\mu_3}\right)^3\frac{G(0)}{G(3)} = \underline{0.660}\,.$$

For the mean response time we use Eq. (7.43):

$$\overline{T}_1 = \frac{\overline{K}_1}{\lambda_1} = \underline{1.366}\,, \quad \overline{T}_2 = \frac{\overline{K}_2}{\lambda_2} = \underline{1.667}\,, \quad \overline{T}_3 = \frac{\overline{K}_3}{\lambda_3} = \underline{3.496}\,.$$

8.1.2 Multiclass Closed Networks

The convolution method introduced in Section 8.1.1 can be extended to the case of multiclass closed product-form queueing networks. According to the BCMP theorem, Eq. (7.80), the state probabilities for closed product-form queueing networks are given by

$$\pi(\mathbf{S}_1,\dots,\mathbf{S}_N) = \frac{1}{G(\mathbf{K})}\prod_{i=1}^{N}F_i(\mathbf{S}_i)\,,$$

where the functions $F_i(\mathbf{S}_i)$ are defined in Eq. (7.82). The determination of the normalization constant of the network

$$G(\mathbf{K}) = \sum_{\substack{\sum\limits_{i=1}^{N}\mathbf{S}_i=\mathbf{K}}}\prod_{i=1}^{N}F_i(\mathbf{S}_i)\,,$$

is analogous to the single class case. For now, we assume that class switching is not allowed and thus the number of jobs in each class is constant. Corresponding to Eq. (8.2), we define for $\mathbf{k} = \mathbf{0},\dots,\mathbf{K}$ the auxiliary functions

$$G_n(\mathbf{k}) = \sum_{\substack{\sum\limits_{i=1}^{n}\mathbf{S}_i=\mathbf{k}}}\prod_{i=1}^{n}F_i(\mathbf{S}_i) \tag{8.18}$$

and determine for $n = 1,\dots,N$ the matrix G_n by convolution:

$$G_n = F_n \otimes G_{n-1}\,,$$

with the initial condition $G_1(.) = F_1(.)$. For $n > 1$ we also have

$$G_n(\mathbf{k}) = \sum_{j_1=0}^{k_1}\cdots\sum_{j_R=0}^{k_R}F_n(\mathbf{j})\cdot G_{n-1}(\mathbf{k}-\mathbf{j})\,. \tag{8.19}$$

The computation of the normalization constant was simplified by [MuWo74b] and [Wong75] by introducing a much simpler iterative formula for $G(\mathbf{K}) =$

$G_N(\mathbf{K})$. For this purpose, we define the vector:

$$\mathbf{k}^{(n)} = \sum_{i=1}^{n} \mathbf{S}_i = \left(k_1^{(n)}, \ldots, k_R^{(n)} \right),$$

where $k_r^{(n)}, 1 \le r \le R$, is the overall number of jobs of class r at the nodes $1, 2, \ldots, n-1, n$. We then have:

$$k_r^{(n)} = \sum_{i=1}^{n} k_{ir} \quad \text{and} \quad \mathbf{k}^{(N)} = \mathbf{K}.$$

With the help of these expressions, it is possible to rewrite Eq. (8.18) in the following manner:

$$G_n(\mathbf{k}^{(n)}) = \sum_{\sum_{i=1}^{n} \mathbf{S}_i = \mathbf{k}^{(n)}} \prod_{i=1}^{n} F_i(\mathbf{S}_i). \tag{8.20}$$

Thus

$$G_n(\mathbf{k}^{(n)}) = \sum_{\mathbf{k}^{(n-1)} + \mathbf{S}_n = \mathbf{k}^{(n)}} \quad \sum_{\sum_{i=1}^{n-1} \mathbf{S}_i = \mathbf{k}^{(n-1)}} \prod_{i=1}^{n-1} F_i(\mathbf{S}_i) \cdot F_n(\mathbf{S}_n)$$

$$= \sum_{\mathbf{k}^{(n-1)} + \mathbf{S}_n = \mathbf{k}^{(n)}} G_{n-1}(\mathbf{k}^{(n-1)}) \cdot F_n(\mathbf{S}_n). \tag{8.21}$$

Equation (8.21), together with the initial conditions $G_1(.) = F_1(.)$, completes the description of the algorithm of [MuWo74b] and [Wong75]. For the normalization constant we have

$$G(\mathbf{K}) = G_N(\mathbf{k}^{(N)}). \tag{8.22}$$

The computation of the performance measures is as follows:

(a) According to Eq. (7.32) the *marginal probability* that there are exactly $\mathbf{S}_i = \mathbf{k}$ jobs at node i is given by

$$\pi_i(\mathbf{k}) = \sum_{\substack{\sum_{j=1}^{N} \mathbf{S}_j = \mathbf{K} \\ \& \mathbf{S}_i = \mathbf{k}}} \pi(\mathbf{S}_1, \ldots, \mathbf{S}_N) = \frac{1}{G(\mathbf{K})} \sum_{\substack{\sum_{j=1}^{N} \mathbf{S}_j = \mathbf{K} \\ \& \mathbf{S}_i = \mathbf{k}}} \prod_{j=1}^{N} F_j(\mathbf{S}_j)$$

$$= \frac{F_i(\mathbf{k})}{G(\mathbf{K})} \sum_{\substack{\sum_{j=1}^{N} \mathbf{S}_j = \mathbf{K} \\ \& \mathbf{S}_i = \mathbf{k}}} \prod_{\substack{j=1 \\ \& j \ne i}}^{N} F_j(\mathbf{S}_j) = \frac{F_i(\mathbf{k})}{G(\mathbf{K})} G_N^{(i)}(\mathbf{K} - \mathbf{k}). \tag{8.23}$$

Then $G_N^{(i)}(\mathbf{K})$ can again be interpreted as the normalization constant of the network without node i:

$$G_N^{(i)}(\mathbf{k}) = \sum_{\substack{\sum_{j=1}^{N} \mathbf{S}_j = \mathbf{K} \\ \&\mathbf{S}_i = \mathbf{K} - \mathbf{k}}} \prod_{\substack{j=1 \\ \&j \neq i}}^{N} F_j(\mathbf{S}_j). \qquad (8.24)$$

In the same way as in the single class case (Eq. (8.11)), the iterative formula for computing the normalization constant $G_N^{(i)}(\mathbf{k})$ is given as follows:

$$G_N^{(i)}(\mathbf{k}) = G(\mathbf{k}) - \sum_{j=0}^{k} \delta(\mathbf{j}) \cdot F_i(\mathbf{j}) \cdot G_N^{(i)}(\mathbf{k} - \mathbf{j}), \qquad (8.25)$$

with $\delta(\mathbf{j})$ defined by

$$\delta(\mathbf{j}) = \begin{cases} 0, & \text{if } \mathbf{j} = \mathbf{0}, \\ 1, & \text{otherwise} \end{cases} \qquad (8.26)$$

and affecting the computation only for $\mathbf{j} = \mathbf{0}$. The initial condition is

$$G_N^{(i)}(\mathbf{0}) = G(\mathbf{0}) = 1, \quad i = 1, \ldots, N. \qquad (8.27)$$

(b) The *throughput* λ_{ir} of class-r jobs at the ith node can be expressed as follows [Wong75]:

$$\lambda_{ir} = e_{ir} \frac{G(K_1, \ldots, K_r - 1, \ldots, K_R)}{G(K_1, \ldots, K_r, \ldots, K_R)},$$

where e_{ir} is the solution of Eq. (7.72). With $(\mathbf{K} - \mathbf{1}_r) = (K_1, \ldots, K_r - 1, \ldots, K_R)$, we have the following simpler form:

$$\lambda_{ir} = e_{ir} \frac{G(\mathbf{K} - \mathbf{1}_r)}{G(\mathbf{K})}. \qquad (8.28)$$

This formula holds for load-dependent and load-independent nodes of all four types. The proof can be found in [BrBa80].

(c) The *utilization* ρ_{ir} is given by Eq. (7.34). If the service rates are constant, then we have the following simplification because of $\lambda_{ir}/(m_i\mu_{ir})$:

$$\rho_{ir} = \frac{\lambda_{ir}}{m_i\mu_{ir}} = \frac{e_{ir}}{m_i\mu_{ir}} \cdot \frac{G(\mathbf{K} - \mathbf{1}_r)}{G(\mathbf{K})}. \qquad (8.29)$$

Example 8.2   Consider the network given in Fig. 8.4 consisting of $N = 4$ nodes and $R = 2$ job classes. In class 1 the number of jobs is $K_1 = 1$ and in

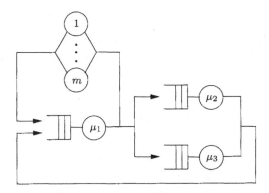

Fig. 8.4   An example of a two-class closed network.

Table 8.2   Computation of the functions $F_i(\mathbf{S}_i)$, $i = 1, 2, 3, 4$,

$\mathbf{S}_i$	$F_1(\mathbf{S}_1)$	$F_2(\mathbf{S}_2)$	$F_3(\mathbf{S}_3)$	$F_4(\mathbf{S}_4)$
(0,0)	1	1	1	1
(1,0)	1	1.6	3.2	2.4
(0,1)	2	2	3	4.8
(1,1)	4	6.4	19.2	11.52
(0,2)	4	4	9	11.52
(1,2)	12	19.2	86.4	27.648

class 2, $K_2 = 2$. It is assumed that class switching is not allowed. The first node is of Type 2, the second and third nodes are of Type 4, and the fourth node is of Type 3 with mean service times:

$$\frac{1}{\mu_{11}} = 1\,\text{sec}, \quad \frac{1}{\mu_{21}} = 4\,\text{sec}, \quad \frac{1}{\mu_{31}} = 8\,\text{sec}, \quad \frac{1}{\mu_{41}} = 12\,\text{sec},$$

$$\frac{1}{\mu_{12}} = 2\,\text{sec}, \quad \frac{1}{\mu_{22}} = 5\,\text{sec}, \quad \frac{1}{\mu_{32}} = 10\,\text{sec}, \quad \frac{1}{\mu_{42}} = 16\,\text{sec}.$$

The visit ratios are given by

$$e_{11} = 1, \quad e_{21} = 0.4, \quad e_{31} = 0.4, \quad e_{41} = 0.2,$$
$$e_{12} = 1, \quad e_{22} = 0.4, \quad e_{32} = 0.3, \quad e_{42} = 0.3.$$

First, with the help of Eq. (7.82), the functions $F_i(\mathbf{S}_i), i = 1, 2, 3, 4$, are computed (Table 8.2). Then for determining the normalization constant we compute the $G_n(\mathbf{k}^{(n)})$ from Eq. (8.21) where $G_1(.) = F_1(.)$. For $n = 2$ we have

$$G_2(0,0) = G_1(0,0)F_2(0,0) \qquad\qquad\qquad = \underline{1},$$
$$G_2(1,0) = G_1(0,0)F_2(1,0) + G_1(1,0)F_2(0,0) \quad = \underline{2.6},$$
$$G_2(0,1) = G_1(0,0)F_2(0,1) + G_1(0,1)F_2(0,0) \quad = \underline{4},$$
$$G_2(1,1) = G_1(0,0)F_2(1,1) + G_1(1,1)F_2(0,0)$$
$$+ G_1(1,0)F_2(0,1) + G_1(0,1)F_2(1,0) = \underline{15.6},$$
$$G_2(0,2) = G_1(0,0)F_2(0,2) + G_1(0,2)F_2(0,0)$$
$$+ G_1(0,1)F_2(0,1) \qquad\qquad\qquad = \underline{12},$$
$$G_2(1,2) = G_1(0,0)F_2(1,2) + G_1(1,2)F_2(0,0)$$
$$+ G_1(1,1)F_2(0,1) + G_1(0,1)F_2(1,1)$$
$$+ G_1(1,0)F_2(0,2) + G_1(0,2)F_2(1,0) = \underline{62.4}.$$

In the same way we can compute the values for $G_3(\mathbf{k}^{(3)})$ and $G_4(\mathbf{k}^{(4)})$ as summarized in Table 8.3. So the normalization constant is $G(\mathbf{K}) = \underline{854.424}$.

Table 8.3   Computation of $G_i(\mathbf{k}^{(i)})$

$\mathbf{k}^{(n)}, 1 \leq n \leq N$	$G_2(\mathbf{k}^{(2)})$	$G_3(\mathbf{k}^{(3)})$	$G_4(\mathbf{k}^{(4)})$
(0,0)	1	1	1
(1,0)	2.6	5.8	8.2
(0,1)	4	7	11.8
(1,1)	15.6	55.4	111.56
(0,2)	12	33	78.12
(1,2)	62.4	334.2	854.424

Now, with the help of Eq. (7.47), we can compute the marginal probability of 0 jobs of class 1 and 2 jobs of class 2 at node 4:

$$\pi_4(0,2) = \frac{F_4(0,2)}{G(\mathbf{K})} G_N^{(4)}(1,0) = \underline{0.0782},$$

where $G_N^{(4)}(1,0)$ can be obtained from Eq. (8.26):

$$G_N^{(4)}(1,0) = G(1,0) - F_4(1,0)\cdot G_N^{(4)}(0,0) = \underline{5}.$$

Similarly, other marginal probabilities at the nodes can be calculated:

$$\pi_4(1,2) = \frac{F_4(1,2)}{G(\mathbf{K})} G_N^{(4)}(0,0) = \underline{0.0324}, \quad \text{with } G_N^{(4)}(0,0) = \underline{1},$$

$$\pi_4(1,1) = \frac{F_4(1,1)}{G(\mathbf{K})} G_N^{(4)}(0,1) = \underline{0.0944}, \quad \text{with } G_N^{(4)}(0,1) = \underline{7},$$

$$\pi_4(0,1) = \frac{F_4(0,1)}{G(\mathbf{K})} G_N^{(4)}(1,1) = \underline{0.3112}, \quad \text{with } G_N^{(4)}(1,1) = \underline{55.4},$$

$$\pi_4(1,0) = \frac{F_4(1,0)}{G(\mathbf{K})}G_N^{(4)}(0,2) = \underline{0.0927}, \quad \text{with } G_N^{(4)}(0,2) = \underline{33},$$

$$\pi_4(0,0) = \frac{F_4(0,0)}{G(\mathbf{K})}G_N^{(4)}(1,2) = \underline{0.3911}, \quad \text{with } G_N^{(4)}(1,2) = \underline{334.2},$$

$$\pi_3(1,2) = \frac{F_3(1,2)}{G(\mathbf{K})}G_N^{(3)}(0,0) = \underline{0.1011}, \quad \text{with } G_N^{(3)}(0,0) = \underline{1},$$

$$\pi_3(1,0) = \frac{F_3(1,0)}{G(\mathbf{K})}G_N^{(3)}(0,2) = \underline{0.1600}, \quad \text{with } G_N^{(3)}(0,2) = \underline{42.72},$$

$$\pi_3(1,1) = \frac{F_3(1,1)}{G(\mathbf{K})}G_N^{(3)}(0,1) = \underline{0.1977}, \quad \text{with } G_N^{(3)}(0,1) = \underline{8.8}.$$

$$\vdots$$

The rest of the marginal probabilities are computed in the same way. The mean number of jobs can be determined with the help of Eq. (7.40). For example, for node 3 the mean number of class 1 and class 2 jobs is respectively given by

$$\overline{K}_{31} = \pi_3(1,0) + \pi_3(1,1) + \pi_3(1,2) \qquad\qquad = \underline{0.4588},$$
$$\overline{K}_{32} = \pi_3(0,1) + \pi_3(1,1) + 2\pi_3(0,2) + 2\pi_3(1,2) = \underline{0.678}.$$

For the IS-node 4 the mean number of jobs by class type is

$$\overline{K}_{41} = \frac{\lambda_{41}}{\mu_{41}} = \underline{0.219}, \quad \overline{K}_{42} = \frac{\lambda_{42}}{\mu_{42}} = \underline{0.627}.$$

The throughputs at each node by class type can be computed with Eq. (8.28):

$$\lambda_{11} = e_{11}\frac{G(0,2)}{G(1,2)} = \underline{0.0914}, \quad \lambda_{12} = e_{12}\frac{G(1,1)}{G(1,2)} = \underline{0.1306},$$

$$\lambda_{21} = e_{21}\frac{G(0,2)}{G(1,2)} = \underline{0.0366}, \quad \lambda_{22} = e_{22}\frac{G(1,1)}{G(1,2)} = \underline{0.0522},$$

$$\lambda_{31} = e_{31}\frac{G(0,2)}{G(1,2)} = \underline{0.0366}, \quad \lambda_{32} = e_{32}\frac{G(1,1)}{G(1,2)} = \underline{0.0392},$$

$$\lambda_{41} = e_{41}\frac{G(0,2)}{G(1,2)} = \underline{0.0183}, \quad \lambda_{42} = e_{42}\frac{G(1,1)}{G(1,2)} = \underline{0.0392}.$$

For the computation of the utilizations of each node by class type, Eq. (8.29) is used to get

$$\rho_{11} = \underline{0.0914}, \quad \rho_{21} = \underline{0.1463}, \quad \rho_{31} = \underline{0.2926},$$
$$\rho_{12} = \underline{0.2611}, \quad \rho_{22} = \underline{0.2611}, \quad \rho_{32} = \underline{0.3917}.$$

The algorithm we presented in the preceding text is applicable only to networks without class switching. For networks with class switching, [Munt72] proved that a closed queueing network with U ergodic chains is equivalent to a closed network with U job classes without class switching. Therefore, the computation of the normalization constant for networks with class switching is an extension of the technique for networks without class switching. More details are given in [BBS77] and [BrBa80] and Section 7.3.6. In [ReKo75] and [Saue83] the convolution method has been extended for analyzing open queueing networks with load-dependent service rates and for analyzing closed queueing networks with load-dependent routing probabilities. The methods can also be extended for networks with *class specific* service rates. These are networks in which the service rates depend not only on the overall number of jobs of each class at the node, but also on the number of jobs at the node. For this case, [LaLi83] modified the convolution algorithm so as to make the best use of storage space. Their algorithm is known as *tree-convolution*.

Because the computation of the normalization constant can cause numerical problems, other techniques were developed that allow the calculation of the performance measures without using the normalization constant. One of the key development in this regard is the *mean value analysis* (MVA), which we discuss next.

8.2 THE MEAN VALUE ANALYSIS

The MVA was developed by Reiser and Lavenberg [ReLa80] for the analysis of closed queueing networks with product-form solution. The advantage of this method is that the performance measures can be computed without explicitly computing the normalization constant. The method is based on two fundamental equations and it allows us to compute the mean values of measures of interest such as the mean waiting time, throughput, and the mean number of jobs at each node. In these computations only mean values are computed (hence the name). For the case of multiserver nodes ($m_i > 1$), it is necessary, however, to compute the marginal probabilities.

The MVA method is based on two simple laws:

1. *Little's theorem*, which is introduced in Eq. (6.9) to express a relation between the mean number of jobs, the throughput, and the mean response time of a node or the overall system:

$$\overline{K} = \lambda \cdot \overline{T}. \tag{8.30}$$

2. *Theorem of the distribution at arrival time* (in short, *arrival theorem*), proven by [LaRe80] and [SeMi81], for all networks that have a product-form solution. The arrival theorem says that in a closed product-form queueing network, the pmf of the number of jobs seen at the time of

arrival to a node i when there are k jobs in the network is equal to the pmf of the number of jobs at this node with one less job in the network ($= (k - 1)$). This property has an intuitive justification [LZGS84]. At the moment a job arrives at a node, it is certain that this job itself is not already in the queue of this node. Thus, there are only $k - 1$ other jobs that could possibly interfere with the new arrival. The number of these at the node is simply the number there when only those $(k - 1)$ jobs are in the network.

At first we introduce the MVA for single class closed queueing networks and explain it in more detail. This algorithm is then extended to multiclass networks, mixed networks, and networks with load-dependent service rates.

8.2.1 Single Class Closed Networks

The fundamental equation of the mean value analysis is based on the *arrival theorem* for closed product-form networks [ReLa80, SeMi81] and it relates the mean response time of a job at the ith node and the mean number of jobs at that node with one job less in the network, that is,

$$\overline{T}_i(K) = \frac{1}{\mu_i} \cdot \left[1 + \overline{K}_i(K - 1) \right], \qquad i = 1, \dots, N. \tag{8.31}$$

For single server stations ($m_i = 1$) with an FCFS strategy, it is easy to give an intuitive explanation for Eq. (8.31) because for each FCFS node i the mean response time $\overline{T}_i(K)$ of a job in a network with K jobs is given by the mean service time ($1/\mu_i$) of that job plus the sum of the mean service times of all jobs that are ahead of this job in the queue. Equation (8.31) can also be derived without using the arrival theorem. For this purposes, we use the formulae for computing the utilization (Eq. (8.15)) and the mean number of jobs (Eq. (8.16)):

$$\rho_i(K) = \frac{e_i}{\mu_i} \cdot \frac{G(K - 1)}{G(K)},$$

and

$$\overline{K}_i(K) = \sum_{k=1}^{K} \left(\frac{e_i}{\mu_i} \right)^k \cdot \frac{G(K - k)}{G(K)}.$$

From Eq. (8.16) it follows that

$$\overline{K}_i(K - 1) = \sum_{k=1}^{K-1} \left(\frac{e_i}{\mu_i} \right)^k \cdot \frac{G(K - k - 1)}{G(K - 1)} = \sum_{k=2}^{K} \left(\frac{e_i}{\mu_i} \right)^{k-1} \cdot \frac{G(K - k)}{G(K - 1)}.$$

If we transform Eq. (8.15) for $G(K-1)$ and insert the result of the preceding equation, then, after rearranging the equation, we get

$$\rho_i(K) \cdot \overline{K}_i(K-1) = \sum_{k=2}^{K} \left(\frac{e_i}{\mu_i} \right)^k \cdot \frac{G(K-k)}{G(K)},$$

and substituting this equation to Eq. (8.16), we have

$$\overline{K}_i(K) = \frac{e_i}{\mu_i} \cdot \frac{G(K-1)}{G(K)} + \rho_i(K) \cdot \overline{K}_i(K-1)$$

$$= \rho_i(K) + \rho_i(K) \cdot \overline{K}_i(K-1) = \rho_i(K) \cdot \left[1 + \overline{K}_i(K-1) \right].$$

If we assume constant service rates and $m_i = 1$, then we get the desired result by using $\rho_i(K) = \lambda_i(K)/\mu_i$ and Little's theorem:

$$\overline{T}_i(K) = \frac{\overline{K}_i(K)}{\lambda_i(K)} = \frac{\rho_i(K)}{\lambda_i(K)} \cdot \left[1 + \overline{K}_i(K-1) \right] = \frac{1}{\mu_i} \cdot \left[1 + \overline{K}_i(K-1) \right].$$

For computing the mean response time at the ith node we need two other equations in addition to Eq. (8.31) to describe the MVA completely. Both equations can be derived from Little's theorem. The first one is for determining the overall throughput of the network,

$$\lambda(k) = \frac{k}{\sum_{i=1}^{N} e_i \cdot \overline{T}_i(k)}, \tag{8.32}$$

whereas the other one determines the mean number of jobs at the ith node:

$$\overline{K}_i(k) = \lambda(k) \cdot \overline{T}_i(k) \cdot e_i, \tag{8.33}$$

where e_i is the visit ratio at the ith node.

The three equations, (8.31), (8.32), and (8.33), allow an iterative computation of the mean response time, mean number of jobs, and throughput of the closed product-form queueing network. The iteration is done over the number of jobs k in the network. Equation (8.31) is valid for FCFS single server nodes, PS nodes, and LCFS PR nodes. The description of the MVA is complete if we extend Eq. (8.31) to the case of IS nodes and FCFS nodes with multiple servers. In the case of IS nodes we have

$$\overline{T}_i(K) = \frac{1}{\mu_i}. \tag{8.34}$$

For the latter case, consider a job that arrives at -/M/m-FCFS node containing $j - 1$ jobs, given that the network population is $k - 1$. This event occurs with probability $\pi_i(j - 1 \mid k - 1)$. Then we have:

$$\overline{T}_i(k) = \sum_{j=1}^{k} \frac{j}{\mu_i \cdot \alpha_i(j)} \cdot \pi_i(j - 1 \mid k - 1),$$

$$\alpha_i(j) = \begin{cases} j, & \text{if } j \le m_i, \\ m_i, & \text{otherwise.} \end{cases}$$

To obtain an expression for the probability $\pi_i(j \mid k)$, we use the formulae presented in the convolution algorithm for computing the performance measures with help of the normalization constant, namely,

$$\pi_i(j \mid k) = F_i(j) \cdot \frac{G_N^{(i)}(k-j)}{G(k)} \quad \text{(see Eq. (8.7))}$$

$$= \frac{e_i}{\mu_i \cdot \alpha_i(j)} \cdot F_i(j-1) \cdot \frac{G_N^{(i)}((k-1)-(j-1))}{G(k)}$$

$$= \frac{1}{\mu_i \cdot \alpha_i(j)} \cdot \underbrace{\frac{e_i \cdot G(k-1)}{G(k)}}_{\lambda_i(k)} \cdot \underbrace{\frac{G_N^{(i)}((k-1)-(j-1))}{G(k-1)} \cdot F_i(j-1)}_{\pi_i(j-1 \mid k-1) \quad \text{see Eq. (8.7)}}$$

$$= \frac{\lambda_i(k)}{\mu_i \cdot \alpha_i(j)} \cdot \pi_i(j-1 \mid k-1),$$

$$\pi_i(0 \mid 0) = 1,$$

$$\pi_i(0 \mid k) = 1 - \sum_{j=1}^{k} \pi_i(j \mid k) = 1 - \sum_{j=1}^{k} \frac{\lambda_i(k)}{\mu_i \cdot \alpha_i(j)} \cdot \pi_i(j-1 \mid k-1).$$

By induction and rearrangement of the equations, the following can be obtained:

$$\overline{T}_i(k) = \sum_{j=1}^{k} \frac{j}{\mu_i \cdot \alpha_i(j)} \cdot \pi_i(j-1 \mid k-1)$$

$$= \frac{1}{m_i \cdot \mu_i} \cdot \left(1 + \overline{K}_i(k-1) + \sum_{j=0}^{m_i-2} (m_i - j - 1) \cdot \pi_i(j \mid k-1) \right),$$

$$\tag{8.35}$$

$$\pi_i(0 \mid k) = 1 - \sum_{j=1}^{k} \frac{\lambda_i(k)}{\mu_i \cdot \alpha_i(j)} \cdot \pi_i(j-1 \mid k-1)$$

$$= 1 - \frac{1}{m_i} \cdot \left(\frac{e_i \cdot \lambda(k)}{\mu_i} + \sum_{j=1}^{m_i-1} (m_i - j) \cdot \pi_i(j \mid k) \right), \tag{8.36}$$

$$\pi_i(j \mid k) = \frac{\lambda_i(k)}{\mu_i \cdot \alpha_i(j)} \cdot \pi_i(j-1 \mid k-1). \tag{8.37}$$

Now the MVA for closed single class product-form queueing networks can be described as follows:

STEP 1 Initialization. For $i = 1, \ldots, N$ and $j = 1, \ldots, (m_i - 1)$:

$$\overline{K}_i(0) = 0, \quad \pi_i(0 \mid 0) = 1, \quad \pi_i(j \mid 0) = 0.$$

STEP 2 Iteration over the number of jobs $k = 1, \ldots, K$.

STEP 2.1 For $i = 1, \ldots, N$, compute the mean response time of a job at the ith node:

$$\overline{T}_i(k) = \begin{cases} \dfrac{1}{\mu_i} \left[1 + \overline{K}_i(k-1) \right], & \text{Type 1,2,4} \\ & (m_i = 1), \\[2mm] \dfrac{1}{\mu_i \cdot m_i} \left[1 + \overline{K}_i(k-1) + \displaystyle\sum_{j=0}^{m_i-2} (m_i - j - 1) \cdot \pi_i(j \mid k-1) \right], & \text{Type 1} \\ & (m_i > 1), \\[2mm] \dfrac{1}{\mu_i}, & \text{Type 3}, \end{cases}$$

(8.38)

where the conditional probabilities are computed using Eqs. (8.37) and (8.36).

STEP 2.2 Compute the overall throughput:

$$\lambda(k) = \frac{k}{\displaystyle\sum_{i=1}^{N} e_i \cdot \overline{T}_i(k)}.$$

(8.39)

The throughput of each node can be computed using the following equation:

$$\lambda_i(k) = e_i \cdot \lambda(k),$$

(8.40)

where the e_i can be determined with Eq. (7.5).

STEP 2.3 For $i = 1, \ldots, N$, compute the mean number of jobs at the ith node:

$$\overline{K}_i(k) = e_i \cdot \lambda(k) \cdot \overline{T}_i(k).$$

(8.41)

The other performance measures, e.g., utilization, mean waiting time, mean queue length, etc., can be derived from the calculated measures using the well-known equations.

The disadvantage of the MVA is its extremely high memory requirement. The memory requirement can be considerably reduced by using approximation techniques (e.g., SCAT, or self-correcting approximation technique), discussed in Chapter 9. Another disadvantage of the MVA is the fact that it is not possible to compute state probabilities. In [AkBo83], the MVA has been extended for computing the normalization constant and for computing the state probabilities. Only Step 2.2 needs to be extended to include the equation:

$$G(k) = \frac{G(k-1)}{\lambda(k)},$$

(8.42)

with the initial condition $G(0) = 1$. This follows immediately from Eq. (8.14). When the iteration stops, we have the normalization constant $G(K)$ that can be used to compute the state probabilities with the help of the BCMP theorem, Eq. (7.80). Two applications of the mean value analysis are now given.

Example 8.3   The central-server model shown in Fig. 8.5 has $N = 4$ nodes and $K = 6$ jobs. The service time of a job at the ith node, $i = 1, 2, 3, 4$, is exponentially distributed with the following mean values:

$$\frac{1}{\mu_1} = 0.02\,\text{sec}, \quad \frac{1}{\mu_2} = 0.2\,\text{sec}, \quad \frac{1}{\mu_3} = 0.4\,\text{sec}, \quad \frac{1}{\mu_4} = 0.6\,\text{sec}.$$

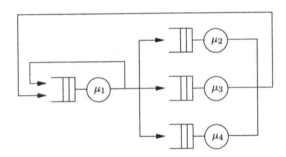

Fig. 8.5   The central-server model.

The visit ratios are given as follows:

$$e_1 = 1, \quad e_2 = 0.4, \quad e_3 = 0.2, \quad e_4 = 0.1.$$

With the help of the MVA, the performance measures and normalization constant of the network are computed in three steps:

STEP 1   Initialization:

$$\overline{K}_1(0) = \overline{K}_2(0) = \overline{K}_3(0) = \overline{K}_4(0) = 0, \quad G(0) = 1.$$

STEP 2   Iteration over the number of jobs in the network starting with $k = 1$:

STEP 2.1   Mean response times, Eq. (8.38):

$$\overline{T}_1(1) = \frac{1}{\mu_1}\left[1 + \overline{K}_1(0)\right] = \underline{0.02}, \quad \overline{T}_2(1) = \frac{1}{\mu_2}\left[1 + \overline{K}_2(0)\right] = \underline{0.2},$$

$$\overline{T}_3(1) = \frac{1}{\mu_3}\left[1 + \overline{K}_3(0)\right] = \underline{0.4}, \quad \overline{T}_4(1) = \frac{1}{\mu_4}\left[1 + \overline{K}_4(0)\right] = \underline{0.6}.$$

STEP 2.2   Throughput, Eq. (8.39), and normalization constant, Eq. (8.42):

$$\lambda(1) = \frac{1}{\sum\limits_{i=1}^{4} e_i \overline{T}_i(1)} = \underline{4.167}, \quad G(1) = \frac{G(0)}{\lambda(1)} = \underline{0.24}.$$

STEP 2.3   Mean number of jobs, Eq. (8.41):

$$\overline{K}_1(1) = \lambda(1)\overline{T}_1(1)e_1 = \underline{0.083}, \quad \overline{K}_2(1) = \lambda(1)\overline{T}_2(1)e_2 = \underline{0.333},$$
$$\overline{K}_3(1) = \lambda(1)\overline{T}_3(1)e_3 = \underline{0.333}, \quad \overline{K}_4(1) = \lambda(1)\overline{T}_4(1)e_4 = \underline{0.25}.$$

Iteration for $k = 2$:

STEP 2.1   Mean response times:

$$\overline{T}_1(2) = \frac{1}{\mu_1} \left[1 + \overline{K}_1(1)\right] = \underline{0.022}, \quad \overline{T}_2(2) = \frac{1}{\mu_2} \left[1 + \overline{K}_2(1)\right] = \underline{0.267},$$
$$\overline{T}_3(2) = \frac{1}{\mu_3} \left[1 + \overline{K}_3(1)\right] = \underline{0.533}, \quad \overline{T}_4(2) = \frac{1}{\mu_4} \left[1 + \overline{K}_4(1)\right] = \underline{0.75}.$$

STEP 2.2   Throughput and normalization constant:

$$\lambda(2) = \frac{2}{\sum\limits_{i=1}^{4} e_i \overline{T}_i(2)} = \underline{6.452}, \quad G(2) = \frac{G(1)}{\lambda(2)} = \underline{3.72 \cdot 10^{-2}}.$$

STEP 2.3   Mean number of jobs:

$$\overline{K}_1(2) = \lambda(2)\overline{T}_1(2)e_1 = \underline{0.140}, \quad \overline{K}_2(2) = \lambda(2)\overline{T}_2(2)e_2 = \underline{0.688},$$
$$\overline{K}_3(2) = \lambda(2)\overline{T}_3(2)e_3 = \underline{0.688}, \quad \overline{K}_4(2) = \lambda(2)\overline{T}_4(2)e_4 = \underline{0.484}.$$

$$\vdots$$

After six steps, the iteration stops and we get the final results as summarized in the Table 8.4. The normalization constant of the network is $G(K) = 5.756 \cdot 10^{-6}$. With Eq. (7.80) we can compute the steady-sate probability that the network is, for example, in the state $(3, 1, 1, 1)$:

$$\pi(3, 1, 1, 1) = \frac{1}{G(K)} \prod_{i=1}^{4} \left(\frac{e_i}{\mu_i}\right)^{k_i} = \underline{5.337 \cdot 10^{-4}}.$$

Example 8.4   As another example, consider the closed queueing network given in Fig. 8.6 with $K = 3$ jobs. At the first node we have $m_1 = 2$ identical processors having exponentially distributed service times with mean $1/\mu_1 = 0.5$ sec. Node 2 and node 3 have exponentially distributed service times with

Table 8.4 Performance measures after completing six iteration steps

Node	1	2	3	4
Mean response time $\overline{T}_i$	0.025	0.570	1.140	1.244
Throughput λ_i	9.920	3.968	1.984	0.992
Mean number of jobs $\overline{K}_i$	0.244	2.261	2.261	1.234
Utilization ρ_i	0.198	0.794	0.794	0.595

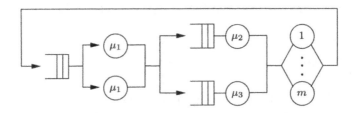

Fig. 8.6 A closed queueing network.

means $1/\mu_2 = 0.6\,\text{sec}$ and $1/\mu_3 = 0.8\,\text{sec}$, respectively. At the fourth node (terminals), the mean service time is $1/\mu_4 = 1\,\text{sec}$. The routing probabilities are as follows:

$$p_{12} = p_{13} = 0.5 \quad \text{and} \quad p_{24} = p_{34} = p_{41} = 1.$$

From Eq. (7.5) we compute the visit ratios:

$$e_1 = 1, \quad e_2 = 0.5, \quad e_3 = 0.5, \quad e_4 = 1.$$

The analysis of the network is carried out in the following steps:

STEP 1 Initialization:

$$\overline{K}_1(0) = \overline{K}_2(0) = \overline{K}_3(0) = 0, \quad \pi_1(0 \mid 0) = 1, \quad \pi_1(1 \mid 0) = 0.$$

STEP 2 Iteration over the number of jobs in the network starting with $k = 1$:

STEP 2.1 Mean response times:

$$\overline{T}_1(1) = \frac{1}{m_1\mu_1}\left[1 + \overline{K}_1(0) + \pi_1(0 \mid 0)\right] = \underline{0.5}, \quad \overline{T}_2(1) = \frac{1}{\mu_2}\left[1 + \overline{K}_2(0)\right] = \underline{0.6},$$

$$\overline{T}_3(1) = \frac{1}{\mu_3}\left[1 + \overline{K}_3(0)\right] = \underline{0.8}, \qquad \overline{T}_4(1) = \frac{1}{\mu_4} = \underline{1}.$$

STEP 2.2 Throughput:

$$\lambda(1) = \frac{1}{\sum\limits_{i=1}^{4} e_i \overline{T}_i(1)} = \underline{0.454}.$$

STEP 2.3 Mean number of jobs:

$$\overline{K}_1(1) = \lambda(1)\overline{T}_1(1)e_1 = \underline{0.227}, \quad \overline{K}_2(1) = \lambda(1)\overline{T}_2(1)e_2 = \underline{0.136},$$
$$\overline{K}_3(1) = \lambda(1)\overline{T}_3(1)e_3 = \underline{0.182}, \quad \overline{K}_4(1) = \lambda(1)\overline{T}_4(1)e_4 = \underline{0.454}.$$

Iteration for $k = 2$:

STEP 2.1 Mean response times:

$$\overline{T}_1(2) = \frac{1}{m_1\mu_1}\left[1 + \overline{K}_1(1) + \pi_1(0 \mid 1)\right] = \underline{0.5},$$

with

$$\pi_1(0 \mid 1) = 1 - \frac{1}{m_1}\left[\frac{e_1}{\mu_1}\lambda(1) + \pi_1(1 \mid 1)\right] = \underline{0.773},$$

and

$$\pi_1(1 \mid 1) = \frac{e_1}{\mu_1}\lambda(1)\pi_1(0 \mid 0) = \underline{0.227}, \quad \overline{T}_2(2) = \frac{1}{\mu_2}\left[1 + \overline{K}_2(1)\right] = \underline{0.682},$$
$$\overline{T}_3(2) = \frac{1}{\mu_3}\left[1 + \overline{K}_3(1)\right] = \underline{0.946}, \quad \overline{T}_4(2) = \underline{1}.$$

STEP 2.2 Throughput:

$$\lambda(2) = \frac{2}{\sum\limits_{i=1}^{4} e_i \overline{T}_i(2)} = \underline{0.864}.$$

STEP 2.3 Mean number of jobs:

$$\overline{K}_1(2) = \lambda(2)\overline{T}_1(2)e_1 = \underline{0.432}, \quad \overline{K}_2(2) = \lambda(2)\overline{T}_2(2)e_2 = \underline{0.295},$$
$$\overline{K}_3(2) = \lambda(2)\overline{T}_3(2)e_3 = \underline{0.409}, \quad \overline{K}_4(2) = \lambda(2)\overline{T}_4(2)e_4 = \underline{0.864}.$$

Iteration for $k = 3$:

STEP 2.1 Mean response times:

$$\overline{T}_1(3) = \frac{1}{m_1\mu_1}\left[1 + \overline{K}_1(2) + \pi_1(0 \mid 2)\right] = \underline{0.512},$$

with

$$\pi_1(0 \mid 2) = 1 - \frac{1}{m_1}\left[\frac{e_1}{\mu_1}\lambda(2) + \pi_1(1 \mid 2)\right] = \underline{0.617}\,,$$

and

$$\pi_1(1 \mid 2) = \frac{e_1}{\mu_1}\lambda(2)\pi_1(0 \mid 1) = \underline{0.334}\,,$$

$$\overline{T}_2(3) = \frac{1}{\mu_2}\left[1 + \overline{K}_2(2)\right] = \underline{0.776}\,,$$

$$\overline{T}_3(3) = \frac{1}{\mu_3}\left[1 + \overline{K}_3(2)\right] = \underline{1.127}\,, \quad \overline{T}_4(3) = \underline{1}\,.$$

STEP 2.2 Throughput:

$$\lambda(3) = \frac{3}{\sum\limits_{i=1}^{4} e_i\overline{T}_i(3)} = \underline{1.217}\,.$$

STEP 2.3 Mean number of jobs:

$$\overline{K}_1(3) = \lambda(3)\overline{T}_1(3)e_1 = \underline{0.624}\,, \quad \overline{K}_2(3) = \lambda(3)\overline{T}_2(3)e_2 = \underline{0.473}\,,$$

$$\overline{K}_3(3) = \lambda(3)\overline{T}_3(3)e_3 = \underline{0.686}\,, \quad \overline{K}_4(3) = \lambda(3)\overline{T}_4(3)e_4 = \underline{1.217}\,.$$

The throughput at each node can be computed with Eq. (8.40):

$$\lambda_1 = \lambda(3)\cdot e_1 = \underline{1.218}\,, \quad \lambda_2 = \lambda(3)\cdot e_2 = \underline{0.609}\,,$$

$$\lambda_3 = \lambda(3)\cdot e_3 = \underline{0.609}\,, \quad \lambda_4 = \lambda(3)\cdot e_4 = \underline{1.218}\,.$$

For determining the utilization of each node, we use Eq. (7.21):

$$\rho_1 = \frac{\lambda_1}{m_1\mu_1} = \underline{0.304}\,, \quad \rho_2 = \frac{\lambda_2}{\mu_2} = \underline{0.365}\,, \quad \rho_3 = \frac{\lambda_3}{\mu_3} = \underline{0.487}\,.$$

8.2.2 Multiclass Closed Networks

The algorithm for computing the performance measures of single class closed queueing networks can easily be extended to the multiclass case in the following way [ReLa80]:

STEP 1 Initialization. For $i = 1,\ldots,N, r = 1,\ldots,R, j = 1,\ldots,(m_i - 1)$:

$$\overline{K}_{ir}(0,0\ldots,0) = 0\,, \quad \pi_i(0 \mid \mathbf{0}) = 1\,, \quad \pi_i(j \mid \mathbf{0}) = 0\,.$$

STEP 2 Iteration: $\mathbf{k} = \mathbf{0},\ldots,\mathbf{K}$:

STEP 2.1 For $i = 1, \ldots, N$ and $r = 1, \ldots, R$, compute the mean response time of class-r jobs at the ith node:

$$
\overline{T}_{ir}(\mathbf{k}) = \begin{cases}
\dfrac{1}{\mu_{ir}}\left[1 + \sum\limits_{s=1}^{R} \overline{K}_{is}(\mathbf{k} - \mathbf{1}_r)\right] & \text{Type 1,2,4} \\[2pt]
& (m_i = 1), \\[8pt]
\dfrac{1}{\mu_{ir} \cdot m_i}\left[1 + \sum\limits_{s=1}^{R} \overline{K}_{is}(\mathbf{k} - \mathbf{1}_r)\right. & \\[8pt]
\left. + \sum\limits_{j=0}^{m_i-2} (m_i - j - 1)\pi_i(j \mid \mathbf{k} - \mathbf{1}_r)\right], & \text{Type 1} \\[4pt]
& (m_i > 1), \\[8pt]
\dfrac{1}{\mu_{ir}}, & \text{Type 3.}
\end{cases}
\tag{8.43}
$$

Here $(\mathbf{k} - \mathbf{1}_r) = (k_1, \ldots, k_r - 1, \ldots, k_R)$ is the population vector with one class-r job less in the system.

The probability that there are j jobs at the ith node ($j = 1, \ldots, (m_i - 1)$) given that the network is in state $\mathbf{k}$ is given by

$$
\pi_i(j \mid \mathbf{k}) = \frac{1}{j}\left[\sum_{r=1}^{R} \frac{e_{ir}}{\mu_{ir}} \lambda_r(\mathbf{k}) \pi_i(j - 1 \mid \mathbf{k} - \mathbf{1}_r)\right],
\tag{8.44}
$$

and for $j = 0$ by:

$$
\pi_i(0 \mid \mathbf{k}) = 1 - \frac{1}{m_i}\left[\sum_{r=1}^{R} \frac{e_{ir}}{\mu_{ir}} \lambda_r(\mathbf{k}) + \sum_{j=1}^{m_i-1} (m_i - j)\pi_i(j \mid \mathbf{k})\right],
\tag{8.45}
$$

where e_{ir} can be computed by Eq. (7.72).

STEP 2.2 For $r = 1, \ldots, R$, compute the throughput:

$$
\lambda_r(\mathbf{k}) = \frac{k_r}{\sum\limits_{i=1}^{N} e_{ir} \overline{T}_{ir}(k)}.
\tag{8.46}
$$

STEP 2.3 For $i = 1, \ldots, N$ and $r = 1, \ldots, R$, compute the mean number of class-r jobs at the ith node:

$$
\overline{K}_{ir}(\mathbf{k}) = \lambda_r(\mathbf{k}) \cdot \overline{T}_{ir}(\mathbf{k}) \cdot e_{ir}.
\tag{8.47}
$$

With these extensions the MVA algorithm for multiclass closed product-form queueing networks is completely specified.

Akyildiz and Bolch [AkBo83] extended the MVA for computing the normalization constant and the state probabilities. For this purpose, Step 2.2 in the preceding algorithm is expanded to include the formula

$$
G(\mathbf{k}) = \frac{G(\mathbf{k} - \mathbf{1}_r)}{\lambda_r(\mathbf{k})},
\tag{8.48}
$$

with the initial condition $G(\mathbf{0}) = 1$. After the iteration stops we get the normalization constant $G(\mathbf{K})$ that can be used to compute the steady-state probabilities with the help of the BCMP theorem, Eq. (7.80).

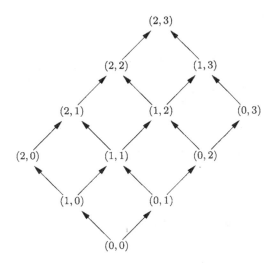

Fig. 8.7 Sequence of intermediate values.

To explain iteration Step 2, we use Fig. 8.7 where the sequence of intermediate values is shown for a network with $R = 2$ job classes. The first job class has $K_1 = 2$ jobs and the second class has $K_2 = 3$ jobs. The mean value algorithm starts with the trivial case where no job is in the system, that is, the population vector $(0,0)$. From this, solutions for all population vectors that consist of exactly one job are computed; in our example these are the population vectors $(1,0)$ and $(0,1)$. Then the solution for all population vectors with exactly two jobs are computed, and so on, until the final result for $\mathbf{K} = (2,3)$ is reached. In general, to compute the solution for a population vector $\mathbf{k}$ we need R intermediate solutions as input, namely the solutions for all population vectors $\mathbf{k} - \mathbf{1}_r$, $r = 1, \ldots, R$.

Now, the algorithm is illustrated by means of an example:

Example 8.5   Consider the queueing network shown in Fig. 8.8 with $N = 3$ nodes and $R = 2$ job classes. Class 1 contains $K_1 = 2$ jobs and class 2 contains $K_2 = 1$ jobs. Class switching of the jobs is not allowed. The service time at the ith, $i = 1, 2, 3$, node is exponentially distributed with mean values:

$$\frac{1}{\mu_{11}} = 0.2\,\text{sec}\,, \quad \frac{1}{\mu_{21}} = 0.4\,\text{sec}\,, \quad \frac{1}{\mu_{31}} = 1\,\text{sec}\,,$$

$$\frac{1}{\mu_{12}} = 0.2\,\text{sec}\,, \quad \frac{1}{\mu_{22}} = 0.6\,\text{sec}\,, \quad \frac{1}{\mu_{32}} = 2\,\text{sec}\,.$$

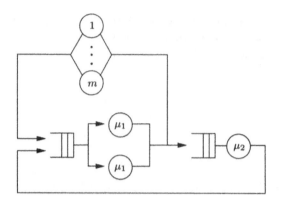

Fig. 8.8 Another closed queueing network.

The visit ratios are given as follows:

$$e_{11} = 1, \quad e_{21} = 0.6, \quad e_{31} = 0.4, \quad e_{12} = 1, \quad e_{22} = 0.3, \quad e_{32} = 0.7.$$

The queueing discipline at node 1 is FCFS and at node 2, processor sharing. The terminal station is modeled as an IS-node. We analyze the network using the MVA in the following three steps:

STEP 1 Initialization:

$$\overline{K}_{ir}(\mathbf{0}) = 0 \quad \text{for } i = 1, 2, 3 \text{ and } r = 1, 2, \quad \pi_1(0 \mid \mathbf{0}) = 1, \quad \pi_1(1 \mid \mathbf{0}) = 0.$$

STEP 2 Iterate over the number of jobs in the network beginning with the population vector $\mathbf{k} = (1, 0)$:

STEP 2.1 Mean response times, Eq. (8.43):

$$\overline{T}_{11}(1, 0) = \frac{1}{2 \cdot \mu_{11}} \left[1 + \overline{K}_{11}(0, 0) + \overline{K}_{12}(0, 0) + \pi_1(0 \mid 0, 0) \right] = \underline{0.2},$$

$$\overline{T}_{21}(1, 0) = \frac{1}{\mu_{21}} \left[1 + \overline{K}_{21}(0, 0) + \overline{K}_{22}(0, 0) \right] = \underline{0.4},$$

$$\overline{T}_{31}(1, 0) = \frac{1}{\mu_{31}} = \underline{1}.$$

STEP 2.2 Throughput for class 1 using Eq. (8.46):

$$\lambda_1(1, 0) = \frac{1}{\sum\limits_{i=1}^{3} e_{i1} \overline{T}_{i1}(1, 0)} = \underline{1.190}.$$

Note that the class 2 throughput for this population vector is 0.

STEP 2.3 Mean number of jobs, Eq. (8.47):

$$\overline{K}_{11}(1,0) = \lambda_1(1,0)\overline{T}_{11}(1,0)e_{11} = \underline{0.238},$$
$$\overline{K}_{21}(1,0) = \lambda_1(1,0)\overline{T}_{21}(1,0)e_{21} = \underline{0.286},$$
$$\overline{K}_{31}(1,0) = \lambda_1(1,0)\overline{T}_{31}(1,0)e_{31} = \underline{0.476}.$$

Iteration for $\mathbf{k} = (0,1)$:

STEP 2.1 Mean response time:

$$\overline{T}_{12}(0,1) = \frac{1}{2 \cdot \mu_{12}} \left[1 + \overline{K}_{11}(0,0) + \overline{K}_{12}(0,0) + \pi_1(0 \mid 0,0)\right] = \underline{0.2},$$
$$\overline{T}_{22}(0,1) = \frac{1}{\mu_{22}} \left[1 + \overline{K}_{21}(0,0) + \overline{K}_{22}(0,0)\right] = \underline{0.6},$$
$$\overline{T}_{32}(0,1) = \frac{1}{\mu_{32}} = \underline{2}.$$

STEP 2.2 Throughput for class 2:

$$\lambda_2(0,1) = \frac{1}{\displaystyle\sum_{i=1}^{3} e_{i2}\overline{T}_{i2}(0,1)} = \underline{0.562}.$$

Note that the class 1 throughput for this population vector is 0.

STEP 2.3 Mean number of jobs:

$$\overline{K}_{12}(0,1) = \lambda_2(0,1)\overline{T}_{12}(0,1)e_{12} = \underline{0.112},$$
$$\overline{K}_{22}(0,1) = \lambda_2(0,1)\overline{T}_{22}(0,1)e_{22} = \underline{0.101},$$
$$\overline{K}_{32}(0,1) = \lambda_2(0,1)\overline{T}_{32}(0,1)e_{32} = \underline{0.787}.$$

Iteration for $\mathbf{k} = (1,1)$:

STEP 2.1 Mean response times:

Class 1:

$$\overline{T}_{11}(1,1) = \frac{1}{2 \cdot \mu_{11}} \left[1 + \overline{K}_{12}(0,1) + \pi_1(0 \mid 0,1)\right] = \underline{0.2},$$

where

$$\pi_1(0 \mid 0,1) = 1 - \frac{1}{m_1} \left[\frac{e_{12}}{\mu_{12}}\lambda_2(0,1) + \pi_1(1 \mid 0,1)\right] = \underline{0.888},$$

and

$$\pi_1(1 \mid 0, 1) = \frac{e_{12}}{\mu_{12}} \lambda_2(0, 1)\pi_1(0 \mid 0, 0) = \underline{0.112},$$

$$\overline{T}_{21}(1, 1) = \frac{1}{\mu_{21}} \left[1 + \overline{K}_{21}(0, 1) + \overline{K}_{22}(0, 1)\right] = \underline{0.44},$$

$$\overline{T}_{31}(1, 1) = \frac{1}{\mu_{31}} = \underline{1}.$$

Class 2:

$$\overline{T}_{12}(1, 1) = \frac{1}{2 \cdot \mu_{12}} \left[1 + \overline{K}_{11}(1, 0) + \pi_1(0 \mid 1, 0)\right] = \underline{0.2},$$

with

$$\pi_1(0 \mid 1, 0) = 1 - \frac{1}{m_1} \left[\frac{e_{11}}{\mu_{11}} \lambda_1(1, 0) + \pi_1(1 \mid 1, 0)\right] = \underline{0.762},$$

and

$$\pi_1(1 \mid 1, 0) = \frac{e_{11}}{\mu_{11}} \lambda_1(1, 0)\pi_1(0 \mid 0, 0) = \underline{0.238},$$

$$\overline{T}_{22}(1, 1) = \frac{1}{\mu_{22}} \left[1 + \overline{K}_{21}(1, 0)\right] = \underline{0.772}, \quad \overline{T}_{32}(1, 1) = \frac{1}{\mu_{32}} = \underline{2}.$$

STEP 2.2 Throughputs:

$$\lambda_1(1, 1) = \frac{1}{\sum\limits_{i=1}^{3} e_{i1}\overline{T}_{i1}(1, 1)} = \underline{1.157},$$

$$\lambda_2(1, 1) = \frac{1}{\sum\limits_{i=1}^{3} e_{i2}\overline{T}_{i2}(1, 1)} = \underline{0.546}.$$

STEP 2.3 Mean number of jobs:

$$\overline{K}_{11}(1, 1) = \lambda_1(1, 1)\overline{T}_{11}(1, 1)e_{11} = \underline{0.231},$$
$$\overline{K}_{12}(1, 1) = \lambda_2(1, 1)\overline{T}_{12}(1, 1)e_{12} = \underline{0.109},$$
$$\overline{K}_{21}(1, 1) = \lambda_1(1, 1)\overline{T}_{21}(1, 1)e_{21} = \underline{0.305},$$
$$\overline{K}_{22}(1, 1) = \lambda_2(1, 1)\overline{T}_{22}(1, 1)e_{22} = \underline{0.126},$$
$$\overline{K}_{31}(1, 1) = \lambda_1(1, 1)\overline{T}_{31}(1, 1)e_{31} = \underline{0.463},$$
$$\overline{K}_{32}(1, 1) = \lambda_2(1, 1)\overline{T}_{32}(1, 1)e_{32} = \underline{0.764}.$$

Iteration for $\mathbf{k} = (2, 0)$:

STEP 2.1 Mean response times:

$$\overline{T}_{11}(2,0) = \frac{1}{2 \cdot \mu_{11}} \left[1 + \overline{K}_{11}(1,0) + \pi_1(0 \mid 1, 0) \right] = \underline{0.2},$$

$$\overline{T}_{21}(2,0) = \frac{1}{\mu_{21}} \left[1 + \overline{K}_{21}(1,0) \right] = \underline{0.514},$$

$$\overline{T}_{31}(2,0) = \frac{1}{\mu_{31}} = \underline{1}.$$

STEP 2.2 Throughput:

$$\lambda_1(2,0) = \frac{2}{\sum\limits_{i=1}^{3} e_{i1} \overline{T}_{i1}(2,0)} = \underline{2.201}.$$

STEP 2.3 Mean number of jobs:

$$\overline{K}_{11}(2,0) = \lambda_1(2,0) \overline{T}_{11}(2,0) e_{11} = \underline{0.440},$$
$$\overline{K}_{21}(2,0) = \lambda_1(2,0) \overline{T}_{21}(2,0) e_{21} = \underline{0.679},$$
$$\overline{K}_{31}(2,0) = \lambda_1(2,0) \overline{T}_{31}(2,0) e_{31} = \underline{0.881}.$$

Iteration for $\mathbf{k} = (2, 1)$:

STEP 2.1 Mean response times:

Class 1:

$$\overline{T}_{11}(2,1) = \frac{1}{2 \cdot \mu_{11}} \left[1 + \overline{K}_{11}(1,1) + \overline{K}_{12}(1,1) + \pi_1(0 \mid 1, 1) \right] = \underline{0.203},$$

where

$$\pi_1(0 \mid 1, 1) = 1 - \frac{1}{m_1} \left[\frac{e_{11}}{\mu_{11}} \lambda_1(1,1) + \frac{e_{12}}{\mu_{12}} \lambda_2(1,1) + \pi_1(1 \mid 1, 1) \right] = \underline{0.685},$$

and

$$\pi_1(1 \mid 1, 1) = \frac{e_{11}}{\mu_{11}} \lambda_1(1,1) \pi_1(0 \mid 0, 1) + \frac{e_{12}}{\mu_{12}} \lambda_2(1,1) \cdot \pi_1(0 \mid 1, 0) = \underline{0.289},$$

$$\overline{T}_{21}(2,1) = \frac{1}{\mu_{21}} \left[1 + \overline{K}_{21}(1,1) + \overline{K}_{22}(1,1) \right] = \underline{0.573}, \quad \overline{T}_{31}(2,1) = \frac{1}{\mu_{31}} = \underline{1}.$$

Class 2:

$$\overline{T}_{12}(2,1) = \frac{1}{2 \cdot \mu_{12}} \left[1 + \overline{K}_{11}(2,0) + \pi_1(0 \mid 2, 0) \right] = \underline{0.205},$$

where

$$\pi_1(0 \mid 2,0) = 1 - \frac{1}{m_1}\left[\frac{e_{11}}{\mu_{11}}\lambda_1(2,0) + \pi_1(1 \mid 2,0)\right] = \underline{0.612}\,,$$

and

$$\pi_1(1 \mid 2,0) = \frac{e_{11}}{\mu_{11}}\lambda_1(2,0)\pi_1(0 \mid 1,0) = \underline{0.335}\,,$$

$$\overline{T}_{22}(2,1) = \frac{1}{\mu_{22}}\left[1 + \overline{K}_{21}(2,0)\right] = \underline{1.008}\,, \quad \overline{T}_{32}(2,1) = \frac{1}{\mu_{32}} = \underline{2}\,.$$

STEP 2.2   Throughputs:

$$\lambda_1(2,1) = \frac{2}{\sum\limits_{i=1}^{3} e_{i1}\overline{T}_{i1}(2,1)} = \underline{2.113}\,, \quad \lambda_2(2,1) = \frac{1}{\sum\limits_{i=1}^{3} e_{i2}\overline{T}_{i2}(2,1)} = \underline{0.524}\,.$$

STEP 2.3   Mean number of jobs:

$$\overline{K}_{11}(2,1) = \lambda_1(2,1)\overline{T}_{11}(2,1)e_{11} = \underline{0.428}\,,$$
$$\overline{K}_{12}(2,1) = \lambda_2(2,1)\overline{T}_{12}(2,1)e_{12} = \underline{0.108}\,,$$
$$\overline{K}_{21}(2,1) = \lambda_1(2,1)\overline{T}_{21}(2,1)e_{21} = \underline{0.726}\,,$$
$$\overline{K}_{22}(2,1) = \lambda_2(2,1)\overline{T}_{22}(2,1)e_{22} = \underline{0.158}\,,$$
$$\overline{K}_{31}(2,1) = \lambda_1(2,1)\overline{T}_{31}(2,1)e_{31} = \underline{0.845}\,,$$
$$\overline{K}_{32}(2,1) = \lambda_2(2,1)\overline{T}_{32}(2,1)e_{32} = \underline{0.734}\,.$$

We can see that the MVA is easy to implement and is very fast.

For networks with multiple server nodes ($m_i > 1$) and several job classes, this technique has some disadvantages such as stability problems, accumulation of roundoff errors, and large storage requirements. The storage requirement is $O\left(N \cdot \prod_{r=1}^{R}(K_r + 1)\right)$. For comparison, the storage requirement of the convolution algorithm is $O\left(\prod_{r=1}^{R}(K_r + 1)\right)$. The time requirement in both cases is approximately the same: $2 \cdot R(N - 1) \cdot \prod_{r=1}^{R}(K_r + 1)$ [CoGe86]. Recall that N is the number of nodes in the network and K_r is the number of jobs of class r.

8.2.3   Mixed Networks

In [ZaWo81] the MVA is extended to the analysis of mixed product-form queueing networks. But the networks are restricted to consist only of single server nodes. Before we introduce the MVA for mixed queueing networks, we consider the arrival theorem for open queueing networks. This theorem says that the probability that a job entering node i will find the network in state $(k_1 \ldots, k_i, \ldots, k_N)$ is equal to the steady-state probability for this state. This theorem is also called PASTA theorem.[1] Therefore, we have for the mean response time of a job in an open network:

$$
\overline{T}_{ir}(\mathbf{k}) = \begin{cases} \dfrac{1}{\mu_{ir}} \left(1 + \sum\limits_{s=1}^{R} \overline{K}_{is}\right), & \text{Type } 1,2,4 \ (m_i = 1), \\[4mm] \dfrac{1}{\mu_{ir}}, & \text{Type } 3. \end{cases}
\tag{8.49}
$$

As from [ZaWo81], we have, for any class r and s, $\overline{K}_{is} = \dfrac{\rho_{is}}{\rho_{ir}}\overline{K}_{ir}$ and, therefore, with the help of the relation $\overline{K}_{ir}(\mathbf{k}) = \lambda_{ir} \cdot \overline{T}_{ir}(\mathbf{k})$, it follows from Eq. (8.49):

$$
\overline{K}_{ir}(\mathbf{k}) = \begin{cases} \rho_{ir}\left(1 + \sum\limits_{s=1}^{R} \dfrac{\rho_{is}}{\rho_{ir}}\overline{K}_{ir}\right) = \dfrac{\rho_{ir}}{1 - \sum\limits_{s=1}^{R} \rho_{is}}, & \text{Type } 1,2,4 \ (m_i = 1), \\[6mm] \rho_{ir}, & \text{Type } 3, \end{cases}
\tag{8.50}
$$

with $\rho_{ir} = \lambda_{ir}/\mu_{ir}$, where $\lambda_{ir} = \lambda_r \cdot e_{ir}$ is given as an input parameter.

These results for open queueing networks are now used for analyzing mixed queueing networks. The arrival theorem for mixed product-form queueing networks says that jobs of the open job classes when arriving at a node see the number of jobs in equilibrium, while jobs of the closed job classes when arriving at that node will see the number of jobs at that node in equilibrium with one job less in its own job class in the network [ReLa80, SeMi81].

If we index the open job classes with $op = 1, \ldots, OP$ and the closed job classes with $cl = 1, \ldots, CL$, then we have for the mean number of jobs in an open class r, taking into account Eq. (8.50):

$$
\begin{aligned}
\overline{K}_{ir}(\mathbf{k}) &= \lambda_{ir} \cdot \dfrac{1}{\mu_{ir}} \left[1 + \sum_{cl=1}^{CL} \overline{K}_{i,cl}(\mathbf{k}) + \sum_{op=1}^{OP} \overline{K}_{i,op}(\mathbf{k})\right] \\[4mm]
&= \dfrac{\rho_{ir}\left[1 + \sum\limits_{cl=1}^{CL} \overline{K}_{i,cl}(\mathbf{k})\right]}{1 - \sum\limits_{op=1}^{OP} \rho_{i,op}},
\end{aligned}
\tag{8.51}
$$

---

[1]PASTA: Poisson arrivals see time averages (see [Wolf82]).

where $\mathbf{k}$ is the population vector of the *closed* job classes. Equation (8.51) is valid for Type 1,2,4 single server nodes, while for Type 3 nodes we have

$$\overline{K}_{ir}(\mathbf{k}) = \rho_{ir}. \tag{8.52}$$

For the mean response time of a job in a *closed* class r at node i we have

$$
\begin{aligned}
\overline{T}_{ir}(\mathbf{k}) &= \frac{1}{\mu_{ir}} \left[1 + \sum_{cl=1}^{CL} \overline{K}_{i,cl}(\mathbf{k} - \mathbf{1}_r) + \sum_{op=1}^{OP} \overline{K}_{i,op}(\mathbf{k} - \mathbf{1}_r) \right] \\
&= \frac{1}{\mu_{ir}} \left[1 + \sum_{cl=1}^{CL} \overline{K}_{i,cl}(\mathbf{k} - \mathbf{1}_r) + \frac{\left[1 + \sum_{cl=1}^{CL} \overline{K}_{i,cl}(\mathbf{k} - \mathbf{1}_r) \right] \sum_{op=1}^{OP} \rho_{i,op}}{1 - \sum_{op=1}^{OP} \rho_{i,op}} \right] \\
&= \frac{1}{\mu_{ir}} \frac{\left[1 + \sum_{cl=1}^{CL} \overline{K}_{i,cl}(\mathbf{k} - \mathbf{1}_r) \right]}{1 - \sum_{op=1}^{OP} \rho_{i,op}}. \tag{8.53}
\end{aligned}
$$

Equation (8.53) is valid for Type 1,2,4 single server nodes. For Type 3 nodes we have:

$$\overline{T}_{ir}(\mathbf{k}) = \frac{1}{\mu_{ir}}. \tag{8.54}$$

With these formulae the MVA for mixed product-form queueing networks can be described completely. We assume for now that all arrival and service rates are load independent.

The algorithm is as follows:

STEP 1 Initialization. For all nodes $i = 1, \ldots, N$, compute the utilization of the open class jobs $op = 1, \ldots, OP$ in the mixed network:

$$\rho_{i,op} = \frac{1}{\mu_{i,op}} \lambda_{op} \cdot e_{i,op}, \tag{8.55}$$

and check for the ergodicity condition ($\rho_{i,op} \le 1$). Set $\overline{K}_{i,cl}(\mathbf{0}) = 0$ for all $i = 1, \ldots, N$ and all closed classes $cl = 1, \ldots, CL$.

STEP 2 Construct a closed queueing network that contains only the jobs of the closed job classes and solve the model with the extended version of the MVA, which considers also the influence of jobs of the open job classes. The results are the performance measures for the closed job classes of the mixed queueing network.

Iteration: $\mathbf{k} = \mathbf{0}, \ldots, \mathbf{K}$:

STEP 2.1 For $i = 1, \ldots, N$ and $r = 1, \ldots, CL$, compute the mean response times with Eqs. (8.53) and (8.54).

STEP 2.2 For $r = 1, \ldots, CL$, compute the throughputs with Eq. (8.46).

STEP 2.3 For $i = 1, \ldots, N$ and $r = 1, \ldots, CL$, evaluate the mean number of jobs with Eq. (8.47).

STEP 3 With the help of the solutions of the closed model and the equations given in Section 7.2, compute the performance measures for the open job classes of the network starting with Eqs. (8.51) and (8.52) for computing the mean number of jobs $\overline{K}_{ir}$, $i = 1, \ldots, N$ and $r = 1, \ldots, OP$.

If we use the iterative formula (8.48) for computing the normalization constant in Step 2, then the steady-state probabilities can be derived from the BCMP theorem, Eq. (7.76) [AkBo83]:

$$\pi(\mathbf{S} = \mathbf{S}_1 \ldots, \mathbf{S}_N) = \frac{1}{G(\mathbf{K})} \prod_{j=0}^{\mathbf{K}(\mathbf{S})-1} \lambda(j) \prod_{i=1}^{N} F_i(\mathbf{S}_i), \tag{8.56}$$

where $\lambda(j)$ is the (possibly state dependent) arrival rate.

Example 8.6   As an example of a mixed queueing network, consider the model given in Fig. 8.9 with $N = 2$ nodes and $R = 4$ job classes. Class 1 and 2 are open, the classes 3 and 4 are closed. Node 1 is of Type 2 and node 2 is of Type 4.

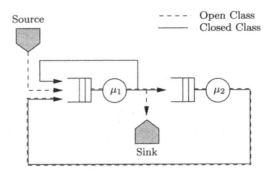

Fig. 8.9 A mixed queueing network.

The mean service times are given as follows:

$$\frac{1}{\mu_{11}} = 0.4\,\mathrm{sec}\,, \quad \frac{1}{\mu_{12}} = 0.8\,\mathrm{sec}\,, \quad \frac{1}{\mu_{13}} = 0.3\,\mathrm{sec}\,, \quad \frac{1}{\mu_{14}} = 0.5\,\mathrm{sec}\,,$$

$$\frac{1}{\mu_{21}} = 0.6\,\mathrm{sec}\,, \quad \frac{1}{\mu_{22}} = 1.6\,\mathrm{sec}\,, \quad \frac{1}{\mu_{23}} = 0.5\,\mathrm{sec}\,, \quad \frac{1}{\mu_{24}} = 0.8\,\mathrm{sec}\,.$$

The arrival rates by class for the open class jobs are $\lambda_1 = 0.5$ jobs/sec and $\lambda_2 = 0.25$ jobs/sec. In each of the closed classes there is $K_3 = K_4 = 1$ job. The routing probabilities are

$$p_{0,11} = 1, \quad p_{11,11} = 0, \quad p_{11,21} = 0.5, \quad p_{21,11} = 1,$$
$$p_{0,12} = 1, \quad p_{12,12} = 0, \quad p_{12,22} = 0.6, \quad p_{22,12} = 1,$$
$$p_{0,13} = 0, \quad p_{13,13} = 0.5, \quad p_{13,23} = 0.5, \quad \cdot p_{23,13} = 1,$$
$$p_{0,14} = 0, \quad p_{14,14} = 0.6, \quad p_{14,24} = 0.4, \quad p_{24,14} = 1.$$

Thus, by Eq. (6.27) the visit ratios are computed to be

$$e_{11} = 2, \quad e_{12} = 2.5, \quad e_{13} = 1, \quad e_{14} = 1,$$
$$e_{21} = 1, \quad e_{22} = 1.5, \quad e_{23} = 0.5, \quad e_{24} = 0.4.$$

The determination of the performance measures of this mixed queueing network is carried out with the MVA in the following three steps:

STEP 1 Initialization. Compute the utilization of both nodes by the open job classes:

$$\rho_{11} = \lambda_1 e_{11} \cdot \frac{1}{\mu_{11}} = \underline{0.4}, \quad \rho_{21} = \lambda_1 e_{21} \cdot \frac{1}{\mu_{21}} = \underline{0.3},$$
$$\rho_{12} = \lambda_2 e_{12} \cdot \frac{1}{\mu_{12}} = \underline{0.5}, \quad \rho_{22} = \lambda_2 e_{22} \cdot \frac{1}{\mu_{22}} = \underline{0.6}.$$

Set $\overline{K}_{11}(\underline{0}) = \overline{K}_{12}(\underline{0}) = \overline{K}_{21}(\underline{0}) = \overline{K}_{22}(\underline{0}) = 0$.

STEP 2 Using MVA, analyze the closed queueing network model, obtained by leaving out the jobs of the open job classes.

Mean value iteration for $\mathbf{k} = (1, 0)$:

Mean response times, Eq. (8.53):

$$\overline{T}_{13}(1,0) = \frac{1}{\mu_{13}} \cdot \frac{1 + \overline{K}_{13}(\underline{0})}{1 - (\rho_{11} + \rho_{12})} = \underline{3}, \quad \overline{T}_{23}(1,0) = \frac{1}{\mu_{23}} \cdot \frac{1 + \overline{K}_{23}(\underline{0})}{1 - (\rho_{21} + \rho_{22})} = \underline{5}.$$

Throughput, Eq. (8.46):

$$\lambda_3(1,0) = \frac{K_3}{\displaystyle\sum_{i=1}^{2} e_{i3} \overline{T}_{i3}(1,0)} = \underline{0.182}.$$

Mean number of jobs, Eq. (8.47):

$$\overline{K}_{13}(1,0) = \lambda_3(1,0) \cdot e_{13} \cdot \overline{T}_{13}(1,0) = \underline{0.545},$$

$$\overline{K}_{23}(1,0) = \lambda_3(1,0) \cdot e_{23} \cdot \overline{T}_{23}(1,0) = \underline{0.454}\,.$$

Mean value iteration for $\mathbf{k} = (0,1)$:

$$\overline{T}_{14}(0,1) = \underline{5}\,, \qquad \overline{T}_{24}(0,1) = \underline{8}\,,$$
$$\lambda_4(0,1) = \underline{0.122}\,,$$
$$\overline{K}_{14}(0,1) = \underline{0.610}\,, \quad \overline{K}_{24}(0,1) = \underline{0.390}\,.$$

Mean value iteration for $\mathbf{k} = (1,1)$:

$$\overline{T}_{13}(1,1) = \underline{4.829}\,, \ \overline{T}_{23}(1,1) = \underline{6.951}\,, \ \overline{T}_{14}(1,1) = \underline{7.727}\,, \ \overline{T}_{24}(1,1) = \underline{11.636}\,.$$
$$\lambda_3(1,1) = \underline{0.120}\,, \quad \lambda_4(1,1) = \underline{0.081}\,.$$
$$\overline{K}_{13}(1,1) = \underline{0.582}\,, \ \overline{K}_{23}(1,1) = \underline{0.418}\,, \ \overline{K}_{14}(1,1) = \underline{0.624}\,, \ \overline{K}_{24}(1,1) = \underline{0.376}\,.$$

STEP 3 With these results and using Eq. (8.51), the performance measures of the open classes can be computed:

$$\overline{K}_{11}(1,1) = \frac{\rho_{11}\left(1 + \overline{K}_{13}(1,1) + \overline{K}_{14}(1,1)\right)}{0.1} = \underline{8.822}\,,$$

$$\overline{K}_{21}(1,1) = \frac{\rho_{21}\left(1 + \overline{K}_{23}(1,1) + \overline{K}_{24}(1,1)\right)}{0.1} = \underline{5.383}\,,$$

$$\overline{K}_{12}(1,1) = \frac{\rho_{12}\left(1 + \overline{K}_{13}(1,1) + \overline{K}_{14}(1,1)\right)}{0.1} = \underline{11.028}\,,$$

$$\overline{K}_{22}(1,1) = \frac{\rho_{22}\left(1 + \overline{K}_{23}(1,1) + \overline{K}_{24}(1,1)\right)}{0.1} = \underline{10.766}\,.$$

For the computation of the mean response times, we use Eq. (7.43):

$$\overline{T}_{11}(1,1) = \underline{8.822}\,, \qquad \overline{T}_{12}(1,1) = \underline{17.645}\,,$$
$$\overline{T}_{21}(1,1) = \underline{10.766}\,, \qquad \overline{T}_{22}(1,1) = \underline{28.710}\,.$$

Now all the other performance measures can be computed.

There are some suggestions to extend this algorithm to mixed queueing networks that are allowed to contain additional multiple server nodes. The most important approaches can be found in [KTK81] and [BBA84].

8.2.4 Networks with Load-Dependent Service

The MVA can be extended to examine product-form queueing networks where the service rates at node depends on the number of jobs at that node [Reis81]. Let $\mu_i(j)$ denote the load-dependent service rate of the ith node when there are j jobs in it.

8.2.4.1 *Closed Networks* The algorithm for single class queueing networks with nodes having load-dependent service rates can be described in the following steps:

STEP 1   Initialization. For $i = 1, \ldots, N$: $\pi_i(0 \mid 0) = 1$.

STEP 2   Iteration: $k = 1, \ldots, K$.

STEP 2.1   For $i = 1, \ldots, N$, compute the mean response time:

$$\overline{T}_i(k) = \sum_{j=1}^{k} \frac{j}{\mu_i(j)} \pi_i(j - 1 \mid k - 1). \tag{8.57}$$

STEP 2.2   Compute the throughput:

$$\lambda(k) = \frac{k}{\sum_{i=1}^{N} e_i \cdot \overline{T}_i(k)}. \tag{8.58}$$

STEP 2.3   For $i = 1, \ldots, N$, compute the conditional probabilities of j jobs at the ith node given that there are k jobs in the network:

$$\pi_i(j \mid k) = \begin{cases} \dfrac{\lambda(k)}{\mu_i(j)} \pi_i(j - 1 \mid k - 1)e_i, & \text{for } j = 1, \ldots, k, \\[2ex] 1 - \displaystyle\sum_{l=1}^{k} \pi_i(l \mid k), & \text{for } j = 0. \end{cases} \tag{8.59}$$

The extension of this algorithm to multiclass queueing networks where the service rate of a node depends on the overall number of jobs at the node can be done very easily. Let $\mu_{ir}(j)$ be the the service rate of class-r jobs at the ith node when there are j jobs in it. Then the MVA can be presented in the following steps:

STEP 1   Initialization. For $i = 1, \ldots, N$: $\pi_i(0 \mid \mathbf{0}) = 1$

STEP 2   Iteration: $\mathbf{k} = 0, \ldots, \mathbf{K}$.

STEP 2.1   For $i = 1, \ldots, N$ and $r = 1, \ldots, R$, compute the mean response times:

$$\overline{T}_{ir}(\mathbf{k}) = \sum_{j=1}^{k} \frac{j}{\mu_{ir}(j)} \pi_i(j - 1 \mid \mathbf{k} - \mathbf{1}_r), \quad \text{where } k = \sum_{r=1}^{R} k_r. \tag{8.60}$$

STEP 2.2 For $r = 1, \ldots, R$, compute the throughputs:

$$\lambda_r(\mathbf{k}) = \frac{k_r}{\displaystyle\sum_{i=1}^{N} e_{ir} \cdot \overline{T}_{ir}(\mathbf{k})}. \tag{8.61}$$

STEP 2.3 For $i = 1, \ldots, N$, determine the marginal probabilities:

$$\pi_i(j \mid \mathbf{k}) = \begin{cases} \displaystyle\sum_{r=1}^{R} \frac{\lambda_r(\mathbf{k})}{\mu_{ir}(j)} \pi_i(j-1 \mid \mathbf{k} - \mathbf{1}_r) e_{ir}, & \text{for } j = 1, \ldots, k, \\[2ex] 1 - \displaystyle\sum_{l=1}^{k} \pi_i(l \mid \mathbf{k}), & \text{for } j = 0. \end{cases} \tag{8.62}$$

Example 8.7   Consider again the example given in Fig. 8.6 where the multiple-server node 1 is replaced by a general load-dependent server. Load-dependent service rates are

$$\mu_1(1) = 2, \quad \mu_2(1) = 1.667, \quad \mu_3(1) = 1.25, \quad \mu_4(1) = 1,$$
$$\mu_1(2) = 4, \quad \mu_2(2) = 1.667, \quad \mu_3(2) = 1.25, \quad \mu_4(2) = 2,$$
$$\mu_1(3) = 4, \quad \mu_2(3) = 1.667, \quad \mu_3(3) = 1.25, \quad \mu_4(3) = 3.$$

STEP 1 Initialization:

$$\pi_i(0 \mid 0) = 1 \quad \text{for } i = 1, 2, 3, 4.$$

STEP 2 Iteration over the number of jobs in the network starting with $k = 1$:

STEP 2.1 Mean response times, Eq. (8.57):

$$\overline{T}_1(1) = \frac{1}{\mu_1(1)} \pi_1(0 \mid 0) = \underline{0.5}, \quad \overline{T}_2(1) = \frac{1}{\mu_2(1)} \pi_2(0 \mid 0) = \underline{0.6},$$
$$\overline{T}_3(1) = \frac{1}{\mu_3(1)} \pi_3(0 \mid 0) = \underline{0.8}, \quad \overline{T}_4(1) = \frac{1}{\mu_4(1)} \pi_4(0 \mid 0) = \underline{1}.$$

STEP 2.2 Throughput, Eq. (8.61) and normalizing constant, Eq. (8.42):

$$\lambda(1) = \frac{1}{\displaystyle\sum_{i=1}^{4} e_i \overline{T}_i(1)} = \underline{0.454}, \quad G(1) = \frac{G(0)}{\lambda(1)} = \underline{2.203}.$$

STEP 2.3 Marginal probabilities, Eq. (8.62):

$$\pi_1(1 \mid 1) = \frac{\lambda(1)}{\mu_1(1)} \pi_1(0 \mid 0) e_1 = \underline{0.227},$$

$$\pi_1(0 \mid 1) = 1 - \pi_1(1 \mid 1) = \underline{0.773}\,,$$
$$\pi_2(0 \mid 1) = \underline{0.864}\,, \quad \pi_2(1 \mid 1) = \underline{0.136}\,,$$
$$\pi_3(0 \mid 1) = \underline{0.818}\,, \quad \pi_3(1 \mid 1) = \underline{0.182}\,,$$
$$\pi_4(0 \mid 1) = \underline{0.545}\,, \quad \pi_4(1 \mid 1) = \underline{0.454}\,.$$

Iteration for $k = 2$:

STEP 2.1   Mean response times:

$$\overline{T}_1(2) = \frac{1}{\mu_1(1)}\pi_1(0 \mid 1) + \frac{2}{\mu_1(2)}\pi_1(1 \mid 1) = \underline{0.5}\,,$$

$$\overline{T}_2(2) = \frac{1}{\mu_2(1)}\pi_2(0 \mid 1) + \frac{2}{\mu_2(2)}\pi_2(1 \mid 1) = \underline{0.682}\,,$$

$$\overline{T}_3(2) = \frac{1}{\mu_3(1)}\pi_3(0 \mid 1) + \frac{2}{\mu_3(2)}\pi_3(1 \mid 1) = \underline{0.946}\,,$$

$$\overline{T}_4(2) = \frac{1}{\mu_4(1)}\pi_4(0 \mid 1) + \frac{2}{\mu_4(2)}\pi_4(1 \mid 1) = \underline{1}\,.$$

STEP 2.2   Throughput and normalizing constant:

$$\lambda(2) = \frac{2}{\displaystyle\sum_{i=1}^{4} e_i\overline{T}_i(2)} = 0.864\,, \quad G(2) = \frac{G(1)}{\lambda(2)} = 2.549\,.$$

STEP 2.3   Marginal probabilities:

$$\pi_1(2 \mid 2) = \frac{\lambda(2)}{\mu_1(2)}\pi_1(1 \mid 1)e_1 = \underline{0.049}\,,$$

$$\pi_1(1 \mid 2) = \frac{\lambda(2)}{\mu_1(1)}\pi_1(0 \mid 1)e_1 = \underline{0.334}\,,$$

$$\pi_1(0 \mid 2) = 1 - \sum_{l=1}^{2} \pi_1(l \mid 2) = \underline{0.617}\,.$$

$$\pi_2(2 \mid 2) = \underline{0.035}\,, \quad \pi_2(1 \mid 2) = \underline{0.224}\,, \quad \pi_2(0 \mid 2) = \underline{0.741}\,,$$
$$\pi_3(2 \mid 2) = \underline{0.063}\,, \quad \pi_3(1 \mid 2) = \underline{0.283}\,, \quad \pi_3(0 \mid 2) = \underline{0.654}\,,$$
$$\pi_4(2 \mid 2) = \underline{0.196}\,, \quad \pi_4(1 \mid 2) = \underline{0.472}\,, \quad \pi_4(0 \mid 2) = \underline{0.332}\,.$$

Iteration for $k = 3$:

STEP 2.1   Mean response times:

$$\overline{T}_1(3) = \frac{1}{\mu_1(1)}\pi_1(0 \mid 2) + \frac{2}{\mu_1(2)}\pi_1(1 \mid 2) + \frac{3}{\mu_1(3)}\pi_1(2 \mid 2) = \underline{0.512}\,,$$
$$\overline{T}_2(3) = \underline{0.776}\,, \quad \overline{T}_3(3) = \underline{1.127}\,, \quad \overline{T}_4(3) = \underline{1}\,.$$

STEP 2.2 Throughput and normalizing constant:

$$\lambda(3) = \underline{1.218}, \quad G(3) = \frac{G(2)}{\lambda(3)} = \underline{2.093}.$$

STEP 2.3 Marginal probabilities:

$$\pi_1(3 \mid 3) = \frac{\lambda(3)}{\mu_1(3)} \pi_1(2 \mid 2) e_1 = \underline{0.015},$$

$$\pi_1(2 \mid 3) = \frac{\lambda(3)}{\mu_1(2)} \pi_1(1 \mid 2) e_1 = \underline{0.102},$$

$$\pi_1(1 \mid 3) = \frac{\lambda(3)}{\mu_1(1)} \pi_1(0 \mid 2) e_1 = \underline{0.375},$$

$$\pi_1(0 \mid 3) = 1 - \sum_{l=1}^{3} \pi_1(l \mid 3) = \underline{0.507}.$$

$$\pi_2(3 \mid 3) = \underline{0.013}, \quad \pi_2(2 \mid 3) = \underline{0.082},$$
$$\pi_2(1 \mid 3) = \underline{0.271}, \quad \pi_2(0 \mid 3) = \underline{0.634},$$
$$\pi_3(3 \mid 3) = \underline{0.031}, \quad \pi_3(2 \mid 3) = \underline{0.138},$$
$$\pi_3(1 \mid 3) = \underline{0.319}, \quad \pi_3(0 \mid 3) = \underline{0.512},$$
$$\pi_4(3 \mid 3) = \underline{0.080}, \quad \pi_4(2 \mid 3) = \underline{0.287},$$
$$\pi_4(1 \mid 3) = \underline{0.404}, \quad \pi_4(0 \mid 3) = \underline{0.229}.$$

Now the iteration stops and all other performance measures can be computed. For the mean number of jobs at the nodes we have, for example,

$$\overline{K}_1(3) = \sum_{k=1}^{3} k \cdot \pi_1(k \mid 3) = \underline{0.624},$$

$$\overline{K}_2(3) = \underline{0.473}, \quad \overline{K}_3(3) = \underline{0.686}, \quad \overline{K}_4(3) = \underline{1.217}.$$

The MVA is extended by [Saue83] to deal with other kinds of load dependence:

- Networks in which the service rate of a node depends on the number of jobs of the different classes at that node (not just on the total number of jobs at the node).

- Networks with load-dependent routing probabilities.

In [TuSa85] and [HBAK86], the MVA is extended to the *tree mean value analysis*, which is well suited for very large networks with only a few jobs. Such networks arise especially while modeling distributed systems. In [BBA84] the *MVALDMX* algorithm (mean value analysis load dependent mixed) is introduced which can be used to analyze mixed BCMP networks with load-dependent service stations.

8.3   FLOW EQUIVALENT SERVER METHOD

The last method for analyzing product-form queueing networks that we consider is called *flow equivalent server (FES) method*. The method is based on *Norton's theorem* from electric circuit theory [CHW75b]. We encounter this method again in Chapter 9 when we deal with approximate solution of non-product-form networks. Here we consider this technique in the context of exact analysis of product-form networks. If we select one or more nodes in the given network and combine all the other nodes into an FES, then Norton's theorem says that the reduced system has the same behavior as the original network.

8.3.1   FES Method for a Single Node

To describe the FES method, we consider the central-server model given in Fig. 8.10 where we suppose that a product-form solution exists. For this network we can construct an equivalent network by choosing one node (e.g., node 1) and combining all other nodes to one *FES node c*. As a result we get a reduced net consisting of two nodes only. This network (shown in Fig. 8.11) is much easier to analyze then the original one, and all computed performance measures are the same as in the original network.

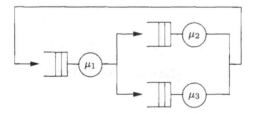

Fig. 8.10   The central-server network.

To determine the service rates $\mu_c(k), k = 1, \ldots, K$ of the FES node, the chosen node 1 in the original network is short-circuited; i.e., the mean service time of this node is set to zero, as shown in Fig. 8.12. The throughput in the short-circuit path with $k = 1, \ldots, K$ jobs in the network is then equated to the load-dependent service rate $\mu_c(k), k = 1, \ldots, K$ of the FES node in the reduced queueing network.

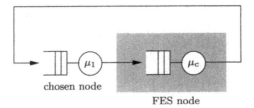

Fig. 8.11 Reduced queueing network.

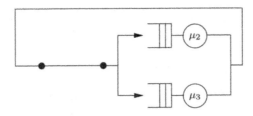

Fig. 8.12 The short-circuited model.

The FES algorithm can now be summarized in the following three steps:

STEP 1 In the given network, choose a node i and short-circuit it by setting the mean service time in that node to zero. Compute the throughputs $\lambda_i^{sc}(k)$ along the short circuit, as a function of the number of jobs $k = 1, \ldots, K$ in the network. For this computation, any of the earlier solution algorithms for product-form queueing networks can be used.

STEP 2 From the given network, construct an equivalent reduced network consisting only of the chosen node i and the FES node c. The visit ratios in both nodes are e_i. The load-dependent service rate of the FES node is the throughput along the short-circuit path when there are k jobs in the network, that is: $\mu_c(k) = \lambda_i^{sc}(k)$ for $k = 1, \ldots, K$.

STEP 3 Compute the performance measures in the reduced network with any suitable algorithm for product-form networks (e.g., convolution or MVA).

The technique just described is very well suited for examining the influence of changes in the parameters of a single node while the rest of the system parameters are kept fixed. To clarify the usage of the FES method, we give the following simple example:

Example 8.8 Consider the closed single class product-form network shown in Fig. 8.13. It consists of $N = 4$ nodes and $K = 2$ jobs. The service times are exponentially distributed with parameters

$$\mu_1 = 1, \quad \mu_2 = 2, \quad \mu_3 = 3, \quad \mu_4 = 4.$$

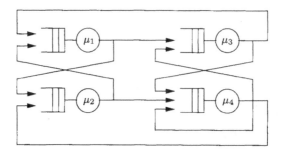

Fig. 8.13 Yet another closed queueing network.

The routing probabilities are given as

$$p_{12} = 0.5, \quad p_{21} = 0.5, \quad p_{31} = 0.5, \quad p_{42} = 0.4,$$
$$p_{13} = 0.5, \quad p_{24} = 0.5, \quad p_{34} = 0.5, \quad p_{43} = 0.4, \quad p_{44} = 0.2.$$

With Eq. (7.5) we compute the visit ratios:

$$e_1 = 1, \quad e_2 = 1, \quad e_3 = 1, \quad e_4 = 1.25.$$

Now, the analysis of the network is carried out in the following three steps:

STEP 1 Choose a node, e.g., node 4 of the network shown in Fig. 8.14, and short circuit it. The throughput of the short circuit is computed for $k = 1, 2$

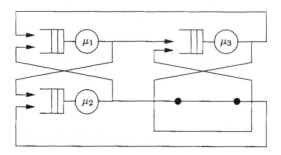

Fig. 8.14 Network after node 4 is short circuit.

by using the MVA, for example. Then we get the following results:

$$\lambda(1) = \underline{0.545}, \qquad \lambda(2) = \underline{0.776}$$

and therefore

$$\lambda_4^{sc}(1) = \underline{0.682}, \quad \lambda_4^{sc}(2) = \underline{0.971}.$$

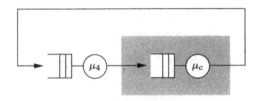

Fig. 8.15 Reduced network.

STEP 2 Construct the reduced network that consists of the selected node 4 and the FES node c as shown in Fig. 8.15. For the service rates of the FES node we have

$$\mu_c(1) = \lambda_4^{\mathrm{sc}}(1) = 0.682, \quad \mu_c(2) = \lambda_4^{\mathrm{sc}}(2) = 0.971.$$

The visit ratios are

$$e_4 = e_c = \underline{1.25}.$$

STEP 3 Analyze the reduced network shown in Fig. 8.15 using, for example, the MVA for load-dependent nodes. The load-dependent service rates of node 4 can be computed by Eq. (7.49) as $\mu_4(1) = \mu_4(2) = 4$.

MVA Iteration for $k = 1$:

$$\overline{T}_4(1) = \frac{1}{\mu_4(1)}\pi_4(0 \mid 0) = \underline{0.25}, \quad \overline{T}_c(1) = \frac{1}{\mu_c(1)}\pi_c(0 \mid 0) = \underline{1.467},$$

$$\lambda(1) = \underline{0.466}, \quad G(1) = \frac{1}{\lambda(1)} = \underline{2.146},$$

$$\pi_4(1 \mid 1) = \underline{0.146}, \quad \pi_c(1|1) = \underline{0.854}, \quad \pi_4(0 \mid 1) = \underline{0.854}, \quad \pi_c(0|1) = \underline{0.146}.$$

MVA Iteration for $k = 2$:

$$\overline{T}_4(2) = \frac{1}{\mu_4(1)}\pi_4(0 \mid 1) + \frac{2}{\mu_4(2)}\pi_4(1 \mid 1) = \underline{0.286},$$

$$\overline{T}_c(2) = \frac{1}{\mu_c(1)}\pi_c(0|1) + \frac{2}{\mu_c(2)}\pi_c(1|1) = \underline{1.974},$$

$$\lambda(2) = \underline{0.708}, \quad G(2) = \frac{G(1)}{\lambda(2)} = \underline{3.032},$$

$$\pi_4(1|2) = \underline{0.189}, \quad \pi_4(2|2) = \underline{0.032}, \quad \pi_4(0|2) = \underline{0.779},$$
$$\pi_c(1|2) = \underline{0.189}, \quad \pi_c(2|2) = \underline{0.779}, \quad \pi_c(0|2) = \underline{0.032}.$$

Now all other performance measures of the network can be computed:

- Mean number of jobs:

For node 4 we get from Eq. (7.26)

$$\overline{K}_4 = \sum_{k=1}^{2} k \cdot \pi_4(k \mid 2) = \underline{0.253} \,.$$

For the other nodes we have by Eq. (8.16):

$$\overline{K}_1 = \sum_{k=1}^{2} \left(\frac{e_1}{\mu_1}\right)^k \cdot \frac{G(2-k)}{G(2)} = \underline{1.038} \,, \quad \overline{K}_2 = \underline{0.436} \,, \quad \overline{K}_3 = \underline{0.273} \,.$$

- Throughputs, Eq. (8.14)

$$\lambda_1 = e_1 \frac{G(1)}{G(2)} = \underline{0.708} \,, \quad \lambda_2 = \underline{0.708} \,, \quad \lambda_3 = \underline{0.708} \,, \quad \lambda_4 = \underline{0.885} \,.$$

- Mean response times, Eq. (7.43):

$$\overline{T}_1 = \frac{\overline{K}_1}{\lambda_1} = \underline{1.466} \,, \quad \overline{T}_2 = \underline{0.617} \,, \quad \overline{T}_3 = \underline{0.385} \,.$$

8.3.2 FES Method for Multiple Nodes

The usage of the FES method, when only one node is short-circuited, has nearly no advantage in reducing the computation time. Therefore, [ABP85] suggest an extension of the concept of [CHW75b]. For this extension, the closed product-form network is partitioned in a number of subnetworks. Each of these subnetworks is analyzed independently from the others, i.e., the whole network is analyzed by short-circuiting the nodes that do not belong to the subnetwork that is to be examined. The computed throughputs of the short-circuited network form the load-dependent service rates of FES node j, which represents the jth subnetwork in the reduced network. When analyzing this reduced network, we get the normalizing constant of the whole network. The normalizing constant can then be used to compute the performance measures of interest. The FES method for this case can be described in the following five steps.

STEP 1 Partition the original network into M disjoint subnetworks $SN-j$ $(1 \leq j \leq M)$. It is even allowed to combine nodes in the subnetwork that do not have a direct connection to this subnetwork.

STEP 2 Analyze each of these subnetworks $j = 1, \ldots, M$ by short-circuiting all nodes that do not belong to the considered subnetwork. The visit ratios of the nodes in the subnetworks are taken from the original network. For analyzing the subnetworks, any product-form algorithm can be used. Determine the throughputs $\lambda_{SN-j}(k)$ and the normalizing constants $G_j(k)$, for $k = 1, \ldots, K$.

STEP 3 Combine the nodes of each subnetwork into one FES node and construct an equivalent reduced network out of the original network by putting the FES nodes (with visit ratio 1) together in a tandem connection. The load-dependent service rates of the FES node j are identical with the throughputs in the corresponding jth subnetwork:

$$\mu_{cj}(k) = \lambda_{SN-j}(k), \quad \text{for } j = 1, \ldots, M \text{ and } k = 1, \ldots, K.$$

STEP 4 The normalizing constant of the reduced network can be computed by convoluting the M normalizing vectors

$$G_j = \begin{pmatrix} G_j(0) \\ \vdots \\ G_j(K) \end{pmatrix}$$

of the subnetworks [ABP85]

$$G = G_1 \otimes G_2 \otimes \ldots \otimes G_M, \tag{8.63}$$

where the convolution operator $\otimes$ is defined as in Eq. (8.1). To determine the normalizing constant, the MVA for networks with load-dependent nodes (see Section 8.2.4) can be used instead.

STEP 5 Compute the performance measures of the original network and the performance measures of the subnetworks with the help of the normalizing constants and the formulae given in Section 8.1.

The FES method is very well suited if the system behavior for different input parameters needs to be examined, because in this case only one new normalizing constant for the corresponding subnetwork needs to be recomputed. The normalizing constants of the other subnetworks remain the same.

Example 8.9   The queueing network from Example 8.8 is now analyzed again.

STEP 1 Partition the whole network into $M = 2$ subnetworks where Subnetwork 1 contains nodes 1 and 2 and Subnetwork 2, nodes 3 and 4 of the original network.

STEP 2 Analyze the two subnetworks: To analyze Subnetwork 1, nodes 3 and 4 are short-circuited (see Fig. 8.16). We use the MVA to compute for $k = 1, 2$ the load-dependent throughputs and normalizing constants of this subnetwork:

$$\lambda_{SN-1}(1) = \underline{0.667}, \quad \lambda_{SN-1}(2) = \underline{0.857},$$

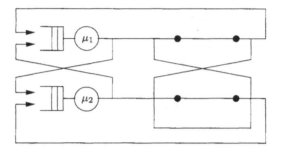

Fig. 8.16   Subnetwork 1 with Subnetwork 2 short-circuited.

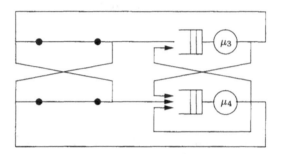

Fig. 8.17   Short-circuit of Subnetwork 1.

$$G_1(1) = \underline{1.5}, \qquad G_1(2) = \underline{1.75}.$$

To analyze Subnetwork 2, nodes 1 and 2 are short-circuited (see Fig. 8.17).

Analogously, we get the following results for Subnetwork 2:

$$\lambda_{SN-2}(1) = \underline{1.548}, \quad \lambda_{SN-2}(2) = \underline{2.064},$$
$$G_2(1) = \underline{0.646}, \qquad G_2(2) = \underline{0.313}.$$

STEP 3   Construct the reduced network shown in Fig. 8.18.

Fig. 8.18   Reduced network.

The first FES node describes the nodes of Subnetwork 1 with service rates $\mu_{c1}(k) = \lambda_{SN-1}(k)$ for $k = 1, 2$, and the second FES node describes the nodes of Subnetwork 2 with service rates $\mu_{c2}(k) = \lambda_{SN-2}(k)$.

STEP 4 The normalization constants of the reduced network are computed using Eq. (8.63):

$$G = G_1 \otimes G_2 = \begin{pmatrix} 1 \\ 1.5 \\ 1.75 \end{pmatrix} \otimes \begin{pmatrix} 1 \\ 0.646 \\ 0.313 \end{pmatrix} = \begin{pmatrix} 1 \\ 2.146 \\ 3.032 \end{pmatrix}.$$

By using Eq. (8.1) we get

$$G(1) = G_1(0) \cdot G_2(1) + G_1(1) \cdot G_2(0) = \underline{2.146},$$
$$G(2) = G_1(0) \cdot G_2(2) + G_1(1) \cdot G_2(1) + G_1(2) \cdot G_2(0) = \underline{3.032}.$$

STEP 5 Compute the performance measures:

Mean number of jobs at each node is computed using, Eq. (8.16):

$$\overline{K}_1 = \sum_{k=1}^{2} \left(\frac{e_1}{\mu_1} \right)^k \cdot \frac{G(2-k)}{G(2)} = \underline{1.038}, \quad \overline{K}_2 = \underline{0.436}, \quad \overline{K}_3 = \underline{0.273},$$
$$\overline{K}_4 = \underline{0.253}.$$

As we can see, the same results are obtained as in Example 8.8.

The FES method can also be applied to the case of multiple class queueing networks. The method is very flexible in the need for resources and lies between the two extremes: MVA and convolution. This flexibility is based on how the subnetworks are chosen.

8.4 SUMMARY

As shown in Table 8.5, the time requirement for convolution, MVA and FES is approximately the same. For larger numbers of nodes, MVA is much worse than FES and convolution. For FES, the storage requirement depends also on the number of subnetworks. The larger the number of subnetworks, the smaller is the storage requirement. In Table 8.6 the main advantages and disadvantages of the convolution algorithm, the MVA, and the FES method are summarized.

Problem 8.1 For the closed queueing network of Problems 7.3 and 7.4, compute the performance measures using the convolution method.

Problem 8.2 For a closed queueing network with $N = 2$ nodes and $R = 2$ job classes, compute the performance measures for each class and node using the convolution method. Node 1 is of Type 3 and node 2 is of Type 2. Each

Table 8.5   Storage requirement in number of storage elements and time requirement in number of operations [CoGe86]

Method	Storage Requirement	Time Requirement
Convolution	$O(\prod_{r=1}^{R} (K_r + 1))$	$O(2 \cdot R(N - 1) \cdot \prod_{r=1}^{R} (K_r + 1))$
MVA	$O(N \cdot \prod_{r=1}^{R} (K_r + 1))$	$\approx O(2 \cdot R(N - 1) \cdot \prod_{r=1}^{R} (K_r + 1))$
FES	$> O(3 \cdot \prod_{r=1}^{R} (K_r + 1))$	$\approx O(2 \cdot R(N - 1) \cdot \prod_{r=1}^{R} (K_r + 1))$

Table 8.6   Comparison of the four solution methods for product-form networks

Method	Advantages	Disadvantages
MVA	Mean values can be computed directly without computing the normalizing constant. But if required, the normalization constant and hence state probabilities can also be computed.	High storage requirement (for multiple classes and multiple servers). Overflow and underflow problems when computing the marginal probabilities for -/M/m nodes.
Convolution	Less storage requirement than MVA.	Normalizing constant (NC) has to be computed. Overflow or underflow problems when computing the NC.
FES	The time and storage requirement is reduced considerably when examining the influence of changing the parameters of an individual or a few nodes while the rest of the system parameters remain unchanged. Basis for solution techniques for non-product-form networks.	Multiple application of convolution or MVA. Network needs to be transformed.

class contains two jobs $K_1 = K_2 = 2$ and class switching is not allowed. The service rates are given as follows:

$$\mu_{11} = 0.4 \sec^{-1}, \quad \mu_{21} = 0.3 \sec^{-1}, \quad \mu_{12} = 0.2 \sec^{-1}, \quad \mu_{22} = 0.4 \sec^{-1}.$$

The routing probabilities are

$$p_{11,11} = p_{11,21} = 0.5, \quad p_{21,11} = p_{22,12} = p_{12,22} = 1.$$

Problem 8.3   Compute performance measures of the network in Problems 7.3 and 7.4 again, with the MVA.

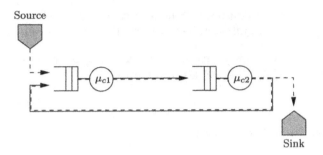

Fig. 8.19 Mixed queueing network.

Problem 8.4   For a mixed queueing network with $N = 2$ Type 2 nodes and $R = 4$ classes (see Fig. 8.19), compute all performance measures per node and class. The system parameters are given as follows:

$$\lambda_1 = 0.2\,\text{sec}^{-1}, \quad \lambda_2 = 0.1\,\text{sec}^{-1}, \quad K_2 = K_3 = 2,$$

$$\mu_{11} = 1.5\,\text{sec}^{-1}, \quad \mu_{12} = 1\,\text{sec}^{-1}, \quad \mu_{13} = 3\,\text{sec}^{-1}, \quad \mu_{14} = 2\,\text{sec}^{-1},$$

$$\mu_{21} = 2\,\text{sec}^{-1}, \quad \mu_{22} = 0.5\,\text{sec}^{-1}, \quad \mu_{23} = 2\,\text{sec}^{-1}, \quad \mu_{24} = 1\,\text{sec}^{-1},$$

$$p_{21,11} = p_{22,12} = p_{21,0} = p_{22,0} = 0.5,$$

$$p_{0,11} = p_{0,12} = p_{11,21} = p_{12,22} = p_{13,23} = p_{14,24} = p_{23,13} = p_{24,14} = 1.$$

Problem 8.5   Consider a closed queueing network, consisting of a Type 3 node and a Type 1 node ($m_2 = 2$). This model represents a simple two-processor system with terminals. The system parameters are given as:

$$K = 3, \quad p_{12} = p_{21} = 1, \quad \frac{1}{\mu_1} = 2\,\text{sec}, \quad \frac{1}{\mu_2} = 1\,\text{sec}.$$

(a) Apply the mean value analysis to determine the performance measures of each node, especially throughput and mean response time of the terminals.

(b) Determine for both nodes the load-dependent service rates and analyze the network again, using the load-dependent MVA.

Problem 8.6   Compute the performance measures for the network given in Problem 8.2 using the RECAL algorithm.

Problem 8.7   Consider a simple central-server model with $N = 3$ Type 1 nodes ($m_i = 1$) and $K = 3$ jobs. The service rates are given as follows:

$$\mu_1 = 1\,\text{sec}^{-1}, \quad \mu_2 = 0.65\,\text{sec}^{-1}, \mu_3 = 0.75\,\text{sec}^{-1},$$

and the routing probabilities are

$$p_{12} = 0.6, \quad p_{21} = p_{31} = 1, \quad p_{13} = 0.4.$$

Combine nodes 2 and 3 into a FES node. Use the FES method and compare the results with the results obtained without the use of FES method.

Problem 8.8   Consider a closed queueing network with $N = 4$ Type 1 nodes and $K = 3$ jobs. The service rates are

$$\mu_1 = 4\,\text{sec}^{-1}, \quad \mu_2 = 3\,\text{sec}^{-1}, \quad \mu_3 = 2\,\text{sec}^{-1}, \quad \mu_4 = 1\,\text{sec}^{-1},$$

and the routing probabilities are

$$p_{23} = p_{24} = 0.5, \quad p_{12} = p_{31} = p_{41} = 1.$$

For computing the performance measures, use the extended FES method by combining nodes 1 and 2 in subsystem 1, and nodes 3 and 4 into subsystem 2.

9

Approximation Algorithms for Product-Form Networks

In Chapter 8, several efficient algorithms for the exact solution of queueing networks are introduced. However, the memory requirements and computation time of these algorithms grows exponentially with the number of job classes in the system. For computationally difficult problems of networks with a large number of job classes, we resort to approximation methods. In Sections 9.1, 9.2, and 9.3 we introduce methods for obtaining such approximate results.

The first group of methods is based on the MVA. The approximate methods that we present need much less memory and computation time than the exact MVA and yet give very accurate results.

The second approach is based on the idea that the mean number of jobs at a network node can be computed approximately if the utilization ρ_i of this node is known. The sum of all the equations for the single nodes leads to the so-called *system equation*. By solving this equation we get approximate performance measures for the whole queueing network.

In some cases it is enough to have upper and lower bounds to make some statements about the network. Therefore, the third approach (Section 9.4) deals with methods on how to get upper and lower bounds on performance measures of a queueing network.

For other techniques that have been developed to analyze very large networks see [ChYu83, SLM86, HsLa89].

9.1   APPROXIMATIONS BASED ON THE MVA

The fundamental equation of the MVA (8.31) describes the relation between the mean response time of a node when there are k jobs at the node and the mean number of jobs in that node with one job less in the network. Therefore, to solve a queueing network it is necessary to iterate over all jobs in the network, starting from 0 to the maximum number K. For this reason, an implementation of the MVA requires a lot of computation time and main memory. For multiclass queueing networks with a large number of jobs in the system, we will very quickly reach the point at which the system runs out of memory, especially if we have multiserver nodes.

An alternative approach is to approximate the fundamental equation of the MVA in a way that the mean response time depends only on K and not on $K - 1$. Then it is not necessary to iterate over the whole population of jobs in the system but we have to improve only the performance measures iteratively, starting with an initial vector. This approach leads to a substantial reduction in computation time and memory requirement. In the following, two techniques based on this idea, are introduced.

9.1.1   Bard Schweitzer Approximation

Bard and Schweitzer [Bard79, Schw79] suggested an approximation of the MVA for single server queueing networks that is based on the following idea: Starting with an initial value $\overline{K}_{ir}(\mathbf{K})$ for the mean number of class-r jobs at node i for a given population vector $\mathbf{K}$, make an estimate of the mean number of jobs for a population vector $(\mathbf{K} - \mathbf{1}_s)$. This estimating needs to be done for all classes s. These $\overline{K}_{ir}(\mathbf{K} - \mathbf{1}_s)$ estimates are then used in the MVA to compute the mean response times, and only one iteration step is now needed to evaluate the performance measures for a given population vector $\mathbf{K}$. In particular, new values for the $\overline{K}_{ir}(\mathbf{K})$ are computed. The values we get are used to estimate the $\overline{K}_{ir}(\mathbf{K} - \mathbf{1}_s)$ again. This iteration stops if the values of the $\overline{K}_{ir}(\mathbf{K})$ in two iteration steps differ by less than the chosen error criterion ϵ.

Obviously, the problem with this MVA approximation is how to estimate the mean value $\overline{K}_{ir}(\mathbf{K} - \mathbf{1}_s)$, given $\overline{K}_{ir}(\mathbf{K})$. An approximate formula for the case of a very large network population was suggested by [Schw79] since in this case a reasonable explanation can be given as to why the mean number of jobs of class r at node i remains nearly the same if the number of jobs in the network is reduced by one. For the mean number of class-r jobs at node i given a population vector $(\mathbf{K} - \mathbf{1}_s)$, they suggest the approximation:

$$\overline{K}_{ir}(\mathbf{K} - \mathbf{1}_s) = \frac{(\mathbf{K} - \mathbf{1}_s)_r}{K_r}\overline{K}_{ir}(\mathbf{K}), \tag{9.1}$$

where

$$(\mathbf{K} - \mathbf{1}_s)_r = \begin{cases} K_r, & r \neq s, \\ K_r - 1, & r = s, \end{cases} \tag{9.2}$$

gives the number of class-r jobs when the network population $\mathbf{K}$ is reduced by one class-s job. If we assume that the jobs are initially equally distributed over the whole network, then the Bard–Schweitzer approximation can be described in the following steps:

STEP 1 Initialization. For $i = 1, \ldots, N$ and $s = 1, \ldots, R$:

$$\overline{K}_{is}(\mathbf{K}) = \frac{K_s}{N}.$$

STEP 2 For all $i = 1, \ldots, N$ and all $r, s = 1, \ldots, R$, compute the estimated values for the mean number of jobs for the population vector $(\mathbf{K} - \mathbf{1}_r)$:

$$\overline{K}_{is}(\mathbf{K} - \mathbf{1}_r) = \frac{(\mathbf{K} - \mathbf{1}_r)_s}{K_s} \cdot \overline{K}_{is}(\mathbf{K}).$$

STEP 3 Analyze the queueing network for a population vector $\mathbf{K}$ by using one iteration step of the MVA.

STEP 3.1 For $i = 1, \ldots, N$ and $r = 1, \ldots, R$, compute the mean response times:

$$\overline{T}_{ir}(\mathbf{K}) = \begin{cases} \dfrac{1}{\mu_{ir}} \left[1 + \displaystyle\sum_{s=1}^{R} \overline{K}_{is}(\mathbf{K} - \mathbf{1}_r) \right], & \text{Type 1,2,4 } (m_i = 1), \\ \dfrac{1}{\mu_{ir}}, & \text{Type 3.} \end{cases}$$

STEP 3.2 For $r = 1, \ldots, R$, compute the throughputs:

$$\lambda_r(\mathbf{K}) = \frac{K_r}{\displaystyle\sum_{i=1}^{N} \overline{T}_{ir}(\mathbf{K}) e_{ir}}.$$

STEP 3.3 For $i = 1, \ldots, N$ and $r = 1, \ldots, R$, compute the mean number of jobs:

$$\overline{K}_{ir}(\mathbf{K}) = \overline{T}_{ir}(\mathbf{K}) \lambda_r(\mathbf{K}) e_{ir}.$$

STEP 4 Check the stopping condition:
If there are no significant changes in the $\overline{K}_{ir}(\mathbf{K})$ values between the nth and $(n-1)$th iteration step, that is, when

$$\max_{i,r} \left| \overline{K}_{ir}^{(n)}(\mathbf{K}) - \overline{K}_{ir}^{(n-1)}(\mathbf{K}) \right| < \varepsilon,$$

for a suitable ε (here the superscript $^{(n)}$ denotes the values for the nth iteration step), then the iteration stops and all other performance values are computed. If the stopping criterion is not fulfilled, then return to Step 2.

This approximation is very easy to program and faster than the exact MVA. The memory requirements are proportional to the product of N and R. Therefore, this approximation needs considerably less memory than the exact MVA or the convolution method, especially for networks with a large number of job classes. A disadvantage of this approximation is that networks with multiple server nodes cannot be solved using it. We give a simple example of the use of the algorithm:

Example 9.1   We revisit the single class network of Example 8.3 (Fig. 8.5) with network parameters as shown in Table 9.1. For ε we choose the value 0.06. The analysis of the network $(K = 6)$ is carried out in the following

Table 9.1 Input parameters for Example 9.1

i	e_i	$1/\mu_i$	m_i
1	1	0.02	1
2	0.4	0.2	1
3	0.2	0.4	1
4	0.1	0.6	1

steps:

STEP 1 Initialization:

$$\overline{K}_1(K) = \overline{K}_2(K) = \overline{K}_3(K) = \overline{K}_4(K) = \frac{K}{N} = \underline{1.5}\,.$$

1. Iteration:

STEP 2 Estimated values, Eq. (9.1):

$$\overline{K}_1(K - 1) = \frac{K - 1}{K} \cdot \overline{K}_1(K) = \underline{1.25}\,,$$
$$\overline{K}_2(K - 1) = \overline{K}_3(K - 1) = \overline{K}_4(K - 1) = \underline{1.25}\,.$$

STEP 3 One MVA iteration:

STEP 3.1 Mean response times:

$$\overline{T}_1(K) = \frac{1}{\mu_1}\left[1 + \overline{K}_1(K - 1)\right] = \underline{0.045}\,, \quad \overline{T}_2(K) = \frac{1}{\mu_2}\left[1 + \overline{K}_2(K - 1)\right] = \underline{0.45}\,,$$

$$\overline{T}_3(K) = \frac{1}{\mu_3}\left[1 + \overline{K}_3(K-1)\right] = \underline{0.9}, \qquad \overline{T}_4(K) = \frac{1}{\mu_4}\left[1 + \overline{K}_4(K-1)\right] = \underline{1.35}.$$

STEP 3.2 Throughput:

$$\lambda(K) = \frac{K}{\sum\limits_{i=1}^{4} e_i \overline{T}_i(K)} = \underline{11.111}.$$

STEP 3.3 Mean number of jobs:

$$\overline{K}_1(K) = \overline{T}_1(K)\lambda(K)e_1 = \underline{0.5}, \quad \overline{K}_2(K) = \overline{T}_2(K)\lambda(K)e_2 = \underline{2},$$
$$\overline{K}_3(K) = \overline{T}_3(K)\lambda(K)e_3 = \underline{2}, \quad \overline{K}_4(K) = \overline{T}_4(K)\lambda(K)e_4 = \underline{1.5}.$$

STEP 4 Check the stopping condition:

$$\max_i \left|\overline{K}_i^{(1)}(K) - \overline{K}_i^{(0)}(K)\right| = \underline{1} \quad > 0.06.$$

2. Iteration:

STEP 2 Estimated values:

$$\overline{K}_1(K-1) = \frac{K-1}{K} \cdot 0.5 = \underline{0.417},$$
$$\overline{K}_2(K-1) = \underline{1.667}, \quad \overline{K}_3(K-1) = \underline{1.667}, \quad \overline{K}_4(K-1) = \underline{1.25}.$$

STEP 3 One MVA iteration:

$$\overline{T}_1(K) = \frac{1}{\mu_1}\left[1 + \overline{K}_1(K-1)\right] = \underline{0.028}$$
$$\overline{T}_2(K) = \underline{0.533}, \quad \overline{T}_3(K) = \underline{1.067}, \quad \overline{T}_4(K) = \underline{1.35},$$
$$\lambda(K) = \frac{K}{\sum\limits_{i=1}^{4} e_i \overline{T}_i(K)} = \underline{10.169},$$
$$\overline{K}_1(K) = \overline{T}_1(K)\lambda(K)e_1 = \underline{0.288},$$
$$\overline{K}_2(K) = \underline{2.169}, \quad \overline{K}_3(K) = \underline{2.169}, \quad \overline{K}_4(K) = \underline{1.373}.$$

STEP 4 Stopping condition:

$$\max_i \left|\overline{K}_i^{(2)}(K) - \overline{K}_i^{(1)}(K)\right| = \underline{0.212} \quad > 0.06.$$

3. Iteration:

STEP 2 Estimated values:

$$\overline{K}_1(K-1) = \frac{K-1}{K} \cdot 0.288 = \underline{0.240},$$
$$\overline{K}_2(K-1) = \underline{1.808}, \quad \overline{K}_3(K-1) = \underline{1.808}, \quad \overline{K}_4(K-1) = \underline{1.144}.$$

STEP 3 MVA iteration:

$$\overline{T}_1(K) = \underline{0.025}, \quad \overline{T}_2(K) = \underline{0.562}, \quad \overline{T}_3(K) = \underline{1.123}, \quad \overline{T}_4(K) = \underline{1.286},$$
$$\lambda(K) = \underline{9.955},$$
$$\overline{K}_1(K) = \underline{0.247}, \quad \overline{K}_2(K) = \underline{2.236}, \quad \overline{K}_3(K) = \underline{2.236}, \quad \overline{K}_4(K) = \underline{1.281}.$$

STEP 4 Stopping condition:

$$\max_i \left| \overline{K}_i^{(3)}(K) - \overline{K}_i^{(2)}(K) \right| = \underline{0.092} \quad > 0.06.$$

4. Iteration:

STEP 2

$$\overline{K}_1(K-1) = \underline{0.206}, \quad \overline{K}_2(K-1) = \underline{1.864},$$
$$\overline{K}_3(K-1) = \underline{1.864}, \quad \overline{K}_4(K-1) = \underline{1.067}.$$

STEP 3

$$\overline{T}_1(K) = \underline{0.024}, \quad \overline{T}_2(K) = \underline{0.573}, \quad \overline{T}_3(K) = \underline{1.145}, \quad \overline{T}_4(K) = \underline{1.240},$$
$$\lambda(K) = \underline{9.896},$$
$$\overline{K}_1(K) = \underline{0.239}, \quad \overline{K}_2(K) = \underline{2.267}, \quad \overline{K}_3(K) = \underline{2.267}, \quad \overline{K}_4(K) = \underline{1.227}.$$

STEP 4

$$\max_i \left| \overline{K}_i^{(4)}(K) - \overline{K}_i^{(3)}(K) \right| = \underline{0.053} \quad < 0.06.$$

Now, the iteration stops and the other performance measures are computed. For the throughputs of the nodes we use Eq. (8.40):

$$\lambda_1 = \lambda(K) \cdot e_1 = \underline{9.896}, \quad \lambda_2 = \underline{3.958}, \quad \lambda_3 = \underline{1.979}, \quad \lambda_4 = \underline{0.986}.$$

For the utilization of the nodes we get

$$\rho_1 = \frac{\lambda_1}{\mu_1} = \underline{0.198}, \quad \rho_2 = \underline{0.729}, \quad \rho_3 = \underline{0.729}, \quad \rho_4 = \underline{0.594}.$$

The final results are summarized in Table 9.2.

Table 9.2 Bard–Schweitzer approximation for Example 9.1

Node:	1	2	3	4
Mean response time $\overline{T}_i$	0.024	0.573	1.145	1.240
Throughput λ_i	9.896	3.958	1.979	0.986
Mean number of jobs $\overline{K}_i$	0.239	2.267	2.267	1.240
Utilization ρ_i	0.198	0.729	0.729	0.594

A comparison with the values from Example 8.2 shows that the Bard–Schweitzer approximation yields results that in this case are very close to the exact ones. The storage requirement of the Bard–Schweitzer approximation depends only on the number of nodes N and the number of classes R $(= O(N \times R))$ and is independent of the number of jobs. This is much smaller than those for convolution and MVA.[1] The accuracy of the Bard–Schweitzer approximation is not very good (the mean deviation from exact results is approximately 6%) but is sufficient for most applications. An essential assumption for the Bard–Schweitzer method is that the removal of a job from a class influences only the mean number of jobs in that class and not the mean number of jobs in other classes (see Eq. (9.1)). This assumption is not very good and we can construct examples for which the computed results deviate considerably from the exact ones. In Section 9.1.2, the *self-correcting approximation technique (SCAT)* is introduced, which does not have this disadvantage. Of course, the algorithm is more complex to understand and implement than the Bard–Schweitzer approximation.

9.1.2 Self-Correcting Approximation Technique

Based on the basic idea of Bard and Schweitzer, Neuse and Chandy [NeCh81, ChNe82] developed techniques that are an improvement over the Bard-Schweitzer approximation [ZES88]. The main idea behind their approach is demonstrated on queueing networks consisting only of single server nodes.

9.1.2.1 Single Server Nodes The SCAT algorithm for single server nodes [NeCh81] gives better results than the Bard–Schweitzer approximation, especially for networks with small populations, because in the SCAT approximation we not only estimate the mean number of jobs but also the change in the mean number of jobs from one iteration step to the next. This estimate is used in the approximate formula for the $\overline{K}_{ir}(\mathbf{K} - \mathbf{1}_s)$. We define $F_{ir}(\mathbf{K})$ as

---

[1] Recall that the storage requirements for convolution and MVA are $O\left(\prod\limits_{r=1}^{R} (K_r + 1) \right)$ and $O\left(N \cdot \prod\limits_{r=1}^{R} (K_r + 1) \right)$, respectively.

the contribution of class-r jobs at the ith node for a given population vector $\mathbf{K}$ as:

$$F_{ir}(\mathbf{K}) = \frac{\overline{K}_{ir}(\mathbf{K})}{K_r}. \tag{9.3}$$

If a class-s job is removed from the network, $D_{irs}(\mathbf{K})$ gives the change in this contribution:

$$\begin{aligned} D_{irs}(\mathbf{K}) &= F_{ir}(\mathbf{K} - \mathbf{1}_s) - F_{ir}(\mathbf{K}) \\ &= \frac{\overline{K}_{ir}(\mathbf{K} - \mathbf{1}_s)}{(\mathbf{K} - \mathbf{1}_s)_r} - \frac{\overline{K}_{ir}(\mathbf{K})}{K_r}, \end{aligned} \tag{9.4}$$

where $(\mathbf{K} - \mathbf{1}_s)_r$ is defined as in Eq. (9.2). Because the values $\overline{K}_{ir}(\mathbf{K} - \mathbf{1}_s)$ are unknown, we are not able to compute the values $D_{irs}(\mathbf{K})$ as well. Therefore, we estimate values for the difference $D_{irs}(\mathbf{K})$ and then we approximate the $\overline{K}_{ir}(\mathbf{K} - \mathbf{1}_s)$ values by the following formula based on Eq. (9.4):

$$\overline{K}_{ir}(\mathbf{K} - \mathbf{1}_s) = (\mathbf{K} - \mathbf{1}_s)_r \cdot [F_{ir}(\mathbf{K}) + D_{irs}(\mathbf{K})]. \tag{9.5}$$

Note that if $D_{irs}(\mathbf{K}) = 0$ we get the Bard–Schweitzer approximation.

The *core algorithm* of SCAT approximation needs the estimated values for $D_{irs}(\mathbf{K})$ as input and, in addition, estimates for the mean values $\overline{K}_{ir}(\mathbf{K})$. The core algorithm can be described in the following three steps:

STEP C1 For $i = 1, \ldots, N$ and $r, s = 1, \ldots, R$, compute with Eq. (9.5) the estimated values for the mean number $\overline{K}_{ir}$ of jobs for the population $(\mathbf{K} - \mathbf{1}_s)$.

STEP C2 Analyze the queueing network by using one iteration step of the MVA as in Step 3 of the Bard–Schweitzer algorithm.

STEP C3 Check the stopping condition. If

$$\max_{i,r} = \frac{\left| \overline{K}_{ir}^{(n)}(\mathbf{K}) - \overline{K}_{ir}^{(n-1)}(\mathbf{K}) \right|}{K_r} < \varepsilon,$$

then stop the iteration, otherwise return to Step C1 (here the superscript (n) denotes the results of the nth iteration step). The use of $\varepsilon = (4000 + 16|K|)^{-1}$ as a suitable value is suggested by [NeCh81].

The SCAT algorithm produces estimates of the differences D_{irs} needed as input parameters of the core algorithm. So the core algorithm is run several times, and with the results of each run the differences $D_{irs}(\mathbf{K})$ are estimated. The estimated differences $D_{irs}(\mathbf{K})$ are computed by the SCAT algorithm, hence the name *self-correcting approximation technique*. To initialize the method, the jobs are distributed equally over all nodes of the network and the differences $D_{irs}(\mathbf{K})$ are set to zero. The SCAT algorithm can be described by the following four steps:

STEP 1 Use the core algorithm for the population vector **K**, and input values $\overline{K}_{ir}(\mathbf{K}) = K_r/N$ and $D_{irs}(\mathbf{K}) = 0$ for all i, r, s. The auxiliary values of the core algorithm are not changed by the SCAT algorithm.

STEP 2 Use the core algorithm for each population vector $(\mathbf{K} - \mathbf{1_j})$ for $j = 1, \ldots, R$, with the input values $\overline{K}_{ir}(\mathbf{K} - \mathbf{1_j}) = (\mathbf{K} - \mathbf{1_j})_r/N$ and $D_{irs}(\mathbf{K} - \mathbf{1_j}) = 0$ for all i, r, s.

STEP 3 For all $i = 1, \ldots, N$ and $r, s = 1, \ldots, R$, compute the estimated values for the $F_{ir}(\mathbf{K})$ and $F_{ir}(\mathbf{K} - \mathbf{1_s})$, respectively, from Eq. (9.3) and the estimated values for the $D_{irs}(\mathbf{K})$ from Eq. (9.4).

STEP 4 Use the core algorithm for the population vector **K** with values $\overline{K}_{ir}(\mathbf{K})$ computed in Step 1, and values $D_{irs}(\mathbf{K})$, computed in Step 3. Finally, compute all other performance measures using Eqs. (7.21)–(7.31).

The memory requirement and computation time of the SCAT algorithm grows as $O(N \cdot R^2)$ and produces, in general, very good results. Now, the algorithm is used on an example:

Example 9.2 Consider the central-server model of Example 8.3 again, but now we use the SCAT algorithm to solve the network.

STEP 1 Use the core algorithm with $K = 6$ jobs in the network and the input data $\overline{K}_i(K) = K/N = 1.5$ and $D_i(K) = 0$, for $i = 1, \ldots, 4$. With these assumptions and $\varepsilon = 0.01$, the usage of the core algorithm will produce the same results as the Bard–Schweitzer algorithm as in Example 9.1. Therefore, we use those results directly:

$$\overline{T}_1(6) = \underline{0.024}, \quad \overline{T}_2(6) = \underline{0.573}, \quad \overline{T}_3(6) = \underline{1.145}, \quad \overline{T}_4(6) = \underline{1.240},$$
$$\lambda(6) = \underline{9.896},$$
$$\overline{K}_1(6) = \underline{0.239}, \quad \overline{K}_2(6) = \underline{2.267}, \quad \overline{K}_3(6) = \underline{2.267}, \quad \overline{K}_4(6) = \underline{1.227}.$$

STEP 2 Use the core algorithm for $(K - 1) = 5$ jobs and the inputs $\overline{K}_i(5) = 5/N = 1.25$ and $D_i(5) = 0$, for $i = 1, \ldots, 4$.

STEP C1 Compute the estimated values for the $\overline{K}_i(K - 1)$ from Eq. (9.5):

$$\overline{K}_i(4) = 4 \cdot \frac{\overline{K}_i(5)}{5} = \underline{1} \quad \text{for } i = 1, \ldots, 4.$$

STEP C2 One step of MVA:

$$\overline{T}_1(5) = \frac{1}{\mu_1} \left[1 + \overline{K}_1(4) \right] = \underline{0.04}, \quad \overline{T}_2(5) = \underline{0.4}, \quad \overline{T}_3(5) = \underline{0.8}, \quad \overline{T}_4(5) = \underline{1.2},$$

$$\lambda(5) = \frac{5}{\sum_{i=1}^{4} e_i \overline{T}_i(5)} = 10.417, \quad \overline{K}_1(5) = \overline{T}_1(5)\lambda(5)e_1 = \underline{0.417},$$

$$\overline{K}_2(5) = \underline{1.667}, \quad \overline{K}_3(5) = \underline{1.667}, \quad \overline{K}_4(5) = \underline{1.25}.$$

STEP C3   Check the stopping condition:

$$\max_i \frac{\left|\overline{K}_i^{(1)}(5) - \overline{K}_i^{(0)}(5)\right|}{5} = \underline{0.166} \quad > \varepsilon.$$

The iteration must be started again with Step C1.

$$\vdots$$

For the sake of brevity, we do not present all the other iterations but give the results after the last (fourth) iteration:

$$\overline{T}_1(5) = \underline{0.024}, \quad \overline{T}_2(5) = \underline{0.495}, \quad \overline{T}_3(5) = \underline{0.989}, \quad \overline{T}_4(5) = \underline{1.123},$$
$$\lambda(5) = \underline{9.405},$$
$$\overline{K}_1(5) = \underline{0.222}, \quad \overline{K}_2(5) = \underline{1.861}, \quad \overline{K}_3(5) = \underline{1.861}, \quad \overline{K}_4(5) = \underline{1.056}.$$

STEP 3   Compute with Eq. (9.4) the estimated values for the differences $D_i(K)$:

$$D_1(6) = \frac{\overline{K}_1(5)}{5} - \frac{\overline{K}_1(6)}{6} = \frac{0.222}{5} - \frac{0.239}{6} = \underline{4.7 \cdot 10^{-3}},$$

$$D_2(6) = \frac{\overline{K}_2(5)}{5} - \frac{\overline{K}_2(6)}{6} = \underline{-5.7 \cdot 10^{-3}},$$

$$D_3(6) = \frac{\overline{K}_3(5)}{5} - \frac{\overline{K}_3(6)}{6} = \underline{-5.7 \cdot 10^{-3}},$$

$$D_4(6) = \frac{\overline{K}_4(5)}{5} - \frac{\overline{K}_4(6)}{6} = \underline{6.72 \cdot 10^{-3}}.$$

STEP 4   Use the core algorithm for $K = 6$ jobs in the network with the $\overline{K}_i(K)$ values from Step 1 and the $D_i(K)$ values from Step 3 as inputs.

STEP C1   Estimated values:

$$\overline{K}_1(5) = 5 \cdot \left[\frac{\overline{K}_1(6)}{6} + D_1(6)\right] = \underline{0.222}, \quad \overline{K}_2(5) = \underline{1.861},$$
$$\overline{K}_3(5) = \underline{1.861}, \qquad\qquad\qquad\qquad \overline{K}_4(5) = \underline{1.056}.$$

STEP C2   One step of MVA:

$$\overline{T}_1(6) = \frac{1}{\mu_1}\left[1 + \overline{K}_1(5)\right] = \underline{0.024}, \quad \overline{T}_2(6) = \underline{0.572}, \quad \overline{T}_3(6) = \underline{1.144},$$
$$\overline{T}_4(6) = \underline{1.234},$$
$$\lambda(6) = \underline{9.908},$$
$$\overline{K}_1(6) = \overline{T}_1(6)\lambda(6)e_1 = \underline{0.242}, \quad \overline{K}_2(6) = \underline{2.267}, \quad \overline{K}_3(6) = \underline{2.267},$$
$$\overline{K}_4(6) = \underline{1.223}.$$

STEP C3   Stopping condition:

$$\max_i \frac{\left|\overline{K}_i^{(1)}(6) - \overline{K}_i^{(0)}(6)\right|}{6} = \underline{8.128 \cdot 10^{-3}} < \varepsilon,$$

is fulfilled and, therefore the iteration stops and the preceding results are the final results from the SCAT algorithm.

The exact results (MVA) are (see Example 8.3):

$$\overline{K}_1(6) = \underline{0.244}, \quad \overline{K}_2(6) = \underline{2.261}, \quad \overline{K}_3(6) = \underline{2.261}, \quad \overline{K}_4(6) = \underline{1.234}.$$

9.1.2.2 Multiple Server Nodes   If we wish to analyze networks with multiple server nodes, then from the MVA analysis, Eq. (8.31), we need not only the $\overline{K}_{ir}$ values but also the conditional marginal probabilities $\pi_i(j \mid \mathbf{K} - \mathbf{1}_r)$. Similarly, for the SCAT algorithm, we need estimates not only for the values $\overline{K}_{ir}(\mathbf{K} - \mathbf{1}_s)$ but also for the probabilities $\pi_i(j \mid (\mathbf{K} - \mathbf{1}_r)$. In [NeCh81] a technique is suggested to estimate the marginal probabilities by putting the probability mass as close as possible to the mean number of jobs in the network. Thus, for example, if $\overline{K}_i(K - 1) = 2.6$, then the probability that there are two jobs at node i is estimated to be $\pi_i(2 \mid K - 1) = 0.4$, and the probability that there are three jobs at node i is estimated to be $\pi_i(3 \mid K - 1) = 0.6$. This condition can be described by the following formulae. We define

$$\text{ceiling}_{ir} = \left\lceil \overline{K}_i(\mathbf{K} - \mathbf{1}_r)\right\rceil, \tag{9.6}$$
$$\text{floor}_{ir} = \left\lfloor \overline{K}_i(\mathbf{K} - \mathbf{1}_r)\right\rfloor, \tag{9.7}$$

with

$$\overline{K}_i = \sum_{s=1}^{R} \overline{K}_{is},$$

and estimate

$$\pi_i(\text{floor}_{ir} \mid \mathbf{K} - \mathbf{1}_r) = \text{ceiling}_{ir} - \overline{K}_i(\mathbf{K} - \mathbf{1}_r), \tag{9.8}$$

$$\pi_i(\text{ceiling}_{ir} \mid \mathbf{K} - \mathbf{1}_r) = 1 - \pi_i(\text{floor}_{ir} \mid \mathbf{K} - \mathbf{1}_r), \tag{9.9}$$

$$\pi_i(j \mid \mathbf{K} - \mathbf{1}_r) = 0 \quad \text{for } j < \text{floor}_{ir} \text{ and } j > \text{ceiling}_{ir}. \tag{9.10}$$

Now we need to modify the core algorithm of SCAT to analyze networks with multiple server nodes. This modified core algorithm, which takes suitable estimates of $D_{irs}(\mathbf{K})$ and $\overline{K}_{ir}$ as inputs, can be described in the following steps:

STEP C1 For $i = 1, \ldots, N$ and $r, s = 1, \ldots, R$, compute estimates for the mean values $\overline{K}_{ir}(\mathbf{K} - \mathbf{1}_s)$ using Eq. (9.5). Compute using Eqs. (9.7)–(9.10), for each multiple server node i, the estimated values of the marginal probabilities $\pi_i(j \mid \mathbf{K} - \mathbf{1}_r), j = 0, \ldots, (m_i - 2)$ and $r = 1, \ldots, R$.

STEP C2 Analyze the queueing network using one step of the MVA via Eqs. (8.38), (8.39), and (8.41).

STEP C3 Check the stopping condition:

$$\max_{i,r} \frac{\left| \overline{K}_{ir}^{(n)}(\mathbf{K}) - \overline{K}_{ir}^{(n-1)}(\mathbf{K}) \right|}{K_r} < \varepsilon.$$

If this condition is not fulfilled, then return to Step C1.

The SCAT algorithm, which computes the necessary estimated input values for the differences $D_{irs}(\mathbf{K})$ and mean values $\overline{K}_{ir}(\mathbf{K})$ for the core algorithm, differs from the algorithm for networks with single server nodes only in the usage of the modified core algorithm, as follows:

STEP 1 Use the modified core algorithm for the population vector $\mathbf{K}$, where all $D_{irs}(\mathbf{K})$ values are set to 0.

STEP 2 Use the modified core algorithm for each population vector $(\mathbf{K} - \mathbf{1}_j)$.

STEP 3 Compute the F_{irs} and D_{irs} values from the previous results (see Eqs. (9.3) and (9.4)).

STEP 4 Use the modified core algorithm with the computed F and D values for the population vector $\mathbf{K}$.

Example 9.3   The closed network from Example 8.4 (Fig. 8.6) is now analyzed with the SCAT algorithm. The network contains $N = 4$ nodes and $K = 3$ jobs. The network parameters are given in Table 9.3. The analysis of the network is carried in the following four steps:

STEP 1 The modified core algorithm for $K = 3$ jobs in the network with the input parameters $\overline{K}_i(K) = K/N = 0.75$ and $D_i(K) = 0$ for $i = 1, \ldots, 4$ is executed as follows:

Table 9.3 Input parameters for the network of Example 9.3

i	e_i	$1/\mu_i$	m_i
1	1	0.5	2
2	0.5	0.6	1
3	0.5	0.8	1
4	1	1	∞

STEP C1 Estimate values for the $\overline{K}_i(K-1)$ from Eq. (9.5):

$$\overline{K}_i(2) = 2 \cdot \frac{\overline{K}_i(3)}{3} = \underline{0.5} \quad \text{for } i = 1, \dots, 4 .$$

Estimate values for $\pi_1(0 \mid K-1)$. Because of $\text{floor}_1 = \lfloor \overline{K}_1(K-1) \rfloor = \underline{0}$ and $\text{ceiling}_1 = \text{floor}_1 + 1 = \underline{1}$, we get

$$\pi_1(0 \mid 2) = \text{ceiling}_1 - \overline{K}_1(2) = \underline{0.5} .$$

STEP C2 One step of MVA:

$$\overline{T}_1(3) = \frac{1}{\mu_1 m_1} \left[1 + \overline{K}_1(2) + \sum_{j=0}^{m_1-2} (m_1 - j - 1) \cdot \pi_1(j \mid 2) \right]$$

$$= \frac{1}{\mu_1 m_1} \left[1 + \overline{K}_1(2) + (m_1 - 1) \cdot \pi_1(0 \mid 2) \right] = \underline{0.5} ,$$

$$\overline{T}_2(3) = \frac{1}{\mu_2} \left[1 + \overline{K}_2(2) \right] = \underline{0.9}, \quad \overline{T}_3(3) = \frac{1}{\mu_3} \left[1 + \overline{K}_3(2) \right] = \underline{1.2} ,$$

$$\overline{T}_4(3) = \frac{1}{\mu_4} = \underline{1}. \quad \lambda(3) = \frac{3}{\sum_{i=1}^{4} e_i \overline{T}_i(3)} = \underline{1.176} ,$$

$$\overline{K}_1(3) = \overline{T}_1(3)\lambda(3)e_1 = \underline{0.588}, \quad \overline{K}_2(3) = \underline{0.529} ,$$

$$\overline{K}_3(3) = \underline{0.706}, \quad \overline{K}_4(3) = \underline{1.176} .$$

STEP C3 Check the stopping condition:

$$\max_i \frac{\left| \overline{K}_i^{(1)}(3) - \overline{K}_i^{(0)}(3) \right|}{3} = \underline{0.142} > \varepsilon = 0.01 .$$

STEP C1 Estimate values for the $\overline{K}_i(K-1)$:

$$\overline{K}_1(2) = 2 \cdot \left(\frac{\overline{K}_1(3)}{3} \right) = \underline{0.392}, \quad \overline{K}_2(2) = \underline{0.353} ,$$

$$\overline{K}_3(2) = \underline{0.471}, \qquad\qquad \overline{K}_4(2) = \underline{0.784}.$$

Estimate values for $\pi_1(0 \mid K - 1)$. With $\text{floor}_1 = \lfloor \overline{K}_1(K - 1) \rfloor = \underline{0}$ and $\text{ceiling}_1 = \text{floor}_1 + 1 = \underline{1}$, we get

$$\pi_1(0 \mid 2) = \text{ceiling}_1 - \overline{K}_1(2) = \underline{0.608}.$$

STEP C2 One step of MVA:

$$\vdots$$

After three iteration steps, we get the following results for the mean number of jobs:

$$\overline{K}_1(3) = \underline{0.603}, \quad \overline{K}_2(3) = \underline{0.480}, \quad \overline{K}_3(3) = \underline{0.710}, \quad \overline{K}_4(3) = \underline{1.207}.$$

STEP 2 The modified core algorithm for $(K - 1) = 2$ jobs in the network with the input parameters $\overline{K}_i(2) = 2/N = 0.5$ and $D_i(2) = 0$ for $i = 1,\ldots,4$ is executed. After three iteration steps, for the mean number of jobs we have

$$\overline{K}_1(2) = \underline{0.430}, \quad \overline{K}_2(2) = \underline{0.296}, \quad \overline{K}_3(2) = \underline{0.415}, \quad \overline{K}_4(2) = \underline{0.859}.$$

STEP 3 Determine the estimates for the F_i and D_i values using Eqs. (9.3) and (9.4), respectively:

$$F_1(3) = \frac{\overline{K}_1(3)}{3} = \underline{0.201}, \quad F_1(2) = \frac{\overline{K}_1(2)}{2} = \underline{0.215},$$

$$F_2(3) = \frac{\overline{K}_2(3)}{3} = \underline{0.160}, \quad F_2(2) = \frac{\overline{K}_2(2)}{2} = \underline{0.148},$$

$$F_3(3) = \frac{\overline{K}_3(3)}{3} = \underline{0.237}, \quad F_3(2) = \frac{\overline{K}_3(2)}{2} = \underline{0.208},$$

$$F_4(3) = \frac{\overline{K}_4(3)}{3} = \underline{0.402}, \quad F_4(2) = \frac{\overline{K}_4(2)}{2} = \underline{0.430}.$$

Therefore, we get

$$D_1(3) = F_1(2) - F_1(3) = \underline{0.014},$$
$$D_2(3) = F_2(2) - F_2(3) = \underline{-0.012},$$
$$D_3(3) = F_3(2) - F_3(3) = \underline{-0.029},$$
$$D_4(3) = F_4(2) - F_4(3) = \underline{0.027}.$$

STEP 4 Execute the modified core algorithm for $K = 3$ jobs in the network with values $\overline{K}_i(K)$ from Step 1 and values $D_i(K)$ from Step 3 as inputs.

STEP C1 Estimate values for the $\overline{K}_i(K - 1)$:

$$\overline{K}_1(2) = 2\left[\frac{\overline{K}_1(3)}{3} + D_1(3)\right] = \underline{0.430}, \quad \overline{K}_2(2) = \underline{0.296},$$

$$\overline{K}_3(2) = \underline{0.415}, \qquad\qquad \overline{K}_4(2) = \underline{0.859}.$$

Estimate value for

$$\text{floor}_1 = \lfloor \overline{K}_1(2) \rfloor = \underline{0}, \quad \text{ceiling}_1 = \text{floor}_1 + 1 = \underline{1},$$
$$\pi_1(0 \mid 2) = \text{ceiling}_1 - \overline{K}_1(2) = \underline{0.570}.$$

$$\vdots$$

After two iterations, the modified core algorithm yields the following final results:

$$\overline{T}_1(3) = \underline{0.5}, \quad \overline{T}_2(3) = \underline{0.776}, \quad \overline{T}_3(3) = \underline{1.122}, \quad \overline{T}_4(3) = \underline{1},$$
$$\lambda(3) = \underline{1.224},$$
$$\overline{K}_1(3) = \underline{0.612}, \quad \overline{K}_2(3) = \underline{0.475}, \quad \overline{K}_3(3) = \underline{0.687}, \quad \overline{K}_4(3) = \underline{1.224}.$$

The exact results are (see Example 8.4):

$$\overline{T}_1(3) = \underline{0.512}, \quad \overline{T}_2(3) = \underline{0.776}, \quad \overline{T}_3(3) = \underline{1.127}, \quad \overline{T}_4(3) = \underline{1},$$
$$\lambda(3) = \underline{1.217},$$
$$\overline{K}_1(3) = \underline{0.624}, \quad \overline{K}_2(3) = \underline{0.473}, \quad \overline{K}_3(3) = \underline{0.686}, \quad \overline{K}_4(3) = \underline{1.217}.$$

9.1.2.3 Extended SCAT Algorithm The accuracy of the SCAT algorithm for networks with multiple server nodes depends not only on the estimates of the $\overline{K}_{ir}(\mathbf{K} - \mathbf{1}_s)$ values, but also on the approximation of the conditional marginal probabilities $\pi_i(j \mid \mathbf{K} - \mathbf{1}_r)$. The suggestion of [NeCh81], to distribute the whole probability mass only between the two values adjacent to the mean number of jobs $\overline{K}_i(\mathbf{K} - \mathbf{1}_r)$, is not very accurate. If, for example, $\overline{K}_i(\mathbf{K} - \mathbf{1}_r) = 2.6$, then the probability of having 0, 1, 4, and 5 jobs at the node cannot be neglected. Especially in the case when values $\overline{K}_i(\mathbf{K} - \mathbf{1}_r)$ are close to m_i (number of identical service units at the ith node), the discrepancy can be large. In this case, all relevant probabilities are zero because the values $\pi_i(j \mid \mathbf{K} - \mathbf{1}_r)$ for computing the mean response time are determined for $j = 0, \ldots, (m_i - 2)$ only. In [AkBo88a] an improvement over this approximation of the conditional probabilities is recommended. In Akyildiz and Bolch's SCAT scheme, the marginal probabilities are spread over the whole range $0, \ldots, \overline{K}_i(\mathbf{K} - \mathbf{1}_r), \ldots, 2\lfloor \overline{K}_i(\mathbf{K}) \rfloor + 1$. They define the following two functions, the first function, PR, being a scaling function with

$$PR(1) = \alpha$$
$$PR(n) = \beta \cdot PR(n - 1) \quad \text{for } n = 2, \ldots, \max_r \{K_r\},$$

where the authors give 45 and 0.7, respectively, as the optimal values for α and β. The second function W is a weighting function that is defined for all $i = 1, \ldots, \max(K_r)$ as follows:

$$W(0, 0) = 1,$$

$$W(i,j) = W(i-1,j) - \frac{W(i-1,j)\cdot PR(i)}{100}, \quad \text{for } j = 1,\ldots,(i-1),$$

$$W(i,i) = 1 - \sum_{j=0}^{i-1} W(i,j).$$

The values of the weighting function $W(i,j)$, $j = 0,\ldots,5$ are given in Table 9.4.

Table 9.4   Values of the weighting function W

i	$W(i,0)$	$W(i,1)$	$W(i,2)$	$W(i,3)$	$W(i,4)$	$W(i,5)$
0	1.0					
1	0.55	0.45				
2	0.377	0.308	0.315			
3	0.294	0.240	0.245	0.221		
4	0.248	0.203	0.208	0.187	0.154	
5	0.222	0.181	0.185	0.166	0.138	0.108

In addition to these two functions, we need the values of floor, ceiling, and maxval, where floor and ceiling are defined as in Eqs. (9.7) and (9.6) and maxval is defined as follows:

$$\text{maxval}_{ir} = \min(2\,\text{floor}_{ir} + 1, m_i). \tag{9.11}$$

The term maxval describes the maximum variance of the probability mass. To compute the conditional probabilities, we divide the probability mass into $(\text{floor}_{ir} + 1)$ pairs, where the sum of each of these pairs has the combined probability mass $W(i,j)$. Formally this computation can be described as follows:

- For $j = 0,\ldots,\text{floor}_{ir}$, compute the conditional probabilities that there are j jobs at the ith node:

$$\pi_i(\text{floor}_{ir} - j \mid \mathbf{K} - \mathbf{1}_r) = W(\text{floor}_{ir}, \text{l\_dist})\frac{\text{upperval} - \overline{K}_i(\mathbf{K} - \mathbf{1}_r)}{\text{upperval} - \text{lowerval}}, \tag{9.12}$$

  with

$$\text{l\_dist} = \text{floor}_{ir} - j,$$
$$\text{upperval} = \text{ceiling}_{ir} + \text{l\_dist},$$
$$\text{lowerval} = \text{floor}_{ir} - \text{l\_dist} = j.$$

- For $j = \text{ceiling}_{ir},\ldots,\text{maxval}_{ir}$, we have

$$\pi_i(j \mid \mathbf{K} - \mathbf{1}_r) = W(\text{floor}_{ir}, \text{u\_dist}) - \pi_i(\text{floor}_{ir} - \text{u\_dist} \mid \mathbf{K} - \mathbf{1}_r), \tag{9.13}$$

  with $\text{u\_dist} = j - \text{ceiling}_{ir}$.

- For $j > \text{maxval}_{ir}$ we get

$$\pi_i(j \mid \mathbf{K} - \mathbf{1}_r) = 0 .$$

This approximation assumes that the $\overline{K}_i(\mathbf{K} - \mathbf{1}_r)$ values are smaller than $K_r/2$. If any of these values is greater than $(K_r - 1)/2$, then we must also make sure that the value of upperval is not outside the range $0, \ldots, (K_r - 1)$. In this case upperval is set to $(K_r - 1)$. More details of this SCAT technique can be found in [AkBo88a].

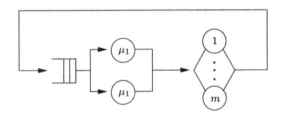

Fig. 9.1   A model of a multiprocessor system.

Example 9.4   Consider the network shown in Fig. 9.1. To compute the marginal probabilities we apply the improved technique of [AkBo88a]. The service time of a job at the ith node, $i = 1, 2$, is exponentially distributed with rates $\mu_1 = 0.5$ and $\mu_2 = 1$. The queueing disciplines are FCFS and IS, respectively. There are $K = 3$ jobs in the network and $e_1 = e_2 = 1$.

STEP 1 Execute the modified core algorithm for $K = 3$ jobs in the network and the input parameters $\overline{K}_i(K) = K/N = 1.5$ and $D_i(K) = 0$ for $i = 1, 2$.

STEP C1 Estimate values for the $\overline{K}_i(2)$, Eq. (9.5):

$$\overline{K}_i(2) = 2 \cdot \left(\frac{\overline{K}_i(3)}{3} \right) = \underline{1}, \quad i = 1, 2 .$$

Estimate value for $\pi_1(0 \mid K - 1)$. Since

$$\text{floor}_1 = \lfloor \overline{K}_1(2) \rfloor = \underline{1},$$
$$\text{ceiling}_1 = \text{floor}_1 + 1 = \underline{2},$$
$$\text{maxval}_1 = \min(2 \cdot \text{floor}_1 + 1, m_1) = \underline{2},$$

$\pi_1(0 \mid 2)$ is determined with

$$\text{l\_dist} = \text{floor}_1 - 0 = \underline{1}, \quad \text{upperval} = \text{ceiling}_1 + 1 = \underline{3},$$
$$\text{lowerval} = \text{floor}_1 - 1 = \underline{0}$$

from Eq. (9.12):

$$\pi_1(0 \mid 2) = W(1, 1) \cdot \frac{3 - \overline{K}_i(2)}{3} = \underline{0.3} .$$

STEP C2 One iteration step of the MVA:

$$\vdots$$

After four iterations, we get the following results:

$$\overline{K}_1(3) = \underline{2.187}, \quad \overline{K}_2(3) = \underline{0.813}.$$

STEP 2 Apply the modified core algorithm for $K = 2$ jobs with the input parameters $\overline{K}_i(2) = 1$ and $D_i(2) = 0$, for $i = 1, 2$.

STEP C1 Estimate values for the $\overline{K}_i(1)$:

$$\overline{K}_i(1) = \left(\frac{\overline{K}_i(2)}{2} \right) = \underline{0.5}, \quad i = 1, 2.$$

Estimate values for $\pi_1(0 \mid 1)$. With

$$\text{floor}_1 = \lfloor \overline{K}_1(1) \rfloor = \underline{0}, \quad \text{ceiling}_1 = \text{floor}_1 + 1 = \underline{1},$$
$$\text{maxval}_1 = \min(2 \cdot \text{floor}_1 + 1, m_1) = \underline{1},$$

we determine

$$\text{l\_dist} = \underline{0}, \quad \text{upperval} = \underline{1}, \quad \text{lowerval} = \underline{0},$$

and use these in Eq. (9.12) to compute

$$\pi_1(0 \mid 1) = W(0,0) \cdot \frac{1 - \overline{K}_1(1)}{1} = \underline{0.5}.$$

STEP C2 One iteration step of the MVA:

$$\vdots$$

After two iterations, we get the following results:

$$\overline{K}_1(2) = \underline{1.333}, \quad \overline{K}_2(3) = \underline{0.667}.$$

STEP C3 Estimate the difference with Eq. (9.4):

$$D_1(3) = F_1(2) - F_1(3) = \underline{-0.062}, \quad D_2(3) = F_2(2) - F_2(3) = \underline{0.062}.$$

STEP 3 Apply the modified core algorithm for $K = 3$ jobs in the network:

STEP C1 Estimate values for the $\overline{K}_i(2)$:

$$\overline{K}_1(2) = \underline{1.333}, \quad \overline{K}_2(2) = \underline{0.667}.$$

Estimate value for $\pi_1(0 \mid 2)$. With:

$$\text{floor}_1 = \lfloor \overline{K}_1(2) \rfloor = 1, \quad \text{ceiling}_1 = 2, \quad \text{maxval}_1 = 2,$$

we determine:

$$\text{l\_dist} = 1, \quad \text{upperval} = 3, \quad \text{lowerval} = 0,$$

and therefore:

$$\pi_1(0 \mid 2) = W(1,1) \cdot \frac{3 - \overline{K}_1(2)}{3} = 0.25.$$

$\boxed{\text{STEP C2}}$ One iteration step of the MVA:

$$\vdots$$

After two iterations, we get the final results:

$$\overline{T}_1(3) = 2.57, \quad \overline{T}_2(3) = 1, \quad \lambda(3) = 0.84, \quad \overline{K}_1(3) = 2.16, \quad \overline{K}_2(3) = 0.84.$$

A comparison with the exact values shows the accuracy of this technique for this example:

$$\overline{T}_1(3) = 2.45, \quad \overline{T}_2(3) = 1, \quad \lambda(3) = 0.87, \quad \overline{K}_1(3) = 2.13, \quad \overline{K}_2(3) = 0.87.$$

If we use the SCAT algorithm from Section 9.1.2.2, we get

$$\overline{T}_1(3) = 3, \quad \overline{T}_2(3) = 1, \quad \lambda(3) = 0.75, \quad \overline{K}_1(3) = 2.25, \quad \overline{K}_2(3) = 0.75.$$

The storage requirement of SCAT is independent of the number of jobs $(= O(N \cdot R^2))$ but higher than for the Bard–Schweitzer approximation, especially for systems with many job classes and many jobs. The storage requirement for SCAT is much less than those of MVA or convolution (see Table 9.5). The accuracy is much better for the SCAT algorithm (the average deviation from exact results is approximately 3%) than for the Bard–Schweitzer approximation and, therefore, in practice SCAT is used more often than MVA, Bard–Schweitzer, or convolution to solve product-form queueing networks.

Problem 9.1 Solve Problem 7.3 with the Bard–Schweitzer and the SCAT algorithms.

Problem 9.2 Solve Problem 7.4 with the SCAT and the extended SCAT algorithms.

Table 9.5 Storage requirement for convolution, MVA, Bard–Schweitzer, and SCAT Algorithms

Methods	Storage Requirement
Convolution	$O(\prod\limits_{r=1}^{R} (K_r + 1))$
MVA	$O(N \cdot \prod\limits_{r=1}^{R} (K_r + 1))$
Bard–Schweitzer	$O(N \cdot R)$
SCAT	$O(N \cdot R^2)$

9.2 SUMMATION METHOD

The summation method (SUM) is based on the simple idea that for each node of a network, the mean number of jobs at the node is a function of the throughput of this node:

$$\overline{K}_i = f_i(\lambda_i) \,. \tag{9.14}$$

For nodes with load-independent service rates, function f_i has the following properties:

- λ_i must be in the range $0 \le \lambda_i \le m_i \mu_i$. For IS nodes we have $0 < \lambda_i \le K \cdot \mu_i$, because $\overline{K}_i = \lambda_i / \mu_i$ and $\overline{K}_i \le K$.

- $f_i(\lambda_i) \le f(\lambda_i + \Delta\lambda_i)$ for $\Delta\lambda_i > 0$, that is, f_i is a monotone nondecreasing function of λ_i.

- $f_i(0) = 0$.

Monotonicity does not hold for nodes with general load-dependent service rates and must therefore be examined for each individual case. For multiple server nodes the monotonicity is maintained.

To analyze product-form queueing networks, [BFS87] suggest the following formulae:

$$f_i(\lambda_i) = \overline{K}_i = \begin{cases} \dfrac{\rho_i}{1 - \dfrac{K-1}{K}\rho_i}, & \text{Type 1,2,4 } (m_i = 1), \\[3mm] m_i\rho_i + \dfrac{\rho_i}{1 - \dfrac{K - m_i - 1}{K - m_i}\rho_i} \cdot P_{m_i}, & \text{Type 1 } (m_i > 1), \\[3mm] \dfrac{\lambda_i}{\mu_i}, & \text{Type 3.} \end{cases}$$

$$\tag{9.15}$$

The utilization ρ_i is computed using Eq. (7.21) and the waiting probability P_{m_i} with Eq. (6.28) or approximately with Eq. (6.100). It should be noted that Eq. (9.15) gives exact values only for Type 3 nodes while the equations for the other node types are approximations. The approximate equations for Type 1 (-/M/1), Type 2, and Type 4 nodes can be derived from Eq. (6.13), $\overline{K}_i = \rho_i/(1 - \rho_i)$, and the introduction of a correction factor $(K - 1)/K$, respectively. The correction factor is motivated by the fact that in the case of $\rho_i = 1$, all K jobs are at node i, hence, for $\rho_i = 1$ we have $\overline{K}_i = K$. The idea of the correction factor can also be found in the equation for Type 1 (-/M/m) nodes by considering Eq. (6.29). If we assume that the functions f_i are given for all nodes i of a closed network, then the following equation for the whole network can be obtained by summing over all functions f_i of the nodes in the network:

$$\sum_{i=1}^{N} \overline{K}_i = \sum_{i=1}^{N} f_i(\lambda_i) = K. \tag{9.16}$$

With the relation $\lambda_i = \lambda \cdot e_i$, an equation to determine the overall throughput λ of the system can be given as

$$\sum_{i=1}^{N} f_i(\lambda \cdot e_i) = g(\lambda) = K. \tag{9.17}$$

In the multiple class case, function (9.14) can be extended:

$$\overline{K}_{ir} = f_{ir}(\lambda_{ir}). \tag{9.18}$$

The function $f_{ir}(\lambda_{ir})$ has the following characteristics:

- f_{ir} is defined in the range $0 \leq \lambda_{ir} \leq \mu_{ir} \cdot m_i$ and, in case of an infinite server node, $0 \leq \lambda_{ir} \leq \mu_{ir} \cdot K_r$.

- f_{ir} is a monotone non decreasing function. $f_{ir}(\lambda_{ir}) \leq f_{ir}(\lambda_{ir} + \epsilon), \epsilon > 0$.

- $f_{ir}(0) = 0$.

For *multiclass product-form queueing networks*, the following formulae, as extensions of Eq. (9.15), are given:

$$\overline{K}_{ir} = \begin{cases} \dfrac{\rho_{ir}}{1 - \dfrac{K-1}{K}\rho_i}, & \text{Type 1,2,4 } (m_i = 1) \\[4mm] m_i\rho_{ir} + \dfrac{\rho_{ir}}{1 - \dfrac{K - m_i - 1}{K - m_i}\rho_i} \cdot P_{m_i}, & \text{Type 1 } (m_i > 1), \\[4mm] \dfrac{\lambda_{ir}}{\mu_{ir}}, & \text{Type 3,} \end{cases} \tag{9.19}$$

with:

$$\rho_i = \sum_{r=1}^{R} \rho_{ir}, \qquad (9.20)$$

$$K = \sum_{r=1}^{R} K_r. \qquad (9.21)$$

For the probability of waiting P_{m_i}, we can use the same formula as for the single class case (Eq. (6.28)) and Eq. (9.20) for the utilization ρ_i. Now we obtain a system of equations for the throughputs λ_r by summing over all functions f_{ir}:

$$\sum_{i=1}^{N} \overline{K}_{ir} = \sum_{i=1}^{N} f_{ir}(\lambda_{ir}) = K_r \quad (r = 1, \ldots, R). \qquad (9.22)$$

With $\lambda_{ir} = \lambda_r e_{ir}$ we obtain the system of equations

$$\sum_{i=1}^{N} f_{ir}(\lambda_r e_{ir}) = g_r(\lambda_r) = K_r \quad r = 1, \ldots, R. \qquad (9.23)$$

In the next section, we introduce solution algorithms to determine the throughput λ in the single class case or the throughputs λ_r in the multiple class case using the preceding fixed-point equations.

9.2.1   Single Class Networks

An algorithm for single class queueing networks to determine λ with the help of Eq. (9.17) and the monotonicity of $f_i(\lambda_i)$ has the following form:

STEP 1 Initialization. Choose $\lambda_l = 0$ as a lower bound for the throughput and $\lambda_u = \min_i \left\{ \dfrac{\mu_i m_i}{e_i} \right\}$ as an upper bound. For $-/G/\infty$ nodes, m_i must be replaced by K.

STEP 2 Use the bisection technique to determine λ:

STEP 2.1 Set $\lambda = \dfrac{\lambda_l + \lambda_u}{2}$.

STEP 2.2 Determine $g(\lambda) = \sum_{i=1}^{N} f_i(\lambda \cdot e_i)$, where the functions $f_i(\lambda \cdot e_i)$ are determined by using Eqs. (9.15) and (9.14).

STEP 2.3 If $g(\lambda) = K \pm \epsilon$, then stop the iteration and compute the performance measures of the network with the equations from Section 7.1. If $g(\lambda) > K$, set $\lambda_u = \lambda$ and go back to Step 2.1.

If $g(\lambda) < K$, set $\lambda_l = \lambda$ and go also back to Step 2.1.

The advantage of this method is that the performance measures can be computed very easily. Furthermore, if we use suitable functions f_i, then the number K of jobs in the network and the number of servers m_i does not affect the computation time. In the case of monotonicity, the convergence is always guaranteed. By using suitable functions $f_i(\lambda_i)$, this method can also be used for an approximate analysis of closed non-product-form networks (see Section 10.1.4). In [BFS87] and [AkBo88b], it is shown that the summation method is also well suited for solving optimization problems in queueing networks. In the following example we show the use of the summation method:

Example 9.5   Consider Example 8.4 from Section 8.2.1. This network is examined again to have a comparison between the exact values and the approximate ones as computed with the summation method. The network parameters are given in Table 9.6. The analysis of the network is carried out

Table 9.6 Input parameters for the network of Example 9.5

i	e_i	$1/\mu_i$	m_i
1	1	0.5	2
2	0.5	0.6	1
3	0.5	0.8	1
4	1	1	∞

in the following steps:

STEP 1 Initialization:

$$\lambda_l = \underline{0} \quad \text{and} \quad \lambda_u = \min_i \left\{ \frac{m_i \mu_i}{e_i} \right\} = \underline{2.5}.$$

STEP 2 Bisection:

STEP 2.1 $\lambda = \dfrac{\lambda_l + \lambda_u}{2} = \underline{1.25}.$

STEP 2.2 Computation of the functions $f_i(\lambda_i)$, $i = 1, 2, 3, 4$ with Eq. (9.15). For the $-/M/2$ node 1 we have

$$\rho_1 = \frac{\lambda \cdot e_1}{\mu_1 \cdot m_1} = \underline{0.3125} \quad \text{and} \quad P_{m_1} = \underline{0.149} \quad \text{(Eq. (6.28))},$$

and therefore

$$f_1(\lambda_1) = 2\rho_1 + \rho_1 P_{m_1} = \underline{0.672}.$$

Furthermore,

$$f_2(\lambda_2) = \frac{\rho_2}{1 - \frac{2}{3}\rho_2} = \underline{0.5} \quad \text{with } \rho_2 = \underline{0.375},$$

$$f_3(\lambda_3) = \frac{\rho_3}{1 - \frac{2}{3}\rho_3} = \underline{0.75} \quad \text{with } \rho_3 = \underline{0.5},$$

$$f_4(\lambda_4) = \frac{\lambda_4}{\mu_4} = \underline{1.25}.$$

Thus, we get

$$g(\lambda) = \sum_{i=1}^{N} f_i(\lambda_i) = \underline{3.172}.$$

STEP 2.3 Check the stopping condition. Because of $g(\lambda) > K$, we set $\lambda_u = \lambda = 1.25$ and go on with the iteration.

STEP 2.1

$$\lambda = \frac{\lambda_l + \lambda_u}{2} = \underline{0.625}.$$

STEP 2.2

$$f_1(\lambda_1) = 2\rho_1 + \rho_1 P_{m_1} = \underline{0.319} \quad \text{with } \rho_1 = \underline{0.156} \text{ and } P_{m_1} = \underline{0.042},$$
$$f_2(\lambda_2) = \underline{0.214} \qquad\qquad\quad \text{with } \rho_2 = \underline{0.1875},$$
$$f_3(\lambda_3) = \underline{0.3} \qquad\qquad\qquad \text{with } \rho_3 = \underline{0.25},$$
$$f_4(\lambda_4) = \underline{0.625}.$$

It follows that

$$g(\lambda) = \sum_{i=1}^{4} f_i(\lambda_i) = \underline{1.458}.$$

STEP 2.3 Because of $g(\lambda) < K$, we set $\lambda_l = \lambda = 0.625$.

STEP 2.1

$$\lambda = \frac{\lambda_l + \lambda_u}{2} = \frac{0.625 + 1.25}{2} = \underline{0.9375}.$$

$$\vdots$$

This iteration is continued until the value of $g(\lambda)$ equals (within $\epsilon = 0.001$) the number of jobs K in the system. To make the bisection clearer, we show the series of λ_l and λ_u values in Table 9.7.

The overall throughput of the network is $\lambda = \underline{1.193}$ as computed by this approximation method. Compare this with the exact value of the overall throughput, $\lambda = \underline{1.217}$. The approximate values of the throughputs and the mean number of jobs calculated using the SUM (Eq. (9.14)) and the exact values for this network calculated using MVA (see Example 8.4) are summarized in Table 9.8.

Table 9.7 Intervals for Example 9.5

Step:	0	1	2	3	4	5	6	7	8	9	10
λ_l	0	0	0.625	0.9375	1.094	1.172	1.172	1.191	1.191	1.191	1.191
λ_u	2.5	1.25	1.25	1.25	1.25	1.25	1.211	1.211	1.201	1.196	1.194

Table 9.8 Exact and approximate values for the throughputs λ_i and the number of jobs $\overline{K}_i$ for Example 9.5

i	1	2	3	4
$\lambda_{i_{\mathrm{SUM}}}$	1.193	0.596	0.596	1.193
$\lambda_{i_{\mathrm{MVA}}}$	1.218	0.609	0.609	1.218
$\overline{K}_{i_{\mathrm{SUM}}}$	0.637	0.470	0.700	1.193
$\overline{K}_{i_{\mathrm{MVA}}}$	0.624	0.473	0.686	1.217

9.2.2 Multiclass Networks

Now we assume R job classes and $\mathbf{K} = (K_1, K_2, \ldots, K_R)$ jobs in the network. Therefore, we get a system of R equations. In general, g_r are nonlinear functions of λ_r: $g_r(\lambda_1, \lambda_2, \ldots, \lambda_R)$. This is a coupled system of nonlinear equations in the throughput of the R job classes and, hence, it is not possible to solve each equation separately from the others. We first transform the system of equations into the fixed-point form:

$$\mathbf{x} = f(\mathbf{x}).$$

Under certain conditions f has only one fixed-point and can then be used to calculate $\mathbf{x}$ iteratively using successive substitution as follows:

$$\mathbf{x}^{(n+1)} = f(\mathbf{x}^{(n)}).$$

This method of solution is also called *fixed-point iteration*. In the case of the SUM, for the calculation of the throughputs λ_r, we get the following system of equations:

$$\sum_{i=1}^{N} \overline{K_{ir}} = \sum_{i=1}^{N} f_{ir}(\lambda_r \cdot e_{ir}) = K_r, \quad r = 1, \ldots, R. \tag{9.24}$$

To apply the fixed-point iteration, this equation can easily be transformed to

$$\lambda_r \sum_{i=1}^{N} \mathrm{fix}_{ir}(\lambda_r \cdot e_{ir}) = K_r,$$

or

$$\lambda_r = \frac{K_r}{\sum\limits_{i=1}^{N} \mathrm{fix}_{ir}(\lambda_r \cdot e_{ir})} = f_r(\lambda_1, \ldots, \lambda_R) = f_r(\boldsymbol{\lambda}), \tag{9.25}$$

see [BaTh94].

For the functions fix_{ir}, we get from Eq. (9.25)

$$\text{fix}_{ir} = \begin{cases} \dfrac{\dfrac{e_{ir}}{\mu_{ir}}}{1 - \dfrac{K-1}{K} \cdot \rho_i}, & \text{Type } 1,2,4 \; m_i = 1, \\[3em] \dfrac{e_{ir}}{\mu_{ir}} + \dfrac{\dfrac{e_{ir}}{m_i \cdot \mu_{ir}}}{1 - \dfrac{K - m_i - 1}{K - m_i} \cdot \rho_i} \cdot P_{m_i}(\rho_i), & \text{Type } 1 \; m_i > 1, \\[3em] \dfrac{e_{ir}}{\mu_{ir}}, & \text{Type } 3, \end{cases} \tag{9.26}$$

and the throughputs λ_r can be obtained by the following iteration:

STEP 1 Initialization:

$$\lambda_1 = \lambda_2 = \cdots = \lambda_R = 0.00001 ; \quad \epsilon = 0.00001 .$$

STEP 2 Compute the throughput λ_r for all $r = 1, \ldots, R$:

$$\lambda_r = \frac{K_r}{\displaystyle\sum_{i=1}^{N} \text{fix}_{ir}(\lambda_1, \ldots, \lambda_R)} .$$

STEP 3 Determine error norm:

$$e = \sqrt{\sum_{r=1}^{R} (\lambda_{r_n} - \lambda_{r_{n+1}})^2} .$$

If $e > \epsilon$, go to Step 2.

STEP 4 Compute other performance measures.

The Newton–Raphson method can also be used to solve the non-linear system of equations for the SUM [Hahn90]. There is no significant difference between fixed-point iteration and the Newton–Raphson method in our experience. The computing time is tolerable for small networks for both methods (below 1 second), but it increases dramatically for the Newton–Raphson method (up to several hours) while it is still tolerable for the fixed-point iteration (several seconds) [BaTh94]. Therefore, the fixed-point iteration should preferably be used. The relative error can be computed with the formula

$$\text{difference} = \frac{|\text{exact value} - \text{computed value}|}{\text{exact value}} \cdot 100,$$

and gives, for the previous example, an error of 2.6% for the overall through-put. Experiments have shown that the relative error for the SUM method is between 5% and 15%. It is clear that the results can be further improved if better functions f_i could be found.

Problem 9.3   Solve Problem 7.3 and Problem 7.4 using the SUM.

9.3   BOTTAPPROX METHOD

We have seen that, compared to the MVA, the SUM approximation needs much less memory and computation time. But if we take a closer look at it, we will see that the number of iterations to obtain reasonably accurate results is still very large for many networks. Now the question arises whether it is possible to choose the initial guess of throughput $\lambda^{(0)}$ so that the number of iterations can be reduced while still maintaining the accuracy of the results. A good initial value of throughput based on the indices of the bottleneck node is suggested by [BoFi93]. Recall that the node with the highest utilization ρ_i is called the bottleneck node. The performance measures of the bottleneck node are characterized by the index bott. The resulting approximation method is called *bottapprox (BOTT)*.

9.3.1   Initial Value of λ

For the throughput λ we have the relation

$$\lambda = \frac{\mu_i \cdot m_i}{e_i} \cdot \rho_i, \quad i = 1, \ldots, N,$$

with

$$0 \leq \rho_i \leq 1, \quad i = 1, \ldots, N.$$

In contrast to the SUM in the BOTT we assume that the utilization ρ_{bott} of the bottleneck node is very close to 1, and therefore we can choose for the initial value of the throughput $\lambda^{(0)}$:

$$\lambda^{(0)} = \min_{i \in \{1, \ldots N\}} \left\{ \frac{\mu_i \cdot m_i}{e_i} \right\}, \quad \text{bott} = \text{argmin} \left\{ \frac{\mu_i \cdot m_i}{e_i} \right\},$$

where argmin is the function that returns the index of the node with the minimum value.

9.3.2   Single Class Networks

The BOTT is an iterative method where for each node i in each iteration step, the mean number of jobs $\overline{K}_i$ is determined as a function of the chosen throughput $\lambda_i^{(k)}$:

$$\overline{K}_i = f_i(\lambda_i). \tag{9.27}$$

For the function f_i, the same approximations are used as in the SUM. By summing over all nodes, we get for the function $g(\lambda)$

$$g(\lambda) = \sum_{i=1}^{N} \overline{K}_i.$$ (9.28)

This sum $g(\lambda)$ is compared to the number of jobs K in the network and a correction factor:

$$\text{corr} = \frac{K}{g(\lambda)},$$ (9.29)

is determined. We assume that in each node i the error in computing the mean number of jobs $\overline{K}_i$ is the same, and therefore for each node the following relation holds:

$$\overline{K}_i^{(k+1)} = \overline{K}_i^{(k)} \cdot \frac{K}{g(\lambda)}.$$ (9.30)

For the bottleneck node the new mean number of jobs in the node is computed as follows:

$$\overline{K}_{\text{bott}}^{(k+1)} = \overline{K}_{\text{bott}}^{(k)} \cdot \frac{K}{g(\lambda)},$$ (9.31)

and the new utilization $\rho_{\text{bott}}^{(k+1)}$ can then be determined as a function of the new mean number of jobs $\overline{K}_{\text{bott}}^{(k+1)}$ at the bottleneck node using Eq. (9.15), that is:

$$\rho_{\text{bott}}^{(k+1)} = h(\overline{K}_{\text{bott}}^{(k+1)}).$$ (9.32)

Now the new value for the throughput λ at the bottleneck node can be computed as a function of the new utilization (Eqs. (7.23) and (7.25)):

$$\lambda^{(k+1)} = \frac{\rho_{\text{bott}}^{(k+1)} \cdot m_{\text{bott}} \cdot \mu_{\text{bott}}}{e_{\text{bott}}}.$$ (9.33)

If none of the following stopping conditions is fulfilled then the iteration is repeated:

- If the correction factor corr is in the interval $[(1 - \epsilon), \ldots, (1 + \epsilon)]$, then stop the iteration. Tests have shown that for $\epsilon = 0.025$ we get good results. For smaller ϵ, only the number of iterations increased but the accuracy of the results did not improve significantly.

- If in the computation the value of $\rho_{\text{bott}} = h(\overline{K}_{\text{bott}})$ is greater than 1, then set $\rho_{\text{bott}} = 1$ and stop the iteration.

- To prevent endless loops, the procedure is stopped when the number of iterations reaches a predefined upper limit.

The function h previously mentioned for different node types is defined as follows (see Eq. (9.15)):

1. For node types -/M/1-FCFS, -/G/1-PS, and -/G/1-LCFS-PR:

$$h(\overline{K}_{\text{bott}}) = \frac{\overline{K}_{\text{bott}}}{1 + \frac{K-1}{K} \cdot \overline{K}_{\text{bott}}}.$$

2. For node type -/G/∞-IS:

$$h(\overline{K}_{\text{bott}}) = \frac{\overline{K}_{\text{bott}}}{K}.$$

3. For node type -/M/m-FCFS it is not possible to give an exact expression for f_i and therefore an appropriate approximation is chosen (see [BoFi93]):

$$h(\overline{K}_{\text{bott}}) = \frac{b - \sqrt{c}}{2 \cdot f \cdot m_{\text{bott}}},$$

with

$$f = \frac{K - m_{\text{bott}} - 1}{K - m_{\text{bott}}},$$
$$b = \overline{K}_{\text{bott}} \cdot f + m_{\text{bott}} + P_{m_{\text{bott}}},$$
$$c = b^2 - 4 \cdot m_{\text{bott}} \cdot \overline{K}_{\text{bott}} \cdot f.$$

Now we give an algorithm for the BOTT:

STEP 1 Initialize:

$$\lambda = \min_{i \in \{1,\dots,N\}} \left\{ \frac{\mu_i \cdot m_i}{e_i} \right\}, \quad \text{bott} = \text{argmin}\left\{ \frac{\mu_i \cdot m_i}{e_i} \right\}.$$

STEP 2 Iteration:

STEP 2.1 Compute the following performance measures for $i = 1, \dots, N$:

$$\lambda_i = \lambda \cdot e_i, \quad \rho_i = \frac{\lambda_i}{m_i \cdot \mu_i}, \quad \overline{K}_i = f_i(\lambda).$$

STEP 2.2 Compute the new throughput λ:

$$g(\lambda) = \sum_{i=1}^{N} \overline{K}_i, \quad \overline{K}_{\text{bott}} = \overline{K}_{\text{bott}} \cdot \frac{K}{g(\lambda)},$$
$$\rho_{\text{bott}} = h(\overline{K}_{\text{bott}}), \quad \lambda = \frac{\rho_{\text{bott}} \cdot m_{\text{bott}} \cdot \mu_{\text{bott}}}{e_{\text{bott}}}.$$

STEP 2.3 Check the stopping condition:

$$\left| \frac{K}{g(\lambda)} \right| \leq \epsilon.$$

If it is fulfilled, then stop the iteration, otherwise return to Step 2.1.

STEP 3 For $i = 1, \ldots, N$ compute

$$\overline{K}_i = \overline{K}_i \cdot \frac{K}{g(\lambda)} \,.$$

To estimate the accuracy of the BOTT and to compare it with the SUM, we investigated 64 different product-form networks. The mean deviation from the exact solution for the overall throughput λ and for the mean response time $\overline{T}$ was 2%. The corresponding value for the summation method was about 3%. The number of iterations of the SUM were about three times the number of iterations of the BOTT. For details see Table 9.9.

Table 9.9 Deviation in percent (%) from the exact values (obtained by MVA)

	Error in λ		Error in $\overline{T}$		Number of Iterations	
	SUM	BOTT	SUM	BOTT	SUM	BOTT
min	0	0	0	0	1	1
mean	2.60	2.10	2.70	2.10	6.60	2.30
max	7.40	13.90	7.80	12.20	13	5

9.3.3   Multiclass Networks

One of the advantages of the BOTT method is that it can easily be extended to multiple class networks. The necessary function $h_r(\overline{K}_{ir})$ is given by Eq. (9.19):

1. For node types $-/M/1$-FCFS, $-/G/1$-PS and $-/G/1$-PR:

$$h_r(\overline{K}_{\text{bott},r}) = \overline{K}_{\text{bott},r} \cdot \left(1 - \frac{K-1}{K} \cdot \rho_{\text{bott}}\right). \tag{9.34}$$

2. For node type $-/G/\infty$-IS, $h_r(\overline{K}_{i,r})$:

$$h_r(\overline{K}_{\text{bott},r}) = \frac{\overline{K}_{\text{bott},r}}{K}. \tag{9.35}$$

3. For the node type $-/M/m$-FCFS:

$$h_r(\overline{K}_{\text{bott},r}) = \frac{\overline{K}_{\text{bott},r}}{m_{\text{bott}} + \dfrac{1}{1 - \dfrac{K - m_{\text{bott}} - 1}{K - m_{\text{bott}}} \cdot \rho_{\text{bott}}} \cdot P_{m_{\text{bott}}}(\rho_{\text{bott}})}. \tag{9.36}$$

Remark: Equations (9.34) and (9.36) are approximations because we used $\rho_{\text{bott},r}$ part of $\rho_{\text{bott}} = \sum_{i=1}^{R} \rho_{\text{bott},i}$ when we created the $h_r(\overline{K}_{i,r})$ function.

Now we give the algorithm for the multiple class BOTT:

STEP 1 Initialize:

$$\lambda_r = \min_{i \in \{1,\ldots,N\}} \left\{ \frac{\mu_{i,r} \cdot m_i}{R \cdot e_{i,r}} \right\}, \text{ for } r = 1,\ldots,R,$$

$$\text{bott}, r = \text{argmin} \left\{ \frac{\mu_{i,r} \cdot m_i}{e_{i,r}} \right\}, r.$$

STEP 2 Iteration:

STEP 2.1 Compute the performance measures:

$$\left. \begin{aligned} \lambda_{ir} &= \lambda_r e_{ir}, \\ \rho_{ir} &= \frac{\lambda_{ir}}{m_i \cdot \mu_{ir}}, \\ \overline{K}_{ir} &= f_{ir}(\lambda_1,\ldots,\lambda_R), \\ \rho_i &= \sum_{r=1}^{R} \rho_{ir}, \end{aligned} \right\} \quad \begin{aligned} r &= 1,\ldots,R, \\ i &= 1,\ldots,N. \end{aligned}$$

STEP 2.2 Compute the new throughputs:

$$\left. \begin{aligned} g(\lambda_r) &= \sum_{i=1}^{N} \overline{K}_{ir}, \\ \overline{K}_{\text{bott},r} &= \overline{K}_{\text{bott},r} \frac{K_r}{g(\lambda_r)}, \\ \rho_{\text{bott},r} &= h_r(\overline{K}_{\text{bott},r}), \\ \lambda_r &= \frac{\rho_{\text{bott},r} \cdot m_{\text{bott}} \cdot \mu_{\text{bott},r}}{e_{\text{bott},r}}, \end{aligned} \right\} \quad r = 1,\ldots,R.$$

STEP 2.3 Stop the iteration for class r. If

$$\left(\left| \frac{K_r}{g(\lambda_r)} \right| > \epsilon \right),$$

go to Step 2.1, else stop iteration for class r.

STEP 3 Final value of $\overline{K}_{ir}$:

$$\overline{K}_{ir} = \overline{K}_{ir} \cdot \frac{K_r}{g(\lambda_r)}, \quad r = 1,\ldots,R, \ i = 1,\ldots,N.$$

The deviation from the exact results is similar to the one in the single class case for the SUM and is slightly worse but tolerable for the BOTT (see Table 9.10). The average number of iterations for the examples of Table 9.10

Table 9.10   Deviation in percent (%) from the exact results for the throughput λ (average of 56 examples)

	BOTT	SUM
min	0	0
mean	6.7	4.68
max	38.62	40.02

used by the bottapprox algorithm is 4.5, and the maximal value is 20, which is much smaller in comparison with the SUM.

Problem 9.4   Solve Problem 7.3 and Problem 7.4 using the BOTT.

9.4 BOUNDS ANALYSIS

Another possibility for reducing the computation time for the analysis of product-form queueing networks is to derive the upper and lower bounds for the performance measures instead of the exact values. In this section we introduce two methods for computing upper and lower bounds, *asymptotic bounds analysis (ABA)* and *balanced-job-bounds analysis (BJB)*. These two methods use simple equations to determine the bounds. Bounds for the system throughput and for the mean number of jobs are calculated. The largest possible throughput and the lowest possible response time of the network are called *optimistic bounds*, and the lowest possible throughput and greatest possible mean response time are called *pessimistic bounds*. So we have

$$\lambda_{\text{pes}} \leq \lambda \leq \lambda_{\text{opt}} \quad \text{and} \quad \overline{T}_{\text{opt}} \leq \overline{T} \leq \overline{T}_{\text{pes}} .$$

With these relations other bounds for the performance measures such as throughputs and utilizations of the nodes can be determined.

In the following we make a distinction between three different network types:

- Type A describes a closed network containing no infinite server nodes.

- Type B describes a closed network with only one IS node.

- Type C describes any kind of open network.

We restrict our consideration to networks with only one job class. Generalizations to networks with several job classes can be found in [EaSe86].

9.4.1 Asymptotic Bounds Analysis

The ABA [MuWo74a, Klei76, DeBu78] gives upper bounds for the system throughput and lower bounds for the mean response time of a queueing model (optimistic bounds). The only condition is that the service rates have to be independent of the number of jobs at the node or in the network. As inputs for the ABA method, the maximum relative utilization of all non-IS nodes, $x_{\max}$, and the sum x_{sum} of the relative utilizations of these nodes are needed:

$$x_{\max} = \max_i(x_i) \quad \text{and} \quad x_{\mathrm{sum}} = \sum_i x_i \,,$$

where $x_i = e_i/\mu_i$ is the relative utilization of node i (Eq. (7.6)). The relative utilization of an IS node (thinking time at the terminals) is called Z.

At first we consider the case of an open network (Type C): Since the utilization $\rho_i = \lambda_i/\mu_i$ with $\lambda_i = \lambda \cdot e_i$, the following relation holds:

$$\rho_i = \lambda \cdot x_i \,. \tag{9.37}$$

Due to the fact that no node can have a utilization greater than 1, the upper bound for the throughput of an open network cannot exceed $\lambda_{\mathrm{sat}} = 1/x_{\max}$ if the network is to remain stable. Therefore, λ_{sat} is the lowest arrival rate where one of the nodes in the system is fully utilized. The node that is utilized most is the node with the highest relative utilization $x_{\max}$. For this *bottleneck* node we have $\rho_{\max} = \lambda \cdot x_{\max} \leq 1$, and the optimistic bound for the throughput is therefore

$$\lambda \leq \frac{1}{x_{\max}} \,. \tag{9.38}$$

To determine the optimistic bounds for the mean response time, we assume that in the best case no job in the network is hindered by another job. With this assumption, the waiting time of a job is zero. Because the relative utilization $x_i = e_i/\mu_i$ of a node is the mean time a job spends being served at this node, the mean system response time in the non-IS nodes is simply given as the sum of the relative utilizations. Therefore, for the optimistic bounds on mean response time we get

$$\overline{T} \geq x_{\mathrm{sum}} \,. \tag{9.39}$$

For closed networks we consider networks with one IS node to model terminals (Type B). From the results obtained for such networks, we can easily derive those for networks of Type A by setting the mean think time Z at the terminals to zero. The bounds are determined by examining the network behavior under both light and heavy loads. We start with the assumption of a heavily loaded system (many jobs are in the system). The greater the number of jobs K in the network, the higher the utilization of the individual nodes. Since

$$\rho_i(K) = \lambda(K)x_i \leq 1 \,,$$

the highest possible overall network throughput is restricted by each node in the network, especially by the bottleneck node. As in open models, we have

$$\lambda(K) \leq \frac{1}{x_{\max}} \,.$$

Now assume that there are only a few jobs in the network (lightly loaded case). In the extreme case $K = 1$, the network throughput is given by $1/(x_{\text{sum}} + Z)$. For $K = 2, 3, \ldots$ jobs in the network, the throughput reaches its maximum when each job in the system is not hindered by other jobs. In this case we have $\lambda(K) = K/(x_{\text{sum}} + Z)$.

These observations can be summarized in the following upper bound for the network throughput:

$$\lambda(K) \leq \min\left\{\frac{1}{x_{\max}}, \frac{K}{x_{\text{sum}} + Z}\right\} . \tag{9.40}$$

To determine the bound for the mean response time $\overline{T}(K)$, we transform Eq. (9.40) with the help of Little's theorem:

$$\frac{K}{\overline{T}(K) + Z} \leq \min\left\{\frac{1}{x_{\max}}, \frac{K}{x_{\text{sum}} + Z}\right\} .$$

Then we get

$$\max\left\{x_{\max}, \frac{x_{\text{sum}} + Z}{K}\right\} \leq \frac{\overline{T}(K) + Z}{K} , \tag{9.41}$$

$$\text{or} \quad \max\left\{x_{\text{sum}}, K \cdot x_{\max} - Z\right\} \leq \overline{T}(K) .$$

The asymptotic bounds for all three network types are listed in Table 9.11.

Table 9.11   Summary of the ABA bounds

	Network Type	ABA Bounds
λ	A	$\lambda(K) \leq \min\left\{\dfrac{K}{x_{\text{sum}}}, \dfrac{1}{x_{\max}}\right\}$
	B	$\lambda(K) \leq \min\left\{\dfrac{K}{x_{\text{sum}} + Z}, \dfrac{1}{x_{\max}}\right\}$
	C	$\lambda \leq \dfrac{1}{x_{\max}}$
$\overline{T}$	A	$\overline{T}(K) \geq \max\left\{x_{\text{sum}}, K \cdot x_{\max}\right\}$
	B	$\overline{T}(K) \geq \max\left\{x_{\text{sum}}, K \cdot x_{\max} - Z\right\}$
	C	$\overline{T} \geq x_{\text{sum}}$

Example 9.6   As an example for the use of ABA, consider the closed product-form queueing network given in Fig. 9.2. There are $K = 20$ jobs in the network and the mean service times and the visit ratios are given as follows:

$$1/\mu_1 = 4.6, \quad 1/\mu_2 = 8, \qquad 1/\mu_3 = 120 = Z,$$
$$e_1 = 2, \qquad e_2 = e_3 = 1.$$

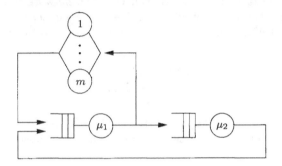

Fig. 9.2   Single class queueing network.

At first we determine

$$x_{\max} = \max\left\{\frac{e_1}{\mu_1}, \frac{e_2}{\mu_2}\right\} = \underline{9.2}, \quad x_{\text{sum}} = \frac{e_1}{\mu_1} + \frac{e_2}{\mu_2} = \underline{17.2},$$
$$Z = \frac{e_3}{\mu_3} = \underline{120}.$$

With these values and the formulae given in Table 9.11, we can determine the asymptotic bounds immediately:

Throughput:

$$\lambda(K) \leq \min\left\{\frac{K}{x_{\text{sum}} + Z}, \frac{1}{x_{\max}}\right\} = \min\left\{\frac{20}{137.2}; \frac{1}{9.2}\right\} = \underline{0.109}.$$

Mean response time:

$$\overline{T}(K) \geq \max\left\{x_{\text{sum}}, K \cdot x_{\max} - Z\right\} = \max(17.2; 64) = \underline{64}.$$

Now the bounds for the other performance measures can be computed. With the formula $\rho_i = \lambda \cdot x_i$, we can, for example, compute the upper bounds for the utilizations $\rho_1 \leq 1$ and $\rho_2 \leq 0.87$. For this example, the exact values for throughput and response time are $\lambda(K) = \underline{0.100}$ and $\overline{T}(K) = \underline{80.28}$, respectively.

9.4.2   Balanced Job Bounds Analysis

With the BJB method [ZSEG82], upper as well as lower bounds can be obtained for the system throughput and mean response time of product-form networks. In general, using this analysis we get better results than with the ABA method. The derivation of bounds by the BJB method is demonstrated only for closed networks of Type A. The derivation of the bounds for the other two network types is similar and is therefore left to the reader as an exercise. In the derivation of BJB, it is assumed that the network is *balanced*: that is, the relative utilizations of all nodes are the same: $x_1 = \ldots = x_N = x$. The throughput of a balanced network can be computed by applying Little's theorem:

$$\lambda(K) = \frac{K}{\sum\limits_{i=1}^{N} \overline{T}_i(K)},$$

where $\overline{T}_i(K)$ is the mean response time in the ith node with K jobs in the network. For product-form networks, Reiser and Lavenberg [ReLa80] have shown that

$$\overline{T}_i(K) = \left[1 + \overline{K}_i(K-1)\right] \cdot x.$$

If we combine the two preceding formulae, we get:

$$\lambda(K) = \frac{K}{\sum\limits_{i=1}^{N} \left[1 + \overline{K}_i(K-1)\right] \cdot x} = \frac{K}{(N+K-1)\cdot x}. \tag{9.42}$$

Equation (9.42) can be used to determine the throughput bounds of any type of product-form queueing network. For this computation, let $x_{\min}$ and $x_{\max}$, respectively, denote the minimum and maximum relative utilization of an individual node in the network. The basic idea of the BJB analysis is to enclose the considered network by two adjoining balanced networks:

1. An "optimistic" network, where the same relative utilization $x_{\min}$ is assumed for all nodes,

2. A "pessimistic" network, where the same relative utilization $x_{\max}$ is assumed for all nodes.

With Eq. (9.42), the BJB bounds for the throughputs can be determined as

$$\frac{K}{(N+K-1)} \cdot \frac{1}{x_{\max}} \leq \lambda(K) \leq \frac{K}{(N+K-1)} \cdot \frac{1}{x_{\min}}. \tag{9.43}$$

Similarly, the bounds for the mean response time can be determined by using Little's theorem:

$$(N+K-1)\cdot x_{\min} \leq \overline{T}(K) \leq (N+K-1)\cdot x_{\max}. \tag{9.44}$$

The optimistic bounds can be improved by observing that among all the networks with relative overall utilization x_{sum}, the one with the relative utilization, $x_i = x_{\text{sum}}/N$ for all i, has the highest throughput. If $x_{\text{ave}} = x_{\text{sum}}/N$ denotes the average relative utilization of the individual nodes in the network, then the improved optimistic bounds are given by

$$\lambda(K) \leq \frac{K}{K+N-1} \cdot \frac{1}{x_{\text{ave}}} = \frac{K}{x_{\text{sum}} + (K-1)x_{\text{ave}}} \qquad (9.45)$$

and

$$x_{\text{sum}} + (K-1)x_{\text{ave}} \leq \overline{T}(K), \qquad (9.46)$$

respectively. The optimistic network, the network with the highest throughput, has, therefore, $N = x_{\text{sum}}/x_{\text{ave}}$ nodes with relative utilization x_{ave}. Analogously, among all networks with relative overall utilization x_{sum} and maximum relative utilization x_{max}, the network with the lowest throughput has altogether $x_{\text{sum}}/x_{\text{max}}$ nodes with relative overall utilization x_{max}. Therefore, the improved pessimistic bounds are given by

$$\frac{K}{K+\frac{x_{\text{sum}}}{x_{\text{max}}}-1} \cdot \frac{1}{x_{\text{max}}} = \frac{K}{x_{\text{sum}} + (K-1)x_{\text{max}}} \leq \lambda(K) \qquad (9.47)$$

and:

$$\overline{T}(K) \leq x_{\text{sum}} + (K-1)x_{\text{max}}, \qquad (9.48)$$

respectively.

The inequalities for open networks can be obtained in a similar way. In Table 9.12, the BJB bounds for all three network types are summarized. The equations for the two closed network types are identical, if the value of Z is set to zero.

Example 9.7   Consider the network of Example 9.6 again, but now we determine the bounds with the BJB method. With $x_{\text{ave}} = x_{\text{sum}}/N = 8.6$, we can obtain the bounds directly from Table 9.12.

Throughput:

$$\frac{20}{17.2 + 120 + 129.5} \leq \lambda(K) \leq \frac{20}{17.2 + 120 + 20.5},$$
$$\underline{0.075} \leq \lambda(K) \leq \underline{0.127}.$$

Response time:

$$17.2 + \frac{163.4}{7.977} \leq \overline{T}(K) \leq 17.2 + \frac{174.8}{1.349},$$
$$\underline{37.70} \leq \overline{T}(K) \leq \underline{146.8}.$$

A comparison of the results of both methods shows that for this example we get better results with the ABA method than with the BJB method. Therefore, the conclusion is irresistible that using both methods in combination will

Table 9.12 Summary of the BJB bounds [LZGS84]

	Network Type	BJB Bounds	
λ	A	$$\frac{K}{x_{\text{sum}} + (K-1)x_{\max}} \le \lambda(K) \le \frac{K}{x_{\text{sum}} + (K-1)x_{\text{ave}}}$$	
	B	$$\frac{K}{x_{\text{sum}} + Z + \frac{(K-1)x_{\max}}{1 + \frac{Z}{K \cdot x_{\text{sum}}}}} \le \lambda(K) \le \frac{K}{x_{\text{sum}} + Z + \frac{(K-1)x_{\text{ave}}}{1 + \frac{Z}{x_{\text{sum}}}}}$$	
	C	$$\lambda \le \frac{1}{x_{\max}}$$	
$\overline{T}$	A	$$x_{\text{sum}} + (K-1)x_{\text{ave}} \le \overline{T}(K) \le x_{\text{sum}} + (K-1)x_{\max}$$	
	B	$$x_{\text{sum}} + \frac{(K-1)x_{\text{ave}}}{1 + \frac{Z}{x_{\text{sum}}}} \le \overline{T}(K) \le x_{\text{sum}} + \frac{(K-1)x_{\max}}{1 + \frac{Z}{K \cdot x_{\text{sum}}}}$$	
	C	$$\frac{x_{\text{sum}}}{1 - \lambda x_{\text{ave}}} \le \overline{T} \le \frac{x_{\text{sum}}}{1 - \lambda x_{\max}}$$	

reduce the area where the performance measures can lie. Figures 9.3 and 9.4 show the results of the ABA and BJB values for the throughput and mean response time as functions of the number of jobs in the network, respectively. The point of intersection for the two ABA bounds is given by

$$K^* = \frac{x_{\text{sum}} + Z}{x_{\max}}.$$

From the point

$$K^+ = \frac{(x_{\text{sum}} + Z)^2 - x_{\text{sum}}x_{\text{ave}}}{(x_{\text{sum}} + Z) \cdot x_{\max} - x_{\text{sum}}x_{\text{ave}}}$$

upwards, the optimistic curve of the ABA method gives better results than the BJB one. These bounds say that the exact values for the throughput and the mean response time reside in the shaded area.

The *performance bound hierarchy method* for determining the bounds of the performance measures was developed by [EaSe83]. In this method we have a hierarchy of upper and lower bounds for the throughput. These bounds converge to the exact solution of the network. The method is based on the MVA. Thus, there is a direct relation between the accuracy of the bounds and the cost of computation. If the number of iterations is high, then the cost is higher but the bounds are closer together. On the other hand, when the number of iterations is lower, the cost is lower but the bounds are farther apart. The extension of this method to multiclass networks can be found in [EaSe86]. Other versions can be found in [McMi84], [Kero86], [HsLa87], and [ShYa88].

Problem 9.5   Solve Problem 7.3 with the ABA and BJB methods and compare the results.

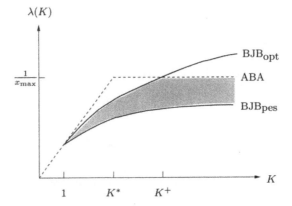

Fig. 9.3   ABA and BJB throughput bounds as a function of K.

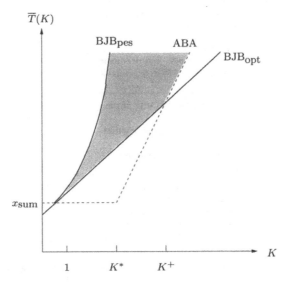

Fig. 9.4   ABA and BJB response time bounds as a function of K.

9.5   SUMMARY

The main advantages and disadvantages of the approximation algorithms for product-form queueing networks described in this chapter are summarized in Table 9.13.

Table 9.13 Comparison of the approximation algorithms for product-form queueing networks

Algorithms	Advantages	Disadvantages
Bard–Schweitzer (BS)	Very low storage and time requirement	No multiple server nodes Low accuracy
SCAT	Good accuracy Very low storage requirement compared with MVA or convolution	Needs more iterations than BS
SUM	Easy to understand and implement Low storage and time requirement Easy to extend to non-product-form networks	Accuracy is not very high (but sufficient for most applications)
BOTT	The same advantages as SUM Fewer iterations than SUM For multiple class networks, easier to implement than SUM Accuracy slightly better than SUM	Accuracy is not very high (but sufficient for most applications)
ABA, BJB	Well-suited for a bottleneck analysis In the design phase, to obtain a rough prediction (insight, understanding) of the performance of a system Extremely low storage and time requirement	Only for single class networks Only upper and lower bounds

10

Algorithms for Non-Product-Form Networks

Although many algorithms are available for solving product-form queueing networks (see Chapters 8 and 9), most practical queueing problems lead to non-product-form networks. If the network is Markovian (or can be Markovized), automated generation and solution of the underlying CTMC via *stochastic Petri nets* (SPNs) is an option provided the number of states is fewer than a million. Instead of the costly alternative of a discrete-event simulation, approximate solution may be considered. Many approximation methods for non-product-form networks are discussed in this chapter. These algorithms and corresponding sections of this chapter are laid out as shown in the flowchart of Fig. 10.1. Networks with nonexponentially distributed service times are treated in the next section, while networks with FCFS nodes having different service times for different classes are treated in Section 10.2. Priority queueing networks and networks with simultaneous resource possession are treated in Sections 10.3 and 10.4, respectively. Network models of programs with internal concurrency are treated in Section 10.5. Fork–join systems and parallel processing are treated in Section 10.6. Networks with

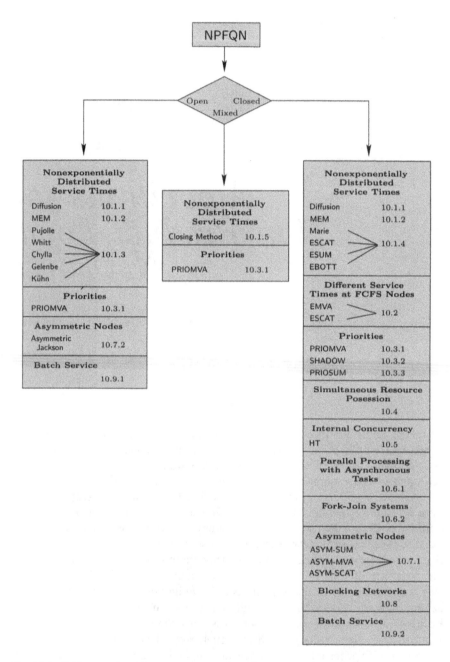

Fig. 10.1 Different algorithms for non-product-form networks and corresponding sections.

asymmetric server nodes and blocking networks are discussed in Section 10.7 and Section 10.8, respectively.

10.1   NONEXPONENTIAL DISTRIBUTIONS

10.1.1   Diffusion Approximation

Although we present diffusion approximation (DA) as a method to deal with networks with nonexponential service time and interarrival time distributions, it is possible to apply this technique to Markovian networks as well. The diffusion approximation is a technique based on an approximate product-form solution. In this approximation, the discrete-state stochastic process $\{K_i(t) \mid t \geq 0\}$ describing the number of jobs at the ith node at time t is approximated by a continuous-state stochastic process (diffusion process) $\{X_i(t) \mid t \geq 0\}$. For the fluctuation of jobs in a time interval we assume a normal distribution. In the steady-state, the density function $f_i(x)$ of the continuous-state process can be shown to satisfy the *Fokker–Planck equation* [Koba74] assuming certain boundary conditions. A discretization of the density function gives the approximated product-form-like state probabilities $\hat{\pi}_i(k_i)$ for node i (see Fig. 10.2).

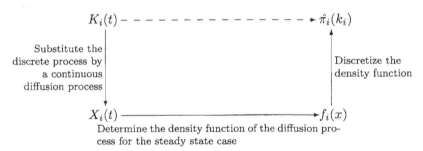

Fig. 10.2 The principle of the diffusion approximation.

Although the derivation of this method is very complex, the method itself is very simple to apply to a given problem. Steady-state behavior of networks with generally distributed service and interarrival times at nodes can be approximately solved with the restriction of a single single server at each node. At present no solutions are available for multiple class networks [Mitz97].

Consider a GI/G/1 queueing system with arrival rate λ, service rate μ, and the coefficients of variation c_A and c_B of the interarrival and service times, respectively. Using the diffusion approximation, the following approximated state probabilities can be obtained [Koba74, ReKo74]:

$$\hat{\pi}(k) = \begin{cases} 1 - \rho, & k = 0, \\ \rho(1 - \hat{\rho})\hat{\rho}^{k-1}, & k > 0, \end{cases} \tag{10.1}$$

with

$$\hat{\rho} = \exp\left(\frac{-2(1-\rho)}{c_B^2 + \rho c_A^2}\right) = \exp(\gamma), \qquad (10.2)$$

and $\rho = \lambda/\mu < 1$. Note that γ is defined in Eq. (10.2) for convenience so that $\hat{\rho} = \exp(\gamma)$. For the mean number of jobs we then have

$$\overline{K} = \sum_{k=1}^{\infty} k \cdot \hat{\pi}(k) = \frac{\rho}{1-\hat{\rho}}. \qquad (10.3)$$

Differential equations underlying the diffusion approximation need boundary conditions for their solution. Different assumptions regarding these boundary conditions lead to different expressions. The preceding expressions are based on the work of [Koba74] and [ReKo74]. Using different boundary conditions, [Gele75] and [Mitz97] get different results that are more accurate than those of [Koba74] and [ReKo74] for larger values of utilization ρ. They [Gele75], [Mitz97] derived the following expression for the approximated state probabilities:

$$\hat{\pi}(k) = \begin{cases} 1-\rho, & k = 0, \\[2mm] \rho\left[1 - \frac{1}{\gamma}(\hat{\rho}-1)\right], & k = 1, \\[2mm] -\frac{\rho}{\gamma\hat{\rho}^2}(1-\hat{\rho})^2\,\hat{\rho}^k, & k \geq 2, \end{cases} \qquad (10.4)$$

with $\hat{\rho}$ and γ as in Eq. (10.2). The mean number of jobs is then

$$\overline{K} = \rho\left[1 + \frac{\rho c_A^2 + c_b^2}{2(1-\rho)}\right]. \qquad (10.5)$$

Next we show how the diffusion approximation can be applied to queueing networks.

10.1.1.1 Open Networks We can make use of the results for the GI/G/1 system for each node in the network, provided that we can approximate the coefficient of variation of interarrival times at each node. We assume the following:

- The external arrival process can be any renewal process with mean interarrival time $1/\lambda$ and coefficient of variation c_A.

- The service times at node i can have any distribution with mean service time $1/\mu_i$ and coefficient of variation c_{B_i}.

- All nodes in the network are single server with FCFS service strategy.

According to [ReKo74] the diffusion approximation for the approximated state probabilities of the network have a product-form solution:

$$\hat{\pi}(k_1, \ldots, k_N) = \prod_{i=1}^{N} \hat{\pi}_i(k_i) \,, \tag{10.6}$$

with the approximate marginal probabilities as in Eq. (10.1):

$$\hat{\pi}_i(k_i) = \begin{cases} 1 - \rho_i \,, & k_i = 0 \,, \\ \rho_i(1 - \hat{\rho}_i)\hat{\rho}_i^{k_i-1} \,, & k \geq 1 \,, \end{cases} \tag{10.7}$$

with

$$\rho_i = \frac{\lambda \cdot e_i}{\mu_i} \,, \tag{10.8}$$

$$\hat{\rho}_i = \exp\left(-\frac{2(1 - \rho_i)}{c_{Ai}^2 \cdot \rho_i + c_{Bi}^2}\right) \,. \tag{10.9}$$

We approximate the squared coefficient of variation of interarrival times at node i using the following expression:

$$c_{Ai}^2 = 1 + \sum_{j=0}^{N}(c_{Bj}^2 - 1) \cdot p_{ji}^2 \cdot e_j \cdot e_i^{-1} \,, \tag{10.10}$$

where we set

$$c_{B0}^2 = c_A^2 \,. \tag{10.11}$$

For the mean number of jobs at node i we get

$$\overline{K}_i = \sum_{k_i=1}^{\infty} k_i \cdot \hat{\pi}(k_i) = \frac{\rho_i}{1 - \hat{\rho}_i} \,. \tag{10.12}$$

Similar results are presented in [Gele75], [GeMi80], and [Koba74].

Example 10.1 In this example (see Fig. 10.3), we show how to use the diffusion approximation for a simple open network with $N = 2$ stations. The external arrival process has mean interarrival time $1/\lambda = 2.0$ and the squared coefficient of variation $c_A^2 = 0.94$. The service times at the two stations have the following parameters:

$$\mu_1 = 1.1 \,, \quad \mu_2 = 1.2 \,, \quad c_{B1}^2 = 0.5 \,, \quad c_{B2}^2 = 0.8 \,.$$

With routing probabilities $p_{10} = p_{12} = p_{21} = p_{22} = 0.5$ and $p_{01} = 1$, we compute the visit ratios, using Eq. (7.4):

$$e_1 = p_{01} + e_2 \cdot p_{21} = 2 \,, \quad e_2 = e_1 \cdot p_{12} + e_2 \cdot p_{22} = 2 \,.$$

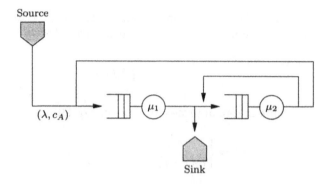

Fig. 10.3 A simple open queueing network.

Node utilizations are determined using Eq. (10.8):

$$\rho_1 = \frac{\lambda \cdot e_1}{\mu_1} = 0.909\,, \quad \rho_2 = \frac{\lambda \cdot e_2}{\mu_2} = 0.833\,.$$

We approximate the coefficients of variation of the interarrival time at both the nodes using Eq. (10.10):

$$
\begin{aligned}
c_{A1}^2 &= 1 + \sum_{j=0}^{2}(c_{Bj}^2 - 1)p_{j1}^2 \cdot \frac{e_j}{e_1} \\
&= 1 + (c_A^2 - 1)p_{01}^2 \cdot \frac{e_0}{e_1} + (c_{B1}^2 - 1)p_{11}^2 \cdot \frac{e_1}{e_1} + (c_{B2}^2 - 1)p_{21}^2 \cdot \frac{e_2}{e_1} \\
&= 0.920\,, \\
c_{A2}^2 &= 1 + \sum_{j=0}^{2}(c_{Bj}^2 - 1)p_{j2}^2 \cdot \frac{e_j}{e_2} = 0.825.
\end{aligned}
$$

With Eq. (10.9) we get

$$\hat{\rho}_1 = \exp\left(-\frac{2(1 - \rho_1)}{c_{A1}^2 \cdot \rho_1 + c_{B1}^2}\right) = 0.873\,,$$

$$\hat{\rho}_2 = \exp\left(-\frac{2(1 - \rho_2)}{c_{A2}^2 \cdot \rho_2 + c_{B2}^2}\right) = 0.799\,.$$

For the mean number of jobs $\overline{K}_i$, we use Eq. (10.12) and obtain

$$\overline{K}_1 = \frac{\rho_1}{1 - \hat{\rho}_1} = 7.147\,, \quad \overline{K}_2 = \frac{\rho_2}{1 - \hat{\rho}_2} = 4.151\,,$$

and the marginal probabilities are given by Eq. (10.7):

$$\hat{\pi}_1(0) = 1 - \rho_1 = 0.091\,, \quad \hat{\pi}_1(k) = \rho_1(1 - \hat{\rho}_1)\hat{\rho}_1^{k-1}$$

$$= 0.116 \cdot 0.873^{k-1} \quad \text{for } k > 0,$$

$$\hat{\pi}_2(0) = 1 - \rho_2 = \underline{0.167}, \quad \hat{\pi}_2(k) = \rho_2(1 - \hat{\rho}_2)\hat{\rho}_2^{k-1}$$

$$= 0.167 \cdot 0.799^{k-1} \quad \text{for } k > 0.$$

10.1.1.2 Closed Networks To apply the diffusion approximation to closed queueing networks with arbitrary service time distributions, we can use Eq. (10.7) to approximate marginal probabilities $\hat{\pi}_i(k_i)$, provided that the throughputs λ_i and utilizations ρ_i are known. In [ReKo74] the following two suggestions are made to estimate λ_i:

1. For large values of K we use the bottleneck analysis: Search for the bottleneck node bott (node with the highest relative utilization), set its utilization to 1, and determine the overall throughput λ of the network using Eq. (10.8). With this throughput and the visit ratios e_i, compute the utilization of each node. The mean number of jobs is determined using Eq. (10.12).

2. If there is no bottleneck in the network and/or the number K of jobs in the network is small, replace the given network by a product-form one with the same service rates μ_i, routing probabilities p_{ij}, and the same K, and compute the utilizations ρ_i. The approximated marginal probabilities given by Eq. (10.7) are then used to determine the approximated state probabilities:

$$\hat{\pi}(k_1, k_2, \ldots, k_N) = \frac{1}{G} \prod_{i=1}^{N} \hat{\pi}(k_i), \tag{10.13}$$

where G is the normalizing constant of the network. Then for the marginal probabilities and the other performance measures improved values can be determined.

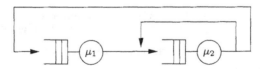

Fig. 10.4 A simple closed queueing network.

Example 10.2 Consider the closed network shown in Fig. 10.4 with the parameters

$$\mu_1 = 1.1, \quad \mu_2 = 1.2, \quad c_{B1}^2 = 0.5, \quad c_{B2}^2 = 0.8,$$

and the routing probabilities

$$p_{12} = 1 \quad \text{and} \quad p_{21} = p_{22} = 0.5.$$

For the visit ratios we use Eq. (7.5) and get

$$e_1 = 1, \quad e_2 = 2.$$

At first we choose $K = 6$ and use the bottleneck analysis to study the network.

Solution with Bottleneck Analysis Let node bott be the bottleneck with the highest relative utilization:

$$\rho_{\text{bott}} = \max_i \{\rho_i\} = \lambda \cdot \max_{1,2} \left\{ \frac{e_1}{\mu_1}, \frac{e_2}{\mu_2} \right\} = \lambda \cdot \max \{0.909, 1.667\},$$

which means that node 2 is the bottleneck and therefore its utilization is set to 1:

$$\rho_2 = 1.$$

Using Eq. (10.8), we determine the overall throughput λ of the network:

$$\lambda = \rho_2 \cdot \frac{\mu_2}{e_2} = \frac{1}{1.667} = \underline{0.6}.$$

For the utilization of node 1 we get

$$\rho_1 = \frac{\lambda \cdot e_1}{\mu_1} = \underline{0.545}.$$

Now we use Eq. (10.10) to determine the coefficients of variation of the inter-arrival times:

$$c_{A1}^2 = 1 + (c_{B1}^2 - 1)p_{11}^2 \cdot e_1 \cdot e_1^{-1} + (c_{B2}^2 - 1)p_{21}^2 \cdot e_2 \cdot e_1^{-1} = \underline{0.9},$$
$$c_{A2}^2 = 1 + (c_{B1}^2 - 1)p_{12}^2 \cdot e_1 \cdot e_2^{-1} + (c_{B2}^2 - 1)p_{22}^2 \cdot e_2 \cdot e_2^{-1} = \underline{0.7}.$$

The $\hat{\rho}_i$ are given by Eq. (10.9):

$$\hat{\rho}_1 = \exp\left(-\frac{2(1 - \rho_1)}{c_{A1}^2 \cdot \rho_1 + c_{B1}^2}\right) = \underline{0.3995}, \quad \hat{\rho}_2 = \exp\left(-\frac{2(1 - \rho_2)}{c_{A2}^2 \cdot \rho_2 + c_{B2}^2}\right) = \underline{1}.$$

Now, by using Eq. (10.12) we can compute the approximate values for the mean number of jobs:

$$\overline{K}_1 = \frac{\rho_1}{1 - \hat{\rho}_1} = \underline{0.908}, \quad \overline{K}_2 = K - \overline{K}_1 = \underline{5.092}.$$

Table 10.1 compares the values obtained by the preceding diffusion approximation (DA) with those obtained via DES. We can see that in this case the results match very well. Such close matching does not always occur, especially when one node in the network is highly utilized and the coefficients of variation are very small.

To show the second method, we assume that there are $K = 3$ jobs in the network.

Table 10.1   Comparison of results for the example ($K = 6$)

	ρ_i (DES)	$\overline{K}_i$ (DES)	ρ_i (DA)	$\overline{K}_i$ (DA)
Node 1	0.544	0.965	0.545	0.908
Node 2	0.996	5.036	1	5.092

Solution Using Product-Form Approximation   Now we substitute the given network by a product-form one. To determine the utilizations of this network, we can use the MVA and get

$$\rho_1 = \underline{0.501} \quad \text{and} \quad \rho_2 = \underline{0.919}\,.$$

The coefficients of variation for the interarrival times remain the same as in the previous method:

$$c_{A1}^2 = \underline{0.9} \quad \text{and} \quad c_{A2}^2 = \underline{0.7}\,.$$

For the $\hat{\rho}_i$ we use Eq. (10.9) and obtain

$$\hat{\rho}_1 = \exp\left(-\frac{2(1 - \rho_1)}{c_{A_1}^2 \cdot \rho_1 + c_{B_1}^2}\right) = \underline{0.35}\,, \quad \hat{\rho}_2 = \underline{0.894}\,.$$

The approximated marginal probabilities can be computed using Eq. (10.7):

$$\hat{\pi}_1(0) = 1 - \rho_1 = \underline{0.499}\,, \qquad \hat{\pi}_2(0) = 1 - \rho_2 = \underline{0.081}\,,$$
$$\hat{\pi}_1(1) = \rho_1(1 - \hat{\rho}_1) = \underline{0.326}\,, \qquad \hat{\pi}_2(1) = \rho_2(1 - \hat{\rho}_2) = \underline{0.0974}\,,$$
$$\hat{\pi}_1(2) = \rho_1(1 - \hat{\rho}_1)\hat{\rho}_1 = \underline{0.114}\,, \quad \hat{\pi}_2(2) = \rho_2(1 - \hat{\rho}_2)\hat{\rho}_2 = \underline{0.0871}\,,$$
$$\hat{\pi}_1(3) = \rho_1(1 - \hat{\rho}_1)\hat{\rho}_1^2 = \underline{0.039}\,, \quad \hat{\pi}_2(3) = \rho_2(1 - \hat{\rho}_2)\hat{\rho}_2^2 = \underline{0.0779}\,.$$

Now the following network states are possible:

$$(3,0)\,, \quad (2,1)\,, \quad (1,2)\,, \quad (0,3)\,,$$

whose probability is computed using Eq. (10.13):

$$\hat{\pi}(3,0) = \hat{\pi}_1(3) \cdot \hat{\pi}_2(0) \cdot \frac{1}{G} = 0.00316 \cdot \frac{1}{G}\,,$$
$$\hat{\pi}(2,1) = 0.01111 \cdot \frac{1}{G}\,,$$
$$\hat{\pi}(1,2) = 0.02839 \cdot \frac{1}{G}\,,$$
$$\hat{\pi}(0,3) = 0.03887 \cdot \frac{1}{G}\,.$$

With the normalizing condition

$$\hat{\pi}(3,0) + \hat{\pi}(2,1) + \hat{\pi}(1,2) + \hat{\pi}(0,3) = 1\,,$$

we determine the normalizing constant G:

$$G = 0.00316 + 0.01111 + 0.02839 + 0.03887 = \underline{0.08153},$$

which is used to compute the final values of the state probabilities:

$$\hat{\pi}(3,0) = \underline{0.039}, \quad \hat{\pi}(2,1) = \underline{0.136}, \quad \hat{\pi}(1,2) = \underline{0.348}, \quad \hat{\pi}(0,3) = \underline{0.477},$$

and with these state probabilities the improved values for the marginal probabilities are immediately derived:

$$\pi_1(3) = \pi_2(0) = \underline{0.039}, \quad \pi_1(2) = \pi_2(1) = \underline{0.136},$$
$$\pi_1(1) = \pi_2(2) = \underline{0.348}, \quad \pi_1(0) = \pi_2(3) = \underline{0.477}.$$

For the mean number of jobs we get

$$\overline{K}_1 = \sum_{k=1}^{3} k \cdot \pi_i(k) = \underline{0.737} \quad \text{and} \quad \overline{K}_2 = \underline{2.263}.$$

In Table 10.2, DES values for this example are compared to values from the preceding approximation.

Table 10.2 Comparison of results for the example ($K = 3$).

	ρ_i (DES)	$\overline{K}_i$ (DES)	ρ_i (DA)	$\overline{K}_i$ (DA)
Node 1	0.519	0.773	0.501	0.737
Node 2	0.944	2.229	0.919	2.263

A detailed investigation of the accuracy of the diffusion method can be found in [ReKo74]. The higher the utilization of the nodes and the closer the coefficient of variation is to 1, the more accurate the results.

10.1.2 Maximum Entropy Method

An iterative method for the approximate analysis of open and closed queueing networks is based on the principle of the maximum entropy. The term *entropy* comes from information theory and is a measure of the uncertainty in the predictability of an event. To explain the principle of the maximum entropy, we consider a simple system that can take on a set of discrete states **S**. The probabilities $\pi(\mathbf{S})$ for different states are unknown; the only information about the probability vector is the number of side conditions given as mean values of suitable functions. Because in general the number of side conditions is smaller than the number of possible states, in most cases there is an infinite number of probability vectors that satisfy the given side conditions. The question now is which of these probability vectors shall be chosen as the one best suited for the information given by the side conditions and which is least prejudiced against the missing information. In this case the principle of maximum entropy says that the best-suited probability vector is the one with the largest entropy.

To solve a queueing network using the maximum entropy method (MEM), the steady-state probabilities $\pi(\mathbf{S})$ are determined so that the entropy function

$$H(\pi) = - \sum_{\mathbf{S}} \pi(\mathbf{S}) \ln \pi(\mathbf{S}) \tag{10.14}$$

is maximized subject to given side conditions. The normalizing condition –the sum of all steady-state probabilities is 1–provides another constraint. In [KGTA88], open and closed networks with one or several job classes and $-/G/1$, $-/G/\infty$ nodes are analyzed using this approach. Queueing disciplines such as FCFS, LCFS, or PS as well as priorities are allowed. An extension to the case of multiple server nodes is given in [KoAl88]. For the sake of simplicity we only consider single class networks with $-/G/1$-FCFS nodes based on the treatment in [Kouv85] and [Wals85].

10.1.2.1   Open Networks   If we maximize the entropy by considering the side conditions for the mean number of jobs $\overline{K}_i$, Eq. (7.18), and the utilization ρ_i, Eq. (7.26), then the following product-form approximation for the steady-state probabilities of open networks can be derived [Kouv85]:

$$\pi(k_1, k_2, \ldots, k_N) = \pi_1(k_1) \cdot \pi_2(k_2) \cdot \ldots \cdot \pi_N(k_N), \tag{10.15}$$

where the marginal probabilities are given by

$$\pi_i(k_i) = \begin{cases} \dfrac{1}{G_i}, & k_i = 0, \\[2mm] \dfrac{1}{G_i} \cdot a_i b_i^{k_i}, & k_i \geq 1, \end{cases} \tag{10.16}$$

and where

$$a_i = \frac{\rho_i^2}{(1 - \rho_i)(\overline{K}_i - \rho_i)}, \tag{10.17}$$

$$b_i = \frac{\overline{K}_i - \rho_i}{\overline{K}_i}, \tag{10.18}$$

$$G_i = \frac{1}{1 - \rho_i}. \tag{10.19}$$

To utilize the MEM, we thus need the utilizations ρ_i and the mean number of jobs $\overline{K}_i$ at each node. The equation $\rho_i = \lambda_i / \mu_i$ can be used to easily determine ρ_i from the given network parameters. The mean number of jobs $\overline{K}_i$ can be approximated as a function of the utilization ρ_i and the squared coefficient of variation of the service and interarrival times [Kouv85]:

$$\overline{K}_i = \frac{\rho_i}{2}\left(1 + \frac{c_{Ai}^2 + \rho_i c_{Bi}^2}{1 - \rho_i}\right),\tag{10.20}$$

if the condition $(1 - c_{Ai}^2)/(1 + c_{Bi}^2) \leq \rho_i < 1$ is fulfilled. The computation of the $\overline{K}_i$ is made possible by approximating the squared coefficient of variation c_{Ai}^2 of interarrival times. The following iterative expressions for computing the squared coefficients of variation is provided by [Kouv85]:

$$c_{Ai}^2 = -1 + \left(\sum_{j=0}^{N} \frac{\lambda_j p_{ji}}{\lambda_i \cdot (c_{ji}^2 + 1)}\right)^{-1},\tag{10.21}$$

$$c_{ji}^2 = 1 + p_{ji}\left(c_{Dj}^2 - 1\right),\ \cdot\tag{10.22}$$

$$c_{Di}^2 = \rho_i(1 - \rho_i) + (1 - \rho_i)c_{Ai}^2 + \rho_i^2 c_{Bi}^2.\tag{10.23}$$

There are other approximations for the computation of the mean number of jobs $\overline{K}_i$ and for the estimation of the necessary squared coefficients of variation. For more information on these equations, see Section 10.1.3.

With this information, the MEM for the analysis of open queueing networks can be summarized in the following three steps:

STEP 1 Determine the arrival rates λ_i using the traffic equations (Eq. (7.1)) and the utilizations $\rho_i = \lambda_i/\mu_i$ for all nodes $i = 1, \ldots, N$.

STEP 2 Determine the squared coefficients of variation.

STEP 2.1 Initialize. The squared coefficients of variations c_{Ai}^2 of the interarrival times are initially set to one for $i = 1, \ldots, N$.

STEP 2.2 Compute the squared coefficients of variation c_{Di}^2 of the interdeparture times of node i for $i = 1, \ldots, N$ using Eq. (10.23). Compute the squared coefficients of variation c_{ij}^2 of node i for $i = 1, \ldots, N$ and $j = 0, \ldots, N$ using Eq. (10.22). From these the new values of the squared coefficients of variation c_{Ai}^2 of the interarrival times are computed using Eq. (10.21).

STEP 2.3 Check the halting condition. If the old and new values for the c_{Ai}^2 differ by more than ϵ, then go back to Step 2.2 with the new c_{Ai}^2 values.

STEP 3 Determine the performance measures of the network beginning with the mean number of jobs $\overline{K}_i$ for $i = 1, \ldots, N$ using Eq. (10.20) and then compute the maximum entropy solutions of the steady-state probabilities using Eq. (10.15).

The MEM normally approximates the coefficients of variation less accurately than the method of Kühn (see Section 10.1.3) and therefore the method of Kühn is generally preferred for open networks with a single job class. The MEM is mainly used for closed networks.

10.1.2.2 *Closed Networks* For closed networks that consist only of -/G/1-FCFS nodes, the following product-form formulation for the approximated steady-state probabilities can be given if we maximize the entropy under the side conditions on the mean number of jobs $\overline{K}_i$ and the utilization ρ_i:

$$\pi(k_1 \ldots, k_N) = \frac{1}{G(K)} \cdot F_1(k_1) \cdot \ldots \cdot F_N(k_N), \qquad (10.24)$$

with

$$F_i(k_i) = \begin{cases} 1, & k_i = 0, \\ a_i b_i^{k_i}, & 1 \le k_i \le K, \end{cases} \qquad (10.25)$$

and

$$G(K) = \sum_{\substack{N \\ \sum\limits_{i=1} k_i = K}} F_1(k_1) \cdot \ldots \cdot F_N(k_N). \qquad (10.26)$$

As in the case of open networks, coefficients a_i and b_i are respectively given by Eqs. (10.17) and (10.18). The computation of the steady-state probabilities of closed networks is more difficult than that for open networks as the ρ_i and $\overline{K}_i$ values of the individual nodes cannot be determined directly. Therefore, we use the following trick: We construct a *pseudo open network* with the same number of nodes and servers as well as identical service time distribution and routing probabilities as the original closed network. The external arrival rates of this open network are determined so that resulting average number of jobs in the pseudo open network equals the given number of jobs K in the closed network:

$$\sum_{i=1}^{N} \overline{K}_i^* = K, \qquad (10.27)$$

where $\overline{K}_i^*$ denotes the mean number of jobs at the ith node in the pseudo open network. The pseudo open network arrival rate λ is determined iteratively by using Eqs. (10.20) and (10.27), assuming that the stability condition

$$\max_i \left\{ \frac{\lambda e_i}{\mu_i} \right\} < 1$$

is satisfied. Here e_i is the relative throughput of node i.

Then the performance measures $\overline{K}_i^*$ and ρ_i^* of the pseudo open network can be computed using the maximum entropy method for open networks as given in the last algorithm. These performance measures are then used to determine the coefficients a_i and b_i and steady-state probabilities of the pseudo open network. To compute the probability vector of the original closed network, we use the convolution method (see Section 8.1). By using the functions $F_i(k_i)$, Eq. (10.25), the normalizing constants as well as the utilizations and mean number of jobs at each node of the closed network can be computed. This approximation uses values for the coefficients a_i and b_i, which are computed for the pseudo open network. To compensate the error in the approximation, we apply the work rate theorem (Eq. (7.7)), which gives us an iterative method for the computation of the coefficients a_i:

$$a_i^{(n+1)} = a_i^{(n)} \cdot K \cdot \rho_i^* \cdot \left(\rho_i \cdot \sum_{j=1}^{N} \frac{\overline{K}_j \cdot \rho_j^*}{\rho_j} \right)^{-1} \tag{10.28}$$

with the initial values

$$a_i^{(0)} = \frac{\rho_i^*}{1 - \rho_i^*} \cdot \frac{\rho_i^*}{\overline{K}_i^* - \rho_i^*}. \tag{10.29}$$

For the coefficients b_i we have

$$b_i = \frac{\overline{K}_i^* - \rho_i^*}{\overline{K}_i^*}. \tag{10.30}$$

The asterisk denotes the performance measures of the pseudo open network.

The algorithm can now be given in the following six steps.

STEP 1 Compute the visit ratios e_i for $i = 1, \ldots, N$, using Eq. (7.2) and the relative utilizations $x_i = e_i / \mu_i$.

STEP 2 Construct and solve the pseudo open network.

STEP 2.1 Initialize. Initial value for the squared coefficients of the interarrival times at node $i = 1, \ldots, N$:

$$c_{Ai}^2 = 1.$$

Initial value for the arrival rate of the pseudo open network:

$$\lambda = \frac{0.99}{x_{\text{bott}}},$$

where $x_{\text{bott}} = \max_i \{x_i\}$ is the relative utilization of the bottleneck node.

STEP 2.2 Compute λ from condition (10.27) by using Eq. (10.20). Thus, λ can be determined as solution of the equation

$$\sum_{i=1}^{N} \frac{\lambda \cdot x_i}{2} \left(1 + \frac{c_{Ai}^2 + \lambda \cdot x_i \cdot c_{Bi}^2}{1 - \lambda \cdot x_i} \right) - K = 0, \tag{10.31}$$

if $(1 - c_{Ai}^2)/(1 + c_{Bi}^2) \leq \lambda \cdot x_i < 1$ for all i is fulfilled. To solve the nonlinear equation, we use the Newton–Raphson method.

STEP 2.3 Compute for $i, j = 1, \ldots, N$ the squared coefficients of variation c_{Di}^2, c_{ij}^2 and c_{Ai}^2 using the Eqs. (10.21)–(10.23). For this computation we need the values $\rho_i = \lambda \cdot x_i$ and $\lambda_i = \lambda \cdot e_i$.

STEP 2.4 Check the halting condition. If the new values for the c_{Ai}^2 differ less than ϵ from the old values, substitute them by the new values and return to Step 2.2.

STEP 3 Determine for $i = 1, \ldots, N$ the utilizations $\rho_i^* = (\lambda \cdot e_i)/\mu_i$, and with Eq. (10.20) the mean number of jobs $\overline{K}_i^*$ in the pseudo open network.

STEP 4 Determine the coefficients a_i and b_i of the pseudo open network using Eqs. (10.29) and (10.30).

STEP 5 Solve the closed network.

STEP 5.1 Use the convolution method to determine the normalizing constant $G(K)$ of the network. The functions $F_i(k_i)$ are given by Eq. (10.25).

STEP 5.2 Determine the performance parameters ρ_i and $\overline{K}_i$ of the closed network. For this we use the normalizing constant $G_N^{(i)}(k)$ as defined in Eq. (8.8). From Eq. (8.7) we have $\pi_i(k) = F_i(k) \cdot G_N^{(i)}(K - k)/G(k)$ and therefore

$$\rho_i = 1 - \pi_i(0) = 1 - \frac{G_N^{(i)}(K)}{G(K)}, \tag{10.32}$$

$$\overline{K}_i = \sum_{k=1}^{K} k \cdot \frac{F_i(k)}{G(K)} \cdot G_N^{(i)}(K - k). \tag{10.33}$$

STEP 5.3 Use Eq. (10.28) to determine a new value for the coefficients a_i. If there is a big difference between the old values and the new values, then return to Step 5.1.

STEP 6 Use Eq. (10.24) to determine the maximum entropy solution for the steady-state probabilities and other performance parameters of the network as, e.g., $\lambda_i = \mu_i \rho_i$ or $\overline{T}_i = \overline{K}_i/\lambda_i$.

Fig. 10.5 Closed network.

Example 10.3   The maximum entropy method is now illustrated on a simple network with $N = 2$ nodes and $K = 3$ jobs (see Fig. 10.5). The queueing discipline at all nodes is FCFS and the service rates and squared coefficients of variation are given as follows:

$$\mu_1 = 1, \quad \mu_2 = 2, \quad \text{and} \quad c_{B1}^2 = 5, \quad c_{B2}^2 = 0.5.$$

The analysis of the network is carried out in the following steps. As condition for the termination of iterations we choose $\epsilon = 0.001$.

STEP 1 Determine the visit ratios and relative utilizations:

$$e_1 = e_2 = \underline{1}, \quad \text{and} \quad x_1 = \underline{1}, \quad x_2 = \underline{0.5}.$$

STEP 2 Construct and solve the pseudo open network.

STEP 2.1 Initialize:

$$c_{A1}^2 = c_{A2}^2 = \underline{1}, \quad \lambda = 0.99/x_1 = \underline{0.99}.$$

STEP 2.2 Determine λ. We use the Newton–Raphson iteration method to solve the equation

$$\sum_{i=1}^{2} \frac{\lambda \cdot x_i}{2} \left(1 + \frac{c_{Ai}^2 + \lambda \cdot x_i \cdot c_{Bi}^2}{1 - \lambda \cdot x_i} \right) - 3 = 0$$

and get $\lambda \approx 0.5558$.

STEP 2.3 Determine the squared coefficients of variation:

$$c_{D1}^2 = \rho_1(1 - \rho_1) + (1 - \rho_1)c_{A1}^2 + \rho_1^2 c_{B1}^2 = \underline{2.236},$$
$$c_{D2}^2 = \rho_2(1 - \rho_2) + (1 - \rho_2)c_{A2}^2 + \rho_2^2 c_{B2}^2 = \underline{0.961},$$
$$c_{12}^2 = 1 + p_{12} \cdot (c_{D1}^2 - 1) = \underline{2.236},$$
$$c_{21}^2 = 1 + p_{21} \cdot (c_{D2}^2 - 1) = \underline{0.961},$$
$$c_{A1}^2 = -1 + \left(\frac{\lambda_2 \cdot p_{21}}{\lambda_1 \cdot (c_{21}^2 + 1)} \right)^{-1} = \underline{0.961},$$
$$c_{A2}^2 = -1 + \left(\frac{\lambda_1 \cdot p_{12}}{\lambda_2 \cdot (c_{12}^2 + 1)} \right)^{-1} = \underline{2.236}.$$

STEP 2.4 Check the halting condition:

$$|c_{A1}^{2(\text{new})} - c_{A1}^{2(\text{old})}| = |0.961 - 1| = 0.039 > \epsilon,$$
$$|c_{A2}^{2(\text{new})} - c_{A2}^{2(\text{old})}| = |2.236 - 1| = 1.236 > \epsilon.$$

As the halting condition is not fulfilled, we return to Step 2.2 and determine a new value for the constant λ.

$$\vdots$$

After 14 iterations, the halting condition is fulfilled and we get the following results:

$$\lambda = \underline{0.490}, \quad c_{A1}^2 = \underline{2.130}, \quad c_{A2}^2 = \underline{2.538}.$$

STEP 3 Determine ρ_i^* and $\overline{K}_i^*$:

$$\rho_1^* = \frac{\lambda \cdot e_1}{\mu_1} = \underline{0.49},\qquad\qquad \rho_2^* = \frac{\lambda \cdot e_2}{\mu_2} = \underline{0.245},$$

$$\overline{K}_1^* = \frac{\rho_1^*}{2}\left(1 + \frac{c_{A1}^2 + \rho_1^* \cdot c_{B1}^2}{1 - \rho_1^*}\right) \qquad \overline{K}_2^* = \frac{\rho_2^*}{2}\left(1 + \frac{c_{A2}^2 + \rho_2^* \cdot c_{B2}^2}{1 - \rho_2^*}\right)$$

$$= \underline{2.446}, \qquad\qquad\qquad\qquad = \underline{0.554}.$$

STEP 4 Determine the coefficients:

$$a_1 = \frac{\rho_1^*}{1 - \rho_1^*} \cdot \frac{\rho_1^*}{\overline{K}_1^* - \rho_1^*} = \underline{0.241}, \qquad a_2 = \frac{\rho_2^*}{1 - \rho_2^*} \cdot \frac{\rho_2^*}{\overline{K}_2^* - \rho_2^*} = \underline{0.257},$$

$$b_1 = \frac{\overline{K}_1^* - \rho_1^*}{\overline{K}_1^*} = \underline{0.8}, \qquad\qquad b_2 = \frac{\overline{K}_2^* - \rho_2^*}{\overline{K}_2^*} = \underline{0.558}.$$

STEP 5 Solve the closed network.

STEP 5.1 Determine the normalizing constant using the convolution method (see Section 8.1). For the functions $F_i(k_i), i = 1, 2$, we use Eq. (10.25) and get

$$F_1(0) = \underline{1}, \qquad F_2(0) = \underline{1},$$
$$F_1(1) = \underline{0.193}, \qquad F_2(1) = \underline{0.144},$$
$$F_1(2) = \underline{0.154}, \qquad F_2(2) = \underline{0.080},$$
$$F_1(3) = \underline{0.123}, \qquad F_2(3) = \underline{0.045}.$$

The procedure to determine the $G(K)$ is similar to the one in Table 8.2 and summarized in the following tabular form:

n \ k	1	2
0	1	1
1	0.193	0.3362
2	0.154	0.2618
3	0.123	0.2054

For the normalizing constant we get $G(K) = \underline{0.2054}$.

STEP 5.2 Determine ρ_i and $\overline{K}_i$. For this we need the normalizing constant $G_N^{(i)}(k)$ of the network with Node i short-circuited. We have only two nodes, and therefore it follows immediately that

$$G_N^{(1)}(k) = F_2(k), \qquad G_N^{(2)}(k) = F_1(k);$$

thus

$$\rho_1 = 1 - \frac{G_N^{(1)}(K)}{G(K)} = \underline{0.782}, \qquad \rho_2 = 1 - \frac{G_N^{(2)}(K)}{G(K)} = \underline{0.400},$$

$$\overline{K}_1 = \sum_{k=1}^{3} k \cdot \frac{F_1(k)}{G(K)} \cdot G_N^{(1)}(K - k) = \underline{2.090},$$

$$\overline{K}_2 = \sum_{k=1}^{3} k \cdot \frac{F_2(k)}{G(K)} \cdot G_N^{(2)}(K - k) = \underline{0.910}.$$

STEP 5.3 Determine the new coefficients a_i. With Eq. (10.28) we get

$$a_1^{(\text{new})} = a_1 \cdot K \cdot \rho_1^* \cdot \left(\rho_1 \cdot \sum_{j=1}^{2} \frac{\overline{K}_j \cdot \rho_j^*}{\rho_j} \right)^{-1} = \underline{0.242},$$

$$a_2^{(\text{new})} = a_2 \cdot K \cdot \rho_2^* \cdot \left(\rho_2 \cdot \sum_{j=1}^{2} \frac{\overline{K}_j \cdot \rho_j^*}{\rho_j} \right)^{-1} = \underline{0.253}.$$

Because the old and new values of the a_i differ more than $\epsilon = 0.001$, the analysis of the closed network has to be started again, beginning with Step 5.1.

$$\vdots$$

After altogether three iteration steps the halting condition is fulfilled and we get the following final results:

$$\rho_1 = \underline{0.787}, \quad \rho_2 = \underline{0.394} \quad \text{and} \quad \overline{K}_1 = \underline{2.105}, \quad \overline{K}_2 = \underline{0.895}.$$

STEP 6 Compute all other performance measures:

$$\lambda_1 = \mu_1 \rho_1 = \underline{0.787}, \qquad \lambda_2 = \mu_2 \rho_2 = \underline{0.787},$$
$$\overline{T}_1 = \overline{K}_1/\lambda_1 = \underline{2.675}, \qquad \overline{T}_2 = \overline{K}_2/\lambda_2 = \underline{1.137}.$$

To compare the results, we want to give the exact values for the mean number of jobs:

$$\overline{K}_1 = \underline{2.206}, \quad \overline{K}_2 = \underline{0.794}.$$

The method of Marie (see Section 10.1.4.2) gives the following results:

$$\overline{K}_1 = \underline{2.200}, \quad \overline{K}_2 = \underline{0.800}.$$

The closer the service time distribution is to an exponential distribution ($c_{Bi}^2 = 1$), the more accurate are the results of the MEM. For exponentially distributed service and arrival times the results are always exact. Another important characteristic of the MEM is the fact that the computation time is relatively independent of the number of jobs in the network.

10.1.3 Decomposition for Open Networks

This section deals with approximate performance analysis of open non-product-form queueing networks, based on the method of decomposition. The first step is the calculation of the arrival rates and the coefficients of variation of the interarrival times for each node. In the second step, performance measures such as the mean queue length and the mean waiting time are calculated using the GI/G/1 and GI/G/m formulae from Sections 6.3.4 and 6.3.6. The methods introduced here are those due to Kühn [Kühn79, Bolc89], Chylla [Chyl86], Pujolle [PuAi86], Whitt [Whit83a, Whit83b] and Gelenbe [GePu87].

Open networks to be analyzed by these techniques must have the following properties:

- The interarrival times and service times are arbitrarily distributed and given by the first and second moments.

- The queueing discipline is FCFS and there is no restriction on the length of the queue.

- The network can have several classes of customers (exception is the method of Kühn).

- The nodes of the network can be of single or multiple server type.

- Class switching is not allowed.

With these prerequisites the method works as follows:

STEP 1 Calculate the arrival rates and the utilizations of the individual nodes using Eqs. (10.35) and (10.36), and (10.37) and (10.38), respectively.

STEP 2 Compute the coefficient of variation of the interarrival times at each node, using Eqs. (10.41)–(10.49).

STEP 3 Compute the mean queue length and other performance measures.

We need the following *fundamental formulae* (Step 1); most of them are from Section 7.1.2:

Arrival rate $\lambda_{ij,r}$ from node i to node j of class r:

$$\lambda_{ij,r} = \lambda_{i,r} \cdot p_{ij,r}. \tag{10.34}$$

Arrival rate $\lambda_{i,r}$ to node i of class r:

$$\lambda_{i,r} = \lambda_{0i,r} + \sum_{j=1}^{N} \lambda_{j,r} \cdot p_{ji,r}. \tag{10.35}$$

Arrival rate λ_i to node i:

$$\lambda_i = \sum_{r=1}^{R} \lambda_{i,r} \,. \tag{10.36}$$

Utilization $\rho_{i,r}$ of node i due to customers of class r:

$$\rho_{i,r} = \frac{\lambda_{i,r}}{m_i \cdot \mu_{i,r}} \,. \tag{10.37}$$

Utilization ρ_i of node i:

$$\rho_i = \sum_{r=1}^{R} \rho_{i,r} \,. \tag{10.38}$$

The mean service rate μ_i of node i:

$$\mu_i = \frac{1}{\displaystyle\sum_{r=1}^{R} \frac{\lambda_{i,r}}{\lambda_i} \cdot \frac{1}{m_i \cdot \mu_{i,r}}} \,. \tag{10.39}$$

Coefficient of variation c_{Bi} of the service time of node i:

$$c_{Bi}^2 = -1 + \sum_{r=1}^{R} \frac{\lambda_{i,r}}{\lambda_i} \cdot \left(\frac{\mu_i}{m_i \cdot \mu_{i,r}} \right)^2 \cdot (c_{i,r}^2 + 1) \,, \tag{10.40}$$

with coefficient of variation $c_{i,r}$ for the service time of customers of class r at node i.

The calculation of the *coefficient of variation* or interarrival times (Step 2) is done iteratively using the following three phases (see Fig. 10.6):

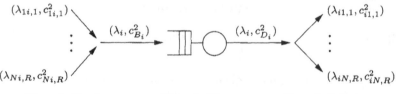

Phase 1: Merging   Phase 2: Flow   Phase 3: Splitting

Fig. 10.6 Phases of the calculation of the coefficient of variation.

- Phase 1: Merging

  Several arrival processes to each node are merged into a single arrival process. The arrival rate λ is the sum of the arrival rates of the individual arrival processes. For the coefficient of variation of the interarrival time, the several authors suggest different approximate formulae.

- Phase 2: Flow

The coefficient of variation c_D of the interdeparture times depends on the coefficients of variation of the interarrival times c_A and the service times c_B. Here again the different authors provide different approximate formulae.

- Phase 3: Splitting

  For the splitting of the departure process, all authors apply the formula

$$c_{ij,r}^2 = 1 + p_{ij,r} \cdot (c_{Di}^2 - 1). \tag{10.41}$$

The iteration starts with phase 1 and the initial values are $c_{ij,r} = 1$.

The corresponding formulae for phases 1 and 2 are introduced in the following:

- Decomposition of Pujolle [PuAi86]:

  Merging:

$$c_{Ai,r}^2 = \frac{1}{\lambda_{i,r}} \cdot \left(\sum_{j=1}^{N} c_{ji,r}^2 \cdot \lambda_{j,r} \cdot p_{ji,r} + c_{0i,r}^2 \cdot \lambda_{0,r} \cdot p_{0i,r} \right), \tag{10.42}$$

$$c_{Ai}^2 = \frac{1}{\lambda_i} \cdot \sum_{r=1}^{R} c_{Ai,r}^2 \cdot \lambda_{i,r}. \tag{10.43}$$

  Flow:

$$c_{Di}^2 = \rho_i^2 \cdot (c_{Bi}^2 + 1) + (1 - \rho_i) \cdot c_{Ai}^2 + \rho_i \cdot (1 - 2\rho_i). \tag{10.44}$$

- Decomposition of Whitt [Whit83b, Whit83a]:

  Merging: see Pujolle.

  Flow:

$$c_{Di}^2 = 1 + \frac{\rho_i^2 \cdot (c_{Bi}^2 - 1)}{\sqrt{m_i}} + (1 - \rho_i^2) \cdot (c_{Ai}^2 - 1). \tag{10.45}$$

- Decomposition of Gelenbe [GePu87]:

  Merging: see Pujolle.

  Flow:

$$c_{Di}^2 = -1 + \lambda_i \sum_{r=1}^{R} \frac{\rho_{i,r}}{\mu_{i,r} \cdot m_i} \cdot (c_{Bi,r}^2 + 1) + (1 - \rho_i) \cdot (c_{Ai}^2 + 1 + 2 \cdot \rho_i). \tag{10.46}$$

- Decomposition of Chylla [Chyl86]:

Merging:

$$c_{Ai,r}^2 = 1 + \sum_{j=0}^{N} \frac{\lambda_{j,r} \cdot p_{ji,r}}{\lambda_{i,r}} \cdot (c_{ji,r}^2 - 1), \qquad (10.47)$$

$$c_{Ai}^2 = 1 + \sum_{r=1}^{R} \frac{\lambda_{i,r}}{\lambda_i} \cdot (c_{Ai,r}^2 - 1). \qquad (10.48)$$

Flow:

$$c_{Di}^2 = 1 + P_{m_i}^2(\rho_i) \cdot (c_{Bi}^2 - 1) + (1 - P_{m_i}(\rho_i)) \cdot (c_{Ai}^2 - 1). \qquad (10.49)$$

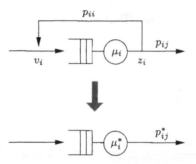

Fig. 10.7   Substitution of a feedback.

- Decomposition of Kühn [Kühn79]:

  Before starting with the iteration in the method of Kühn, nodes with direct feedback are replaced by nodes without feedback (see Fig. 10.7 and Eq. (10.50)).

$$\mu_i^* = \mu_i(1 - p_{ii}), \quad \lambda_i^* = \lambda_i(1 - p_{ii}), \quad c_{Bi}^{2*} = p_{ii} + (1 - p_{ii})c_{Bi}^2,$$
$$p_{ij}^* = \frac{p_{ij}}{1 - p_{ii}} \text{ for } j \neq i, \quad p_{ii}^* = 0.$$

$$(10.50)$$

Merging:

$$c_{Ai}^2 = 2\lambda_{1i} \cdot \lambda_{2i} \cdot (\lambda_{1i} + \lambda_{2i}) \cdot (I_1 + I_2 + I_3 + I_4) - 1. \qquad (10.51)$$

Only two processes can be merged by this formula. If there are more processes to merge, then the formula must be used several times. The terms $I_l, l = 1, 2, 3, 4$ are functions of the λ_{ji} and the c_{ji}. For details see [Kühn79] or [Bolc89].

Flow:

$$c_{Di}^2 = c_{Ai}^2 + 2\rho_i^2 c_{Bi}^2 - \rho_i^2(c_{Ai}^2 + c_{Bi}^2) \cdot G_{\text{KLB}}. \qquad (10.52)$$

For G_{KLB} see Eqs. (6.92), (6.112) and (10.56).

To calculate the *mean queue length* (Step 3) and the other performance measures, GI/G/1 and GI/G/m formulae from Sections 6.3.4 and 6.3.6 can be used in the case of a single class of customers. Here we give the extension of the most important formulae to the case of multiple classes of customers:

- M/M/m-FCFS:

$$\overline{Q}_{i,r\mathrm{M/M/m}} = \frac{\rho_{i,r}}{1 - \rho_i} \cdot P_{mi}, \tag{10.53}$$

with the probability of waiting, P_{mi}, given by Eq. (6.28)

- Allen–Cunneen [Alle90]:

$$\overline{Q}_{i,r\mathrm{AC}} \approx \overline{Q}_{i,r\mathrm{M/M/m}} \cdot \frac{(c_{Ai,r}^2 + c_{Bi,r}^2)}{2}. \tag{10.54}$$

- Krämer–Langenbach-Belz (KLB)[KrLa76]:

$$\overline{Q}_{i,r\mathrm{KLB}} \approx \overline{Q}_{i,r\mathrm{AC}} \cdot G_{\mathrm{KLB}}, \tag{10.55}$$

with the correction factor:

$$G_{\mathrm{KLB}} = \begin{cases} e^{\left(-\frac{2}{3} \cdot \frac{(1 - \rho_i)}{P_{mi}} \cdot \frac{(1 - c_{Ai,r}^2)^2}{(c_{Ai,r}^2 + c_{Bi,r}^2)}\right)}, & c_{Ai,r}^2 \leq 1, \\ e^{-\left((1 - \rho_i) \cdot \frac{(c_{Ai,r}^2 - 1)}{(c_{Ai,r}^2 + c_{Bi,r}^2)}\right)}, & c_{Ai,r}^2 > 1. \end{cases} \tag{10.56}$$

- Kimura [Kimu85]:

$$\overline{Q}_{i,r\mathrm{KIM}} \approx \frac{\overline{Q}_{i,r\mathrm{M/M/m}}}{2}$$
$$\cdot \frac{(c_{Ai,r}^2 + c_{Bi,r}^2)}{\frac{1 - c_{Ai,r}^2}{1 - 4C(m_i,\rho_i)} \cdot e^{\left(-\frac{2}{3}\frac{(1 - \rho_i)}{\rho_i}\right)} + \frac{1 - c_{Bi,r}^2}{1 + C(m_i,\rho_i)} + c_{Ai,r}^2 + c_{Bi,r}^2 - 1}, \tag{10.57}$$

with the coefficient of cooperation:

$$C(m_i, \rho_i) = (1 - \rho_i)(m_i - 1)\frac{\sqrt{4 + 5m_i} - 2}{16 m_i \rho_i}. \tag{10.58}$$

- Marchal [Marc78]:

$$\overline{Q}_{i,r\mathrm{MAR}} \approx \overline{Q}_{i,r\mathrm{M/M/m}} \frac{(1 + c_{Bi,r}^2) \cdot (c_{Ai,r}^2 + \rho_i^2 c_{Bi,r}^2)}{2(1 + \rho_i^2 c_{Bi,r}^2)} \cdot G_{\mathrm{KLB}}. \tag{10.59}$$

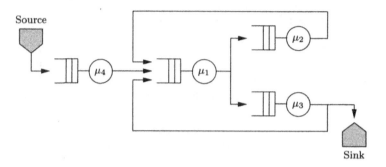

Fig. 10.8 Open non-product-form queueing network.

Example 10.4   Consider the open non-product-form queueing network of Fig. 10.8. The routing probabilities of the network are given by

$$p_{12} = 0.5, \quad p_{13} = 0.5, \quad p_{31} = 0.6, \quad p_{21} = p_{41} = 1.$$

Jobs are served at different nodes with rates

$$\mu_1 = 25, \quad \mu_2 = 33.333 \quad \mu_3 = 16.666, \quad \mu_4 = 20,$$

and arrive at node 4 with rate $\lambda_{04} = 4$. The coefficients of variation of the service times at different nodes as well as of the interarrival times are given by

$$c_{B_1}^2 = 2.0, \quad c_{B_2}^2 = 6.0, \quad c_{B_3}^2 = 0.5, \quad c_{B_4}^2 = 0.2, \quad c_{04}^2 = 4.0.$$

With Eq. (10.35) we can determine the arrival rate at each node:

$$\lambda_1 = \underline{20}, \quad \lambda_2 = \underline{10}, \quad \lambda_3 = \underline{10}, \quad \lambda_4 = \underline{4},$$

and using Eq. (10.37) we determine the utilizations of different nodes:

$$\rho_1 = \underline{0.8}, \quad \rho_2 = \underline{0.3}, \quad \rho_3 = \underline{0.6}, \quad \rho_4 = \underline{0.2}.$$

The initial values for the coefficients of variation c_{ij} are

$$c_{12}^2 = c_{13}^2 = c_{21}^2 = c_{31}^2 = c_{41}^2 = 1.$$

Now we start the iteration using the decomposition of Pujolle to obtain the different coefficients of variation of the interarrival times $c_{A_i}^2, i = 1, 2, 3, 4$.

<u>Iteration 1:</u>
Merging (Eq. (10.42)):

$$
\begin{aligned}
c_{A_1}^2 &= \frac{1}{\lambda_1}(c_{21}^2 \lambda_2 p_{21} + c_{31}^2 \lambda_3 p_{31} + c_{41}^2 \lambda_4 p_{41}) \\
&= \frac{1}{20}(1 \cdot 10 \cdot 1 + 1 \cdot 10 \cdot 0.6 + 1 \cdot 4 \cdot 1)
\end{aligned}
$$

$$= \underline{1}.$$

Similarly, we obtain

$$c_{A_2}^2 = \underline{1}, \quad c_{A_3}^2 = \underline{1}, \quad c_{A_4}^2 = \underline{4}.$$

Flow (Eq. (10.44)):

$$
\begin{aligned}
c_{D_1}^2 &= \rho_1^2(c_{B_1}^2 + 1) + (1 - \rho_1)c_{A_1}^2 + \rho_1(1 - 2\rho_1) \\
&= 0.8^2(2 + 1) + (1 - 0.8) \cdot 1 + 0.8 \cdot (1 - 2 \cdot 0.8) \\
&= \underline{1.64}.
\end{aligned}
$$

Similarly, we get

$$c_{D_2}^2 = \underline{1.45}, \quad c_{D_3}^2 = \underline{0.82}, \quad c_{D_4}^2 = \underline{3.368}.$$

Splitting (Eq. (10.41)):

$$
\begin{aligned}
c_{12}^2 &= 1 + p_{12} \cdot (c_{D_1}^2 - 1) \\
&= 1 + 0.5 \cdot (1.64 - 1) \\
&= \underline{1.32},
\end{aligned}
$$

$$c_{13}^2 = \underline{1.32}, \quad c_{21}^2 = \underline{1.45}, \quad c_{31} = \underline{0.892}, \quad c_{41}^2 = \underline{3.368}.$$

Iteration 2:
Merging (Eq. (10.42)):

$$c_{A_1}^2 = \underline{1.666}, \quad c_{A_2}^2 = \underline{1.32}, \quad c_{A_3}^2 = \underline{1.32}, \quad c_{A_4}^2 = \underline{4.0}.$$

Flow (Eq. (10.44)):

$$c_{D_1}^2 = \underline{1.773}, \quad c_{D_2}^2 = \underline{1.674}, \quad c_{D_3}^2 = \underline{0.948}, \quad c_{D_4}^2 = \underline{3.368}.$$

Splitting (Eq. (10.41)):

$$c_{12}^2 = \underline{1.387}, \quad c_{13}^2 = \underline{1.387}, \quad c_{21}^2 = \underline{1.674}, \quad c_{31}^2 = \underline{0.969}, \quad c_{41}^2 = \underline{3.368}.$$

$$\vdots$$

After six iteration steps we get the values for the coefficients of variation of the interarrival times $c_{A_i}^2$ at the nodes, shown in Table 10.3. In Table 10.4 the mean number of jobs at different nodes are given. We use the input parameters and values for the coefficients of variation for the interarrival times at the nodes after six iterations (given in Table 10.3). As we can see in Table 10.4, we get better results if we use the Krämer–Langenbach-Belz formula instead of the Allen–Cunneen formula.

Example 10.5   In this second example we have two servers at node 1. The queueing network model is shown in Fig. 10.9.

Table 10.3 Coefficients of variation $c_{A_i}^2$, $i = 1, 2, 3, 4$ of the inter-arrival times at the nodes

Iteration	$c_{A_1}^2$	$c_{A_2}^2$	$c_{A_3}^2$	$c_{A_4}^2$
1	1.0	1.0	1.0	4.0
2	1.666	1.320	1.320	4.0
3	1.801	1.387	1.387	4.0
4	1.829	1.400	1.400	4.0
5	1.835	1.403	1.403	4.0
6	1.836	1.404	1.404	4.0

Table 10.4 Mean number of jobs $\overline{K}_i$ at node i using the formulae of Allen–Cunneen (AC), Krämer–Langenbach-Belz (KLB), and discrete-event simulation (DES)

Methods	$\overline{K}_1$	$\overline{K}_2$	$\overline{K}_3$	$\overline{K}_4$
AC	6.94	0.78	1.46	0.31
KLB	5.97	0.77	1.42	0.27
DES	4.36	0.58	1.42	0.23

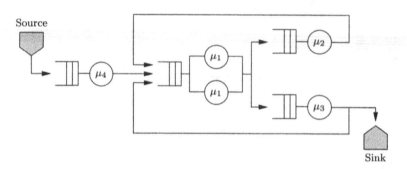

Fig. 10.9 Open non-product-form queueing network with two servers at node 1.

The service rate of each server at node 1 is $\mu_1 = 12.5$. All other input parameters are the same as in Example 10.4. We also have the same values for the arrival rates λ_i and the utilizations ρ_i. The initial values for the coefficients of variation c_{ij} are again

$$c_{12}^2 = c_{13}^2 = c_{21}^2 = c_{31}^2 = c_{41}^2 = 1.0 \,.$$

This time we use the formula of Whitt to determine the coefficients of variation c_{A_i} iteratively.

Iteration 1:
Merging: Whitt uses the same merging formula as Pujolle (Eq. (10.42)) and

we get

$$c_{A_1}^2 = c_{A_2}^2 = c_{A_3}^2 = \underline{1}, \quad c_{A_4}^2 = \underline{4.0}.$$

Flow (Eq. (10.45)):

$$c_{D_1}^2 = 1 + \frac{\rho_1^2(c_{B_1}^2 - 1)}{\sqrt{m_1}} + (1 - \rho_1^2)(c_{A_1}^2 - 1)$$

$$= 1 + \frac{0.64(2 - 1)}{\sqrt{2}} + (1 - 0.64)(1 - 1)$$

$$= \underline{1.453}$$

$$c_{D_2}^2 = \underline{1.45}, \quad c_{D_3}^2 = \underline{0.82}, \quad c_{D_4}^2 = \underline{3.848}.$$

Splitting (Eq. (10.41)):

$$c_{12}^2 = \underline{1.226}, \quad c_{13}^2 = \underline{1.226}, \quad c_{21}^2 = \underline{1.450}, \quad c_{31}^2 = \underline{0.892}, \quad c_{41}^2 = \underline{3.848}.$$

Iteration 2:
Merging (Eq. (10.42)):

$$c_{A_1}^2 = \underline{1.762}, \quad c_{A_2}^2 = \underline{1.226}, \quad c_{A_3}^2 = \underline{1.226}, \quad c_{A_4}^2 = \underline{4.0}.$$

Flow (Eq. (10.45)):

$$c_{D_1}^2 = \underline{1.727}, \quad c_{D_2}^2 = \underline{1.656}, \quad c_{D_3}^2 = \underline{0.965}, \quad c_{D_4}^2 = \underline{3.848}.$$

Splitting (Eq. (10.41)):

$$c_{12}^2 = \underline{1.363}, \quad c_{13}^2 = \underline{1.363}, \quad c_{21}^2 = \underline{1.656}, \quad c_{31}^2 = \underline{0.979}, \quad c_{41}^2 = \underline{3.848}.$$

$$\vdots$$

After seven iterations, we get the values for the coefficients of variation of the interarrival times $c_{A_i}^2$ at the nodes shown in Table 10.5.

Table 10.5 Coefficients of variation $c_{A_i}^2$ of the interarrival times at the nodes

Iteration	$c_{A_1}^2$	$c_{A_2}^2$	$c_{A_3}^2$	$c_{A_4}^2$
1	1.0	1.0	1.0	4.0
2	1.762	1.226	1.226	4.0
3	1.891	1.363	1.363	4.0
4	1.969	1.387	1.387	4.0
5	1.983	1.401	1.401	4.0
6	1.991	1.403	1.403	4.0
7	1.992	1.405	1.405	4.0

In Table 10.6, the mean number of jobs at the different nodes are given. We use the input parameters and values for the coefficients of variation for the interarrival times at the nodes after seven iterations (given in Table 10.5). These results are similar to the results of Example 10.4, and the differences with discrete-event simulation results are slightly smaller than in Example 10.4.

Table 10.6 Mean number of jobs $\overline{K}_i$ at the nodes using the formula of Allen–Cunneen (AC), Krämer–Langenbach-Belz (KLB), and discrete-event simulation (DES)

Methods	$\overline{K}_1$	$\overline{K}_2$	$\overline{K}_3$	$\overline{K}_4$
AC	6.48	0.78	1.46	0.31
KLB	6.21	0.77	1.42	0.27
DES	4.62	0.57	1.38	0.23

10.1.4 Methods for Closed Networks

10.1.4.1 Robustness for Closed Networks In case of closed non-product-form queueing networks with -/G/1 and -/G/m FCFS nodes, the easiest way to analyze them is to replace the -/G/1 and -/G/m FCFS nodes by -/M/1 and -/M/m FCFS nodes and to use a product-form method such as MVA, convolution, or the SCAT algorithm. This substitution can be done due to the property of *robustness* of such closed non-product-form queueing networks. Robustness, in general, means that a major change in system parameters generates only minor changes in the calculated performance measures. In our case it means that if we replace the values of the coefficients of variation of the service times c_{Bi} $(i = 1, \ldots, N)$ by 1 (i.e., we assume exponentially distributed service times instead of arbitrarily distributed service times), we obtain a tolerable deviation in the calculated performance measures.

We have investigated more than 100 very different closed non-product-form queueing networks with -/G/1 and -/G/m FCFS nodes. The maximum number of nodes was 10 and the coefficient of variation varied from 0.1 to 15. For a network with only -/G/1 FCFS nodes, the mean deviation is about 6% for both single and multiple class networks. In the case of only -/G/m FCFS nodes, the deviation decreases to about 2%. If we have only -/G/∞ nodes, the deviation is zero. The difference between single and multiple class networks is that the maximum deviation is greater for multiple class networks (40% instead of 20% for single class networks).

For open non-product-form queueing networks, the influence of the coefficients of variation is much greater, especially if the network is a tandem one without any feedback. So we cannot suggest to use the property of robustness to obtain approximate results for open non-product-form queueing networks. It is better to use one of the approximation methods introduced in Section 10.1.2.

Example 10.6   In this example we study the robustness property of several closed networks with different coefficients of variation. Consider the three network topologies given in Fig. 10.10 together with four combinations of the coefficients of variation (see Table 10.7). Note that the first combination (a) corresponds the case of all service times being exponentially distributed and the corresponding network being a product-form one. The analysis of the resulting models is done with $K = 5$ and $K = 10$ jobs in the system.

Table 10.7 Squared coefficients of variation for the networks in Fig. 10.10

Combinations	Node 1 $c_{B_1}^2$	Node 2 $c_{B_2}^2$	Node 3 $c_{B_3}^2$
a	1.0	1.0	1.0
b	0.2	0.4	0.8
c	4.0	1.0	0.2
d	2.0	4.0	8.0

We use two combinations of service rates. The first combination leads to a more balanced network, while the second combination results in a network with a bottleneck at node 1 (see Table 10.8).

Table 10.8 Two combinations of service rates

	μ_1	μ_2	μ_3
Balanced network	0.5	0.333	0.666
Network with bottleneck	0.5	1.0	2.0

In Tables 10.9 and 10.10, the throughputs λ for the three network topologies together with the different combinations of the coefficient of variation are given. If we compare the product-form approximation results (case a, Table 10.7) with the results for non-product-form networks (cases b, c, and d, Table 10.7), given in Tables 10.9 and 10.10, we see that the corresponding product-form network is a good approximation to the non-product-form one provided:

- The coefficients of variation are < 1 (case b) and/or not too big (case c).

- The network has a bottleneck.

Network 1

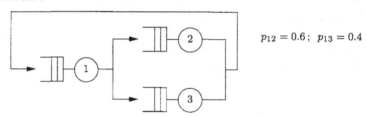

$p_{12} = 0.6$; $p_{13} = 0.4$

Network 2

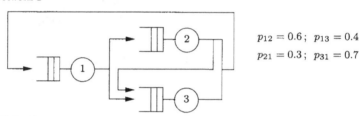

$p_{12} = 0.6$; $p_{13} = 0.4$
$p_{21} = 0.3$; $p_{31} = 0.7$

Network 3

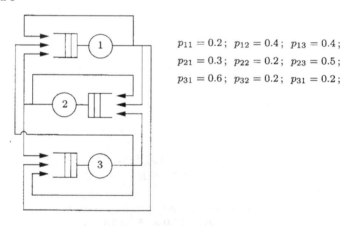

$p_{11} = 0.2$; $p_{12} = 0.4$; $p_{13} = 0.4$;
$p_{21} = 0.3$; $p_{22} = 0.2$; $p_{23} = 0.5$;
$p_{31} = 0.6$; $p_{32} = 0.2$; $p_{31} = 0.2$;

Fig. 10.10 Different closed networks.

Table 10.9 Throughputs λ for the networks in Fig. 10.10, for the balanced case

Squared Coefficient	Network 1		Network 2		Network 3	
of Variations	$K = 5$	$K = 10$	$K = 5$	$K = 10$	$K = 5$	$K = 10$
a	0.43	0.47	0.41	0.47	0.37	0.42
b	0.47	0.50	0.45	0.49	0.40	0.44
c	0.39	0.44	0.48	0.47	0.36	0.41
d	0.37	0.42	0.33	0.39	0.29	0.34

Table 10.10 Throughputs λ for the networks in Fig. 10.10, for the case with a bottleneck.

Squared Coefficient	Network 1		Network 2		Network 3	
of Variations	$K = 5$	$K = 10$	$K = 5$	$K = 10$	$K = 5$	$K = 10$
a	0.50	0.50	0.51	0.50	0.50	0.52
b	0.50	0.50	0.52	0.50	0.50	0.52
c	0.49	0.50	0.52	0.50	0.49	0.52
d	0.47	0.50	0.47	0.50	0.48	0.50

- The number of jobs in the network is high.

The larger the coefficients of variation ($c_{B_i}^2 > 1$), the larger are the deviations.

Example 10.7   In this example we consider in more detail the dependency of the approximation accuracy on the number of jobs K in the network. The examination is based on the first network given in Fig. 10.10 together with the same input parameters except that $\mu_i = 1$, for $i = 1, 2, 3$. The results for the throughputs λ for different numbers of jobs K are shown in Table 10.11. As we can see in Table 10.11, the larger the number of jobs K, the better the approximation.

Table 10.11 Throughput λ for the Network 1 shown in Fig. 10.10, for different values of K and combinations of the $c_{B_i}^2$

K	3	4	5	10	20	50
a	0.842	0.907	0.940	0.996	1.00	1.00
b	–	0.970	0.991	1.00	1.00	1.00
c	–	0.856	0.894	0.972	0.998	1.00
d	0.716	0.766	0.805	0.917	0.984	1.00

10.1.4.2 Marie's Method If the network contains single or multiple server FCFS nodes with generally distributed service times, then an approximate iterative method due to Marie [Mari79, Mari80] can be used. In this method each node of the network is considered in isolation with a Poisson arrival stream having load-dependent arrival rates $\lambda_i(k)$. To determine the unknown arrival rates $\lambda_i(k)$, we consider a product-form network corresponding to the given non-product-form network. The product-form network is derived from the original network by simply substituting the FCFS nodes with generally distributed service times by FCFS nodes with exponentially distributed service times (and load-dependent service rates). Because the service rates $\mu_i(k)$ for the nodes of the substitute network are determined in the isolated analysis of each node, the method is iterative where the substitute network is initialized with the service rates of the original network. The load-dependent arrival rates $\lambda_i(k)$ of the nodes $1, \ldots, N$ can be computed by short-circuiting node i in the substitute network (see Fig. 10.11). All other nodes of the network are combined into the composite node c (see Section 8.3).

Fig. 10.11 Short-circuiting node i in the substitute network.

If there are k jobs at node i, then there are exactly $K - k$ jobs at the composite node c. The load-dependent service rate through the substitute network with $K - k$ jobs and short-circuited node i is, therefore, the arrival rate at node i with k jobs at node i, which is

$$\lambda_i(k) = \lambda_c^{(i)}(K - k), \quad \text{for } k = 0, \ldots, (K - 1). \tag{10.60}$$

Here $\lambda_c^{(i)}(K - k)$ denotes the throughput of the network with node i short-circuited. This throughput can be computed using any algorithm for a product-form network.

The arrival rates $\lambda_i(k)$ can now be used for the isolated analysis of the single node of the network. The state probabilities $\pi_i(k)$ of the nodes in the substitute network and the values of the load-dependent service rates $\mu_i(k)$ for the next iteration step have to be computed. It is shown in [MaSt77] that in the considered network for each isolated node, a partition of the state space can be found so that the following condition holds:

$$\nu_i(k) \cdot \pi_i(k) = \lambda_i(k - 1) \cdot \pi_i(k - 1). \tag{10.61}$$

According to this equation the probability that a job leaves a node in state k is equal to the probability that a job arrives at the same node when this node is in state $k - 1$. The stochastic process, which can be assigned to the node under consideration, behaves exactly like a birth–death process with birth rates $\lambda_i(k)$ and death rate $\nu_i(k)$ (see Section 3.1). The state probabilities, which we are looking for, can therefore be computed as follows:

$$\pi_i(k) = \pi_i(0) \prod_{j=0}^{k-1} \frac{\lambda_i(j)}{\nu_i(j+1)}, \quad k = 1, \ldots, K, \tag{10.62}$$

$$\pi_i(0) = \frac{1}{1 + \sum_{j=1}^{K} \prod_{k=0}^{j-1} \frac{\lambda_i(k)}{\nu_i(k+1)}}. \tag{10.63}$$

The new load-dependent service rates for the nodes in the substitute network for the next iteration step are then chosen as follows:

$$\mu_i(k) = \nu_i(k) \quad \text{for } i = 1, \ldots, N. \tag{10.64}$$

Now the main problem is to determine the rates $\nu_i(k)$ for the individual nodes in the network. The fact that in closed queueing networks the number of jobs is constant allows us to consider the individual nodes in the network as elementary $\lambda(k)/G/m/K$ stations in the isolated analysis. The notation describes a more specific case of a $GI/G/m$ node where the arrival process is Poisson with state-dependent arrival rates and for which an additional number K exists, so that $\lambda(k) > 0$ for all $k < K$ and $\lambda(k) = 0$ for all $k \geq K$.

If we do the isolated analysis, we have to differentiate between the following:

• Product-form node

• Single server node with FCFS service strategy and generally distributed service time

In the first case, a comparison of the Eqs. (10.62) and (10.63) with the corresponding equations to compute the state probabilities of product-form nodes with load-dependent arrival rates [Lave83] shows that the rates $\nu_i(k)$ correspond to the service rates $\mu_i(k)$. For nodes with exponentially distributed service times and service discipline PS, IS, LCFS PR, respectively, it follows immediately from Eq. (10.64), that the service rates of the substitute network remain unchanged in each iteration step.

Next, consider FCFS nodes with generally distributed service times, which can be analyzed as elementary $\lambda(k)/G/1/K$-FCFS systems. Under the assumption that the distribution of the service time has a rational Laplace transform, we use a Cox distribution to approximate the general service times. Depending on the value of the squared coefficient of variation c_{Bi}^2 of the service time, different Cox models are used: If $c_{Bi}^2 = 1/m \pm \varepsilon$ for $m = 3, 4, \ldots$, where ε is a suitable tolerance area, then we use an Erlang model with m exponential phases. If $c_{Bi}^2 \geq 0.5$, then a Cox-2 model is chosen. For all other cases a special Cox model with m phases is chosen.

According to [Mari80], the parameters for the models are determined as follows:

- For a *Cox-2 model* (Fig. 10.12):

$$\hat{\mu}_{i1} = 2\mu_i, \quad \hat{\mu}_{i2} = \mu_i \cdot \frac{1}{c_{Bi}^2}, \quad a_i = \frac{1}{2c_{Bi}^2}.$$

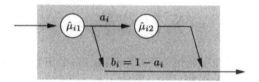

Fig. 10.12   Cox-2 model.

- The special *Cox-m model* is shown in Fig. 10.13. A job leaves this model with probability b_i after completing service at the first phase or runs through the rest of the $(m-1)$ exponential phases before it is completed. The number m of exponential phases is given by

$$m = \left\lceil \frac{1}{c_{Bi}^2} \right\rceil.$$

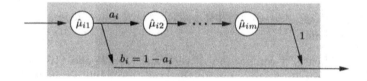

Fig. 10.13   Special Cox-m model.

From Section 1.3.1, we have

$$b_i = \frac{2mc_{Bi}^2 + (m-2) - \sqrt{m^2 + 4 - 4mc_{Bi}^2}}{2(m-1)(c_{Bi}^2 + 1)},$$

$$\hat{\mu}_i = [m - b_i(m-1)]\,\mu_i.$$

Here b_i denotes the probability that a job leaves node i after the first phase.

- For the *Erlang-m model* we have:

$$\hat{\mu}_i = m \cdot \mu_i.$$

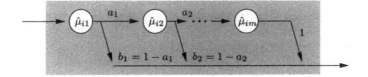

Fig. 10.14 General Cox-m model.

Next consider the general Cox-m model, shown in Fig. 10.14. Each of the elementary $\lambda(k)/C_m/1/K$ nodes can be assigned a CTMC whose states (k,j) indicate that the service process is in the jth phase and k jobs are at node i with corresponding steady-state probability $\pi(k,j)$. The underlying CTMC is shown in Fig. 10.15.

The corresponding balance equations for each node i are given by

$$\lambda_i(0)\pi_i(0,0) = \sum_{j=1}^{m} b_{i,j}\hat{\mu}_{i,j}\pi_i(1,j)\,, \tag{10.65}$$

$$\lambda_i(k-1)\pi_i(k-1,1) + \sum_{j=1}^{m} b_{i,j}\hat{\mu}_{i,j}\pi_i(k+1,j)$$
$$= \lambda_i(k)\pi_i(k,1) + \hat{\mu}_{i,1}\pi_i(k,1)\,, \tag{10.66}$$

$$\lambda_i(k-1)\pi_i(k-1,j) + a_{i,j-1}\hat{\mu}_{i,j-1}\pi_i(k,j-1)$$
$$= \lambda_i(k)\pi_i(k,j) + \hat{\mu}_{i,j}\pi_i(k,j)\,. \tag{10.67}$$

Let $\pi_i(k)$ denote the probability that there are k jobs at node i and $\nu_i(k)$ the conditional throughput of node i with k jobs:

$$\pi_i(k) = \sum_{j=1}^{m} \pi_i(k,j)\,, \tag{10.68}$$

$$\nu_i(k) = \frac{\displaystyle\sum_{j=1}^{m} b_{i,j}\hat{\mu}_{ij}\pi_i(k,j)}{\displaystyle\sum_{j=1}^{m} \pi_i(k,j)}\,. \tag{10.69}$$

Here $b_{i,j}$ denotes the probability that a job at node i leaves the node after the jth phase. If we add Eq. (10.66) to Eq. (10.67) (for $j = 2, \ldots, m$), we obtain the following for node i:

$$\lambda_i(k-1)\cdot\sum_{j=1}^{m} \pi_i(k-1,j) + \sum_{j=1}^{m} b_{i,j}\hat{\mu}_{ij}\pi_i(k+1,j)$$
$$= \lambda_i(k)\sum_{j=1}^{m} \pi_i(k,j) + \sum_{j=1}^{m} b_{i,j}\hat{\mu}_{ij}\pi_i(k,j)\,. \tag{10.70}$$

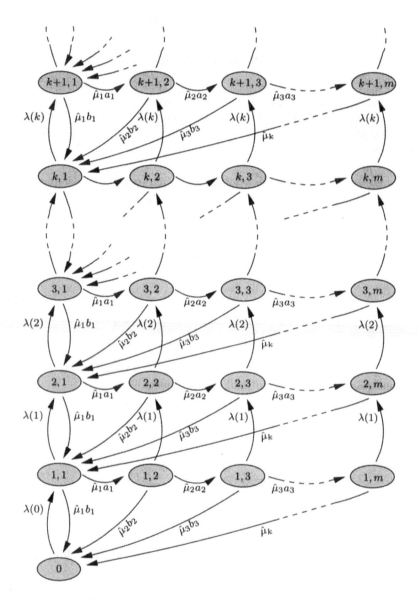

Fig. 10.15 CTMC for the general Cox-*m* model.

Now we can apply Eq. (10.69) to Eq. (10.70) and obtain

$$\lambda_i(k-1) \cdot \sum_{j=1}^{m} \pi_i(k-1,j) + \nu_i(k+1) \cdot \sum_{j=1}^{m} \pi_i(k+1,j)$$

$$= \lambda_i(k) \cdot \sum_{j=1}^{m} \pi_i(k,j) + \nu_i(k) \sum_{j=1}^{m} \pi_i(k,j),$$

$$\lambda_i(k-1)\pi_i(k-1) + \nu_i(k+1)\pi_i(k+1) = \lambda_i(k)\pi_i(k) + \nu_i(k)\pi_i(k).$$

If we now consider Eq. (10.65), we finally get for the solution of an elementary $\lambda(k)/C_m/1/K$ system in steady-state [Mari78]:

$$\nu_i(k)\pi_i(k) = \lambda_i(k-1)\pi_i(k-1).$$

But this is exactly Eq. (10.61). For this reason the state probabilities of an FCFS node with generally distributed service time can be determined using Eqs. (10.62) and (10.63) where Eq. (10.69) is used to compute the conditional throughputs $\nu_i(k)$ of the single nodes.

To determine the conditional throughputs, efficient algorithms are given in [Mari80]. In the case of an exponential distribution, $\nu_k(k)$ is just given by

$$\nu_i(k) = \mu_i. \tag{10.71}$$

In the case of an Erlang-m model, $\nu_i(k)$ is given by

$$\nu_i(k) = \frac{\hat{\mu}_i}{E_{i,k}(0) + \sum_{j=1}^{y-1} (-1)^j E_{i,k}(j) \left(\prod_{l=1}^{j} \frac{\nu_i(k-l)}{\hat{\mu}_i} \right)}, \tag{10.72}$$

with

$$E_{i,k}(j) = \sum_{n=j}^{y-1} T_{ik,n}(j), \qquad T_{ik,n}(j) = \sum_{\tau_{n,j}^k} \prod_{h=k-j}^{k} u_{i,h}^{V_h},$$

$$\tau_{n,j}^k = \left\{ (V_k, V_{k-1}, \dots, V_{k-j}) : \forall k \in \{k, \dots, k-j\}, \right.$$

$$\left. 0 \le V_h \le n-j, \quad \sum_{h=k-j}^{k} V_h = n-j \right\},$$

$$u_{i,h} = 1 + \phi_{i,h}, \qquad \phi_{i,h} = \frac{\lambda_i(h)}{\hat{\mu}_i}, \qquad y = \min\{m,k\}.$$

For the Cox-2 model ($c_{Bi}^2 \ge 0.5$), an even simpler equation exists:

$$\nu_i(1) = \frac{\lambda_i(1) \cdot \hat{\mu}_{i1} \cdot b_{i,1} + \hat{\mu}_{i1} \cdot \hat{\mu}_{i2}}{\lambda_i(1) + \hat{\mu}_{i2} + a_{i1} \cdot \hat{\mu}_{i1}}, \tag{10.73}$$

$$\nu_i(k) = \frac{\lambda_i(k) \cdot \hat{\mu}_{i1} \cdot b_{i,1} + \hat{\mu}_{i1} \cdot \hat{\mu}_{i2}}{(\lambda_i(k) + \hat{\mu}_{i1} + \hat{\mu}_{i2}) - \nu_i(k-1)}, \quad \text{for } k > 1. \tag{10.74}$$

We do not discuss the algorithms for the computation of the conditional throughputs of the other model type ($c_{Bi}^2 < 0.5$). The interested reader is referred to the original literature [Mari80].

The computed $\nu_i(k)$ are used in the next iteration step as load-dependent service rates for the nodes in the substitute network, Eq. (10.64). To stop the iteration, two conditions must be fulfilled. First, after each iteration we check whether the sum of the mean number of jobs equals the number of jobs in the network:

$$\left| \frac{K - \sum_{i=1}^{N} \overline{K}_i}{K} \right| < \varepsilon, \tag{10.75}$$

where ε is a suitable tolerance and

$$\overline{K}_i = \sum_{k=1}^{K} k \cdot \pi_i(k)$$

is the mean number of jobs at the ith node.

Second, we check whether the conditional throughputs of each node are consistent with the topology of the network:

$$\left| \frac{r_j - \frac{1}{N} \sum_{i=1}^{N} r_i}{\frac{1}{N} \sum_{i=1}^{N} r_i} \right| < \varepsilon \quad \text{for } j = 1, \dots, N, \tag{10.76}$$

where

$$r_j = \frac{\lambda_j}{e_j} = \frac{1}{e_j} \left(\sum_{k=1}^{K} \pi_j(k) \cdot \nu_j(k) \right)$$

is the normalized throughput of node j and e_j is the visit ratio. Usual values for ε are 10^{-3} or 10^{-4}.

Thus, the method of Marie can be summarized in the following four steps:

STEP 1 Initialize. Construct a substitute network with the same topology as the original one. The load-dependent service rates $\mu_i(k)$ of the substitute network for $i = 1, \dots, N$ and $k = 0, \dots, K$ are directly given by the original network. A multiple server node with service rate μ_i and m_i servers can be replaced by a single server node with the following load-dependent service rate (see also Section 6.2.3):

$$\mu_i(k) = \begin{cases} 0, & \text{for } k = 0, \\ \min\{k, m_i\} \cdot \mu_i, & \text{for } k > 0. \end{cases}$$

STEP 2 With Eq. (10.60), determine for $i = 1, \ldots, N$ and $k = 0, \ldots, (K-1)$ the load-dependent arrival rates $\lambda_i(k)$ by analyzing the substitute network. Here any product-form analysis algorithm from Chapter 8 can be used. For $k = K$ we have $\lambda_i(K) = 0$.

STEP 3 Analyze each individual node of the non-product-form network with the load-dependent arrival rate $\lambda_i(k)$. Determine the values $\nu_i(k)$ and the state probabilities $\pi_i(k)$. The computation method to be used for the $\nu_i(k)$ depends on the type of node being considered (for example, Eqs. (10.73) and (10.74) for Cox-2 nodes), while the $\pi_i(k)$ can all be determined using Eqs. (10.62) and (10.63).

STEP 4 Check the stopping conditions (10.75) and (10.76). If the conditions are fulfilled, then the iteration stops and the performance measures of the network can be determined with the equations given in Section 7.2. Otherwise use Eq. (10.64), $\mu_i(k) = \nu_i(k)$, to compute the new service rates for the substitute network and return to Step 2.

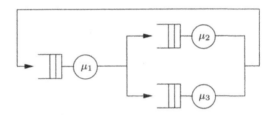

Fig. 10.16   Closed queueing network.

Example 10.8   In this example we show how to apply the method of Marie to a simple closed queueing network with $N = 3$ nodes and $K = 2$ jobs (see Fig. 10.16). The service strategy at all nodes is FCFS and the service times are generally distributed with the mean service times

$$\frac{1}{\mu_1} = 2\,\mathrm{sec}\,, \quad \frac{1}{\mu_2} = 4\,\mathrm{sec}\,, \quad \frac{1}{\mu_3} = 2\,\mathrm{sec}\,,$$

and the squared coefficients of variation are

$$c_{B1}^2 = 1\,, \quad c_{B2}^2 = 4\,, \quad c_{B3}^2 = 0.5\,.$$

The visit ratios are given as

$$e_1 = 1\,, \quad e_2 = 0.5\,, \quad e_3 = 0.5\,.$$

We assume a tolerance of $\varepsilon = 0.001$. The analysis of the network can then be done in the given four steps.

STEP 1 Initialize. The service rates of the substitute network can directly be taken from the given network:

$$\mu_1(1) = \mu_1(2) = 0.5, \quad \mu_2(1) = \mu_2(2) = 0.25, \quad \mu_3(1) = \mu_3(2) = 0.5.$$

We obtain the substitute network from the original model by replacing the FCFS nodes with generally distributed service times by FCFS nodes with exponentially distributed service times.

STEP 2 Compute the load-dependent arrival rates $\lambda_i(k)$ for $i = 1, 2, 3$, and for $k = 0, 1, 2$: When using the convolution method (see Section 8.1), we first need to determine the normalizing constants of the substitute network:

$$G(0) = 1, \quad G(1) = 5, \quad G(2) = 17,$$

and the normalizing constants $G_N^{(i)}$ for the network with the short-circuited node i with Eq. (8.11) are

$$G_N^{(1)}(0) = \underline{1}, \quad G_N^{(2)}0) = \underline{1}, \quad G_N^{(3)}0) = \underline{1},$$
$$G_N^{(1)}1) = \underline{3}, \quad G_N^{(2)}1) = \underline{3}, \quad G_N^{(3)}1) = \underline{4},$$
$$G_N^{(1)}2) = \underline{7}, \quad G_N^{(2)}2) = \underline{7}, \quad G_N^{(3)}2) = \underline{12}.$$

The throughput in the short-circuited node with $(K - k)$ jobs in the substitute network is then given by Eq. (8.14):

$$\lambda_c^{(i)}(K - k) = e_i \frac{G_N^{(i)}(K - k - 1)}{G_N^{(i)}(K - k)}.$$

Now the load-dependent arrival rates can be determined using Eq. (10.60):

$$\lambda_1(0) = \lambda_c^{(1)}(2) = e_1 \frac{G_N^{(1)}(1)}{G_N^{(1)}(2)} = \underline{0.429}.$$

Similarly, the values for the other arrival rates can be determined:

$$\lambda_2(0) = \underline{0.214}, \quad \lambda_3(0) = \underline{0.167},$$
$$\lambda_1(1) = \underline{0.333}, \quad \lambda_2(1) = \underline{0.167}, \quad \lambda_3(1) = \underline{0.125},$$
$$\lambda_1(2) = \underline{0}, \quad \lambda_2(2) = \underline{0}, \quad \lambda_3(2) = \underline{0}.$$

STEP 3 Perform the isolated analysis of the nodes of the given network. Because $c_{B1}^2 = 1$, the service time of node 1 is exponentially distributed (product-form node) and we have to use Eq. (10.64) $\nu_1(k) = \mu_1(k)$. The state probabilities for node 1 can be determined using Eqs. (10.62) and (10.63):

$$\pi_1(0) = \cfrac{1}{1 + \sum_{j=1}^{2} \prod_{k=0}^{j-1} \frac{\lambda_1(k)}{\mu_1(k+1)}} = \underline{0.412},$$

$$\pi_1(1) = \pi_1(0) \cdot \frac{\lambda_1(0)}{\mu_1(1)} = \underline{0.353},$$

$$\pi_1(2) = \pi_1(0) \cdot \prod_{j=0}^{1} \frac{\lambda_1(j)}{\mu_1(j+1)} = \underline{0.235}.$$

Nodes 2 and 3 are non-product-form nodes. Because $c_{Bi}^2 \geq 0.5$, for $i = 2, 3$, we use the Cox-2 distribution to model the nonexponentially distributed service time. The corresponding parameters for node 2 are given as follows:

$$\hat{\mu}_{21} = 2\mu_2 = \underline{0.5}, \quad \hat{\mu}_{22} = \mu_2 \cdot \frac{1}{c_{B2}^2} = \underline{0.0625}, \quad a_2 = \frac{1}{2c_{B2}^2} = \underline{0.125}.$$

For node 3 we get

$$\hat{\mu}_{31} = \hat{\mu}_{32} = a_3 = 1.$$

Using Eqs. (10.73) and (10.74), we can determine the conditional throughputs:

$$\nu_2(1) = \frac{\lambda_2(1)\hat{\mu}_{21}b_2 + \hat{\mu}_{21}\hat{\mu}_{22}}{\lambda_2(1) + \hat{\mu}_{22} + a_{21}\hat{\mu}_{21}} = \underline{0.357},$$

$$\nu_2(2) = \frac{\lambda_2(2)\hat{\mu}_{21}b_2 + \hat{\mu}_{21}\hat{\mu}_{22}}{(\lambda_2(2) + \hat{\mu}_{21} + \hat{\mu}_{22}) - \nu_2(1)} = \underline{0.152}.$$

Then

$$\nu_3(1) = \underline{0.471}, \quad \nu_3(2) = \underline{0.654}.$$

For the state probabilities we use Eqs. (10.62) and (10.63) and get

$$\pi_2(0) = \underline{0.443}, \quad \pi_2(1) = \underline{0.266}, \quad \pi_2(2) = \underline{0.291},$$
$$\pi_3(0) = \underline{0.703}, \quad \pi_3(1) = \underline{0.249}, \quad \pi_3(2) = \underline{0.048}.$$

STEP 4 Check the stopping condition, Eqs. (10.75) and (10.76):

$$\left| \frac{K - \sum_{i=1}^{N} \sum_{k=1}^{K} k \cdot \pi_i(k)}{K} \right| = 7.976 \cdot 10^{-3}.$$

Because the first stopping condition is not fulfilled, it is not necessary to check the second one.

Now we compute the new service rates of the substitute network using Eq. (10.64):

$$\mu_i(k) = \nu_i(k), \quad \text{for } i = 1, 2, 3,$$

and go back to Step 2.

STEP 2 Determine the load-dependent arrival rates using Eq. (10.60) and analyze the nodes of the network isolated from each other:

$$\vdots$$

After four iterations we get the following final results:

Marginal probabilities:

$$\pi_1(0) = \underline{0.438}, \quad \pi_2(0) = \underline{0.438}, \quad \pi_3(0) = \underline{0.719},$$
$$\pi_1(1) = \underline{0.311}, \quad \pi_2(1) = \underline{0.271}, \quad \pi_3(1) = \underline{0.229},$$
$$\pi_1(2) = \underline{0.251}, \quad \pi_2(2) = \underline{0.291}, \quad \pi_3(2) = \underline{0.052}.$$

Conditional throughputs:

$$\nu_2(1) = \underline{0.356}, \quad \nu_3(1) = \underline{0.466}, \quad \nu_2(2) = \underline{0.151}, \quad \nu_3(2) = \underline{0.652}.$$

Utilizations, Eq. (7.19):

$$\rho_1 = 1 - \pi_1(0) = \underline{0.562}, \quad \rho_2 = \underline{0.562}, \quad \rho_3 = \underline{0.281}.$$

Mean number of jobs, Eq. (7.26):

$$\overline{K}_1 = \pi_1(1) + 2 \cdot \pi_1(2) = \underline{0.813}, \quad \overline{K}_2 = \underline{0.853}, \quad \overline{K}_3 = \underline{0.333}.$$

In most cases the accuracy of Marie's method has proven to be very high. The predecessor of Marie's method was the iterative method of [CHW75a], which is a combination of the FES method and a recursive numerical method. A network to be analyzed is approximated by K substitute networks where the kth substitute network consists of node k of the original network and one composite node representing the rest of the network. The service rates of the composite nodes are determined using the FES method under the assumption that the network has a product-form solution. Then, the K substitute networks are analyzed using the recursive numerical method. The performance measures obtained from this analysis are approximate values for the whole network. If the results are not precise enough, the substitute networks are analyzed again with appropriate modifications of the service rates. These results are then the approximate values for the given network. The results thus obtained are normally less accurate than the ones derived by Marie's method.

10.1.4.3 Extended SCAT (ESCAT)

The core of the SCAT algorithm (see Section 9.1.2) is a single step of the MVA in which the formula for the mean response time, Eq. (10.186), is approximated so that it depends only on K and not on $(K-1)$. However, it is necessary to estimate the marginal probabilities $\pi_i(j \mid K-1)$ that need a lot of computation time. For this reason, another formula for the mean response time is derived from Eq. (6.29). This derivation depends only on $\overline{K}_i$ and μ_i. For $-/M/1$ nodes we apply Little's theorem to the equation

$$\overline{K}_i = \frac{\rho_i}{1 - \rho_i},$$

and after reorganizing we get

$$\overline{T}_i = \frac{1}{\mu_i}(1 + \overline{K}_i),$$

which is the basic equation of the core of the SCAT algorithm for $-/M/1$ nodes. Similarly, from Eq. (6.29), namely

$$\overline{K}_i = m_i \rho_i + \frac{\rho_i}{1 - \rho_i} \cdot P_{m_i}(\rho_i),$$

along with Little's theorem, we obtain for $-/M/m$ nodes

$$\overline{T}_i = \frac{1}{\mu_i \cdot m_i} \cdot (m_i + \overline{K}_i + P_{m_i}(\rho_i) - m_i \cdot \rho_i). \tag{10.77}$$

From Eq. (6.53), namely

$$\overline{K}_i = \rho_i + \frac{\rho_i^2}{1 - \rho_i} \cdot a_i, \quad \text{with } a_i = \frac{c_{B_i}^2 + 1}{2}, \tag{10.78}$$

we obtain for $-/G/1$ nodes

$$\overline{T}_i = \frac{1}{\mu_i} \cdot (1 + \overline{K}_i + \rho_i \cdot (a_i - 1)), \tag{10.79}$$

and from Eq. (6.99) and Little's theorem, namely

$$\overline{K}_i \approx m_i \rho_i + \frac{\rho_i}{1 - \rho_i} \cdot a_i \cdot P_{m_i}(\rho_i), \tag{10.80}$$

we obtain for $-/G/m$ nodes

$$\overline{T}_i = \frac{1}{\mu_i \cdot m_i} \cdot (m_i + \overline{K}_i + a_i \cdot P_{m_i}(\rho_i) - m_i \cdot \rho_i). \tag{10.81}$$

And finally we get as core equations for ESCAT

$$\overline{T}_i(K) = \frac{1}{\mu_i} \cdot (1 + \overline{K}_i(K - 1) + \rho_i(K - 1) \cdot (a_i - 1)), \tag{10.82}$$

for $-/G/1$ nodes and

$$\overline{T}_i(K) = \frac{1}{\mu_i \cdot m_i} \cdot (m_i + \overline{K}_i(K-1) + a_i \cdot P_{m_i}(\rho_i(K-1)) - m_i \cdot \rho_i(K-1)), \tag{10.83}$$

for $-/G/m$ nodes. For P_m see Eq. (6.28).

ESCAT can easily be extended to multiclass queueing networks in the same way as SCAT (9.1.2).

Table 10.12 Input parameters for the ESCAT example

Nodes	μ_i	c_{B_i}	m_i	Type
1	0.20	0.3	4	M/G/m FCFS
2	0.08	2.4	7	M/G/m FCFS
3	0.80	1.0	1	M/M/1 FCFS
4	0.12	3.9	10	M/G/m FCFS
5	0.05	-	-	M/G/∞ IS

Example 10.9   In this example the use of ESCAT is shown for a closed non-product-form queueing network with generally distributed service times, $N = 5$ nodes and $K = 10$ jobs in the system. The routing probabilities are $p_{12} = 0.1$, $p_{14} = 0.9$, $p_{24} = 1$, $p_{32} = 0.1$, $p_{34} = 0.9$, $p_{41} = 0.3$, $p_{42} = 0.3$, $p_{44} = 0.1$, $p_{45} = 0.2$, and $p_{54} = 1$. The other input parameters are listed in Table 10.12.

The results determined using ESCAT are listed in Table 10.13 together with results obtained by discrete-event simulation (DES) using PEPSY (see Section 12.1).

Table 10.13 Utilizations ρ and mean number of jobs from discrete-event simulation (DES) and ESCAT

Nodes	$\rho_{i_{\text{DES}}}$	$\rho_{i_{\text{ESCAT}}}$	$\overline{K}_{i_{\text{DES}}}$	$\overline{K}_{i_{\text{ESCAT}}}$
1	0.21	0.22	0.85	0.85
2	0.33	0.34	2.29	2.29
3	0.07	0.07	0.07	0.08
4	0.47	0.49	4.67	4.57
5	-	-	2.13	2.21

In this example where the utilizations are low the accuracy is very high. In fact the accuracy of the approximation is always high in case of low traffic. In the heavy traffic case the deviations are higher. This was found by examining 30 different test cases. The mean deviations for these examples were around 10%. ESCAT has been further improved to get a higher accuracy for the heavy traffic case [ESV00].

The accuracy of ESCAT can be further improved by incorporating the coefficient of variation of the interarrival time c_{A_i} in a_i in Eq. (10.78):

$$a_i = \frac{c_{B_i}^2 + c_{A_i}^2}{2}. \tag{10.84}$$

The coefficient of variation of the interarrival time c_{A_i} is calculated iteratively in a way similar to that done in Section 10.1.3 for open networks. The only difference is that the utilization ρ_i and the arrival rate λ_i need to be calculated in each iteration step while in the case of open networks ρ_i and λ_i need to be calculated only once. For the iteration the formulae of Section 10.1.3 can be used, e.g., the formulae of Pujolle/Ai [PuAi86]:

- Phase 1: Merging (Eq. (10.42)):

$$c_{Ai}^2 = \frac{1}{\lambda_i} \cdot \sum_{j=1}^{N} c_{ji}^2 \cdot \lambda_j \cdot p_{ji} \qquad (10.85)$$

- Phase 2: Flow (Eq. (10.44)):

$$c_{Di}^2 = \rho_i^2 \cdot (c_{Bi}^2 + 1) + (1 - \rho_i) \cdot c_{Ai}^2 + \rho_i \cdot (1 - 2\rho_i). \qquad (10.86)$$

- Phase 3: Splitting (Eq. (10.41)):

$$c_{ij}^2 = 1 + p_{ij} \cdot (c_{Di}^2 - 1). \qquad (10.87)$$

10.1.4.4 Extended SUM (ESUM) and BOTT (EBOTT)

The summation method and the bottapprox method can easily be extended to non-product-form queueing networks with generally distributed service times. For the mean number of jobs $\overline{K}_i$ in a $-/G/1$-FCFS node, we use again Eq. (10.78).

We also introduce a correction factor in Eq. (10.78) as we did in the product-form case to obtain

$$\overline{K}_i = \rho_i + \frac{\rho_i^2 \cdot a_i}{1 - \frac{K-1-a_i}{K-1} \rho_i}. \qquad (10.88)$$

In a $-/G/m$-FCFS node the mean number of jobs $\overline{K}_i$ is given by Eq. (10.80). The introduction of the correction factor in this equation yields

$$\overline{K}_i = m_i \rho_i + \frac{\rho_i}{1 - \frac{K-m_i-a}{K-m_i} \cdot \rho_i} \cdot a_i \cdot P_{m_i}. \qquad (10.89)$$

Now we can use the SUM method for networks with these two additional node types by introducing Eqs. (10.88) and/or (10.89) in the system equation (9.23) and calculate the throughput λ and other performance measures in the same way as in Section 9.2.

We can also use the BOTT method if we use the corresponding formulae for the function $h(\overline{K}_i)$ for the bottleneck $\overline{K}_i = \overline{K}_{\text{bott}}$. For the node type $-/G/1$-FCFS, $h(\overline{K}_{\text{bott}})$ has the form

$$h(\overline{K}_{\text{bott}}) = \frac{-(1 + b \cdot \overline{K}_{\text{bott}}) + \sqrt{(1 + b \cdot \overline{K}_{\text{bott}})^2 + 4 \cdot \overline{K}_{\text{bott}} \cdot (a - b)}}{2 \cdot (a - b)}, \qquad (10.90)$$

with

$$b = (K - 1 - a)/(K - 1). \qquad (10.91)$$

For the node type $-/G/m$-FCFS, the function $h(\overline{K}_{\mathrm{bott}})$ cannot be given in an exact form and therefore an approximation is used:

$$h(\overline{K}_{\mathrm{bott}}) = \frac{b - \sqrt{c}}{2 \cdot f \cdot m_{\mathrm{bott}}}, \qquad (10.92)$$

with

$$f = \frac{K - m_{\mathrm{bott}} - a}{K - m_{\mathrm{bott}}}, \quad a = \frac{1 + c_B^2}{2},$$

$$b = \overline{K}_{\mathrm{bott}} \cdot f + m_{\mathrm{bott}} + P_{m_{\mathrm{bott}}} \cdot a, \quad c = b^2 - (4 \cdot m_{\mathrm{bott}} \cdot \overline{K}_{\mathrm{bott}} \cdot f).$$

Example 10.10   In this example results of the application of ESUM and EBOTT are shown for a closed non-product-form queueing network with generally distributed service times, $N = 5$ nodes and $K = 17$ jobs in the system. The routing probabilities are: $p_{12} = 0.2$, $p_{13} = 0.4$, $p_{14} = 0.3$, $p_{15} = 0.1$, $p_{21} = 1$, $p_{31} = 1$, $p_{41} = 1$, $p_{51} = 1$. The other input parameters are listed in Table 10.14.

Table 10.14   Input parameters for the ESUM and EBOTT example

Nodes	μ_i	c_{B_i}	m_i	Type
1	13.50	1.0	1	M/M/1 FCFS
2	1.15	0.8	3	M/G/m FCFS
3	1.20	1.0	-	M/G/∞ IS
4	1.20	3.0	4	M/G/m FCFS
5	1.70	1.6	1	M/G/1 FCFS

The results determined using ESUM and EBOTT are listed in Table 10.15 together with results obtained by discrete-event simulation (DES) using PEP-SY (see Section 12.1).

Table 10.15   Overall throughput λ and overall response time $\overline{T}$ from DES, ESUM, and EBOTT

Method	λ	$\Delta\lambda/\%$	$\overline{T}$	$\Delta\overline{T}/\%$
DES	11.74	-	1.45	-
ESUM	11.34	3.4	1.50	3.5
EBOTT	11.93	1.5	1.43	1.5

The mean deviations of the overall throughput λ and the overall response time $\overline{T}$ for 96 different parameter sets which are solved using ESUM and EBOTT are listed in Table 10.16 together with the number of iterations. As can be seen from this table EBOTT is around 4 times faster than ESUM and also the accuracy of EBOTT is slightly better.

Table 10.16 Mean deviations of the overall throughput λ and overall response time $\overline{T}$ for 96 different non-product-form networks determined by ESUM and EBOTT

Method	$\Delta\lambda/\%$	$\Delta\overline{T}/\%$	Iterations
ESUM	6.08	9.40	10.44
EBOTT	3.78	7.41	2.25

The accuracy of the ESUM and EBOTT methods for non-product-form queueing networks can also be improved by introduction of the coefficient of variation of the interarrival time c_{A_i} via a_i in Eq. (10.78) as was done for ESCAT (Section 10.1.4.3). They can also easily be extended to the multiclass case as was done in Sections 9.2 and 9.3 for product-form queueing networks.

Problem 10.1 Analyze the closed queueing networks of Fig. 10.4 and Fig. 10.5 using ESCAT, ESUM, and EBOTT and compare the results.

10.1.5 Closing Method for Open and Mixed Networks

In [BGJ92], a technique is presented that allows every open queueing network to be replaced by a suitably constructed closed network. The principle of the *closing method* is quite simple: The external world of the open network is replaced by a $-/G/1$ node with the following characteristics:

1. The service rate μ_∞ of the new node is equal to the arrival rate λ_0 of the open network.

2. The coefficient of variation c_{B_∞} of service time at the new node is equal to the coefficient of variation of the interarrival time of the open network.

3. If the routing behavior of the open network is specified by visit ratios, then the visit ratio e_∞ of the new node is equal to 1. Otherwise the routing probabilities are assigned so that the external world is directly replaced by the new node.

The idea behind this technique is shown in Figs. 10.17 and 10.18.

A very high utilization of the new node is necessary to reproduce the behavior of the open network with adequate accuracy. This utilization is achieved when there is a large number of jobs K in the closed network. The performance measures are sufficiently accurate after the number of jobs in the

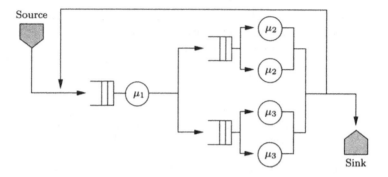

Fig. 10.17   Open network before using the closing method.

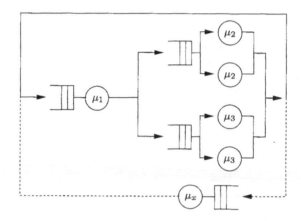

Fig. 10.18   Closed queueing network with the additional -/G/1 node for the closing method.

network has passed a certain threshold value K_i (see Fig. 10.19). Depending on the chosen solution algorithm, the following values for the number of jobs are proposed (see [BGJ92]):

$$K = 100 \quad \text{mean value analysis,}$$
$$K = 5000 \quad \text{summation method,}$$
$$K = 10000 \quad \text{SCAT algorithm.}$$

To check whether the number of jobs K in the closed network is sufficiently large, we can repeat the closing method with a larger number of jobs K and compare the results.

The closing method can be extended to multiple class queueing networks and mixed queueing networks. In the case of multiple class networks, the service rate $\mu_{\infty r}$ of the new node is given by $R \cdot \lambda_{0r}$, with $r = 1, \ldots, R$, where R is the number of classes. In the case of mixed queueing networks, the

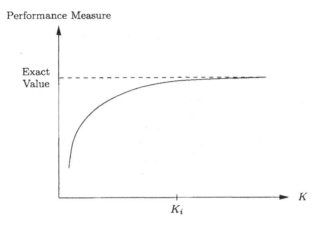

Fig. 10.19 Threshold in the closing method.

new node is added to the open classes only and then we can use any method for closed queueing networks also for mixed queueing networks. Open non-product-form networks are easier to analyze using the methods introduced in Section 10.1.3. But the closing method in combination with the SCAT algorithm or the SUM method is an interesting alternative to the MVA (see Section 8.2) for mixed queueing networks in the product-form case. The main advantage of the closing method is the ability to solve mixed non-product-form-queueing networks.

Example 10.11   In this first example, the closing method is shown for an open product-form queueing network with $\lambda_0 = 5$, $p_{12} = 0.3$, $p_{13} = 0.7$, and $p_{21} = p_{31} = 0.1$ (see Fig. 10.17). The service rates are $\mu_1 = 7$, $\mu_2 = 4$, and $\mu_3 = 3$. Using Jackson's method for this network, we obtain the following values for the mean response time $\overline{T}$ and the mean number of jobs in the open network:

$$\overline{T} = \underline{1.30}, \quad \overline{K} = \underline{6.52}.$$

In Table 10.17 the values for these performance measures are shown as a function of K_{closed}, the number of jobs in the closed network. The results are obtained using MVA. From the table we see that we obtain exact results if K_{closed} is large enough and that, in this case, $K_{\text{closed}} = 50$ is the threshold.

Example 10.12   In this second example, the closing method is used for an open non-product-form network (see Fig. 10.20) with $\lambda_0 = 3$ and $c_0^2 = 1.5$. The routing probabilities are given by

$$p_{12} = 0.3, \quad p_{13} = 0.7, \quad p_{22} = 0.3, \quad p_{24} = 0.7,$$
$$p_{33} = 0.1, \quad p_{34} = 0.9, \quad p_{41} = 0.4.$$

Table 10.17 Mean response time $\overline{T}$ and mean number of jobs $\overline{K}$ for several values of the number of jobs in the closed network

K_{closed}	$\overline{T}$	$\overline{K}$
10	0.95	4.76
20	1.24	6.17
30	1.29	6.47
50	1.30	6.52
60	1.30	6.52
100	1.30	6.52

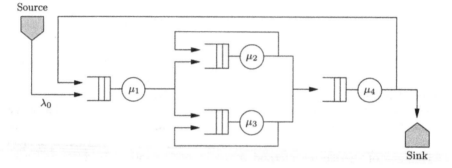

Fig. 10.20 Open queueing network.

Table 10.18 Input parameters for the network of Fig. 10.20

Node	μ_i	c_{Bi}^2
1	9	0.5
2	10	0.8
3	12	2.4
4	4	4

The other parameters are listed in Table 10.18.

With the method of Pujolle (see Section 10.1.3), we can analyze the open network directly to obtain

$$\overline{T} = \underline{4.43} \quad \text{and} \quad \overline{K} = \underline{13.47}.$$

The alternative method we use to solve the network of Fig. 10.20 is to first use the closing method and then use the SUM method (see Section 10.1.4.4) to approximately solve the resulting closed non-product-form network. The results are shown in Table 10.19. From this table we see that the threshold for

Table 10.19 Mean response time $\overline{T}$ and mean number of jobs $\overline{K}$ as a function of K_{closed}

K_{closed}	$\overline{T}$	$\overline{K}$
100	3.94	11.83
200	4.16	12.49
500	4.29	12.88
1000	4.34	13.02
5000	4.37	13.12
10000	4.37	13.12

K_{closed} is about 5000 and the results of the closing method and the method of Pujolle differ less than 3%.

Example 10.13   The open networks of Example 10.11 and Example 10.12 are much easier to analyze using the Jackson's method and the Pujolle method, respectively, rather than by the closing method. Mixed product-form networks are easy to analyze using the MVA for mixed networks but also using SCAT or SUM in combination with the closing method, especially for networks with multiple server nodes. But the main advantage of the closing method is the solution of mixed non-product-form queueing networks. As an example we use a mixed network with $N = 5$ nodes and $R = 2$ classes. Class 1 is closed and contains $K_1 = 9$ jobs (see Fig. 10.21). Class 2 (see Fig. 10.22) is open with arrival rate $\lambda_{01} = 5$ and the squared coefficient of variation $c_{01}^2 = 0.7$. There is no class switching in this network. The routing probabilities of the

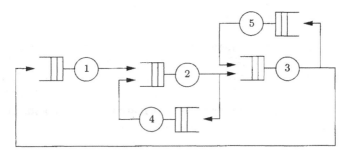

Fig. 10.21   Closed class 1 of the mixed network.

two classes are as follows:

Class 1: $p_{12,1} = 1.0$,   Class 2: $p_{12,2} = 1.0$,

$p_{23,1} = 0.7$, $p_{24,1} = 0.3$, $p_{23,2} = 0.8$, $p_{24,2} = 0.2$,

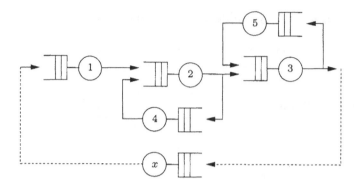

Fig. 10.22 Open class 2 of the mixed network with an additional node.

$$p_{31,1} = 0.7, \quad p_{35,1} = 0.3, \qquad\qquad p_{30,2} = 0.7, \quad p_{35,2} = 0.3,$$
$$p_{42,1} = 1.0, \qquad\qquad\qquad\qquad\quad p_{42,2} = 1.0,$$
$$p_{53,1} = 1.0, \qquad\qquad\qquad\qquad\quad p_{53,2} = 1.0.$$

The other input parameters are listed in Table 10.20. Now we solve the

Table 10.20 Input parameters for the mixed queueing network

Nodes	$\mu_{i,1} = \mu_{i,2}$	$c_{i1}^2 = c_{i2}^2$	m_i
1	4	0.4	3
2	4	0.3	4
3	6	0.3	3
4	4	0.4	2
5	5	0.5	2

equivalent closed queueing network with an additional node ($\mu_\infty = 5$, $c_\infty^2 = 0.7$) for class 2 using the SUM method (see Section 10.1.4.4) and $K_2 = 500$. The results for the utilization ρ_i and the mean number of jobs $\overline{K}_i$ are listed in Table 10.21. Since SUM is an approximation method, we obtain deviations from discrete-event simulation results, which are about 6% for the utilizations ρ_i and about 5% for the mean number of jobs $\overline{K}_i$.

10.2   DIFFERENT SERVICE TIMES AT FCFS NODES

If the given non-product-form network is a "small-distance" away from a product-form network, we may be able to modify the MVA algorithm to approximately solve the network. Based on this idea [Bard79] extended the MVA to priority scheduling as well as to networks with FCFS nodes and with

Table 10.21 Utilizations ρ and mean number of jobs from discrete-event simulation (DES) and the closing method (CLS)

Nodes	$\rho_{i_{\text{DES}}}$	$\rho_{i_{\text{CLS}}}$	$\overline{K}_{i_{\text{DES}}}$	$\overline{K}_{i_{\text{CLS}}}$
1	0.89	0.84	4.9	5.2
2	0.90	0.85	5.5	5.8
3	0.85	0.80	4.0	4.1
4	0.46	0.43	1.1	1.0
5	0.46	0.43	1.1	1.0

different service rates for different job classes. The key idea for treating the latter case is to replace the MVA Eq. (8.31) by

$$\overline{T}_{ir}(\mathbf{k}) = \frac{1}{\mu_{ir}} + \sum_{s=1}^{R} \frac{1}{\mu_{is}} \cdot \overline{K}_{is}(\mathbf{k} - \mathbf{1}_r). \tag{10.93}$$

This equation can also be extended for multiple server nodes [Hahn88] as

$$\begin{aligned}
\overline{T}_{ir}(\mathbf{k}) = \frac{1}{m_i} &\left(\frac{1}{\mu_{ir}} + \sum_{s=1}^{R} \frac{1}{\mu_{is}} \cdot \overline{K}_{is}(\mathbf{k} - \mathbf{1}_r) \right. \\
&\left. + \frac{1}{\mu_i} \sum_{j=0}^{m_i-2} (m_i - 1 - j) \cdot \pi_i(j \mid \mathbf{k} - \mathbf{1}_r) \right),
\end{aligned} \tag{10.94}$$

with

$$\frac{1}{\mu_i} = \sum_{s=1}^{R} \frac{\lambda_{is}}{\lambda_i} \cdot \frac{1}{\mu_{is}}.$$

In order to demonstrate the use of this method, consider a queueing network consisting of $N = 6$ nodes and $R = 2$ job classes. Nodes 1 to 5 are FCFS nodes with one server each and node 6 is an IS node. Class 1 contains $K_1 = 10$ jobs and class 2 has $K_2 = 6$ jobs. All service times are exponentially distributed. Table 10.22 shows all service rates and visit ratios for the network. Service rate μ_{11}, marked with an $*$ in the table, is varied and the mean overall response time for all class-1 jobs is determined. We compare the results of the extended MVA with DES.

The values given in Table 10.23 show that the extension of the MVA by Eq. (10.93) gives good results for queueing networks with FCFS nodes with different service rates for different job classes. SCAT and Bard–Schweitzer algorithm can also be extended [Förs89] in a similar fashion.

Table 10.22 Parameters of the network described in Fig. 10.22

i	μ_{i1}	μ_{i2}	e_{i1}	e_{i2}
1	*	0.5	8	20
2	1	1	2	2
3	1	1	2	4
4	1	1	2	6
5	1	1	2	8
6	0.1	10^7	1	1

Table 10.23 Mean overall response time of a class-1 job for different service rates μ_{11}

μ_{11}	0.5	2	128
MVA-Ext.	260.1	143.1	105.4
DES	260.1	141.1	102.0

10.3   PRIORITY NETWORKS

We will consider three different methods for networks with priority classes: extended MVA, shadow server technique, and extended SUM method.

10.3.1   PRIOMVA

If a queueing network contains nodes where some job classes have priority over other job classes, then the product-form condition is not fulfilled and the techniques presented in Chapters 8 and 9 cannot be used. For the approximate analysis of such priority networks [BKLC84] suggest an extension of the MVA. The MVA for mixed queueing networks is thus extended to formulae for the following priority node types:

- -/M/1-FCFS PRS (preemptive resume):

  As per this strategy, as soon as a high priority job arrives at the node it will be serviced if no other job with higher priority is in service at the moment. If a job with lower priority is in service when the higher priority one arrives, then this low priority job is preempted. When all jobs with higher priority are serviced, then the preempted job is resumed starting at the point where it was stopped. It is assumed that the preemption does not cause any overhead. The service strategy within a priority class is FCFS.

- -/M/1-FCFS HOL (non-preemptive head-of-line):

Here a job with higher priority must wait until the job that is processed at the moment is ready to leave the node even if its priority is lower than the priority of the arriving job. The service strategy within a class is again FCFS.

The priority classes are ordered linearly where class R has lowest priority and class 1 has the highest one.

At first we consider $-/M/1$-FCFS PRS nodes and derive a formula for the mean response time $\overline{T}_r$. In this case the mean response time of a class-r job consists of

- The mean service time:
$$\frac{1}{\mu_r}.$$

- The time it has to wait until all jobs with higher or the same priority are serviced:
$$\sum_{s=1}^{r} \frac{\overline{K}_s}{\mu_s}.$$

- The time for serving all jobs with a higher priority that arrive during the response time $\overline{T}_r$:
$$\sum_{s=1}^{r-1} \frac{\overline{T}_r \cdot \lambda_s}{\mu_s}.$$

If we combine these three expressions, then for the mean response time of a priority class-r job, $1 \leq r \leq R$, we get the following implicit equation:

$$\overline{T}_r = \sum_{s=1}^{r} \frac{\overline{K}_s}{\mu_s} + \sum_{s=1}^{r-1} \frac{\overline{T}_r \cdot \lambda_s}{\mu_s} + \frac{1}{\mu_r},$$

where $\overline{K}_s$ denotes the mean number of class-s jobs at the node. With the relation $\rho_s = \lambda_s/\mu_s$ it follows that:

$$\overline{T}_r = \frac{\dfrac{1}{\mu_r} + \sum\limits_{s=1}^{r} \dfrac{\overline{K}_s}{\mu_s}}{1 - \sum\limits_{s=1}^{r-1} \rho_s}. \tag{10.95}$$

If, on the other hand, we consider an $-/M/1$-FCFS HOL node, then the considered job can not be preempted any more. In this case the mean response time consists of

- The mean service time:
$$\frac{1}{\mu_r}.$$

- The time needed to serve higher priority jobs that arrive during the waiting time with the mean $\overline{W} = \left(\overline{T}_r - \frac{1}{\mu_r}\right)$:

$$\sum_{s=1}^{r-1} \left(\overline{T}_r - \frac{1}{\mu_r}\right) \cdot \frac{\lambda_s}{\mu_s}.$$

- The mean time needed for the job that is being served at the moment:

$$\sum_{s=1}^{R} \frac{\rho_s}{\mu_s}.$$

- The mean service time of jobs with the same or higher priority that are still in the queue:

$$\sum_{s=1}^{r} \frac{\overline{K}_s - \rho_s}{\mu_s}.$$

If we combine these three terms for the mean response time of a priority class-r job, $1 \le r \le R$, we get

$$\overline{T}_r = \sum_{s=1}^{r} \frac{\overline{K}_s - \rho_s}{\mu_s} + \sum_{s=1}^{R} \frac{\rho_s}{\mu_s} + \sum_{s=1}^{r-1} \left(\overline{T}_r - \frac{1}{\mu_r}\right) \cdot \frac{\lambda_s}{\mu_s} + \frac{1}{\mu_r}.$$

If we use the relation $\rho_s = \lambda_s/\mu_s$, then we get

$$\overline{T}_r = \frac{1}{\mu_r} + \frac{\sum_{s=1}^{r} \frac{\overline{K}_s}{\mu_s} + \sum_{s=r+1}^{R} \frac{\rho_s}{\mu_s}}{1 - \sum_{s=1}^{r-1} \rho_s}. \tag{10.96}$$

The equations just developed give the exact $\overline{T}_r$ values for M/M/1- priority queues. But in a priority queueing network the arrival process at a node is normally not Poisson. This condition means that if we apply the equations given to a priority network, we do not get exact results. Furthermore, these equations allow class-dependent service time distributions, while in product-form queueing networks a class-independent service time distribution is required at FCFS nodes. Nevertheless, the approximation of [BKLC84] is based on these assumptions.

Let $\overline{K}_{is}^{(r)}$ denote the mean number of class-s jobs at node i that an arriving class-r job sees given the network population vector $\mathbf{k}$. If Eq. (10.95) is applied to node i of a priority network, then this equation can be generalized as follows:

$$\overline{T}_{ir}(\mathbf{k}) = \frac{\frac{1}{\mu_{ir}} + \sum_{s=1}^{r} \frac{\overline{K}_{is}^{(r)}(\mathbf{k})}{\mu_{is}}}{1 - \sum_{s=1}^{r-1} \rho_{is}}. \tag{10.97}$$

Similarly, Eq. (10.96) is of form

$$\overline{T}_{ir}(\mathbf{k}) = \frac{1}{\mu_{ir}} + \frac{\sum\limits_{s=1}^{r} \dfrac{\overline{K}_{is}^{(r)}(\mathbf{k})}{\mu_{is}} + \sum\limits_{s=r+1}^{R} \dfrac{\rho_{is}}{\mu_{is}}}{1 - \sum\limits_{s=1}^{r-1} \rho_{is}}. \tag{10.98}$$

For the computation of $\overline{K}_{is}^{(r)}(\mathbf{k})$, the arrival theorem (see [BKLC84]) is assumed to hold:

$$\overline{K}_{is}^{(r)}(\mathbf{k}) = \begin{cases} \overline{K}_{is}(\mathbf{k} - \mathbf{1}_r), & \text{for closed classes } r, \\ \overline{K}_{is}(\mathbf{k}), & \text{for open classes } r. \end{cases} \tag{10.99}$$

For open class s, the equation

$$\overline{K}_{is}(\mathbf{k}) = \lambda_s \cdot e_{is} \cdot \overline{T}_{is}(\mathbf{k}) \tag{10.100}$$

can be used to eliminate the unknown $\overline{K}_{is}(\mathbf{k})$. Then in Eqs. (10.97) and (10.98) we only need to determine the utilizations ρ_{is}. For open classes we have $\rho_{is} = \lambda_{is}/\mu_{is}$, independent of the number of jobs in the closed classes. But for closed class s the utilization ρ_{is} depends on the population vector and it is not obvious which $\rho_{is}(\mathbf{k})$, $\mathbf{0} \leq \mathbf{k} \leq \mathbf{K}$, should be used. However, [ChLa83] suggests

$$\rho_{is} = \rho_{is}(\mathbf{k} - \overline{K}_{is}), \tag{10.101}$$

where $(\mathbf{k} - \overline{K}_{is})$ is the population vector of the closed classes with $\overline{K}_{is}$ jobs fewer in class s. If $\overline{K}_{is}$ is not an integer, then linear interpolation is used to get a convenient value for ρ_{is}. Equation (10.101) is based on the assumption that if there are already $\overline{K}_{is}$ jobs at node i, then the arrival rate of class-s jobs at node i is given by the remaining $\mathbf{k} - \overline{K}_{is}$ jobs in the network, and hence

$$\rho_{is} = \frac{\lambda_{is}(\mathbf{k} - \overline{K}_{is})}{\mu_{is}} = \rho_{is}(\mathbf{k} - \overline{K}_{is}).$$

If we insert Eqs. (10.99)–(10.101) into Eqs. (10.97) and (10.98), respectively, then the mean response time of the *closed* class r at the priority node i when there are $\mathbf{k}$ jobs in the closed classes of the network:

$$\text{PRS:} \quad \overline{T}_{ir}(\mathbf{k}) = \frac{\dfrac{1}{\mu_{ir}} + \sum\limits_{s=1}^{r} \dfrac{\overline{K}_{is}(\mathbf{k} - \mathbf{1}_r)}{\mu_{is}}}{1 - \sum\limits_{s=1}^{r-1} \rho_{is}'}, \tag{10.102}$$

$$\text{HOL:} \quad \overline{T}_{ir}(\mathbf{k}) = \frac{1}{\mu_{ir}} + \frac{\sum\limits_{s=1}^{r} \dfrac{\overline{K}_{is}(\mathbf{k} - \mathbf{1}_r)}{\mu_{is}} + \sum\limits_{s=r+1}^{R} \dfrac{\rho_{is}(\mathbf{k} - \mathbf{1}_r)}{\mu_{is}}}{1 - \sum\limits_{s=1}^{r-1} \rho_{is}'}, \tag{10.103}$$

where

$$\rho'_{is} = \begin{cases} \rho_{is}(\mathbf{k} - \overline{K}_{is}), & \text{for closed class } s, \\ \rho_{is}, & \text{for open class } s. \end{cases}$$

Analogously, for the mean response time of the *open* class r at the priority node i we get

$$\text{PRS: } \overline{T}_{ir}(\mathbf{k}) = \frac{\dfrac{1}{\mu_{ir}} + \sum\limits_{s=1}^{r} \dfrac{\overline{K}_{is}(\mathbf{k})}{\mu_{is}}}{1 - \sum\limits_{s=1}^{r-1} \rho'_{is}}, \tag{10.104}$$

$$\text{HOL: } \overline{T}_{ir}(\mathbf{k}) = \frac{\dfrac{1}{\mu_{ir}}\left(1 - \sum\limits_{s=1}^{r-1} \rho'_{is}\right) + \sum\limits_{s=1}^{r-1} \dfrac{\overline{K}_{is}(\mathbf{k})}{\mu_{is}} + \sum\limits_{s=r+1}^{R} \dfrac{\rho_{is}(\mathbf{k})}{\mu_{is}}}{1 - \sum\limits_{s=1}^{r-1} \rho'_{is}}. \tag{10.105}$$

Now the modified MVA with the additional priority node types can be given. The closed classes are denoted by $1, \ldots, CL$ and the open classes by $1, \ldots, OP$. Because of the fact that the values of $\overline{K}_{is}(\mathbf{k})$ are also needed for the open class s, it is necessary to determine the performance measures for the open classes in each iteration step of the algorithm.

STEP 1 Initialize. For each node $i = 1, \ldots, N$ determine the utilizations of the open classes $op = 1, \ldots, OP$ using Eq. (8.55), that is

$$\rho_{i,op} = \frac{1}{\mu_{i,op}} \lambda_o p e_{i,op},$$

and check for stability ($\rho_{i,op} \leq 1$).
Set $\overline{K}_{i,cl}(0) = 0$ for all nodes $i = 1, \ldots, N$ and all closed classes $cl = 1, \ldots, CL$.

STEP 2 Construct a closed model that contains only jobs of the closed classes and solve this model using the MVA.
Iterate for $\mathbf{k} = 0, \ldots, \mathbf{K}$:

STEP 2.1 Determine $\overline{T}_{ir}(\mathbf{k})$ for all nodes $i = 1, \ldots, N$ and all closed classes $r = 1, \ldots, CL$ using Eqs. (8.53), (8.54), and (10.102) and (10.103), respectively.

STEP 2.2 Determine $\lambda_r(\mathbf{k})$ for all closed classes $r = 1, \ldots, CL$ (Eq. (8.46)).

STEP 2.3 Determine $\overline{K}_{ir}(\mathbf{k})$ for all nodes $i = 1, \ldots, N$ and for all closed classes $r = 1, \ldots, CL$ using Eq. (8.47).

STEP 2.4 Determine $\overline{K}_{ir}(\mathbf{k})$ for all nodes $i = 1, \ldots, N$ and all open classes $r = 1, \ldots, OP$ using Eqs. (8.51) and (8.52), and Eqs. (10.100), (10.104), and (10.105), respectively.

STEP 3 Determine all other performance measures using the computed results.

The computation sequenced from the highest priority down to the lowest one.
The algorithm is demonstrated in the following example:

Example 10.14   Consider the queueing network in Fig. 10.23. The network consists of $N = 2$ nodes and $R = 3$ job classes. Class 1 has highest priority and is open, while the classes 2 and 3 are closed. Node 1 is of type

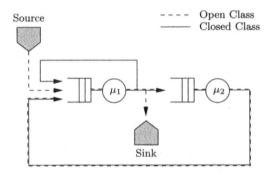

Fig. 10.23 Mixed network with priority nodes.

$-/G/1$-PS and node 2 is of type $M/M/1$-FCFS PRS. The arrival rate for class-1 jobs is $\lambda_1 = 1$. Each closed class contains one job $(K_2 = K_3 = 1)$. The mean service times are given as

$$\frac{1}{\mu_{11}} = 0.4\,, \quad \frac{1}{\mu_{12}} = 0.3\,, \quad \frac{1}{\mu_{13}} = 0.5\,,$$

$$\frac{1}{\mu_{21}} = 0.6\,, \quad \frac{1}{\mu_{22}} = 0.5\,, \quad \frac{1}{\mu_{23}} = 0.8\,.$$

The visit ratios are

$$e_{11} = 2\,, \quad e_{12} = 1\,, \quad e_{13} = 1\,, \quad e_{21} = 1\,, \quad e_{22} = 0.5\,, \quad e_{23} = 0.4\,.$$

STEP 1 Initialize. Determine the utilizations of the open class:

$$\rho_{11} = \lambda_1 e_{11}\cdot \frac{1}{\mu_{11}} = \underline{0.8}\,, \quad \rho_{21} = \lambda_1 e_{21}\cdot \frac{1}{\mu_{21}} = \underline{0.6}\,.$$

Set $\overline{K}_{12}(\mathbf{0}) = \overline{K}_{13}(\mathbf{0}) = \overline{K}_{22}(\mathbf{0}) = \overline{K}_{23}(\mathbf{0}) = 0$.

STEP 2 Analyze the model using the MVA.

MVA-iteration for $\mathbf{k} = (0,0)$:

STEP 2.4 Mean number of jobs in the open class 1:
Node 1, Eq. (8.51):

$$\overline{K}_{11}(0,0) = \frac{\rho_{11}\left[1 + \overline{K}_{12}(0,0) + \overline{K}_{13}(0,0)\right]}{1 - \rho_{11}} = \underline{4}\,.$$

Node 2, Eqs. (10.100) and (10.104):

$$\overline{K}_{21}(0,0) = \lambda_1 \cdot e_{21} \cdot \frac{1/\mu_{21}}{1 - \rho_{21}} = \underline{1.5}\,.$$

MVA-iteration for $\mathbf{k} = (0,1)$:

STEP 2.1 Mean response time for the closed class 3: We use Eqs. (8.53)
and (10.102) to get

$$\overline{T}_{13}(0,1) = \frac{1}{\mu_{13}} \cdot \frac{1}{1 - \rho_{11}} = \underline{2.5}\,, \quad \overline{T}_{23}(0,1) = \frac{\frac{1}{\mu_{23}} + \frac{\overline{K}_{21}(0,0)}{\mu_{21}}}{1 - \rho_{21}} = \underline{4.25}\,.$$

STEP 2.2 Throughput of class 3, from Eq. (8.46):

$$\lambda_3(0,1) = \underline{0.238}\,.$$

STEP 2.3 Mean number of jobs in class 3, from Eq. (8.47):

$$\overline{K}_{13}(0,1) = \underline{0.595}\,, \quad \overline{K}_{23}(0,1) = \underline{0.405}\,.$$

STEP 2.4 Mean number of jobs in the open class 1, using Eqs. (8.51),
(10.100), and (10.104), respectively:

$$\overline{K}_{11}(0,1) = \underline{6.38}\,, \quad \overline{K}_{21}(0,1) = \underline{1.5}\,.$$

MVA-iteration for $\mathbf{k} = (1,0)$:

STEP 2.1 $\overline{T}_{12}(1,0) = \underline{1.5}\,, \quad \overline{T}_{22}(1,0) = \underline{3.5}\,,$

STEP 2.2 $\lambda_2(1,0) = \underline{0.308}\,,$

STEP 2.3 $\overline{K}_{12}(1,0) = \underline{0.462}\,, \quad \overline{K}_{22}(1,0) = \underline{0.538}\,,$

STEP 2.4 $\overline{K}_{11}(1,0) = \underline{5.85}\,, \quad \overline{K}_{21}(1,0) = \underline{1.5}\,.$

MVA-iteration for $\mathbf{k} = (1,1)$:

STEP 2.1 $\overline{T}_{12}(1,1) = \underline{2.393}\,, \quad \overline{T}_{22}(1,1) = \underline{3.5}\,,$

STEP 2.2 $\lambda_2(1,1) = \underline{0.241}\,,$

STEP 2.3 $\overline{K}_{12}(1,1) = \underline{0.578}\,, \quad \overline{K}_{22}(1,1) = \underline{0.422}\,,$

| STEP 2.1 | $\overline{T}_{13}(1,1) = \underline{3.654}, \quad \overline{T}_{23}(1,1) = \underline{4.923},$ |

| STEP 2.2 | $\lambda_3(1,1) = \underline{0.178},$ |

| STEP 2.3 | $\overline{K}_{13}(1,1) = \underline{0.65}, \quad \overline{K}_{23}(1,1) = \underline{0.35},$ |

| STEP 2.4 | $\overline{K}_{11}(1,1) = \underline{8.91}, \quad \overline{K}_{21}(1,1) = \underline{1.5}.$ |

These are the final results for the given mixed priority network.

Up to now we considered a pure HOL service strategy or a pure PRS service strategy at a node. In many cases we may need to combine these two priority strategies. Let us therefore consider a combination of PRS and HOL:

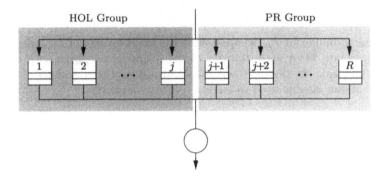

Fig. 10.24 Splitting into HOL and PRS classes.

In this combination, some job classes are processed by HOL and some job classes are processed by PRS. It is assumed that HOL jobs always have higher priority than PRS jobs (see Fig. 10.24). In this case:

- An arriving HOL job cannot preempt a HOL job that is still in progress, but it can preempt a PRS job that is still in progress. Therefore, the response time of a HOL job is not influenced by the arrival of PRS jobs.

$$\overline{T}_r = \sum_{s=1}^{r} \frac{\overline{K}_s - \rho_s}{\mu_s} + \sum_{s=1}^{j} \frac{\rho_s}{\mu_s} + \sum_{s=1}^{r-1} \left(\overline{T}_r - \frac{1}{\mu_r} \right) \cdot \frac{\lambda_s}{\mu_s} + \frac{1}{\mu_r}.$$

By simplifying the above expression we get, in analogy to Eq. (10.103)

$$\overline{T}_{ir}(\mathbf{k}) = \frac{1}{\mu_{ir}} + \frac{\displaystyle\sum_{s=1}^{r} \frac{\overline{K}_{is}(\mathbf{k}-\mathbf{1}_r)}{\mu_{is}} + \sum_{s=r+1}^{j} \frac{\rho_{is}(\mathbf{k}-\mathbf{1}_r)}{\mu_{is}}}{1 - \displaystyle\sum_{s=1}^{r-1} \tilde{\rho}_{is}}, \quad (10.106)$$

$$\tilde{\rho}_{is} = \rho_{is}(\mathbf{k} - \overline{K}_{is}).$$

- A PRS job can always be preempted by a job with higher priority. For priority class r we therefore get

$$\overline{T}_r = \sum_{s=1}^{r} \frac{\overline{K}_s}{\mu_s} + \sum_{s=1}^{r-1} \frac{\overline{T}_r \cdot \lambda_s}{\mu_s} + \frac{1}{\mu_r}.$$

In analogy to Eq. (10.102), after simplification we get

$$\overline{T}_{ir}(\mathbf{k}) = \frac{\frac{1}{\mu_{ir}} + \sum_{s=1}^{r} \frac{\overline{K}_{is}(\mathbf{k}-1_r)}{\mu_{is}}}{1 - \sum_{s=1}^{r-1} \tilde{\rho}_{is}}, \qquad (10.107)$$

$$\tilde{\rho}_{is} = \rho_{is}(\mathbf{k} - \overline{K}_{is}).$$

If these formulae for $\overline{T}_{ir}(\mathbf{k})$ are used in the MVA instead of $\overline{T}_{ir}$, then we get an approximate MVA algorithm for closed queueing networks with a mixed priority strategy. In the next section we introduce the technique of shadow server that can also be used to analyze priority networks.

10.3.2   The Method of Shadow Server

10.3.2.1   The Original Shadow Technique   This technique, shown in Fig 10.25, was introduced by [Sevc77] for networks with a pure PRS priority strategy. The idea behind this technique is to split each node in the network into R parallel nodes, one for each priority class. The problem then is that in this extended node, jobs can run in parallel whereas this is not possible in the original node. To deal with this problem and to simulate the effect of preemption, we iteratively increase the service time for each job class in the shadow node. The lower the priority of a job, the higher is the service time for the job. After all iterations we have

$$\tilde{s}_{i,r} \geq s_{i,r}.$$

Here $\tilde{s}_{i,r} = 1/\mu_{ir}$ denotes the approximate service time of class-r jobs at node i in the shadow node.

Shadow Algorithm

STEP 1   Transform the original model into the shadow model.

STEP 2   Set $\lambda_{i,r} = 0$.

STEP 3   Iterate:

STEP 3.1   Compute the utilization for each shadow node:

$$\tilde{\rho}_{i,r} = \lambda_{i,r} \cdot \tilde{s}_{i,r}. \qquad (10.108)$$

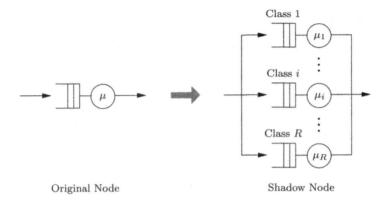

Class 1

Class i

Class R

Original Node   Shadow Node

Fig. 10.25 The idea behind the shadow technique.

STEP 3.2 Compute the shadow service times:

$$\tilde{s}_{i,r} = \frac{s_{i,r}}{1 - \sum\limits_{s=1}^{r-1} \tilde{\rho}_{i,s}} . \tag{10.109}$$

Here $s_{i,r}$ denotes the original service time of a class-r job at node i in the original network while $\tilde{s}_{i,r}$ is the approximated service time in the shadow network.

STEP 3.3 Compute the new values of the throughput of class r at node i, $\lambda_{i,r}$, of the shadow model using any standard analysis technique from Chapters 8 and 9. If the $\lambda_{i,r}$ differ less than ε in two successive iterations, then stop the iteration. Otherwise go back to Step 3.1.

10.3.2.2 Extensions of the Shadow Technique

Extension of Kaufmann Measurements on real systems and comparisons with discrete-event simulation results have shown that the technique of shadow server is not very accurate. Because of the lack of accuracy, [Kauf84] suggested an extension of the original shadow technique. This extended method differs from the original method in that an improved expression $\hat{\rho}_{i,r}$ is used instead of $\tilde{\rho}_{i,r}$. To derive this improved utilization, we define

c = class of the job under consideration,

$N_{i,r}$ = number of class-r jobs at node i in the shadow node,

$s_{i,r}^{pt}$ = approximated mean service position time,

$M_{i,r} = \sum\limits_{l=1}^{r-1} N_{i,l}$.

For a pure PR strategy the following relations hold:

- The improved utilization $\hat{\rho}_{ir}$ is expressed using $s_{i,r}^{pt}$:

$$\hat{\rho}_{i,r} = \lambda_{i,r} \cdot s_{i,r}^{pt},$$
$$\hat{\rho}_{i,r} = P(N_{i,r} > 0),$$
$$\lambda_{i,r} = P(N_{i,1} = 0 \wedge N_{i,2} = 0 \wedge \cdots \wedge N_{i,r-1} = 0$$
$$\wedge N_{i,r} > 0) \cdot \mu_{i,r}$$
$$= P(M_{i,r} = 0 \wedge N_{i,r} > 0) \cdot \mu_{i,r},$$
$$P(N_{i,r} > 0) = P(M_{i,r} = 0 \wedge N_{i,r} > 0) \cdot \mu_{i,r} \cdot s_{i,r}^{pt}.$$

- The last equation can now be used to derive an expression for s_{ir}^{pt}:

$$s_{i,r}^{pt} = \frac{P(N_{i,r} > 0)}{P(M_{i,r} = 0 \wedge N_{i,r} > 0)} \cdot \mu_{i,r}^{-1}$$
$$= \frac{1}{\dfrac{P(M_{i,r} = 0 \wedge N_{i,r} > 0)}{P(N_{i,r} > 0)}} \cdot \mu_{i,r}^{-1}.$$

- By using conditional probabilities the last expression can be rewritten into

$$s_{i,r}^{pt} = \frac{1}{P(M_{i,r} = 0 \mid N_{i,r} > 0)} \cdot \mu_{i,r}^{-1}.$$

- Now we rewrite the numerator of the last expression and define the correction factor $\delta_{i,r}$ which we use to express $\hat{\rho}_{i,r}$:

$$s_{i,r}^{pt} = \frac{1}{1 - P(M_{i,r} > 0 \wedge c \neq r \mid N_{i,r} > 0)} \cdot \mu_{i,r}^{-1}.$$

With

$$\delta_{i,r} = 1 - P(M_{i,r} > 0 \wedge c \neq r \mid N_{i,r} > 0)$$

and

$$\tilde{\rho}_{i,r} = \lambda_{i,r} \cdot \tilde{s}_{i,r}, \quad \hat{\rho}_{i,r} = \lambda_{i,r} \cdot s_{i,r}^{pt},$$

we obtain

$$\hat{\rho}_{i,r} = \lambda_{i,r} \cdot \tilde{s}_{i,r} \cdot \frac{1}{\delta_{i,r}} = \tilde{\rho}_{i,r} \cdot \frac{1}{\delta_{i,r}}. \tag{10.110}$$

As we see, the improved utilization $\hat{\rho}_{i,r}$ is just given by dividing the utilization $\tilde{\rho}_{i,r}$ by the correction factor $\delta_{i,r}$. In [Kauf84] it is shown that the following approximation holds:

$$\delta_{i,r} = \frac{\rho(r)[1 - \rho(r)] + \alpha(r) \cdot \rho(r+1)}{\rho(r)[1 - \rho(r)]^2 + \alpha(r) \cdot \rho_{i,r}} \quad \text{for } r = 2, 3, \ldots, R, \quad (10.111)$$

with

$$\alpha(r) = \sum_{k=1}^{r-1} \frac{\tilde{\rho}_{i,r}}{\omega_{r,k}}, \quad \omega_{r,k} = \frac{s_{i,k}}{s_{i,r}}, \quad \rho(r) = \sum_{k=1}^{r-1} \tilde{\rho}_{i,k}.$$

This correction factor δ changes only Step 3.2 in the original shadow algorithm:

$$\tilde{s}_{i,r} = \frac{s_{i,r}}{1 - \sum\limits_{s=1}^{r-1} \frac{1}{\delta_{i,s}} \cdot \tilde{\rho}_{i,s}}. \quad (10.112)$$

If the utilizations of the shadow nodes are very close to 1, then the expression

$$1 - \sum_{s=1}^{r-1} \frac{1}{\delta_{i,s}} \cdot \tilde{\rho}_{i,s}$$

can be negative because of the approximate technique. In this case the iteration must be stopped or the extended shadow technique with bisection can be used.

In the original and extended shadow method the expression

$$\sum_{s=1}^{r-1} \tilde{\rho}_{i,s}$$

can be very close to 1. Then the service time $s_{i,r}$ gets very large and the throughput through this node is very low. In addition, the value of

$$\sum_{s=1}^{r-1} \tilde{\rho}_{i,s}$$

will be very small in the next iteration step and in some cases the shadow technique does not converge but the computed values move back and forth between the two values. Whenever such swinging occurs, we simply use the average of the last two values of throughput in Step 3.3. Thus, we have used bisection.

Class Switching If class switching is allowed in the network then we use the shadow technique in combination with the concept of chains (see Section 7.3.6.1). The resulting algorithm can now be given in the following steps:

STEP 1 Determine the chains in the network.

STEP 2 Transform the original model into the shadow model.

STEP 3 Set $\lambda_{i,r} = 0$.

STEP 4 Iterate:

STEP 4.1

$$\tilde{\rho}_{i,r} = \lambda_{i,r} \cdot \tilde{s}_{i,r} \, .$$

STEP 4.2

$$\tilde{s}_{i,r} = \frac{s_{i,r}}{1 - \sum_{s=1}^{r-1} \frac{1}{\delta_{i,s}} \cdot \tilde{\rho}_{i,s}} \, .$$

Here $s_{i,r}$ is the original service time of a class-r job at node i in the original network.

STEP 4.3 Transform the class values into chain values (to differentiate between class values and chain values, the chain values are marked with *).

STEP 4.4 Compute the performance parameters of the transformed shadow model with the MVA. If the model parameters swing, then set

$$\overline{T}_{i,r} = \frac{\overline{T}_{i,r}^{\text{old}} + \overline{T}_{i,r}^{\text{new}}}{2}$$

and compute the performance measures with this new response time.

STEP 4.5 Transform chain values into class values. If the $\lambda_{i,r}^*$ (chain value) in two successive steps differs less than ε, stop the iteration. Otherwise go back to Step 4.1.

Class Switching and Mixed Priorities The original shadow technique [Sevc77] can only be used for networks with a pure PRS-strategy, but in the models of a UNIX-based operating system (see Section 13.1.5), for example, or in models of cellular mobile networks [GBB98], a mixed priority strategy is used. For this reason the original shadow technique is extended to deal with mixed priority strategies.

Consider the situation where the classes $1, 2, \ldots, u$ are pure HOL classes and $u+1, u+2, \ldots, R$ are pure PRS classes, as shown in Fig. 10.26. We define

$$y_{i,r} = \sum_{l=1,l\neq r}^{u} N_{i,l}$$

and get the following relations:

- The utilizations can be expressed as

$$\hat{\rho}_{i,r} = \lambda_{i,r} \cdot s_{i,r}^{pt} \, ,$$

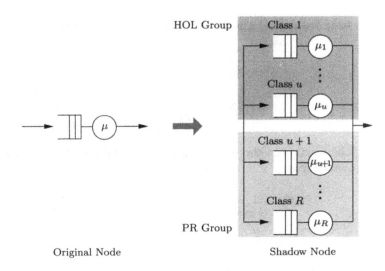

Fig. 10.26 Shadow technique and mixed strategy.

$$\hat{\rho}_{i,r} = P(N_{i,r} > 0)\,.$$

- A PRS job can only be served if no other job with higher priority is waiting:
$$\lambda_{i,r} = P(M_{i,r} = 0 \wedge N_{i,r} > 0) \cdot \mu_{i,r}\,.$$

- For HOL jobs, two cases have to be distinguished:

  1. Class-r jobs as well as HOL jobs are at the node and a class-r job that is just served. Then we get
$$N_{i,r} > 0 \quad \wedge \quad c = r \quad \wedge \quad y_{i,r} > 0$$

  2. Class-r jobs but no HOL jobs are waiting for service at a node. Then we get
$$N_{i,r} > 0 \quad \wedge \quad y_{i,r} = 0\,.$$

For the HOL classes we therefore get the following:

  - The utilization of node i by class-r HOL jobs is given by the probability that at least one class-r job is at node i (and is served) and other HOL jobs are waiting, or at least one class-r job is at node i and no other HOL jobs are waiting for service:

$$\begin{aligned}
\lambda_{i,r} &= \rho_{i,r} \cdot \mu_{i,r} \\
&= [P(N_{i,r} > 0 \wedge c = r \wedge y_{i,r} > 0) + P(N_{i,r} > 0 \wedge y_{i,r} = 0)] \cdot \mu_{i,r} \\
&= [P(N_{i,r} > 0) - P(N_{i,r} > 0 \wedge y_{i,r} > 0 \wedge c \neq r)] \cdot \mu_{i,r}\,.
\end{aligned}$$

- As in the case of a pure PRS strategy, we get analogously:

$$s_{i,r}^{pt} = \frac{P(N_{i,r} > 0)}{P(N_{i,r} > 0) - P(N_{i,r} > 0 \wedge y_{i,r} > 0 \wedge c \neq r)} \cdot \mu_{i,r}^{-1}$$

$$= \frac{1}{1 - \dfrac{P((y_{i,r} > 0 \wedge c \neq r) \wedge N_{i,r} > 0)}{P(N_{i,r} > 0)}} \cdot \mu_{i,r}^{-1} .$$

- The fraction in the numerator is now rewritten using conditional probabilities:

$$s_{i,r}^{pt} = \frac{1}{1 - P(y_{i,r} > 0 \wedge c \neq r) \mid N_{i,r} > 0)} \cdot \mu_{i,r}^{-1} .$$

- As before, we now define the correction factor $\psi_{i,r}$ as

$$\psi_{i,r} = 1 - P(y_{i,r} > 0 \wedge c \neq r \mid N_{i,r} > 0) ,$$

and get

$$\hat{\rho}_{i,r} = \frac{1}{\psi_{i,r}} \cdot \tilde{\rho}_{i,r} . \qquad (10.113)$$

As we see, the correction factor for the HOL case differs from the PRS case only in the summation index. We then get to the following equations:

$$\psi_{i_j,r} = \frac{\varrho(r)[1 - \varrho(r)] + \beta(r) \cdot \tilde{\varrho}(r)}{\varrho(r)[1 - \varrho(r)]^2 + \beta(r) \cdot \varrho_{i,r}} , \qquad (10.114)$$

$$\beta(r) = \begin{cases} \displaystyle\sum_{k=1, k \neq r}^{u} \frac{\tilde{\rho}_{i,r}}{\omega_{r,k}} , & \text{if } r \in \text{HOL} , \\ \displaystyle\sum_{k=1}^{r-1} \frac{\tilde{\rho}_{i,r}}{\omega_{r,k}} , & \text{if } r \in \text{PR} , \end{cases}$$

$$\omega_{r,k} = \frac{s_{i,k}}{s_{i,r}} ,$$

$$\varrho(r) = \begin{cases} \displaystyle\sum_{k=1, k \neq r}^{u} \tilde{\rho}_{i,k} , & \text{if } r \in \text{HOL}, \\ \displaystyle\sum_{k=1}^{r-1} \tilde{\rho}_{i,k} , & \text{if } r \in \text{PR} , \end{cases}$$

$$\tilde{\varrho}(r) = \varrho(r) + \tilde{\rho}_{i,r} .$$

Class Switching and Arbitrary Mixed Priorities   Up to now we assumed that the classes $1, \ldots, u$ are pure HOL classes and the classes $u + 1, \ldots, R$ are pure PRS classes. In the derived formulae we can see that in the HOL case, the summation in the correction factor is over all HOL classes excluding the actually considered class. In the PRS case, the summation is over all classes with higher priority. With this relation in mind it is possible to extend the formulae in such a way that every mixing of PRS and HOL classes is possible. Let ξ_r be the set of jobs that cannot be preempted by a class-r job, $r \notin \xi_r$. For the computation of the correction factor, this mixed priority strategy has the consequence that:

- For an HOL class r the summation is not from 1 to u but over all priority classes $s \in \xi_r$.

- For a PRS class r the summation is not from 1 to $r - 1$ but over all priority classes $s \in \xi_r$.

For any mixed priority strategy we get the following formulae:

$$\tilde{s}_{i,r} = \frac{s_{i,r}}{1 - \sum\limits_{s \in \xi_r} \frac{1}{\psi_{i,s}} \cdot \tilde{\rho}_{i,s}},$$

$$\psi_{i,r} = \frac{\varrho(r)[1 - \varrho(r)] + \beta(r) \cdot \tilde{\varrho}(r)}{\varrho(r)[1 - \varrho(r)]^2 + \beta(r) \cdot \varrho_{i,r}}, \qquad (10.115)$$

$$\beta(r) = \sum_{k \in \xi_r} \frac{\tilde{\rho}_{i_j,r}}{\omega_{r,k}}, \qquad \omega_{r,k} = \frac{s_{ik}}{s_{ir}},$$

$$\varrho(r) = \sum_{k \in \xi_r} \tilde{\rho}_{i,k}, \qquad \tilde{\varrho}(r) = \sum_{k \in \{\xi_r \cup r\}} \tilde{\rho}_{i,k}.$$

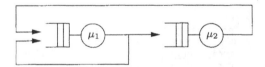

Fig. 10.27 Simple network for Example 10.15.

Example 10.15 Consider the network, given in Fig. 10.27. Node 1 is a -/M/1-PRIORITY node and node 2 is a -/M/1-FCFS node. The network contains two priority job classes, Class 1: PRS, and Class 2: HOL, and class switching is allowed. The routing probabilities are given by

$$p_{11,11} = 0.5, \quad p_{11,12} = 0.4, \quad p_{11,21} = 0.1,$$
$$p_{12,11} = 0.5, \quad p_{12,12} = 0.4, \quad p_{12,22} = 0.1,$$
$$p_{21,11} = 1.0, \quad p_{22,11} = 1.0.$$

The other parameters are

$$s_{11} = 0.1, \quad s_{12} = 0.1, \quad s_{21} = 1.0, \quad s_{22} = 1.0,$$

$$K = 3, \quad \epsilon = 0.001.$$

Compute the visit ratios e_{ir} in the network. For this we use the formula

$$e_{ir} = \sum_{j=1}^{2} \sum_{s=1}^{2} e_{js} p_{js,ir}$$

and get $e_{11} = 1.5$, $e_{12} = 1.0$, $e_{21} = 0.15$, $e_{22} = 0.1$.

Determine the chains in the network and compute the number of jobs in each chain (refer to Section 7.3.6). In Fig. 10.28 the DTMC state diagram for the routing matrix for the chains is shown, and it can easily be seen that we can reach each state from every other state (in a state (i, r), i denotes the node number and r denotes the class number). Therefore, we have only one chain in the network.

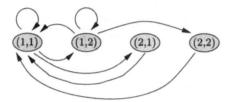

Fig. 10.28 DTMC state diagram of the routing matrix.

Transform the network into the shadow network, as it is shown in Fig. 10.29.

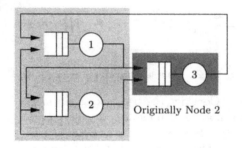

Originally Node 1

Fig. 10.29 The Shadow model for the network in Example 10.15.

The corresponding parameters are

$$s_{11} = 0.1, \quad e_{11} = 1.5, \quad e_{31} = 0.15,$$
$$s_{12} = 0,$$

$$s_{21} = 0,$$
$$s_{22} = 0.1, \quad e_{12} = 0.0, \quad e_{32} = 0.1,$$
$$s_{31} = 1.0, \quad e_{21} = 0.0,$$
$$s_{32} = 1.0, \quad e_{22} = 1.0.$$

STEP 4 Compute the visit ratios e_{iq}^* per chain and remember that we have only one chain:

$$e_{iq}^* = \frac{\sum\limits_{r \in q} e_{ir}}{\sum\limits_{r \in q} e_{1r}}.$$

Then we get

$$e_{11}^* = \frac{e_{11} + e_{12}}{e_{11} + e_{12}} = 1, \quad e_{21}^* = \frac{e_{21} + e_{22}}{e_{11} + e_{12}} = \frac{2}{3}, \quad e_{31}^* = \frac{e_{31} + e_{32}}{e_{11} + e_{12}} = \frac{1}{6}.$$

STEP 5 Compute the scale factors α_{ir}. For this we use the equation

$$\alpha_{ir} = \frac{e_{ir}}{\sum\limits_{s \in q} e_{is}}$$

and get

$$\alpha_{11} = \frac{e_{11}}{e_{11} + e_{12}} = 1.0, \quad \alpha_{12} = \frac{e_{12}}{e_{11} + e_{12}} = 0,$$
$$\alpha_{21} = \frac{e_{21}}{e_{21} + e_{22}} = 0, \quad \alpha_{22} = \frac{e_{22}}{e_{21} + e_{22}} = 1.0,$$
$$\alpha_{31} = \frac{e_{31}}{e_{31} + e_{32}} = 0.6, \quad \alpha_{32} = \frac{e_{32}}{e_{31} + e_{32}} = 0.4.$$

STEP 6 Compute the service times per chain that are given by the equation

$$s_{iq}^* = \frac{1}{\mu_{iq}} = \sum_{r \in q} s_{ir} \cdot \alpha_{ir},$$

which gives

$$s_{11}^* = s_{11} \cdot \alpha_{11} + s_{12} \cdot \alpha_{12} = 0.1,$$
$$s_{21}^* = s_{21} \cdot \alpha_{21} + s_{22} \cdot \alpha_{22} = 0.1,$$
$$s_{31}^* = s_{31} \cdot \alpha_{31} + s_{32} \cdot \alpha_{32} = 1.0.$$

Now we can analyze the shadow network given in Fig. 10.29 using the following parameters:

$$e_{11}^* = 1.0, \quad e_{21}^* = \frac{2}{3}, \quad e_{31}^* = \frac{1}{6},$$
$$s_{11}^* = 0.1, \quad s_{21}^* = 0.1, \quad s_{31}^* = 1.0,$$
$$\alpha_{11} = 1.0, \quad \alpha_{21} = 0.0, \quad \alpha_{31} = 0.6,$$
$$\alpha_{12} = 0.0, \quad \alpha_{22} = 1.0 \quad \alpha_{32} = 0.4,$$

$$K = 3, \qquad N = 3, \qquad \epsilon = 0.001,$$
$$\xi_1 = \{2\}, \qquad \xi_2 = \{1\}.$$

STEP 7   Start the shadow iterations.

1. Iteration:

$$\rho_{11}^* = 0.505, \quad \rho_{21}^* = 0.337, \quad \rho_{31}^* = 0.841,$$
$$\lambda_{11}^* = 5.049, \quad \lambda_{21}^* = 3.366, \quad \lambda_{31}^* = 0.841,$$
$$\tilde{s}_{11}^* = 0.117, \quad \tilde{s}_{21}^* = 0.145, \quad \tilde{s}_{31}^* = 1.000.$$

2. Iteration:

$$\rho_{11}^* = 0.541, \quad \rho_{21}^* = 0.447, \quad \rho_{31}^* = 0.766,$$
$$\lambda_{11}^* = 4.597, \quad \lambda_{21}^* = 3.065 \quad \lambda_{31}^* = 0.766,$$
$$\tilde{s}_{11}^* = 0.125, \quad \tilde{s}_{21}^* = 0.142, \quad \tilde{s}_{31}^* = 1.000.$$

3. Iteration:

$$\rho_{11}^* = 0.569, \quad \rho_{21}^* = 0.430, \quad \rho_{31}^* = 0.756,$$
$$\lambda_{11}^* = 4.533, \quad \lambda_{21}^* = 3.022, \quad \lambda_{31}^* = 0.756,$$
$$\tilde{s}_{11}^* = 0.122, \quad \tilde{s}_{21}^* = 0.147, \quad \tilde{s}_{31}^* = 1.000.$$

4. Iteration:

$$\rho_{11}^* = 0.556, \quad \rho_{21}^* = 0.447, \quad \rho_{31}^* = 0.755,$$
$$\lambda_{11}^* = 4.531, \quad \lambda_{21}^* = 3.012, \quad \lambda_{31}^* = 0.755,$$
$$\tilde{s}_{11}^* = 0.124, \quad \tilde{s}_{21}^* = 0.144, \quad \tilde{s}_{31}^* = 1.000.$$

5. Iteration:

$$\rho_{11}^* = 0.565, \quad \rho_{21}^* = 0.437, \quad \rho_{31}^* = 0.754,$$
$$\lambda_{11}^* = 4.528, \quad \lambda_{21}^* = 3.018, \quad \lambda_{31}^* = 0.754.$$

Now we stop the iteration because the difference between the λ^* in the last two iteration steps is smaller than ϵ.

STEP 8   Retransform the chain values into class values:

$$\lambda_{11} = \alpha_{11} * \lambda_{11}^* = 4.528,$$
$$\lambda_{22} = \alpha_{22} * \lambda_{21}^* = 3.018,$$
$$\lambda_{31} = \alpha_{31} * \lambda_{31}^* = 0.452,$$
$$\lambda_{32} = \alpha_{32} * \lambda_{31}^* = 0.301.$$

If we rewrite the results of the shadow network in terms of the original network, we get the following final results:

$$\lambda_{11}^{(\text{approx})} = \lambda_{11} = 4.528, \quad \lambda_{12}^{(\text{approx})} = \lambda_{22} = 3.018,$$
$$\lambda_{21}^{(\text{approx})} = \lambda_{31} = 0.452, \quad \lambda_{22}^{(\text{approx})} = \lambda_{32} = 0.301.$$

STEP 9 Verify the results. To verify these results we generate and solve the underlying CTMC via MOSEL-2 [BoHe96] (see Chapter 12). The exact results for this network are given by

$$\lambda_{11}^{(\text{exact})} = 4.52, \quad \lambda_{12}^{(\text{exact})} = 3.01,$$
$$\lambda_{21}^{(\text{exact})} = 0.45, \quad \lambda_{22}^{(\text{exact})} = 0.30.$$

As we see, the differences between the exact MOSEL-2 results $\lambda_{ij}^{(\text{exact})}$ and the approximated results $\lambda_{ij}^{(\text{approx})}$ are very small.

Quota Nodes The idea of quota is to assign to each user in the system the percentage of CPU power he or she can use. If, for example, we have three job classes and the quota are given by $q_1 = 0.3$, $q_2 = 0.2$ and $q_3 = 0.5$, then class 1 customers get 30% of the overall CPU time, class 2 customers get 20% CPU time, and class 3 customers get 50% of the overall CPU time. In an attempt to enforce this quota, we give priority to jobs so that the priority of job class r, pri_r, is

$$\text{pri}_r = \frac{q_r}{\rho_{i,r}}. \tag{10.116}$$

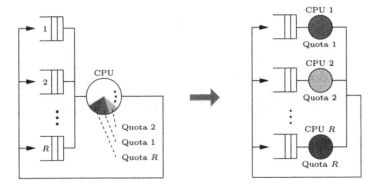

Fig. 10.30 Transformation of a quota node into the corresponding shadow node.

The higher the pri_r, the higher is the priority of class r. As we can see, the priority of class-r jobs depends on the CPU utilization at node i and the

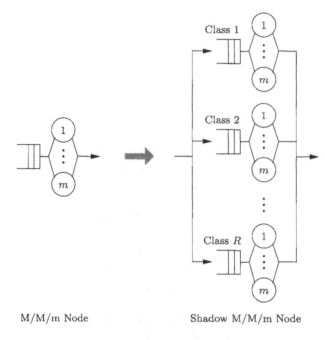

M/M/m Node   Shadow M/M/m Node

Fig. 10.31 Shadow transformation of a -/M/m node.

quota q_r of class r. A high-priority job class is influenced only a little by lower-priority job classes, while lower-priority job classes are influenced much more by higher-priority job classes. A measure for this influence is the weight $\tilde{\omega}_{r,j}$, which is defined as

$$\tilde{\omega}_{r,j} = \begin{cases} \dfrac{\mathrm{pri}_j^2}{\mathrm{pri}_j^2 + \mathrm{pri}_r^2}, & \text{if } r \neq j, \\ 1, & \text{if } r = j. \end{cases} \tag{10.117}$$

The weights $\tilde{\omega}_{r,j}$ are a measure for how much the rth job class is influenced by the jth job class. The closer these weights are to one, the higher is the influence. Since an active job influences itself directly ($r = j$), we have $\tilde{\omega}_{r,r} = 1$. The mean virtual service time of a class-r job can approximately be given as

$$s_{(i,r)}^{\mathrm{virt}} = \frac{s_{i,r}}{1 - \displaystyle\sum_{k=1}^{R} \rho_{i,r} \cdot \tilde{\omega}_{r,k}}. \tag{10.118}$$

If we compare this equation with the shadow equation to compute the shadow service time, then the basic structure of the equations are the same. The idea, therefore, is to use the weights $\tilde{\omega}_{r,j}$ to include quota into the shadow technique. In order to transform a quota node with R job classes into a shadow node, we apply the shadow technique as shown in Fig. 10.30. The resulting equation for computing the service time at each shadow node can now be given as

$$
\tilde{s}_{i,r} = \begin{cases} \dfrac{s_{i,r}}{1 - \sum\limits_{\substack{k=1 \\ k \neq r}}^{R} \dfrac{\tilde{\omega}_{r,k}}{\psi_{i,k}} \cdot \tilde{\rho}_{i,k}} & \text{if } r \in \text{HOL}, \\[4ex] \dfrac{s_{i,r}}{1 - \sum\limits_{k=1}^{r-1} \dfrac{\tilde{\omega}_{r,k}}{\psi_{i,k}} \cdot \tilde{\rho}_{i,k}} & \text{if } r \in \text{PRS}, \end{cases} \qquad (10.119)
$$

$$
\tilde{\omega}_{i,k} = \frac{\text{pri}_k^2}{\text{pri}_k^2 + \text{pri}_i^2}, \qquad \text{pri}_k = \frac{q_k}{\rho_{i,k}}.
$$

By using the weights $\tilde{\omega}_{i,k}$, we make sure that each class gets the amount of CPU time as specified by the CPU quota. In the case of an arbitrary mixed priority strategy, we use the summation index ξ as defined before.

Extended Nodes We now introduce a new node type, the so-called *extended -/M/m node*. It is very easy to analyze a regular -/M/m node since we only have to transform it into the corresponding shadow form and use the equations for -/M/m nodes when analyzing the node. The transformation of a -/M/m node into its corresponding shadow form is shown in Fig. 10.31.

The node shown in Fig. 10.32 is called *extended -/M/m node*. This node contains R priority classes that are priority ordered (job class 1 has highest priority, job class R has lowest priority). The CPU can process all job classes, while at the *APU* (associate processing unit), only the job classes $u, \ldots, R$ $(1 \leq u \leq R)$ can be processed.   For a better overview, a queue

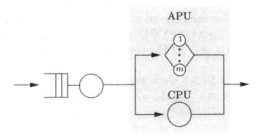

Fig. 10.32 Extended -/M/m node (original form).

is introduced for each job class. The result of this transformation is shown in Fig. 10.33. For example, let $u = R = 3$, then the job classes 1, 2, and 3 can enter the CPU, while the APU can only be entered by class-3 jobs. Priority class-3 jobs can always be preempted at the CPU, but in this case they cannot be preempted at the APU because this job class is the only one entering

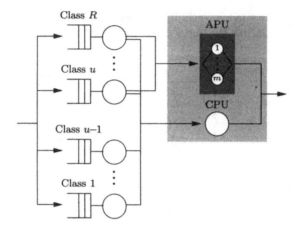

Fig. 10.33 Extended -/M/m node with priority queues.

the APU. In general, the actual priority of a job also depends on the node it enters. If it enters the CPU, the priority order remains unchanged (priority class 1 has highest priority and priority class R has lowest priority), while at the APU priority, class-u jobs have highest priority. Since preemption is also possible at the APUs, if we have more than one priority class, the shadow transformation can be applied to each node, from which a switch to the APU is possible. The result of this transformation is shown in Fig. 10.34.

The components $APU_u, \ldots, APU_R$ form the original APU in the network. Now it is possible to apply the shadow algorithm to the network transformed in this way. Because the priority order at the APU can change, we have to separate the APU iterations from the CPU iterations meaning that in one shadow step we first iterate over all CPUs and then over all APUs. Because of the separated iterations over CPU and APU, we usually get different final service times for CPU_i and APU_i, resulting in a so-called asymmetric node [BoRo94].

At the CPU, class-1 jobs have highest priority, while at the APUs, class-u jobs have highest priority. In the shadow iteration, only Step 3.2 is to be changed as follows:

$$\forall c: \tilde{s}_{i,r} = \frac{s_{i,r}}{1 - \sum\limits_{s \geq c, s \in \xi_r} \frac{1}{\psi_{i,s}} \cdot \tilde{\rho}_{i,s}}, \quad c = 1, \ldots, R.$$

An application for this node type can be found in Section 13.1.5, where it plays an important role.

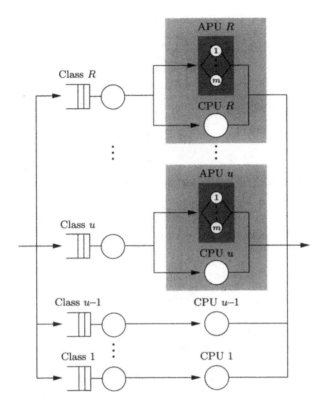

Fig. 10.34 Extended -/M/m node (complete shadow form).

10.3.3 PRIOSUM

The SUM method (see Section 9.2) can be applied to priority networks if we extend the formula for the mean number of jobs in the system, $\overline{K}_{ir}$ of an individual priority node as we did for an FCFS node with multiple job classes. The extended method is referred to as PRIOSUM. Here we use the formulae for single station queueing systems with priorities (see Section 6.4). In this context it is convenient to consider class 1 as the class with the highest priority and class R as the one with the lowest priority. Of course, this convention leads to minor changes in the formulae.

- -/M/1-FCFS-HOL node
  After we have introduced multiple classes and considered the reverse order of priorities, we derive the following formulae from Eqs. (6.126) and (6.127):

$$\overline{W}_{ir} = \frac{\overline{W}_{i0}}{(1 - \sigma_{ir})(1 - \sigma_{ir+1})}, \qquad (10.120)$$

with

$$\sigma_{ir} = \begin{cases} \sum\limits_{j=1}^{r-1} \rho_{ij}, & r > 1, \\ 0, & \text{otherwise} \end{cases} \tag{10.121}$$

and the remaining service time:

$$\overline{W}_{i0} = \sum_{j=1}^{R} \frac{\rho_{ij}}{\mu_{ij}}. \tag{10.122}$$

We obtain the final result for the mean number of jobs $\overline{K}_{ir}$ in node i with

$$\overline{K}_{ir} = \rho_{ir} + \overline{Q}_{ir}, \tag{10.123}$$

and the introduction of correction factors as in the FCFS case:

$$\overline{K}_{ir} = \rho_{ir} + \frac{\lambda_{ir} \sum\limits_{j=1}^{R} \frac{\rho_{ij}}{\mu_{ij}}}{\left(1 - \frac{K-2}{K-1}\sigma_{ir}\right)\left(1 - \frac{K-2}{K-1}\sigma_{ir+1}\right)}. \tag{10.124}$$

- -/M/m-FCFS-HOL node
  In this case we also can use Eqs. (10.120) and (10.121). Together with

$$\overline{W}_{i0} = \frac{P_{m_i}(\rho_i)}{m_i\rho_i} \sum_{j=1}^{R} \frac{\rho_{ij}}{\mu_{ij}} \tag{10.125}$$

and

$$\overline{K}_{ir} = m_i\rho_{ir} + \overline{Q}_{ir}, \tag{10.126}$$

we obtain

$$\overline{K}_{ir} = m_i\rho_{ir} + \frac{\lambda_{ir} \sum\limits_{j=1}^{R} \frac{\rho_{ij}}{\mu_{ij}}}{\left(1 - \frac{K-m_i-1}{K-m_i}\sigma_{ir}\right)\left(1 - \frac{K-m_i-1}{K-m_i}\sigma_{ir+1}\right)} \cdot \frac{P_{m_i}(\rho_i)}{m_i\rho_i}. \tag{10.127}$$

- -/G/1-FCFS-HOL node
  Again with Eqs. (10.120) and (10.121) and the remaining service time

$$\overline{W}_{i0} = \frac{1}{2} \sum_{j=1}^{R} \lambda_{ij}\alpha_2(B_{ij}), \tag{10.128}$$

we obtain

$$\overline{K}_{ir} = \rho_{ir} + \frac{\frac{1}{2}\lambda_{ir} \sum\limits_{j=1}^{R} \lambda_{ij}\alpha_2(B_{ij})}{\left(1 - \frac{K-1-a}{K-1}\sigma_{ir}\right)\left(1 - \frac{K-1-a}{K-1}\sigma_{ir+1}\right)}, \tag{10.129}$$

with the second moment of the service time distribution:

$$\alpha_2(B_{ij}) = \frac{c_{ij}^2 + 1}{(\mu_{ij})^2} , \tag{10.130}$$

$$a = \frac{c_{ir}^2 + 1}{2} . \tag{10.131}$$

- -/G/m-FCFS-HOL node

$$\overline{W}_{i0} = \frac{P_{m_i}(\rho_i)}{2m_i\rho_i} \sum_{j=1}^{R} \rho_{ij}\alpha_2(B_{ij})\mu_{ij} , \tag{10.132}$$

$$\overline{K}_{ir} = m_i\rho_{ir} + \frac{\frac{1}{2}\lambda_{ir} \sum_{j=1}^{R} \rho_{ij}\alpha_2(B_{ij})\mu_{ij}}{\left(1 - \frac{K-m_i-a}{K-m_i}\sigma_{ir}\right)\left(1 - \frac{K-m_i-a}{K-m_i}\sigma_{ir+1}\right)} \cdot \frac{P_{m_i}(\rho_i)}{m_i\rho_i} . \tag{10.133}$$

If *preemption* is allowed, we get

$$\overline{W}_{ir} = \frac{1}{\rho_{ir}}\left(\sigma_{ir+1}w_{ir}^{(r)} - \sigma_{ir}w_{ir}^{(r-1)}\right) . \tag{10.134}$$

Let $w_{ir}^{(r)}$ be the mean waiting time of a job of class r in node i if we consider only the classes with the r highest priorities. Then we get

$$w_{ir}^{(r)} = \frac{\overline{W}_{i0}^{(r)}}{(1 - \sigma_{ir+1})} . \tag{10.135}$$

- -/M/1-FCFS-PRS node

$$\overline{W}_{i0}^{(r)} = \sum_{i=1}^{r} \frac{\rho_{ij}}{\mu_{ij}} . \tag{10.136}$$

Using Eq. (10.123) and Little's theorem, we obtain

$$\overline{K}_{ir} = \rho_{ir} + \lambda_{ir}\overline{W}_{ir} . \tag{10.137}$$

For $\overline{W}_{ir}$ we use Eqs. (10.134), (10.135), and (10.136) and introduce the correction factor:

$$\overline{K}_{ir} = \rho_{ir} + \frac{\lambda_{ir}}{\rho_{ir}}\left[\frac{\sigma_{ir+1}\sum_{j=1}^{r}\frac{\rho_{ij}}{\mu_{ij}}}{\left(1 - \frac{K-2}{K-1}\sigma_{ir+1}\right)} - \frac{\sigma_{ir}\sum_{j=1}^{r-1}\frac{\rho_{ir}}{\mu_{ir}}}{\left(1 - \frac{K-2}{K-1}\sigma_{ir}\right)}\right] . \tag{10.138}$$

- -/M/m-FCFS-PRS node

$$\overline{W}_{i0}^{(r)} = \frac{P_{mi}^r}{m_i \sigma_{ir+1}} \cdot \sum_{j=1}^{r} \frac{\rho_{ij}}{\mu_{ij}}, \qquad (10.139)$$

with probability of waiting P_{mi}^r, which we get from the probability of waiting P_{m_i} by replacing ρ_i by σ_{ir+1}. With Eq. (10.126) and Little's theorem we get

$$\overline{K}_{ir} = m_i \rho_{ir} + \lambda_{ir} \overline{W}_{ir}, \qquad (10.140)$$

and using Eqs. (10.134), (10.135), and (10.139) we obtain

$$\overline{K}_{ir} = m_i \rho_{ir} + \frac{\lambda_{ir}}{\rho_{ir}} \left[\frac{\frac{P_{m_i}^r}{m_i} \sum_{j=1}^{r} \frac{\rho_{ij}}{\mu_{ij}}}{\left(1 - \frac{K-m_i-1}{K-m_i}\sigma_{ir+1}\right)} - \frac{\frac{P_{m_i}^{r-1}}{m_i} \sum_{j=1}^{r-1} \frac{\rho_{ij}}{\mu_{ij}}}{\left(1 - \frac{K-m_i-1}{K-m_i}\sigma_{ir}\right)} \right]. \qquad (10.141)$$

- -/G/1-FCFS-PRS node
  With the remaining service time

$$\overline{W}_{i0}^{(r)} = \frac{1}{2} \sum_{j=1}^{r} \lambda_{ij} \alpha_2(B_{ij}), \qquad (10.142)$$

we obtain Eqs. (10.130), (10.131), (10.134), and (10.135) and the correction factors:

$$\overline{K}_{ir} = m_i \rho_{ir} + \frac{\lambda_{ir}}{\rho_{ir}} \left[\frac{\frac{\sigma_{ir+1}}{2} \sum_{j=1}^{r} \lambda_{ij} \alpha_2(B_{ij})}{\left(1 - \frac{K-m_i-a}{K-m_i}\sigma_{ir+1}\right)} - \frac{\frac{\sigma_{ir}}{2} \sum_{j=1}^{r-1} \lambda_{ij} \alpha_2(B_{ij})}{\left(1 - \frac{K-m_i-a}{K-m_i}\sigma_{ir}\right)} \right]. \qquad (10.143)$$

- -/G/m-FCFS-PRS node
  Using the corresponding equations as before, we get for this node type:

$$\overline{W}_{i0}^{(r)} = \frac{P_{m_i}^r}{2m_i \sigma_{ir+1}} \sum_{j=1}^{r} \rho_{ij} \alpha_2(B_{ij}) \mu_{ij}, \qquad (10.144)$$

$$\overline{K}_{ir} = m_i \rho_{ir} + \frac{\lambda_{ir}}{\rho_{ir}} \left[\frac{\frac{P_{m_i}^r}{2m_i} \sum_{j=1}^{r} \rho_{ij} \alpha_2(B_{ij}) \mu_{ij}}{\left(1 - \frac{K-m_i-a}{K-m_i}\sigma_{ir+1}\right)} - \frac{\frac{P_{m_i}^{r-1}}{2m_i} \sum_{j=1}^{r-1} \rho_{ij} \alpha_2(B_{ij}) \mu_{ij}}{\left(1 - \frac{K-m_i-a}{K-m_i}\sigma_{ir}\right)} \right]. \qquad (10.145)$$

10.4   SIMULTANEOUS RESOURCE POSSESSION

In a product-form queueing network, a job can only occupy one service station at a time. But if we consider a computer system, then a job at times uses two or more resources at once. For instance, a computer program first reserves memory before the CPU or disks can be used to run the program. Thus, memory (a passive resource) and CPU (or another active resource such as a disk) have to be reserved at the same time. To take the simultaneous resource possession into account in performance modeling, *extended queueing networks* were introduced [SMK82]. In addition to the normal (active) nodes, these networks also contain *passive nodes* that consist of a set of *tokens* and a number of *allocate* queues that request these tokens. These passive nodes can also contain additional nodes for special actions such as the release or generation of tokens. The tokens of a passive node correspond to the service units of a passive node: A job that arrives at an allocate queue requests a number of tokens from the passive node. If it gets them, it can visit all other nodes of the network, otherwise the job waits at the passive node. When the job arrives at the release node that corresponds to the passive node, the job releases all its tokens. These tokens are then available for other jobs. We consider two important cases where we have simultaneous resource possession. These two cases are queueing systems with memory constraints and I/O subsystems.

10.4.1   Memory Constraints

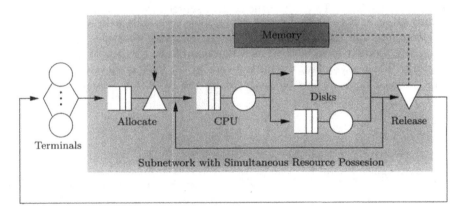

Fig. 10.35 Central-server model with memory constraint.

In Fig. 10.35 we see the typical case of a queueing network with memory constraints where a job first needs to reserve some memory (passive resource) before it can use an active resource such as the CPU or the disks. The memory queue is shown as an allocate node. The job occupies two resources simultaneously: the memory and the CPU or memory and disks. The maximum number of jobs that can compete for CPU and disks equals the number of tokens in the passive node, representing available memory units. The service strategy is assumed to be FCFS. Such models can be solved by generating and solving the underlying CTMC (via SPNP for example). To avoid the generation and solution of resulting large CTMC, we use an approximation technique.

The most common approximation technique is the flow-equivalent server method (FES) (see Section 8.3). We assume that the network without considering the simultaneous resource possession would have product-form solution. The subsystem that contains the simultaneous resource possession is replaced by an FES node and analyzed using product-form methods. For simple systems with only one job class, the load-dependent service rates $\mu_i(k)$ of the FES node can be determined by an isolated analysis of the active nodes of the subsystem (CPU, disks), where the rest of the network is replaced by a short circuit path, and determination of the throughputs $\lambda(k)$ for each possible job population along the short circuit. If K denotes the number of memory units, then the service rates of the FES node are determined as follows [SaCh81]:

$$\mu(k) = \begin{cases} \lambda(k), & \text{for } k = 1, \ldots, K, \\ \lambda(K), & \text{for } k > K. \end{cases}$$

If there are no more than K jobs in the network, then there is no memory constraint, and the job never waits at the allocate node.

Example 10.16   To analyze the central-server network with memory constraint given in Fig. 10.35, we first replace the terminals with a short circuit and compute the throughputs $\lambda(k)$ of the resulting central-server network using any product-form method as a function of the number of jobs k in the network. Then we replace the subnetwork by an FES node (see Fig. 10.36). This reduced network can be analyzed by the MVA for queueing networks

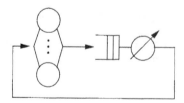

Fig. 10.36   Reduced network.

with load-dependent service rates or as we do it here using a birth–death type CTMC (see Section 3.1). With the mean think time $1/\mu$ at the terminals and the number of terminals M, we obtain for the birth rates

$$\lambda_k = (M - k)\mu, \quad k = 0, 1, \ldots, K \tag{10.146}$$

and for the death rates

$$\nu_k = \begin{cases} \lambda(k), & i = 1, \ldots, K, \\ \lambda(K), & i > K. \end{cases} \tag{10.147}$$

Marginal probabilities of the FES node are (see Eq. (3.11))

$$\pi(k) = \pi(0) \frac{M!}{(M-1)!} \cdot \frac{\mu^k}{\prod\limits_{i=1}^{k} \nu_i}, \tag{10.148}$$

and with Eq. (3.12) we obtain

$$\pi(0) = \frac{1}{1 + \sum\limits_{k=1}^{M} \frac{M!}{(M-1)!} \cdot \frac{\mu^k}{\prod\limits_{i=1}^{k} \nu_i}}. \tag{10.149}$$

Now the throughput λ of the FES node that is also the throughput of terminal requests is obtained by

$$\lambda = \sum_{i=1}^{M} \pi(i)\nu_i = \sum_{i=1}^{K} \pi(i)\nu_i + \nu_K \sum_{i=K+1}^{M} \pi(i). \tag{10.150}$$

As a numerical example, assume that the mean think time $1/\mu$ is $15\,\text{sec}$, the maximum number of programs in the subnetwork is $K = 4$ and the other parameters are as shown in Table 10.24. We calculate the mean response time $\overline{T}$ for terminal requests as a function of the number of terminals M and, in comparison, the mean response time $\hat{T}$ for the case that the main memory is large enough so that no waiting for memory space is necessary, i.e., $K \geq M$. In the case with ample memory, the overall network is of product-form type and, hence, $\hat{T}$ can be determined using algorithms from Chapter 8. From

Table 10.24 Parameters for the central-server network

Node	CPU	Disk 1	Disk 2	Disk 3
μ_j	89.3	44.6	26.8	13.4
p_{1j}	0.05	0.5	0.3	0.15

Table 10.25 we see that $\hat{T}$ is only a good approximation if the number of terminals M is not too large.

For other related examples see [Triv01], and for a **SHARPE** implementation of this technique see [STP96, HSZT00]. An extension of this technique to

Table 10.25 Response time for the central-server system with memory constraint

M	10	20	30	40	50
$\overline{T}$	1.023	1.23	1.64	2.62	7.03
$\hat{T}$	1.023	1.21	1.46	1.82	3.11

multiple class networks can be found in [Saue81], [Bran82], and [LaZa82]. This extension is simple in the case that jobs of different job classes reserve different amounts of memory. But if on the other side, jobs of different job classes use the same amount of memory, then the analysis is very costly and complex. Extensions to deal with complex memory strategies like paging or swapping also exist [BBC77, SaCh81, LZGS84]. For the solution of multiple class networks with memory constraints see also the ASPA algorithm (**average subsystem population algorithm**) of [JaLa83]. This algorithm can also be extended to other cases of simultaneous resource possession.

Problem 10.2   Verify the results shown in Table 10.25 using **PEPSY** and **SHARPE**.

Problem 10.3   Solve the network of Fig. 10.35 by formulating it as a GSPN and solve it using **SHARPE** or **SPNP**. Compare the results obtained by means of FES approximation.

10.4.2 I/O Subsystems

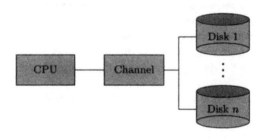

Fig. 10.37 I/O subsystem.

The other very important case of simultaneous resource possession occurs in modeling an I/O subsystem consisting of several disks that are connected by a common channel and a controller with the CPU. In the model of Fig. 10.37 channel and controller can be considered as one single service station because they are either simultaneously busy or simultaneously idle. An I/O operation from/to a certain disk can be subdivided into several phases: The first phase is the seek phase when the read/write head is moved to the right cylinder.

The second phase is the latency phase, which is the disk rotation time needed for the read/write head to move to the right sector. Finally, we have the data transfer phase. During the last two phases the job simultaneously needs the disk and the channel. In the special cases of rotational position sensing systems (RPS) the channel is needed only in the last phase. Delays due to simultaneous resource possession occur if a disk is blocked and no data can be transmitted because another disk is using the channel.

For the analysis of I/O subsystem models, different suggestions have been published. In the formulation of [Wilh77], an increased service time for the disks is computed approximately. This increased service time takes into account the additional delay for the competition of the channel. Each disk is then modeled as an simple M/G/1 queueing system. To determine the disk service time, another model for the probability that the channel is free when needed is used. Hereby it is assumed that only one access path for each disk exists, and a disk cannot be used by several processors simultaneously. This limitation is removed by [Bard80] and [Bard82] where the free path probability is determined using the maximum entropy principle.

We consider a method for analyzing disk I/O systems without RPS, introduced in [LZGS84]. The influence of the channel is considered by an additional term in the mean service time s_i of the disk. The mean service time of the $disk_i$ is given then as

$$s_i = s_{si} + s_{li} + s_{ti} + s_{ci}, \tag{10.151}$$

with

$$s_{si} = \text{mean seek time}, \qquad s_{li} = \text{mean latency time},$$
$$s_{ti} = \text{mean transfer time}, \qquad s_{ci} = \text{mean contention time}.$$

The contention time s_{ci} is calculated iteratively. To obtain an approximate formula for s_{ci} , we first use the formula for the mean number of customers in a $-/M/1$ FCFS node in an open queueing network (see Section 6.2.1):

$$\overline{K}_i = \frac{\rho_i}{1 - \rho_i}. \tag{10.152}$$

In the case of the channel, any requests ahead of a $disk_i$ request at the channel must be associated with some disk other than i, so the equation is slightly modified to

$$\overline{K}_{ch} = \frac{\rho_{ch} - \rho_{ch}(i)}{1 - \rho_{ch}}, \tag{10.153}$$

where

$$\rho_{ch} = \sum_{i=1}^{U} \rho_{ch}(i) \tag{10.154}$$

is the utilization of the channel and $\rho_{ch}(i)$ is the utilization of the channel due to requests from $disk_i$. Here we have assumed that the number of disks is equal to U. Because the mean channel service time is the sum of the mean latency time s_{si} and the mean transfer time s_{ti}, we obtain for the contention time

$$s_{ci} = \frac{\rho_{ch} - \rho_{ch}(i)}{1 - \rho_{ch}} \cdot (s_{li} + s_{ti}), \qquad (10.155)$$

and, hence, for the mean disk service time

$$s_i = s_{si} + \frac{(s_{li} + s_{ti})(1 - \rho_{ch}(i))}{1 - \rho_{ch}}. \qquad (10.156)$$

The measures $\rho_{ch}(i)$ and ρ_{ch}, and other interesting performance measures can be calculated using the following iterative algorithm. We assume that the overall model is a closed queueing network that will be of product-form type if we have no contention at the channel.

STEP 1 Initialization. System throughput $\lambda = 0$.

STEP 2 Iteration:

STEP 2.1 For each $disk_i$, the contribution of $disk_i$ to the utilization of the channel:

$$\rho_{ch}(i) = \lambda \cdot e_i(s_{li} + s_{ti}). \qquad (10.157)$$

STEP 2.2 Channel utilization:

$$\rho_{ch} = \sum_{i=1}^{U} \rho_{ch}(i). \qquad (10.158)$$

STEP 2.3 Compute the mean service time for each $disk_i$ using Eq. (10.156).

STEP 2.4 Use MVA or any other product-form method method to calculate system throughput λ. Return to Step 2.1 until successive iterates of λ are sufficiently close.

STEP 3 Obtain the performance measures after the final iteration.

Example 10.17   We consider a batch computer system with multiprogramming level $K = 10$, the mean CPU service time of 15 sec, one channel, and five disks [LZGS84], each with the mean seek time $s_{si} = 8$ sec, the mean latency time $s_{li} = 1$ sec, and the mean transfer time $s_{ti} = 2$ sec. We assume $e_i = 1$ for each node. With a queueing network having six nodes, we can solve the problem iteratively using the preceding method. The first 12 steps of the iteration are given in Table 10.26. This iterative algorithm can be easily extended to disk I/O systems with

- Multiple classes

Table 10.26 Iterative calculation of the throughput and the channel utilization

	λ_{in}	$\rho_{ch}(i)$	ρ_{ch}	$s_i \cdot e_i$	λ_{out}
1	0	0	0	11.00	0.056
2	0.056	0.167	0.836	23.24	0.030
3	0.030	0.090	0.449	12.96	0.050
4	0.050	0.150	0.749	18.16	0.038
5	0.038	0.113	0.564	14.10	0.047
6	0.047	0.140	0.701	16.63	0.041
7	0.041	0.123	0.611	14.77	0.045
8	0.045	0.135	0.674	15.96	0.042
9	0.042	0.127	0.635	15.18	0.044
10	0.044	0.132	0.659	15.63	0.043
11	0.043	0.129	0.645	15.36	0.043
12	0.043	0.130	0.651	15.48	0.043

- RPS

- Additional channels

- Controllers

For details see [LZGS84]. For related examples of the use of **SHARPE**, see [STP96, HSZT00].

Problem 10.4   Verify the results shown in Table 10.26 using the iterative algorithm of this section and implement it in **SHARPE**.

Problem 10.5   Develop a stochastic reward net model for Example 10.17 and solve using **SPNP**.

10.4.3 Method of Surrogate Delays

Apart from special algorithms for the analysis of I/O subsystem models and queueing networks with memory constraints, more general methods for the analysis of simultaneous resource possession in queueing networks are also known. Two general methods with which a satisfactory accuracy can be achieved in most cases are given in [FrBe83] and [JaLa82]. Both methods are based on the idea that simultaneously occupied service stations can be subdivided into two classes: the *primary stations* that are occupied before an access to a *secondary station* can be made. In the case of I/O subsystem models, the disks constitute the primary stations and the channel the secondary station. In the case of models with memory constraints, the memory constitutes the primary stations and the CPU and disks constitute the *secondary stations*. The time while a job occupies a primary station without trying to get service from the secondary station is called *non-overlapping service time*. This non-overlapping service time can also be zero. The time while a job occupies both the primary and the secondary station is called *overlapping service time*. Characteristic for the method of surrogates [JaLa82] is that the additional delay time can be subdivided into:

- The delay that is caused when the secondary station is heavily loaded. Heavy load means that more jobs arrive at a station than can immediately be served.

- The delays due to an overload at the primary station. This is exactly the time one has to wait for a primary station while the secondary station is not heavily loaded.

To estimate the delay times, two models are generated, where each model provides the parameters for the other model. In this way we have an iterative procedure for the approximate determination of the performance measures of the considered queueing model. This method can be applied to single class closed queueing networks that would otherwise have product-form solution if we neglect the simultaneous resource possession. Studies have shown that the method of surrogate delays tends to overestimate the delay and therefore underestimate the throughput. In all examples we considered, the deviation compared to the exact values was found to be less than 5%.

10.4.4   Serialization

Up to now we only modeled contention for the hardware resources of a computer system. But delays can also be caused by the competition for software resources. Delay that is caused because a process has to wait for software resources and that makes a serialization in the system necessary is called *serialization delay*. Examples for such software resources are critical areas that have to be controlled by semaphores or non-reentrant subprograms. Serialization can be considered as a special case of simultaneous resource possession because jobs simultaneously occupy a special (passive) serialization node and an active resource such as CPU or disk.

The phrase *serialization phase* is used here for the processing phase in which at most one job can be active at any time. A job that wishes to enter the serialization phase or that is in the serialization phase is called a *serialized job*. The processing phase in which a job is not serialized is called the *non-serialized phase*, and a job that is in a non-serialization phase is called a *non-serialized job*. There can always be more than one job in the non-serialization phase. All nodes that can be visited by serialized jobs are called *serialized nodes*, all other nodes are called *non-serialized nodes*. Each serialized node can be visited by serialized jobs as well as non-serialized jobs.

An iterative algorithm for the analysis of queueing networks with serialization delay is suggested by [JaLa83]. This algorithm can also be used for multiple class networks where the jobs can be at most in one serialization phase at one time. The starting point of this method is the idea that the entrance of a job in a serialization phase shall lead to a class switch of that job. Therefore, a class-r job that intends to enter the serialization phase s is denoted as a class-(s, r) job. When the job leaves the serialization phase, it goes back to its former class. In each serialization phase, there can be at most one job at a time. Non-serialization phases are not limited in this way. The method uses a two layer model and the iteration takes place between these two layers. For more information see [JaLa83].

There are also other formulations for the analysis of queueing networks with serialization delays, but these methods are all restricted to single class queueing networks. In the aggregate server method [AgBu83], an additional node for each serialization phase is introduced and the mean service time of the node is set equal to the mean time that a job spends in the serialization phase. The mean service time of the original node and the additional node have to be increased in a proper way to take into consideration the fact that at the original node the service of non-serialized jobs can be hindered by the service of serialized jobs and vice versa. Therefore, iteration is needed to solve the model.

Two different techniques to model the deleterious effect of the serialization delays to the performance of a computer system are proposed by [Thom83]: an iterative technique and a decomposition technique. The decomposition technique also uses two different model layers. At the lower layer the mean system throughputs for each possible system state is computed. These throughputs are then used in the higher-level model to determine the transition rates between the states from which the steady-state probabilities can be determined by solving the global balance equations of the underlying CTMC. More details and solution techniques can be found in [AgTr82], [MüRo87], and [SmBr80]. A SHARPE implementation can be found in [STP96, HSZT00].

Another complex extension of the models with serialization delays are models for the examination of synchronization mechanism in database systems (concurrency control). These mechanisms make sure that the data in a database are consistent when requests and changes to these data can be done concurrently. These models can be found in [BeGo81], [CGM83], [MeNa82], [Tay87], or [ThRy85].

10.5   PROGRAMS WITH INTERNAL CONCURRENCY

A two-level hierarchical model for the performance prediction of programs with internal concurrency was developed by Heidelberger and Trivedi (see [HeTr83]). Consider a system consisting of M servers (processors and I/O devices) and a pseudo server labeled 0. A job consists of a primary task and a set of secondary tasks. Each primary task spawns a fixed number of secondary tasks whenever it arrives at node 0. The primary task is then suspended until all the secondary tasks it has spawned have finished execution.

Upon spawning, each of the secondary tasks is serviced by network nodes and finally on completion return to node 0. Each secondary task may have to wait for its siblings to complete execution and return to node 0. When all its siblings complete execution, the primary task is reactivated. It should be clear that a product-form queueing model cannot be directly used here as the waiting of the siblings at node 0 violates the assumptions. Of course, under the assumption of service times being exponentially distributed, a CTMC can be used. In such a CTMC, each state will have to include the number of tasks of each type (primary, secondary1, secondary2, etc.) at each node. The resulting large state space makes it infeasible for us to use the one-level CTMC model even if it can be automatically generated starting from an SPN model.

The two-level decomposition we use here is based on the following idea: Assume that the service requirements of the primary and secondary tasks are substantial so that on the average, queue lengths in the network will have time to reach steady state before any spawned task actually completes. We could then use a product-form queueing network model to compute throughputs and other measures for a fixed number of tasks of each type. This is assumed to be a closed product-form network and, hence, efficient algorithms such as MVA can be used to compute steady-state measures of this lower level queueing network.

A CTMC is constructed as the outer model where the states are described just by a vector containing the numbers or each type of tasks currently in the system. To describe the generator matrix of the CTMC, we introduce the following notation. For simplicity, we assume that the parent task spawns exactly two children after arriving at node 0. The parent tasks are labeled 0, while the children tasks are labeled 1 and 2. Let a_i denote the number of tasks of type i. Then a generic CTMC state is $\mathbf{a} = (a_0, a_1, a_2)$. The state space of the CTMC will then be

$$S = \{\mathbf{a} : 0 \leq a_0 \leq N, 0 \leq a_1 \leq N - a_0, 0 \leq a_2 \leq N - a_0, N \leq a_0 + a_1 + a_2\}$$

Let w_i be the number of children of type i waiting for their siblings at node 0. Let $r_i(\mathbf{a})$ be the probability that a task of type i, on arrival at node 0, finds its sibling waiting for it there. This probability is w_2/a_1 for a Type 1 task and w_1/a_2 for a Type 2 task. Denote the throughputs obtained from the solution of the lower level PFQN by $\lambda_i(\mathbf{a})$ as a function of the task vector $\mathbf{a}$. Then the generator matrix $\mathbf{Q}$ of the CTMC can be derived as shown in Table 10.27. The entries listed are the off-diagonal entries $q(\mathbf{a}, \mathbf{b})$. For vectors $\mathbf{b}$ not in the table, the convention is that $q(\mathbf{a}, \mathbf{b}) = 0$. For a detailed SHARPE implementation of this algorithm see [STP96, HSZT00].

10.6   PARALLEL PROCESSING

Two different types of models of networks with parallel processing of jobs will be considered in this section. The first subsection deals with a network in

Table 10.27 Generator matrix

b	$q(\mathbf{a}, \mathbf{b})$	Transition Explanation
$(a_0 - 1, a_1 + 1, a_2 + 1)$	$\lambda_0(\mathbf{a})$	Task 0 completion, tasks 1 and 2 spawned
$(a_0, a_1 - 1, a_2)$	$\lambda_1(\mathbf{a})(1 - r_1(\mathbf{a}))$	Task 1 completion, sibling active
$(a_0, a_1, a_2 - 1)$	$\lambda_2(\mathbf{a})(1 - r_2(\mathbf{a}))$	Task 2 completion, sibling active
$(a_0 + 1, a_1 - 1, a_2)$	$\lambda_1(\mathbf{a})(r_1(\mathbf{a}))$	Task 1 completion, sibling waiting
$(a_0 + 1, a_1, a_2 - 1)$	$\lambda_2(\mathbf{a})(r_2(\mathbf{a}))$	Task 2 completion, sibling waiting

which jobs can spawn tasks that do not need any synchronization with the spawning task. In the second subsection we consider fork-join networks in which synchronization is required.

10.6.1 Asynchronous Tasks

In this section we consider a system that consists of m active resources (e.g., processors and channels). The workload model is given by a set of statistically independent tasks. Each job again consists of one primary and zero or more statistically identical secondary tasks. The primary task is labeled 1 and the secondary tasks are labeled 2. A secondary task is created by a primary task during the execution time and proceeds concurrently with it, competing for system resources. It is assumed that a secondary task always runs independently from the primary task. This condition in particular means that we do not account for any synchronization between the tasks. Our treatment here is based on [HeTr82].

The creation of a secondary task takes place whenever the primary task enters a specially designated node, labeled 0; the service time at this node is assumed to be 0. Let $e_{i1}, i = 1, 2, \ldots, m$, denote the average number of visits to node i per visit to node 0 by a primary task. Furthermore $e_{i2}, i = 1, 2, \ldots, m$, denotes the average number of visits to node i by a secondary task. After completing execution, a secondary task leaves the network and $e_{i1}, e_{i2} < \infty$. Let $1/\mu_{ij}$ denote the average service time of a type j task at node i. Each task type does not hold more than one resource at any time. The number of primary jobs in the system is assumed to be constant K and concurrency within a job is allowed only through multitasking, while several independent jobs are allowed to execute concurrently and share system resources.

In the case that the scheduling strategy at a node is FCFS, we require each task to have an exponentially distributed service time with common mean. In the case of PS or LCFS-PRS scheduling, or when the node is an IS node, an arbitrary differentiable service time distribution is allowed and each task can have a distinct service time distribution. Since concurrency within a job is allowed, the conditions for a product-form network are not fulfilled (see

[BCMP75]). Therefore, we use an iterative technique where in each step a product-form queueing network is solved (the product-form solution of each network in the sequence is guaranteed due to the assumptions on the service time distributions and the queueing disciplines).

The queueing network model of the system consists of two job classes, an open class and a closed class. The open classes are used to model the behavior of the secondary tasks and, therefore, the arrival rate of the open class is equal to the throughput of the primary task at node 0. The closed class models the behavior of the primary tasks. Because there are K primary tasks in the network, the closed class population is K. Notice that the approximation assumes the arrival process at node 0 is a Poisson process that is independent of the network state. Due to these two assumptions (Poisson process and independence of the network state), our solution method is approximate. A closed form solution is not possible because the throughput of the closed chain itself is a nonlinear function of the arrival rate of the open chain. For the solution of this nonlinear function, a simple algorithm can be used. In [HeTr82] *regula falsi* was used and on the average only a small number of iteration steps was necessary to obtain an accurate solution of the non-linear function.

Let λ_{02} denote the arrival rate and λ_2 the throughput of the open class, respectively. If any of the queues is saturated, then $\lambda_2 \leq \lambda_{02}$. Furthermore, let λ_1 denote the throughput of the closed class at node 0. We require $\lambda_2 = \lambda_1$ and for the stability of the network, we must have $\lambda_2 = \lambda_{02}$. Therefore, the following nonlinear fixed-point equation needs to be solved:

$$\lambda_1(\lambda_{02}) = \lambda_{02} . \tag{10.159}$$

Here $\lambda_1(\lambda_{02})$ is the throughput of the primary task at Node 0, given that the throughput of the secondary task is λ_{02}. Equation (10.159) is a nonlinear function in λ_{02} and can be evaluated by solving the product-form queueing network with one open and one closed job class for any fixed value of λ_{02}. Let λ_1^* and λ_{02}^* denote the solution of Eq. (10.159), i.e., $\lambda_1^* = \lambda_{02}^* = \lambda_1(\lambda_{02}^*)$. Here λ_{02}^* is the approximated arrival rate of a secondary task. At this point, the arrival rate of a secondary task is equal to the throughput of a primary task at node 0. To get the approximate arrival rate λ_{02}, we can use any algorithm for the solution of a nonlinear function in a single variable. Since the two classes share system resources, an increase of λ_{02} does not imply an increase in λ_1; we conclude that λ_1 is a monotone non-increasing function of λ_{02}. If $\lambda_2^{\max}$ denotes the maximum throughput of the open class, then the stability condition is given by

$$\lambda_{02} < \lambda_2^{\max} . \tag{10.160}$$

The condition for the existence of a stable solution (i.e., a solution where no queue is saturated), for Eq. (10.159) is given by

$$\lim_{\lambda_{02} \to \lambda_2^{\max}} \lambda_1(\lambda_{02}) < \lambda_2^{\max} . \tag{10.161}$$

If a stable solution exists, it is also a unique one (monotonicity property). If the condition is not fulfilled, primary tasks can generate secondary tasks at a rate that exceeds the system capacity. The node that presents a bottleneck determines the maximum possible throughput of the open class. The index of this node is given by

$$\text{bott} = \text{argmax}\left\{\frac{e_{i2}}{\mu_{i2}}\right\}. \tag{10.162}$$

Here argmax is the function that determines the index of the largest element in a set. Therefore the maximum throughput is given by

$$\lambda_2^{\max} = \frac{\mu_{\text{bott},2}}{e_{\text{bott},2}}. \tag{10.163}$$

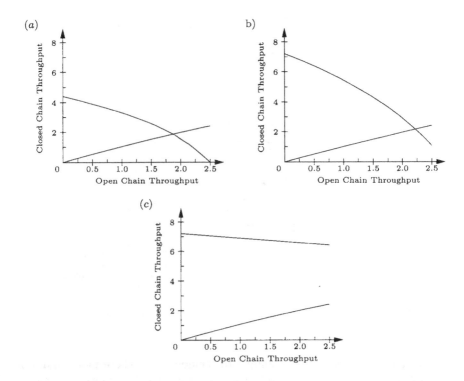

Fig. 10.38   Three types of possible behavior of the approximation method (a) Contention at bottleneck (b) No contention at bottleneck, moderate contention at other devices (c) No contention at bottleneck, little contention at other devices.

In Fig. 10.38 the three types of possible behaviors, depending on the system parameters, are shown. In case (a), node bott is utilized by the primary task, i.e.,

$$\frac{e_{\text{bott},1}}{\mu_{\text{bott},1}} > 0. \tag{10.164}$$

When the queue of node bott grows without bound due to an excessive arrival of jobs of the open job class, the throughput of the closed job class will approach zero. If condition (10.164) is not satisfied, but some other network nodes are shared by two job classes, then either case (*b*) or case (*c*) results depending on the degree of sharing. Because case (*b*) yields a unique solution to (10.159), we can conclude that (10.164) is a sufficient but not necessary condition for convergence, while (10.162) is both necessary and sufficient. Compared to condition (10.162), condition (10.164) has the advantage that it is testable prior to the solution of the network.

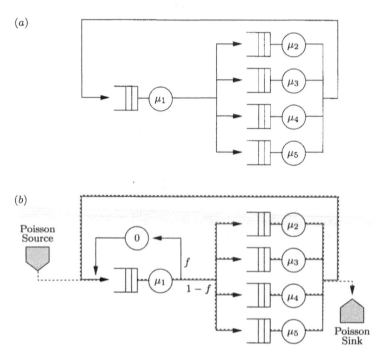

Fig. 10.39 The central-server model without overlapped I/O (*a*) and with overlapped I/O (*b*).

We can also extend this technique to include cases where more than one type of secondary task is created. Assume, whenever a primary task passes node 0, a secondary task of type k, $(k = 2, \ldots, C)$ is created with probability p_k. To model the secondary tasks, $C - 1$ open job classes are used with arrival rate λ_{0k} each. Let $\lambda_0 = \sum_{k=2}^{C} \lambda_{0k}$ denote the total arrival rate of the secondary tasks. The individual arrival rates are constrained so that $\lambda_{0k} = p_k \lambda_0$, $k \geq 2$. The total throughput of secondary tasks is set equal to the throughput of the primary tasks at node 0. This condition again defines a nonlinear equation in the single variable λ. This equation is similar to Eq. (10.159) and must be solved. As a concrete example we consider the central-server model. The standard model without overlap is shown in Fig. 10.39a.

Node 1 represents the CPU and nodes 2, 3, 4, and 5 represent I/O devices such as disks. In the case without overlap, the number of primary tasks is constant C. The processing time of the primary tasks at the CPU is assumed to be random with mean $1/\mu_p$. As soon as a primary task finishes processing, a secondary task is created with probability f. Once a secondary task is generated, the primary task returns to the CPU for processing (with mean $1/\mu_1$), while the secondary task starts at the CPU queue (with mean service time $1/\mu_{12}$). With probability p_{12}, the secondary task moves to an I/O device. The secondary tasks leaves the system with probability $1 - p_2$ and returns to the CPU with probability p_2. In the case that a primary task completes CPU processing and does not create a new secondary task (with probability $1 - f$), it begins an overhead period of CPU processing with mean $1/\mu_0$. On completion of this overhead processing, the primary task moves to I/O device i with probability p_i and then returns to the CPU. As scheduling strategies, we assume PS at the CPU and FCFS at all I/O devices with mean service times $1/\mu_i$.

A primary task can issue I/O requests that are processed by the secondary task. Then the task continues processing without waiting for these I/O requests to complete. If we assume $1/\mu_0 = 1/\mu_{12}$, then this is the average time to initiate an I/O request. In the notation of Section 10.5 we have $1/\mu_{11} = 1/\mu_p + (1 - f)1/\mu_0$. On the average, each secondary task generates $1/(1 - p_2)$ I/O requests and thus the fraction of all I/Os that overlap with a primary task, is given by

$$f_{ol} = \frac{f}{(1 - p_2)} \cdot \frac{1}{(1 - f) + \frac{f}{(1-p_2)}} . \tag{10.165}$$

In the case of $p_2 = 0$, we get $f_{ol} = f$.

In Fig. 10.39b, the central-server model with overlap is shown. Here the additional node 0 with service time 0 is introduced. Once a primary task leaves the CPU, it moves to node 0 with probability f and to I/O device i with probability $(1 - f)p_{i1}$. The rate at which secondary tasks are generated is equal to the throughput of primary tasks at node 0. Secondary tasks are modeled by an open class of customers where all arrivals are routed to the CPU. We will assume that the arrival process is Poisson. The arrival rate of the open class customers λ_{02} is set equal to the throughput of the closed class (primary tasks) at node 0. The routing of the secondary task is as described earlier. A secondary task leaving the CPU visits the I/O device labeled i with probability p_{i2} and returns then to the CPU with probability p_2 or leaves the system with $1 - p_2$, respectively. For this mixed queueing network, the stability condition is $e_{i2}\lambda_{02} < \mu_{i2}$ for all i where $e_{12} = 1/(1 - p_2)$ and $e_{i2} = p_{i2}/(1 - p_2)$, $i > 1$.

Six sets of central-server models with 60 models for each set were analyzed by [HeTr82]. Each set contains models with a moderate utilization

Table 10.28   Input parameters for the central-server model

Model Set	$\frac{1}{\mu_2} = \frac{1}{\mu_3} = \frac{1}{\mu_4}$	$\frac{1}{\mu_5}$	$p_{21} = p_{22}$ $p_{31} = p_{32}$ $p_{41} = p_{42}$	$p_{51} = p_{52}$	p_{02}
I	0.04	0.04	0.25	0.25	0.00
II	0.04	0.04	0.25	0.25	0.50
III	0.04	0.04	0.30	0.10	0.00
IV	0.04	0.10	0.30	0.10	0.00
V	0.04	0.20	0.30	0.10	0.00
VI	0.04	0.40	0.30	0.10	0.00

of the devices as well as heavily CPU and/or I/O bound cases. Each set also contains a range of values of multiprogramming levels, K, the mean service time at the CPU, $1/\mu_1$, and an overlapping factor f. A model within a set is then characterized by the triple $(K, 1/\mu_1, f)$ where $K = 1, 3, 5$, $1/\mu_1 = 0.002, 0.01, 0.02, 0.10, 0.2$ and $f = 0.1, 0.25, 0.5, 0.75$. The differences between the sets are the I/O service times and the branching probabilities p_{ij} and p_{02}, respectively. Furthermore, they assume for all models that the overhead processing times are equal, $1/\mu_0 = 1/\mu_{12} = 0.0008$. The mean service times of the least active I/O device is varied in an interval around the mean service time of the other I/O devices. In addition the I/O access pattern ranges from an even distribution to a highly skewed distribution. The input parameters for each set of 60 central-server models are shown in Table 10.28.

As a second example, we consider the terminal-oriented timesharing systems model shown in Fig. 10.40a. Basically it is the same model as the central-server model except that it has an extra IS node (node 6) that represents a finite number of terminals. The mean service time at node 6, $1/\mu_{61}$, can be considered as the mean thinking time. The number of terminals submitting primary tasks to the computer system is denoted by K. Each primary task first uses the CPU and then moves to I/O device i with probability p_{i1}. As soon as a primary task completes service at the I/O, it returns to the CPU with probability p_1 or enters the terminals with probability $1 - p_1$. We assume that during the execution of a primary task a secondary task is generated which executes independently from the primary task (except for queueing effects). In particular, the secondary task can execute during the thinking time of its corresponding primary task. Routing and service times are the same as in the central-server model. The approximate model of this overlap situation is shown in Fig. 10.40b. Between the I/O devices and the terminals, node 0 (with service time 0) is inserted. Therefore, the throughputs at node 0 and the terminals are equal. The secondary tasks are modeled as open job classes and arrive at the CPU. The corresponding arrival rate λ_{02} is set equal to the throughput of primary tasks at node 0. The stability condi-

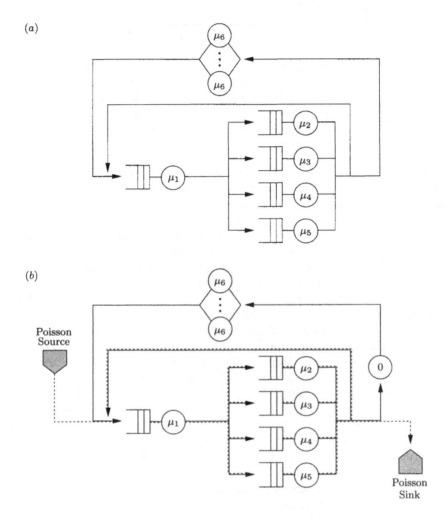

Fig. 10.40 Terminal model without overlapped tasks (*a*) and with overlapped tasks (*b*).

tion for this model is the same as in the model before: $\lambda_{02}/(\mu_{12} \cdot (1 - p_2)) < 1$ and $\lambda_{02} \cdot p_{i2}/(\mu_{i2} \cdot (1 - p_2)) < 1$ for all $i > 2$.

Table 10.29 Input parameters for the terminal system

$\frac{1}{\mu_{61}}$	$\frac{1}{\mu_{11}} = \frac{1}{\mu_{12}}$	$\frac{1}{\mu_i}$	p_{ij}	$p_1 = p_2$	K (no. of terminals)
10	0.010	0.04	0.25	0.10	1, 5, 10, 20, 30, 40, 50,60, 70, 80, 90, 100
10	0.015	0.04	0.25	0.10	1, 5, 10, 20, 30, 40, 50,60, 70, 80, 90, 100
15	0.002	0.04	0.25	0.10	25, 50, 100
15	0.010	0.04	0.25	0.10	25, 50, 100
15	0.020	0.04	0.25	0.10	25, 50, 100
15	0.020	0.08	0.25	0.02	1, 5, 10, 20, 30, 40, 50
15	0.030	0.08	0.25	0.02	1, 5, 10, 20, 30, 40, 50

This model can be interpreted as follows: Each terminal interaction can be split into two tasks. A terminal user has to wait for completion of a primary task but not for completion of a secondary task. In [HeTr82], 47 models of this type were considered. The corresponding parameter settings for this model are described in Table 10.29. Each of the models was simulated using the IBM Research Queueing package **RESQ**, which offers a so-called *split node* that splits a job into two tasks that proceed independently through the network. This node type was used to model the overlap of jobs. In Table 10.30, 10.32, and 10.31, the results of the comparison of discrete-event simulation (DES) with the approximate analytic method are shown.

Both absolute errors as well as relative errors are given. For each set of models, the mean and maximum absolute and relative errors are listed for utilizations, queue lengths, throughputs, and response times (for the terminal models). As we can see from the results, the approximation is particularly accurate for estimating utilizations and throughputs. For these quantities, the relative error is about 1.3%, and the maximum relative error is 17.4%. We can also see that the estimates for the mean queue length are somewhat less accurate, the mean relative error is 4.2%. The maximum relative error is very high (up to 89%). These high errors always occur in very imbalanced systems such as the central-server model sets V and VI (in these sets the overlap factor is very high). If, on the other hand, the model is better balanced (lower values of f_{ol}), we get quite accurate approximations. In this case the mean relative error for all performance measures is 1.9%, and the maximum relative error is 2.9%.

10.6.2   Fork–Join Systems

In this section we consider computer systems that are able to run programs in parallel by using the *fork–join* or *parbegin–parend* constructs. The parallel programs (jobs) consist of *tasks* that have to be processed in a certain order.

Table 10.30 Comparison of DES and analytic approximations for the central-server models with balanced I/O subsystem

Models		CPU ρ	Disk ρ	CPU $\overline{Q}$	Disk $\overline{Q}$	λ
Set I				*Absolute error*		
	Mean	0.003	0.006	0.011	0.044	0.056
	Maximum	0.026	0.071	0.110	0.457	0.710
				Relative error		
	Mean	1.2	1.5	2.8	3.2	1.3
	Maximum	15.9	15.4	26.2	16.2	15.5
Set II				*Absolute error*		
	Mean	0.003	0.005	0.009	0.072	0.034
	Maximum	0.020	0.054	0.073	0.670	0.317
				Relative error		
	Mean	1.0	1.1	2.8	3.1	1.0
	Maximum	9.2	8.7	25.0	15.1	9.0

Table 10.31 Comparison of simulation and analytic approximations for the terminal models with balanced I/O subsystem

Models		CPU ρ	Disk ρ	CPU $\overline{Q}$	Disk $\overline{Q}$	Terminal $\overline{Q}$	λ	$\overline{T}$
Terminal					*Absolute error*			
Models	Mean	0.002	0.002	0.429	0.056	0.115	0.010	0.228
	Maximum	0.010	0.008	2.261	0.421	0.674	0.036	1.141
					Relative error			
	Mean	0.3	0.4	3.0	1.4	0.6	0.7	1.7
	Maximum	1.4	1.5	13.4	4.5	2.7	2.2	5.4

Table 10.32 Comparison of simulation and analytic approximations for the central server models with imbalance I/O

Models		CPU ρ	Disk 2 ρ	Disk 3-5 ρ	CPU $\overline{Q}$	Disk 2 $\overline{Q}$	Disk 3-5 $\overline{Q}$	λ
Set III				*Absolute error*				
	Mean	0.003	0.007	0.003	0.010	0.060	0.004	0.050
	Maximum	0.025	0.076	0.025	0.101	0.799	0.065	0.631
				Relative error				
	Mean	1.2	1.4	1.9	2.9	3.8	2.0	1.3
	Maximum	14.7	14.4	13.9	27.7	18.3	13.1	14.3
Set IV				*Absolute error*				
	Mean	0.003	0.007	0.007	0.010	0.044	0.030	0.055
	Maximum	0.029	0.078	0.069	0.095	0.555	0.323	0.636
				Relative error				
	Mean	1.4	1.6	2.1	2.9	3.1	4.2	1.4
	Maximum	15.8	16.2	17.4	27.0	13.6	29.2	15.9
Set V				*Absolute error*				
	Mean	0.003	0.005	0.008	0.009	0.043	0.394	0.039
	Maximum	0.024	0.059	0.102	0.065	0.878	5.499	0.490
				Relative error				
	Mean	1.4	1.5	1.6	3.1	3.2	11.4	1.4
	Maximum	15.6	15.3	15.9	28.7	37.4	73.2	15.3
Set VI				*Absolute error*				
	Mean	0.002	0.002	0.007	0.008	0.041	0.592	0.019
	Maximum	0.011	0.180	0.067	0.032	0.394	4.224	0.149
				Relative error				
	Mean	1.0	1.2	1.3	4.0	7.4	13.7	1.1
	Maximum	7.2	7.2	8.1	28.9	47.9	89.0	7.1

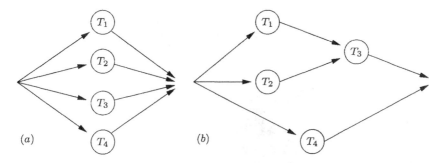

Fig. 10.41 Task precedence graph with (*a*) four parallel tasks and (*b*) four tasks.

This order can be given by a *task precedence graph*. In Fig. 10.41*a* the task precedence graph for a program consisting of four parallel tasks T_1, T_2, T_3, T_4 is shown, and in Fig. 10.41*b* another task precedence graph for a program consisting of four tasks T_1, T_2, T_3, T_4 is shown. In case *b*, task T_3 can start after both tasks T_1 and T_2 are finished while task T_4 can be executed in parallel with task T_1, T_2, and T_3. The corresponding *parbegin–parend* constructs look as shown in Fig.10.42.

10.6.2.1 Modeling

Consider a model of a system in which a series of external parallel program execution requests arrive at an open queueing network where nodes indicate the processors and external arrivals indicate the jobs [Duda87]. It is assumed that each processor can execute exactly one special task. Thus, task T_i always needs to be served by processor P_i. Furthermore it is assumed that processors are always available when needed. The interarrival times of jobs are exponentially distributed with mean value $1/\lambda$, and the service times of the tasks are generally distributed with mean value $1/\mu_i$, $\quad i = 1, \ldots, N$. Figure 10.43 shows the queueing network model for jobs with the structure shown in Fig. 10.41*a*. Figure 10.44 shows the queueing network model of a parallel system, where jobs have the structure shown in Fig. 10.41*b*. The fork and join operators are shown as triangles.

The basic structure of an *elementary fork–join system* is shown in Fig. 10.43. The fork–join operation splits an arriving job into N tasks that arrive simultaneously at the N parallel processors $P_1, \ldots, P_N$. As soon as job i is served, it enters the join queue Q_i and waits until all N tasks are done. Then the job can leave the fork–join system.

A special case of the elementary fork–join system is the *fission–fusion system* (see Fig. 10.45), where all tasks are considered identical. A job can leave the system, as soon as any N tasks are finished. These tasks do not necessarily have to belong to the same job.

Another special case is the *split–merge system* (see Fig. 10.46) where the N tasks of a job occupy all N processors. Only when all the N tasks have completed, can a new job occupy the processors. In this case there is a queue

(a)

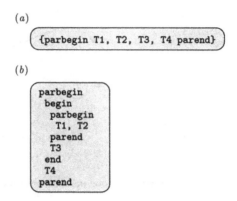

(b)

Fig. 10.42 *Parbegin–parend* construct for (a) four parallel tasks (Fig. 10.41a), (b) four parallel-sequential tasks (Fig. 10.41b).

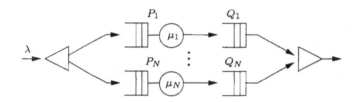

Fig. 10.43 Fork–join system.

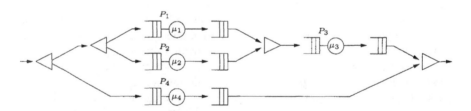

Fig. 10.44 Queueing network model for the execution of a job with the task precedence graph shown in Fig. 10.41b.

in front of the split station and there are no processor queues. This variant corresponds to a blocking situation: A completed task blocks the processor until all other tasks are finished.

10.6.2.2 Performance Measures In this section some performance measures of fork–join systems are defined. They are based on some basic measures of queueing systems that are defined as follows (see Fig. 10.47 and Section 7.2.1):

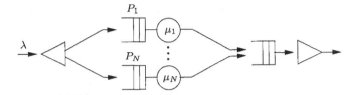

Fig. 10.45 Fission–fusion system.

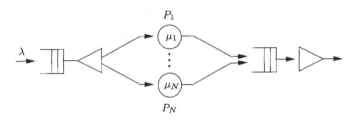

Fig. 10.46 Split–merge system.

$\overline{T}_{S,i}$ = The mean response time of task T_i belonging to a purely sequential job

$\overline{T}_S$ = The mean response time of a purely sequential job, equal to the sum of all $\overline{T}_{S,i}$

$\overline{T}_{\text{FJ}}$ = The mean response time of parallel-sequential jobs with fork–join synchronization

$\overline{T}_{J,i}$ = The mean time spent by task T_i waiting for the join synchronization in queue Q_i

$\overline{Q}_i$ = The mean number of tasks waiting for the join synchronization in queue Q_i

$\overline{T}_J$ = The mean join time

$\overline{K}_{\text{FJ}}$ = The mean number of parallel-sequential jobs with fork–join synchronization

In addition to these measures, the following other measures are of interest for parallel systems:

Speedup: Ratio of the mean response time in the system with N sequential tasks to the mean response time in a fork–join system with N parallel tasks:

$$G_N = \frac{\overline{T}_S}{\overline{T}_{\text{FJ}}}.$$

(10.166)

Normalized Speedup:

$$G = \frac{G_N}{N}, \quad G \in \left[\tfrac{1}{N}, 1\right].$$

(10.167)

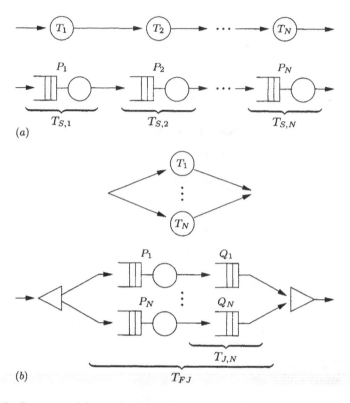

(a)

(b)

Fig. 10.47 System used for performance comparison: (a) sequential system, (b) parallel system.

Synchronization Overhead: Ratio of the mean time that tasks have to wait altogether in the join queues until all tasks are finished to the mean response time of the fork–join system:

$$S_N = \frac{\sum\limits_{i=1}^{N} \overline{T}_{J,i}}{\overline{T}_{\mathrm{FJ}}}. \tag{10.168}$$

Normalized Synchronization Overhead:

$$S = \frac{S_N}{N}, S \in [0, 1]. \tag{10.169}$$

Blocking Factor: The blocking factor is the average total number of blocked tasks in the join queues:

$$B_N = \sum_{i=1}^{N} \overline{Q}_i. \tag{10.170}$$

Normalized Blocking Factor. The normalized blocking factor is given by

$$B = \frac{B_N}{N}.\qquad(10.171)$$

10.6.2.3 Analysis The method of [DuCz87] for the approximate analysis of fork-join systems is based on an application of the decomposition principle: We consider fork-join systems as *open* networks and replace the subnetwork that contains the fork-join construct by a composite load-dependent node. The service rates of the composite node are determined by analyzing the isolated (closed and) short-circuited subnetwork. This analysis is done for every number of tasks $(1\ldots\infty)$. For practical reasons this number is, of course, limited. During the analysis, the mean number of tasks in the join queues is determined by considering the CTMC and computing the corresponding probabilities using a numerical solution method.

Consider an elementary fork-join system with N processors as shown in Fig. 10.43. To analyze this system, we consider the closed queueing network shown in Fig. 10.48. This network contains $N \cdot k$ tasks, $k = 1, 2, \ldots$.

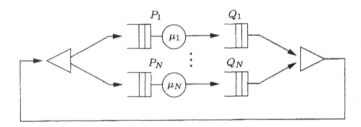

Fig. 10.48   Closed network with $N \cdot k$ tasks.

The analysis of the network, via the numerical solution of the underlying CTMC, gives the throughput rates $\lambda(j)$, which are used as the load-dependent service rates $\mu(j)$ of a FES node that replaces the fork–join system (see Fig. 10.49). The analysis of the FES node finally gives the performance measures of the overall system.

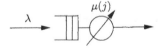

Fig. 10.49   Composite service station.

In the similar fashion, the performance measures of more complex fork–join systems can be computed because a job can have task precedence graphs with several nested fork and join constructs. The model for the service of a job has then the form of a series–parallel network with subnetworks that again contain fork–join systems. In Fig. 10.44 we have already given an example of

Fig. 10.49 Composite service station.

such a model. The described decomposition principle is applied in this case several times in order to solve the fork–join subnetworks numerically so that they can be replaced by FES service stations.

Example 10.18   Consider a two-server fork–join queue. Here we assume that the service times at both servers are exponentially distributed with mean $1/\mu$. This system is shown in Fig. 10.43 (with $N = 2$). In order to solve the fork–join queue, we consider the FES queue with state-dependent service rate, shown in Fig. 10.49. The decomposition step consists of analyzing the closed subsystem with $2k$ tasks, $k = 1, 2, \ldots$ (Fig. 10.48). The service rate of the FES in Fig. 10.49 is set equal to the throughput of the closed subsystem. The solution of the FES gives overall performance measures. At the level of the closed subsystem we deal with forked tasks and delays caused by the join primitive, while at the FES level we obtain the performance measures for jobs.

Let $X(t)$ denote the total number of tasks in collectively both the join queues at time t. The stochastic process is a CTMC with the state space $\{i = 1, 2, \ldots, k\}$.

The state diagrams of the CTMCs for various values of k are shown in Fig. 10.50. Each of these CTMCs is a finite birth–death process, and thus the

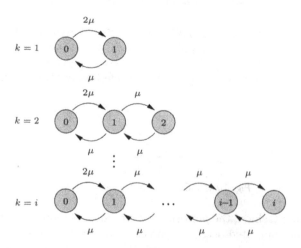

Fig. 10.50 CTMC for the total number of tasks waiting in the join queues.

steady-state probabilities can be derived using Eqs. (3.11) and (3.12):

$$\pi_0 = \frac{1}{1 + 2k}, \quad \pi_i = 2\pi_0, \quad i = 1, 2, \ldots, k, \quad k = 1, 2, \ldots. \quad (10.172)$$

In order to calculate the throughput of jobs $\lambda(k)$, we observe that a successful join corresponds to a transition at rate μ from state i to state $(i-1)$ in the CTMCs of Fig. 10.50. Thus, assigning reward rate μ to each of the states $i = 1, 2, \ldots, k$, we obtain the needed throughput as the expected reward rate in the steady states.

$$\lambda(k) = \mu \sum_{i=1}^{k} \pi_i = \frac{2k}{2k+1}\mu, \quad k = 1, 2, \ldots. \quad (10.173)$$

State-dependent service rates of the FES are then set equal to the throughput of the closed subsystem: $\mu(k) = \lambda(k)$, $k = 1, 2, \ldots$. The solution of the FES queue then gives the steady-state probabilities of the number of jobs in the fork–join queue and is given by

$$\nu_k = \nu_0 \rho^k \frac{\displaystyle\prod_{i=1}^{k} 2i + 1}{\displaystyle\prod_{i=1}^{k} 2i}, \quad k = 1, 2, \ldots,$$

and with Eq. (3.12) it is expressed as

$$\nu_0 = \frac{1}{1 + \displaystyle\sum_{k=1}^{\infty} \rho^k \frac{\displaystyle\prod_{i=1}^{k} 2i + 1}{\displaystyle\prod_{i=1}^{k} 2i}} = (1 - \rho)^{\frac{3}{2}} \quad (10.174)$$

(see [BrSe91], p. 32), where $\rho = \lambda/\mu$. These results hold for $\rho < 1$, which is just the ergodicity condition for fork–join queues. Knowing the probabilities of Eq. (10.174), the expected number of jobs in the system is

$$\overline{K}_{\mathrm{FJ}} = \sum_{k=1}^{\infty} k\nu_k = \frac{3}{2} \cdot \frac{\rho}{1 - \rho}; \quad (10.175)$$

using Little's theorem, we obtain

$$\overline{T}_{\mathrm{FJ}} = \frac{\overline{K}_{\mathrm{FJ}}}{\lambda} = \frac{3}{2} \cdot \frac{1}{\mu(1 - \rho)}. \quad (10.176)$$

If these values are compared to the corresponding ones for the M/M/1 queue, it can be seen that the mean number of jobs and the mean response time of the fork–join queue are both 3/2 times larger than those of the M/M/1 queue.

The performance indices of the synchronization primitives are derived in the following: With the mean response time $\overline{T}_{S,i}$ of task i of a purely sequential job (= mean response time of an M/M/1 queue, Eq. (6.15))

$$\overline{T}_{S,i} = \frac{1}{\mu(1 - \rho)}, \quad i = 1, 2,$$

we obtain the mean response time of a purely sequential job:

$$\overline{T}_S = \sum_{i=1}^{2} \overline{T}_{S,i} = \frac{2}{\mu(1 - \rho)}.$$

Equations (10.166) and (10.176) are used to derive the speedup:

$$G_N = \frac{\overline{T}_S}{\overline{T}_{FJ}} = \frac{4}{3}.$$

The normalized speedup is computed using Eq. (10.167). With $N = 2$ we get

$$G = \frac{G_N}{N} = \frac{2}{3}.$$

To compute the synchronization overhead S_N, we need the average time a task spends in the join queues. This time is just given by the difference between $\overline{T}_{FJ}$ and $\overline{T}_{M/M/1}$ (where $\overline{T}_{M/M/1}$ is the time a task spends in the servers). If

$$\overline{T}_{J,i} = \overline{T}_{FJ} - \overline{T}_{M/M/1}, \quad i = 1, 2, \qquad (10.177)$$

then the synchronized overhead (Eq. (10.168)) is given by

$$S_N = \frac{\sum_{i=1}^{2} \overline{T}_{J,i}}{\overline{T}_{FJ}} = \frac{3}{2},$$

and the normalized (Eq. (10.169)) synchronization overhead

$$S = \frac{S_N}{N} = \frac{1}{3}.$$

In order to obtain the blocking factor B_N, we need the mean number of tasks in the join queues. This number can be determined from Eq. (10.177), using Little's theorem:

$$B_N = \sum_{i=1}^{2} \overline{Q}_{J,i} = \sum_{i=1}^{2} \lambda \overline{T}_{J,i} = \lambda \sum_{i=1}^{2} \left(\frac{3}{2} \frac{1}{\mu(1 - \rho)} - \frac{1}{\mu(1 - \rho)} \right)$$

$$= \frac{\rho}{1 - \rho};$$

the normalized blocking factor is given by

$$B = \frac{1}{2} \frac{\rho}{1 - \rho}.$$

Note that the speedup as well as the synchronization overhead do not depend on the utilization ρ and that the speedup is rather poor. It can be shown that the speedup G_2 increases from $4/3$ to 2 if the coefficient of variation of the service time decreases from 1 (exponential distribution) to 0 (deterministic distribution). In Tables 10.33, 10.34, and 10.35, the speedup is given for different distribution functions of the service times. It is assumed that jobs arrive from outside with arrival rate $\lambda = 0.1$. The service rate is $\mu = 10$.

For the case $c_X < 1$, we use an Erlang-k distribution and get the results shown in Table 10.33. For $c_X > 1$, we use a hyperexponential distribution and

Table 10.33 System speedup with Erlang-k distributed service times

var(X)	Exact				Erlang-k				
var(X)	0.0	$\frac{1}{1000}$	$\frac{1}{800}$	$\frac{1}{600}$	$\frac{1}{500}$	$\frac{1}{400}$	$\frac{1}{300}$	$\frac{1}{200}$	$\frac{1}{100}$
c_X	0.0	$\frac{1}{\sqrt{10}}$	$\frac{1}{\sqrt{8}}$	$\frac{1}{\sqrt{6}}$	$\frac{1}{\sqrt{5}}$	$\frac{1}{\sqrt{4}}$	$\frac{1}{\sqrt{3}}$	$\frac{1}{\sqrt{2}}$	1.0
Speedup	2.0	1.760	1.671	1.630	1.605	1.569	1.524	1.453	1.333

get the results shown in Table 10.34. Because of the phase-type service time

Table 10.34 System speedup with hyperexponentially distributed service times

var(X)	2.0	4.0	6.0	8.0	10.0	20.0
Speedup	1.060	1.043	1.029	1.028	1.027	1.026

distribution, the method discussed so far can easily be extended. Underlying CTMCs will now be more complex than those shown in Fig. 10.50. We chose to generate and solve these CTMCs using the MOSEL-2 tool (see Chapter 12).

Finally, we show in Table 10.35 the system speedup if the service times are normally distributed with standard deviation σ_X and mean value $\mu = 10$. The results in this case were obtained using DES. Using a similar approach,

Table 10.35 System speedup with normally distributed service times

var(X)	0.001	0.005	0.01	0.02	0.08	0.1	0.2	0.4
Speedup	1.697	1.525	1.487	1.463	1.439	1.437	1.433	1.431

performance measures for the fission fusion system (see Fig. 10.45) and the split merge system (see Fig. 10.46) can be computed [DuCz87].

For fork–join systems with a large number of processors, [Duda87] developed an approximate method for the analysis of the subnetworks. To demonstrate this method we consider a series–parallel closed fork–join network as shown in Fig. 10.51. In this method it is assumed that all service times of

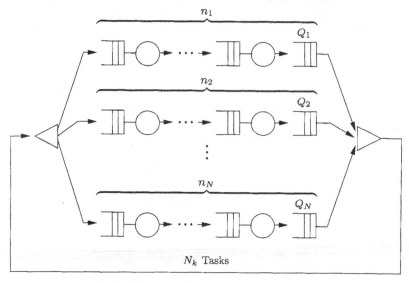

Fig. 10.51   Series–parallel closed fork–join network.

the tasks are exponentially distributed. The method is based on construction of a product-form network with the same topology as the original fork–join subnetwork and approximately the same number of states in the corresponding CTMC. The main difficulty when following this procedure is to determine the number of states in the fork–join subnetwork. If N denotes the number of parallel branches where the ith branch contains exactly n_i nodes and k tasks, then the number of ways to distribute k tasks to $n_i + 1$ nodes (including the join queue Q_i) is given by

$$z_i(k) = \binom{n_i + k}{n_i}.$$

(10.178)

Then the number of ways to distribute $N \cdot k$ tasks to all nodes is given by the following expression:

$$\prod_{i=1}^{N} z_i(k).$$

(10.179)

This number can be limited further. If there is exactly one job in each join queue Q_i, then the jobs are passed immediately to the first station. Thus,

$\prod_{i=1}^{N} z_i(k-1)$ of the given combinations are not possible because this is the number of possible ways to distribute $k-1$ tasks over n_i stations, given that there is one task in each join queue Q_i. The number of states in the fork–join subnetwork is therefore given by

$$Z_{\mathrm{FJ}}(k) = \prod_{i=1}^{N} z_i(k) - \prod_{i=1}^{N} z_i(k-1). \qquad (10.180)$$

In the following steps, the algorithm for the analysis of fork–join networks with possibly several nested fork–join subnetworks is given. The replacement of these fork–join subsystems by FES nodes is done iteratively.

STEP 1 Choose one subsystem that does not contain any other fork–join structures and construct a closed network with $N \cdot k$ tasks by using a short circuit.

STEP 2 Compute the number of states $Z_{\mathrm{FJ}}(k)$ of this closed subnetwork for $k = 1, 2, \ldots$ (Eq. (10.180)).

STEP 3 Compute the throughput of the subnetwork.

STEP 3.1 Construct a product-form network with exactly the same $M = \sum_{i=1}^{N} n_i$ number of nodes as the subnetwork. The number of jobs K in the product-form network is chosen so that the number of states in the product-form network is approximately equal to the number of states in the fork–join subnetwork. Since the state-space structures of both networks are almost identical, the throughput of the product-form network $\lambda_{\mathrm{PF}}(K)$ will be approximately equal to the throughput of the fork–join network $\lambda_{\mathrm{FJ}}(k)$. Thus, we need to determine K by using Eqs. (10.178) and (10.180) so that for $k = 1, 2, \ldots$ we have

$$K : |Z_{\mathrm{FJ}}(k) - Z_{\mathrm{PF}}(K)| = \min_l |Z_{\mathrm{FJ}}(k) - Z_{\mathrm{PF}}(l)|, \qquad (10.181)$$

with $Z_{\mathrm{PF}}(K) = z_{M-1}(K)$ holding.

STEP 3.2 Solve the product-form network for all possible numbers of jobs K using an algorithm such as MVA and determine the throughput as a function of K. This throughput $\lambda_{\mathrm{PF}}(K)$ of the product-form network is approximately equal to the throughput $\lambda_{\mathrm{FJ}}(k)$ of the fork–join subnetwork for the value of K corresponding to k.

STEP 4 Replace the subnetwork in the given system by a composite node whose load-dependent service rates are equal to the determined throughput rates:

$$\mu(k) = \lambda_{\mathrm{FJ}}(k) \quad \text{for } k = 1, 2 \ldots .$$

STEP 5 If the network under consideration contains only one FES node, then the analysis of this node gives the performance measures for the whole network. Otherwise go back to Step 1.

STEP 6   From these performance measures, special characteristic values of parallel systems such as the speedup G_N and the synchronization overhead S_N can be computed.

The algorithm is now demonstrated on a simple example.

Example 10.19   Consider a parallel program with the precedence graph depicted in Fig. 10.41b. The model in the form of a series–parallel fork–join network is presented in Fig. 10.44. The arrival rate of jobs is $\lambda = 0.04$. All service times of the tasks are exponentially distributed with the mean values:

$$\frac{1}{\mu_1} = 10, \quad \frac{1}{\mu_2} = 5, \quad \frac{1}{\mu_3} = 2, \quad \frac{1}{\mu_4} = 10.$$

The analysis of the model is carried out in the steps that follow:

STEP 1   At first the inner fork–join system, consisting of the processors P_1 and P_2, is chosen and short-circuited. This short circuited model consists of $2 \cdot k$ tasks, $k = 1, 2, \ldots$ (Fig. 10.52).

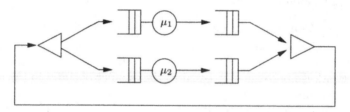

Fig. 10.52   Short-circuited model.

STEP 2   The number of states Z_{FJ} of this subsystem is determined using Eq. (10.180). For $k = 1, \ldots, 10$, the results are shown in Table 10.36.

Table 10.36   Number of states in the subsystem

k	1	2	3	4	5	6	7	8	9	10
Z_{FJ}	3	5	7	9	11	13	15	17	19	21
K	3	5	7	9	11	13	15	17	19	21
Z_{PF}	3	5	7	9	11	13	15	17	19	21

STEP 3   Determine the throughput of the subnetwork.

STEP 3.1   Construct a product-form network with $N = 2$ nodes. The number of jobs K in this network are determined using Eq. (10.181). For $k = 1, 2 \ldots, 10$ the values for K and the number of states $Z_{PF}(K)$ are given in Table 10.36.

$\boxed{\text{STEP 3.2}}$ Using MVA we solve the product-form network, constructed in Step 3.1, for all possible number of jobs K. The throughputs are given by

$$\lambda_{\mathrm{PF}}(3) = \underline{0.093}, \quad \lambda_{\mathrm{PF}}(5) = \underline{0.098},$$
$$\lambda_{\mathrm{PF}}(7) = \underline{0.100}, \quad \lambda_{\mathrm{PF}}(9) = \underline{0.100},$$

$$\vdots$$

These values approximately correspond to the throughputs $\lambda_{\mathrm{FJ}}(k)$ of the fork–join subnetwork:

$$\lambda_{\mathrm{FJ}}(1) = \lambda_{\mathrm{PF}}(3) = \underline{0.093},$$
$$\lambda_{\mathrm{FJ}}(2) = \lambda_{\mathrm{PF}}(5) = \underline{0.098},$$
$$\lambda_{\mathrm{FJ}}(3) = \lambda_{\mathrm{PF}}(7) = \underline{0.100},$$
$$\lambda_{\mathrm{FJ}}(4) = \lambda_{\mathrm{PF}}(9) = \underline{0.100},$$

$$\vdots$$

$\boxed{\text{STEP 4}}$ The analyzed fork-join subnetwork, consisting of the processors P_1 and P_2, is replaced by an FES node (see Fig. 10.53) with the load-dependent service rates $\mu(k) = \lambda_{\mathrm{FJ}}(k)$ for $k = 1, 2, \ldots$.

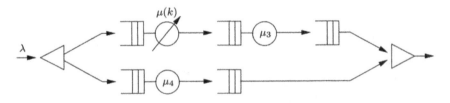

Fig. 10.53 Fork–join subnetwork with one FES node.

$\boxed{\text{STEP 5}}$ This network contains more than one node and therefore we return to Step 1.

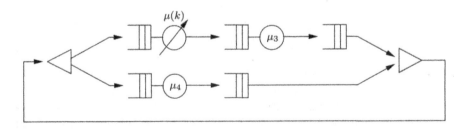

Fig. 10.54 Short-circuited network.

STEP 1 The short circuit of the network shown in Fig. 10.53 results in a closed network with $2 \cdot k$ tasks, $k = 1, 2, \ldots$ (Fig. 10.54).

STEP 2 The possible number of states Z_{FJ} in this network is given by Eq. (10.180). For $k = 1, \ldots, 10$ the values are shown in the Table 10.37.

Table 10.37 The possible number of states Z_{FJ} as a function of $k = 1, \ldots, 10$

k	1	2	3	4	5	6	7	8	9	10
Z_{FJ}	5	12	22	35	51	70	92	117	145	176
K	2	3	5	7	9	10	12	14	16	17
Z_{PF}	6	10	21	36	55	66	91	120	153	171

STEP 3 Determine the throughputs.

STEP 3.1 Construct a product-form network consisting of $N = 3$ nodes and determine the number of jobs K in the network using Eq. (10.181). For $k = 1, \ldots, 10$ the values of K and $Z_{PF}(K)$ are given in Table 10.37.

STEP 3.2 Solve the network for all number of jobs K. For the computed throughputs, the following relation approximately holds: $\lambda_{PF}(K) = \lambda_{FJ}(k)$. Using the MVA we get

$$\lambda_{FJ}(1) \approx \lambda_{PF}(2) = \underline{0.062}, \qquad \lambda_{FJ}(2) \approx \lambda_{PF}(3) = \underline{0.072},$$
$$\lambda_{FJ}(3) \approx \lambda_{PF}(5) = \underline{0.082}, \qquad \lambda_{FJ}(4) \approx \lambda_{PF}(7) = \underline{0.087},$$
$$\lambda_{FJ}(5) \approx \lambda_{PF}(9) = \underline{0.089}, \qquad \lambda_{FJ}(6) \approx \lambda_{PF}(10) = \underline{0.090},$$
$$\lambda_{FJ}(7) \approx \lambda_{PF}(12) = \underline{0.092}, \qquad \lambda_{FJ}(8) \approx \lambda_{PF}(14) = \underline{0.093},$$
$$\lambda_{FJ}(9) \approx \lambda_{PF}(16) = \underline{0.094}, \quad \lambda_{FJ}(10) \approx \lambda_{PF}(17) = \underline{0.094}.$$

STEP 4 The fork–join subnetwork is now an FES node (see Fig. 10.55) with the load-dependent service rates $\mu'(k) = \lambda_{FJ}(k)$, $k = 1, 2, \ldots$.

Fig. 10.55 FES queue for Example 10.19.

STEP 5 The fork–join network is now reduced to only one node and therefore the iteration stops and the performance measures of the system can be determined by analyzing the FES node of Fig. 10.55. For this analysis we determine at first the steady-state probabilities of the node using Eqs. (3.11) and (3.12):

$$\pi_0 = \underline{0.428}, \quad \pi_1 = \underline{0.287}, \quad \pi_2 = \underline{0.155},$$

$$\pi_3 = \underline{0.076}, \quad \pi_4 = \underline{0.034}, \quad \pi_5 = \underline{0.015},$$
$$\pi_6 = \underline{0.007}, \quad \pi_7 = \underline{0.003}, \quad \pi_8 = \underline{0.001},$$
$$\pi_9 = \underline{0.001}, \quad \pi_{10} = \underline{0.000}.$$

These probabilities are then used to determine the mean number of jobs in the fork–join model, Eq. (6.8):

$$\overline{K}_{\mathrm{FJ}} = \underline{1.116}$$

and the mean response time, Eq. (6.9):

$$\overline{T}_{\mathrm{FJ}} = \underline{27.89}.$$

With Eq. (6.15) and

$$\overline{T}_S = \sum_{i=1}^{4} \overline{T}_{S,i} = \sum_{i=1}^{4} \frac{1}{\mu_1(1 - \rho_i)},$$

we obtain

$$\overline{T}_S = \underline{41.76}.$$

To obtain the speedup, we use Eq. (10.166) and get

$$G_N = \underline{1.497}.$$

From this result we can compute the normalized speedup (Eq. (10.167)):

$$G = \frac{G_N}{N} = \underline{0.374}.$$

To obtain the synchronization overhead S_N, we need the mean waiting time $\overline{T}_{J,i}$, $(i = 1, 2, 3, 4)$ in synchronization queues. First we consider the fork–join queue of node 1 and node 2:

$$\overline{T}_{J,i} = \overline{T}_{FJ_{12}} - \overline{T}_i, \tag{10.182}$$

where $\overline{T}_{FJ_{12}}$ is the mean response time in the fork-join construct consisting of node 1 and node 2. It is obtained from the state-dependent service rate $\mu(k)$ for the composite node of the inner fork–join construct. At first we need the state probabilities, using Eqs. (3.11) and (3.12):

$$\pi_0 = \underline{0.580}, \quad \pi_1 = \underline{0.249}, \quad \pi_2 = \underline{0.102},$$
$$\pi_3 = \underline{0.041}, \quad \pi_4 = \underline{0.016}, \quad \pi_5 = \underline{0.007},$$
$$\pi_6 = \underline{0.003}, \quad \pi_7 = \underline{0.001}, \quad \pi_8 = \underline{0.000},$$

and the mean number of jobs:

$$\overline{K}_{FJ_{12}} = \sum_{k=1}^{\infty} k\pi_k = \underline{0.701}.$$

If we apply Little's theorem, we obtain for the mean response time of a parallel–sequential job with fork–join synchronizations:

$$\overline{T}_{FJ_{12}} = \frac{\overline{K}_{FJ_{12}}}{\lambda} = \underline{17.514}\,.$$

In order to obtain the mean response time, we apply Eq. (6.15) and get

$$\overline{T}_1 = \frac{1}{\mu_1(1 - \rho_1)} = \underline{16.667}\,, \quad \overline{T}_2 = \underline{6.25}\,.$$

Using Eq. (10.182), we finally get the results:

$$\overline{T}_{J,1} = \overline{T}_{FJ_{12}} - \overline{T}_1 = \underline{0.874}\,, \quad \overline{T}_{J,2} = \underline{11.264}\,.$$

From Fig. 10.53 we have

$$\overline{T}_{J,3} = \overline{T}_{FJ} - \overline{T}_{FJ_{12}} - \overline{T}_3\,, \qquad \overline{T}_{J,4} = \overline{T}_{FJ} - \overline{T}_4\,.$$

Applying Eq. (6.15) again, we obtain

$$\overline{T}_3 = \underline{2.174}\,, \qquad \overline{T}_4 = \underline{16.667}\,.$$

Together with $\overline{T}_{FJ}$ and $\overline{T}_{FJ_{12}}$ from the preceding, we get

$$\overline{T}_{J,3} = \underline{8.202}\,, \qquad \overline{T}_{J,4} = \underline{11.233}\,,$$

which finally gives us the synchronization overhead (see Eq. (10.168)):

$$S_N = \frac{\sum_{i=1}^{4} \overline{T}_{J,i}}{\overline{T}_{FJ}} = \underline{1.132}\,.$$

The normalized synchronization overhead (Eq. (10.169)) is given by

$$S = \frac{S_N}{N} = \underline{0.292}\,.$$

Now it is easy to calculate the blocking factor. Using Little's theorem, we obtain via Eq. (10.170)

$$B_N = \sum_{i=1}^{4} \overline{Q}_i = \underline{1.263}\,,$$

and with Eq. (10.171) the normalized blocking factor is given by

$$B = \frac{B_N}{4} = \underline{0.316}\,.$$

A comparison of the mean response time computed with the preceding approximation with DES shows that the differences between the DES results ($\overline{T} = 26.27$) and the approximation results are small in our example. For fork–join systems with higher arrival rates, the differences become larger. It can be shown in general that the power of the fork–join operations is reduced considerably by join synchronization operations. Also the speedup is reduced considerably by using synchronization operations. Other approaches to the analysis of parallel programs can be found in [Flei89], [MüRo87], [NeTa88], [SaTr87], and [ThBa86].

Problem 10.6   Consider a parallel program with the precedence graph shown in Fig. 10.56. The arrival rate is $\lambda = 1$ and the service rates of the tasks are $\mu_1 = 2$, $\mu_2 = 4$, $\mu_3 = 4$, $\mu_4 = 5$. All service times are exponentially distributed. Determine the mean response time and the speedup.

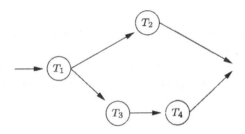

Fig. 10.56   Precedence graph for Problem 10.6.

10.7   NETWORKS WITH ASYMMETRIC NODES

Whenever we have considered multiserver nodes in a network, we have always assumed that all the servers at the node are statistically identical. We now consider the case of networks where servers at a multiserver node have different service times. We first consider closed networks with asymmetric nodes followed by open networks with asymmetric nodes. Note that in each of these cases we will only obtain an approximative solution.

10.7.1   Closed Networks

In this section we show how to modify the summation method, the mean value analysis, and the SCAT algorithm to deal with asymmetric nodes [BoRo94] as introduced in Section 9.1.2.

10.7.1.1   Asymmetric SUM (ASYM-SUM)   To extend the SUM (see Section 9.2) to queueing networks with asymmetric nodes, we need to adapt the formula for computing the mean number of jobs (see Eqs. (9.15) and (6.28)):

$$\overline{K}_i = m_i \cdot \rho_i + \frac{\rho_i}{1 - \dfrac{K - m_i - 1}{K - m_i} \cdot \rho_i} \cdot P_{m_i}, \quad \text{Type 1,}$$

with

$$P_{m_i}(\rho_i) = \frac{\dfrac{(m_i \cdot \rho_i)^{m_i}}{m_i! \cdot (1 - \rho_i)}}{\displaystyle\sum_{k=0}^{m_i-1} \dfrac{(m_i \cdot \rho_i)^k}{k!} + \dfrac{(m_i \cdot \rho_i)^{m_i}}{m_i \cdot (1 - \rho_i)}}.$$

For symmetric nodes, ρ_i is given by

$$\rho_i = \frac{\lambda_i}{m_i \cdot \mu_i},$$

and for asymmetric nodes we can use Eq. (6.156) or Eqs. (6.170) and (6.166), respectively. Since the SUM is an approximate method, so is the extended SUM. By replacing ρ_i and P_{m_i} by the corresponding formulae, the BOTT can be extended to queueing networks with asymmetric nodes in the same way.

10.7.1.2 Asymmetric MVA (ASYM-MVA)

The MVA was developed by Reiser and Lavenberg [ReLa80] for the exact analysis of product-form networks and is based on Little's theorem and the arrival theorem of Reiser, Lavenberg, Sevcik, and Mitrani (see Section 8.2) [Litt61, ReLa80, SeMi81]. To extend the MVA to networks with asymmetric nodes, the equation for the calculation of the mean response time for symmetric -/M/m nodes,

$$\overline{T}_i(k) = \frac{1}{\mu_i \cdot m_i} \left(1 + \overline{K}_i(k-1) + \sum_{j=0}^{m_i-2} (m_i - j - 1) \cdot \pi_i(j \mid k - 1) \right),$$
(10.183)

needs to be extended. Here $\pi_i(j \mid k)$ is the probability that there are j jobs at the ith node, given that there are k jobs in the network. This condition probability $\pi_i(j \mid k)$ is given by

$$\pi_i(j \mid k) = \frac{e_i \cdot \lambda(k)}{\mu_i \cdot j} \cdot \pi_i(j - 1 \mid k - 1), \quad j = 1, \ldots, m_i - 1, \qquad (10.184)$$

and $\pi_i(0 \mid k)$ is obtained by

$$\pi_i(0 \mid k) = 1 - \frac{1}{m_i} \left[\frac{e_i}{\mu_i} \cdot \lambda(k) + \sum_{j=1}^{m_i-1} (m_i - j) \cdot \pi_i(j \mid k) \right]. \qquad (10.185)$$

In Eq. (10.183), the factor $\mu_i \cdot m_i$ is the overall service rate of the symmetric $-/M/m$ node. For asymmetric nodes, $m_i \cdot \mu_i$ is replaced by

$$\sum_{j=1}^{m_i} \mu_{ij} \cdot$$

Thus, the mean response time of an asymmetric node can be computed as follows:

$$\overline{T}_i(k) = \frac{1}{\displaystyle\sum_{j=1}^{m_i} \mu_{ij}} \left(1 + \overline{K}_i(k-1) + \sum_{j=0}^{m_i-2} (m_i - j - 1) \cdot \pi_i(j \mid k-1) \right).$$

$$(10.186)$$

Because of the different service rates, it is necessary to take into account which servers are occupied as well as the number of servers when calculating the marginal probabilities $\pi_i(j \mid k)$. If we assume that the free server with the highest service rate is selected and the service rates are arranged in descending order, then the marginal probabilities are given by

$$\pi_i(j \mid k) = \frac{e_i \cdot \lambda(k)}{\displaystyle\sum_{l=1}^{j} \mu_{il}} \cdot \pi_i(j-1 \mid k-1), \tag{10.187}$$

$$\pi_i(0 \mid k) = 1 - \frac{e_i}{\displaystyle\sum_{l=1}^{m_i} \mu_{il}} \cdot \lambda(k) - \frac{1}{m_i} \cdot \sum_{j=1}^{m_i-1} (m_i - j) \cdot \pi_i(j \mid k). \tag{10.188}$$

If we consider Eq. (10.185) in more detail, we see that the factor

$$\frac{e_i \cdot \lambda(k)}{\displaystyle\sum_{l=1}^{m_i} \mu_{il}}$$

is the utilization ρ_i; therefore, Eq. (10.188) can be rewritten as

$$\pi_i(0 \mid k) = 1 - \rho_i(k) - \frac{1}{m_i} \cdot \sum_{j=1}^{m_i-1} (m_i - j) \cdot \pi_i(j \mid k).$$

10.7.1.3 Asymmetric SCAT (ASYM-SCAT)

The core of the SCAT algorithm (see Section 9.1.2) is a single step of the MVA in which the formula for the mean response time, Eq. (10.186), is approximated so that it depends only on K and not on $(K-1)$. However, it is necessary to estimate the marginal probabilities $\pi_i(j \mid K-1)$ that need a lot of computation time. For this reason, another formula for the mean response time is derived from Eq. (6.29). This derivation depends only on $\overline{K}_i$ and μ_i. For -/M/1 nodes we apply Little's theorem to the following equation:

$$\overline{K}_i = \frac{\rho_i}{1 - \rho_i},$$

and after reorganizing we get

$$\overline{T}_i = \frac{1}{\mu_i}(1 + \overline{K}_i),$$

which is the basic equation of the core of the SCAT algorithm for -/M/1 nodes. Similarly, from

$$\overline{K}_i = m_i\rho_i + \frac{\rho_i}{1 - \rho_i} \cdot P_{m_i}(\rho_i)$$

(Eq. (6.29)) and Little's theorem we obtain for -/M/m nodes

$$\overline{T}_i = \frac{1}{\mu_i \cdot m_i} \cdot (m_i + \overline{K}_i + P_{m_i}(\rho_i) - m_i \cdot \rho_i). \tag{10.189}$$

This formula can easily be extended to asymmetric nodes as follows:

$$\overline{T}_i = \frac{1}{\sum\limits_{j=1}^{m_i} \mu_{ij}} \cdot (m_i + \overline{K}_i + P_{m_i}(\rho_i) - m_i \cdot \rho_i). \tag{10.190}$$

And finally we get the following as core equation for ASYM-SCAT:

$$\overline{T}_i(K) = \frac{1}{\sum\limits_{j=1}^{m_i} \mu_{ij}} \cdot \left(m_i + \overline{K}_i(K-1) + P_{m_i}(\rho_i(K-1)) - m_i \cdot \rho_i(K-1)\right). \tag{10.191}$$

At this point we proceed analogously to the SUM by using the equations belonging to the selected method to calculate ρ_i and $P_{m_i}(\rho_i)$. Our experience with these methods suggests that the mean deviation for ASYM-SUM and ASYM-MVA is in the order of 2–3% for the random selection of a server as well as for the selection of the fastest free server. For ASYM-SCAT, the situation is somewhat different. For random selection the deviation is considerably high (7%), while for the selection of the fastest free server the mean deviation is only of the order of 1%. Therefore, this method can be recommended. ASYM-SUM, ASYM-MVA, and ASYM-SCAT can easily be extended to multiclass queueing networks [AbBo97].

10.7.2   Open Networks

Jackson's method for analyzing open queueing networks with symmetric nodes (see Section 7.3.4) can also be extended to analyze open queueing networks with asymmetric nodes. Note that this will result in an approximate solution. Jackson's theorem can be applied to open single class queueing networks with symmetric $-/M/m$-FCFS nodes.

The analysis of an open Jackson network is done in two steps:

STEP 1 Calculate the arrival rates λ_i using traffic Eq. (7.1) for all nodes $i = 1 \ldots N$.

STEP 2 Consider each node i as an elementary $-/M/m$-FCFS queueing system. Check whether the ergodicity ($\rho < 1$) is fulfilled and calculate the performance measures using the known formulae.

Step 1 remains unchanged when applied to asymmetric open networks. In Step 2, formulae given in Section 6.5 for asymmetric $-/M/m$-FCFS queueing networks should be used instead of formulae for symmetric $-/M/m$-FCFS queueing systems. Similarly, the BCMP theorem for open networks (see Section 7.3.6) can be extended to the approximate analysis of asymmetric queueing networks. Using the closing method (see Section 10.1.5) ASYM-MVA, ASYM-SCAT and ASYM-SUM can also be applied to open asymmetric queueing networks. Furthermore, it is also possible to apply the closing method to open and mixed queueing networks with asymmetric nodes.

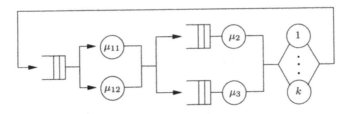

Fig. 10.57   Closed network with one asymmetric node.

Example 10.20   To demonstrate different methods for networks with asymmetric nodes we use the example network shown in Fig. 10.57. Node 1 is an asymmetric $-/M/2$, nodes 2 and 3 are of $-/M/1$ type, and node 4 is $-/G/\infty$. The routing probabilities are

$$p_{12} = 0.5, \quad p_{13} = 0.5, \quad p_{24} = p_{34} = p_{41} = 1,$$

and the service rates are

$$\mu_{11} = 3.333, \quad \mu_{12} = 0.666, \quad \mu_2 = 1.666, \quad \mu_3 = 1.25, \quad \mu_4 = 1.$$

The total number of jobs in the network is $K = 3$. From the routing probabilities we obtain the following visit ratios:

$$e_1 = 1, \quad e_2 = 0.5, \quad e_3 = 0.5, \quad e_4 = 1.$$

We start with the ASYM-SUM method, which consists of the same steps
as the SUM algorithm (see Section 9.2), but with different formulae for the
utilization ρ_1 and the probability of waiting P_{m_1}. For the stopping condition
we use $\epsilon = 0.001$.

STEP 1 Initialization:

$$\lambda_l = 0 \quad \text{and} \quad \lambda_u = \min_i \left\{ \frac{\sum\limits_{l=1}^{m_i} \mu_{il}}{e_i} \right\} = \underline{2.5}.$$

STEP 2 Bisection:

STEP 2.1 $\lambda = \dfrac{\lambda_l + \lambda_u}{2} = \underline{1.25}.$

STEP 2.2 Calculation of the values $f_i(\lambda_i)$, $i = 1, \ldots, 4$. In order to get the
result for node 1, Eqs. (6.166) and (6.170) for the exact analysis of networks
with asymmetric nodes are used (selection of the fastest free server). With

$$\rho_1 = \frac{e_1 \cdot \lambda}{\mu_{11} + \mu_{12}} = \frac{1.25}{4} = \underline{0.3125}$$

and

$$P_{m_1} = \frac{1}{1-c} \cdot \frac{P_m^{(L)}}{N} = \frac{1}{1 - 0.3125} \cdot \frac{0.111}{1.050} = \underline{0.153},$$

we get

$$\overline{K}_1(\lambda_1) = 2\rho_1 + \rho_1 \cdot P_{m_1} = \underline{0.673}.$$

Accordingly, we obtain

$$\overline{K}_2(\lambda_2) = \frac{\rho_2}{1 - \frac{2}{3}\rho_2} = 0.500 \quad \text{with } \rho_2 = \underline{0.375},$$

$$\overline{K}_3(\lambda_3) = \frac{\rho_3}{1 - \frac{2}{3}\rho_3} = 0.750 \quad \text{with } \rho_3 = \underline{0.500},$$

$$\overline{K}_4(\lambda_4) = \frac{\lambda_4}{\mu_4} = \underline{1.250},$$

which yields

$$g(\lambda) = \sum_{i=1}^{N} \overline{K}_i(\lambda_i) = \underline{3.173}.$$

STEP 2.3 Check the stopping condition. Because of $g(\lambda) > K$, we set $\lambda_u = \lambda = 1.25$.

STEP 2.1 $\lambda = \dfrac{\lambda_l + \lambda_u}{2} = \underline{0.625}$.

STEP 2.2 Calculation of the values $f_i(\lambda_i)$, $i = 1, \ldots, 4$. With

$$\overline{K}_1(\lambda_1) = 2\rho_1 + \rho_1 \cdot P_{m_1} = \underline{0.319}. \quad \text{with } \rho_1 = \underline{0.156} \text{ and } P_{m_1} = \underline{0.0386},$$

$$\overline{K}_2(\lambda_2) = \frac{\rho_2}{1 - \frac{2}{3}\rho_2} = \underline{0.214} \quad \text{with } \rho_2 = \underline{0.188},$$

$$\overline{K}_3(\lambda_3) = \frac{\rho_3}{1 - \frac{2}{3}\rho_3} = \underline{0.300} \quad \text{with } \rho_3 = 0.250,$$

$$\overline{K}_4(\lambda_4) = \frac{\lambda_4}{\mu_4} = \underline{0.625},$$

we get

$$g(\lambda) = \sum_{i=1}^{N} \overline{K}_i(\lambda_i) = \underline{1.458}.$$

STEP 2.3 Check the stopping condition. Because of $g(\lambda) < K$, we set $\lambda_l = \lambda = 0.625$.

STEP 2.1 $\lambda = \dfrac{\lambda_l + \lambda_u}{2} = \underline{0.938}$.

STEP 2.2 Calculation of the values $f_i(\lambda_i)$, $i = 1, \ldots, 4$. With

$$\overline{K}_1(\lambda_1) = 2\rho_1 + \rho_1 \cdot P_{m_1} = \underline{0.489}. \quad \text{with } \rho_1 = \underline{0.234} \text{ and } P_{m_1} = \underline{0.0881},$$

$$\overline{K}_2(\lambda_2) = \frac{\rho_2}{1 - \frac{2}{3}\rho_2} = \underline{0.346} \quad \text{with } \rho_2 = \underline{0.281},$$

$$\overline{K}_3(\lambda_3) = \frac{\rho_3}{1 - \frac{2}{3}\rho_3} = \underline{0.500} \quad \text{with } \rho_3 = \underline{0.375},$$

$$\overline{K}_4(\lambda_4) = \frac{\lambda_4}{\mu_4} = \underline{0.938},$$

we get

$$g(\lambda) = \sum_{i=1}^{N} \overline{K}_i(\lambda_i) = \underline{2.273}.$$

STEP 2.3 Check the stopping condition. Because of $g(\lambda) < K$, we set $\lambda_l = \lambda = 0.9375$.

$$\vdots$$

After 11 iterations we have $g(\lambda) = 3.000$ and the stopping condition is fulfilled. For the overall throughput we obtain $\lambda = 1.193$, which yields

$$\lambda_1 = \underline{1.193}, \quad \lambda_2 = \underline{0.596}, \quad \lambda_3 = \underline{0.596}, \quad \lambda_4 = \underline{1.193}.$$

For the utilizations of the nodes we get

$$\rho_{11} = \underline{0.294}, \quad \rho_{12} = \underline{0.317}, \quad \rho_1 = \underline{0.298}, \quad \rho_2 = \underline{0.358}, \quad \rho_3 = \underline{0.477},$$

and for the mean number of jobs in a node we obtain

$$\overline{K}_1 = \underline{0.638}, \quad \overline{K}_2 = \underline{0.470}, \quad \overline{K}_3 = \underline{0.700}, \quad \overline{K}_4 = \underline{1.193}.$$

Now we analyze the network using the ASYM-MVA algorithm.

STEP 1 Initialization:

$$\overline{K}_1(0) = \overline{K}_2(0) = \overline{K}_3(0) = 0, \quad p_1(0 \mid 0) = 1, \quad p_1(1 \mid 0) = 0.$$

STEP 2 Iteration over the number of jobs in the network starting with $k = 1$.

STEP 2.1 Mean response times:

$$\overline{T}_1(1) = \frac{1}{\mu_{11} + \mu_{12}} \left(1 + \overline{K}_1(0) + p_1(0 \mid 0)\right) = \underline{0.5},$$

$$\overline{T}_2(1) = \frac{1}{\mu_2} \left(1 + \overline{K}_2(0)\right) = \underline{0.6},$$

$$\overline{T}_3(1) = \frac{1}{\mu_3} \left(1 + \overline{K}_3(0)\right) = \underline{0.8},$$

$$\overline{T}_4(1) = \frac{1}{\mu_4} = \underline{1}.$$

STEP 2.2 Throughput:

$$\lambda(1) = \frac{1}{\sum\limits_{i=1}^{4} e_i \overline{T}_i(1)} = \underline{0.455}.$$

STEP 2.3 Mean number of jobs:

$$\overline{K}_1(1) = \lambda(1)\overline{T}_1(1)e_1 = \underline{0.227}, \quad \overline{K}_2(1) = \lambda(1)\overline{T}_2(1)e_2 = \underline{0.136},$$

$$\overline{K}_3(1) = \lambda(1)\overline{T}_3(1)e_3 = \underline{0.182}, \quad \overline{K}_4(1) = \lambda(1)\overline{T}_4(1)e_4 = \underline{0.455}.$$

Iteration for $k = 2$:

STEP 2.1 Mean response times:

$$\overline{T}_1(2) = \frac{1}{\mu_{11} + \mu_{12}} \left(1 + \overline{K}_1(1) + p_1(0 \mid 1)\right) = \underline{0.511},$$

where

$$p_1(0 \mid 1) = 1 - \frac{e_1\lambda(1)}{\mu_{11} + \mu_{12}} - \frac{1}{m_1}p_1(1 \mid 1) = \underline{0.818} \quad \text{and}$$

$$p_1(1 \mid 1) = \frac{e_1\lambda(1)}{\mu_{11}}p_1(0 \mid 0) = \underline{0.136},$$

$$\overline{T}_2(2) = \frac{1}{\mu_2}\left(1 + \overline{K}_2(1)\right) = \underline{0.682},$$

$$\overline{T}_3(2) = \frac{1}{\mu_3}\left(1 + \overline{K}_3(1)\right) = \underline{0.945},$$

$$\overline{T}_4(2) = \frac{1}{\mu_4} = \underline{1}.$$

STEP 2.2 Throughput:

$$\lambda(2) = \frac{2}{\sum\limits_{i=1}^{4} e_i\overline{T}_i(2)} = \underline{0.860}.$$

STEP 2.3 Mean number of jobs:

$$\overline{K}_1(2) = \lambda(2)\overline{T}_1(2)e_1 = \underline{0.440}, \quad \overline{K}_2(2) = \lambda(2)\overline{T}_2(2)e_2 = \underline{0.293},$$

$$\overline{K}_3(2) = \lambda(2)\overline{T}_3(2)e_3 = \underline{0.407}, \quad \overline{K}_4(2) = \lambda(2)\overline{T}_4(2)e_4 = \underline{0.860}.$$

Iteration for $k = 3$:

STEP 2.1 Mean response times:

$$\overline{T}_1(3) = \frac{1}{\mu_{11} + \mu_{12}}\left(1 + \overline{K}_1(2) + p_1(0 \mid 2)\right) = \underline{0.530},$$

where

$$p_1(0 \mid 2) = 1 - \frac{e_1\lambda(2)}{\mu_{11} + \mu_{12}} - \frac{1}{m_1}p_1(1 \mid 2) = \underline{0.679} \quad \text{and}$$

$$p_1(1 \mid 2) = \frac{e_1\lambda(2)}{\mu_{11}}p_1(0 \mid 1) = \underline{0.211},$$

$$\overline{T}_2(3) = \frac{1}{\mu_2}\left(1 + \overline{K}_2(2)\right) = \underline{0.776},$$

$$\overline{T}_3(3) = \frac{1}{\mu_3}\left(1 + \overline{K}_3(2)\right) = \underline{1.125},$$

$$\overline{T}_4(3) = \frac{1}{\mu_4} = \underline{1}.$$

STEP 2.2   Throughput:

$$\lambda(3) = \frac{3}{\sum\limits_{i=1}^{4} e_i \overline{T}_i(3)} = \underline{1.209}\,.$$

STEP 2.3   Mean number of jobs:

$$\overline{K}_1(3) = \lambda(3)\overline{T}_1(3)e_1 = \underline{0.641}\,, \quad \overline{K}_2(3) = \lambda(3)\overline{T}_2(3)e_2 = \underline{0.469}\,,$$
$$\overline{K}_3(3) = \lambda(3)\overline{T}_3(3)e_3 = \underline{0.681}\,, \quad \overline{K}_4(3) = \lambda(3)\overline{T}_4(3)e_4 = \underline{1.209}\,.$$

The algorithm stops after three iteration steps. For the throughput we get

$$\lambda_1 = \underline{1.209}\,, \quad \lambda_2 = \underline{0.605}\,, \quad \lambda_3 = \underline{0.605}\,, \quad \lambda_4 = \underline{1.209}\,.$$

For the utilizations of the nodes we obtain

$$\rho_1 = \underline{0.302}\,, \quad \rho_2 = \underline{0.363}\,, \quad \rho_3 = \underline{0.484}\,,$$

and for the mean number of jobs in a node we get

$$\overline{K}_1 = \underline{0.641}\,, \quad \overline{K}_2 = \underline{0.469}\,, \quad \overline{K}_3 = \underline{0.681}\,, \quad \overline{K}_4 = \underline{1.209}\,.$$

Finally, the analysis of the network is performed using the ASYM-SCAT algorithm, where we use the approximate Eqs. (6.155) and (6.156) and Eqs. (6.27) and (6.28) for asymmetric multiple server nodes.

STEP 1   Modified core algorithm for $K = 3$ jobs in the network with the input parameters:

$$\overline{K}_i(K) = \frac{K}{N} = \underline{0.75} \quad \text{and} \quad D_i(K) = 0 \quad \text{for } i = 1,\dots,4\,.$$

STEP C1   Estimate values for the $\overline{K}_i(K-1)$ from Eq. (9.5):

$$\overline{K}_i(2) = 2\frac{\overline{K}_i(3)}{3} = 0.5 \quad \text{for } i = 1,\dots,4\,.$$

For the next step we need an initial value for the throughput:

$$\lambda(2) = \mu_4 \cdot \overline{K}_4(2) = \underline{0.5}\,.$$

STEP C2   One step of MVA:

$$\overline{T}_1(3) = \frac{1}{\mu_{11} + \mu_{12}} \cdot \left(m_1 + \overline{K}_1(2) + P_{m_1}(\rho_1) - m_1\rho_1\right) = \underline{0.569}\,,$$

with

$$\rho_1 = \frac{e_1 \lambda(2)}{\mu_{11} + \mu_{12}} = \frac{0.5}{4} = 0.125 \quad \text{and} \quad P_{m_1}(\rho_1) = 0.0278,$$

$$\overline{T}_2(3) = \frac{1}{\mu_2}\left(1 + \overline{K}_2(2)\right) = 0.9,$$

$$\overline{T}_3(3) = \frac{1}{\mu_3}\left(1 + \overline{K}_3(2)\right) = 1.2,$$

$$\overline{T}_4(3) = \frac{1}{\mu_4} = 1.$$

Throughput:

$$\lambda(3) = \frac{3}{\displaystyle\sum_{i=1}^{4} e_i \overline{T}_i(3)} = 1.145.$$

Mean number of jobs:

$$\overline{K}_1(3) = \lambda(3)\overline{T}_1(3)e_1 = 0.652, \quad \overline{K}_2(3) = \lambda(3)\overline{T}_2(3)e_2 = 0.515,$$

$$\overline{K}_3(3) = \lambda(3)\overline{T}_3(3)e_3 = 0.687, \quad \overline{K}_4(3) = \lambda(3)\overline{T}_4(3)e_4 = 1.145.$$

STEP C3 Check the stopping condition:

$$\max_i \left\{ \frac{\left|\overline{K}_i^{(1)}(3) - \overline{K}_i^{(0)}(3)\right|}{3} \right\} = 0.132 > \epsilon = 0.01.$$

STEP C1 Estimate values for the $k_i(K-1)$ from Eq. (9.5):

$$\overline{K}_1(2) = 0.435, \quad \overline{K}_2(2) = 0.344, \quad \overline{K}_3(2) = 0.458, \quad \overline{K}_4(2) = 0.764.$$

For the next step we need an initial value for the throughput:

$$\lambda(2) = \mu_4 \overline{K}_4(2) = 0.764.$$

STEP C2 One step of MVA:

$$\vdots$$

After three iterations we get the following results for the mean number of jobs:

$$\overline{K}_1(3) = 0.626, \quad \overline{K}_2(3) = 0.475, \quad \overline{K}_3(3) = 0.702, \quad \overline{K}_4(3) = 1.198.$$

STEP 2 Modified core algorithm for $(K-1) = 2$ jobs in the network with the input parameters:

$$\overline{K}_i(2) = \frac{2}{N} = 0.5 \quad \text{and} \quad D_i(2) = 0 \quad \text{for } i = 1, \dots, 4.$$

STEP C1 Estimate values for the $\overline{K}_i(1)$ from Eq. (9.5):

$$\overline{K}_i(1) = 2\frac{\overline{K}_i(2)}{2} = \underline{0.25} \quad \text{for } i = 1, \dots, 4.$$

For the next step we need an initial value for the throughput:

$$\lambda(1) = \mu_4 \cdot \overline{K}_4(1) = \underline{0.25}.$$

STEP C2 One step of MVA:

$$\overline{T}_1(2) = \frac{1}{\mu_{11} + \mu_{12}} \cdot (m_1 + \overline{K}_1(1) + P_{m_1}(\rho_1) - m_1\rho_1) = \underline{0.533},$$

with

$$\rho_1 = \frac{e_1\lambda(1)}{\mu_{11} + \mu_{12}} = \frac{0.25}{4} = \underline{0.0625} \quad \text{and} \quad P_{m_1}(\rho_1) = \underline{0.00735},$$

$$\overline{T}_2(2) = \frac{1}{\mu_2}\left(1 + \overline{K}_2(1)\right) = \underline{0.75},$$

$$\overline{T}_3(2) = \frac{1}{\mu_3}\left(1 + \overline{K}_3(1)\right) = \underline{1.0},$$

$$\overline{T}_4(2) = \frac{1}{\mu_4} = \underline{1}.$$

Throughput:

$$\lambda(2) = \frac{2}{\sum\limits_{i=1}^{4} e_i\overline{T}_i(2)} = \underline{0.831}.$$

Mean number of jobs:

$$\overline{K}_1(2) = \lambda(2)\overline{T}_1(2)e_1 = \underline{0.443}, \quad \overline{K}_2(2) = \lambda(2)\overline{T}_2(2)e_2 = \underline{0.311},$$
$$\overline{K}_3(2) = \lambda(2)\overline{T}_3(2)e_3 = \underline{0.415}, \quad \overline{K}_4(2) = \lambda(2)\overline{T}_4(2)e_4 = \underline{0.831}.$$

STEP C3 Check the stopping condition:

$$\max_i \left\{ \frac{\left|\overline{K}_i^{(1)}(2) - \overline{K}_i^{(0)}(2)\right|}{2} \right\} = \underline{0.165} > \epsilon = 0.01.$$

STEP C1 Estimate values for the $\overline{K}_i(1)$ from Eq. (9.5):

$$\overline{K}_1(1) = \underline{0.221}, \quad \overline{K}_2(1) = \underline{0.156}, \quad \overline{K}_3(1) = \underline{0.208}, \quad \overline{K}_4(1) = \underline{0.415}.$$

For the next step we need an initial value for the throughput:

$$\lambda(1) = \mu_4 \overline{K}_4(1) = \underline{0.415}.$$

STEP C2 One step of MVA:

$$\vdots$$

After three iterations we get the following results for the mean number of jobs:

$$\overline{K}_1(2) = \underline{0.434}, \quad \overline{K}_2(2) = \underline{0.295}, \quad \overline{K}_3(2) = \underline{0.414}, \quad \overline{K}_4(2) = \underline{0.857}.$$

STEP 3 Estimate the F_i and D_i values with Eqs. (9.3) and (9.4), respectively:

$$F_1(3) = \frac{\overline{K}_1(3)}{3} = \underline{0.209}, \quad F_1(2) = \frac{\overline{K}_1(2)}{2} = \underline{0.217},$$

$$F_2(3) = \frac{\overline{K}_2(3)}{3} = \underline{0.158}, \quad F_2(2) = \frac{\overline{K}_2(2)}{2} = \underline{0.148},$$

$$F_3(3) = \frac{\overline{K}_3(3)}{3} = \underline{0.234}, \quad F_3(2) = \frac{\overline{K}_3(2)}{2} = \underline{0.207},$$

$$F_4(3) = \frac{\overline{K}_4(3)}{3} = \underline{0.399}, \quad F_4(2) = \frac{\overline{K}_4(2)}{2} = \underline{0.429}.$$

Therefore, we get

$$D_1(3) = F_1(2) - F_1(3) = \underline{0.00835}, \quad D_2(3) = F_2(2) - F_2(3) = \underline{-0.0106},$$
$$D_3(3) = F_3(2) - F_3(3) = \underline{-0.0270}, \quad D_4(3) = F_4(2) - F_4(3) = \underline{0.0292}.$$

STEP 4 Modified core algorithm for $K = 3$ jobs in the network with the computed $\overline{K}_i(K)$ values from Step 1 and the $D_i(K)$ values from Step 3 as inputs.

STEP C1 Estimate values for the $\overline{K}_i(K-1)$ from Eq. (9.5):

$$\overline{K}_1(2) = 2\left(\frac{\overline{K}_1(3)}{3} + D_1(3)\right) = \underline{0.434},$$

$$\overline{K}_2(2) = \underline{0.295}, \quad \overline{K}_3(2) = \underline{0.414}, \quad \overline{K}_4(2) = \underline{0.857}.$$

For the next step we need an initial value for the throughput:

$$\lambda(2) = \mu_4 \cdot \overline{K}_4(2) = \underline{0.857}.$$

STEP C2 One step of MVA:

$$\overline{T}_1(3) = \frac{1}{\mu_{11} + \mu_{12}} \cdot \left(m_1 + \overline{K}_1(2) + P_{m_1}(\rho_1) - m_1\rho_1\right) = \underline{0.520},$$

with

$$\rho_1 = \underline{0.24} \quad \text{and} \quad P_{m_1}(\rho_1) = \underline{0.0756},$$

$$\overline{T}_2(3) = \frac{1}{\mu_2}\left(1 + \overline{K}_2(2)\right) = \underline{0.777},$$

$$\overline{T}_3(3) = \frac{1}{\mu_3}\left(1 + \overline{K}_3(2)\right) = \underline{1.131},$$

$$\overline{T}_4(3) = \frac{1}{\mu_4} = \underline{1}.$$

Throughput:

$$\lambda(3) = \frac{3}{\sum\limits_{i=1}^{4} e_i \overline{T}_i(3)} = \underline{1.212}.$$

Mean number of jobs:

$$\overline{K}_1(3) = \lambda(3)\overline{T}_1(3)e_1 = \underline{0.631}, \quad \overline{K}_2(3) = \lambda(3)\overline{T}_2(3)e_2 = \underline{0.471},$$

$$\overline{K}_3(3) = \lambda(3)\overline{T}_3(3)e_3 = \underline{0.686}, \quad \overline{K}_4(3) = \lambda(3)\overline{T}_4(3)e_4 = \underline{1.212}.$$

STEP C3   Check the stopping condition:

$$\max_i \left\{ \frac{\left|\overline{K}_i^{(1)}(3) - \overline{K}_i^{(0)}(3)\right|}{3} \right\} = \underline{0.00531} < \epsilon = 0.01.$$

The stopping condition is fulfilled. Therefore, the algorithm stops and we get the following results:

Throughputs:

$$\lambda_1 = \underline{1.212}, \quad \lambda_2 = \underline{0.606}, \quad \lambda_3 = \underline{0.606}, \quad \lambda_4 = \underline{1.212}.$$

Mean number of jobs:

$$\overline{K}_1 = \underline{0.631}, \quad \overline{K}_2 = \underline{0.471}, \quad \overline{K}_3 = \underline{0.686}, \quad \overline{K}_4 = \underline{1.212}.$$

Utilizations:

$$\rho_1 = \underline{0.303}, \quad \rho_2 = \underline{0.364}, \quad \rho_3 = \underline{0.485}.$$

In Table 10.38 we compare the results for the throughputs λ_i and the mean number of jobs $\overline{K}_i$. For this example, the results of all methods are close together and close to the DES results. As in the case of asymmetric single station queueing networks, the preceding approximations are accurate if the values of the service rates in a station do not differ too much. As a rule of thumb, we consider the approximations to be satisfactory if $\mu_{\min}/\mu_{\max} < 10$ at each asymmetric node.

Table 10.38 Throughput λ_i and mean number of jobs $\overline{K}_i$ for different methods compared with the DES results

Node		1	2	3	4
	ASYM-SUM	1.19	0.60	0.60	1.19
	ASYM-MVA	1.21	0.61	0.61	1.21
λ_i	ASYM-SCAT	1.21	0.61	0.61	1.21
	DES	1.21	0.61	0.61	1.21
	ASYM-SUM	0.64	0.47	0.70	1.19
	ASYM-MVA	0.64	0.47	0.68	1.21
$\overline{K}_i$	ASYM-SCAT	0.63	0.47	0.69	1.21
	DES	0.64	0.47	0.68	1.21

Problem 10.7   Formulate the closed network of Fig. 10.57 as a GSPN and solve it exactly using **SHARPE** or **SPNP**. Compare the exact results with the approximative ones obtained in this section using ASYM-SUM, ASYM-MVA and ASYM-SCAT.

Problem 10.8   Change the closed network of Fig. 10.57 to an open network by removing the feedback from node 3 to node 1. Jobs arrive from outside to node 1 with $\lambda_1 = 1.2$. Analyze this open network using the extended Jackson's method for networks with asymmetric nodes.

Problem 10.9   Introduce in the open network of Fig. 7.14 an asymmetric node by replacing node 1 by an asymmetric multi server with two servers and the service rates $\mu_{11} = 10$ and $\mu_{12} = 15$ and alternatively with $\mu_{11} = 5$ and $\mu_{12} = 20$. Analyze this open network using the extended Jackson's method for networks with asymmetric nodes. Discuss the results.

10.8 NETWORKS WITH BLOCKING

Up to now we have generally assumed that all nodes of a queueing network have queues with infinite buffers. Thus, a job that leaves a node will always find an empty place in the queue of the next node. But in the case of finite capacity queues, blocking can occur and jobs can be rejected at a node if the queue is full. Queueing networks that consist of nodes with finite buffer capacity are called *blocking networks*. Blocking networks are not only important for modeling and performance evaluation of computer systems and computer networks, but also in the field of production line systems. Exact results for general blocking networks can only be obtained by using the generation and numerical solution of the underlying CTMC (either directly or via SPNs) that is possible for medium-sized networks. For small networks, closed-form results may be derived. The most common techniques presented in the literature for the analysis of large blocking networks are, therefore, approximate techniques.

10.8.1   Different Blocking Types

In [Perr94] three different types of blocking are introduced:

1. *Blocking after service* (see Fig. 10.58): With this type of blocking, node i is blocked when a job completing service at node i wishes to join node j that is already full. This job then has to stay in the server of node i and block until a job leaves node j. This type of blocking is used to model production line systems or external I/O devices [Akyi87, Akyi88b, Akyi88a, Alti82, AlPe86, OnPe89, PeAl86, Perr81, PeSn89, SuDi86]. In the literature, this type of blocking has been referred to by a variety of different terms such as Type 1 blocking, transfer blocking, production blocking, and non-immediate blocking.

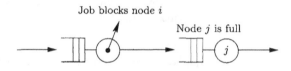

Job blocks node i

Node j is full

Fig. 10.58   Blocking after service.

2. *Blocking before service* (see Fig. 10.59): The job to be served next at node i determines its source node j before it enters service at node i. If node j is already full, the job cannot enter the server of node i and node i is therefore blocked. Only if another job leaves node j will node i be deblocked and the first job in the queue served [BoKo81, GoNe67b, KoRe76, KoRe78]. In the literature, this type of blocking has also been referred to as Type 2 blocking, service blocking, communication blocking, and immediate blocking.

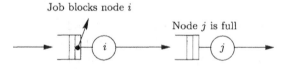

Job blocks node i

Node j is full

Fig. 10.59   Blocking before service.

3. *Repetitive blocking* (see Fig. 10.60): If a job that has been served at node i wishes to go to node j whose queue is full, it is rejected at node j and goes to the rear of node i queue. This procedure is repeated until a job leaves node j, thereby creating one empty slot in the queue of node j. This type of blocking is used to model communications networks

or flexible production line systems [Akvo89, BaIa83, Hova81, Pitt79]. Other terms for this type of blocking are Type 3 blocking or rejection blocking.

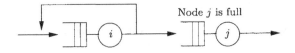

Fig. 10.60 Repetitive blocking.

In [OnPe86], different kinds of blocking are compared with each other. In [YaBu86], a queueing network model is presented that has finite buffer capacity and yet is not a blocking network. Here a job that would block node i because node j is full is buffered in a central buffer until there is an empty spot at node j. Such models are especially interesting for production line systems.

10.8.2   Product-Form Solution for Networks with Two Nodes

Many different techniques for the solution of blocking networks are available. We only consider a simple exact solution technique in detail that can be applied to closed blocking networks with only two nodes and blocking after service [Akyi87]. This method can also be extended to queueing networks with several nodes, but then it is an approximate technique that is very costly and it can only be used for throughput analysis [Akyi88b].

We consider queueing networks with one job class, exponentially distributed service times, and FCFS service strategy. Transitions back to the same node are not possible ($p_{ii} = 0$). Each node has a fixed capacity M_i that equals the capacity of the queue plus the number m_i of servers. Nodes with an infinity capacity can be taken into consideration by setting $M_i > K$ so that the capacity of the node is bigger than the number of jobs in the network. The overall capacity of the network must, of course, be bigger than the number of jobs in the network for it to be a blocking network:

$$K < M_1 + M_2 \,. \tag{10.192}$$

In the case of equality in the preceding expression, deadlocks can occur [Akyi87]. The possible number of states in the underlying CTMC of a closed network with two nodes and unlimited storage capacity is

$$Z = K + 1 \,. \tag{10.193}$$

At first we consider the simple case where the number of servers at each node is $m_1 = m_2 = 1$. The state diagram of the CTMC for such a two-node network *without blocking* is shown in Fig. 10.61.

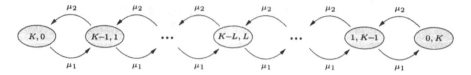

Fig. 10.61 CTMC for a two-node network without blocking ($m_1 = m_2 = 1$).

As can be seen, the minimum capacity at each node is $M_i = K$ (the queue must contain at least $K - 1$ spots) so that all states are possible. If the nodes have a finite capacity $M_i < K$, then not all $K + 1$ states of Fig. 10.61 are possible. Possible states of the blocking network are then given by the condition that the number of jobs at a node cannot be bigger than the capacity of that node. In the case of a transition to a state where the capacity of a node is exceeded, blocking occurs. The corresponding state is called a blocking state. Recall that due to Type 1 blocking, the job does not switch to the other node but stays in the original node. The state diagram of the blocking network is shown in Fig. 10.62.

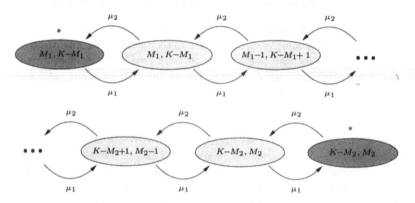

Fig. 10.62 CTMC for the two-node blocking network (states marked with * are blocking states) $m_1 = m_2 = 1$ (blocking after service).

The state diagram of the blocking network shown in Fig. 10.62 differs from the state diagram shown in Fig. 10.61 in two aspects. First, all states where the capacity is exceeded are no longer necessary and, second, Fig. 10.62 contains the blocking states marked with an asterisk. The number of states $\hat{Z}$ in the blocking network is the sum of all possible nonblocking states plus all the blocking states:

$$\hat{Z} = \min\{K, M_1 + 1\} + \min\{K, M_2 + 1\} - K + 1. \qquad (10.194)$$

It can be shown [Akyi87] that for this closed network with two nodes and blocking after service, an equivalent closed network with two nodes without blocking exists. Apart from the number of jobs $\hat{K}$ in the equivalent network, all other parameter values remain the same. The population $\hat{K}$ in the equivalent network can be determined using Eqs. (10.193) and (10.194):

$$\hat{K} = \min\{K, M_1 + 1\} + \min\{K, M_2 + 1\} - K. \tag{10.195}$$

The CTMC of the equivalent network is shown in Fig. 10.63. The equivalent network is a product-form network with $\hat{k}_i$ jobs at node i, $0 \leq \hat{k}_i \leq \hat{K}$.

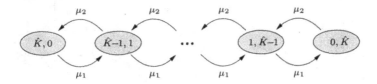

Fig. 10.63 CTMC of the equivalent network corresponding to the CTMC in Fig. 10.61.

The steady-state probabilities of the equivalent network can be determined using the following equation:

$$\pi(\hat{k}_1, \hat{k}_2) = \frac{1}{G(\hat{K})} \cdot \prod_{i=1}^{2} \left(\frac{1}{\mu_i}\right)^{\hat{k}_i}. \tag{10.196}$$

By equating the steady-state probabilities of the blocking network with the corresponding steady-state probabilities of the equivalent nonblocking network, the performance measures of the blocking network can be determined. To demonstrate how this technique works, the following simple example is used:

Example 10.21 Consider a closed queueing network with two nodes, $K = 10$ jobs, and the routing probabilities:

$$p_{12} = 1, \quad p_{21} = 1.$$

All other system parameters are given in Table 10.39.

Table 10.39 System parameters for Example 10.21

i	e_i	$1/\mu_i$	m_i	M_i
1	1	2	1	7
2	1	0.9	1	5

All service times are exponentially distributed and the service discipline at each node is FCFS. Using Eq. (10.193), the number of states in the network without blocking is determined to be $Z = K + 1 = 11$. The CTMC of the

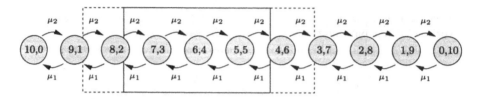

Fig. 10.64   CTMC of the Example 10.21 network without blocking.

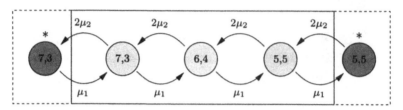

Fig. 10.65   CTMC of the Example 10.21 network with blocking.

network without blocking is shown in Fig. 10.64, while the state diagram of
the blocking network (see Fig. 10.65) contains all possible nonblocking states
from the original network as well as the blocking states (marked with asterisk).
The number of states $\hat{Z}$ of this blocking network is given by Eq. (10.194) as the
sum of all possible nonblocking states ($= 3$) and the blocking states ($= 2$):
$\hat{Z} = 5$. The CTMC of the equivalent network (see Fig. 10.66) therefore
contains $\hat{Z} = 5$ states and the number of jobs in the network is given by
Eq. (10.195) as $\hat{K} = 4$.

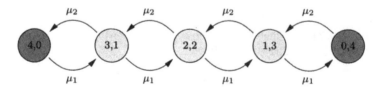

Fig. 10.66   CTMC of the equivalent network for Example 10.21.

The throughput $\lambda_{\mathrm{NB}}(4)$ of the equivalent network can be determined directly using the MVA and is equal to the throughput $\lambda_B(10)$ of the blocking
network:

$$\lambda_{\mathrm{B}}(10) = \lambda_{\mathrm{NB}}(4) = \underline{0.500}\,.$$

The steady-state probabilities of the blocking network are obtained by equating with the corresponding steady-state probabilities of the equivalent nonblocking network:

$$\pi(6,4) = \pi(2,2) = \underline{0.120}\,,$$
$$\pi(7,3)^* = \pi(4,0) = \underline{0.550}\,, \qquad \pi(5,5) = \pi(1,3) = \underline{0.054}\,,$$

$$\pi(7,3) = \pi(3,1) = \underline{0.248}, \quad \pi(5,5)^* = \pi(0,4) = \underline{0.026}.$$

The mean number of jobs at both nodes can be determined using the steady-state probabilities:

$$\overline{K}_1 = 7\left[\pi(7,3)^* + \pi(7,3)\right] + 6\pi(6,4) + 5\left[\pi(5,5) + \pi(5,5)^*\right] = \underline{6.73},$$
$$\overline{K}_2 = 3\left[\pi(7,3)^* + \pi(7,3)\right] + 4\pi(6,4) + 5\left[\pi(5,5) + \pi(5,5)^*\right] = \underline{3.27}.$$

With these results all other performance measures can be derived using the well-known formulae. Of special interest for blocking networks are the blocking probabilities:

$$P_{B_1} = \pi(5,5)^* = \underline{0.026},$$
$$P_{B_2} = \pi(7,3)^* = \underline{0.0550}.$$

Next consider the case where each node has multiple servers ($m_i > 1$). The state diagram of the CTMC underlying such a two-node network without blocking is shown in Fig. 10.67, while the state diagram of the blocking

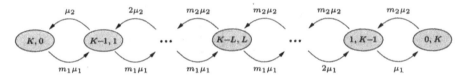

Fig. 10.67   CTMC of a two-node network without blocking ($m_i > 1$).

network (with Type 1 blocking) is shown in Fig. 10.68. Here a state marked

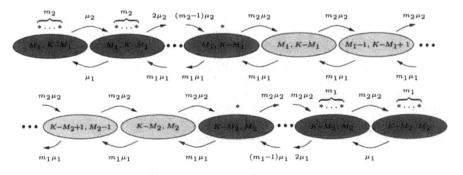

Fig. 10.68   CTMC for the two-node network with blocking. Blocking states are marked with $*$ ($m_i > 1$).

with asterisks shows the number of blocked servers in that state. Therefore, we get for the number $\hat{Z}$ of states in the blocking network, the sum of all possible nonblocking states plus the blocking states ($m_i > 1$):

$$\hat{Z} = \min\{K, M_1 + m_2\} + \min\{K, M_2 + m_1\} - K + 1. \tag{10.197}$$

$\hat{K}$ is given by Eqs. (10.193) and (10.197):

$$\hat{K} = \min\{K, M_1 + m_2\} + \min\{K, M_2 + m_1\} - K .\qquad(10.198)$$

The state diagram of the equivalent network is shown in Fig. 10.69. The equivalent network is again a product-form network with $\hat{k}_i$ jobs at node i, $0 \leq \hat{k}_i \leq \hat{K}$.

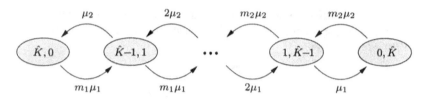

Fig. 10.69 CTMC of the equivalent product-form network $(m_i > 1)$.

The steady-state probabilities of the equivalent network can be determined using Eq. (7.59):

$$\pi(\hat{k}_1, \hat{k}_2) = \frac{1}{G(\hat{K})} \cdot \prod_{i=1}^{2} \left(\frac{1}{\mu_i}\right)^{\hat{k}_i} \cdot \frac{1}{\beta_i(\hat{k}_i)} ,\qquad(10.199)$$

with $\beta_i(\hat{k}_i)$, Eq. (7.62).

The following example is used to demonstrate this technique:

Example 10.22   Consider again the network given in Example 10.21, but change the number of servers, $m_2 = 2$. All other network parameters remain the same. From Eq. (10.193), the number of states in the network without blocking is $Z = K + 1 = 11$. The corresponding state diagram is shown in Fig. 10.70. The state diagram of the blocking network (see Fig. 10.71)

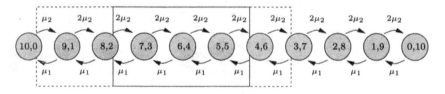

Fig. 10.70 CTMC of the Example 10.22 network without blocking.

contains all possible nonblocking states from the original network plus all the blocking states (marked with ∗). The number of states $\hat{Z}$ of the blocking network is given by Eq. (10.197) as the sum of all nonblocking states ($= 3$) and the blocking states ($= 3$): $\hat{Z} = 6$.

The state diagram of the equivalent network (see Fig. 10.72) contains, therefore, also $\hat{Z} = 6$ states, and for the number of jobs in the network,

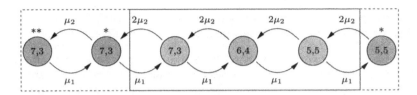

Fig. 10.71   CTMC of the Example 10.22 blocking network.

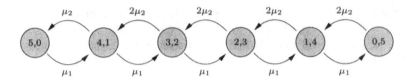

Fig. 10.72   CTMC of the Example 10.22 equivalent product-form network.

Eq. (10.195) is used: $\hat{K} = 5$. The throughput $\lambda_{NB}(5)$ of the equivalent network can be determined using the MVA and, because of the equivalence of both networks, is identical with the throughput $\lambda_B(10)$ of the blocking network:

$$\lambda_B(10) = \lambda_{NB}(5) = \underline{0.499}.$$

The state probabilities of the blocking network are given by equating the corresponding steady-state probabilities of the equivalent nonblocking network. The steady-state probabilities are determined using Eq. (10.196):

$$\pi(7,3)^{**} = \pi(5,0) = \underline{0.633}, \qquad \pi(6,4) = \pi(2,3) = \underline{0.014},$$
$$\pi(7,3)^{*} = \pi(4,1) = \underline{0.284}, \qquad \pi(5,5) = \pi(1,4) = \underline{0.003},$$
$$\pi(7,3) = \pi(3,2) = \underline{0.064}, \qquad \pi(5,5)^{*} = \pi(0,5) = \underline{0.001}.$$

The mean number of jobs at each node can be determined using the steady-state probabilities:

$$\overline{K}_1 = 7\left[\pi(7,3)^{**} + \pi(7,3)^{*} + \pi(7,3)\right] + 6\pi(6,4) + 5\left[\pi(5,5) + \pi(5,5)^{*}\right]$$
$$= \underline{6.978},$$
$$\overline{K}_2 = 3\left[\pi(7,3)^{**} + \pi(7,3)^{*} + \pi(7,3)\right] + 4\pi(6,4) + 5\left[\pi(5,5) + \pi(5,5)^{*}\right]$$
$$= \underline{3.022}.$$

All other performance measures can be determined using the well-known equations. Of special interest are blocking probabilities:

$$P_{B_1} = \pi(5,5)^{*} = \underline{0.001},$$
$$P_{B_2} = \pi(7,3)^{**} + \frac{1}{2}\,\pi(7,3)^{*} = \underline{0.775}.$$

For computing P_{B_2}, the steady-state probability $\pi(7,3)^*$ can only be counted half because in state $(7,3)^*$ only one of the two servers of node 2 is blocked.

For blocking networks with more than two nodes, no equivalent nonblocking network can be found. For these cases only approximate results exist, which are often inaccurate or can only be used for very special types of networks such as tandem networks [SuDi84, SuDi86] or open networks [PeSn89]. A detailed consideration of networks with blocking can be found in [Perr94, BNO01].

Problem 10.10   Consider a closed blocking network with $N = 2$ nodes and $K = 9$ jobs. The service times are exponentially distributed and the service strategy at all nodes is FCFS. All other network parameters are given in the Table 10.40.

Table 10.40 Network input parameters for Problem 10.10

i	e_i	$1/\mu_i$	m_i	M_i
1	1	1	1	6
2	1	1	1	4

(a) Draw the state diagram for the network without blocking as well as the state diagram of the blocking network and the equivalent non-blocking network.

(b) Determine the throughput λ_B, the mean number of jobs $\overline{K}_i$, and the blocking probabilities $P_{B_i} (i = 1, 2)$ for the blocking network.

Problem 10.11   Directly solve the CTMC of the blocking network in Problem 10.10 by using **SHARPE**. Note that you can easily obtain transient probabilities and cumulative transients with **SHARPE** in addition to the steady-state probabilities.

Problem 10.12   For Problem 10.11, construct an SRN and solve it using **SPNP**. Compare the results obtained from SPNP with those obtained by hand computation. Note that SRNs enable computation of transients and cumulative transients. Furthermore, easy definition of rewards at the net levels enables straightforward computation of all desired performance measures.

10.9 NETWORKS WITH BATCH SERVICE

10.9.1 Open Networks with Batch Service

In this section the decomposition method for open networks (see Section 10.1.3) is extended to open networks with batch service and FB policy. To extend the decomposition method to open queueing networks with batch service we first need to choose an approximation for the mean number of jobs for service stations with batch service and batch input. This approximation can be derived from Eq. (6.184) in Section 6.6.2:

$$\overline{K}_{\text{batch}} \approx b \cdot \frac{\rho P_m}{1 - \rho} \cdot \frac{c_{A^b}^2 + c_B^2}{2} + \frac{b - 1}{2} + \rho m b, \tag{10.200}$$

with the coefficient of variation $c_{A^b}^2$ of the interarrival time of batches with batch size b given by Eq. (6.183):

$$c_{A^b}^2 = \frac{\overline{X}}{b}(c_X^2 + c_{A^x}^2), \tag{10.201}$$

with the mean value of the size of the arriving batches $\overline{X}$, the coefficient of variation of the size of the arriving batches c_X^2 and the coefficient of variation of the interarrival time of the arriving batches $c_{A^x}^2$. Recall that the utilization ρ is

$$\rho = \frac{\lambda}{\mu m b}. \tag{10.202}$$

The mean input batch size $\overline{X}_i$ for every node i, $i = 1, \ldots, N$, is given by

$$\overline{X}_i = \frac{\sum_{j=1}^N \lambda e_j p_{ji}}{\sum_{j=1}^N \lambda \frac{e_j}{b_j} p_{ji}} \tag{10.203}$$

where b_j is the service batch size at node j [Fran03]. With

$$\overline{X^2}_i = \frac{\sum_{j=1}^N \lambda e_j b_j p_{ji}}{\sum_{j=1}^N \lambda \frac{e_j}{b_j} p_{ji}}, \tag{10.204}$$

we can calculate the coefficient of variation $c_X^2 = \overline{X^2}/\overline{X}^2 - 1$.

To calculate performance measures for open networks with batch service Zisgen [Zisg99] extended the decomposition method. For details see Section 10.1.3 or Pujolle/Ai [PuAi86], Gelenbe [Gele75], and Whitt [Whit83b].

There an approximation of the coefficient of variation of output process D_i^b of node i ($i = 1, \ldots, N$) is derived using renewal theory:

$$c_{D_i^b}^2 = \rho_i^2 c_{B_i}^2 + (1 - \rho_i)\left(\frac{\overline{X}_i}{b_i}(c_{X_i}^2 + c_{A_i^x}^2)\right) + (1 - \rho_i)\rho_i, \tag{10.205}$$

with the mean value of the size of the arriving batches $\overline{X}_i$ (Eq. (10.203)), the coefficient of variation of the size of the arriving batches $c_{X_i}^2$ (Eq. (10.204)), the coefficient of variation of the interarrival time of the arriving batches $c_{A_i^x}^2$, and the coefficient of variation of the service time of the jobs $c_{B_i}^2$ at node i.

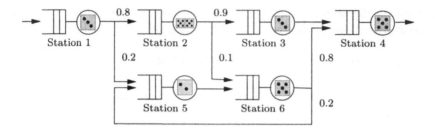

Fig. 10.73 Open network with batch service (dice denote the batch sizes).

With probability p_{ij} a batch will go from station i to station j. This leads to the characteristics of the interarrival times from station i to station j:

$$c^2_{A_{ij}^X} = p_{ij}c^2_{D_i^b} + 1 - p_{ij}.\tag{10.206}$$

The superposition of the different processes leads to a system of equations that in turn allows us to determine the coefficients of variation:

$$c^2_{A_i^X} = \frac{1}{\lambda_i^X}\sum_{j=1}^{N}\lambda_j^X p_{ji}c^2_{A_{ji}^X}.\tag{10.207}$$

Using these equations the coefficients of variation of the interarrival time of the arriving batches $c^2_{A_i^X}$ can be calculated iteratively with the initial values $c^2_{A_{ij}^X} = 1$.

Numerical Results   To assess numerical efficiency of the approximation, we consider an example queueing network shown in Fig. 10.73. The results are shown in Table 10.41.

10.9.2 Closed Networks with Batch Service

The summation method of Section 9.2 can be extended for closed networks with batch service in the same way as in the decomposition method for open networks with batch service. To calculate the mean number of jobs for service stations with batch service and batch input the same formulae as for open networks can be used (Eqs. (10.200) – (10.204)).

To use Eq. (10.200) in closed queueing networks, we again need to include a factor C to limit the mean number of jobs to K for $\rho \to 1$. Using

$$C = 1 - \frac{1}{2}\frac{b\cdot(c^2_{A^b}+c^2_B)}{K - \frac{b-1}{2} - bm}\tag{10.208}$$

Table 10.41 Analytical vs. DES results for $\overline{Q}_{\text{batch}}$ for model 10.73

Node	b	m	ρ	c_B	DES	Decomp.
1	3	3	0.80	1.0	12.740 ± 0.161	13.378
2	10	1	0.72	1.0	4.421 ± 0.034	4.127
3	3	1	0.65	1.0	16.408 ± 0.211	17.793
4	5	2	0.80	1.0	22.716 ± 0.278	27.267
5	2	4	0.60	1.0	6.513 ± 0.086	6.362
6	5	3	0.89	1.0	24.350 ± 1.098	24.474
1	3	3	0.80	0.5	10.033 ± 0.038	10.465
2	10	1	0.72	0.5	2.925 ± 0.014	2.366
3	3	1	0.65	0.5	12.097 ± 0.041	12.938
4	5	2	0.80	0.5	14.540 ± 0.085	18.323
5	2	4	0.60	0.5	6.169 ± 0.044	5.973
6	5	3	0.89	0.5	18.207 ± 0.677	16.304
1	3	3	0.80	1.2	14.027 ± 0.136	15.086
2	10	1	0.72	0.3	4.054 ± 0.044	3.910
3	3	1	0.65	1.2	18.525 ± 0.396	20.266
4	5	2	0.80	0.3	19.146 ± 0.135	18.697
5	2	4	0.60	1.2	6.701 ± 0.057	6.555
6	5	3	0.89	0.3	17.219 ± 0.570	16.458

leads to

$$\overline{K}_{\text{batch}} \approx b \cdot \frac{\rho P_m}{1 - C \cdot \rho} \cdot \frac{c_{A^b}^2 + c_B^2}{2} + \frac{b-1}{2} + \rho m b. \tag{10.209}$$

In case the capacity of the node is higher than the number of jobs to be served, there can only be jobs in the queue, that wait for the next batch to be completed. But there can be no waiting for other reasons. The mean number of jobs actually waiting to be part of a batch and start service is $(b-1)/2$. This finally leads to

$$\overline{K}_{\text{batch}} \approx \begin{cases} \frac{\rho}{1-C\cdot\rho} \cdot P_m \cdot \frac{b\cdot(c_{A^b}^2+c_B^2)}{2} + \frac{b-1}{2} + \\ \quad + b\cdot m \cdot \rho, \qquad \frac{b-1}{2} + bm < K, \\ \frac{b-1}{2} + b\cdot m \cdot \rho, \qquad \frac{b-1}{2} + bm \geq K. \end{cases} \tag{10.210}$$

To calculate performance parameters for closed networks with batch service Frank [Fran03] used a decomposition approach. Decomposition suits the summation method very well, since the input and output streams of every station are considered separately and approximated to be renewal processes. Therefore, the same formulas (Eqs. (10.205) – (10.207)) as for the decomposition method can be used to determine iteratively the coefficients of variation of the interarrival time of the arriving batches $c_{A_i}^2 x$.

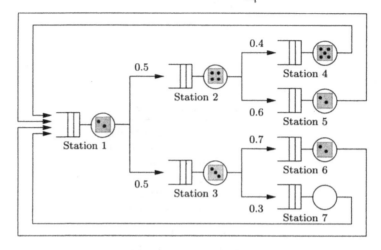

Fig. 10.74   Model with splitting (dice denote the batch sizes).

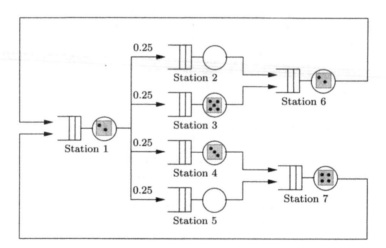

Fig. 10.75   Model with superposition (dice denote the batch sizes).

Numerical Results   To assess the numerical accuracy of the approximation we consider two different queueing networks shown in Fig. 10.74 and Fig. 10.75. The results are shown in Table 10.42. A crucial part of the approximation is the decomposition. We have therefore examined two models, where many superpositions and many branches occur. As we can clearly see, the results for networks with higher workload are better. This is due to the fact that the accuracy of Eq. (10.200) increases with the traffic volume.

Table 10.42 Analytical vs. DES results for models 10.74 and 10.75

					DES		Approx	
Node	b	m	$\overline{x}$	c_B^2	ρ	$\overline{Q}_{\text{batch}}$	ρ	$\overline{Q}_{\text{batch}}$
					$K = 15$			
1	2	4	2	0.8	0.1757	0.5098	0.1673	0.5021
2	4	1	6	0.3	0.5259	1.3538	0.5020	1.7709
3	3	1	4	0.5	0.4694	1.4334	0.4462	1.6135
4	5	3	7	0.6	0.0650	2.1210	0.0625	2.0000
5	2	1	1	0.2	0.1051	0.1067	0.1004	0.0359
6	2	2	8	0.0	0.4932	0.8343	0.4685	0.9355
7	1	2	4	1.2	0.2116	0.1423	0.2008	0.0434
					$K = 40$			
1	2	4	2	0.8	0.3132	0.5433	0.3013	0.5291
2	4	1	6	0.3	0.9391	10.1266	0.9038	10.6527
3	3	1	4	0.5	0.8358	6.0043	0.8034	5.8737
4	5	3	7	0.6	0.1165	2.0173	0.1125	2.0029
5	2	1	1	0.2	0.1878	0.1999	0.1808	0.1067
6	2	2	8	0.0	0.8775	5.4907	0.8436	5.9818
7	1	2	4	1.2	0.3726	0.4696	0.3615	0.2712

					DES		Approx	
Node	b	m	$\overline{x}$	c_B^2	ρ	$\overline{Q}_{\text{batch}}$	ρ	$\overline{Q}_{\text{batch}}$
					$K = 15$			
1	2	5	2	1.0	0.1703	1.0001	0.1907	0.0019
2	1	2	4	0.2	0.4268	0.2505	0.4767	0.2599
3	5	1	5	1.2	0.2155	2.1025	0.2384	2.2519
4	3	2	1	0.5	0.0355	1.0863	0.0397	1.0002
5	1	1	2	0.3	0.4241	0.4881	0.4767	0.4235
6	2	6	7	0.8	0.2494	0.4600	0.2781	0.5047
7	4	4	2	1.0	0.0536	1.4917	0.0596	1.5000
					$K = 40$			
1	2	5	2	1.0	0.3719	0.0497	0.3690	0.0644
2	1	2	4	0.2	0.9347	7.7527	0.9225	7.5398
3	5	1	5	1.2	0.4627	2.8910	0.4613	3.3732
4	3	2	1	0.5	0.0776	1.0404	0.0769	1.0014
5	1	1	2	0.3	0.9297	8.4001	0.9225	8.2324
6	2	6	7	0.8	0.5431	0.7179	0.5381	0.7604
7	4	4	2	1.0	0.1163	1.4788	0.1153	1.5004

11

Discrete-Event Simulation

11.1 INTRODUCTION TO SIMULATION

In many fields of engineering and science, we can use a computer to simulate natural or man-made phenomena rather than to experiment with the real system. Examples of such computer experiments are simulation studies of congestion control in a network and contention for resources in a computer operating system. A simulation is an experiment to determine characteristics of a system empirically. It is a modeling method that mimics or emulates the behavior of a system over time. It involves generation and observation of artificial history of the system under study, which leads to drawing inferences concerning the dynamic behavior of the real system. In earlier chapters, we have discussed the essential role of analytic models in solving performance problems. In this chapter, we focus on simulation models for performance evaluation.

Computer simulation is a discipline of designing a model of an actual or theoretical system, executing the model (an experiment) on a digital computer, and statistically analyzing the execution output (see Fig. 11.1). The current state of the physical system is represented by state variables (program variables). A simulation program modifies state variables to reproduce the evolution of the physical system over time. Therefore, at any point in time the simulation algorithm has only to consider the current state and the set of possible next events to happen but never needs an overall view over the state space of the system under investigation. As a consequence, one of the advantages of simulation is its low memory consumption. However, the

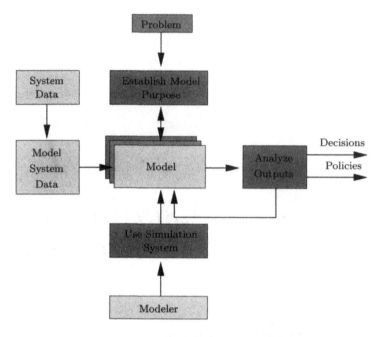

Fig. 11.1   Simulation-based problem solving.

most appealing advantage of simulation is its relative simplicity and therefore wide applicability. In Section 1.2 we have discussed the process of going from system description to model in general and in particular to simulation model. This chapter provides an introductory treatment of various concepts related to simulation. In Section 11.2 we provide some guidelines for choice between simulative and analytical solutions. In Section 11.3, we provide a broad classification of simulation models followed by a classification of simulation modeling tools/languages in Section 11.4. In Section 11.5 the role of probability and statistics in simulation is discussed while in Section 11.6 some networking applications using simulation tools: CSIM-19, OPNET Modeler, and ns-2 is developed.

11.2   SIMULATIVE OR ANALYTIC SOLUTION?

While modeling has proved to be a viable and reliable alternative to experiments on the real system, the choice between analytic and simulative solutions is still a matter of importance. For large and complex systems, analytic model formulation and/or solution may require making unrealistic assumptions and approximations. For such systems, simulation models can be created and solved to study the whole system more accurately. Clear advantage of simulation over all analytic approaches is that it can handle systems with extremely large or even infinite state spaces without any necessity of approximations.

Thus, one of the advantages of simulation is its relatively low memory consumption since large matrices containing information regarding the model state space need not be generated and stored in memory. However, the most appealing advantage of simulation is its relative simple approach for defining and setting up the underlying system which leads to its wide applicability. For example one does not need to solve large systems equations but instead the simulative solution follows an intuitive approach of exploring paths through the state space of the modeled system.

In principle, simulative solution can be applied to all classes of systems but this flexibility comes at the cost of two disadvantages: The first disadvantage is the fact that simulation sometimes requires very long execution times. For example, if we are simulating a model with rare events, we need a large number of simulation runs to get such a rare event at least once and many more replications may be needed to get statistically significant results. This disadvantage can be somewhat alleviated by the use of special rare event simulation techniques (see Section 11.5.4.1). The second disadvantage relates to safety-critical systems. Since simulation only explores paths through the state space, it is quite likely that some rare events or improbable states are never encountered by the simulation runs. In other words, simulation can only show the existence of critical states but can never prove the absence of such states.

Many users often employ simulation where a faster analytic solution would have served the purpose. A user should be aware of the following issues when applying simulation:

1. Model building requires special training. Frequently, simulation languages like Simula [Sim05c], Simscript [Sim05a] and Automod [AUT05] are used. Users need some programming expertise before using these languages. There are several formal paradigms available for developing a simulation model (e.g., event graphs [LaKe00] and DEVS [ZKP00]), but quite often simulation models are developed in an ad-hoc fashion.

2. Simulation results are difficult to interpret, since most simulation outputs are samples of random variables. However, some of the recent simulation packages have built-in capabilities to statistically analyze the outputs of simulation experiments.

3. The proper use of these tools requires a deep understanding of statistical methods and necessary assumptions to assert the credibility of obtained results. Due to a lack of understanding of statistical techniques, frequently simulation results are wrongly interpreted [PJL02].

4. Simulation modeling and analysis are time consuming and expensive. With availability of faster machines, development in parallel and dis-

tributed simulation [ChMi81, Fuji99] and in variance reduction techniques such as importance sampling [Fish97, GlIg89, TuTr01], importance splitting/RESTART [HaTo99, Heid95, TuTr00] and regenerative simulation [LaSl75], this difficulty is being alleviated.

In spite of some of the difficulties, simulation is widely used in practice and the use of simulation will surely increase manifold as experimenting with real systems gets increasingly difficult due to cost and other reasons. Simulation is one of the most popular methods to analyze state-based systems. The discrete-event simulation (DES) for the study of queueing systems and stochastic Petri nets is widely applied. Hence it is important for every computer scientist/engineer (in fact, any engineer) to be familiar with the basics of simulation.

11.3 CLASSIFICATION OF SIMULATION MODELS

Simulation models can be classified according to several criteria [BCNN01]:

1. *Continuous vs. discrete:* Depending upon the way in which state variables of the modeled system change over time. For example concentration of a substance in a chemical reactor changes in a smooth, continuous fashion like in a fluid flow whereas changes in the length of a queue in a packet switching network can be tracked at discrete points in time. Continuous system simulation may be used either because the system contains such a phenomena or as fluid approximation to a discrete phenomena [KuMi01].

2. *Deterministic vs. stochastic:* This classification refers to the type of timing associated with the events in the model being simulated. A simulation with all deterministic event durations will have no random variability. By contrast, stochastic simulation is experimental in nature and hence necessitates statistical analysis of results.

3. *Terminating vs. steady-state:* A terminating simulation is used to study the behavior of a system over a well-defined period of time, for example for the reliability analysis of a flight control system over a designated mission time. This corresponds to transient analysis put in the context of Chapter 5. Whereas steady-state simulation corresponds to the steady-state analysis in the context of Chapter 3 and Chapters 6-10. As such, we have to wait for the simulation system output variables to reach steady-state values. For example, the performance evaluation of a computer or networking system is normally (but not always) done using steady-state simulation. Likewise, the availability analysis is typically carried out for steady-state behavior.

4. *Synthetic or distribution-driven vs. trace-driven:* A time-stamped sequence of input events is required to drive a simulation model. Such an event trace may already be available to drive the simulation hence making it a trace-driven simulation. Examples are CPU cache simulations for which many traces are available [GHPS93, Los, SmSa95]. Similarly, traces of packet arrival events (packet size, etc.) are first captured by using a performance measurement tool such as tcpdump. Then these traces are used as input traffic to the simulation. A lot of traces are freely available on the Web (e.g., http://ita.ee.lbl.gov). Alternatively, event traces can be synthetically generated. For the synthetic generation, if the distributions of all inter-event times are assumed to be known or given, then random deviates of the corresponding distributions are used as the time to next event of that type. We show how to generate random deviates of important distributions later in this chapter. The distributions needed to drive such distribution-driven simulations may have been obtained by statistical inference based on real measurement data.

5. *Sequential vs. distributed simulation:* Sequential simulation processes events in a nondecreasing time order. In distributed simulation a primary model is distributed over heterogeneous computers, which independently perform simulations locally. Distributed simulation has the potential to speed up simulation runs of large scale systems. The challenge is to produce an overall order of events, which is identical with the order that would be generated if the primary model was sequentially simulated on a single computer. There is extensive research in parallel and distributed simulation [ChMi81, Fuji99].

6. *Symbolic simulation:* Symbolic simulation involves evaluating system behavior using special symbolic values to encode a range of operating conditions. In one simulation run, a symbolic simulator can compute what would require many runs of a traditional simulator. Symbolic simulation has applications in both logic and timing verification, as well as sequential test generation. This type of simulation is used to formally verify high-level descriptions of electric circuits (see [Brya90, LMR01]).

7. *Event-oriented vs. process-oriented:* In an event-oriented simulation a system is modeled by identifying its characteristic events and then writing a set of event routines that give a detailed description of the state changes taking place at the time of each event. The simulation evolves over time by executing the events in increasing order of time. By contrast, in process oriented simulation the behavior of the system is represented by a set of interacting processes. A process is a time-ordered sequence of interrelated events which describe the entire experience of an entity as it flows through the system. However, the internal structure of a process-oriented simulation is based on scheduling events as

well [BCNN01, LaKe00]. The difference can be viewed as storing and searching through a single event list vs. a set of event lists. Most visual simulation systems provide a process-oriented view to the user since the model creation in such systems is easier.

The rest of this chapter refers to sequential, stochastic, distribution-driven DES. Both terminating and steady-state simulation is discussed.

11.4   CLASSIFICATION OF TOOLS IN DES

Simulation tools can be broadly divided into four basic categories (Fig. 11.2):

1. *General-Purpose Programming Language (GPPL):* C, C^{++}, Java are some of the languages which have the advantage of being readily available. They also provide a total control over the software development process. But the disadvantage is that the model construction takes considerable time. Also it does not provide any basic support for DES. Furthermore, generation of random deviates for various needed distributions and the statistical analysis of output needs to be learned and programmed.

2. *Plain Simulation Language (PSL):* SIMULA [Sim05c], SIMSCRIPT II.5 [Sim05a], SIMAN [PSSP95], GPSS [GPS], JSIM [JSI], SILK [HeKi97] are some of the examples. Almost all of them have basic support for DES. One drawback is that they are not readily available. One also needs to get programming expertise in a new language before executing simulation models.

3. *Software libraries for simulation:* CSIM-19 [CSI05], C^{++}SIM [C++] and its Java counterpart JavaSim [Jav], baseSim [bas], TomasWeb [Tom] are some examples. They provide extensive libraries written in GPPL (C, C^{++}, Java) which aid in simulation model development and execution. They provide various monitoring and statistical gathering functions with very little or no need to learn any new language.

4. *Simulation Packages (SPs):* OPNET Modeler [OPN05], ns-2 [Ns05], OMNET^{++} [OMN], Arena [Are], Qualnet [Qua], and SPNP [HTT00] are some examples. They facilitate the model building process with most of them providing graphical user interfaces along with flexibility of working at the textual level. They provide basic support for DES and statistical analysis. They also cover several application domains like TCP/IP networks. This ensures that model construction time is shorter. Some simulation tools like SPNP, MOSEL-2, and OPNET Modeler also provide the user with an option of analytic solutions of the model. Like PSL, SPs require expertise in new language/environment, and they tend to be less flexible than the PSLs.

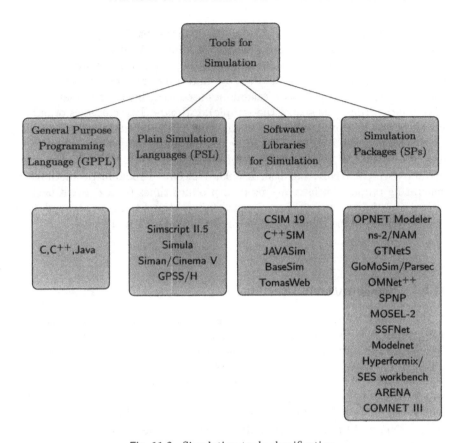

Fig. 11.2 Simulation tools classification.

Information about a variety of available simulation tools can be found at [Sim05b].

11.5 THE ROLE OF PROBABILITY AND STATISTICS IN SIMULATION

Models discussed in this book are stochastic in nature and hence their simulations will naturally be distribution-driven. There are two different uses of statistical methods and one use of probabilistic methods in such distribution-driven simulations. First, the distributions and parameters of input random variables such as interarrival times, times to failure, service times and times to repair need to be estimated from real measurement data. Statistical inference techniques for parameter estimation and fitting distributions were covered in Chapter 1 of this book. Second, models treated in this book have random outputs containing sequences of random variables with unknown distributions.

Simulation runs are performed as computer experiments in order to determine the characteristics of its output random variables. A single simulation run produces a sample of values of an output variable over time. Statistical techniques are employed to examine the data and to get meaningful output from the experiment. Also statistical techniques used to define the necessary length of simulation runs (the size of the sample), estimate characteristics of output variables such as a probability or a mean value and some assessments regarding an accuracy of the results. Two principal methods, independent replication and the method of batch means, are discussed in Section 11.5.3. For other methods see [BCNN01, LaKe00, Klei95, Welc83]. Probabilistic methods of generating random variates are used to produce times to next event such as interarrival times, service times, times to failure, times to repair etc. and drive the simulation. This topic is covered in Sections 11.5.1 and 11.5.2.

11.5.1 Random Variate Generation

In this section we describe methods of generating random variates (or deviates) of any arbitrary distribution, assuming a routine to generate uniformly distributed random numbers is available. Such a routine can be obtained from standard-packages that are freely available, as for example [MaNi98]. Note that the choice of a good random number generator is important for the accuracy of the simulation [BCNN01, LaKe00]. The distribution can be either continuous or discrete. Most of the simulation packages like OPNET Modeler, ns-2, SPNP, and CSIM-19 have built-in routines for generating some random variates. But still knowledge of random variate generation is necessary to more accurately model the real-world problem especially when available built-in generators in simulation packages do not support the needed distribution. We examine here some popular methods for generating random variates. For further information see [BCNN01, Devr86, LaKe00, Triv01]:

Inverse transform: With this method the following property is used: If X is a random variable with the CDF F, then the new random variable $Y = F(X)$ is uniformly distributed over the interval [0, 1]. Thus, to generate a random variate x of X first a random number u from a uniform distribution over [0, 1] is generated and then the function F is inverted. Thus, $F^{-1}(u)$ gives the required value of x. The principal idea of this method is shown in Figure 11.3.

This method can be used to sample from the exponential, the uniform, the Weibull, the Pareto, as well as several empirical and discrete distributions. It is most useful when the inverse of the CDF, $F(.)$, can be easily computed. Taking the example of the *exponential* distribution:

$$F(x) = 1 - e^{-\lambda x},$$

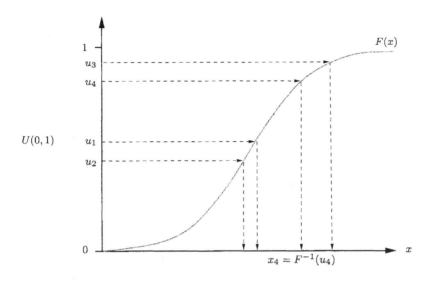

Fig. 11.3   Idea of the inverse transform method. Here $U(0,1)$ refers to uniform random variable.

given u drawn from $U(0,1)$, generate x drawn from exponential distribution using the formula

$$x = -\frac{\ln(1-u)}{\lambda}. \tag{11.1}$$

Noting further that $(1-u)$ is also a random number, the formula above can be simplified to

$$x = -\frac{\ln(u)}{\lambda}. \tag{11.2}$$

In order to convince ourselves that Eq. (11.2) provides a random variate of the exponential distribution, we conduct an experiment. We repeatedly generate random numbers and correspondingly generate variates as per above formula. Three cases are taken with 10, 100, and 1000 random variates and histograms from these observations are plotted in Fig. 11.4. We over-plot the pdf of the exponential distribution in each case. From the figure it becomes clear that increasing the number of random variates in the sample decreases mean-square error between theoretical pdf and empirical plots.

Next we calculate random variate for the *Weibull* distribution which is frequently used to model mechanic component failure [Aber00]. It is given by

$$F(x) = 1 - e^{-\lambda x^{\alpha}}$$

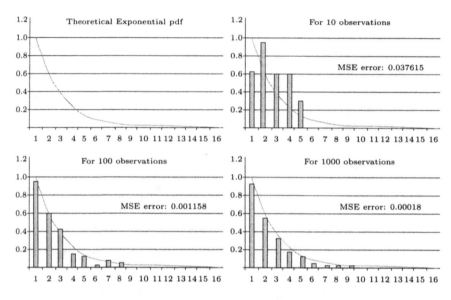

Fig. 11.4   pdf of exponential distribution and histograms of random variates.

Its random variate is generated using

$$x = \left(\frac{- \ln(1 - u)}{\lambda} \right)^{\frac{1}{\alpha}}.$$

The *uniform* distribution $U(a, b)$ is given by

$$F(x) = \begin{cases} 0, & x < 0, \\ \frac{x-a}{b-a}, & a \le x \le b, \\ 1, & x > b. \end{cases}$$

Given u from $U(0, 1)$, the random variate u' of $U(a, b)$ is calculated using

$$u' = (b - a)u + a. \tag{11.3}$$

The *Pareto* distribution is frequently used to specify Web file size on the Internet servers and thinking time of the Web browser [CrBe96, Deng96]. It is given by

$$F(x) = \begin{cases} 1 - \left(\frac{k}{x} \right)^{\alpha}, & x \ge k, \ \alpha, \ k > 0, \\ 0, & x < k. \end{cases}$$

The random variate is generated using

$$x = \frac{k}{(1 - u)^{\frac{1}{\alpha}}}.$$

The *triangular* distribution is given by

$$F(x) = \frac{2}{a}\left(x - \frac{x^2}{2a}\right), \qquad a > 0.$$

The random variate is given by

$$a(1 - \sqrt{1 - u}).$$

The *Rayleigh* distribution is given by

$$F(x) = \begin{cases} 1 - e^{\frac{-x^2}{2\sigma^2}}, & x \geq 0, \\ 0, & \text{otherwise.} \end{cases}$$

The random variate can be generated using

$$x = \sqrt{-2\sigma^2 \ln(1 - u)}. \tag{11.4}$$

Similarly the *log-logistic* distribution, useful in software reliability models [GoTr99], is given by

$$F(x) = 1 - \frac{1}{1 + (\lambda x)^\kappa}, \qquad \lambda > 0 \text{ and } \kappa > 0.$$

The random variate is generated using

$$x = \frac{1}{\lambda}\left(\frac{u}{1 - u}\right)^{\frac{1}{\kappa}}.$$

The random variate of a (discrete) *Bernoulli* distribution with parameter $(1 - q)$ can also be generated by the inverse transform technique. The CDF (see Fig. 11.5) is given for $0 \leq q \leq 1$ by

$$F(x) = \begin{cases} 0, & x < 0, \\ q, & 0 \leq x < 1, \\ 1, & x \geq 1. \end{cases}$$

The inverse function for a Bernoulli distribution becomes

$$F^{-1}(u) = \begin{cases} 0, & 0 < u \leq q, \\ 1, & q < u \leq 1. \end{cases} \tag{11.5}$$

We can thus write the random variate of the Bernoulli distribution as an indicator function of the event $\{u > q\}$:

$$x = 1_{\{u > q\}}$$

Table 11.1   Random variate table

Name	Density $f(x)$	$F(x)$	$x = F^{-1}(u)$	Simplified Formula
Exponential	$\lambda e^{-\lambda x}$	$1 - e^{-\lambda x}$	$-\dfrac{\ln(1-u)}{\lambda}$	$-\dfrac{\ln(u)}{\lambda}$
Weibull	$\lambda e^{-\lambda x^{\alpha}}$	$1 - e^{-\lambda x^{\alpha}}$	$\left(\dfrac{-\ln(1-u)}{\lambda}\right)^{\frac{1}{\alpha}}$	$\left(\dfrac{-\ln(u)}{\lambda}\right)^{\frac{1}{\alpha}}$
Uniform	$\left(\dfrac{1}{b-a}\right)$	$\left(\dfrac{x-a}{b-a}\right)$	$(b-a)u + a$	$(b-a)u + a$
Pareto	$\alpha k^{\alpha} x^{-\alpha-1}$	$1 - \left(\dfrac{k}{x}\right)^{\alpha}$	$\dfrac{k}{(1-u)^{\frac{1}{\alpha}}}$	$\dfrac{k}{(u)^{\frac{1}{\alpha}}}$
Rayleigh	$\dfrac{x}{\sigma^2} \exp\left[-\dfrac{1}{2}\left(\dfrac{x}{\sigma}\right)^2\right]$	$1 - e^{\frac{-x^2}{2\sigma^2}}$	$\sqrt{-2\sigma^2 \ln(1-u)}$	$\sqrt{-2\sigma^2 \ln(u)}$
Log-logistic	$\dfrac{\lambda\kappa(\lambda x)^{\kappa-1}}{[1+(\lambda x)^{\kappa}]^2}$	$1 - \dfrac{1}{1+(\lambda x)^{\kappa}}$	$\dfrac{1}{\lambda}\left(\dfrac{u}{1-u}\right)^{\frac{1}{\kappa}}$	$\dfrac{1}{\lambda}\left(\dfrac{u}{1-u}\right)^{\frac{1}{\kappa}}$
Triangular	$\dfrac{2}{a}\left(1-\dfrac{x}{a}\right)$	$\dfrac{2}{a}\left(x-\dfrac{x^2}{2a}\right)$	$a(1-\sqrt{1-u})$	$a(1-\sqrt{u})$

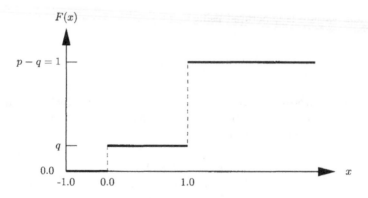

Fig. 11.5   Bernoulli distribution. Note that due to right continuity $F(x) = q$ at $x = 0$ and $F(x) = 1$ at $x = 1$

For a *general* discrete distribution, given by

$$F(x) = (X \le x) = \sum_{x_i \le x} p(x_i),$$

a general algorithm for calculating a discrete random variate consists of the following steps:

STEP 1 Draw u from $U(0,1)$.

STEP 2 Assume that $x_1, x_2, \ldots$ are arranged in increasing order, i.e., $x_1 < x_2 < \ldots < x_k$. Choose the smallest integer i, such that $u \leq F(x_i)$, then the random variate is given by $x = x_i$.

Let $p(x_i) = \alpha_i$ for $i = 1, 2, \ldots, k$ and further let $\alpha_0 = 0$ and $\alpha_{k+1} = 1$, then in terms of the indicator function $\mathbf{1}_{\ldots}$ it is given by

$$x = \sum_{i=1}^{k} \mathbf{1}_{\left\{ \sum_{j=0}^{i-1} \alpha_j < u \leq \sum_{j=0}^{i} \alpha_j \right\}} x_i \,.$$

The *geometric* pmf is given by

$$p(1-p)^{j-1}, \qquad j \geq 1 \,,$$

and the corresponding distribution is given by

$$F(j) = 1 - (1-p)^j \,.$$

So a geometric random variate j satisfies

$$1 - (1-p)^{j-1} < u \leq 1 - (1-p)^j \,,$$
$$(1-p)^{j-1} < 1 - u \leq (1-p)^j \,.$$

Hence a geometric random variable x becomes

$$x = \min\{i \mid (1-p)^i < 1 - u\} = \min \left\{ i \mid i > \frac{\ln(1-u)}{\ln(1-p)} \right\} = \left\lceil \frac{\ln(1-u)}{\ln(1-p)} \right\rceil \,.$$

Since $(1-u)$ is also a uniform random number, the above formula simplifies to

$$x = \left\lceil \frac{\ln(u)}{\ln(1-p)} \right\rceil \,.$$

Convolution method: This is very helpful in such cases when the random variable Y can be expressed as a sum of other random variables that are independent and easier to generate than Y. Let

$$Y = X_1 + X_2 + \ldots + X_k \,.$$

Taking an example of *hypoexponential* case, a random variable Y with parameters $(\lambda_1, \ldots, \lambda_k)$ is the sum of k independent exponential RV's X_i with mean $1/\lambda_i$. From the inverse transform technique, each X_i is generated using Eq. (11.1) and their sum is the required result. Note that *Erlang* is a special case of the hypoexponential distribution when all the k sequential phases have identical distribution. A random variate for hypoexponential distribution is given by

$$y = \sum_{i=1}^{k} -\frac{1}{\lambda_i} \ln(1 - u_i).$$

The *binomial* random variable is known to be the sum of n independent and identically distributed Bernoulli random variables. Hence generating n Bernoulli random variates and adding them, this sum will result in a random variate of the binomial. Let $(x_1, x_2, \ldots, x_n,)$ be the Bernoulli random variates given by Eq. (11.5) and let y be a binomial random variate. Then,

$$y = \sum_{i=1}^{n} x_i = \sum_{i=1}^{n} \mathbf{1}_{\{u_i > q\}},$$

where $u_1, u_2, \ldots, u_n$ are n uniform random numbers derived from $U(0,1)$.

To obtain a geometric pmf, we observe the number of trials until the first success. Now let us observe the number of trials until the rth success, and let T_r be the random variable denoting this number. This distribution is called *negative binomial* distribution. The density is given by

$$p_{T_r}(n) = \binom{n-1}{r-1} p^r (1-p)^{n-r}, \qquad n = r, r+1, r+2, \ldots.$$

Negative binomial, also called *Pascal* random variable, can also be expressed as a sum of r independent and identically distributed geometric random variables assuming r success is desired. Hence for generating a negative binomial random variate r geometric random variates are generated and added together.

The convolution method can also be used for generating *triangular* random variates from two uniform variates and *Chi-Squared(n)* random variates from n iid normal(0,1) variates.

Mixing or Composition Method: Suppose that the distribution $F_Y(y)$ or the density function $f_Y(y)$ can be represented by either of the following forms:

1. $F_Y(y) = \alpha_1 F_{X_1}(y) + \alpha_2 F_{X_2}(y) + \ldots + \alpha_k F_{X_k}(y)$,

2. $f_Y(y) = \alpha_1 f_{X_1}(y) + \alpha_2 f_{X_2}(y) + \ldots + \alpha_k f_{X_k}(y)$,

where $\alpha_1, \ldots, \alpha_k$ are non-negative and sum to one. Given that the $X_i's$ are relatively easy to generate, these kind of variates can be generated using the composition method. One example is that of the hyperexponential distribution.

The *hyperexponential* distribution is given by

$$F(x) = \sum_{i} \alpha_i (1 - e^{-\lambda_i x}) \qquad x \geq 0, \lambda_i, \alpha_i > 0 \text{ and } \sum \alpha_i = 1.$$

The random variate for the hyperexponential can be generated in two steps. Consider, for example, a three-stage hyperexponential distribution with parameters α_1, α_2, α_3 and λ_1, λ_2, λ_3. First a uniform random number u is generated and like Eq. (11.5) the following inverse function is used in the first step:

$$\Phi(u) = \begin{cases} 1, & 0 < u \leq \alpha_1, \\ 2, & \alpha_1 < u \leq \alpha_1 + \alpha_2, \\ 3, & \alpha_1 + \alpha_2 < u \leq 1. \end{cases}$$

The desired variate is then given by

$$x = \mathbf{1}_{\{u \leq \alpha_1\}} \left(-\frac{\ln(1 - u_1)}{\lambda_1} \right) + \mathbf{1}_{\{\alpha_1 < u \leq \alpha_1 + \alpha_2\}} \left(-\frac{\ln(1 - u_1)}{\lambda_2} \right)$$
$$+ \mathbf{1}_{\{\alpha_1 + \alpha_2 < u \leq 1\}} \left(-\frac{\ln(1 - u_1)}{\lambda_3} \right).$$

Here u_1 denotes different uniform variate than u. Note that this example was for 3 stages but it can be easily extended to k stages. The variate for the general case is given by

$$x = \sum_{j=0}^{k-1} \left(\mathbf{1}_{\left\{ \sum_{i=0}^{j} \alpha_i < u \leq \sum_{i=0}^{j+1} \alpha_i \right\}} \left(-\frac{\ln(1 - u_1)}{\lambda_{j+1}} \right) \right).$$

where $\alpha_0 = 0$ and $\alpha_{k+1} = 1$.

Direct Transformation Method: This method uses convenient properties of some distributions. We apply it to the *normal* distribution. Since the inverse of a normal distribution cannot be expressed in closed form, we cannot apply the inverse transform method. The corresponding CDF is given by

$$F(x) = \int_{-\infty}^{x} \frac{1}{\sqrt{2\pi}} e^{-\frac{t^2}{2}} \, dt.$$

In order to derive a method of generating a random variate of this distribution, we use a property of the normal distribution that relates it to the Rayleigh distribution. Assume that Z_1 and Z_2 are independent standard normal random variables $N(0, 1)$. Then the square root of their sum of squares,

$$B^2 = Z_1^2 + Z_2^2,$$

is known to have the Rayleigh distribution [Triv01] for which we know how to generate its random variate. The original normal random variables can be written in polar coordinates as (see Fig. 11.6):

$$Z_1 = B \cos \theta, \qquad Z_2 = B \sin \theta,$$

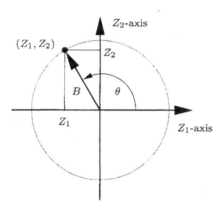

Fig. 11.6 Polar representation.

where θ is uniformly distributed between 0 and 2π. Using the inverse trans-form technique (see Eq. (11.4)), we have

$$b = \sqrt{-2\ln(1 - u_1)}.$$

Next we generate a random value of θ to finally get two random variates of the standard normal:

$$z_1 = \sqrt{-2\ln(1 - u_1)}\cos(2\pi u_2), \qquad z_2 = \sqrt{-2\ln(1 - u_1)}\sin(2\pi u_2)$$

where u_1 and u_2 are two uniform random numbers.

Acceptance–Rejection Method: This is a universally applicable method for generating random variates [Devr86, LaKe00]. Our goal is to generate a random variate x having a probability density $f_X(.)$. Assume that we can generate a random variate y with density $f_Y(.)$. Assume that for a constant $c > 0$ the two densities satisfy (see Fig. 11.7):

$$\frac{f_X(x)}{f_Y(x)} \le c, \qquad \text{for all } x.$$

A random variable Y satisfying the above property is called a majorizing random variable and the density $f_Y(.)$ is called majorizing density for $f_X(.)$.

Then the acceptance–rejection algorithm is given as follows:

STEP 1 Generate y according to density $f_Y(.)$ and generate u from $U(0,1)$.

STEP 2 If $u < \frac{f_X(y)}{cf_Y(y)}$, then $x = y$; else return to previous step.

This technique is used when other methods have no straightforward solu-tion (e.g., no closed form solution). Efficiency of this method depends on

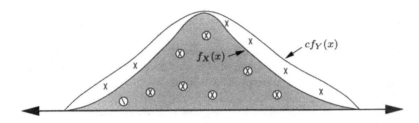

Fig. 11.7 The density function $f_X(x)$ and $cf_Y(x)$

the fraction of generated random numbers that are rejected. The acceptance-rejection technique is used in generation of Poisson, hypergeometric and negative binomial random variates. The *Poisson* pmf is given by

$$\alpha_i = P(X = i) = e^{-\lambda} \lambda^i / i! \qquad i = 0, 1, \ldots .$$

Using the identity, we obtain

$$\alpha_{i+1} = \frac{\lambda}{i+1} \alpha_i, \qquad i \geq 0 .$$

An algorithm for generating a Poisson variate is as follows:

STEP 1 Generate a uniform random number u.

STEP 2 $i := 0$, $\alpha := e^{-\lambda}$, $F := \alpha$.

STEP 3 if $u < F$, then $x := i$ and STOP.

STEP 4 $\alpha := \frac{\lambda \alpha}{i+1}$; $F := F + \alpha$; $i : i + 1$. and go to step 3.

The *hypergeometric* pmf is given by

$$h(i; m, d, n) = \frac{\binom{d}{i} \cdot \binom{n-d}{m-i}}{\binom{n}{m}} , \qquad i = 0, 1, 2, \ldots, \min\{d, m\}.$$

Using the identity, we obtain

$$\alpha_{i+1} = \alpha_i \frac{(d - i + 1)(m - i + 1)}{i(L + i)} , \qquad L = N - d - m .$$

An algorithm for generating a hypergeometric variate is as follows:

STEP 1 Generate a uniform random number u.

STEP 2 $i := 0$, $\alpha := \frac{\binom{n-d}{m}}{\binom{n}{m}}$, $F := \alpha$.

STEP 3 If $u < F$, then $x := i$ and STOP.

STEP 4 $\alpha := \alpha \frac{(d-i+1)(m-i+1)}{i(L+i)}$; $F := F + \alpha;$. $i : i + 1$ and go to step 3.

Earlier the convolution method was used to generate the random variates of the *negative binomial* distribution. Alternatively, random variates from the negative binomial can be generated as follows:

- First draw a sufficiently large sequence of uniform random variates u_i and record the count of which uniform random variates are greater than p and the count of which ones are less than p.

- When the count of uniform random variates greater than p first reaches r then the count of variates less than p is the negative binomial random variate (i.e., the number of trials n to get the rth success).

11.5.2 Generating Events from an Arrival Process

Frequently in simulations, we need to generate a sequence of random points in time $\{T_i\}$ where $i = 1, 2, 3, \ldots$ for different types of stochastic arrival processes. Let $N(t) = \max\{i : T_i \leq t\}$ the number of arrivals and $W_i = T_i - T_{i-1}$ be the interarrival time between the ith and $(i-1)$st arrivals. In this section we deal with the variate generation for various common arrival processes.

Bernoulli Process: In a slotted system, an arrival occurs in each slot with probability $(1 - q)$ and no arrival occurs with probability q. This relates to the discrete Bernoulli distribution for which the variate generation has been described earlier.

Poisson Process: In an unslotted system, an arrival occurs in an interval $(t, t + dt)$ with probability (λdt). In this case interarrival times $\{W_i\}$ are known to be iid exponentially distributed with rate λ and mean $(1/\lambda)$. Thus, t_i can be generated recursively as follows:

STEP 1 Set $t_0 = 0$, $i = 0$.

STEP 2 Increment $i = i + 1$.

STEP 3 Generate u_i from $U(0, 1)$.

STEP 4 $t_i = t_{i-1} + (-\frac{1}{\lambda} \ln u_i)$.

The above algorithm can be applied for the generation of event times of any renewal arrival process in which interarrival times are an iid, with an appropriate modification of step 4. For example consider the case of a renewal arrival process with Weibull interarrival times. In this case, step 4 would be modified as $t_i = t_{i-1} + (-\frac{1}{\lambda} \ln u_i)^{\frac{1}{\alpha}}$.

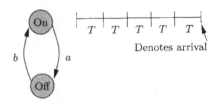

Fig. 11.8 Graphical representation of an On–Off process.

Nonhomogeneous Poisson Process (NHPP): This is a generalization of the Poisson process in which the rate of arrival λ is allowed to be a function of t. Note that the algorithm for the generation of the Poisson process cannot directly be applied to generate variates for NHPP with λ replaced by $\lambda(t_i)$.

Many methods for generating NHPP variates are discussed in [Lave83, LaKe00]. We discuss one of these methods known as the *Thinning Method:* Assume that there is a constant $\bar{\lambda}$ such that $\lambda(t) \leq \bar{\lambda}$ for all t. Let $t_1^*, t_2^*, \ldots$ be the arrival times of a homogeneous Poisson process with intensity $\bar{\lambda}$. In this method, we accept the ith arrival with probability $\lambda(t_i^*)/\bar{\lambda}$, independent of all other arrivals. This sequence $t_1, t_2, \ldots$, of the accepted arrival times forms a sequence of arrival times of an NHPP with arrival rate $\lambda(t)$. The algorithm amounts to either rejecting or accepting a particular arrival as part of thinned process. The resulting algorithm is as follows:

STEP 1 Set $t_0 = 0$ and $t^* = 0$.

STEP 2 Generate an exponential random variable e with rate $\bar{\lambda}$.

STEP 3 Set $t^* = t^* + e$.

STEP 4 Generate a random number $u \sim U(0, 1)$.

STEP 5 If $u > \lambda(t^*)/\bar{\lambda}$, then return to step 2 (reject the arrival time) else set $t_i = t^*$ (accept the arrival time).

The thinning method can be considered as an application of the acceptance rejection method for variate generation. Thus, a t_i^* will be more likely to be accepted if an arrival rate $\lambda(t_i^*)$ is high.

On–Off Process [PaWi00]: Packet stream from a single source is often modeled as a sequence of alternating burst periods and silence periods. The duration of each burst is generally distributed. During such periods packets are generated with a constant interarrival time T. After that a generally distributed silence period follows (see Fig. 11.8). Assuming the Of periods are Weibull distributed and the Off periods are Pareto distributed the algorithm for the generation of sequence of message times of an On–Off process is shown in Fig. 11.9.

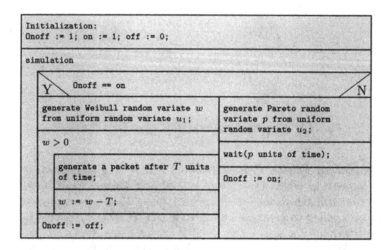

Fig. 11.9   Algorithm for the generation of On–Off packet arrivals.

Markov-Modulated Poisson Process (MMPP) (see Section 6.8.2):   In this model, source changes state according to a multistate homogenous continuous Markov chain. In each state of the source, the arrival process is Poisson with possibly state-dependent arrival rates. Thus, the CTMC modulates the Poisson process which introduces correlations between successive interarrival times. Recall that for an ordinary Poisson arrival process the interarrival times are independent. MMPP is extensively used to model various ISDN sources, such as voice and video and to characterize superposed traffic. An MMPP with two states and two (different) intensities λ_1 and λ_2 is also known as the Switched Poisson process (SPP) (see Fig. 11.10).

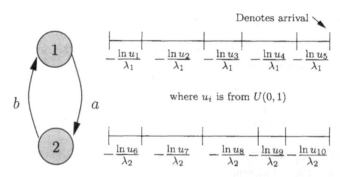

Fig. 11.10   Graphical representation of a SPP.

MMPP variates can be easily obtained by extending the algorithm for variate generation for On–Off processes. One noticeable difference is that in MMPP states with nonzero arrival rates, the traffic is Poisson unlike the case of the On–Off process where traffic is generated at deterministic intervals during the On periods. Also there are more than two states that the MMPP source can take unlike the On–Off process.

Interrupted Poisson process (IPP):   This is a special case of an MMPP with two states. No packets are generated in one of the states and traffic based on a Poisson process is generated in the other state (see Fig. 11.11). A routine which can be used by a discrete-event simulator to generate packet arrivals according to an IPP is shown in Fig. 11.12.

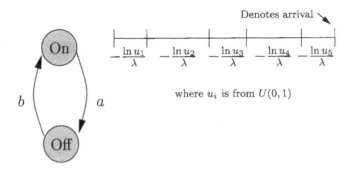

Fig. 11.11   Graphical representation of an IPP.

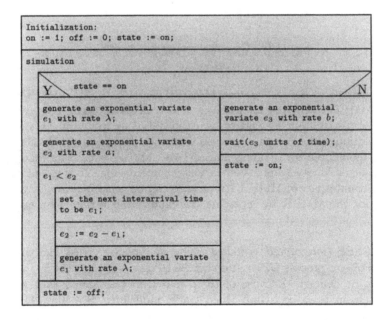

Fig. 11.12   Algorithm for the generation of IPP packet arrivals.

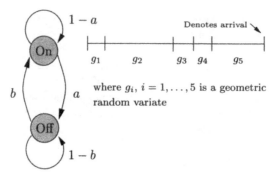

Fig. 11.13 Graphical representation of an IBP.

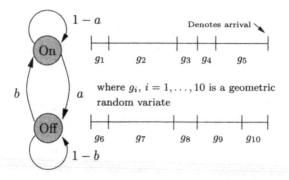

Fig. 11.14 Graphical representation of SBP.

Markov-modulated Bernoulli process (MMBP) [Oeze97, Triv01]: MMBP is similar to MMPP where the interarrivals and transitions occur according to a geometric distribution instead of an exponential distribution. MMBP is a generalization of the Bernoulli process where the parameters of Bernoulli process are modulated by a homogenous DTMC. MMBP is used for modeling of self-similar or bursty traffic such as video conferencing. It is essentially the discrete counterpart of MMPP. Its variates can be generated in a way similar to that for the MMPP, by replacing the exponential variate by a geometric variate.

IBP and SBP (Interrupted/Switched Bernoulli Process): Just like the interrupted Poisson process (IPP) and switched Poisson process (SPP), IBP and SBP can be defined. Because of their analytical tractability they are used extensively in voice data and video traffic modeling. See Figs. 11.13 and 11.14 for a graphical illustration. Figure 11.15 shows an algorithm for the generation of IBP packet arrivals.

Constant Bit Rate (CBR) Traffic: Two different definitions of CBR are in use:

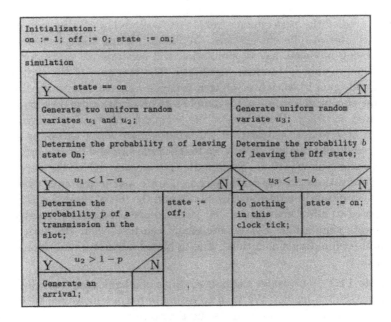

Fig. 11.15   Algorithm for the generation of IBP packet arrivals.

- Def. 1: This says that packets are generated at deterministic intervals of time. The size of the generated packets is also fixed. This definition is used in the ns-2 simulator.

- Def. 2: This assumes that data arrive in a continuous stream of fluid flow at a constant deterministic rate. This definition is primarily used when the number of bits or packets is approximated to be a continuous quantity.

In both of the above cases, there is no randomness; hence the generation of traffic is straightforward.

Markov-Modulated Fluid Source (MMFS) [McQi98]:   The traffic is generated via a fluid source modulated by a homogeneous CTMC. The (fluid) arrival rate c_1 is constant when the CTMC is in state i. MMFS are used in communications and computer systems performance modeling. MMFS is the continuous-state counterpart of MMPP. Variates can be generated by extending the MMPP variate generation method.

11.5.3 Output Analysis

DES takes random numbers as inputs so that different set of outputs are produced in general in each run. Output analysis utilizes statistical inference techniques to examine the data generated by a simulation. It can be used to predict the performance/reliability/availability of a system or compare attributes of different systems. While estimating some measure of the system being modeled, say θ, simulation will generate an estimator $\hat{\Theta}$ of θ due to the presence of random variability. The precision of the estimator $\hat{\Theta}$ will depend upon its variance. Output analysis helps in estimating this variance and also in determining the number of observations needed to achieve the desired accuracy. Phenomenon like sampling error and systematic error influence how well an estimate $\hat{\theta}$ will approximate θ. A sampling error is introduced due to random inputs, limited sample size and dependence or correlation among observations. Systematic errors occur due to the dependence of the observations on the initially chosen state and initial condition of the system. Note that in a terminating simulation the initial condition is normally supposed to affect the results, but its influence needs to be removed from steady-state simulation. Systematic errors can also occur (in both steady-state and terminating simulations) due to biased estimator and wrong model assumptions.

Example 11.1   Consider a simple example of a bank with five tellers modeled as a GI/G/5 queue. We wish to estimate output measures such as the mean response time, the throughput and the average waiting time. We assume that the interarrival times as well as the service times are Weibull distributed. We carry out this simulation in CSIM-19 which has a built-in facility for multiple servers and provides a routine to generate random variates of the Weibull distribution. We repeat this simulation 5 times for a duration of 1000 minutes each. We used a different seed for the random number generator in each replication. We notice that there is variation in the results from different runs (see Table 11.2). Statistical inference techniques can help in estimating the extent of the variability in simulation outputs.

Table 11.2 Outputs for the bank teller example

Repli-cation	Average Service Time	Utili-zation	Through-put	Average Queue Length	Average Response Time	Mean Waiting time	Custo-mers served
1	3.55115	3.453	0.97225	6.82096	7.01566	3.405532	990
2	3.59328	3.657	1.01785	6.77477	6.65599	3.059713	1023
3	3.79645	3.642	0.95919	6.85135	6.75921	3.156353	967
4	3.62218	3.266	0.90162	5.88733	6.52973	2.901117	904
5	3.45522	3.643	1.05429	7.71457	7.31732	3.858472	1064

11.5.3.1 Point and Interval Estimates   Estimation of a parameter by a single number from the output of a simulation is called a point estimate. Let the random variables $Y_1, Y_2, \ldots, Y_n$, be the set of observations of an output random variable Y obtained from simulation. We are interested in estimating the characteristics of Y, say its mean value θ where $\theta = E[Y]$. Then a common point estimator for parameter θ is the sample mean:

$$\widehat{\Theta} = \frac{1}{n} \sum_{i=1}^{n} Y_i \,.$$

The point estimator $\widehat{\Theta}$ is also a random variable and is said to be unbiased if its expected value is θ, i.e.,

$$E[\widehat{\Theta}] = \theta \,.$$

In general if $\widehat{\Theta}$ is a biased estimator and

$$E[\widehat{\Theta}] = \theta + b \,,$$

then b is called the bias of the point estimator.

The confidence interval provides an interval or a range of values around the point estimate [Triv01]. A confidence interval (A, B) for the parameter θ is defined as

$$P(A < \theta < B) = 1 - \alpha \,,$$

where A and B are functions of the point estimate $\widehat{\theta}$ generated from the output sequences. The confidence interval is a random interval, centered around $\widehat{\theta}$, which covers the true value of θ with probability $1 - \alpha$. For a single parameter, such as the mean, the standard deviation, or probability level, the most common intervals are two-sided (i.e., the parameter is between the lower and upper limit) and one-sided (i.e., the parameter is smaller or larger than the end point). For the simultaneous estimation of two or more parameters, a confidence region, which is a generalization of a confidence interval, can be defined [Nels82, YST01].

Terminating simulation and steady-state simulation require different methods of output analysis, hence they are separately discussed below.

11.5.3.2 Terminating Simulation This applies to the situation where we are interested in the transient value of some measure, e.g., channel utilization after 10 minutes of system operation or the transient availability of the system after 10 hours of operation. In these cases each simulation run is conducted until the required simulated time and from each run a single sample value of the measure is collected. By making m independent simulation runs, point and interval estimates of the required measure are obtained using standard statistical techniques.

In the above-cited examples, each simulation run will provide a binary value of the measure and hence we use the inference procedure based on sampling from the Bernoulli random variable [Triv01].

Yet another situation for terminating simulation arises when the system being modeled has some absorbing states. For instance, we are interested in estimating the mean time to failure of a system. Then from each simulation run a single value is obtained and multiple independent runs are used to get the required estimate. In this case, we could use inference procedures assuming sampling from the exponential or the Weibull distribution [Triv01].

In the general case the distribution of the output variable is unknown. In that case we use statistical inference based on normal distribution because of the *Central Limit Theorem*. But when the sample size is small we use *Student's t distribution* to approximately derive the confidence interval [LaKe00].

11.5.3.3 Sampling from Normal Distribution Suppose a sample, $\{Y_i\}$, of size n is taken from a normal population $N(\mu, \sigma^2)$ then the sample mean is $N(\mu, \sigma^2/n)$, where n is the sample size [Triv01]. Now define

$$Z = \frac{(\bar{Y} - \mu)}{\sigma/\sqrt{n}},$$

such that it becomes standard normal, $N(0,1)$. Here $\bar{Y}$ is defined as

$$\bar{Y} = \frac{\sum_{i=1}^{n} Y_i}{n}.$$

To find the $100\gamma\%$ confidence interval for the population mean, where γ is a value between 0 and 1, we find numbers a and b from $N(0,1)$ such that $P(a < Z < b) = \gamma$. Once the numbers a and b are determined, we obtain the required confidence interval as

$$a < \frac{(\bar{Y} - \mu)}{\sigma/\sqrt{n}} < b, \quad \bar{y} - \frac{b\sigma}{\sqrt{n}} < \mu < \bar{y} - \frac{a\sigma}{\sqrt{n}};$$

therefore $(\bar{y} - \frac{b\sigma}{\sqrt{n}}, \bar{y} - \frac{a\sigma}{\sqrt{n}})$ is a $100\gamma\%$ confidence interval for μ.

For a more general case, we do not know the distribution of variables in an output sequence. However, when $\{Y_i\}$ are obtained from independent runs, we can assume that they are independent. Let $Y_1, Y_2, \ldots Y_n$ are iid random variables with finite mean μ and variance σ^2. Let $\bar{Y} = \sum_{i=1}^{n} Y_i/n$ be the sample mean, then from the central limit theorem: $\frac{\bar{Y}-\mu}{\sigma/\sqrt{n}}$ converges to a standard normal distribution $N(0,1)$ as $n \to \infty$. The theorem remains valid when instead of σ^2 we use its unbiased and consistent estimator S^2, which is

$$S^2 = \frac{\sum_{i=1}^{n} Y_i^2 - n\bar{Y}^2}{n-1}.$$

Therefore, for a sufficiently large n, above confidence interval based on the normal population remains valid. When n is small we replace the normal distribution with Student's t distribution with $(n-1)$ degrees of freedom. Thus, we obtain the $100(1-\alpha)\%$ confidence interval of μ as

$$\bar{y} - t_{n-1;\alpha/2}\frac{s}{\sqrt{n}} < \mu < \bar{y} + t_{n-1;\alpha/2}\frac{s}{\sqrt{n}}.$$

We can obtain a point estimate and a confidence interval for the bank example mentioned in the previous section. First we get an overall point estimate for the response time as

$$\widehat{r} = \frac{7.01566 + 6.5599 + 6.75921 + 6.52973 + 7.31732}{5} = 6.8364\,.$$

The estimated variance becomes

$$s^2(\widehat{R}) = \frac{(6.8364 - 7.01566)^2 + \ldots + (6.8364 - 7.31372)^2}{4} = (0.3317)^2\,.$$

To compute the 95% confidence interval, we obtain $t_{0.025,4} = 2.776$ from standard tables. The confidence interval can now be written as

$$\widehat{r} \pm t_{0.025,4}\widehat{r}s\,,$$
$$7.475082 \pm (2.776) * 0.3317\,,$$
$$7.475082 \pm 0.9208\,.$$

11.5.3.4 Sampling from the Bernoulli Distribution

Suppose the random variable, denoting the experimental observation, Y_i, be Bernoulli distributed with parameter p. The estimated success probability is

$$\hat{P} = \frac{\sum_{i=1}^{n} Y_i}{n} = \overline{Y} = \frac{S_n}{n}\,.$$

Recall that the statistic $\sum_{i=1}^{n} Y_i$ is binomially distributed with parameters n and p. Derivation of the $100(1-\alpha)\%$ confidence interval becomes: Let k_0 be the largest integer such that

$$\sum_{k=0}^{k_0} b(k; n, p) = B(k_0; n, p) \le \frac{\alpha}{2}\,,$$

and let k_1 be the smallest integer such that

$$\sum_{k=k_1}^{n} b(k; n, p) = -B(k_1 - 1; n, p) \le \frac{\alpha}{2}\,,$$

$$P(k_0(p) < S_n < k_1(p)) \approx 1 - \alpha\,.$$

Then the confidence interval for p can be obtained by inverting [Triv01]:

$$k_0(p) < s_n < k_1(p).$$

When $np \ge 5$ and $nq \ge 5$, S_n may be approximated by the normal distribution. In such cases, S_n is approximately normal with $\mu = np$ and $\sigma^2 = np(1-p)$. When p is close to 0 or 1, Poisson approximation may be used provided sample size is large enough [Triv01].

11.5.3.5 Steady-State Simulation In steady-state simulation we are interested in estimating characteristics such as the mean or the variance of some output in steady state. This situation is basically different from terminating simulation due to the unknown length of the transient phase and due to correlations in the output sequence. In this case, we can in principle use the method of independent replications but since the transient phase needs to be thrown away and since it can be long, this approach is wasteful. An attempt is therefore made to get the required statistics from a single long run. The first problem encountered then is to estimate the length of the transient phase. The second problem is the dependence in the resulting sequence. Section 11.5.3.7 describes how to estimate the correlation in the sequence, first using independent runs.

The estimator random variable of the mean measure, θ, to be estimated is given by Eq. (11.6), where n is the number of observations:

$$\widehat{\Theta} = \lim_{n \to \infty} \frac{1}{n} \sum_{i=1}^{n} Y_i. \qquad (11.6)$$

This value should be independent of the initial conditions. Under real conditions, simulation is stopped after some number of observations n have been collected. The simulation run length is determined by how large the bias in the point estimator is, the desired precision, or resource constraint for computing.

11.5.3.6 Initialization Bias Initial conditions may be artificial or unrealistic. There are two methods that reduce the point-estimator bias in steady-state simulation. One method is called *intelligent initialization* that involves initialization of the simulation in a state that is more representative of long-run conditions. In cases where the system does not exist or it is very difficult to obtain data directly from the system, data on similar systems or a simplified model is collected. The collected data are then used to initialize the simulation.

The second method involves dividing the simulation into two phases. One of them is called the *initialization phase* from time 0 to T_0 and the other is called the *data-collection phase* from T_0 to $T_0 + T_E$. The choice of T_0 [LaCa79, Shru81, Welc83] is important as system state at time T_0 will be more representative of steady-state behavior than at the time of original initial conditions (i.e., at time $t = 0$). Unfortunately, there is no widely accepted and proven technique to determine how much data to delete to reduce the initialization bias. Plots of ensemble averages (average across batch means or across independent replications) and cumulative averages (average over a single run) are used to generally decide how much data to delete [BCNN01].

- Ensemble average: This will become smoother and more precise as the number of replications n is increased. Since each ensemble average is the sample mean of iid observations, a confidence interval based on t-distribution can be placed around each point and these intervals can be used to decide whether or not the plot is precise enough to remove the bias.

- Cumulative average: This will become less variable as more and more data are averaged. So as time progresses the plot becomes smoother. Hence this method should only be used if ensemble averages are difficult to plot.

Apart from this, some of the methods include testing for bias [GSS94], modeling the bias [SnSc85] and randomly sampling the initial conditions on multiple replications [Kelt89]. Generally T_E is taken to be more than five times T_0 [Kelt86].

11.5.3.7 Dealing with Dependency   Successive values of variables monitored from a simulation run exhibit dependencies, such as high correlation between the response times of consecutive requests to a file server. Assume that the observed quantities are dependent random variables, $\{Y_1, Y_2, \ldots, Y_m\}$ having index invariant mean μ and variance σ^2. The sample mean is given by

$$\bar{Y} = \sum_{i=1}^{m} \frac{Y_i}{m} .$$

The sample mean is an unbiased point estimator of the population mean but the variance of the sample mean is not equal to σ^2/m. Assuming the sequence to be wide sense stationary, the variance is given by

$$\mathrm{var}[\bar{Y}] = \frac{\sigma^2}{m} + \frac{2}{m} \sum_{i=1}^{m-1} \left(1 - \frac{i}{m}\right) K_i ,$$

where K_i is the covariance of the wide-sense stationary sequence $\{Y_i\}$ which depends only on the distances between Y_j and Y_{i+j} for every j [Triv01]. So

$$m\,\mathrm{var}[\bar{Y}] = \sigma^2 a, \quad \text{where} \quad a = 1 + 2 \sum_{i=1}^{\infty} \frac{K_i}{\sigma^2} .$$

The statistic

$$\frac{\bar{Y} - \mu}{\sigma\sqrt{\frac{a}{m}}}$$

approaches a standard normal distribution as m approaches infinity. Therefore, an approximate $100(1 - \alpha)\%$ confidence interval becomes

$$\bar{Y} \pm \sigma z_{\alpha/2} \sqrt{\frac{a}{m}} .$$

The value of $\sigma^2 a$ is unknown and should be estimated from data. This need to estimate $\sigma^2 a$ can be avoided using the *the method of independent replications* [LaKe00]. It is used to estimate the point-estimator variability. In this method, the simulation experiment is replicated m times with n observations each. If the initial state is chosen randomly for all m observations, the result will be independent of each other. But the n observations within each experiment will be dependent. Let the sample mean and sample variance of the jth experiment be given by $\bar{Y}(j)$ and $\bar{S}^2(j)$, respectively. From individual sample means, the point estimator of the population mean is given by

$$\bar{Y} = \frac{1}{m} \sum_{j=1}^{m} \bar{Y}(j) = \frac{1}{mn} \sum_{j=1}^{m} \sum_{i=1}^{n} Y_i(j). \tag{11.7}$$

Here $\bar{Y}(1), \bar{Y}(2), \ldots, \bar{Y}(m)$ are independent and identically distributed (iid) random variables. Assume that the common variance of $\bar{Y}(j)$ is denoted by v^2. The estimator of the variance v^2 is given by

$$\bar{V}^2 = \frac{1}{m-1} \sum_{j=1}^{m} [\bar{Y}(j) - \bar{Y}]^2, \tag{11.8}$$

and the $100(1-\alpha)\%$ confidence interval for θ is approximately given by

$$\bar{y} - t_{m-1;\alpha/2} \frac{v}{\sqrt{m}} \leq \theta \leq \bar{y} + t_{m-1;\alpha/2} \frac{v}{\sqrt{m}}. \tag{11.9}$$

11.5.3.8 Method of Batch Means One major disadvantage of the independent replication method is that initialization phase data from each replication is wasted. To address the issue, we use a design based on a single, long simulation run divided into m contiguous segments (or batches), each having length n. The sample mean $\bar{Y}(j)$ of the jth segment is then treated as an individual observation. This method, called the method of batch means [LaKe00], reduces the unproductive portion of simulation time to just one initial stabilization period. But the disadvantage with this is that the set of sample means are not statistically independent and usually the estimator is biased. Estimation of the confidence interval for a single run method can be done following the same procedure as done for replication method. We just replace the jth replication in independent replication by the jth batch. The method of batch means is also called the *single run method.*

11.5.4 Speedup Techniques

One of the most important issues in DES is the length of the simulation run needed to achieve a desired level of accuracy. In order to reduce the time needed to carry out simulation while maintaining accuracy, two approaches are in use: variance reduction techniques and parallel/distributed simulation.

11.5.4.1 Variance Reduction Techniques
Variance reduction techniques (VRTs) help in obtaining greater precision of simulation results (smaller confidence interval) for the same number of

simulation runs, or in reducing the number of runs required for the desired precision. They are used to improve the efficiency and accuracy of the simulation process. VRTs are extensively researched topic in literature [LaKe00]. The most common are *antithetic variates* and *control variates* which take advantage of correlation between random variables to obtain variance reduction. The *conditioning method* replaces an estimate of a quantity by its exact analytical value in order to remove the source of variability. The most popular technique of *common random number* applies to situations when we compare alternative system configurations. Different simulation runs are driven by the same random number sequence in order to be confident that observed fluctuations in performance are due to differences in the system configuration rather than in variability of experimental conditions. In this section we describe the methods of importance sampling and importance splitting/RESTART which are used specially in rare event simulation.

Importance Sampling   One of the frequently used techniques is importance sampling [Fish97, Gllg89, Shah94, TuTr01]. In this approach, the stochastic behavior of the system is modified in such a way that some events occur more often. This helps in dealing with rare events scenarios. Assume $f(y)$ be the original sampling density and Y is the random variable. Let $g(y)$ be the property of interest. The problem now is that the probability of observing $g(y) > 0$ is very small when taking samples from $f(y)$. To solve this problem, importance sampling changes the sampling density to a new density $f^*(y)$. Taking samples from $f^*(y)$ should increase the probability of observing $g(y) > 0$ significantly. Let the property of interest be $\theta = E[g(Y)]$. This can now be rewritten as

$$\theta = E_f[g(Y)] = \int g(y)f(y)\,\mathrm{d}x = \int g(y)\frac{f(y)}{f^*(y)}f^*(y)\,\mathrm{d}y, \qquad (11.10)$$

$$\theta = E_{f^*}[g(Y)f(Y)], \qquad (11.11)$$

where $L(y) = \frac{f(y)}{f^*(y)}$ is called by likelihood ratio. But this modification causes the model to be biased, which can be removed using the above likelihood ratio function. Note that the expected value of observations under f is equal to the expected value of observations under f^* corrected for bias by the likelihood ratio. An unbiased estimator of θ, taking samples y from $f^*(y)$ is given by

$$\widehat{\Theta} = \frac{1}{n}\sum_{j=1}^{n} g(Y_j)L(Y_j), \qquad (11.12)$$

with variance

$$\mathrm{var}[\widehat{\Theta}] = \frac{1}{n}\,\mathrm{var}_{f^*}[g(Y)L(Y)] = \frac{1}{n}E_{f^*}[(g(Y)L(Y) - \theta)^2]. \qquad (11.13)$$

From Eq. (11.10), the only restriction to the probability density $f^*(y)$ is that $f^*(y) > 0$ holds for all samples y where $g(y)f(y) \neq 0$. This means that the samples y with a positive value $f(y) > 0$ and a nonzero $g(y)$ must have a positive value also in the new density, $f^*(y)$. This is a necessary condition which serves as a guideline for choosing $f^*(y)$. A wrong choice of $f^*(y)$ may cause the variance of $\hat{\theta}$ to be larger than the variance of the estimates obtained by direct simulation. The variance is minimized to $\text{var}[\hat{\gamma}] = 0$ when $g(y)f(y) = f^*(y)$. This observation serves as a guideline suggesting $g(y)f(y) \propto f^*(y)$. Thus, if carefully done, the variance of the estimator of the simulated variable is smaller than the original one implying a reduction in the size of the confidence interval.

Importance Splitting/RESTART Another important technique for variance reduction is known as importance splitting [Heid95, HaTo99, TuTr00, NSH93]. Suppose that we wish to estimate by simulation the probability of a rare event A. We define multiple thresholds where we split the standard simulation path. Consider $k + 1$ sets B_i, such that $A = B_{k+1} \subset \ldots \subset B_1$ and use the formula

$$P(A) = P(A|B_k)P(B_k|B_k - 1)\ldots P(B_2|B_1)P(B_1), \qquad (11.14)$$

where each conditioning event on the right side is assumed to be less rare. The idea of importance splitting is to make a Bernoulli trial to see if the (not rare) set B_1 is hit. If this is the case, we split this trial into n_1 trials and look (still by a Bernoulli simulation) for each new trial if B_2 is hit. We repeat this splitting procedure at each level if a threshold is hit; i.e., we make n_i retrials each time B_i is hit by a previous trial. If a threshold is not hit, neither is A, so we stop the current retrial. By this procedure we have then considered $n_1, \ldots, n_k$ (dependent) trials, considering for example that if we have failed to reach B_i at the ith step then $n_i \ldots n_k$ possible retrials have failed. This constitutes saving in computational time. Using n_0 independent replications of this procedure the unbiased estimator of $P(A)$ is given as

$$\widehat{P} = \frac{1}{n_0 \ldots n_k} \sum_{i_0=1}^{n_0} \ldots \sum_{i_k=1}^{n_k} \mathbf{1}_{i_0} \mathbf{1}_{i_0 i_1} \ldots \mathbf{1}_{i_0 i_1 \ldots i_k}, \qquad (11.15)$$

where $\mathbf{1}_{i_0 i_1 \ldots i_k}$ is the result of the ith Bernoulli retrial at stage j (its value is 1 if it is successful, and 0 otherwise). An estimator of the variance can also be easily derived, as usually done in simulation using independent replications. One another method which derives itself from importance splitting is called RESTART (*RE*petitive *S*imulation *T*rials *A*fter *R*eaching *T*hreshold). The method [GöSc96, HoKl79] is based on restarting the simulation in valid states of the system that provokes the events of interest more often than in a straightforward simulation, thus leading to a shorter simulation run time. The results of the simulation have to be modified to reflect the correct weights of the events.

11.5.4.2 Parallel and Distributed Simulation (PADS) Other methods used to speed up simulations are parallel and distributed simulation [Fuji99, Zoma96].

It refers to the execution of single DES on multiple computers which can be distributed or locally placed. It has attracted a considerable amount of interest as large simulations in engineering, computer science, economics and military applications consume enormous amounts of time on sequential machines. PADS mechanisms broadly fall into two categories: *conservative* and *optimistic*. *Conservative* approaches strictly avoid the possibility of any causality error [Fuji99] ever occurring. These approaches rely on some strategy to determine when it is safe to process an event (i.e., they must determine when all events that could affect the event in question have been processed). Several methods including deadlocks avoidance, synchronization, conservative time windows and lookahead improvement have been proposed for PADS using conservative methods. On the other hand, optimistic approaches use a detection and recovery approach: Causality errors are detected, and a rollback mechanism is invoked to recover. This includes methods like time-warp mechanism, lazy cancellation and evaluation and several others. One distinct advantage of parallel simulation is in case of terminating simulations. Several runs of simulation can be run on multiple machines generating totally independent output results [Heid86]. This is ideally desired in case of independent replications described in Section 11.5.3.2.

11.5.5 Summary of Output Analysis

To summarize, before generating any sound conclusions on the basis of the simulation-generated output data, a proper statistical analysis is required. The simulation experiment helps in estimating different measures of the system. The statistical analysis helps in acquiring some assurance that these estimates are sufficiently accurate for the proposed use of the model. Depending on the initial conditions and choice of run length, terminating simulations or steady-state simulations can be performed. A confidence interval can be used to measure the accuracy of point estimates.

11.6 APPLICATIONS

In this section we present the application of some simulation packages and software libraries like CSIM-19 [CSI05], OPNET Modeler [OPN05], and ns-2 [Ns05]. CSIM-19 is a process-oriented general-purpose discrete-event simulator. OPNET Modeler and ns-2 are application-oriented simulation packages. While OPNET Modeler uses GUI extensively for configuring network, ns-2 is an OTcl interpreter and uses code in OTcl and C++ to specify the network to be simulated. In the Web cache example we compare results from an analytic solution with those from simulation and then show advantages of simulation in a case where analytical solution is not feasible. In the RED example we point out that when using special-purpose simulation packages like OPNET or ns2, one must be very careful in interpretation of the results due to hidden details of the built-in behavior of elements provided by the simulator for the construction of the model, which are usually not accessible to a user.

11.6.1   CSIM-19

CSIM-19 is a process-oriented discrete-event simulator [CSI05]. It is implemented in C/C^{++} as a library of routines which implement basic simulation operations. It models a system as a collection of CSIM processes that interact with each other by using the CSIM structures. It is patterned after the simulation language ASPOL [ASP72, MaMc73].

11.6.1.1   Features of CSIM-19   The simplicity of CSIM-19 comes from its use of C/C^{++} for simulation development. Simulation features such as random variate generation, traffic generating source, output analysis, facilities, etc., are implemented as calls to procedures (functions) in a runtime library. In CSIM-19 models are compiled, not interpreted (unlike ns-2 [Ns05]). With a compiled language, code entered is translated into a set of machine-specific instructions before being saved as an executable file. With interpreted languages, the code is saved in the same format that is entered by the user. Compiled programs generally run faster than interpreted ones because interpreted programs must be translated to machine instructions at runtime. CSIM-19 is a commercial simulation tool like **OPNET Modeler**. Confidence intervals and run length control can be performed. Each process has both a private data store (memory) as well as access to global data. Data gathering is partially automated and very easy to extend in CSIM-19. Provisions for tracing execution of the simulation model and for logging simulated events are also present. CSIM-19 is available on Windows, Linux, Solaris, AIX, and Macintosh platforms.

11.6.1.2   CSIM-19 Objects   For each CSIM-19 structure, the program must have a declaration, which is a pointer to an object. Some of the common objects extensively used for modeling in CSIM-19 are:

- *Processes* are the active entities that request services at facilities, wait for events, etc.

- *Facilities* consist of queues and servers reserved or used by processes. These are also known as *active* resources.

- *Storage blocks* are resources which can be partially allocated to processes. These are also known as *passive* resources.

- *Events* are used to synchronize process activities.

- *Mailboxes* are used for inter process communications.

- *Table structures* are used to collect data during execution of a model.

- *Process classes* are used to segregate statistics for reporting purpose.

- *Streams* are used to generate streams of random deviates for more than twenty distributions.

11.6.2 Web Cache Example in CSIM-19

We discuss the effect of Web caching on network planning in the sense of determining the bandwidth of the access link interconnecting the ISP's subnet with the Internet. This example is a slightly modified version of the model presented in [SZT99]. The latency of a browser retrieving files is studied for given traffic characteristics, number of users, bandwidth of access link, and cache hit ratio.

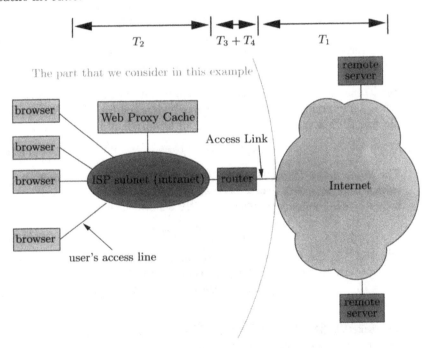

Fig. 11.16 A typical network infrastructure interconnecting an ISP's subnet (intranet) with the Internet.

11.6.2.1 System description The system we model represents a typical network infrastructure interconnecting a subnet with the Internet. Browsers access the subnet via Modem, ISDN Basic Rate interface (BRI), Ethernet, or Token Ring. When a WWW browser clicks on a hyperlink, several URL requests may be sent from the browser to the Web proxy, which may be just one cache or a collection of caches. If the proxy has a copy of the requested file and is consistent with the original copy on the remote server, the proxy sends the copy to the browser. This is called a cache hit. If the proxy does not have a copy of the file that the browser is looking for, the proxy retrieves an original copy from the remote server, sends it to the browser and keeps a copy in the cache. This is called a cache miss.

In this example we are interested in *file delivering latency* or *mean response time*, which is defined as the time interval from the browser clicking an object to the requested object being displayed on the monitor. Excluding some trivial terms such as the delay of monitor display and cache retrieval and the HTTP interaction within the subnet, the response time consists of the following components:

- Delay on the Internet side T_1: The delay related with HTTP interaction in the Internet, which consists of a request sent from the proxy to the server, followed by a response sent back from the server to the proxy. This also includes transmission and queueing delay for a round trip on the Internet side. In general T_1 is a random variable but for simplicity we assume that it is constant.

- T_2: This consists of two delays. Transmission delays while the browser sends a file request packet to the server is denoted by T_2^{out}. The request file size Z^{out} is typically of the order of hundreds of bytes. Thus,

$$T_2^{\text{out}} = 8Z^{\text{out}}/\text{speed} . \qquad (11.16)$$

The other delay T_2^{in} constitutes the time of the files sojourning on the subnet and browser's access line, when the requested file has arrived. If browsers use modems to access the ISP's subnet, then

$$T_2^{\text{in}} = 8Z^{\text{in}}/\text{speed}, \qquad (11.17)$$

where Z^{in} is the file length in bytes and "speed" denotes the speed of the modem in terms of bits per second. So the total delay becomes $T_2 = T_2^{\text{out}} + T_2^{\text{in}}$. Since the file size is treated as a random variable, T_2 is also a random variable.

- T_3: Like T_2, this has two parts. The transmission delays involved in sending a request packet at the entrance of subnet forms T_3^{out}. The other part is the transmission delay of the file sojourning at the entrance of the subnet. It depends on the speed of the access link as well as the size of the file requested. Total delay $T_3 = T_3^{\text{in}} + T_3^{\text{out}}$ is given by

$$T_3 = \frac{8Z^{\text{in}}}{\text{bandwidth}} + \frac{8Z^{\text{out}}}{\text{bandwidth}}, \qquad (11.18)$$

where "bandwidth" denotes the capacity of the access link in terms of bits per second.

- T_4: This constitutes the queueing and servicing delay due to the presence of a queue at both ends of the access link. Here we note that two queues will be formed; one when Z^{out} packets are being sent from browsers and the other while files are fetched from the Internet. We assume the buffer size at the entrance to be infinite.

The total latency or response time is given by

$$R_{\text{overall}} = \begin{cases} T_1 + T_2 + T_3 + T_4 & \text{in case of a cache miss;} \\ T_2 & \text{in case of a cache hit.} \end{cases} \quad (11.19)$$

We divide our study into two cases. In the first case we assume the arrival process to be Poisson and in the second case we assume that the arrival process is an On–Off process, with sojourn time in On state being Weibull distributed and in Off state being Pareto distributed.

Case 1: Poisson Arrival Process Since the arrival process is Poisson and assuming that Z^{out} is fixed at x bytes which is typically in hundreds of bytes, T_4^{out} is then the delay through an M/D/1 queue. We assume that Z^{in} is Pareto distributed and we assume that traffic from the server back to the browser is Poisson, hence T_4^{in} is the delay through an M/G/1 queue. Three scenarios will be studied:

1. Bandwidth=256 Mb/sec with no Web cache.

2. Bandwidth=256 Mb/sec with Web cache having 50% hit ratio.

3. Bandwidth=512 Mb/sec with no Web cache.

The overall mean response time is given by

$$E[R_{\text{overall}}] = (1 - h)(E[T_1] + E[T_2] + E[T_3] + E[T_4]) + h(E[T_2]), \quad (11.20)$$

where h is the cache hit ratio. Also

$$E[R_{\text{overall}}] = (1 - h)(E[T_1] + E[T_2] + E[T_3] + E[R^{\text{in}}]_{M/G/1} \\ + E[R^{\text{out}}]_{M/D/1}) + h(E[T_2]). \quad (11.21)$$

Since the typical file size distribution on the Internet is observed to be Pareto, the mean length of files is given by

$$E[Z^{in}] = \frac{\alpha k}{\alpha - 1}.$$

Here α denotes the shape parameter and k the location parameter. We take the values $\alpha = 2.5$ and $k = 0.25$, which yields $E[T_2]^{\text{in}} = E[Z^{\text{in}}](\text{Mb})/\text{speed} = 41.67\,\text{msec}$ where the speed was set to 10 Mbps. Similarly, the mean transmission delay on the router side, $E[T_3]^{\text{in}} = E[Z^{\text{in}}](\text{Mb})/\text{bandwidth}$ becomes

1.6 msec for a bandwidth of 256 Mbps and assuming the delay on the Internet side $E[T_1]$ to be constant 100 msec. Also the queue on the access link (router) is regarded to be M/G/1 for incoming and M/D/1 for outgoing requests. Then the mean response time for the two cases are given by

$$E[R^{\text{out}}]_{M/D/1} = \frac{1}{\lambda_{\text{eff}}}\left(\rho + \frac{\rho^2}{2(1-\rho)}\right), \tag{11.22}$$

$$E[R^{\text{in}}]_{M/G/1} = \frac{1}{\lambda_{\text{eff}}}\left(\rho + \frac{\lambda_{\text{eff}}^2 E[B^2]}{2(1-\rho)}\right), \tag{11.23}$$

where $\lambda_{\text{eff}} = (1-h)\lambda$ is the effective arrival rate and $\rho = \lambda_{\text{eff}} E[B]$, $E[B]$ is the average service time. For outgoing requests, $E[B] = 8x/c$ where x is the size of request typically in hundreds of bytes and c is the speed of the router (256 Mb/sec in our case) while for the incoming files, $E[B] = 8E[Z^{\text{in}}]/c$. Also for the incoming requests, $E[B^2]$ is the second moment of the service time which in our case can be obtained easily using the following formula for the Pareto distribution:

$$E[B^2] = \frac{k^2\alpha}{c^2(\alpha-2)}(\text{Mb})^2 \quad \alpha > 2.$$

Thus, the mean response time can be calculated analytically using Eq. (11.20). Note that if there is no Web cache, it becomes special case of Eq. (11.20) where $h = 0$. Using this formula and simulation in CSIM-19 we calculated the mean response time and found that they are quite close (see Table 11.3). In general, there can be many cases where simulation results can be much closer to the analytical results. We assume that the mean interarrival time for each URL request is 1000/500 msec, which by superposition principle of 500 independent Poisson processes [Triv01] gives an overall λ of 1/1000 msec^{-1}. Service time for incoming files is assumed to be generally distributed (specifically Pareto distributed) with mean of 1.6 msec. From the above equation $E[R^{\text{in}}]_{M/G/1}$ becomes 44.87 msec excluding $E[T_1]$. For in the outgoing direction we can ignore $E[T_2]^{\text{out}}$ and $E[T_3]^{\text{out}}$ as they are in order of 10^{-6} msec. We use Eq. (11.22) to calculate $E[R^{\text{out}}]_{M/G/1}$ which gives time of the order of 10^{-4} msec. Finally, combining $E[T_2]^{\text{out}}$ and $E[T_2]^{\text{in}}$ we get 144.87 msec.

When assuming a 50% hit rate $E[R^{\text{out}}]_{M/G/1}$ becomes 6.4 msec and $E[T_2]^{\text{in}}$ and $E[T_3]^{\text{in}}$ remains same as above. From the above equation $E[R^{\text{in}}]_{M/G/1}$ becomes 49.67 msec excluding $E[T_1]$. Again, in the outgoing direction individual times are in the order of 10^{-6} msec. Using Eq. (11.23) we obtain

$$0.5 \cdot 149.67\,\text{msec} + 0.5 \cdot 41.67\,\text{msec} = 95.67\,\text{msec}.$$

The mean response time in this case could be derived by analytic methods but we need to use simulation to obtain the CDF of the overall response time. Figure 11.17 compares the cumulative distribution function (CDF) for the response time R_{overall}. In this figure we see that the two cases "bandwidth 512 Mb/sec with no Web cache" and '256 Mb/sec with Web cache and 50%

Table 11.3 Comparison of analytical calculation vs. simulation

	Analytical (Mean Response Time)	Simulation (Mean Response Time) ± (95% confidence interval)
No Web cache	144.87 msec	151.7 ± 9.4 msec
Web cache 50% hit ratio	95.67 msec	98.24 ± 7.1 msec

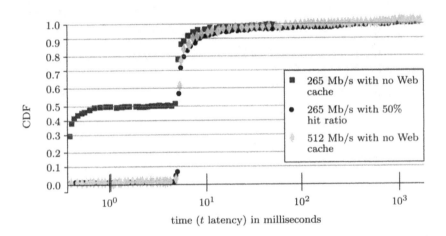

Fig. 11.17 CDF of response time for the three cases with Poisson arrivals.

hit ratio" have nearly the same CDF. In fact doubling the bandwidth is not so effective as compared to having a Web cache with respect to decreasing retrieval latency. Note that in this model 500 users simultaneously access the Internet via their browsers which is a common size in many real-world scenarios.

Case 2: On–Off Arrival Process In this case we will assume the arrival process to be an On–Off process. The On periods are initiated by user's clicks on the hypertext links. The On period is found to follow Weibull distribution whose pdf is given by

$$f_{on}(x) = \frac{k}{\theta} \left(\frac{x}{\theta}\right)^{k-1} \exp\left(-\left(\frac{x}{\theta}\right)^k\right),$$

with $k = 0.77$ to 0.91 and $\theta = e^{4.4}$ to $e^{4.6}$.

During the On period, requests are generated at deterministic intervals. During the Off period, no request is generated. The duration of the Off period

Table 11.4   Parameters for simulation model for second case

On period:	Weibull ($k = 0.9$, $\theta = e^{4.4}$)
Off period:	Pareto ($k = 60$, $\alpha = 0.5$)
Interarrival during On period:	Weibull ($k = 0.5$, $\theta = 1.5$)
File size:	Pareto ($k = 300$, $\alpha = 1.3$)

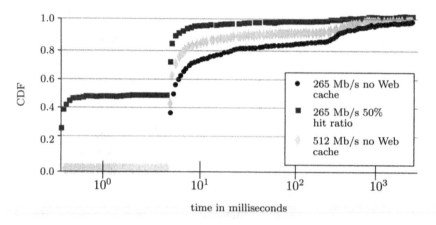

Fig. 11.18   CDF of the response time for the three cases when arrival is On–Off process.

follows a Pareto distribution whose pdf is given by

$$f_{off}(x) = \frac{\alpha k^\alpha}{x^{\alpha+1}} \text{with } \alpha = 0.58 \text{ to } 0.9, \, k = 60 \,.$$

The file size of the request is also Pareto distributed with $\alpha = 1.1$ to 1.3 and k is determined by mean length of files:

$$k = (\alpha - 1)\frac{E[Z^{in}]}{\alpha} \,.$$

With the non-Poisson arrival process even the mean response time cannot be obtained analytically and we use CSIM-19 simulation for this purpose. We use Eq. (11.19) to determine the mean response time of the system. The model parameters we use are given in Table 11.4.

The CDF of the response time was determined by first calculating delays for all the simulated requests. Since CSIM-19 doesn't provide support for generating CDF from data, a routine was written to compute the empirical CDF from the set of all observed delays [LaKe00]. From Fig. 11.18 we can see the CDF of the response time and conclude that a Web cache with 50% hit ratio is more effective than doubling the bandwidth.

CSIM-19 was helpful in solving the Web cache problem as it provides random variates from several distributions including Weibull and Pareto used in this example. Apart from this it also calculates confidence interval of estimated parameters by using the batch means method. We saw in this example that when the arrival process was Poisson we were able to calculate the mean response time using both analytical and simulative methods. but when arrival process was On–Off analytical techniques were not feasible in solving even the mean response time. The response time CDF estimation for both cases — i.e., Poisson and On–Off process — was only possible by simulative techniques.

11.6.3　OPNET Modeler

OPNET Modeler is a commercial simulation environment with a GUI support (originally developed at MIT). It is designed to support the modeling, design and validation of communication protocols and architectures. It has been used in areas like:

1. Network (LAN/WAN) planning. It has built-in libraries for all the standard TCP/IP protocols and applications including IP Quality of Service (QoS), Resource Reservation Protocol (RSVP), etc.

2. Wireless and satellite communication schemes and protocols.

3. Microwave and fiber-optic-based network management.

4. Evaluation of new routing algorithms for routers, switches and other connecting devices, before plugging them physically in the network.

OPNET provides an object oriented approach in building a model. Menu driven graphical user interface helps a user in designing a model. OPNET Modeler gives an option of importing traffic patterns from an external source like an XML file, spreadsheet or an OPNET VNE server. It has animation capabilities that can help in understanding and debugging the network. It runs on different flavors of Windows, Linux, and Unix platforms. Parallel simulation is also possible.

OPNET Modeler allows the user to model network topologies using three hierarchical levels [HaJa03]:

1. *Network level:* It is the highest level of modeling in OPNET Modeler. Topologies are modeled using network level components like routers, hosts and links. These network models can be dragged and dropped from an object palette, can be chosen from the OPNET Modeler menu which contains numerous topologies like star, bus, ring, mesh or can be imported from a real network by collecting network topology information.

2. *Node level:* This level models the internal structure of a network level component. It captures the architecture of a network device or system by depicting the interactions between functional elements called modules. Modules have the capability of generating, sending and receiving packets from other modules to perform their functions within the node. They typically represent applications, protocol layers and physical resources ports, buses and buffers. Modules are connected by *"streams"* that can be a *packet stream*, a *statistic stream* or an *association stream*. As the name suggests packet stream represents packet flows between modules, a statistic stream is used to convey the statistics of the stream between modules. An association stream is used for logically associating different modules and it does not carry any information.

3. *Process level:* This level uses a finite-state machine (FSM) description to support specification of protocols, resources, applications, algorithms and queueing policies. States and transitions graphically define the evolution of a process in response to events. Each state of the process model contains C/C++ code, supported by an extensive library for protocol programming.

Example 11.2   We consider two distinct management approaches to IP router queue management in this example. The classical behavior of router queues on the Internet is called FIFO with tail-drop. Tail-drop works by queueing the incoming messages up to its queue length and then dropping all traffic that arrives when the queue is full. This could be unfair, and may lead to many retransmissions. The sudden burst of drops from a router that has reached its buffer size will cause a delayed burst of retransmits, which will over fill the congested router again.

Random early detection (RED) [FlJa93, ChCl03] is an active queue management scheme proposed for IP routers. It is a router based congestion avoidance mechanism. RED is effective in preventing congestion collapse when the TCP window size is configured to exceed the network storage capacity. RED distributes losses in time and maintains a normally low queue length while absorbing spikes. When enabled on an interface, RED begins dropping packets when congestion occurs at a rate selected during configuration. It reduces congestion and end-to-end delay by controlling the average queue size. Packets are dropped randomly with certain probability even before the queue gets full (see Fig. 11.19).

The measure for queue length at any point in time at the RED buffer is estimated by the first-order weighted running average:

$$\text{AvgQLen}(t) = (1-\text{weight}) * \text{AvgQLen}(t-1) + \text{weight} * \text{InstQLen}(t), \quad (11.24)$$

where weight is usually much less than 1 and $\text{InstQLen}(t)$ is the value of current queue length. The packet drop probability is given by

$$p(t) = \text{Max}_{\text{drop}} \frac{\text{AvQLen}(t) - \text{min}_{th}}{\text{max}_{th} - \text{min}_{th}}, \quad \text{AvQLen}(t) \in [\text{min}_{th}, \text{max}_{th}],$$

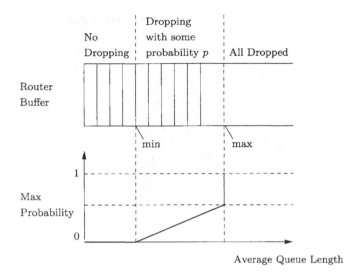

Fig. 11.19   Active queue management by RED.

where no packets are dropped till the queue length is less than min_{th} and all packets are dropped when the queue length become greater than max_{th}. Max_{drop} is the packet drop probability at max_{th}.

In this example we compare the performance of RED and FIFO with tail-drop. The network for the example consists of two routers A and B and three clients with their corresponding servers. The capacity of the link between two routers is assumed to be 2.048 Mbps. All other links have a capacity of 100 Mbps fast Ethernet. Clearly the link between routers is the bottleneck. Our goal is to estimate the buffer occupancy at router A for the two schemes using the OPNET Modeler.

Model is constructed using network level editor of the OPNET Modeler (see Fig. 11.20). Hosts and servers are joined together with the help of routers and switches that are dragged and dropped from the object palette. Attributes are assigned for various components. Configuration parameters are assigned with the help of utility objects. The chosen application is video conferencing with each of the clients having different parameter sets: heavy, streaming multimedia, best effort, standard and with background traffic. The incoming and outgoing frame sizes are set to 1500 bytes. The incoming rate is set to 30 frames per second with deterministic interarrival times. Instead of deterministic start time for the sources, we have chosen the exponential distribution to randomize the start of each application.

For modeling the RED policy, we have used the convenient feature of OPNET Modeler for duplicating scenarios. The applications and profile configuration for RED remains the same as in the FIFO case. Only the QoS attributes configuration needs to be changed. RED parameters are set as

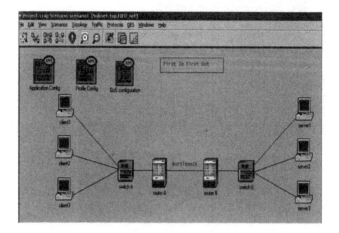

Fig. 11.20 Network level modeling for FIFO queue management. Three clients are connected to 3 servers via 2 routers.

Table 11.5 RED parameters

Parameters	Values
$\min_{th}$	25
$\max_{th}$	90
Exponential weight factor	100
weight (w_q)	1/100
Mark probability denominator	15
Max_{drop} at $\max_{th}$	1/15

in Table 11.5. *Mark probability denominator* is set to 15 which means that when the average queue size reaches the maximum threshold then, every 15th packet is marked to be deleted. So this implies Max_{drop} to be 1/15 when the average queue reaches the maximum threshold. The *exponential weight factor* is used to calculate the average queue size as per Eq. (11.24). After the model is completely specified, a DES is run and the statistics like the buffer occupancy according to Eq. (11.24) for the router A are collected.

Figure 11.21 shows the simulation results where the buffer occupancy for the two cases are plotted as functions of time. Notice that both buffers using RED and FIFO tail-drop behave similarly when the link utilization is low. After 40 sec, when the utilization jumps to almost 100%, congestion starts to build at the router buffer that uses FIFO tail-drop. In case of an active queue management (RED case), the buffer occupancy remains low and it never saturates. In fact the buffer occupancy is much smaller than that of FIFO during the congestion period.

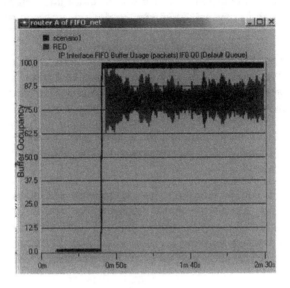

Fig. 11.21 RED vs. FIFO for buffer occupancy (running average queue length).

11.6.4 ns-2

Network simulator (ns) is a discrete-event simulator for applications in networking. ns started as a variant of REAL network simulator [REA] with the support of DARPA and universities like UCB, CMU, and USC. Its current version, ns-2, is written in C^{++} and object-oriented TCL [OTC05]. ns-2 is an object oriented OTcl script interpreter containing a simulation event scheduler, network components, object libraries and network setup module libraries. It is a public domain simulation package in contrast to OPNET Modeler. Like OPNET Modeler, it also uses an object oriented approach towards problem solving. All network components and characteristics are represented by classes. ns-2 provides a substantial support for simulation of TCP, routing and multicast protocols over wired and wireless networks. Details about ns-2 can be found from its official Web site [Ns05, ChCl05].

11.6.4.1 ns-2 Features

- It provides canned sub-models for several network protocols like TCP and UDP, router queue management mechanism like tail-drop, RED, CBQ routing algorithms like Dijkstra [Dijk59].

- It also provides several traffic sources such as Telnet, FTP, or CBR. It implements some MAC protocols for LAN and multicast protocols.

- It contains a simulation event scheduler and a large number of network objects, such as routers, links, etc., which are interconnected to form a network.

- The user needs to write an OTcl script that initiates an event scheduler, sets up the network topology using network objects, and tells traffic sources when to start and stop transmitting packets through the event scheduler.

- It is available under public license and runs under several flavors of Unix and Linux. A Cygwin version of ns-2 is also available for different Windows versions.

Being an open source simulator, ns-2 lacks code and documentation consistency across its different versions. Apart from this, it does not provide any built-in output analysis graphing tools. Unlike OPNET modeler, ns-2 does not have a built-in function for confidence interval calculation. ns-2 does have a large and active user-base and thus has extensive resources for solving problems.

11.6.5   Model Construction in ns-2

11.6.5.1   Network Components (ns Objects)   Objects are built from a hierarchical C^{++} class structure. As shown in Fig. 11.22, all objects are derived from class NsObject. It consists of two classes: *connectors* and *classifiers*. Connector is an NsObject from which links like queue and delay are derived. Classifiers examine packets and forward them to appropriate destinations. Some of the most frequently used objects are:

1. Nodes: This represents clients, hosts, routers and switches. Two type of nodes which can be defined are unicast and multicast.

2. Classifiers: It determines the outgoing interface object based on source address and packet destination address. Some of the classifiers are address classifier, multicast classifier, multipath classifier, and replicators.

3. Links: These are used for the connection of nodes to form a network topology. A link is defined by its head which becomes its entry point, a reference to main queue element and a queue to process packets dropped at the link. Network activities are traces around simplex links.

4. Agents: these are the transport end-points where packets originate or are destined. Types of agents are TCP and UDP. ns-2 supports wide

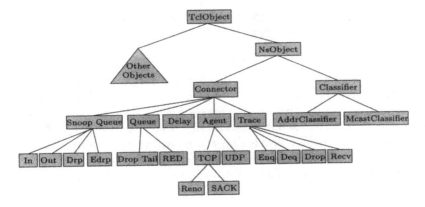

Fig. 11.22 Hierarchy. (Taken from "NS by example" [ChCl05].)

variants of TCP and it gives an option for setting ECN bit specification, congestion control mechanism, and window settings.

5. Application: The major types of applications that ns-2 supports are traffic generators and simulated applications. Some of the TCP-based applications supported by ns-2 are Telnet and FTP.

6. Traffic generators: In cases of a distribution driven simulation automated traffic generation with desired shape and pattern is required. Some traffic generators which ns-2 provides are Poisson, On–Off, CBR, and Pareto On–Off.

11.6.5.2 Event Schedulers Event scheduler is used by network components that simulate packet-handling delay or components that need timers. The network object that issues an event will handle that event later at a scheduled time. Event scheduler is also used to schedule simulated events, such as when to start a Telnet application or when to finish a simulation. ns-2 has two event schedulers namely real-time which is for emulation and non-real-time. Non-real-time scheduler can be implemented either by a list, heap or a calendar.

11.6.5.3 Data Collection and Execution ns-2 uses tracing and monitoring for data collection. Events such as a packet arrival, packet departure or a packet drop from a link/queue are recorded by tracing. Since the tracing module does not collect data for any specific performance metrics, it is only useful for debugging and verification purposes.

Monitoring is a better alternative to tracing where we need to monitor a specific link or node. Several trace objects are created which are then inserted into a network topology at desired places. These trace objects collect different performance metrics. Monitoring objects can also be written in C^{++} (tracing can be written in OTcl only) and inserted into source or sink functions.

After constructing a network model and setting different parameters, the ns-2 model is executed by using run command.

11.6.5.4   Network Animator (NAM)   NAM is an animation tool that is used extensively along with ns-2. It was developed in LBL and is used for viewing network simulation packet traces and real-world packet traces. It supports packet level animation that shows packets flowing through the link, packets being accumulated in the buffer and packets dropping when the buffer is full. It also supports topology layout that can be rearranged to suit user's needs. It has various data inspection tools that help in better understanding of the output. More information about NAM can be found at [Nam].

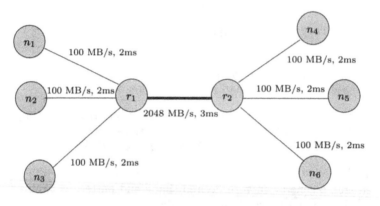

Fig. 11.23   Network connection for a RED configuration.

Example 11.3   In this example we shall study the dynamics of an average queue size in a RED and FIFO queue in an eight node network similar to the example described in Section 11.2. All links are duplex in nature with their speed and delay shown in Fig. 11.23. CBR traffic is chosen (given in Section 11.5.2) for all the source nodes n_1, n_2, and n_3. Nodes n_4, n_5, and n_6 are the sink nodes. The remaining two nodes are the router nodes r_1 and r_2. The window size for TCP application is chosen to be 15. The CBR traffic is generated for each of the sources from 0 to 100 sec. Instead of the deterministic start time for the sources we have chosen the exponential distribution to randomize the start time. The CBR sink is defined for handling traffic from source nodes. For output data collection, the monitoring feature of ns-2 is used. XGraph [Xgr05] is used to display the graph of the buffer occupancy as per Eq. (11.24) vs. time.

The buffer size is taken to be 100 packets and the router parameters are given in Table 11.6. Implementation of nodes in ns-2 can be different from the implementation of similar nodes in **OPNET Modeler**. Therefore, we do not expect identical results. The output shows average buffer occupancy at the router r_1, for the FIFO and the RED case (see Figs. 11.24 and 11.25).

Table 11.6   RED parameters for ns-2 case

Parameters	Values
$\min_{th}$	25
$\max_{th}$	90
Buffer length	100
Weight (w_q)	0.01
Max_{drop}	1/15

This simple example and the similar example using **OPNET** Modeler, provides an overview to reader explaining how to define a problem, how to construct models, how to setup model parameters and how to analyze the results obtained from these simulators. These simulators are very popular and have been extensively used to perform large-scale and more complex problem than described here. The purpose of these examples is to provide an overview of basic capabilities of these simulators.

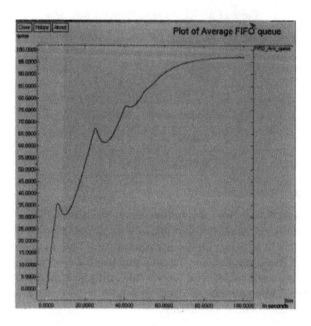

Fig. 11.24   Plot of average FIFO queue.

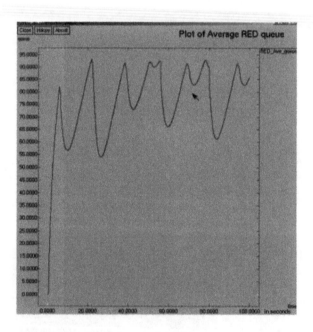

Fig. 11.25   Plot of average RED queue trace path.

12

Performance Analysis Tools

Performance analysis tools have acquired increased importance due to increased complexity of modern systems. It is often the case that system measurements are not available or are very difficult to get. In such cases the development and the solution of a system model is an effective method of performance assessment. Software tools that support performance modeling studies provide one or more of the following solution methods:

- Discrete-event simulation.

- Generation and (steady-state and/or transient) solution of CTMC and DTMC.

- Exact and/or approximate solution of product-form queueing networks.

- Approximate solution of non-product-form queueing networks.

- Hierarchical (multilevel) models combining one or more of the preceding methods.

If we use DES, then the system behavior can be described very accurately, but computation time and resource needs are usually extremely high. In this book queueing network solutions as well as Markov chain analysis methods have been introduced to analyze system models. Queueing networks are very easy to understand and allow a very compact system description. For a limited class of queueing networks (see Chapters 8 and 9), so-called product-form queueing networks, efficient solution algorithms (such as convolution, MVA,

SCAT) are available. But many queueing networks do not fulfill the product-form requirements. In this case approximation methods can be used (see Chapter 10). It is also possible to develop a multilevel model to approximately solve a non-product-form queueing network. If the approximation methods are not accurate enough or cannot be applied to a certain problem, as it can be the case when we have nonexponentially distributed service times or when blocking is allowed in the network, then DES or CTMC is used. As we have seen in earlier chapters, the state space for even simple systems can be huge and grows exponentially with the number of nodes and the number of jobs in the network. Nevertheless, due to increasing computational power and better solution algorithms, CTMCs have acquired greater importance.

For the solution of our models we realize that:

- The application of exact or approximate solution algorithms for queueing networks is very cumbersome, error prone, and time consuming and hence not feasible to carry out by hand calculations.

- The generation and solution of the CTMC/DTMC for even small systems by hand is nearly impossible.

We therefore need tools that can automatically generate the state space of the underlying CTMC and apply exact or approximate solution algorithms and/or that have implemented exact and approximate algorithms for queueing networks. In this chapter we introduce four representative performance modeling and analysis tools: PEPSY, SPNP, MOSEL-2, and SHARPE. As mentioned previously, tools differ not only in their solution algorithms (DES, exact and approximate solution algorithms for queueing networks, generation and transient and steady-state solution of CTMCs) but also in their input/output facilities. There are tools that have a graphical user interface (GUI) and/or a special input language (batch mode or interactive) or both. Some tools are based on special model types such as queueing networks, SPNs, CTMC models, or precedence graphs (and others) or a combination of these model types.

12.1 PEPSY

PEPSY (**p**erformance **e**valuation and **p**rediction **sy**stem) [BoKi92] has been developed at the University of Erlangen-Nürnberg. Using this tool, it is possible to describe and solve PFQNs and NPFQNs. More than 30 solution algorithms are incorporated. It is possible to specify open, closed, and mixed networks where jobs cannot switch to another job class. Closed job classes are described by the number of jobs in each class, while open job classes are described by the arrival rate of jobs at the class. To compute performance measures such as throughput, utilization, or mean response time of a given network, different exact as well as approximation algorithms are provided for

PFQN and NPFQN. The network is described in textual and/or graphical form. The X11-windows version of PEPSY is called XPEPSY [Kirs93]. PEPSY can be used on almost all UNIX machines, whereas XPEPSY needs the X11 windows system. A Windows98, WindowsNT and WindowsXP version Win-PEPSY (see Section 12.1.5) with similar features is also available but it has a restricted set of solution algorithms.

12.1.1 Structure of PEPSY

The Input File The PEPSY input file contains the specification of the system to be modeled. The specification file has a name of the form e_name and is divided into a standard block and possibly additional blocks that contain additional information specific to solution algorithms. The standard block contains all necessary descriptions for most of the solution methods, for example, the number of nodes and job classes, the service time distribution at each node, the routing behavior of the jobs, and a description of the classes. The input file is usually generated using the input program, which is provided by PEPSY. Information that is not asked for by the input program is provided using the addition program.

The type of each node (and its characteristics) needs to be specified: the $-/M/m$-FCFS, $-/G/1$-PS, $-/G/$-IS, $-/G/1$-LCFS and several other node types. For example, two types of multiple server nodes with different service time distributions $-/M/m$-FCFS-ASYM and $-/G/m$-FCFS-ASYM can be specified. The service time distribution is defined by its first and second moment, or by the service rate and the squared coefficient of variation.

For the solution of the specified models, many different algorithms are implemented in PEPSY. These algorithms can be divided into six groups:

1. Convolution algorithm for product-form networks.

2. MVA for product-form networks.

3. Approximation algorithms for product-form networks.

4. Approximation methods for non-product-form networks.

5. Automated generation and steady-state solution of the underlying CTMC.

6. DES.

In addition to these six groups, there are some methods and techniques that do not fit in any of these groups. For example, the bounds method performs a calculation of the upper bounds of the throughput and lower bounds of the average response time.

The Output File   For each file name e_name, an output file with the computed performance measures is generated. It is called xx_name where "name" is the same as in the input file and "xx" is an abbreviation for the used method. The output file consists of a short header with the name of the model, the corresponding input file, and the solution method used to calculate the performance measures. The performance measures of all nodes are separated by classes.

The output file contains the performance measures for each node (throughput, visit ratio, utilization, average response time, average number of jobs, average waiting times, and average queue length) and the performance measures for the entire network (throughput, average response time and average number of jobs).

Control Files   As already mentioned, PEPSY offers more than 30 solution methods to compute performance measures of queueing networks. In general, each method has certain limitations and restrictions. Because it would be impossible for the user to know all these restrictions and limitations, control files are used in PEPSY. Each control file contains the limitations and restrictions for each solution method, namely the network type (open, closed, or mixed), the maximum number of nodes and classes allowed, whether the method can deal with routing probabilities or visit ratios, whether the service rates at $-/M/m$-FCFS nodes have to be the same for different job classes, the range of the coefficients of variation, and the maximum number of servers at a $-/M/m$ node.

12.1.2   Different Programs in PEPSY

The following lists the most important programs of PEPSY. For a more detailed reference see [Kirs94].

input: This program is used for an interactive description of a queueing network.

addition: For some solution algorithms, additional information is needed. This information is provided by invoking the addition program.

selection: Using this program, all possible solution algorithms can be listed that are applicable to the system description (the file that was just created). The output of this program consists of two columns. The left-hand column contains all solution algorithms that can be applied directly, while the right-hand column contains the methods that can only be applied after specifying an additional block (using the addition program).

analysis: After a complete system description, the queueing network desired solution algorithm can be applied by invoking the analysis program. The results are printed on the screen as well as written to a file.

pinf: This program is invoked to obtain information about the solution method utilized.

transform: As already mentioned before, it is possible to specify the network using either routing probabilities or visit ratios. By invoking the transform program, visit ratios can be transformed into routing probabilities.

pdiff: By invoking pdiff, the results of two different solution methods can be compared.

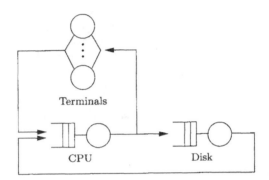

Fig. 12.1   Simple queueing network example.

12.1.3   Example of Using PEPSY

The main steps in working with PEPSY are described herein. Starting with the model, first the model file is created, then the solution method is selected, and the analysis is carried out. The model to be analyzed is shown in Fig. 12.1.

At first the input program is invoked by means the command input. The user will then be asked to enter the type and number of job classes (a closed network with only one job class in our case), the number of nodes (three in our case), the type of each node, and its service rate. After the basic network information is specified, the routing information is entered. In PEPSY we need to consider one special point when talking about routing: open job classes contain a (combined) source/sink-node that is used in PEPSY as a reference node (outside node) for relative performance values. For closed queueing networks the outside node also needs to be specified. Finally, the number of jobs in the network needs to be specified.

Now the file is saved in, say, e_first_example. The corresponding model description, generated by input, is shown in Fig. 12.2. The next step is the selection of the solution method. PEPSY supports the user with a built-in database with the limitations and restrictions of each algorithm and returns a list of applicable solution methods. In our example the result of applying the selection program to the model description is shown in Fig. 12.3.

```
#
# filename  e_simple_example
#

NUMBER NODES: 3
NUMBER CLASSES: 1

NODE SPECIFICATION

    node   |      name        |      type
  ---------+------------------+-------------------
       1   |     cpu          |    -/G/1-PS
       2   |     disk         |    -/M/1-FCFS
       3   |     terminal     |    -/G/0-IS

CLASS SPECIFICATION

    class  | arrival rate   number of jobs
  ---------+-----------------------------------
       1   |     -              10

CLASS SPECIFIC PARAMETERS

CLASS  1

    node   | service_rate  squared_coeff._of_variation
  ---------+------------------------------------------------
    cpu    |  '10                    1
    disk   |  3                      1
    terminal | 0.2                   1

ROUTING PROBABILITIES

  from/to  | outside    cpu        disk        terminal
  ---------+----------------------------------------------------
  outside  | 0.000000  1.000000   0.000000    0.000000
  cpu      | 0.000000  0.000000   0.700000    0.300000
  disk     | 1.000000  0.000000   0.000000    0.000000
  terminal | 1.000000  0.000000   0.000000    0.000000
```

Fig. 12.2   Model description file e_simple_example.

```
simple_example:
    applicable solution method |  need further specification
  ----------------------------------------------------------------
       ammva                   |
       bol_aky                 |
       bounds                  |
       cmva                    |
                               |  hm
       marie                   |
       mmva                    |
       monosum                 |
       multisum                |
       num_app                 |
       num_single              |
       pm_2                    |
       pmapp_1                 |
       priomva2c               |
       priomva2m               |
       recal                   |
                               |  sim2
```

Fig. 12.3   The list of applicable solution methods presented by the selection program.

```
PERFORMANCE_INDICES FOR NET:  simple_example
description of the network is in file 'e_simple_example'
the closed network was solved using the method 'marie'
jobclass 1
marie      |  lambda     e    1/mu    rho    mvz     maa     mwz    mwsl
-----------+------------------------------------------------------------
cpu        |   3.922   1.000  0.100  0.392  0.156   0.613   0.056  0.221
disk       |   2.746   0.700  0.333  0.915  1.276   3.503   0.943  2.588
terminal   |   1.177   0.300  5.000  0.000  5.000   5.883   0.000  0.000

characteristic indices:
marie      |  lambda    mvz     maa
-----------+--------------------------
           |   3.922   2.549  10.000

legend
e    : average number of visits        mu    : service rate
rho  : utilisation                     lambda: mean throughput
mvz  : average response time
maa  : average number of jobs
mwz  : average waiting time
mwsl: average queue-length
```

Fig. 12.4   Performance measures of the network e_simple_example.

This list contains all the methods that can be used to analyze the network e_simple_example. In addition to the standard network description that has been entered in the beginning, some methods need further parameters.

For example, in order to run the sim2 method (DES), some information about the maximum simulation time and accuracy needs to be specified. If the user is not familiar with a particular method, PEPSY assists the user with online help.

Suppose the user selects the method name marie and invokes the solution method by entering the command analysis marie simple_example. The resulting performance measures are shown on the screen and are also written into the file a_c_simple_example. The contents of this file are shown in Fig. 12.4.

12.1.4   Graphical User Interface XPEPSY

To demonstrate how to work with XPEPSY, the same queueing model as in Section 12.1.2 is used. After the program is started, the user can graphically draw the network as shown in Fig. 12.5. Nodes or connections can be created, moved, or deleted simply by mouse clicks. The necessary parameters are specified in the corresponding dialog boxes. Consider, for example, the node data dialog box, shown in Fig. 12.6a that appears when clicking on the service station in the drawing area and that is used to specify node parameters.

When the queueing network is fully specified, XPEPSY switches from the edit mode to the analysis mode and a new set of menus will appear. Now PEPSY assists you in selecting the solution method. Four possibilities are offered:

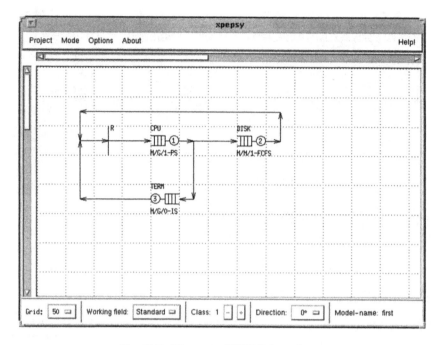

Fig. 12.5 The main XPEPSY window.

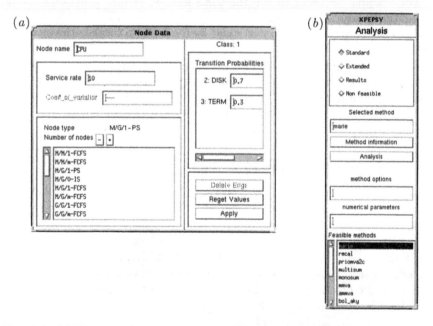

Fig. 12.6 (*a*) The node data dialog box and (*b*) the menu of solution methods corresponding to Fig. 12.5.

1. Select from a limited set of methods, i.e., the ones that are very well suited for the problem.

2. Select from all possible methods.

3. Show only methods that have already been used and for which valid results can be found without resolving the model.

4. Show all methods that cannot be used due to inherent restrictions of the solution method.

Now the solution can be carried out by simply activating the **Analysis** button from the analysis menu, shown in Fig. 12.6*b*, and the results are presented on the screen (see Fig. 12.4).

Some additional programs from PEPSY are integrated into XPEPSY as well. Using the **Method Information** button, the PEPSY internal database can be searched for information about the selected method.

12.1.5 WinPEPSY

Motivation The reason for developing WinPEPSY at the department of computer science of the University of Erlangen where also the queueing network tool PEPSY was developed was that there is a lack of queueing network tools with a modern GUI and facilities for extensive experiments which can be run on Windows PCs. The feedback from more than 50 installations of PEPSY was used when developing and implementing WinPEPSY [BBG04]. From among 40 methods and algorithms of PEPSY the most important six are included in WinPEPSY besides a discrete-event simulation component. In addition a CTMC solver is implemented. The user interface with a network editor (graphical and textual), a scenario editor and a graphical and tabular representation of the results are completely redesigned.

Description and Architecture The Tool WinPEPSY can analyze queueing networks with one or more job-classes. Routing of jobs is described by visit ratios or routing probabilities. The arrival process of open networks is given by its mean arrival rate and the squared coefficient of variation of the interarrival time. The same queueing network types (open, closed and mixed) and the same node types as in PEPSY can be handled. The service time distribution of a node is defined by its mean service time and squared coefficient of variation.

Analysis of the network can be done by

- MVA (Section 8.2) for closed product-form networks

- Open product-form network methods (Sections 7.3.4, 7.3.6.2)

- Marie's method (Section 10.1.4.2) for closed non-product-form networks

- Decomposition (Section 10.1.3) for open non-product-form networks

- Steady-state CTMC solution (Chapter 3)

- Discrete-event simulation (Chapter 11)

The following results (steady state) are calculated:

- Mean waiting time

- Mean response time

- Mean number of jobs

- Mean number of jobs waiting

- Utilization

- Throughput

The user of WinPEPSY enters the network details with the graphical network editor and chooses an appropriate analysis method. Then he can enter up to two parameters, which will be changed for each analysis, or start the analysis run directly. After the calculation is finished, the performance results are presented. Table views as well as graphical representations are possible.

Model of a Web Server with content Replication   In this section we present a model of a Web server with content replication (Fig. 12.7) which is described in detail in [MeAl98]. In contrast to the model in the reference we define all nodes with different service rates for each class as M/G/1-PS nodes. The node Incoming Link remains unchanged.

The system is modelled as an open network with several classes. Each class describes the service demands for a different type of HTTP request. The network model includes the incoming link, the router, the LAN and two Web servers (Fig. 12.8). The clients and the internet are described implicitly by different arrival rates of HTTP requests for each class.

The WinPEPSY method for open product-form networks (OPFN) is used to calculate the performance measures. A three-dimensional bar graph is selected to show the obtained results for the mean response time (Fig. 12.9). Note that the class 4 produces the largest response times at the OUTgoing link and the server discs. This is because jobs of this class request not only HTML pages but also large files, in contrast to the other classes. The jobs of class 5 lead to the execution of CGI scripts on the server CPUs. This corresponds to the response times of the server CPUs as can be seen from the diagram.

12.2 SPNP

The stochastic Petri net package (SPNP) developed by Ciardo, Muppala, and Trivedi [CTM89, HTT00], is a versatile modeling package for performance,

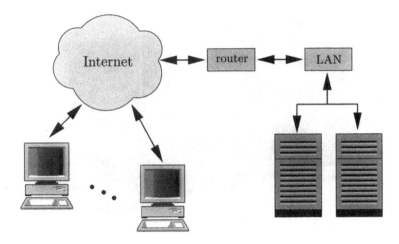

Fig. 12.7 Web server with content replication.

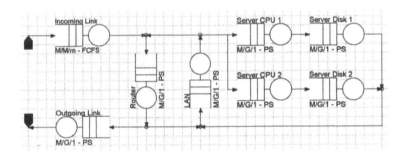

Fig. 12.8 WinPEPSY model of the Web server with content replication.

dependability, and performability analysis of complex systems. Models developed based on the theory of stochastic Petri nets are solved by efficient and numerically stable algorithms or by simulation. Steady-state, transient, cumulative transient, time-averaged, and up-to-absorption measures can be computed. Parametric sensitivity analysis of these measures is efficiently carried out. Some degree of logical analysis capabilities is also available in the form of assertion checking and the number and types of markings in the reachability graph. Advanced constructs available — such as marking-dependent arc multiplicities, guards, arrays of places and transitions, and subnets — reduce modeling complexity and enhance the power of expressiveness of the package. The most powerful feature is the capability to assign reward rates at the net level and subsequently compute the desired measures of the system being modeled. The model description language is CSPL, a C-like language, although no previous knowledge of the C language is necessary to use SPNP.

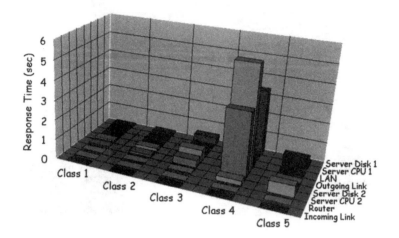

Fig. 12.9   Response times of the WinPEPSY model of the Web server.

In the current version (V6.0), the user can specify non-Markovian SPNs and Fluid Stochastic Petri Nets (FSPNs). Such SPNs are solved using discrete-event simulation rather than by analytic-numeric methods. Several types of simulation methods are available: *standard discrete-event simulation with independent replications or batch means, importance splitting techniques (splitting and Restart), importance sampling, regenerative simulation without or with importance sampling,* among many others.

12.2.1   SPNP Features

The input language of SPNP is CSPL (C-based Stochastic Petri net Language). A CSPL file is compiled using the C compiler and then linked with the precompiled files that are provided with SPNP. The full power of the C programming language can be used to increase the flexibility of the net description. An important feature of CSPL is that it is a superset of C. Thus, a CSPL user can exploit C language constructs to represent a large class of SPNs within a single CSPL file. Most applications will only require a limited knowledge of the C syntax, as predefined functions are available to define SPNP objects.

Another important characteristic of SPNP is the provision of a function to input values at run time, before reading the specification of the SPN. These input values can be used in SPNP to modify the values of scalar parameters or the structure of the SPN itself. Arrays of places and transitions and subnets are two features of SPNP that are extremely useful in specifying large structured nets. A single CSPL file is sufficient to describe any legal SPN, since the user of SPNP can input at run time the number of places and transitions, the arcs among them, and any other required parameters.

The SPNP package allows the user to perform *steady-state, transient, cumulative transient*, and *sensitivity analysis* of SPNs. Steady-state analysis is often adequate to study the performance of a system, but time-dependent behavior (transient analysis) is sometimes of greater interest: instantaneous availability, interval availability, and reliability (for a fault-tolerant system); response time distribution of a program (for performance evaluation of software); computation availability (for a degradable system) are some examples. Sensitivity analysis is useful to estimate how the output measures are affected by variations in the value of input parameters, allowing the detection of system performance bottlenecks or aiding in design optimization.

Sophisticated steady-state and transient solvers are available in SPNP. In addition, the user is not limited to a predefined set of measures: Detailed expressions reflecting exactly the measures sought can be easily specified. The measures are defined in terms of reward rates associated with the markings of the SPN. The numerical solution methods provided in the package address the *stiffness* problems often encountered in reliability and performance models.

A number of important Petri net constructs such as marking dependency, variable cardinality arc, guards, arrays of places and transitions, subnets, and assertions facilitate the construction and debugging of models for complex systems. A detailed description of SPNP and the input language CSPL can be found in [Triv99]. SPNP has been installed at more than 250 sites and has been used to solve many practical problems at Digital Equipment Corporation, Hewlett Packard, EMC, Motorola, GE, IBM, Boeing and other corporations.

New features have been added to include the capability of discrete-event simulation of non-Markovian nets and fluid stochastic Petri nets [CNT99].

```
/* begin of CSPL file */   |
                           |
parameters(){              |   ac_init(){
    . . .                  |       . . .
}                          |   }
                           |
net(){                     |   ac_reach(){
    . . .                  |       . . .
}                          |   }
                           |
assert(){                  |   ac_final(){
    . . .                  |       . . .
}                          |   }
                           |
                           |   /* end of CSPL file */
```

Fig. 12.10 Basic structure of a CSPL input file.

12.2.2 The CSPL Language

Modeling with SPNP implies that an input file describing the system structure and behavior must be prepared. Alternatively, the user may decide to prepare such a file himself. The language designed to do so is named **CSPL**, a superset of the C language [KeRi78]. What distinguishes **CSPL** from C is a set of predefined functions specially developed for the description of SPN entities. Any legal C construct can be used anywhere in the **CSPL** file. All the C library functions, such as `fprintf`, `fscanf`, `log`, `exp`, etc., are available and perform as expected. The only restriction to this generic rule is that the file should not have a `main` function. In spite of being a programming language, **CSPL** enables the user to describe SPN models very easily. There is no need to be a programmer to fully exploit all the built-in features of **SPNP**. Just a basic knowledge of C is sufficient to describe SPNs effectively; although, for experienced programmers, **CSPL** brings the full power and generality of the C language.

Figure 12.10 shows the basic structure of a **CSPL** input file with six different segments.

```
net(){                              |   /* output arcs,
    ...                             |       multiple output arcs */
    /* places */                    |   ...
    place("p1");                     |   oarc("t2","p1");
    place("p2");                     |   oarc("t1","p2");
    ...                             |   moarc("t3","p6",2);
    place("pn");                     |   ...
    /* initial markings */           |   /* inhibitor arcs,
    ...                             |       multiple inhibitor arcs */
    init("p1",n);                    |   ...
    ...                             |   harc("t4","p5");
    init("pn",2);                    |   mharc("t5","p3");
                                    |   ...
    /* transitions */                |   /* priorities */
    ...                             |   ...
    trans("t1");                     |   priority("t3",10);
    trans("t2");                     |   priority("t6",5);
    ...                             |   ...
    trans("tm");                     |   /* firing probabilities
    ...                             |       of immediate transitions */
                                    |   ...
    /* input arcs,                   |   probval("t7",0.4);
        multiple input arcs */       |   probval("t8",0.6);
    ...                             |   ...
    iarc("t1","p1");                 |   /* firing rates of timed transitions */
    iarc("t3","p2");                 |   ...
    miarc("t5","p6",3);              |   rateval("t1",2.0);
    ...                             |   rateval("t2",5.0);
                                    |   ...
                                    |   }
```

Fig. 12.11 The CSPL net segment.

12.2.2.1 Parameters Segment The function `parameters` allows the user to customize the package. Several parameters establishing a specific behavior can be selected. The function `iopt` (`fopt`) enables the user to set `option` to have the integer (double-precision floating point) `value`. Any of the available options can be selected and modified. For example

```
parameters(){
  ...
  iopt(IOP_METHOD,VAL_TSUNIF);
  iopt(IOP_PR_RGRAPH,VAL_YES);
  ...
}
```

specifies that the solution method to be used is transient solution using uniformization and that the reachability graph is to be printed. The function **input** permits the input of parameter values at run time.

12.2.2.2 Net Segment The function **net** allows the user to define the structure and parameters of an SPN model. Some examples for the built-in functions that can be used inside the **net** segment are illustrated in Fig. 12.11.

A set of additional functions which serve different purposes are available in CSPL: **guard** functions allow the user to define the enabling condition of a transition as a function of tokens in various places. Moreover, there exist functions that can be used to specify arrays of places and transitions and functions that facilitate a sensitivity analysis of a CSPL system description.

12.2.2.3 Assert Segment The **assert** function allows the evaluation of a logical condition on a marking of the SPN. For example, the **assert** definition

```
assert(){
  if(mark("p2")+mark("p3") != 4 || enabled("t11") && enabled("t7"))
    return(RES_ERROR);
  else
    return(RES_NOERR);
}
```

will stop the execution in a marking where the sum of the number of tokens in places p2 and p3 is not 4, or where t11 and t7 are both enabled.

12.2.2.4 Ac_init and Ac_reach Segment The function **ac_init** is called just before starting the reachability graph construction. It can be used to output data about the SPN in the ". out" file. This is especially useful when the number of places or transitions is defined at run time (otherwise it is merely a summary of the CSPL file). The function **ac_reach** is called after the reachability graph construction is completed. It can be used to output data about the reachability graph in the ". out" file.

12.2.2.5 Discrete-Event Simulation in SPNP The firing distributions in case of simulation can be other than the exponential distribution. Three functions are available to define one distribution with different kind of parameterizations: marking independent value (with xxxval()), marking dependent (with xxxdep()), and function dependent (with xxxfun()). So for Weibull the function definitions beomes: weibval(), weibdep(), weibfun(). Some of the distributions currently supported by SPNP are exponential, constant, uniform, geometric, Weibull, truncated normal, lognormal, Erlang, gamma, beta, truncated cauchy, Binomial, Poisson, Pareto, Hyperexponential (2-stage), Hypoexponential (2 or 3 stage).

The following simulation related information can be specified in a CSPL file, using calls to iopt() or fopt().

IOP_SIMULATION specifies if the simulative solution will be used. Default value is VAL_NO.

IOP_SIM_RUNMETHOD specifies the simulation method. Value VAL_REPL is specified if independent replications are used and value VAL_BATCH if the method of batch means is used.

FOP_SIM_LENGTH is the length of each simulation run, in simulated time, to be specified with a call to fopt. In the case where batches are used, it represents the length of each batch. If no value is specified, it is possible to use calls to at_time or cum_time in function ac_final instead.

FOP_SIM_CONFIDENCE specifies the confidence level to be used when computing the confidence intervals. Possible values are 90%, 95% and 99% with default being 99%.

IOP_SIM_RUNS specifies the maximum number of simulation runs to be performed, to obtain meaningful statistics. FOP_SIM_ERROR specifies the target half-width of the confidence interval, relative to the point estimate.

IOP_SIM_SEED allows the user to change the seed of the pseudo random number generator.

IOP_SIM_STD_REPORT specifies that the results will be displayed in the .out file and the call of pr_message(char *msg) in ac_final() allows to print a message in the .out file.

Importance Splitting Techniques are specified by the option IOP_SIM_RUNMETHOD. It is set to VAL_RESTART if we use RESTART, estimating $P(\#p \geq x \in [0, T])$, and to VAL_SPLIT if we use splitting, with estimate $P(\tau_{F,x} < \min(T, \tau_0))$.

Importance Sampling is invoked by setting IOP_SIM_RUNMETHOD to VAL_IS. Presently the sampling distribution can be changed only for the exponential, uniform, Weibull, Erlang, truncated Cauchy, Pareto and Hyperexponential (2-stage) distributions.

Regenerative Simulation is used to estimate steady-state measures. It is called by setting IOP_SIM_RUNMETHOD to VAL_REG. The number of used regenerative cycles may be specified by IOP_SIM_RUNS or the desired precision by FOP_SIM_ERROR.

Regenerative Simulation with Importance Sampling is also used to estimate steady-state measures. It is called by setting IOP_SIM_RUNMETHOD to VAL_ISREG. It combines regenerative simulation with importance sampling to speed up the simulation.

12.2.2.6 Ac_final Segment The function ac_final is called after the solution of the CTMC or the simulation runs are completed, to carry out the computation and printing of user-requested outputs. For example, the function

```
ac_final(){
    pr_std_average();
    pr_std_average_der();
    pr_message(goodbye);
}
```

writes in the ".out" file for each place the probability that it is not empty and its average number of tokens, and for each transition the probability that it is enabled and its average throughput, respectively. The derivatives of all the preceding standard measures with respect to a set of previously chosen parameters are computed and printed. In the end it writes the message goodbye. Before applying the ac_final function, additional functions often have to be provided. This can be done, for instance, using the construct reward_type as it is shown in the following example:

```
reward_type ep1() { return(mark("p1")); } reward_type ep3()
{return(mark("p3")); } reward_type ep7() { return(mark("p7")); }
... ac_final(){ x = expected(ep1)*expected(ep7)+expected(ep3)*1.2;
  printf("%f",x);
}
```

12.2.2.7 Example Using the example of Fig. 12.12, we show that the application of SPNP to analyze systems that can be modeled by an SPN is a straightforward manner. The inhibitor arcs avoid a capacity overflow in places p1 and p2. The CSPL file for this SPN is shown in Fig. 12.13. A short description of this specification is given in Table 12.1.

12.2.3 iSPN

Input to SPNP is specified using CSPL, but iSPN (integrated environment for modeling using stochastic Petri nets) removes this burden from the user by providing an interface for graphical representation of the model. The use

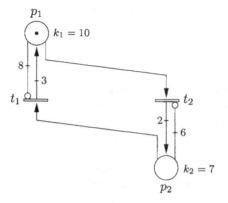

Fig. 12.12   Petri net example.

```
 1  #include "user.h"                          32  ac_init(){
 2                                             33    pr_net_info();
 3  parameter(){                               34  }
 4    iopt(IOP_PR_RGRAPH,VAL_YES);             35
 5    iopt(IOP_PR_MC,VAL_YES);                 36  ac_reach(){
 6    iopt(IOP_PR_PROB,VAL_YES);               37    pr_rg_info();
 7  }                                          38  }
 8                                             39
 9  net(){                                     40  reward_type s1(){ return(mark("p1")); }
10    place("p1");                             41  reward_type s2(){ return(mark("p2")); }
11    init("p1",1);                            42  reward_type p1_empty(){ if(mark("p1")==0)
12    place("p2");                             43                           return(1.0);
13    trans("t1");                             44                         else
14    trans("t2");                             45                           return(0.0);}
15    rateval("t1",1.0);                       46  reward_type p2_empty(){ if(mark("p2")==0)
16    rateval("t2",1.0);                       47                           return(1.0);
17    iarc("t1","p2");                         48                         else
18    iarc("t2","p1");                         49                           return(0.0);}
19    moarc("t1","p1",3);                      50
20    moarc("t2","p2",2);                      51  ac_final(){
21    mharc("t1","p1",8);                      52    pr_mc_info();
22    mharc("t2","p2",6);                      53    pr_expected("Mean number of tokens in p1",s1);
23  }                                          54    pr_expected("Mean number of tokens in p2",s2);
24                                             55    pr_expected("Prob. that p1 is empty",p1_empty);
25  assert(){                                  56    pr_expected("Prob. that p2 is empty",p2_empty);
26    if ((mark("p1")>10)||(mark("p2")>7))     57    pr_std_average();
27      return(RES_ERROR);                     58  }
28    else
29      return(RES_NOERR);
30  }
31
```

Fig. 12.13   CSPL file for the SPN in Fig. 12.12.

of Tcl/Tk [Welc95] in designing iSPN makes this application portable to all platforms [HWFT97].

The major components of the iSPN interface (see Fig. 12.14) are a Petri net editor, which allows graphical input of the stochastic Petri net, and an extensive collection of visualization routines to display results of SPNP and aid in debugging. Each module in iSPN is briefly described in the following:

Table 12.1 CSPL code description

Line	Description
4 – 6	The reachability graph, the CTMC, and the state probabilities of the CTMC are to be printed.
10 – 22	The places, the initial marking of the places, the transitions, their firing rates, and the input, output, and inhibitor arcs are defined.
26 – 29	Check whether a capacity overflow in either place p1 or place p2 has occurred.
33 – 37	Data about the SPN and the reachability graph of the SPN are to be printed.
40	The number of tokens in place p1 in each marking is the reward rate s1 in that marking.
41	The number of tokens in place p2 in each marking is the reward rate s2 in that marking.
42 – 49	Self-explanatory.
52	Data about the CTMC and its solution that are to be printed.
53 – 54	The mean number of tokens in the places p1 and p2 are to be computed and printed.
55 – 57	The probabilities that the places p1 and p2 are empty are to be computed and printed.

Input Data: iSPN provides a higher level input format to CSPL, which provides great flexibility to users. iSPN is capable of executing SPNP with two different file formats: Files created directly using the CSPL and files created using iSPN's Petri net editor.

The Petri net editor (see Fig. 12.15), the software module of iSPN that allows users to graphically design the input models, introduces another way of programming SPNP: The user can draw the SPN model and establish all the necessary additional functions (i.e., rewards rates, guards, etc.) through a common environment. The Petri net editor provides several characteristics normally available only in sophisticated two-dimensional graphical editors and a lot of features designed specifically for the SPNP environment.

iSPN also provides a textural interface, which is necessary if we wish to accommodate several categories of users. Beginners may feel more comfortable using the Petri net editor, whereas experienced SPNP users may wish to input their models using CSPL directly. Even if the textual input is the option of choice, many facilities are offered through the integrated environment. In both cases, the "SPNP control" module provides everything a user needs to run and control the execution of SPNP without having to switch back and forth among different environments.

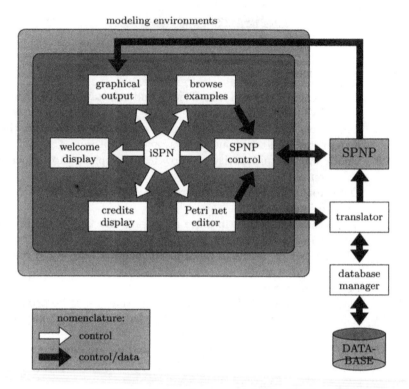

Fig. 12.14 Main software modules for iSPN.

Output Data: The main goal of most GUIs is only to facilitate the creation of input data for its underlying package. Usually, the return communication between the software output and the GUI is neglected. One of the advantages of iSPN is the incorporation of displaying SPNP results in the GUI application. iSPN's own graphing capability allows the results of experiments to be graphically displayed in the same environment (see Fig. 12.16). Different combinations of input data may be compared against each other on one plot or viewed simultaneously. The graphical output format is created in such a way that it may be viewed by other visualization packages such as gnuplot of xvgr.

12.3   MOSEL-2

12.3.1   Introduction

In this section the modeling language MOSEL-2 (MOdeling, Specification and Evaluation Language, 2nd revision) is introduced using examples from queueing networks with finite capacity, retrial systems, and unreliable multiproces-

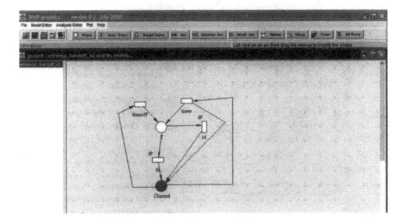

Fig. 12.15   The Petri net editor.

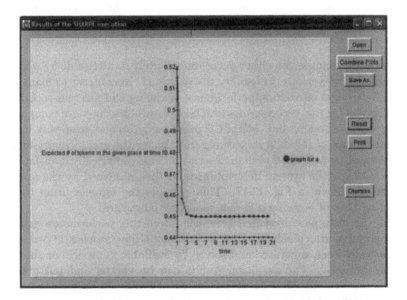

Fig. 12.16   The iSPN output page.

sor systems. MOSEL-2 is an substantially enhanced and redesigned version of its predecessor MOSEL which is described in detail in [BBH01]. The core of the MOSEL-2 formal description technique (FDT) consists of a set of language constructs to specify the possible states and state transitions of the model. Conditions under which a transition is en- or disabled can also be specified.

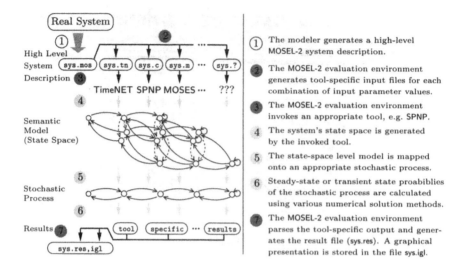

The following descriptions appear to the right of the diagram:

1. The modeler generates a high-level MOSEL-2 system description.

2. The MOSEL-2 evaluation environment generates tool-specific input files for each combination of input parameter values.

3. The MOSEL-2 evaluation environment invokes an appropriate tool, e.g. SPNP.

4. The system's state space is generated by the invoked tool.

5. The state-space level model is mapped onto an appropriate stochastic process.

6. Steady-state or transient state proabiblies of the stochastic process are calculated using various numerical solution methods.

7. The MOSEL-2 evaluation environment parses the tool-specific output and generates the result file (sys.res). A graphical presentation is stored in the file sys.igl.

Fig. 12.17 The modeling and analysis process in the MOSEL-2 evaluation environment.

Possible state changes can either occur immediately or after a delay which can be sampled from various probability distributions. In contrast to many specification languages of existing performance modeling and analysis tools, which often tend to be too verbose, most MOSEL-2 specifications are compact but anyhow easy to understand. MOSEL-2 provides constructs to specify interesting performance or reliability measures and also for the graphical presentation of the results. A MOSEL-2 system description is automatically translated into various tool specific system descriptions and then analyzed by the appropriate tools as shown in Fig. 12.17. This exempts the modeler from the time consuming task of learning different modeling languages.

In all the examples presented here, the interesting performance and reliability measures are calculated after a continuous-time stochastic process has automatically been generated and solved by available numerical or simulative solution methods. The graphical results can be viewed and postprocessed using the utility IGL (Intermediate Graphical Language) which is provided with the MOSEL-2 evaluation environment (see [BBH01]).

A variety of (freeware and commercial) tools which are based on one of the high-level modeling formalisms: SPNP [Triv99] (see Section 12.2), GreatSPN [CFGR95] and WebSPN [BPST98] are based on GSPNs or SRNs. Other packages, for example PEPSY [Kirs94] (see Section 12.1) and QNAP2 [VePo85] favor the queueing theoretical approach, whereas SHARPE [STP96, HSZT00] (see Section 12.4) and MOSES [BGMT98] use a mixture of methods for the generation and solution of the underlying stochastic models.

These packages usually have their own textual or graphical specification language which depends largely on the underlying modeling formalism. The different syntax of the tool specific modeling languages implies that once a tool is chosen it will be difficult to switch to another one as the model needs to be rewritten using a different syntax. On the other hand, the solution of the underlying stochastic process is performed in most tools by the same algorithms from numerical mathematics.

Starting from these observations the development of MOSEL-2 [BoHe95, GrBo96b, BBH01, BBBZ03] is based on the following idea: Instead of creating another tool with all the components needed for system description, state-space generation, stochastic process derivation, and numerical solution, we focus on the formal system description part and exploit the power of various exisiting and well-tested packages the subsequent stages. The MOSEL-2 evaluation environment is equipped with a set of translators which transform the model specification into various tool specific system descriptions. Figure 12.17 gives an overview of performance and reliability modeling and analysis using the MOSEL-2 evaluation environment.

At present the MOSEL-2 evaluation environment is able to translate models into system descriptions for the GSPN-based packages SPNP and TimeNET [ZFGH00]. MOSEL-2 models containing nonexponential transitions can now be analyzed by the discrete-event simulation components of either SPNP or TimeNET. TimeNET also offers numerical solution methods for a restricted class of systems with nonexponentially distributed transitions.

12.3.2   The MOSEL-2 Formal Description Technique

In the following the structure of a MOSEL-2 system description and the apploication of some important language constructs are described and exemplified via a simple queueing network model.

12.3.2.1   The Model Structure   A MOSEL-2 model consists of up to six parts. Each part is dedicated to the description of a different aspect of the performance and reliability model by means of appropriate language constructs [BBH01]:

1. Parameter declaration part

2. Component definition part

3. Function and condition part

4. Transition definition part

5. Result part (optional)

6. Picture part (optional)

Parameter Declaration Part: In this part constants and variable system parameters can be declared. The ENUM construct allows the definition of a set of
constants. The possibility of declaring a list of values in the PARAMETER construct facilitates experimenting with the model. The MOSEL-environment
automatically analyzes the specified system for any combination of system
parameters.

```
CONST pi = 3.14159265358979;
PARAMETER lambda = 0.25, 0.30, 0.35;
ENUM cpu_states = {idle, user, kernel, driver};
```

Component Definition Part: This part consists of the NODE part and the
ASSERT part.

NODE Part: This part specifies the components (nodes) of the model. Each
node can hold an integer-valued number of jobs (or tokens) up to its capacity.
For every component in the model a name and the capacity is specified. As
an example we consider

```
NODE N1[k] = k;        /* range: 0..k */
NODE cpu[cpu_states]; /* range: 0..3 */
```

ASSERT Part: This construct can be used to specify the prohibited system
states. For a closed system, we can assure that the total number of jobs
summed over all the nodes is a constant. This can be expressed as

```
ASSERT N1 + N2 + N3 == K;
```

Function and Condition Part: In this optional part of a MOSEL-2 description,
complex enabling conditions for the rules defined in the subsequent transition
definition part and functions which can be used to model arbitraty properties
of the target system can be specified. An illustrative example for the use-
fulness of the MOSEL-2 functions and conditions is given in the study of a
UMTS cell in Section 13.2.6.

Transition Definition Part: In this part the state transitions are specified. The
transitions are called rules in MOSEL. Each rule consists of a global part and,
optionally, a local part.

```
FROM node_1 TO node_2 RATE mue_1;

IF (cpu_control == idle) FROM node_1 RATE mue_1
   THEN { TO node_2 WEIGHT p_12 ;
          TO node_3 WEIGHT p_13 ; }
```

The first rule consists of a local part only and specifies a transition from
node_1 to node_2 with rate mue_1. The keyword RATE indicates that the
transition is exponentially timed. The second example shows how rules containing a local part can be used to model probabilistic branching. The timed

Fig. 12.18 Open tandem queueing network.

transition with rate `mue_1` emanating from the component `node_1` is split up into two branches which lead to the components `node_2` and `node_3`. The branches have weights `p_12` and `p_13` which of course have to be defined in the parameter declaration part of the model. Each branch is probabilistically selected according to its weight and the selection is assumed to happen immediately. The condition `IF (cpu_control == idle)` in the global rule part is used to model synchronization. Jobs can leave `node_1` only if node `cpu_control` is in state `idle`. Otherwise the transition is blocked.

Result Part: In this section the modeller defines the performance and reliability measures. In the following example the mean value `MEAN`, utilization `UTIL` and the distribution of jobs `DIST` for `node_1` are specified as measures of interest:

```
RESULT N_1 = MEAN (node_1);
RESULT rho_1 = UTIL (node_1);
RESULT DIST node_1;
```

If the probability that a component of the system is in a certain state is required the keyword `PROB` can be used:

```
RESULT p_working = PROB (working > 0);
RESULT rho_cpu_kernel = PROB (cpu == kernel
                              AND cpu_state == up);
```

Picture Part: In this optional part the user specifies the quantities to be plotted. With the following construct the mean queue length defined in the Result part is plotted as a function of `lambda` with the values of `lambda` specified in the declaration part:

```
PICTURE "Mean Queue Length"
    PARAMETER lambda
    CURVE mean_queue_length
```

12.3.2.2 A Tandem Network As a first example for a complete model specification in MOSEL-2, we consider an open tandem queueing network with two M/M/1-FCFS nodes:

and get the following simple MOSEL-2 model:

```
 1 // Tandem network
 2
 3 // Parameter declaration part
 4
 5 PARAMETER K := 1, 2, 3, 4, 5, 6, 7, 8, 9, 10;
 6 CONST lambda := 0.25;
 7 CONST mue1 := 0.28;
 8 CONST mue2 := 0.22;
 9
10 // System component part
11
12 NODE N1[K] = 0;
13 NODE N2[K] = 0;
14 NODE num[K];
15
16 // Transition part
17
18 FROM EXTERN TO N1, num  RATE lambda;
19 FROM N1 TO  N2  RATE mue1;
20 FROM N2, num TO EXTERN RATE mue2;
21
22 // Result part
23
24 PRINT rho1 = UTIL (N1);
25 PRINT rho2 = UTIL (N2);
26 PRINT throughput = rho2 * mue2;
27 PRINT WIP = MEAN (num);
28
29 // Picture part
30
31 PICTURE "utilization"
32 PARAMETER K
33 CURVE rho1
34 CURVE rho2;
```

In the parameter declaration part the constant arrival rate lambda as well as the two rates mue1 and mue2 for the servers are defined. The system is analyzed assuming that the maximum overall number of jobs in the System K varies from 1 to 10. This implies that during the performance analysis the MOSEL-2 evaluation environment automatically generates a sequence of ten high-level system descriptions (e.g., CSPL-files) and as well invokes the appropriate tool (i.e., SPNP) for each variant. The results of the experiment (the set of analysis runs performed by the tool) are collected in a *single* file by the MOSEL-2 evaluation environment. The same applies to the graphical presentation of the results.

In the component definition part the two nodes N1, N2 representing the queues and the auxiliary node num are defined. Each node is capable of holding up to K jobs. The num-node is needed for keeping track of the overall number of jobs in the system which is important to derive the WIP (Work In Progress) measure as specified in the result part. In the transition part the possible transitions of jobs between the nodes are defined using the rates and values specified in the parameter declaration part. The special node EXTERN is used to model arrivals and departures of jobs from and to the environment. In the result part the utilizations (probability that the queue is not empty) of the

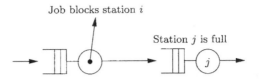

Fig. 12.19 Tandem network with blocking after service.

nodes N1 (rho1) and N2 (rho2) are defined using the keyword UTIL. Moreover, the throughput of the second server is specified as the product of rho2 and the service rate mue2. Using the keyword MEAN the mean total number of jobs in the system (WIP) is specified. The keyword PRINT at the beginning of each line of the result part indicates, that all specified results are going to be saved in MOSEL-2 result file. Intermediate results, which often are used to compute complex measures and which need not be stored in the result file, can be specified by using the keyword RESULT at the beginning of the line.

The picture part contains the definition of a plot showing the utilizations rho1 and rho2 as a function of the varying system parameter K.

12.3.3 Tandem Network with Blocking after Service

We consider an open tandem queueing network with blocking after service (Fig. 12.19) where the second station of type M/M/1/N-FCFS has a finite capacity queue. In order to express the blocking behavior adequately in the MOSEL-2 model we append an additional node BLOCK to the node N1, which models the first station in the network. Node N2 specifies the finite capacity queue of the second station. For technical reasons we introduce an auxiliary node num to the model, which can be used to obtain global performance measures and assures the finiteness of the resulting system state space. The complete MOSEL-2 model is listed below:

```
 1 // Blocking after service              21 PRINT throughput = rho2*mue2;
 2                                         22 PRINT blockpr = PROB(BLOCK==1);
 3 CONST K := 20;                          23 PRINT WIP = MEAN (num);
 4 PARAMETER capacity := 1,2,3,5,7,10;     24
 5 CONST lambda := 1.0;                    25 PICTURE "utilizations"
 6 CONST mue1 := 1.2;                      26 PARAMETER capacity
 7 CONST mue2 := 1.4;                      27 CURVE rho1
 8                                         28 CURVE rho2;
 9 NODE N1[K] = 0;                         29
10 NODE N2[capacity] = 0;                  30 PICTURE "throughput"
11 NODE BLOCK[1];                          31 PARAMETER capacity
12 NODE num[K];                            32 CURVE throughput;
13                                         33
14 FROM EXTERN TO N1, num  RATE lambda;    34 PICTURE "blockprob"
15 FROM N1 TO BLOCK RATE mue1;             35 PARAMETER capacity
16 IF (N2 < capacity) FROM BLOCK TO N2;    36 CURVE blockpr;
17 FROM N2, num TO EXTERN RATE mue2;       37
18                                         38 PICTURE "WIP"
19 PRINT rho1 = PROB(N1>0 OR BLOCK == 1);  39 PARAMETER capacity
20 PRINT rho2 = PROB(N2 > 0);              40 CURVE WIP;
```

In the parameter declaration part the maximum number of jobs in the system K, the `capacity` of node N2, the arrival rate and the service rates are specified. In the component definition part four nodes are declared as explained above. Lines 14–17 of the MOSEL-2-model contain the rules which describe the flow of jobs through the queueing network. It is interesting to note that the explicit declaration of the condition (IF (N2 < capacity)) is unnecessary, since MOSEL-2 automatically ensures that a node cannot be populated by more jobs than specified by its capacity. In the result part five performance measures are specified. A graphical presentation of the performance measures depending on the variable parameter `capacity` is specified in the picture part.

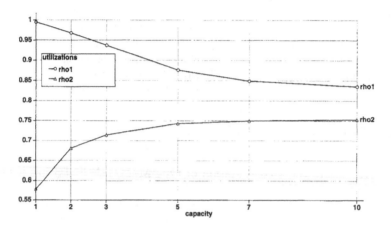

Fig. 12.20 Server utilizations in the tandem network with blocking after service.

Analysis of the Blocking After Service Model Suppose that we have saved our MOSEL-2-model in a file named `blocking_after.msl`. We are now ready to invoke the MOSEL-2 evaluation environment on the command line via

```
>> mosel2 -cs blocking_after.msl
```

The option-c indicates that we requested that our MOSEL-2 specification should be translated into a set of CSPL-files (C based Stochastic Petri net Language), which serve as input for the tool SPNP. Because we also used the s option (s ≡ "start appropriate tool automatically") the MOSEL-2 evaluation environment will now perform the rest of the model analysis and result-postprocessing steps automatically using the SPNP package for state-space generation, derivation of the underlying CTMC and numerical solution (cf. Fig. 12.17).

Upon completion of the analysis, the MOSEL-2 evaluation environment creates two files named `blocking_after.res` and `blocking_after.igl` which contain the requested performance measures of the system in numerical and

graphical form. Figure 12.21 shows a compressed form of the result file
`blocking_after.res`.

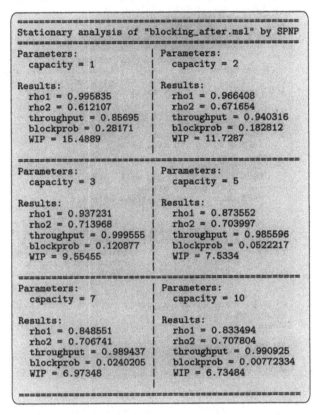

```
============================================================
Stationary analysis of "blocking_after.msl" by SPNP
============================================================
Parameters:              | Parameters:
  capacity = 1           |   capacity = 2
                         |
Results:                 | Results:
  rho1 = 0.995835        |   rho1 = 0.966408
  rho2 = 0.612107        |   rho2 = 0.671654
  throughput = 0.85695   |   throughput = 0.940316
  blockprob = 0.28171    |   blockprob = 0.182812
  WIP = 15.4889          |   WIP = 11.7287
                         |
============================================================
Parameters:              | Parameters:
  capacity = 3           |   capacity = 5
                         |
Results:                 | Results:
  rho1 = 0.937231        |   rho1 = 0.873552
  rho2 = 0.713968        |   rho2 = 0.703997
  throughput = 0.999555  |   throughput = 0.985596
  blockprob = 0.120877   |   blockprob = 0.0522217
  WIP = 9.55455          |   WIP = 7.5334
                         |
============================================================
Parameters:              | Parameters:
  capacity = 7           |   capacity = 10
                         |
Results:                 | Results:
  rho1 = 0.848551        |   rho1 = 0.833494
  rho2 = 0.706741        |   rho2 = 0.707804
  throughput = 0.989437  |   throughput = 0.990925
  blockprob = 0.0240205  |   blockprob = 0.00772334
  WIP = 6.97348          |   WIP = 6.73484
                         |
============================================================
```

Fig. 12.21 `blocking_after.res` result file.

As expected the blocking probability, the utilization of station 1 and the
work in progress (WIP) decrease if the capacity of station 2 increases while
the utilization of node 2 and the throughput increase. The graphical result
file can be viewed and postprocessed with the IGL-utility. In Fig. 12.20 we
show an IGL-generated plot of the utilizations rho1 and rho2 as a function of
the capacity of station N2. It is interesting that the utilization of station N1
decreases if the capacity of station N2 increases.

12.3.4 A Retrial Queue

Retrial queues were introduced in Section 6.7. Here we consider a simple
M/M/1 retrial queue with exponentially distributed repeating or delay times.

To get the MOSEL-2 model for this simple type of retrial system we introduce the delay node D (see Fig. 6.23) for the rejected requests. This node is of

```
//                    M|M|1_Retrial System

//              Declaration part
PARAMETER lambda := 0.01, 1, 2, 5, 8, 9, 9.5, 9.9;
PARAMETER mue := 0.01, 0.1, 1, 1000000;
CONST nue := 10;
ENUM status := { idle, busy };
CONST N := 100;      // Maximum number of requests

//            Component definition part
NODE server[status] = idle; // server initially idle
NODE D[N-1]; // delay node contains maximal N-1 req.
NODE num[N]; // counts total number of requests

//              Transition part
IF (server == idle) {
    FROM EXTERN TO server, num RATE lambda;
    FROM D TO server RATE D*mue;
    }
IF (server == busy) {
    FROM EXTERN TO D, num RATE lambda;
    FROM D TO D RATE D*mue;
    FROM server, num TO EXTERN RATE nue;
    }

//              Result part
PRINT rho = UTIL (server); // Server utilization
PRINT K = MEAN (num);      // Mean number of requests
PRINT Q = MEAN (D);        // Mean queue length
PRINT T = K/lambda;        // Mean system time

//              Picture part
PICTURE "Mean Queue Length"
PARAMETER lambda
CURVE Q;

PICTURE "Utilization"
PARAMETER lambda
CURVE rho;
```

Fig. 12.22   MOSEL-2 file for the M/M/1 retrial system.

the type $M/M/\infty$-IS. This means that there is no queue and the service rate of this node is the repeating rate. The MOSEL-2 model is shown in Fig. 12.22.

Because the component which represents the single server can assume the two states idle and busy, we explicitly define them as members of an ENUM-list. Although it would be possible to specify the model without the explicit naming of the server states, the possibility to refer to the status of a system component by a name improves the readability. Recall that the M/M/1 retrial queue has an infinite number of states. Here we limit the the maximum number of jobs to a finite value N. In the component definition part the three nodes for the server (server), the delay node (D), and the auxiliary node num are defined. The capacity of node server is 1, since the the ENUM-list status contains the two elements idle $\equiv 0$ and busy $\equiv 1$. Since the behavior of the system depends mainly on the status of the server we define two sets of transitions. The first block of transitions is enabled when the server node is in state idle: Primary requests arrive from outside with rate lambda and

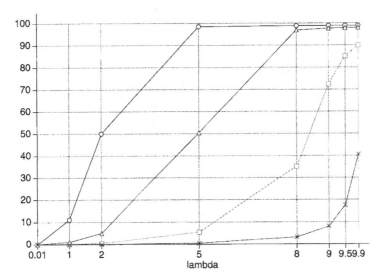

Fig. 12.23   Mean queue length $\overline{Q}$ as function of the arrival rate μ.

occupy the server, repeated requests try to enter the **server** node from the delay node D at rate D * **mue**. Note that a node name in the rate definition part of a rule stands for the actual number of jobs in the node. Each primary request which enters the system is also counted in the auxiliary **num** node. When the **server** node is in state **busy** the following transitions are possible: Primary requests are directly routed to the delay node with rate **lambda** since the server is not free. Repeating requests find the server busy and are rerouted to the delay node D.

The results obtained by the SPNP runs are collected in the result file **mm1.retrial.res** and is partly shown in Fig. 12.24. Figure 12.23 shows one of the plots after it was postprocessed by the IGL-utility.

12.3.5 Conclusions

We kept the presented examples simple due to the introductory nature of this section. Nevertheless, the basic models can easily be extended to express the structure and behavior of real systems in more detail. It is possible to specify systems with generally distributed state transitions. In particular, there are constructs to model deterministic and uniformly distributed transitions. Another, more realistic example where not only exponential distributions are used can be found in the applications chapter (see Section 13.2.6).

```
====================================================
Steady state analysis of "mm1.retrial.msl" by SPNP
====================================================

                             . . .

Parameters:               | Parameters:
  lambda = 0.01           |   lambda = 9.9
  mue = 0.01              |   mue = 1
                          |
Rsults:                   | Results:
  rho = 0.001             |   rho = 0.908255
  K = 0.002002            |   K = 90.806
  Q = 0.001002            |   Q = 89.8977
  T = 0.2002              |   T = 9.17232
=====================================================
Parameters:               | Parameters:
  lambda = 0.01           |   lambda = 9.9
  mue = 0.1               |   mue = 1e+06
                          |
Results:                  | Results:
  rho = 0.001             |   rho = 0.984317
  K = 0.0011011           |   K = 41.6014
  Q = 0.000101101         |   Q = 40.617
  T = 0.11011             |   T = 4.20216
=====================================================
  . . .
```

Fig. 12.24   Partial content of the MOSEL-2 result file.

12.4   SHARPE

The SHARPE (Symbolic Hierarchical Automated Reliability Performance Evaluator) tool was originally developed in 1986 by Sahner and Trivedi at Duke University [STP96, HSZT00]. It is implemented in ANSI standard C and runs on virtually all platforms. The advantage of SHARPE is that it is not restricted to one model type unlike PEPSY or SPNP. It offers nine different model types including product-form queueing networks, stochastic Petri nets, CTMCs, semi-Markov and Markov regenerative models, task precedence graphs, reliability block diagrams, reliability graphs and fault trees. The user can choose the model type that is most convenient for the problem. The models can also be hierarchical, which means the output of a sub-model can be used as the input for another sub-model. The user can also define variables and functions, write conditional statements and loops within SHARPE. Loops can be used for repeated evaluation of models or for the specification of CTMCs.

A graphical user interface for SHARPE is now available but the original version used a textual, line-oriented ASCII input file. An interactive or batch oriented input is possible. The syntax for both cases is the same. As an added feature, the SHARPE GUI provides a way to plot the results of SHARPE, as it allows also the creation of Excel spreadsheets containing these data. The possibilities that **SHARPE** offers depend on the chosen model type. If the user wishes to analyze a product-form queueing network, the commands for the performance measures such as throughput, utilization, mean response time at a station, and mean queue length are provided by **SHARPE**. MVA is used to solve closed, single or multiple chain, product-form queueing networks. For a continuous-time Markov chain, reward rates and initial state probability vector can be specified. Steady-state, transient, and cumulative transient measures can be computed. The tool offers two steady-state solution methods: Successive Overrelaxation (SOR) and Gauss–Seidel. At first it starts with the SOR method and after a certain period of time, it prints the number of iteration steps and the tolerance if no solution is found. In this case the user is asked whether he or she wants to go on with the SOR method or switch to Gauss Seidel. Three different variants of uniformization for the transient CTMC solution are available.

Because of space limitations, it is not possible to describe the full power of **SHARPE**. Instead of describing the syntax and semantics in detail, we give examples that use the most important model types to show the reader that **SHARPE** is very powerful and also easy to use. **SHARPE** has been installed at more than 350 sites. For more detailed information the reader can consult [STP96, HSZT00] and for recent additions to **SHARPE** see [Triv02].

12.4.1 Central-Server Queueing Network

As the first example we consider a central-server queueing network consisting of a CPU and two disks with the following input parameters:

$$\mu_1 = 1000/20, \quad \mu_2 = 1000/30, \quad \mu_3 = 1000/42.9,$$
$$p_{11} = 0.1, \quad p_{12} = 0.667, \quad p_{13} = 0.233, \quad p_{21} = p_{31} = 0.1.$$

This model file is created by **SHARPE** GUI with the above parameters as shown in Fig. 12.25. By right clicking each node, information such as name of the node, scheduling discipline as FCFS and service times as exponential distribution are specified. Similarly, the routing probabilities are specified for the arcs. Fig. 12.26 shows a **SHARPE** input file for this model which is automatically created after the GUI model description is finished. Lines with comments begin with an asterisk. The description of the **SHARPE** specification is given in Fig. 12.26.

Table 12.2 contains a description of some important code lines of the central-server network in code in Fig. 12.26.

In the **SHARPE** GUI, the user can select various performance measures such as throughput, response time, queue length and utilization to be calculated for the pfqn model type. By assigning the number of jobs in the system, the output will be created. Figure 12.27 shows how to select the outputs and specification of number of jobs. The output produced by **SHARPE** GUI for this model is shown in Fig. 12.3. To demonstrate the plot features in

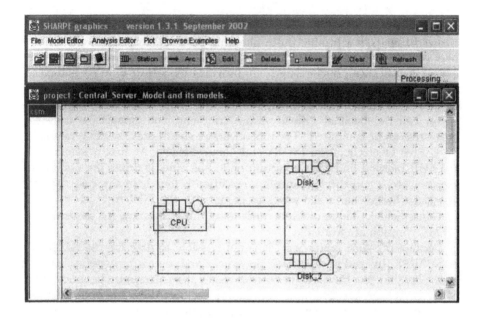

Fig. 12.25 A central-server model.

```
 1 * central-server model     18
 2                            19 * servers
 3 bind                       20 CPU FCFS mu_1
 4 mu_3   1000/42.918         21 Disk_1 FCFS mu_2
 5 mu_2   1000/30             22 Disk_2 FCFS mu_3
 6 mu_1   1000/20             23 end
 7 p13    0.223               24 * number of jobs
 8 p12    0.667               25 chain1 jobs
 9 end                        26 end
10                            27
11 pfqn csm(jobs)             28 expr tput(csm, CPU; 2)
12 * routing probabilities... 29 expr util(csm, CPU; 2)
13 CPU Disk_1 p12             30 expr qlength(csm, CPU; 2)
14 CPU Disk_2 p13             31 expr rtime(csm, CPU; 2)
15 Disk_1 CPU 1               32
16 Disk_2 CPU 1               33 end
17 end
```

Fig. 12.26 SHARPE input for the central-server network.

SHARPE GUI, we have graphically shown the throughput for the number of jobs varying from 2 to 40 in Fig. 12.28. We see that the throughput grows from 29 to 49 as the number of jobs increases.

Table 12.2 SHARPE code description

Line	Description
3 – 9	The values of the parameters and routing probabilities are specified.
11	Specifies the model type pfqn (product-form queueing network), model name csm, and model parameter jobs.
12 – 17	Specification of the routing probabilities.
19 – 23	Specification of the nodes and the service parameters of the nodes (name, queueing discipline, service rate).
24 – 26	Gives the number of jobs in the network.
28 – 31	Requests the computation and printing of the throughput, the utilization, the mean queue length, and the mean response time at the cpu with 2 jobs in the network.

Table 12.3 SHARPE results for the central-server model

tput(csm, CPU; 2):	2.95982540e+001
util(csm, CPU; 2):	5.91965079e-001
qlength(csm, CPU; 2):	8.30753524e-001
rtime(csm, CPU; 2):	2.80676531e-002

12.4.2   M/M/m/K System

As a simple example for the CTMC model type, we consider an M/M/5/10 queueing system. To demonstrate how to use the **SHARPE** GUI, this queueing model is drawn choosing the CTMC model type. After the program is started, the user can graphically draw the state diagram as shown in Fig. 12.29. Nodes can be created, moved, or deleted simply by mouse. The necessary parameters are specified adjacent to the arcs. Each state name gives the number of jobs in the system. This is an example of an irreducible CTMC of birth–death type.

The **SHARPE** textual input file is automatically created by the GUI and is shown in Fig. 12.30.

In the first two lines the values of the arrival and the service rates are bound. Then the model type (CTMC) and the name (mm5) of the model are

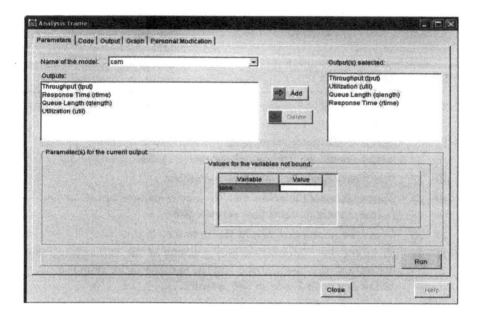

Fig. 12.27 **SHARPE** GUI Analysis frame for the central-server model.

given. Subsequent lines specify the transitions between the states together with their transition rates. Then five variables are defined.

In the **SHARPE** GUI, this queueing model can be solved by clicking Analysis Editor in the main screen and the following results can be obtained (see Fig. 12.31). *Pidle* is the steady-state probability of being in state 0. This is the probability that there are no jobs being served, i.e., that the station is idle. *Pfull* is the steady-state probability of being in state 10. This is the probability that the queue is full. *Lreject* is the rate at which jobs are rejected. *Mqueue* is the mean number of jobs in the system. The above two variables are calculated by assigning proper reward rates. These also can be calculated by the built-in function sum. *Mresp* is the mean response time of accepted jobs computed using Little's theorem as $Mqueue/(\lambda - Lreject)$, *expr expression* prints the value of the expression in the output file (Fig. 12.32).

Note that an extension to **SHARPE** includes a loop specification in the definition of a CTMC that can be used to make concise specification of a large, structured CTMC. For example, we consider the above queueing system with buffer size 100. In this case, the above Markov chain model type shown in Fig. 12.30 can be updated with the following specifications, as shown in Fig. 12.33, so that arbitrary size can be enabled using loop specification. For an example of how to use this feature, see also Section 13.3.2.

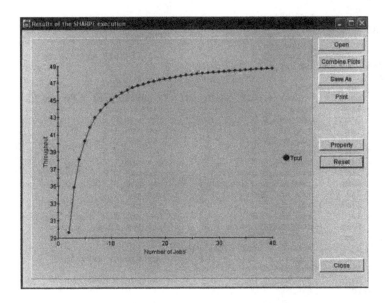

Fig. 12.28 SHARPE GUI graph of throughput for the central-server model.

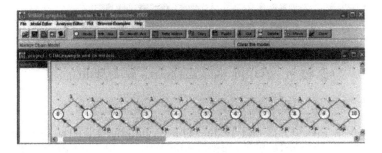

Fig. 12.29 Markov chain for the M/M/5/10 system.

12.4.3 M/M/1/K System with Server Failure and Repair

Now we extend the M/M/1/K queueing model by allowing for the possibility that a server could fail and could be repaired. Let the job arrival rate be λ and the job service rate be μ. The processor failure rate is γ and the processor repair rate is τ. This system can be modeled using an irreducible CTMC, as shown in Fig. 12.34 for the case where $m = 1$ (one server) and $K = 10$ (the maximum number of jobs in the server and queue is 10). Each state is named with a two-digit number ij, where $i \in \{0, 1, \ldots, 9, a\}$ is the number of jobs in the system (a is 10 in hexadecimal) and $j \in \{0, 1\}$ denotes the number of operational processors. SHARPE input and output files for this example

```
bind lam 5.3      7 rew_mm5_7            end
bind mu 1.2       8 rew_mm5_8
                  9 rew_mm5_9            var Lreject exrss(mm5)
markov mm5        10 rew_mm5_10
0 1     lam       end
1 2     lam       end
1 0     mu        * Reward conf. assigned:   * REWARD conf.: Mqueue
2 3     lam       bind                   bind
2 1     2*mu      rew_mm5_0 0            rew_mm5_0 0
3 4     lam       rew_mm5_1 0            rew_mm5_1 1
3 2     3*mu      rew_mm5_2 0            rew_mm5_9 9
4 5     lam       rew_mm5_3 0            rew_mm5_8 8
4 3     4*mu      rew_mm5_4 0            rew_mm5_7 7
5 6     lam       rew_mm5_5 0            rew_mm5_6 6
5 4     5*mu      rew_mm5_6 0            rew_mm5_5 5
6 7     lam       rew_mm5_7 0            rew_mm5_4 4
6 5     5*mu      rew_mm5_8 0            rew_mm5_3 3
7 8     lam       rew_mm5_9 0            rew_mm5_2 2
7 6     5*mu      rew_mm5_10 0           rew_mm5_10 10
8 9     lam       end                    end
8 7     5*mu      * REWARD conf.: Lreject
9 10    lam       bind                   var Mqueue exrss(mm5)
9 8     5*mu      rew_mm5_0 0
10 9    5*mu      rew_mm5_1 0            var Pidle prob(mm5,0)
* Reward:         rew_mm5_9 0            var Pfull prob(mm5,10)
reward            rew_mm5_8 0            var Mresp Mqueue/(lam-Lreject)
0 rew_mm5_0       rew_mm5_7 0            expr Pidle
1 rew_mm5_1       rew_mm5_6 0            expr Pfull
2 rew_mm5_2       rew_mm5_5 0            expr Lreject
3 rew_mm5_3       rew_mm5_4 0            expr Lreject
4 rew_mm5_4       rew_mm5_3 0            expr Mresp
5 rew_mm5_5       rew_mm5_2 0
6 rew_mm5_6       rew_mm5_10 lam         end
```

Fig. 12.30 SHARPE input file for the $M/M/5/10$ queue.

are shown in Figs. 12.35 and 12.36. The probability that the system is idle is given by **prob**($mm1k, 00$) + **prob**($mm1k, 01$), and the rate at which jobs are rejected because the system is full or the server is down is given by λ (**prob**($mm1k, a0$) + **prob**($mm1k, a1$)).

The generalized stochastic Petri net (GSPN) in Fig. 12.37 is equivalent to the CTMC model and will let us vary the value of K without changing the model structure. We use this example to show how to handle GSPNs using **SHARPE**. The loop in the upper part of the GSPN is a representation of an M/M/1/K queue. The lower loop models a server that can fail and be repaired. The inhibitor arc from place **server-down** to transition **service** shows that customers cannot be served while the server is not functioning. The number within each place is the initial number of tokens in the place. All of the transitions are timed, and their firing rates are shown below the transitions. The input file for this model is shown in Fig. 12.38, and a description of this file is given in Table 12.4.

This GSPN is irreducible. For GSPNs that are nonirreducible, **SHARPE** can compute the expected number of tokens in a place at a particular time t,

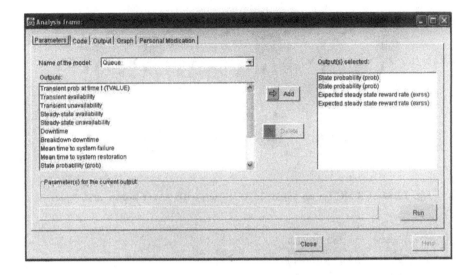

Fig. 12.31   SHARPE GUI analysis editor screen for the M/M/5/10 system.

```
Pidle: 9.22499e-03
Pfull: 6.9483e-02
Lreject: 3.6826e-01
Mqueue: 5.3547e+00
Mresp: 7.4671e-02
```

Fig. 12.32   SHARPE GUI output file for the M/M/5/10 system.

```
markov mm5 loop i,0,99,1 $i $(i+1) lam end loop i,0,3,1 $(i+1) $i
(i+1)*mu end loop i,4,99,1 $(i+1) $i 5*mu end
```

Fig. 12.33   The loop specification for the M/M/5/100 system.

the probability that a place is empty at time t, the throughput and utilization of a transition at time t, the time-average number of tokens in a place during the interval $(0, t)$, and the time average throughput of a transition during $(0, t)$. See [STP96, HSZT00] for the syntax for these functions. Furthermore, SHARPE now supports the Stochastic Reward Net model type with all the relevant extensions.

12.4.4   GSPN Model of a Polling System

Consider an example of a polling system that is discussed in Section 2.3.5. To demonstrate the GSPN model in **SHARPE** GUI, we consider a three-station single-buffer polling system as shown in Fig. 12.40. The nodes are shown in dark color with their initial number of tokens.

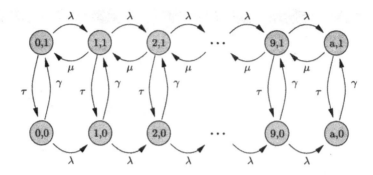

Fig. 12.34 CTMC for an M/M/1/10 system with server failure and repair.

```
markov mm1k        11 10 GAM        30 40 LAM
01 11 LAM          10 11 TAU        40 50 LAM
11 01 MU           21 10 GAM        60 70 LAM
11 21 LAM          20 21 TAU        70 80 LAM
21 11 MU           31 30 GAM        80 90 LAM
21 31 LAM          30 31 TAU        90 a0 LAM
31 21 MU           41 40 GAM        end
31 41 LAM          40 41 TAU
41 31 MU           51 50 GAM        bind
41 51 LAM          50 51 TAU        LAM 1
51 41 MU           61 60 GAM        MU 2
51 61 LAM          60 61 TAU        GAM 0.0001
61 51 MU           71 70 GAM        TAU 0.1
61 71 LAM          70 71 TAU        end
71 61 MU           81 80 GAM        var Pidle prob(mm1k,00)+prob(mm1k,01)
71 81 LAM          80 81 TAU        var Pfull prob(mm1k,a0)+prob(mm1k,a1)
81 71 MU           91 90 GAM        var Lreject LAM*PFULL
81 91 LAM          90 91 TAU        expr Pidle
91 81 MU           a1 a0 GAM        expr Lreject
91 a1 LAM          a0 a1 TAU        end
a1 91 MU           00 10 LAM
01 00 GAM          10 20 LAM
00 01 TAU          20 30 LAM
```

Fig. 12.35 SHARPE input for the M/M/1/10 system with server failure and repair.

```
Pidle:   4.9953e-01
Lreject: 9.6239e-04
```

Fig. 12.36 SHARPE output for the M/M/1/10 system with server failure and repair.

For a numerical computation, we use the parameters $\mu_i = \mu = 1$, $1/\gamma_i = 1/\gamma = 0.005$ and $\lambda_i = \lambda$ for all stations $1 \leq i \leq 3$. From Eq. (2.85), we can obtain mean response time at station 1. We obtain this result using function command in the **SHARPE GUI**. The analyze frame is shown in Fig. 12.41. In this figure, we can see that the steady-state probability that the place $P1$ is empty is given by **prempty** (PollingSystem,P1;0.1). Using loop specification

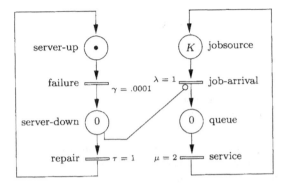

Fig. 12.37 GSPN model for queue with server failure and repair.

```
1   bind                              28  * arcs, transitions-places
2   lambda 1.0                        29  job-arrival queue 1
3   mu 2.0                            30  service jobsource 1
4   gamma 0.0001                      31  failure serverdown 1
5   tau 0.1                          32  repair serverup 1
6   K 10                             33  end
7   end                              34  * inhibitor arcs
8                                    35  serverdown service 1
9   gspn mm1k-fail                   36  end
10  * places                        37
11  jobsource K                     38  var Pidle prempty(mm1k-fail,queue)
12  queue 0                         39
13  serverup 1                      40  var Preject prempty(mm1k-fail,jobsource)
14  serverdown 0                    41
15  end                             42  var Lreject prempty(mm1k-fail,jobsource)
16  * transitions                   43
17  job-arrival ind lambda          44  var avquelength etok(mm1k-fail,queue)
18  service ind mu                  45
19  failure ind gamma               46  var thruput tput(mm1k-fail,service)
20  repair ind tau                  47
21  end                             48  var utilization uitl(mm1k-fail,service)
22  * arcs, places-transitions      49
23  jobsource job-arrival 1         50  expr Pidle
24  queue service 1                 51  expr Lreject, Preject
25  serverup failure 1              52  expr avquelength
26  serverdown repair 1             53  expr thruput, utilization
27  end                             54  end
```

Fig. 12.38 Input for GSPN model of the system with server failure and repair.

for different values of offered load ρ, we obtain the mean response time at station 1. Then the result is plotted by clicking 'Graph' in 'Analyze frame:' window, and is shown in Fig. 12.42. We observe that the mean response time increases as the offered load increases. Other model types provided by SHARPE are multiple-chain product-form queueing networks, semi-Markov chains, reliability block diagrams, fault trees, reliability graphs, series–parallel task graphs, and Markov regenerative processes. For a detailed description see [STP96].

Table 12.4 CSPL code description

Line	Description
1 – 7	The input parameters are bound to specific values.
9	Model type and name of the model.
11 – 15	Places and initial numbers of the tokens.
17 – 21	Timed transitions.
23 – 27	Arcs from places to transitions and arc multiplicities.
29 – 33	Arcs from transitions to places and arc multiplicities.
35 – 36	Inhibitor arcs and their multiplicities.
38	Define a variable for the probability that the server is idle.
40	Define a variable for the probability that a job is rejected.
42	Define a variable for the rejection rate.
44	Define a variable for the average number of jobs in the system.
46	Define a variable for the throughput of server.
48	Define a variable for the utilization of server.
50 – 54	Values requested to be printed (Fig. 12.39).

```
Pidle:  4.9953e-01
Lreject:  9.6239e-04
Preject:  9.6239e-04
avquelength:  1.0028e+00
thruput:  9.9904e-01
utilization:  4.9952e-01
```

Fig. 12.39 Output for the GSPN model.

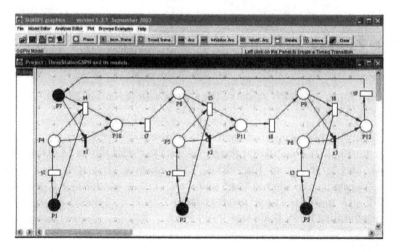

Fig. 12.40 GSPN model for a three-station polling system.

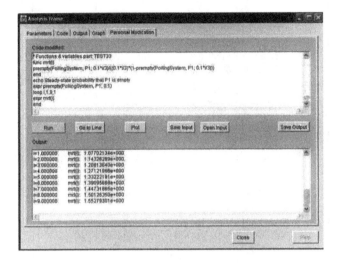

Fig. 12.41 **SHARPE** GUI Analysis Editor screen for the polling system.

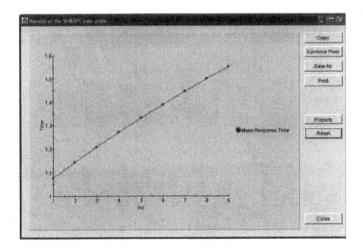

Fig. 12.42 **SHARPE** GUI plot screen for the polling system.

Table 12.5 Performance Evaluation Tools

Tool	Location	Reference	Modeltype			Solution Method					GUI	Language	Hierarch.
			QN	SPN	other	DES	PFQN	NPFQN	m-tr	m-ss			
QNAP2	SIMULOG	[VePo85]	*	-	-	*	*	-	-	*	*	*	-
MAOS	TU München	[FeJo90, Jobm91]	*	-	-	*	*	-	-	*	-	SIMULA-Interface	*
PEPSY	Univ. Erlangen	[Kirs94]	*	-	-	*	*	*	-	*	XPEPSY	Menue	-
WinPEPSY	Univ. Erlangen	[BBG04]	*	-	-	*	*	*	-	*	*	Menue	-
HIT	Univ. Dortmund	[BMW94]	*	-	-	*	*	-	-	*	HITGRAPHIC	HISLANG	*
VITO	Univ. Duisburg-Essen	[MFV99]	*	-	-	-	*	-	-	-	*	-	-
SPNP	Duke Univ.	[HTT00]	-	*	-	*	-	-	*	*	iSPN	CSPL	-
GreatSPN	Univ. Torino	[ABC+95]	-	*	-	*	-	-	*	*	*	*	-
TIMENET	TU Berlin	[GKZH94]	-	*	-	*	-	-	*	*	*	(*)	-
DSPNexpress	Univ. Dortmund	[LTK+00]	-	*	-	*	-	-	*	*	*	-	-
Ultra SAN	Univ. Illinois	[SOQW95]	-	*	-	*	-	-	*	*	*	-	-
Moebius	Univ. Illinois	[DCC+02]	-	*	*	*	-	-	*	*	*	-	*
CPN Tools	Univ. of Aarhus	[CPN05]	-	*	*	*	-	-	-	-	*	*	*
APNN	Univ. Dortmund	[BFKT01]	-	*	*	*	-	-	*	*	*	*	*
QPN	Univ. Dortmund	[BaKe94]	*	*	-	-	-	-	-	*	*	*	*
SHARPE	Duke Univ.	[HSZT00]	*	*	*	-	*	*	*	*	*	*	*
MOSEL	Univ. Erlangen	[BBBZ03]	*	*	*	*	-	-	*	*	-	MOSEL-2	*
MARCA	NC State Univ.	[Stew90, Stew94]	*	-	*	-	-	-	*	*	XMARCA	*	-
MACOM	Univ. Dortmund	[KMS90]	*	-	-	-	-	-	*	*	*	(USENUM)	-
CSIM 19	Mesquite Software	[CSI05]	*	-	-	*	-	-	-	-	-	C++	-
OPNET Modeler	OPNET Technologies	[OPN05]	*	-	-	*	-	-	-	-	*	-	-
ns-2	DARPA	[Ns05]	*	-	-	*	-	-	-	-	-	C++	-

Notes: m-tr: CTMC solver transient, m-ss: CTMC solver steady-state

12.5 CHARACTERISTICS OF SOME TOOLS

In Table 12.5 other important tools are listed together with their main features. Many of them provide a GUI and/or a textual input language. The GUI is very convenient for getting accustomed to the tools, but the experienced user often prefers the use of a textual input language.

13

Applications

The concluding chapter of this textbook is devoted to the application of queueing network and Markov chain methodology in larger real-world performance and reliability case studies. The applications presented in Section 13.1 are based on queueing networks as the modeling formalism and employ the exact or approximate solution algorithms of Chapter 8, 9, and 10 for the calculation of performance and reliability measures. In Section 13.2 we present case studies in which a real-world system is formalized either directly as a Markov chain or alternatively in a high-level modeling formalism if more complex properties of the real-world system have to be captured. The system analysis is performed either by application of the numerical solution methods for Markov chains presented in Chapters 3, 4, and 5 or by discete event simulation (see Chapter 11). In Section 13.3, case studies of hierarchical models are presented.

13.1 CASE STUDIES OF QUEUEING NETWORKS

Eight different case studies are presented in this section. These range from a multiprocessor system model, several networking applications one operating system model and a flexible production system model.

13.1.1 Multiprocessor Systems

Models of tightly coupled multiprocessor systems will be discussed first followed by models of loosely coupled systems.

13.1.1.1 Tightly Coupled Systems Consider a tightly coupled multiprocessor system with caches at each processor and a common memory, connected to the processors via a common bus (see Fig. 13.1). The system consists of m

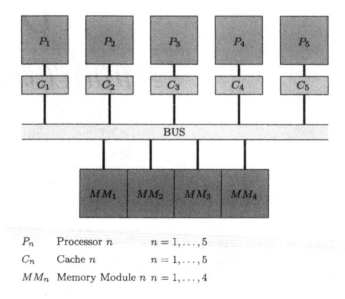

P_n	Processor n	$n = 1, \ldots, 5$
C_n	Cache n	$n = 1, \ldots, 5$
MM_n	Memory Module n	$n = 1, \ldots, 4$

Fig. 13.1 A multiprocessor system with caches, common memory, and common bus.

processors and a common memory with n memory modules. A processor sends a request via the common bus to one of the n memory modules when a cache miss occurs, whereupon the requested data is loaded into the cache via the common bus. Figure 13.2 shows a product-form queueing network model of such a multiprocessor system.

The m processors are modeled by an IS node and the bus and the memory modules by single server nodes. The number of requests in the system is m since a processor is either working or waiting until a memory request is finished. Using this model we can calculate the utilization of the bus or the mean response time of a memory request from a processor. The mean time between two cache misses is modeled by the mean thinking time of the IS node. Other parameters that we need are the mean bus service time, the mean memory request time, and, finally, p_i, the probability of a request to memory module i. In the absence of additional information, we assume $p_i = 1/n$ $(i = 1, 2, \ldots, n)$. Note that for each cache miss, there are two service requests to the bus. It is for this reason that with probability 0.5, a completed bus

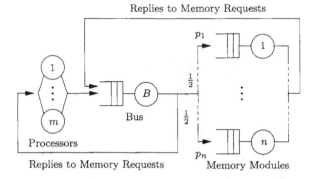

Fig. 13.2 Queueing network model of the multiprocessor system shown in Fig. 13.1.

request returns to the processor station. Similarly, the probability of visiting memory module i subsequent to the completion of a bus request is $p_i/2$.

We assume that the service times for the two types of bus requests have the same mean. This assumption can be easily relaxed. If we have explicit values for these parameters, then we can calculate interesting performance measures such as bus utilization and mean response time as functions of the number of processors. Another interesting measure is the increase in the mean response time assuming a fixed number of processors and memory modules, while changing the mean bus service time and the mean time between two cache misses. In Fig. 13.3 the mean response time is shown as a function of the mean bus service time and the mean time between two cache misses [Bolc91].

We can see that there is a wide area where the increase in the mean response time is tolerable (15%) compared to the case with no waiting time at the bus or a memory queue (this is the minimum value of the mean response time). On the other hand, there is also a wide area with an intolerable increase in the mean response time. Thus, the analysis of such a queueing network model can help the system designer to choose parameter values in a tolerable area.

Problem 13.1   Verify the results shown in Fig. 13.3 by hand computation and using any software package available (e.g., **SHARPE** or **WinPEPSY**).

Problem 13.2   Improve the model described in Section 13.1.1.1 by considering different mean service times for bus requests from processor to memory and from memory to cache. (Hint: Use a multiclass queueing network model.)

13.1.1.2 Loosely Coupled Systems In a loosely coupled multiprocessor system, the processors have only local memory and local I/O devices, and communicate via an interconnection network by exchanging messages. A simple queueing network model of such a system is shown in Fig. 13.4. The mean service time at the processors is the mean time between the messages, and the mean service time at the network node N is the mean message delay at

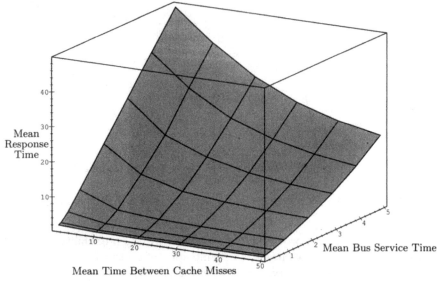

Fig. 13.3   Mean response time as a function of the bus service time, and the mean time between cache misses $(m = 5, n = 4)$.

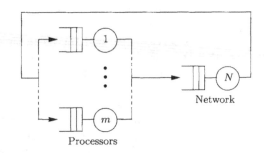

Fig. 13.4   A simple queueing network model of a loosely coupled system.

the network. The routing probability p_i is the probability that a message is directed to processor i.

A more complex and detailed model of a loosely coupled multiprocessor system is shown in Fig. 13.5 [MAD94]. Here we assume that we have n separate I/O processors and m "computing" processors. The computing processors send I/O requests via the network N to the I/O processors and get the replies to these requests also via the network N. We assume that a job that begins at a computing processor is also completed at this processor and that the processors are not multiprogrammed and are heavily loaded (for each job that leaves the system after being processed by a processor, a new job arrives immediately at the processor). This system can be modeled by a closed product-form

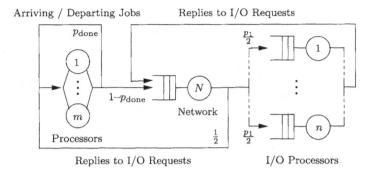

Fig. 13.5 A complex queueing network model of a loosely coupled system.

queueing network with m job classes (one for each computing processor) with the population of each class equal to 1 [MAD94].

As an example, we consider a loosely coupled multiprocessor system with eight computing processors and $n = 2, 3, 4$ I/O processors. The mean computing processor service time is 30 msec, the mean I/O time is 50 msec, and the mean message delay at the network is 1 msec. The probability that a job leaves the system after it has been processed at a computing processor is $p_{\mathrm{done}} = 0.05$. We assume that $p_i = 1/n$ for all i. Some interesting results from this model are listed in Table 13.1.

Table 13.1 Performance measures for the loosely coupled multiprocessor system for different numbers of I/O processors

Number of I/O Processors	2	3	4
Mean response time	4.15 sec	3.18 sec	2.719 sec
Throughput	$1.93\,\mathrm{sec}^{-1}$	$2.51\,\mathrm{sec}^{-1}$	$2.944\,\mathrm{sec}^{-1}$
$\rho_{\mathrm{computingprocessor}}$	0.145	0.189	0.220
ρ_{network}	0.070	0.090	0.106
$\rho_{\mathrm{I/Oprocessor}}$	0.867	0.754	0.662

Problem 13.3 Verify the results shown in Table 13.1 by hand computation and using any software package available (e.g., WinPEPSY or SHARPE).

Problem 13.4 Extend the complex model described in Section 13.1.1.2 so as to allow distinct mean network service times for request from a computing processor to I/O processor and vice versa.

13.1.2 Client–Server Systems

A client–server system consists of client and server processes and some method of interprocess communication. Usually the client and the server processes are executing on different machines and are connected by a LAN. The client interacts with the user, generates requests for a server, transmits the request to the server, receives the results from the server, and presents the results to the user. The server responds to requests from the clients and controls access to resources such as file systems, databases, wide area networks, or printers. As an example, we consider a client–server system with a fixed number m of client workstations that are connected by an Ethernet network to a database server. The server consists of a single disk (node number 4) and a single CPU (node number 3). This leads to a closed product-form queueing network model shown in Fig. 13.6.

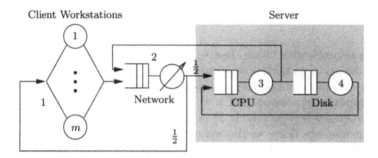

Fig. 13.6 Closed queueing network model of a client–server system.

The client workstations are modeled as an IS node (node number 1) [MAD94], and the number of jobs in the closed system is equal to the number of workstations m. The Ethernet network (carrier sense multiple access with collision detection, or CSMA/CD, network) can be modeled as a server (node number 2) with the load-dependent service rate [LZGS84, MAD94, HLM96]:

$$\mu_{\text{net}}(k) = \begin{cases} \left(\frac{1}{N_p} \cdot \frac{\overline{L_p}}{B} + S \cdot C(1)\right)^{-1}, & k = 1, \\ \left(\frac{1}{N_p} \cdot \frac{\overline{L_p}}{B} + S \cdot C(k+1)\right)^{-1}, & k > 1, \end{cases} \tag{13.1}$$

where $C(k) = (1 - A(k))/A(k)$ is the average number of collisions per request and $A(k) = (1 - 1/k)^{k-1}$ is the probability of a successful transmission and k the number of workstations that desire the use of the network.

Other parameters are described in Table 13.2 and shown in Fig. 13.6.

We compute the throughput λ and other interesting performance measures using the load-dependent MVA (see Section 8.2). As an example we use the parameters from Table 13.3 and determine the throughput as a function of the number of client workstations m (see Fig. 13.7). A more detailed example for client–server systems can be found in [Mena02].

Problem 13.5   Verify the results shown in Fig. 13.7 by hand computation and using an available modeling package such as **SHARPE** or **WinPEPSY**.

Table 13.2 Parameters for the client–server example

Parameter	Description
N_p	Average number of packets generated per request
B	Network bandwidth in bits per second
S	Slot duration (i.e., time for collision detection)
$\overline{L}_p$	Average packet length in bits

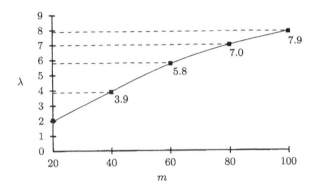

Fig. 13.7 Throughput as a function of the number of workstations.

Table 13.3 Numerical parameter values for the client–server example

$N_p = 7$	$\mu_1 = \mu_{CL} = 0.1/\sec$	
$B = 10\,\mathrm{Mb}/\sec$	$\mu_2 = \mu_{Net}(k)$	
$S = 51.2\,\mu\sec$	$\mu_3 = \mu_{CPU} = 16.7/\sec$	
$\overline{L}_p = 1518\,\text{bits}$	$\mu_4 = \mu_{Disk} = 18.5/\sec$	
$p_{12} = 1$	$p_{21} = 0.5$	$p_{32} = 0.5$	$p_{43} = 1$
	$p_{23} = 0.5$	$p_{34} = 0.5$	

13.1.3 Communication Systems

13.1.3.1 Description of the System As an example of a more complex queueing network and the performance evaluation of communication systems, we consider the LAN of a medium-sized enterprise [Ehre96]. The LAN connects several buildings that are close together and it is divided into several network sections. In each network section the devices of a building are connected to each other. Three other sections are located in one building and are used to connect the servers to the LAN. As a backbone for the sections, a fiber distributed data interface (FDDI) ring is used. Eleven bridges for the sections, a WAN router, and a LAN analyzer that measures the utilization of the LAN

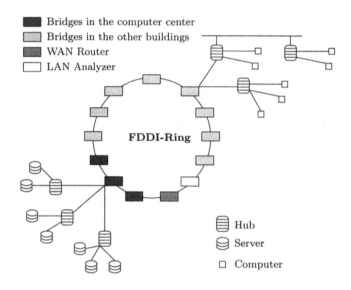

Fig. 13.8   The FDDI backbone with 13 stations.

constitute the stations of this ring. Figure 13.8 shows the FDDI ring with the stations schematically, wherein the structure of two stations is shown in more detail.

The typical structure of a section in which all computers within a building are connected is shown in Fig. 13.9. The building has four floors. The computers on the second and third floor are connected to different segments, and the computers on the fourth and fifth floor are connected to the same segment. These three segments are connected to each other and to the FDDI ring via a multiport bridge. The computers are connected to a stack of four hubs that are connected to each other via Cheapernet, and there is a connection of the hubs to the multiport bridge via optical fibers. The CSMA/CD principle is used with a transmission rate of 10 Mb/sec which is equal for all segments. Each segment is a different collision domain.

13.1.3.2 Queueing Network Model   The nodes of the queueing network model of the LAN are the FDDI ring, the Ethernets that connect the devices in the segment, the computers, the servers in the computer center, and the bridges. The FDDI ring can be considered as a server that serves the stations in the ring (12 stations without the LAN analyzer) in a fixed sequence. A station (or section) sends a request to another station via the FDDI ring, and the reply to the request is sent back to the station again via the FDDI ring. Thus, we obtain a closed queueing network model for the LAN as shown in Fig. 13.10 with a simplified representation of the individual sections. The FDDI ring is modeled by a single server node. The LAN analyzer does not influence the other sections and for this reason it does not appear in the model. The WAN

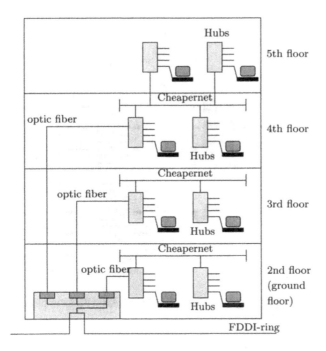

Fig. 13.9 A section of a communication system with three segments.

router can be modeled as a single server node and the WAN itself as an IS
node (see Fig. 13.11). There are arrivals at the WAN router from the WAN
and from the ring. The WAN router sends requests to the ring and to the
WAN as well. A similar situation occurs at all the bridges/routers.

In the sections, the segments are connected to the bridge via Ethernet.
The Ethernet segments are modeled by a multiple server with only one queue
where the number of servers is the number of segments in the section. Because
of the CSMA/CD strategy, the frames are either transferred successfully to
the computers of the sections to the bridge or in the case of a collision, sent
back to the Ethernet queue with collision probability q_i. Each computer sends
requests to a server via the LAN and waits for the reply before it sends another
request. Therefore, the number of requests in the LAN equals the number of
active computers, and the requests do not have to wait at the computers. For
this reason the computers at a section can be modeled by an IS node.

A queueing network model of a section of the LAN is shown in Fig. 13.12
and the queueing network model of a computer center section with the servers
is shown in Fig. 13.13. Servers are modeled by a multiple server node with one
queue where the number of servers at the node is equal to the number of real
servers. The queue is necessary because there are more computers that can
send requests to the servers than the number of available servers. The whole
LAN model consists of 1 node for the FDDI ring, 12 nodes for the bridges,

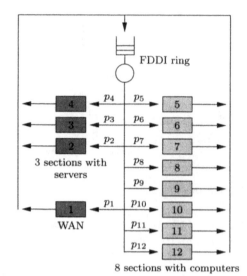

Fig. 13.10 Closed queueing network model of the LAN with a simplified representation of the individual sections.

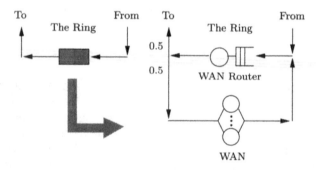

Fig. 13.11 Detailed model of the WAN and the WAN router.

11 nodes for the Ethernet segments, 8 nodes for the computers, 3 nodes for the servers, and 1 node for the WAN. Because we will allow nonexponential service time distributions, it is a non-product-form network.

13.1.3.3 Model Parameters As parameters we need the total number of requests, the routing probabilities p_i, the mean service times $1/\mu_i$, and the coefficient of variation c_i of the service times of the nodes. Furthermore, we need the number of service units m_i for the multiple server nodes. The total number of requests in the network is the mean number of active computers $K = 170$ (measured value). We estimate the routing probabilities p_i from the FDDI ring to the bridges by measuring the amount of data transferred from the ring to the bridges, which can be measured at the bridges. This estimate

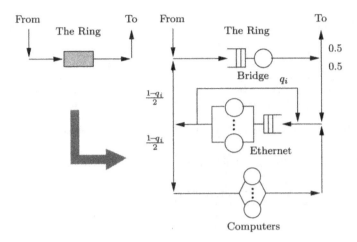

Fig. 13.12 Detailed model of a section of the LAN.

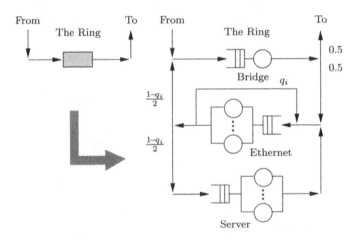

Fig. 13.13 Detailed model of a section with servers.

was done based on measurements over a two-day period (see Table 13.4). The other necessary routing probabilities are shown in Figs. 13.11, 13.12, and 13.13 and in Table 13.5. The number of servers m_i of the multiple server nodes is given by the number of servers for the server nodes (Table 13.6) or by the number of segments for the Ethernet node (Table 13.7). To obtain the mean service times of the nodes, we use a LAN analyzer which can measure the interarrival times and the length of the frames transferred by the FDDI ring. The results of the measurement are shown in Tables 13.8 and 13.9. From these tables we obtain the mean values and the square coefficients of variation (Table 13.10) of the interarrival times and frame lengths. Given the measured throughput $\lambda = 1/346\,\mu\text{sec} = 2890$ per sec and the routing probabilities (see

Table 13.4 Total data transferred from FDDI ring to the bridges for a two-day period and estimated routing probabilities p_i

Section	1	2	3	4	5	6	7	8	9	10	11	12
Data/Mb	2655	1690	2800	1652	2840	1500	3000	200	1940	1180	4360	4380
p_i/%	9.5	6.2	10	5.9	10.1	5.4	10.7	0.7	6.9	4.2	14.8	15.6

Notes: 1: WAN; 2, 3, 4: section with servers; 5, ..., 12: section with computers.

Table 13.5 Collision probabilities q_i

Section	1	2	3	4	5	6	7	8	9	10	11	12
q_i/%	1	5	1	1	1	1	1	1	1	1	3	2

Table 13.6 Number of service units m_i in server nodes

Section	2	3	4
m_i	14	23	13

Table 13.7 Number of service units m_i (= number of segments) in Ethernet nodes

Section	2	3	4	5	6	7	8	9	10	11	12
m_i	4	7	2	4	5	3	1	4	2	5	5

Table 13.8 Empirical pmf of the interarrival times of the frames.

Interarrival Time (μs)	%
≤ 5	3.0
5 – 20	0.9
20 – 82	15.3
82 – 328	42.7
329 – 1300	27.1
1300 – 5200	1.0

Table 13.9 Empirical pmf of the frame length L

Length (Bytes)	%
≤ 32	11.1
32 – 63	10.0
64 – 95	33.0
96 – 127	7.9
128 – 191	6.8
192 – 511	4.1
512 – 1023	5.8
1024 – 1526	21.3

Table 13.4), the traffic equations are solved to produce the arrival rates of the individual sections (see Table 13.11). Normally in a closed network, relative throughputs are computed from routing probabilities or visit ratios. However, in this case we have measured throughput at one node, and hence the relative throughputs at each node are also the actual throughputs.

The mean service time of the FDDI ring is given by the sum of the mean transfer time of a frame and the mean waiting time until the token arrives at a station [MSW88]. We assume *one-limited service*; i.e., a station transmits at most one frame when the token arrives. An upper limit of the service time of

Table 13.10 Characteristic values of the interarrival times and the frame length

	Interarrival Times of the Frames	Length of the Frames
Mean value	346 μ sec	382.5 Byte
Squared coefficient of variation	1.48	1.67

Table 13.11 Arrival rates of frames at the stations

Section	1	2	3	4	5	6	7	8	9	10	11	12
λ_i	275	179	289	171	292	156	309	20	199	121	428	451

the ring is the token rotation time. The utilization $\rho_i = \lambda_i/\mu$ is the probability that a station wishes to transmit a frame, and $1 - \rho_i$ is the probability that the token is transferred to the next station without transmitting a frame. In this case the transfer of the token can be considered as the transfer of a frame with length 0. Accordingly, the mean token rotation time $\overline{T}_r$ can be calculated as follows:

$$\overline{T}_r = U + R^{-1} \cdot \overline{L} \cdot \sum \rho_i \,. \tag{13.2}$$

Here U denotes the free token rotation time (22μsec), R denotes the transfer rate (100 Mb/sec), and $\overline{L}$ denotes the mean frame length (see Table 13.10). With

$$\sum \rho_i = \frac{\sum \lambda_i}{\mu} = \frac{\lambda}{\mu}, \tag{13.3}$$

the approximation

$$\overline{T}_r \approx \frac{1}{\mu},$$

and Eq. (13.2), it follows that the service rate of the FDDI ring is given by

$$\mu = \frac{R - \lambda \overline{L}}{U \cdot R}. \tag{13.4}$$

With $\lambda = 2890/\text{sec}$, $\overline{L} = 382.5$ Bytes (see Table 13.10), we obtain the mean service rate

$$\mu = 41,435/\sec,$$

and the mean service time at the FDDI ring,

$$\overline{T}_r \approx \frac{1}{\mu} = 24\,\mu\text{sec}.$$

The variance of the token rotation time T_r is given by

$$\sigma_{T_r}^2 = R^{-2} \left(\rho \cdot \text{var}\,(L) + \rho \cdot \left(1 - \rho \sum p_i^2 \right) \overline{L}^2 \right), \tag{13.5}$$

where $\sigma_{T_r}^2$ and $\overline{L}^2$ can be calculated using the values from Table 13.9, the values of p_i are given in Table 13.4, and $\rho = \lambda/\mu = 0.070$ is given by the values of λ and μ [MSW88]. Then we obtain the squared coefficient of variation:

$$c_{T_r}^2 = 0.3.$$

The service time at each bridge is deterministic with the forwarding rate of $10,000$ frames/sec. And, of course, the coefficient of variation is 0 in this case.

The service time of the Ethernet is given by the sum of the transfer time T_t and the delay time T_d of a frame. To obtain the transfer time T_t we need the mean frame length, which we get from Table 13.9 given that the minimum frame size is 72 bytes in the CSMA/CD case. Thus, 54.1% of the frames have a size between 72 and 95 bytes, and it follows that

$$\overline{L}_{eth} = 395 \text{ bytes} = 3160 \text{ bit}, \quad c_{L_{eth}}^2 = 1.51.$$

Given a transfer rate of 10 Mb/sec (see Table 13.12), we finally have

$$\overline{T}_t = \frac{3160 \text{ bit}}{10 \text{ Mb/sec}} = 316 \text{ } \mu\text{sec} \quad c_{T_t}^2 = 1.51.$$

The mean delay time $\overline{T}_d$ of the Ethernet can be calculated using Fig. 13.9 and Table 13.12:

$$\overline{T}_d = 0.011 \text{ km} \cdot 5 \text{ } \mu\text{sec/km} + 0.005 \text{ km} \cdot 4.3 \text{ } \mu\text{sec/km} + 0.05 \text{ km} \cdot 4.8 \text{ } \mu\text{sec/km}$$
$$= 0.3 \text{ } \mu\text{sec}.$$

Table 13.12   Characteristics of the optical fiber, the Cheapernet, and the twisted pair

	Transfer Rate	Mean Length	Signal Time
Optical fiber	10 Mb/sec	11 m	5 μsec/km
Cheapernet	10 Mb/sec	5 m	4.3 μsec/km
Twisted pair	10 Mb/sec	50 m	4.8 μsec/km

In this case, $\overline{T}_d$ can be neglected compared to $\overline{T}_t$ and we have

$$\frac{1}{\mu_{eth}} = 316 \text{ } \mu\text{sec}.$$

To obtain the service rates of the IS node (WAN or computers in a section), we use the formula

$$\overline{K}_i = \frac{\lambda_i}{\mu_i},$$

and have

$$\mu_i = \frac{\lambda_i}{\overline{K}_i}. \tag{13.6}$$

The values of the λ_i are listed in Table 13.11. In order to get the $\overline{K}_i$ we need the mean number of active computers. This value cannot be measured directly, but we have the mean total number of active computers which is 170. To get an approximation, for the $\overline{K}_i$, we divide 170 by the number of sections, which is 12 (see Table 13.4 or Fig. 13.18). This yields

$$\overline{K}_i \approx \frac{170}{12} = 14.17 \, .$$

For sections with servers (multiple server node) we use

$$\rho_i = \frac{\lambda_i}{m_i \mu_i}$$

and obtain

$$\mu_i = \frac{\lambda_i}{m_i \rho_i} \tag{13.7}$$

and can calculate the service rates using the utilization ρ_i of the servers, which was measured approximately as 90%, and the number of servers m_i from Table 13.6. The values of the service rates are listed in Table 13.13.

Table 13.13 Service rates of the computers, servers, and the WAN

Section	1	2	3	4	5	6	7	8	9	10	11	12
μ_i	19.4	14.2	14.0	14.6	20.6	11.0	21.8	14.1	14.8	8.5	30.2	31.8

Now the closed non-product-form queueing network model of the considered communication system is completely defined. We solve it using Marie's method (see Section 10.1.4.2).

13.1.3.4 Results The closed queueing network model from Section 13.1.3.2 together with the model parameters (as derived in Section 13.1.3.3) is solved using the queueing network package PEPSY (see Chapter 12). The PEPSY input file for this example is given in Fig. 13.14. In PEPSY, either routing probabilities p_{ij} or visit ratios e_i can be used in the input file. Here we use the visit ratios, which are calculated from the originally given routing probabilities using Eq. (7.5). We have a closed queueing network; therefore we have the total number of jobs $K = 170$ as input. From the input file, PEPSY produces an output file, shown in Fig 13.15, with all performance measures. Since we have a non-product-form network, Marie's method was used for the analysis.

In the output file we see that the computed utilization of the servers is $\rho_{\text{serv}} = 0.9$, which matches with the estimated value from actual measurement, and that the utilization of the ring is $\rho_r = 0.07$, which also matches with the measured value. The mean queue length of the ring is negligible and queue length at the servers is $\overline{Q}_{cc2} = 3.1$, $\overline{Q}_{cc3} = 1.9$, and $\overline{Q}_{cc4} = 3.2$. The mean response time $\overline{T}$ for the LAN is 0.59 sec.

```
#
# filename e_lan170
#

NUMBER NODES: 36
NUMBER CLASSES: 1

NODE SPECIFICATION

  node |    name     |    type     |  node |    name     |    type
  -----+-------------+-------------+-------+-------------+-------------
    1  |  ring       |  -/G/1-FCFS  |  19  |  pc-b7      |  -/G/0-IS
    2  |  bridge-cc2 |  -/G/1-FCFS  |  20  |  bridge-b8  |  -/G/1-FCFS
    3  |  eth-cc2    |  -/G/4-FCFS  |  21  |  eth-b8     |  -/G/1-FCFS
    4  |  serv-cc2   |  -/G/14-FCFS |  22  |  pc-b8      |  -/G/0-IS
    5  |  bridge-cc3 |  -/G/1-FCFS  |  23  |  bridge-b9  |  -/G/1-FCFS
    6  |  eth-cc3    |  -/G/7-FCFS  |  24  |  eth-b9     |  -/G/4-FCFS
    7  |  serv-cc3   |  -/G/23-FCFS |  25  |  pc-b9      |  -/G/0-IS
    8  |  bridge-cc4 |  -/G/1-FCFS  |  26  |  bridge-b10 |  -/G/1-FCFS
    9  |  eth-cc4    |  -/G/2-FCFS  |  27  |  eth-b10    |  -/G/2-FCFS
   10  |  serv-cc4   |  -/G/13-FCFS |  28  |  pc-b10     |  -/G/0-IS
   11  |  bridge-b5  |  -/G/1-FCFS  |  29  |  bridge-b11 |  -/G/1-FCFS
   12  |  eth-b5     |  -/G/4-FCFS  |  30  |  eth-b11    |  -/G/5-FCFS
   13  |  pc-b5      |  -/G/0-IS    |  31  |  pc-b11     |  -/G/0-IS
   14  |  bridge-b6  |  -/G/1-FCFS  |  32  |  bridge-b12 |  -/G/1-FCFS
   15  |  eth-b6     |  -/G/5-FCFS  |  33  |  eth-b12    |  -/G/5-FCFS
   16  |  pc-b6      |  -/G/0-IS    |  34  |  pc-b12     |  -/G/0-IS
   17  |  bridge-b7  |  -/G/1-FCFS  |  35  |  wanrouter  |  -/G/1-FCFS
   18  |  eth-b7     |  -/G/3-FCFS  |  36  |  wan        |  -/G/0-IS

CLASS SPECIFICATION

  class | arrival rate   number of jobs
  ------+------------------------------------
    1   |     -              170

CLASS SPECIFIC PARAMETERS

CLASS 1               ( sc_o_v = squared coeffecient of variation )

  node       | service_rate | sc_o_v | visit_rat |  node       | service_rate | sc_o_v | visit_rat
  -----------+--------------+--------+-----------+-------------+--------------+--------+----------
  ring       |    41345     |  0.3   |  9.901    |  pc-b7      |    21.81     |   1    |  1.059
  bridge-cc2 |    9999      |  0.1   |  1.228    |  bridge-b8  |    9999      |  0.1   |  0.139
  eth-cc2    |    3164      |  1.51  |  1.292    |  eth-b8     |    3164      |  1.5   |  0.14
  serv-cc2   |    14.2      |  1     |  0.614    |  pc-b8      |    14.11     |   1    |  0.069
  bridge-cc3 |    9999      |  0.1   |  1.98     |  bridge-b9  |    9999      |  0.1   |  1.366
  eth-cc3    |    3164      |  1.51  |  2        |  eth-b9     |    3164      |  1.51  |  1.38
  serv-cc3   |    14.0      |  1     |  0.99     |  pc-b9      |    14.11     |   1    |  0.683
  bridge-cc4 |    9999      |  0.1   |  1.168    |  bridge-b10 |    9999      |  0.1   |  0.832
  eth-cc4    |    3164      |  1.51  |  1.18     |  eth-b10    |    3164      |  1.51  |  0.84
  serv-cc4   |    14.6      |  1     |  0.584    |  pc-b10     |    8.54      |   1    |  0.416
  bridge-b5  |    9999      |  0.1   |  2        |  bridge-b11 |    9999      |  0.1   |  2.93
  eth-b5     |    3164      |  1.51  |  2.02     |  eth-b11    |    3164      |  1.51  |  3.021
  pc-b5      |    20.61     |  1     |  1        |  pc-b11     |    30.21     |   1    |  1.465
  bridge-b6  |    9999      |  0.1   |  1.069    |  bridge-b12 |    9999      |  0.1   |  3.089
  eth-b6     |    3164      |  1.51  |  1.08     |  eth-b12    |    3164      |  1.51  |  3.152
  pc-b6      |    11.01     |  1     |  0.535    |  pc-b12     |    31.83     |   1    |  1.545
  bridge-b7  |    9999      |  0.1   |  2.119    |  wanrouter  |    9999      |  0.1   |  1.881
  eth-b7     |    3164      |  1.51  |  2.1      |  wan        |    19.41     |   1    |  0.941
```

Fig. 13.14   PEPSY input file for the LAN example.

Now we can use this model for some experiments. For example, we can change the number of active computers K. As we can then see in Table 13.14, the server nodes are the bottleneck and the number of active computers should not exceed 250 because the queue lengths become very large. The mean response time $\overline{T}$ is not influenced much by the number of active computers K in our example. Table 13.15 demonstrates the influence of changing the number of servers at server nodes. If the number of servers is reduced to $m_2 = 10$, $m_3 = 16$, and $m_4 = 9$, then we have a bottleneck at the server nodes. The situation deteriorates if we further reduce the number of servers. The utilization of the ring decreases when the number of servers decreases, whereas the utilization of the server nodes increases.

```
PERFORMANCE_MEASURE FOR NETWORK:  lan170
description of the network is in file 'e_lan170'
the closed network was solved using the method 'marie'
jobclass 1
```

marie	lambda	e	1/mu	rho	mvz	maa	mwz	mwsl
ring	2879.034	9.901	0.000	0.070	0.000	0.377	0.000	0.089
bridge-cc2	357.080	1.228	0.000	0.036	0.000	0.037	0.000	0.001
eth-cc2	375.690	1.292	0.000	0.030	0.000	0.119	0.000	0.000
serv-cc2	178.540	0.614	0.070	0.898	0.088	15.669	0.017	3.096
bridge-cc3	575.749	1.980	0.000	0.058	0.000	0.060	0.000	0.002
eth-cc3	581.564	2.000	0.000	0.026	0.000	0.184	0.000	0.000
serv-cc3	287.874	0.990	0.071	0.894	0.078	22.479	0.007	1.916
bridge-cc4	339.634	1.168	0.000	0.034	0.000	0.035	0.000	0.001
eth-cc4	343.123	1.180	0.000	0.054	0.000	0.113	0.000	0.004
serv-cc4	169.817	0.584	0.069	0.895	0.087	14.858	0.019	3.227
bridge-b5	581.564	2.000	0.000	0.058	0.000	0.060	0.000	0.002
eth-b5	587.380	2.020	0.000	0.046	0.000	0.186	0.000	0.000
pc-b5	290.782	1.000	0.049	0.000	0.049	14.109	0.000	0.000
bridge-b6	310.846	1.069	0.000	0.031	0.000	0.032	0.000	0.001
eth-b6	314.045	1.080	0.000	0.020	0.000	0.099	0.000	0.000
pc-b6	155.568	0.535	0.091	0.000	0.091	14.130	0.000	0.000
bridge-b7	616.167	2.119	0.000	0.062	0.000	0.064	0.000	0.003
eth-b7	622.274	2.140	0.000	0.066	0.000	0.199	0.000	0.003
pc-b7	307.938	1.059	0.046	0.000	0.046	14.119	0.000	0.000
bridge-b8	40.419	0.139	0.000	0.004	0.000	0.004	0.000	0.000
eth-b8	40.709	0.140	0.000	0.013	0.000	0.013	0.000	0.000
pc-b8	20.064	0.069	0.071	0.000	0.071	1.422	0.000	0.000
bridge-b9	397.208	1.366	0.000	0.040	0.000	0.041	0.000	0.001
eth-b9	401.279	1.380	0.000	0.032	0.000	0.127	0.000	0.000
pc-b9	198.604	0.683	0.071	0.000	0.071	14.075	0.000	0.000
bridge-b10	241.931	0.832	0.000	0.024	0.000	0.025	0.000	0.000
eth-b10	244.257	0.840	0.000	0.039	0.000	0.078	0.000	0.001
pc-b10	120.965	0.416	0.117	0.000	0.117	14.165	0.000	0.000
bridge-b11	852.282	2.931	0.000	0.085	0.000	0.090	0.000	0.005
eth-b11	878.453	3.021	0.000	0.056	0.000	0.278	0.000	0.000
pc-b11	425.996	1.465	0.033	0.000	0.033	14.101	0.000	0.000
bridge-b12	898.226	3.089	0.000	0.090	0.000	0.096	0.000	0.006
eth-b12	916.545	3.152	0.000	0.058	0.000	0.290	0.000	0.000
pc-b12	449.258	1.545	0.031	0.000	0.031	14.114	0.000	0.000
wanrouter	546.961	1.881	0.000	0.055	0.000	0.057	0.000	0.002
wan	273.626	0.941	0.051	0.000	0.051	14.097	0.000	0.000

```
characteristic indices:
```

marie	lambda	mvz	maa
	290.782	0.585	170.000

```
legend

e    : average number of visits        mu    : service rate
rho  : utilisation                     lambda: mean throughput
mvz  : average response time
maa  : average number of jobs
mvz  : average waiting time
mwsl : average queue-length
```

Fig. 13.15 PEPSY output file for the queueing network model of the communication system (cc_i = computing center$_i$, b_i = building$_i$).

These two experiments show that it is easy to study the influence of the variation in the parameters or the structure of the LAN once we have constructed and parameterized the queueing network model. We can use PEPSY, or any other queueing network package such as QNAP2 [VePo85] or RESQ [SaMa85], for analysis. To obtain the mean response time for a computer, its service time has to be subtracted from mean system response time $\overline{T} = 0.585$ and we obtain for the computers in building 5, for example, the mean response time 0.54 sec.

Problem 13.6   Verify the results of the communication system model just described using another queueing network package such as RESQ or QNAP2.

Table 13.14 Performance measures for the communication system for different numbers of active computers K

K	100	130	150	170	180	200	300
ρ_2	0.55	0.71	0.81	0.90	0.93	0.97	0.99
ρ_3	0.55	0.71	0.81	0.89	0.92	0.97	0.99
ρ_4	0.55	0.71	0.81	0.90	0.92	0.97	0.99
$\overline{Q}_2$	0.05	0.6	1.4	3.1	4.8	7.9	50
$\overline{Q}_3$	0.02	0.3	0.8	1.9	2.9	8.3	30
$\overline{Q}_4$	0.11	0.6	1.4	3.2	4.8	9.3	43
$\overline{T}$	0.56	0.56	0.57	0.59	0.60	0.64	0.94
ρ_{ring}	0.043	0.055	0.063	0.070	0.072	0.075	0.077

Notes: $m_2 = 14$, $m_3 = 23$, $m_4 = 13$.

Table 13.15 Performance measures for the communication system for different number of servers m_2, m_3, and m_4 of section 1, section 2, and section 3, respectively

m_2	7	10	14	16	18	28
m_3	12	16	23	18	26	46
m_4	7	9	13	16	17	26
ρ_2	0.999	0.96	0.90	0.69	0.73	0.47
ρ_3	0.95	0.98	0.89	0.998	0.88	0.47
ρ_4	0.92	0.98	0.90	0.64	0.71	0.47
$\overline{Q}_2$	63	10	3.1	0.4	0.5	0
$\overline{Q}_3$	9.6	13	1.9	28	0.7	0
$\overline{Q}_4$	7.0	23	3.2	0.2	0.5	0
$\overline{T}$	1.05	0.77	0.59	0.67	0.56	0.56
ρ_{ring}	0.039	0.053	0.070	0.061	0.072	0.073

Note: $K = 170$.

13.1.4 Proportional Differentiated Services

13.1.4.1 Introduction Two quality of service architectures in the Internet context have been proposed: Integrated Services (IntServ) and Differentiated Services (DiffServ). IntServ was proposed as a model to introduce flow-based quality of service into the Internet, which has traditionally been limited to best-effort type services [Whit97].

The differentiated services (DS) architecture is based on a model where traffic entering a DS domain is classified and conditioned at the edges, then marked and assigned to different behavior aggregates (BA) by setting the value of the DiffServ Code Point (DSCP). Within the core of the network, packets are forwarded according to the per-hop behavior (PHB) associated with the DS codepoint [BBC+98, NBBB98].

Two approaches to service differentiation have been defined: *absolute differentiated services* and *relative differentiated services*. The first relies on admission control and resource reservation mechanisms in order to provide guarantees and statistical assurances for performance measures, such as minimum service rate and end-to-end delay [DoRa99]. The approach of relative differentiated services groups network traffic in a small number of classes. These classes are ordered based on their packet forwarding quality, in terms of queueing delays and packet losses [DSR99, DoRa99]. This ordering should guarantee that higher classes $(j > i, \forall i, j \in \{1, \ldots, R\}$, where R denotes the number of traffic classes) experience a better forwarding behavior.

The *proportional differentiation model* is a further refinement of the relative differentiated services approach. It is thought to control the quality spacing between classes at each hop, where the packet forwarding performance for each class are ratioed proportionally to certain differentiation parameters set by the network operator [DoRa99].

13.1.4.2 Dynamic Priority Scheduling

Proportional Delay Differentiation The proportional differentiation model states that certain traffic class performance metrics should be proportional to the differentiation parameters the network operator chooses [DoRa99].

The proportional differentiation model can be applied in the context of queueing delays where $E[W_i)]$ is the average queueing delay of the class i packets. The proportional delay differentiation states that for all pairs (i, j) of traffic classes:

$$\frac{E[W_i]}{E[W_j]} = \frac{\delta_i}{\delta_j} . \qquad (13.8)$$

The parameters δ_i are known as the *Delay Differentiation Parameters (DDPs)*. As higher classes $(j > i, \forall i, j \in \{1, \ldots, R\})$ should experience shorter delays than lower classes, the DDPs satisfy the relation $\delta_1 > \delta_2 > \ldots > \delta_R$.

Dynamic Scheduling Model The waiting time priority scheduler used to implement the proportional delay differentiation is a non-preemptive packet scheduling discipline, characterized by a set of parameters b_r, $1 \leq r \leq R$. The priority of a class-r packet at time t, which arrives to the system at time t_0, increases with the elapsed time in the queueing system linearly with the rate b_r. The priority of a class-r packet, $1 \leq r \leq R$, denoted by $q_r(t)$, is given by (see Eq. (6.146)):

$$q_r(t) = (t - t_0)b_r, \qquad 1 \leq r \leq R. \qquad (13.9)$$

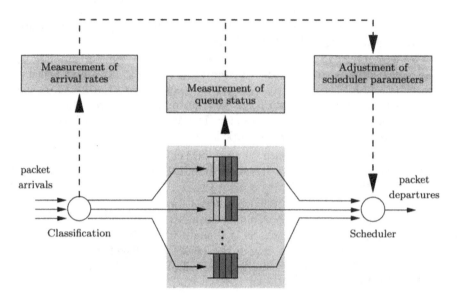

Fig. 13.16 Dynamic priority scheduling.

The mean queueing delay of a class-r packet in an M/G/1 queueing system is given by (see Eq. (6.147))

$$E[W_r] = \frac{E[W_{FIFO}] - \sum_{i=1}^{r-1} \rho_i E[W_i](1 - \frac{b_i}{b_r})}{1 - \sum_{i=r+1}^{R} \rho_i(1 - \frac{b_r}{b_i})}, \tag{13.10}$$

where ρ_i denotes the utilization of class i, $1 \leq i \leq R$. The queueing delay given in Eq. (13.10) refers to the long-time average queueing delay and is therefore not time-dependent, as apposed to queueing delays of Eq. (13.8). The characterization of the queueing system and the use of Eq. (13.10) enables us to analytically derive interesting properties of the scheduler in the particular case of the proportional differentiation model.

The proposed dynamic priority scheduling is depicted by Fig. 13.16. The diagram shows the router classification, queueing and scheduling functions. The router is furthermore extended by a weight adjustment module, which when triggered activates a new set of scheduler parameters based on collected measurement data. The measurements yield estimates of the packet arrival rates in different queues, service rates of different traffic classes and the backlog in the queues. As we assume infinite buffers in this paper, only arrival and service rates and the derived utilizations are considered.

For the dynamic priority scheduling, we extend the priority function in Eq. (13.9) as follows. Instead of fixing the rate of increase b_r for the priorities of class-r packets, the increase rate b_r is dynamically adjusted depending on

- the current system conditions, defined by the utilizations of the R traffic classes $\rho_1, \rho_2, \ldots, \rho_R$, and

- the differentiation requirements, defined by the delay differentiation parameters $\delta_1, \delta_2, \ldots, \delta_R$.

The rate of increase b_r for the priorities of class-r packets is regarded as a function of the parameters $\rho_1, \rho_2, \ldots, \rho_R$ and $\delta_1, \delta_2, \ldots, \delta_R$. We define the dynamic priority scheduler as a scheduler which applies the priority function:

$$q_r(t) = (t - t_0)b_r(\rho_1, \ldots, \rho_R, \delta_1, \ldots, \delta_R), \qquad 1 \le r \le R. \tag{13.11}$$

Note that the rates of increase for **all** classes are adjusted simultaneously, in order to avoid the case that higher classes get lower rates of increase than lower classes ($b_j > b_i, \forall i, j \in \{1, \ldots, R\}$ and $j > i$).

Optimization Problem Determining the rates of increase $b_1, b_2, \ldots, b_R$ of the priority functions $q_1, q_2, \ldots, q_R$ can in fact be considered as an optimization problem which can be stated as follows: given the delay differentiation parameters $\delta_1, \delta_2, \ldots, \delta_R$ and the utilizations of the traffic classes $\rho_1, \rho_2, \ldots, \rho_R$, determine the parameters $b_1, b_2, \ldots, b_R$ so that the scheduler achieves the required ratios of the average queueing delay (see Eq. (13.8)). The ratios of the experienced queueing delays are ideally equal to the required delay differentiation ratios for all pairs of consecutive classes, which in terms of the sum of error squares is given by

$$\sum_{i=1}^{R-1} \left(\frac{E[W_i]}{E[W_{i+1}]} - \frac{\delta_i}{\delta_{i+1}} \right)^2 = 0, \tag{13.12}$$

where the mean queueing delay $E[W_i]$ of a class-i packets is given by Eq.(13.10). In case the average delay differentiation is not feasible, we minimize the cost function

$$\sum_{i=1}^{R-1} g_i \cdot \left(\frac{E[W_i]}{E[W_{i+1}]} - \frac{\delta_i}{\delta_{i+1}} \right)^2 \to min, \tag{13.13}$$

which means that the ratios of the average queueing delay have to be as close as possible to the required DDP ratios. In Eq. (13.13) the sum of error squares is to be minimized. The positive factor g_i in Eq. (13.13) is used to weight specific delay ratios, e.g., higher weights for the delay ratios of the higher priority classes. Here all weights $g_i, 1 \le i < R$ are set to 1.

Exact Solution for two Traffic Classes For two traffic classes, a set of results can be derived ([BEM01, EBA01, LLY00]):

1. A specific delay differentiation ratio $\frac{\delta_2}{\delta_1}$ is feasible, if and only if the system utilization ρ satisfies the condition

$$\rho > 1 - \frac{\delta_2}{\delta_1}. \tag{13.14}$$

For example a differentiation ratio of 4 ($\frac{\delta_1}{\delta_2} = 4$), cannot be achieved at a load less than or equal to 75%.

2. In order to meet predefined delay differentiation requirements specified by δ_1 and δ_2, the rates of increase of the priority functions are set to

$$b_1 = 1 \text{ and} \tag{13.15}$$

$$\frac{b_1}{b_2} = 1 - \frac{1}{\rho}(1 - \frac{\delta_2}{\delta_1}). \tag{13.16}$$

3. As the load approaches 100%, the ratio of the scheduler parameters b_1/b_2 approaches the inverse of the corresponding DDPs (consider limit of Eq. (13.16) as the utilization ρ tends to 1).

4. The scheduler parameters do not depend on the class load distribution. These parameters depend only on the *total* utilization in the queueing system, which is in line with the simulation results presented in [DoRa99].

For further details on multiclass case and numerical results see [EsBo03].

13.1.5   UNIX Kernel

We now develop a performance model of the UNIX operating system [GrBo96a]. A job in the considered UNIX system is in the *user* context when it executes user code, in the *kern* context when it executes code of the operating system kernel, or in the *driv* context when it executes driver code to set up or perform an I/O operation. In Fig. 13.17 the life cycle of a UNIX job is shown.

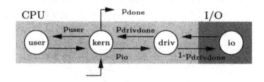

Fig. 13.17   Model of a UNIX job.

The job always starts in the *kern* context from where it switches to the *user* context with probability p_{user} or to the *driv* context with probability p_{io}. After a mean service time s_{user}, the job returns from the *user* context to the *kern* context. From the *driv* context, where the job remains s_{drive} time units, the job returns to the *kern* context with probability p_{drivdone} or starts a I/O operation with probability $1 - p_{\text{drivdone}}$. After finishing the I/O, the job returns to the context *driv* and again remains there (with mean time s_{driv}) and returns to the *kern* context with probability p_{drivdone}. A job can only leave the system when it is in the *kern* context (with probability p_{done}). The mean service time in the *kern* context is s_{kern}. The I/O operation is carried

Table 13.16 Routing probabilities of the monoprocessor model

	(1,1)	(1,2)	(1,3)	(2,1)	(2,2)	(2,3)
(1,1)	0	$p_{drivdone}$	0	$1 - p_{drivdone}$	0	0
(1,2)	p_{io}	p_{done}	p_{user}	0	0	0
(1,3)	0	1	0	0	0	0
(2,1)	1	0	0	0	0	0
(2,2)	0	0	0	0	0	0
(2,3)	0	0	0	0	0	0

out on an I/O device, whereas the remaining three activities are serviced by the CPU.

13.1.5.1 Model of the UNIX Kernel The UNIX kernel just described can be modeled by a closed non-product-form queueing network with two nodes (CPU, I/O) and three job classes, priorities, and class-switching (see Fig. 13.18).

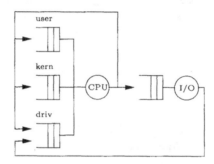

Fig. 13.18 The monoprocessor model.

Since we have only one processor, we call this the *monoprocessor model*. The three job classes correspond to the *user*, *kern*, and *driv* context. To obtain a closed network, we assume that when a job leaves the system a new job immediately arrives at the *kern* queue.

A state in the routing DTMC of this model is described as a pair (node number, class number) with

node number 1 : CPU,   class number 1 : driv,

node number 2 : I/O,   class number 2 : kern,

class number 3 : user.

The routing probabilities are shown in Table 13.16. In the following we summarize the assumptions we make:

- The network is closed and there are $K = 10$ jobs in the system.

- The service times are exponentially distributed.

- The system has three job classes *user, kern, driv* with the priority order driv > kern > user.

- Class switching is allowed.

- Jobs in the context *user* can always be preempted by *driv* and *kern* jobs. Other preemptions are not possible.

- The time unit used in this section is msec.

- The following parameter values are used:

$$p_{io} = 0.05, \qquad s_{user} = \text{varied from } 0.25 \text{ to } 20.0,$$
$$p_{done} = 0.01, \qquad s_{kern} = 1.0,$$
$$p_{drivdone} = 0.4, \quad s_{driv} = 0.5.$$

- The I/O system consists of several devices that are assumed to form a composite node (see Section 8.3) with load-dependent service times (which were measured from the real system):

$$s_{io}(1) = 28.00, \qquad s_{io}(6) \; = 12.444,$$
$$s_{io}(2) = 18.667, \qquad s_{io}(7) \; = 12.000,$$
$$s_{io}(3) = 15.555, \qquad s_{io}(8) \; = 11.667,$$
$$s_{io}(4) = 14.000, \qquad s_{io}(9) \; = 11.407,$$
$$s_{io}(5) = 13.067, \qquad s_{io}(10) = 11.200.$$

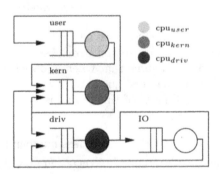

Fig. 13.19 The transformed monoprocessor model.

The parameters s_{user}, s_{kern}, and s_{driv} are the mean service times of class i ($i = 1, 2, 3$) at the CPU (node 1), and $s_{io}(k)$ is the mean load-dependent service time of class 1 jobs at the I/O devices (node 2). At the CPU (node 1) we have a mixture of preemptive and non-preemptive service strategy, while at the I/O devices (node 2) we have only one job class with the FCFS service discipline. The following concepts are used to solve this model:

- Chain concept (see Section 7.3.6)

- Extended shadow technique with mixed priority strategy and class switching (see Section 10.3.2.2).

Since we have three job classes, we split the CPU into three CPUs, one for each job class (see Fig. 13.19). Approximate iterative solution of the resulting model can be carried out by PEPSY or SHARPE.

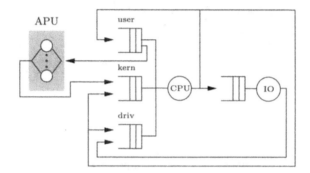

Fig. 13.20   The master–slave model.

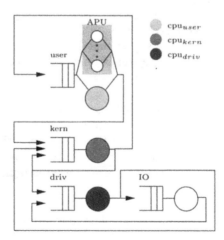

Fig. 13.21   The transformed master–slave model.

13.1.5.1.1   The Master–Slave Model   For this model, shown in Fig. 13.20, we have the same basic assumptions as for the monoprocessor model but now we also have two APUs. An APU is an additional processor (associate processing unit) that works with the main CPU. It is assumed that only jobs in the *user* context can be processed at the APU. Since this is the only job class that can be processed at the APU, no preemption can take place there. If the APU as well as the CPU are free and there are jobs in the user queue, then the processor (CPU, APU) on which the job is served is decided randomly. If we apply the shadow transformation to the original model, we get the transformed master–slave model, shown in Fig. 13.21.

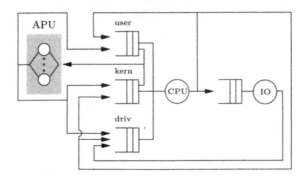

Fig. 13.22   The associated processor model.

13.1.5.1.2 The Associated Processor Model   The assumptions for the associated processor model, shown in Fig. 13.22, are the same as for the master slave model but now jobs in the *kern* context can also be processed on the APU. Since jobs in *kern* context have higher priority than jobs in *user* context, preemption can also take place at the APU. The model needs to be transformed so that it can be solved approximately using the shadow technique in combination with the MVA. The transformed associated processor model is shown in Fig. 13.23.

13.1.5.2 Analysis   We now solve the three non-product-form network models using approximate techniques and compare the results to the ones obtained from the (exact) numerical solution of the underlying CTMC. We show the computation of the performance measures for the monoprocessor model step by step. The procedure is analogous for the other two models and therefore only the results are given.

13.1.5.2.1 The Monoprocessor Model   The model, given in Fig. 13.18, contains three job classes and two nodes:

$$
\begin{aligned}
&\text{class 1} : driv, \\
&\text{class 2} : kern, \\
&\text{class 3} : user,
\end{aligned}
\qquad
\begin{aligned}
&\text{node 1} : \text{CPU}, \\
&\text{node 2} : \text{I/O}.
\end{aligned}
$$

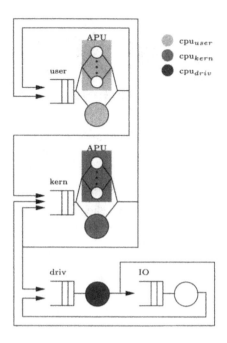

Fig. 13.23 The transformed associated processor model.

The priority order of the job classes is $driv > kern > user$. Jobs in the context *user* can always be preempted by *driv* and *kern* jobs. Other preemptions are not possible. The chosen numerical parameter values are

$$p_{11,12} = p_{\text{drivdone}} = 0.4, \qquad\qquad p_{11,21} = 1 - p_{\text{drivdone}} = 0.6,$$
$$p_{12,11} = p_{\text{io}} = 0.05, \qquad\qquad p_{12,12} = p_{\text{done}} = 0.01,$$
$$p_{12,13} = p_{\text{user}} = 0.94, \qquad\qquad p_{13,12} = 1.0,$$
$$p_{21,11} = 1.0,$$
$$s_{11} = s_{\text{driv}} = 0.5, \qquad\qquad \epsilon = 0.001,$$
$$s_{12} = s_{\text{kern}} = 1.0, \qquad\qquad N = 5 \text{ jobs},$$
$$s_{13} = s_{\text{user}} = 1.5,$$
$$s_{21} = s_{\text{io}} = \text{load dep.}$$

Now the analysis of the monoprocessor model can be done in the following nine steps:

STEP 1 Compute the visit ratios e_{ir} using Eq. (7.15):

$$e_{11} = e_{\text{driv}} = 12.5, \qquad e_{12} = e_{\text{kern}} = 100,$$
$$e_{13} = e_{\text{user}} = 94, \qquad e_{21} = e_{\text{io}} = 7.5.$$

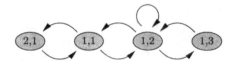

Fig. 13.24 The routing DTMC for the monoprocessor model.

STEP 2 Determine the chains in the network. The DTMC state diagram for the routing matrix in the monoprocessor model is given in Fig. 13.24. As we can see, every state can be reached from every other state, which means that we have only one chain in the system.

STEP 3 Transform the network into the shadow network (see Fig. 13.19). The corresponding parameter values for the transformed model are

$$s_{11} = 0.5, \qquad s_{22} = 1.0, \qquad s_{33} = 1.0, \qquad s_{41} = \text{load dep.}$$
$$e_{11} = 12.5, \qquad e_{22} = 100, \qquad e_{33} = 94, \qquad e_{41} = 7.5.$$

STEP 4 Compute the visit ratios e_{iq}^* per chain and recall that we have only one chain:

$$e_{iq}^* = \frac{\sum\limits_{r \in \pi_q} e_{ir}}{\sum\limits_{r \in \pi_q} e_{1r}}.$$

Then we get

$$e_{11}^* = \frac{e_{11}}{e_{11}} = 1, \qquad e_{21}^* = \frac{e_{22}}{e_{11}} = 8,$$
$$e_{31}^* = \frac{e_{33}}{e_{11}} = 7.52, \qquad e_{41}^* = \frac{e_{41}}{e_{11}} = 0.6.$$

STEP 5 Compute the scale factors α_{ir}, using the equation

$$\alpha_{ir} = \frac{e_{ir}}{\sum\limits_{s \in \pi_q} e_{is}},$$

and get

$$\alpha_{11} = \frac{e_{11}}{e_{11} + e_{12} + e_{13}} = 1.0, \qquad \alpha_{31} = \frac{e_{31}}{e_{31} + e_{32} + e_{33}} = 0,$$
$$\alpha_{12} = \frac{e_{12}}{e_{11} + e_{12} + e_{13}} = 0, \qquad \alpha_{32} = \frac{e_{32}}{e_{31} + e_{32} + e_{33}} = 0,$$
$$\alpha_{13} = \frac{e_{13}}{e_{11} + e_{12} + e_{13}} = 0, \qquad \alpha_{33} = \frac{e_{33}}{e_{31} + e_{32} + e_{33}} = 1.0,$$
$$\alpha_{21} = \frac{e_{21}}{e_{21} + e_{22} + e_{23}} = 0, \qquad \alpha_{41} = \frac{e_{41}}{e_{41} + e_{42} + e_{43}} = 1.0,$$

$$\alpha_{22} = \frac{e_{22}}{e_{21} + e_{22} + e_{23}} = 1.0, \qquad \alpha_{42} = \frac{e_{42}}{e_{41} + e_{42} + e_{43}} = 0,$$

$$\alpha_{23} = \frac{e_{23}}{e_{21} + e_{22} + e_{23}} = 0, \qquad \alpha_{43} = \frac{e_{43}}{e_{41} + e_{42} + e_{43}} = 0.$$

STEP 6 Compute the mean service times for the chain:

$$s_{iq}^* = \frac{1}{\mu_{iq}} = \sum_{r \in \pi_q} s_{ir} \cdot \alpha_{ir},$$

$$s_{11}^* = 0.5, \qquad s_{21}^* = 1.0, \qquad s_{31}^* = 1.5, \qquad s_{31}^* = \text{load dep.}$$

STEP 7 Start the shadow iterations:

Iteration 1:

$$\rho_{11}^* = 0.037, \qquad \rho_{21}^* = 0.594, \qquad \tilde{s}_{11}^{(1)*} = 0.5,$$

$$\rho_{31}^* = 0.838, \qquad \rho_{41}^* = 0.659, \qquad \tilde{s}_{21}^{(1)*} = 1.03842,$$

$$\lambda_{11}^* = 0.075, \qquad \lambda_{21}^* = 0.594, \qquad \tilde{s}_{31}^{(1)*} = 4.06504,$$

$$\lambda_{31}^* = 0.558, \qquad \lambda_{41}^* = 0.045, \qquad \tilde{s}_{41}^{(1)*} = \text{load dep.}$$

Iteration 2:

$$\rho_{11}^* = 0.016, \qquad \rho_{21}^* = 0.27, \qquad \tilde{s}_{11}^{(2)*} = 0.5,$$

$$\rho_{31}^* = 0.994, \qquad \rho_{41}^* = 0.289, \qquad \tilde{s}_{21}^{(2)*} = 1.01626,$$

$$\lambda_{11}^* = 0.033, \qquad \lambda_{21}^* = 0.26, \qquad \tilde{s}_{31}^{(2)*} = 2.10084,$$

$$\lambda_{31}^* = 0.245, \qquad \lambda_{41}^* = 0.02, \qquad \tilde{s}_{41}^{(2)*} = \text{load dep.}$$

Iteration 3:

$$\rho_{11}^* = 0.030, \qquad \rho_{21}^* = 0.481, \qquad \tilde{s}_{11}^{(3)*} = 0.5,$$

$$\rho_{31}^* = 0.935, \qquad \rho_{41}^* = 0.525, \qquad \tilde{s}_{21}^{(3)*} = 1.03092,$$

$$\lambda_{11}^* = 0.033, \qquad \lambda_{21}^* = 0.473, \qquad \tilde{s}_{31}^{(3)*} = 3.16373,$$

$$\lambda_{31}^* = 0.245, \qquad \lambda_{41}^* = 0.02, \qquad \tilde{s}_{41}^{(3)*} = \text{load dep.}$$

$$\vdots$$

Iteration 15: Finally, after 15 iterations, the solution converges and we have the following results.

$$\rho_{11}^* = 0.025, \qquad \rho_{21}^* = 0.402, \qquad \tilde{s}_{11}^{(15)*} = 0.5,$$

$$\rho_{31}^* = 0.969, \qquad \rho_{41}^* = 0.435, \qquad \tilde{s}_{21}^{(15)*} = 1.0256,$$

$$\lambda_{11}^* = 0.049, \qquad \lambda_{21}^* = 0.392, \qquad \tilde{s}_{31}^{(15)*} = 2.61780,$$

$$\lambda_{31}^* = 0.368, \qquad \lambda_{41}^* = 0.029, \qquad \tilde{s}_{41}^{(15)*} = \text{load dep.}$$

STEP 8 Retransform the chain values into class values:

$$\lambda_{11} = \alpha_{11}\lambda_{11}^* = 0.049,$$
$$\lambda_{22} = \alpha_{22}\lambda_{21}^* = 0.392,$$
$$\lambda_{33} = \alpha_{33}\lambda_{31}^* = 0.368,$$
$$\lambda_{41} = \alpha_{41}\lambda_{41}^* = 0.029.$$

If we rewrite the results of the shadow network in terms of the original network, we get the following final results:

$$\lambda_{11}^{(\text{approx})} = 0.049, \quad \lambda_{12}^{(\text{approx})} = 0.392,$$
$$\lambda_{13}^{(\text{approx})} = 0.368, \quad \lambda_{21}^{(\text{approx})} = 0.029.$$

The overall throughput of the network is then given by

$$\lambda^{(\text{approx})} = 1000 \cdot p_{\text{done}} \cdot \lambda_{12}^{(\text{approx})} = 3.92.$$

STEP 9 Verify the results: To verify the results, we construct and solve the underlying CTMC using **MOSEL-2**. The exact overall throughput is given by

$$\lambda^{(\text{exact})} = 3.98.$$

The difference between the exact and the approximate throughput is small.

The other two models were also analyzed using the extended shadow technique with class switching in a similar manner.

The overall system throughput of the UNIX models is plotted as a function of the user service time in Fig. 13.25. There is nearly no difference between the approximate values and the exact values obtained from the numerical solution of the CTMC for the chosen parameter set. As expected, the associated processor model has the highest system throughput. In a more detailed study it was found that for the master slave model the throughput cannot be increased any more by adding additional APUs because the CPU is the bottleneck. For systems that spend a lot of time doing system jobs (UNIX is said to spend approximately 50% of its time doing system work) it is worth having a parallel system kernel. For more I/O-intensive jobs, the number of peripheral devices must also be increased, otherwise the processors have to wait too long for I/O jobs to be completed. In this case it is not worth having a multiprocessor system. Another very important result from this study is that for $s_{\text{user}} \geq 3$ msec, the master slave and associated processor model behave nearly the same and only for $s_{\text{user}} < 3$ msec does the associated processor model gives better results than the master slave model.[1] However, it does not seem to matter

---

[1]This result means that if the system will be used for very time consuming jobs, it is worthwhile to have a parallel kernel.

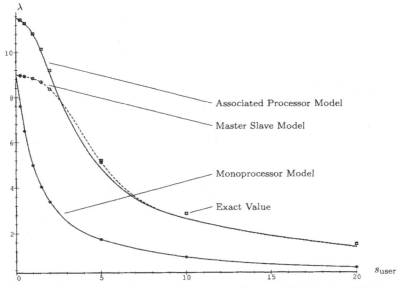

Fig. 13.25   Throughput for the different models covered in Section 13.1.5.

as to which of the two (an associated processor configuration or a master–slave one) configurations is used. On the other hand, if the system is used in a real-time environment where response times are required to be extremely short, the associated processor model delivers better performance than the master–slave configuration and it is worthwhile to have a completely parallel kernel.

The computation time to solve the model was approximately 45 minutes if the CTMC method is used. Using the shadow technique, the results were generated within seconds for the entire parameter set. By contrast, when this non-product-form network was solved using DES, it took nearly 120 hours of computing time. The impact of priorities is shown for the master–slave model in Table 13.17. Here p_{io} is varied while p_{done} is constant. On the right-hand side of the table the throughput for the network without priorities (nonpre), the original values with priorities (orig = obtained from numerical solution of the CTMC), and approximate values (approx = obtained using the shadow approximation) are given. As can be seen in the table, for some parameter values the impact of priorities cannot be neglected.

13.1.6   J2EE Applications

13.1.6.1   Introduction   Over the past several years, the Java 2 Enterprise Edition Platform (J2EE) has established itself as major technology for developing modern e-business solutions.

Table 13.17   Impact of the priorities ($p_{\text{done}} = 0.01$)

S_{user}	p_{io}	Throughput		
		Nonpre	Approx	Orig
0.50	0.9	0.6593	0.6594	0.6593
0.50	0.01	9.0049	9.8765	9.8765
0.50	0.001	9.0897	9.9875	9.9875
1.00	0.9	0.6593	0.6594	0.6593
1.00	0.01	8.2745	9.8736	9.8716
1.00	0.001	8.3398	9.9859	9.9832
1.50	0.9	0.6593	0.6594	0.6592
1.50	0.01	7.6517	9.8194	9.7666
1.50	0.001	7.7027	9.9395	9.8983
2.00	0.9	0.6593	0.6594	0.6592
2.00	0.01	7.1078	9.5178	9.4700
2.00	0.001	7.1489	9.6326	9.4000
2.50	0.9	0.6592	0.6593	0.6592
2.50	0.01	6.6170	8.8123	8.6144
2.50	0.001	6.6485	8.8802	8.5100

The aim of J2EE is to enable developers to quickly and easily build scalable, reliable and secure applications without having to develop their own complex *middleware services*. One of the key services within this infrastructure, provided by most J2EE application servers, is clustering. When a J2EE application is deployed in a clustered environment, its components are transparently replicated on the servers participating in the cluster. Client requests are then load-balanced across cluster nodes and in this way scalability can be achieved.

Modern J2EE applications are typically based on highly distributed, multi-tiered architectures comprising multiple components deployed in a clustered environment. The inherent complexity of the latter makes it difficult for system deployers to estimate the size and capacity of the deployment environment needed to guarantee that Service Level Agreements (SLAs) are met. Deployers are often faced with questions such as the following:

- What hardware and software resources are needed to guarantee that SLAs are met? More specifically, how many application servers need to be included in the clusters used and how fast should they be?

- What are the maximum load levels that the system will be able to handle in the production environment?

- What would the average response time, throughput, and resource utilization be under the expected workload?

- Which components have the largest effect on the overall system performance and are they potential bottlenecks?

The application studied here is the SPECjAppServer 2002 J2EE benchmark, a real-world application which is an adapted and shortened version of the contribution [KoBu03] by Samuel Kounev and Alejandro Buchmann.

13.1.6.2 The SPECjAppServer 2002 Benchmark

SPECjAppServer 2002 is a new industry standard benchmark for measuring the performance and scalability of J2EE-based application servers. Server vendors can use the SPECjAppServer benchmarks to measure, optimize and showcase their product's performance and scalability. Their customers, on the other hand, can use them to gain a better understanding and insight into the tuning and optimization issues surrounding the development of modern J2EE applications.

SPECjAppServer 2002 *Business Model* The SPECjAppServer 2002 workload is based on a distributed application claimed to be large enough and complex enough to represent a real-world e-business system [Stan02].

SPECjAppServer 2002 models businesses using four domains. Figure 13.26 illustrates these domains and gives some examples of typical transactions run in each of them.

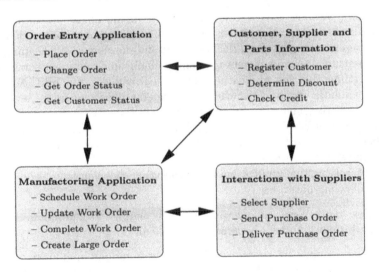

Fig. 13.26 SPECjAppServer 2002 business domains.

The *customer domain* models customer interactions using an order entry application, which provides some typical online ordering functionality. Orders can be placed by individual customers as well as by distributors. Orders placed by distributors are called *large orders*.

The *manufacturing domain* models the activity of production lines in a manufacturing plant. Products manufactured by the plant are called *widgets*. There are two types of production lines, namely *planned lines* and *large order lines*. Planned lines run on schedule and produce a predefined number of widgets. Large order lines run only when a large order is received in the customer domain. The unit of work in the manufacturing domain is a *work order*. Each work order is for a specific quantity of a particular type of widget. When a work order is created, the bill of materials for the corresponding type of widget is retrieved and the required parts are taken out of inventory. As the widgets move through the assembly line, the work order status is updated to reflect progress. Once the work order is complete, it is marked as completed and inventory is updated. When inventory of parts gets depleted, suppliers need to be located and *purchase orders (POs)* need to be sent out. This is done by contacting the *supplier domain*, which is responsible for interactions with external suppliers. The *corporate domain* is managing all customer, product, and supplier information.

Modeling **SPECjAppServer 2002** This section discusses how to build and validate a queueing network model of SPECjAppServer 2002 and, then, how to exploit this model for the purposes of performance prediction in the capacity planning process.

Motivation Assume that a company wishes to automate its internal and external business operations with the help of e-business technology. The company chooses to employ the J2EE platform and develops a J2EE application for supporting its order-inventory, supply-chain and manufacturing operations. Assume that this application is the one provided in the SPECjAppServer 2002 benchmark. Assume also that the company plans to deploy the application in the deployment environment depicted in Fig. 13.27. This environment uses a cluster of WebLogic servers (WLS) as a J2EE container and an Oracle database server (DBS) for persistence. We assume that all machines in the WLS cluster are identical.

Before putting the application into production the company needs answers to the following questions:

- How many WebLogic servers would be needed to guarantee adequate performance under the expected workload?

- What would the average transaction throughput and mean response time be? How utilized (CPU/Disk utilization) would the WebLogic servers and the database server be?

- Will the capacity of the database server suffice to handle the incoming load?

- Does the system scale or are there any other potential system bottlenecks?

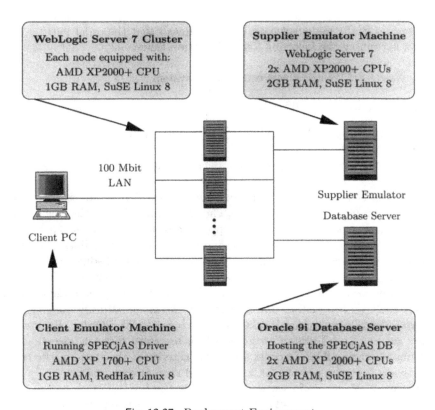

Fig. 13.27 Deployment Environment

These issues can be approached with the help of queueing network-based performance models.

Workload Characterization The first step in the capacity planning process is to describe the workload of the system under study in a qualitative and quantitative manner. This includes four major steps:

1. Describe the types of requests that are processed by the system.

2. Identify the hardware and software resources used by each request class.

3. Measure the total amount of service time for each request class at each resource.

4. Give an indication of the number of requests of each class that the system will be exposed to.

The **SPECjAppServer 2002** workload is made up of two major components: the *order entry application* in the customer domain and the *manufacturing* *application* in the manufacturing domain. Recall that the order entry application is running the following four transaction types:

1. *NewOrder*: places a new order in the system

2. *ChangeOrder*: modifies an existing order

3. *OrderStatus*: retrieves the status of a given order

4. *CustStatus*: lists all orders of a given customer

We map each of them to a separate *request class*. The manufacturing application, is running production lines. The main unit of work there is a *work order*. Each work order produces a specific quantity of a particular type of widget. There are two types of production lines: planned lines and large order lines. While planned lines run on a predefined schedule, large order lines run only when a large order arrives in the customer domain. Each large order results in a separate work order. During the processing of work orders multiple transactions are executed in the manufacturing domain, i.e., scheduleWorkOrder, updateWorkOrder and completeWorkOrder. Each work order moves along 3 virtual *stations*, which represent distinct operations in the manufacturing flow. In order to imitate activity at the stations, the manufacturing application waits for a designated time at each station.

We model the manufacturing workload at the level of work orders. We define the additional request class *WorkOrder*, which represents a request for processing a work order. The following resources are used during the processing of the 5 request classes:

- The CPU of a WebLogic server (WLS-CPU)

- The local area network

- The CPUs of the database server (DBS-CPU)

- The disk drives of the database server (DBS-I/O)

In order to determine the service demands at these resources, a separate experiment for each of the 5 request classes is conducted. In each case, we deployed the benchmark in a configuration with a single WebLogic server and then injected requests of the respective class into the system. During the experiment, we monitored the system resources and measured the time requests spend at each resource during their processing. For the database server, we used the Oracle 9i Intelligent Agent, which provides exhaustive information about CPU consumption, as well as I/O wait times. For the application server, we monitored the CPU utilization using operating system tools to derive the CPU service demand: the service demand $\overline{x}$ of requests at a given resource is equal to the average resource utilization ρ divided by the average request throughput λ, during the measurement interval, i.e. (Eq.(6.3)),

$$\overline{x} = \frac{\rho}{\lambda} \tag{13.17}$$

We can safely ignore network service demands, since all communications were taking place over a 100 MBit LAN and communication times were negligible. Table 13.18 reports the service demand measurements for the 5 request classes in our workload model.

Table 13.18   Workload service demands

TX-Type	WLS-CPU	DBS-CPU	DBS-I/O
NewOrder	12.98 msec	10.64 msec	1.12 msec
ChangeOrder	13.64 msec	10.36 msec	1.27 msec
OrderStatus	2.64 msec	2.48 msec	0.58 msec
CustStatus	2.54 msec	2.08 msec	0.3 msec
WorkOrder	24.22 msec	34.14 msec	1.68 msec

As we can see from Table 13.18, database I/O service demands are much lower than CPU service demands. This stems from the fact that data are cached in the database buffer and disks are usually accessed only when updating or inserting new data.

For each request class, we also must specify the rates at which requests arrive. We should also be able to vary these rates so that we can consider different scenarios. To this end, we modified the SPECjAppServer 2002 driver to allow more flexibility in configuring the intensity of the workload generated. Specifically, the new driver allows us to set the number of concurrent order entry clients simulated, as well as their average *think time*. In addition to this, we can specify the number of planned production lines run in the manufacturing domain and the time they wait after processing a work order before starting a new one.

Building a Performance Model   In this section, we build a queueing network model of our SPECjAppServer 2002 deployment environment. We first define the model in a general fashion and then customize it to concrete workload scenarios. sending requests to the system is fixed. Figure 13.28 shows a high-level view of our closed queueing network model.

Following is a brief description of the nodes used:

C : IS node used to model the client machine which runs the SPECjAppServer 2002 driver and emulates virtual clients sending requests to the system. The service time of order entry requests at this node is equal to the average client think time, while the service time of WorkOrder requests is equal to the average time a production line waits after processing a

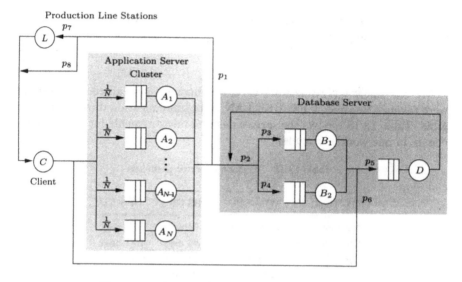

Fig. 13.28   Queueing network model of the system.

work order before starting a new one. Note that times spent on this
node are not part of system response times.

$A_1..A_N$: PS nodes used to model the CPUs of the N WebLogic servers.

B_1, B_2 : PS nodes used to model the two CPUs of the database server.

D : FCFS node used to model the disk subsystem (made up of a single 100
GB disk drive) of the database server.

L : IS node (delay resource) used to model the virtual production line stations
in the manufacturing domain. Only WorkOrder requests ever visit this
node. Their service time at the node corresponds to the average delay at
the production line stations simulated by the manufacturing application
during work order processing.

The model is a closed queueing network model with the 5 classes of requests
(jobs) defined in the previous section. The behavior of requests in the model
is defined by specifying their respective routing probabilities p_i and service
demands at each node which they visit. We discussed the service demands
in the previous section. To set the routing probabilities we examine the life
cycle of client requests in the queueing network. Every request is initially at
the client node C, where it waits for a user-specified think time. After the
think time elapses, the request is routed with probability $1/N$ to a randomly
chosen node A_i. Processing at the CPU may be interrupted multiple times
if the request requires some database accesses. Each time this happens, the
request is routed to the database server where it receives service at one of the

two CPU nodes B_1 or B_2 (each chosen equally likely so that $p_3 = p_4 = 0.5$). Processing at the database CPUs may be interrupted in case I/O accesses are needed. For each I/O access the request is sent to the disk subsystem node D and after receiving service there, is routed back to the database CPUs. This may be repeated multiple times depending on routing probabilities p_5 and p_6. Having completed their service at the database server, requests are sent back to the application server. Requests may visit the database server multiple times during their processing, depending on routing probabilities p_1 and p_2. After completing service at the application server requests are sent back to the client node C. Order entry requests are sent directly to the client node (for them $p_8 = 1$, $p_7 = 0$), while WorkOrder requests are routed through node L (for them $p_8 = 0$, $p_7 = 1$), where they are additionally delayed for 1 second. This delay corresponds to the 1 second delay at the 3 production line stations imposed by the manufacturing application during work order processing.

In order to set routing probabilities p_1, p_2, p_5, and p_6 we need to know how many times a request visits the database server during its processing and for each visit how many times I/O access is needed. Since we only know the total service demands over all visits to the database, we assume that requests visit the database just once and need a single I/O access during this visit. This allows us to drop routing probabilities p_1, p_2, p_5, and p_6 and leads us to the following simplified model depicted in Fig. 13.29.

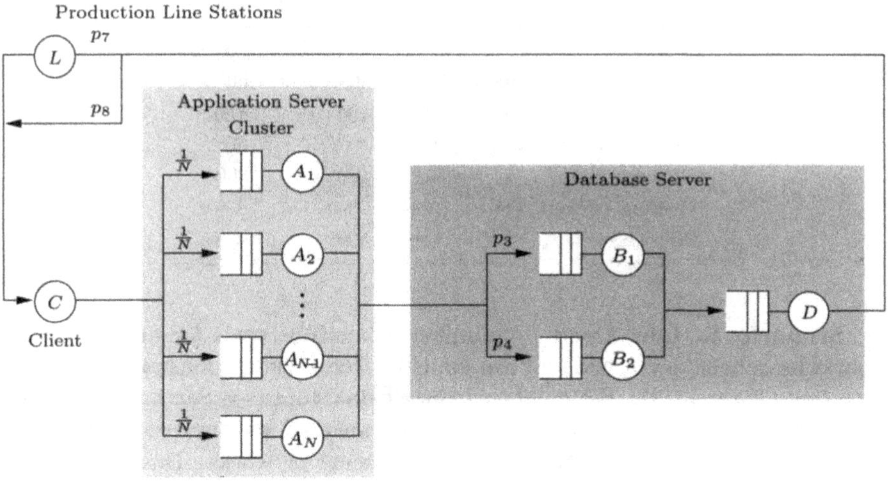

Fig. 13.29 Simplified QN Model of the System

The following input parameters need to be supplied before the model can be analyzed:

- Number of order entry clients (NewOrder, ChangeOrder, OrderStatus and CustStatus).

- Average think time of order entry clients - *Customer Think Time.*

- Number of planned production lines generating WorkOrder requests.

- Average time production lines wait after processing a work order, before starting a new one — *Manufacturing (Mfg) Think Time.*

- Service demands of the 5 request classes at nodes A_i, B_j, and D (see Table 13.18).

Model Analysis and Validation We now proceed to analyze several different instances of the model introduced in the previous section and then validate them by comparing results from the analysis with measured data. We first consider only the case without large order lines and study the system in three scenarios representing low, moderate and heavy load, respectively. The case with large order lines is described in [KoBu03]. In each case, we examine deployments with different number of application servers — from 1 to 9.

Table 13.19 summarizes the input parameters for the three scenarios that we consider.

Table 13.19 Model input parameters for the 3 scenarios

Parameter	Low	Moderate	Heavy
NewOrder Clients	30	50	100
ChangeOrder Clients	10	40	50
OrderStatus Clients	50	100	150
CustStatus Clients	40	70	50
Planned Lines	50	100	200
Customer Think Time	2 sec	2 sec	3 sec
Mfg Think Time	3 sec	3 sec	5 sec

Scenario 1: Low Load A number of analysis tools for queueing networks have been developed and are available free of charge for noncommercial use (see Chapter 12). We employed the PEPSY tool (see Section 12.1). We chose PEPSY because it supports a wide range of solution methods (over 30) for product- and non-product-form queueing networks. Both exact and approximate methods are provided, which are applicable to models of considerable size and complexity. For the most part, we have applied the *multisum method* (see Section 10.1.4.4) for solution of the queueing network models in this section. However, to ensure plausibility of the results, we cross-verified them with results obtained from other methods. In all cases the difference was negligible.

Table 13.20 summarizes the results we obtained for our first scenario. We studied two different configurations: the first one with 1 application server, the second with 2. The table reports throughput (TP) and response time (RT) for the 5 request classes, as well as CPU utilization (UT) of the application server and the database server. Results obtained from the model analysis are compared against results obtained through measurements.

Table 13.20 Analysis results for scenario 1 — low load (TP = throughput, RT = response time, UT = utilization)

METRIC	1 Application Server			2 Application Servers		
	Model	Measured	Error	Model	Measured	Error
NewOrder TP	14.59	14.37	1.5%	14.72	14.49	1.6%
ChangeOrder TP	4.85	4.76	1.9%	4.90	4.82	1.7%
OrderStatus TP	24.84	24.76	0.3%	24.89	24.88	0.0%
CustStatus TP	19.89	19.85	0.2%	19.92	19.99	0.4%
WorkOrder TP	12.11	12.19	0.7%	12.20	12.02	1.5%
NewOrder RT	56ms	68ms	17.6%	37ms	47ms	21.3%
ChangeOrder RT	58ms	67ms	13.4%	38ms	46ms	17.4%
OrderStatus RT	12ms	16ms	25.0%	8ms	10ms	20.0%
CustStatus RT	11ms	17ms	35.2%	7ms	10ms	30.0%
WorkOrder RT	1127ms	1141ms	1.2%	1092ms	1103ms	1.0%
Web Server CPU UT	66%	70%	5.7%	33%	37%	10.8%
DB Server CPU UT	36%	40%	10%	36%	38%	5.2%

As we can see from the table, while throughput and utilization results are extremely accurate, the same does not hold to this extent for response time results.

From Table 13.20 we see that the response time error for requests with very low service demands (e.g., OrderStatus and CustStatus) is much higher than average. This is because the processing times for such requests are very low (around 10ms) and the additional delays from software contention, while not that high as absolute values, are high relative to the overall response times. The results show that the higher the service demand of a request type, the lower the response time error. Indeed, the requests with the highest service demand (WorkOrder) always have the lowest response time error.

Scenario 2: Moderate Load   We study 2 deployments: the first with 3 application servers, the second with 6. Table 13.21 summarizes the results from the model analysis. Again we obtain very accurate throughputs and utilizations, and accurate response times. The response time error does not exceed 35%, which is considered acceptable in most capacity planning studies [MeAl00].

Table 13.21 Analysis results for scenario 2 — moderate load (TP = throughput, RT = response time, UT = utilization)

METRIC	3 Application Servers			6 Application Servers		
	Model	Measured	Error	Model	Measured	Error
NewOrder TP	24.21	24.08	0.5%	24.29	24.01	1.2%
ChangeOrder TP	19.36	18.77	3.1%	19.43	19.32	0.6%
OrderStatus TP	49.63	49.48	0.3%	49.66	49.01	1.3%
CustStatus TP	34.77	34.24	1.5%	34.80	34.58	0.6%
WorkOrder TP	23.95	23.99	0.2%	24.02	24.03	0.0%
NewOrder RT	65ms	75ms	13.3%	58ms	68ms	14.7%
ChangeOrder RT	66ms	73ms	9.6%	58ms	70ms	17.1%
OrderStatus RT	15ms	20ms	25.0%	13ms	18ms	27.8%
CustStatus RT	13ms	20ms	35.0%	11ms	17ms	35.3%
WorkOrder RT	1175ms	1164ms	0.9%	1163ms	1162ms	0.0%
Web Server CPU UT	46%	49%	6.1%	23%	25%	8.0%
DB Server CPU UT	74%	76%	2.6%	73%	78%	6.4%

Scenario 3: Heavy Load We consider three configurations: with 4, 6 and 9 application servers, respectively. However, we slightly increase the think times in order to make sure that our single machine database server is able to handle the load. Table 13.22 summarizes the results for this scenario. For models of this size, the available algorithms do not produce reliable results for response time and therefore we only consider throughput and utilization in this scenario.

Table 13.22 Analysis results for scenario 3 — heavy load (TP = throughput, UT = utilization)

METRIC	4 App. Servers		6 App. Servers		9 App. Servers	
	Model	Error	Model	Error	Model	Error
NewOrder TP	32.19	0.3%	32.22	1.3%	32.24	0.7%
ChangeOrder TP	16.10	0.9%	16.11	0.5%	16.12	0.4%
OrderStatus TP	49.59	1.4%	49.60	0.8%	49.61	0.7%
CustStatus TP	16.55	1.8%	16.55	1.9%	16.55	0.5%
WorkOrder TP	31.69	0.2%	31.72	1.1%	31.73	1.8%
Web Server CPU UT	40%	4.8%	26%	10.3%	18%	10.0%
DB Server CPU UT	87%	2.2%	88%	3.3%	88%	3.3%

Conclusions from the Analysis   The results enable us to give answers to the questions we started with in Section 13.1.6.2. For each configuration, we obtained approximations for the average request throughput, response time and server utilization. Depending on the service level agreements (SLAs) and the expected workload intensity, we can now determine how many application servers we need in order to guarantee adequate performance. We can also see for each configuration which component is mostly utilized and thus could become a potential bottleneck. In scenario 1, we saw that by using a single application server, the latter could easily turn into bottleneck since its utilization would be twice as high as that of the database server. The problem is solved by adding an extra application server. In scenarios 2 and 3, we saw that with more than 3 application servers as we increase the load, the database CPU utilization approaches 90%, while the application servers remain in all cases less than 50% utilized. This clearly indicates that, in this case, our database server is the bottleneck.

13.1.7   Flexible Production Systems

Queueing network models combined with numerical solution methods are also a very powerful paradigm for the performance evaluation of production systems. In this section we demonstrate how to model and analyze such systems using open and closed queueing networks.

13.1.7.1 An Open Network Model   Consider a simple production system that can be modeled as an open queueing network. The system contains the following stations (Fig. 13.30):

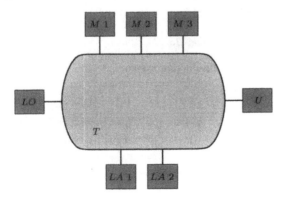

Fig. 13.30 Layout of a simple production system.

- A load station where the workpieces are mounted on to the pallet (*LO*).

- Two identical lathes (*LA*).

- Three identical milling machines (M).

- A transfer system (T) that does the transfer between the stations and consists of two automatically controlled vehicles.

- A station to unload the pallet (U), which removes the workpieces from the production system.

The identical milling machines and lathes are modeled by multiple server nodes. With these assumptions we obtain the queueing network model shown in Fig. 13.31. The production system produces different products and there-

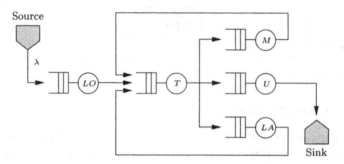

Fig. 13.31 Open network model of the production system shown in Fig. 13.30.

fore the machines are used in several different ways. For example, 60% of the workpieces that leave the milling machine are passed to the station that unloads the pallet, the rest are transferred to the lathes. Table 13.23 contains the probabilities q_{ij}, $i = LO, M, LA$, $j = M, LA, U$ that the transfer system T moves workpieces from a station i to station j.

Table 13.23 Routing table for the transfer system of the production system

i \ j	M	LA	U
LO	0.5	0.5	0
M	0	0.4	0.6
LA	0.7	0	0.3

To obtain from the transfer probabilities q_{ij} the probabilities of routing from the transfer system T to the lathes $p_{T,LA}$, milling machines $p_{T,M}$, or the unload station $p_{T,U}$, we have to weight the individual probabilities q_{ij} with the individual arrival rates λ_i, $i = LO, M, LA$. The arrival rate λ_T to the transfer system can be easily obtained from Fig. 13.31:

$$\lambda_T = \lambda_{LO} + \lambda_M + \lambda_{LA}, \tag{13.18}$$

and it follows that

$$p_{T,M} = \frac{\lambda_{LO}}{\lambda_T} \cdot q_{LO,M} + \frac{\lambda_{LA}}{\lambda_T} \cdot q_{LA,M} \,,$$

$$p_{T,LA} = \frac{\lambda_{LO}}{\lambda_T} \cdot q_{LO,LA} + \frac{\lambda_M}{\lambda_T} \cdot q_{M,LA} \,,$$ (13.19)

$$p_{T,U} = \frac{\lambda_{LA}}{\lambda_T} \cdot q_{LA,U} + \frac{\lambda_M}{\lambda_T} \cdot q_{M,U} \,.$$

Then, using the values from Table 13.23:

$$p_{T,M} = \frac{1}{\lambda_T} \left(\lambda_{LO} \cdot 0.5 + \lambda_{LA} \cdot 0.7 \right) \,,$$

$$p_{T,LA} = \frac{1}{\lambda_T} \left(\lambda_{LO} \cdot 0.5 + \lambda_M \cdot 0.4 \right) \,,$$

$$p_{T,U} = \frac{1}{\lambda_T} \left(\lambda_{LA} \cdot 0.3 + \lambda_M \cdot 0.6 \right) \,,$$

we finally obtain the matrix of routing probabilities p_{ij} for the queueing network model (see Fig. 13.31 and Table 13.24). In order to apply Jackson's

Table 13.24   Routing matrix for the queueing
model of the production system

i \ j	outside	LO	LA	M	U	T
outside	0	1	0	0	0	0
LO	0	0	0	0	0	1
LA	0	0	0	0	0	1
M	0	0	0	0	0	1
U	1	0	0	0	0	0
T	0	0	p_{LA}	p_M	p_U	0

theorem for open product-form networks (see Section 7.3.4), we make the following assumptions:

- The service times at the stations and the interarrival times at the load station LO are exponentially distributed.

- The service discipline at each station is FCFS.

- The system is stable.

The service rates μ_i, the arrival rates λ_{0i}, and number of servers m_i are listed in Table 13.25. Now we use Eq. (7.1) and the values of the routing probabilities from Table 13.24 to determine the arrival rates λ_i,

$$\lambda_i = \lambda_{0i} + \sum_j \lambda_j \cdot p_{ji}, \quad i,j = LO, LA, M, U, T, \quad (13.20)$$

Table 13.25 Arrival rates λ_{0i}, service rates μ_i (in 1/h), and number of servers m_i for the model of Fig. 13.31

i	λ_{0i}	μ_i	m_i
LO	15	20	1
LA	0	10	2
M	0	7	3
U	0	20	1
T	0	24	2

and we obtain

$$\lambda_{LO} = \lambda_{0\,LO} = 15\,,$$
$$\lambda_{LA} = \lambda_T \cdot p_{T,LA} = \lambda_{LO} \cdot 0.5 + \lambda_M \cdot 0.4 = 14.58\,,$$
$$\lambda_M = \lambda_T \cdot p_{T,M} = \lambda_{LO} \cdot 0.5 + \lambda_{LA} \cdot 0.7 = 17.71\,,$$
$$\lambda_U = \lambda_T \cdot p_{T,U} = \lambda_{LO} = 15\,,$$
$$\lambda_T = \lambda_{LO} + \lambda_{LA} + \lambda_M = 47.29\,.$$

Performance measures, which were calculated using Jackson's theorem, are listed in Table 13.26.

Table 13.26 Performance measures for the production system

i	$\overline{Q}_i$	$\overline{W}_i$	ρ_i
LO	2.25	0.15	0.75
LA	1.66	0.11	0.73
M	3.86	0.22	0.84
U	2.25	0.15	0.75
T	65.02	1.38	0.985

We see that the transfer system T is heavily loaded and has a very long queue, and that its utilization is nearly 100%. If we use an additional vehicle, the utilization and the queue length are reduced substantially to

$$\rho_T = 0.66\,, \quad \overline{Q}_T = 0.82\,, \quad \overline{W}_T = 0.02\,.$$

The work in progress (WIP, the mean number of workpieces in the system) and the mean response time $\overline{T}$ for both cases are listed in Table 13.27. To improve the performance, we could increase the number of milling machines because they have a very high utilization, $\rho_M = 0.84$, or the number of lathes. The results for these changes are shown in Table 13.28. A further increase in

Table 13.27 Work in progress
(WIP) and mean response time $\overline{T}$
for 2 and 3 vehicles in the transfer
system

m_T	WIP	$\overline{T}$
2	82.5	5.5
3	18.3	1.2

Table 13.28 WIP and $\overline{T}$ for different numbers of vehicles, milling
machines, and lathes

m_T	m_{LA}	m_M	WIP	$\overline{T}$
3	2	3	18.3	1.22
4	2	3	17.6	1.18
3	3	3	16.9	1.12
3	2	4	15.0	1.00
3	3	4	13.6	0.90

the number of the vehicles in the transfer system has a negligible influence on
the performance. Maximum improvement can be achieved by increasing the
number of milling machines, because that station is the bottleneck ($\rho_i = 0.84$,
see Table 13.26). Although we have used a very simple model to approximate
the behavior of the production system, it does provide insight into the system
behavior and enables us to identify bottlenecks.

13.1.7.2 A Closed Network Model Now we consider the example of a production system that can be modeled as a closed multiclass queueing network
[SuHi84]. The production system consists of (see Fig. 13.32)

- Two load/unload stations (LU)

- Two lathes with identical tools (LA)

- Three machine centers with different tools (M)

- A central transfer system with eight transporters (T)

In the production system, workpieces for three different product classes are
manufactured. A fixed number K_i of pallets for each class circulates through
the system. In the first class the workpieces are mounted on the pallets in the
LU station then they are moved by the transfer system either to the lathes LA
or to the third M_3. Finally, they are shipped back to the LU station where
they are unmounted. For workpieces of the other two classes, the procedure
is similar, but they use machine M_4 and M_5, respectively. The load/unload

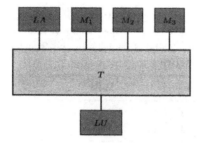

Fig. 13.32 Layout of the production system for a closed multiclass network model.

station LU and the lathes LA can be modeled as a multiple server node and the machines M_3, M_4, and M_5 as single server nodes. Since there are more transporters in the transfer system T than circulating pallets, it is modeled by an IS node. The parameters for the model are summarized in Table 13.29.

Table 13.29 Parameter values for the stations

Station$_i$	m_i	$1/\mu_i$	e_{i1}	e_{i2}	e_{i3}
LU	2	7.5	2	2	2
LA	2	28	0.5	0	0
M_3	1	12	0.5	0	0
M_4	1	30	0	1	0
M_5	1	15	0	0	1
T	8	2.5	2	2	2

The numbers of pallets K_i, which are the numbers of workpieces in class i, are given by

$$K_1 = 4, \quad K_2 = 1, \quad K_3 = 2.$$

With these parameters we obtain the queueing network models for these three classes, given in Fig. 13.33. Furthermore, we assume that each network fulfills the product-form assumptions. Now we have all the necessary input parameters and can use any solution method for closed product-form queueing networks such as MVA, convolution, or SCAT to obtain performance measures. In Table 13.30, the throughput and the mean system response time for the different classes are shown. The total throughput is 0.122 workpieces/min or 7.32 workpieces/hr. For other interesting performance measures of the machines, see Table 13.31.

Using the queueing network model, the system can be optimized. In Table 13.31 we can see that the load/unload stations are the bottleneck and, therefore, we increase the number of these stations to three and four, respectively. The results for the modified system models are given in Table 13.32, columns 2, 3, and 4. In column 1, the number of load/unload stations is

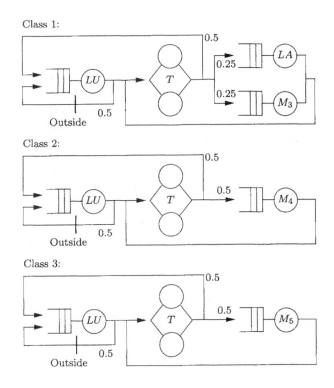

Fig. 13.33 Closed queueing model of the production system shown in Fig. 13.32.

Table 13.30 Performance measures for workpieces

Class	Throughput	Mean Response Time
1	0.069	58.35
2	0.016	63.88
3	0.037	53.77

reduced to one. In this case the system works, but with a severe bottleneck at LU, the mean response time is greater than 100 units, and the mean queue length at LU is 3.23. For $m_{LU} = 3$, the utilization of the LU station is still $\rho_{LU} = 0.73$, and the system is still unbalanced, which means that the utilization of the different stations differ considerably, as seen in the last column of Table 13.32. For $m_{LU} = 4$, the system is more balanced, but now $\rho_{M_5} = 0.72$ is too high. To get a more balanced system, a fraction of the workpieces of class 2 are serviced at station M_3 (see Fig. 13.34). Now the system is better balanced especially in case 4c, but the utilizations are relatively low ($\rho_i \approx 0.6$). Therefore, we double the number of pallets in the system ($K_1 = 8$, $K_2 = 2$,

Table 13.31   Performance measures for the stations

Station$_i$	m_i	ρ_i	Mean number in Service	Mean Queue Length $\overline{Q}_i$
LU	2	0.91	1.82	1.68
LA	2	0.48	0.96	0.12
M_3	1	0.41	0.41	0.19
M_4	1	0.47	0.47	0.00
M_5	1	0.56	0.56	0.16
T	8	0.08	0.61	0.00

Table 13.32   Utilizations of the stations for different number of servers at the LU stations and routing probabilities of Table 13.33

m_{LU} Station$_i$	1	2	3	4	4a	4b	4c	4d
LU	0.997	0.91	0.73	0.58	0.55	0.56	0.56	0.77
LA	0.27	0.48	0.58	0.61	0.60	0.60	0.60	0.82
M_3	0.23	0.41	0.49	0.52	0.56	0.54	0.56	0.77
M_4	0.26	0.47	0.57	0.59	0.68	0.64	0.59	0.82
M_5	0.30	0.56	0.68	0.72	0.52	0.62	0.62	0.85
T	0.04	0.08	0.092	0.096	0.09	0.09	0.08	0.129
$\rho_{max} - \rho_{min}$	0.77	0.50	0.240	0.200	0.16	0.10	0.06	0.08

Table 13.33   Routing probabilities

	p_{32}	p_{42}	p_{43}	p_{53}
a	0.2	0.8	0.2	0.8
b	0.1	0.9	0.1	0.9
c	0.2	0.8	0.1	0.9
d	$K_1 = 8$, $K_2 = 2$, $K_3 = 4$			

and $K_3 = 4$) and obtain the values of the utilization in column 4d. The system is now balanced at a higher level of utilization ($\rho_i \approx 0.8$). Increasing the number of LU servers leads to a higher throughput and a lower mean response time (see Table 13.34). If we increase the number of pallets to $K_1 = 8$, $K_2 = 2$ and $K_3 = 4$, the mean response time increases along with the throughput (see Table 13.35).

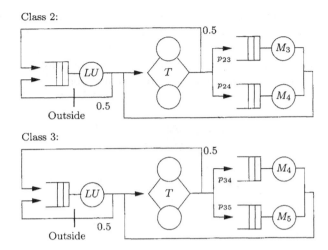

Fig. 13.34 Queueing network model of the production system shown in Fig. 13.33, with different routing for class 2 and 3.

Table 13.34 Performance measures with four *LU* servers

Class	Throughput	Mean Response Time
1	0.086	46.29
2	0.020	50.61
3	0.048	41.97

Table 13.35 Performance measures with four *LU* servers and a higher number of pallets

Class	Throughput	Mean Response Time
1	0.117	68.21
2	0.026	76.56
3	0.063	63.54

Notes: $K_1 = 8$, $K_2 = 2$, $K_3 = 4$.

13.1.8   Kanban Control

In order to limit the work in progress and minimize the lead time many manufactures use Kanban systems, an idea developed in the Japanese automobile industry. Consider a production line which consists of some machines and a storage for finished parts. The Kanban control organizes the production facility so that each part is tagged with a card (Japanese: kanban). Every time a finished part is removed from the storage its card is removed and tagged to a new part for production. Parts in the line may only be processed, when tagged with a card. In this way the work in progress is effectively limited to the number of Kanbans. See Fig. 13.35 for visualization.

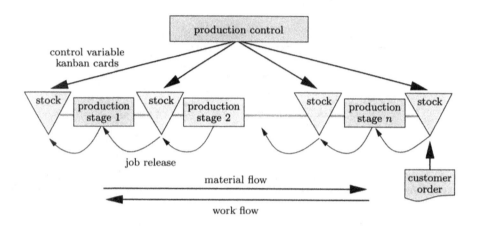

Fig. 13.35 Principle of a Kanban system.

Machines are grouped by sectors, in which the production is controlled by a Kanban system. The material flow then extends over many sectors. In an environment like this the number of Kanban cards in the different sectors is crucial in determing performance. Choosing the number of Kanban cards too low means wasting capacity and creating a bottleneck due to insufficient planning. The other way around means the work in progress is too high resulting in excessive waiting times and lead times.

By representing the Kanban cards as jobs we can model a Kanban controlled production line as closed queueing network. The impact of the number of Kanban cards on the performance measures can then be estimated with analytical-numeric methods.

Figure 13.36 shows the final sector of a mass production line, an assembly sector with rework cycles. Given this configuration and a target for the throughput of more than one part each 20 time units in every class we have to determine the number of Kanban cards. By using the summation method (see Section 10.1.4.4) to solve this non-product-form network we can approximately calculate the work in progress needed to reach this throughput (see Table 13.37). This sector of a production line has therefore to be configured with exactly 6 Kanban cards, 2 in each class. The requirements can also be met with an additional card for class 3. But this means unnecessarily higher cost. For more details see [Fran03].

Table 13.36 Mean values $\bar{x}_i$ and coefficients of variations $c^2_{B_i}$ of the service times in the assembly line of Fig. 13.36

Station	$\bar{x}_1$	$\bar{x}_2$	$\bar{x}_3$	$c^2_{B_1}$	$c^2_{B_2}$	$c^2_{B_3}$
Assembly 1	2.2	3	2	0.8	0	1
Inspection	1.5	2.5	1	0.8	1	1.1
Assembly 2	1.5	1	2	0.7	1.5	0.3
Assembly 3	–	5	4	–	0.5	0
Test	5	10	1	1	0	0.8
Package	3	3	3	0.7	0.7	1
Rework 1	–	–	5	–	–	1.2
Rework 2	–	–	2	–	–	0

Table 13.37 Throughput λ_i for a given number of Kanban cards K_i

$K_1 = 1$	$K_2 = 1$	$K_2 = 2$	$K_2 = 3$
	$\lambda_1 = 0.0517$	$\lambda_1 = 0.0379$	$\lambda_1 = 0.0290$
$K_3 = 1$	$\lambda_2 = 0.0408$	$\lambda_2 = 0.0631$	$\lambda_2 = 0.0752$
	$\lambda_3 = 0.0451$	$\lambda_3 = 0.0357$	$\lambda_3 = 0.0289$
	$\lambda_1 = 0.0514$	$\lambda_1 = 0.0371$	$\lambda_1 = 0.0283$
$K_3 = 2$	$\lambda_2 = 0.0409$	$\lambda_2 = 0.0624$	$\lambda_2 = 0.0740$
	$\lambda_3 = 0.0894$	$\lambda_3 = 0.0702$	$\lambda_3 = 0.0567$
	$\lambda_1 = 0.0506$	$\lambda_1 = 0.0363$	$\lambda_1 = 0.0277$
$K_3 = 3$	$\lambda_2 = 0.0405$	$\lambda_2 = 0.0614$	$\lambda_2 = 0.0727$
	$\lambda_3 = 0.1316$	$\lambda_3 = 0.1033$	$\lambda_3 = 0.0834$

$K_1 = 2$	$K_2 = 1$	$K_2 = 2$	$K_2 = 3$
	$\lambda_1 = 0.0870$	$\lambda_1 = 0.0650$	$\lambda_1 = 0.0508$
$K_3 = 1$	$\lambda_2 = 0.0356$	$\lambda_2 = 0.0554$	$\lambda_2 = 0.0670$
	$\lambda_3 = 0.0399$	$\lambda_3 = 0.0318$	$\lambda_3 = 0.0260$
	$\lambda_1 = 0.0851$	$\lambda_1 = 0.0634$	$\lambda_1 = 0.0497$
$K_3 = 2$	$\lambda_2 = 0.0351$	$\lambda_2 = 0.0546$	$\lambda_2 = 0.0659$
	$\lambda_3 = 0.0782$	$\lambda_3 = 0.0622$	$\lambda_3 = 0.0510$
	$\lambda_1 = 0.0831$	$\lambda_1 = 0.0620$	$\lambda_1 = 0.0486$
$K_3 = 3$	$\lambda_2 = 0.0345$	$\lambda_2 = 0.0536$	$\lambda_2 = 0.0648$
	$\lambda_3 = 0.1147$	$\lambda_3 = 0.0915$	$\lambda_3 = 0.0751$

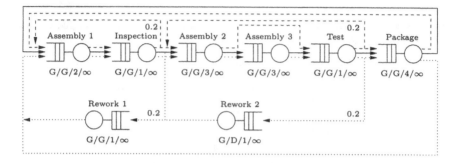

Fig. 13.36   QN Model of an assembly line.

Problem 13.7   Verify the results of the closed queueing network model of the production system by hand computation.

13.2   CASE STUDIES OF MARKOV CHAINS

In this section we show the use of CTMC models either constructed directly or via a higher-level specification.

13.2.1   Wafer Production System

In this section we formulate and solve a CTMC model of the photolithographic process of a wafer production system. The system is specified using a queueing network formalism, although for this product-form network we use the package MOSEL-2 to automatically generate and solve the underlying CTMC. In [Rand86] the photolithographic process is divided into the following five steps:

1. Clean the wafer.

2. Put on photoresist.

3. Bake wafer and remove solvent by evaporation.

4. Align mask, and illuminate and develop the wafer.

5. Etch the wafer and remove photoresist.

The photolithographic processing is carried out by three machine types:

- The spinning machine does the first three steps (machine type 1).

- The masking machine does step four (machine type 2).

- The etching machine does step five (machine type 3).

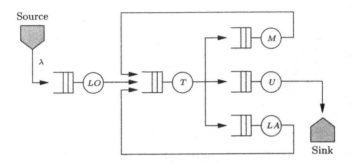

Fig. 13.37 Open network model of the wafer production system discussed in Section 13.2.1.

From the first machine type, three machines are available. From the second machine type, we have machines from different producers. These machines are called masking 1 and masking 2. The third machine type does the steps of etching and removing the photoresist. The job of removing the photoresist is done by machine five in case the wafer has not been produced correctly. Since three layers of photoresist have to be put on, the wafer needs to be processed again after a successful masking and spinning. In Fig. 13.37, the open queueing network model of the wafer production system is shown and in Table 13.38 its parameter values are given.

Table 13.38 Parameter values for the wafer production system discussed in Section 13.2.1

m_i	p_{ij}	μ_i	λ
$m_1 = 3 \ldots 10$	$p_{12} = 0.5$	$\mu_1 = 1$	$\lambda = 1$
$m_2 = 2 \ldots 4$	$p_{13} = 0.5$	$\mu_2 = 1$	
$m_3 = 2 \ldots 4$	$p_{24} = 0.9$	$\mu_3 = 1$	
$m_4 = 2$	$p_{25} = 0.1$	$\mu_4 = 2$	
$m_5 = 1$	$p_{34} = 0.9$	$\mu_5 = 1$	
	$p_{35} = 0.1$		
	$p_{40} = 0.1$		
	$p_{46} = 0.9$		
	$p_{60} = 2/3$		
	$p_{61} = 1/3$		

In the representation of the queueing network model we use node 6 as a dummy node with zero service time. We use node 6 in order to easily determine the routing probabilities, but it is, of course, also possible to determine the routing probabilities without the introduction of node 6. Since it is an open network, we limit the number of wafers in the system to K so that the underlying CTMC is not infinite.

```
/* Nodes */
<1..5> NODE Buffer_#[K];
<1..5> NODE Active_#[m_#];
       NODE Station_6 [K];
       NODE num  [K];

/* Rules */
       FROME TO Buffer_1, num W lambda;
<1..5>    FROM Buffer_# TO Active_#;

<1..m_1> FROM Active_1 TO Buffer_2  W #*mu_1 P p12 IF Active_1 == #;
<1..m_1> FROM Active_1 TO Buffer_3  W #*mu_1 P p13 IF Active_1 == #;

<1..m_2> FROM Active_2 TO Buffer_4  W #*mu_2 P p24 IF Active_2 == #;
<1..m_2> FROM Active_2 TO Buffer_5  W #*mu_2 P p25 IF Active_2 == #;

<1..m_3> FROM Active_3 TO Buffer_4  W #*mu_3 P p34 IF Active_3 == #;
<1..m_3> FROM Active_3 TO Buffer_5  W #*mu_3 P p35 IF Active_3 == #;

<1..m_4> FROM Active_4, num TOE     W #*mu_4 P p40 IF Active_4 == #;
<1..m_4> FROM Active_4 TO node_6    W #*mu_4 P p46 IF Active_3 == #;

       FROM Active_5 TO Buffer_1  W mu_5;
       FROM Station_6, num TOE    P p60;
       FROM Station_6 TO Buffer_1 P p61;

/* Results */
/* Mean number of active machines */
<1..5>    RESULT >> A_# = MEAN Active_#;

/* Utilization of the machines */
<1..5>    RESULT >> rho_# = A_#/m_#;

/* Mean buffer length */
<1..5>    RESULT >> Q_# = MEAN Buffer_#;

/* Mean number of wafers (WIP) */
       RESULT >> WIP = MEAN num;

/* Mean system time */
       RESULTS >> T = WIP / lambda;
```

Fig. 13.38 MOSEL-2 specification of the wafer production system.

The MOSEL-2 specification of the network is shown in Fig. 13.38; note the use of the loop construct to reduce the size of the specification. WIP is the work in progress, i.e., the mean number of wafers in the system. This value needs to be as small as possible for good system performance.

Results of the analysis are shown in Fig. 13.39. Here we plot the mean response time as a function of the number of machines at station 1. There are several things that we can learn from the plot:

- Having more than five machines at station 1 does not reduce the mean response time very much. Therefore, it is not cost effective to have more than five machines at station 1.

- Increasing the number of machines at stations 2 and 3 from two to three reduces the mean response time considerably, while adding more machines is not any more cost effective.

- We get a large decrease in the mean response time if we increase the number of machines at station 1 from three to four and if we increase the number of machines at stations 2 and 3 from two to three.

It is possible to extend this model to allow for machine failures and repairs.

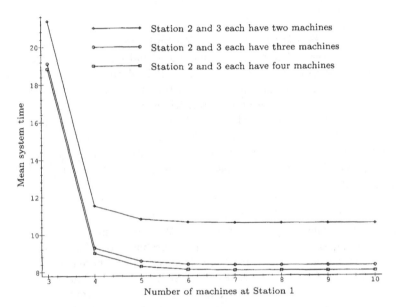

Fig. 13.39 Results for different number of machines.

Problem 13.8   Formulate the wafer production problem as a stochastic reward net and solve it using SPNP.

13.2.2 Polling Systems

Polling systems are very important in modeling computer and communication systems. The generic *polling model* is a system of multiple queues visited by a single server in cyclic order (see Fig. 13.40). For an exposition and performance models of polling systems, we refer the reader to [Taka90]. In this subsection we wish to illustrate the use of SPNs in automatically generating and solving the underlying CTMC for polling system performance models. Our treatment is based on the paper by Ibe and Trivedi [IbTr90].

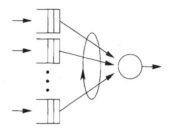

Fig. 13.40 A polling queue.

In Fig. 13.41, the GSPN for a three-station finite population single service polling system is shown. The potential customers that can arrive at station j are indicated by tokens in place P_{jI}, $j = 1, 2, 3$. This place contains M_j tokens where M_j denotes the population size of customers that potentially arrive at the station. The firing rate of the transitions t_{lj} is marking-dependent, meaning that when the number of tokens in place P_{jI} is k, then the firing rate of t_{lj} is given by $\lambda_j k$, $0 \leq k \leq M_j$. A marking-dependent firing rate is represented by the # symbol and placed next to transition t_{lj}. A token in place P_{jP} represents the condition that the server is polling station j; a token in place P_{jB} represents the condition that a customer has arrived at station j, and P_{jS} represents the condition that the server has arrived at station j.

The server commences polling station $j+1$ after finishing service at station j or found no customer there when it arrived. Whenever the timed transition t_{lj} fires, a customer arrives at station j. The server has finished polling station j when the timed transition t_{rj} fires. Such a GSPN can be easily input to any of the SPN tools such as **SHARPE** or **SPNP**, whereby the underlying CTMC will be automatically generated and solved for steady-state or transient behavior, and resulting performance measures calculated. We refer the reader to [STP96] and [IbTr90].

Other kinds of polling schemes have also been considered by [IbTr90]. In the gated service polling system, only those customers are served by the server that arrive prior to the server's arrival at station j. The SRN model of station 1 in a finite population gated service polling system is shown in Fig. 13.42. The difference between the GSPN model of the single service polling system is that now a variable multiplicity arc (indicated by the zigzag sign) is needed. Therefore, the new model is a stochastic reward net model that cannot be solved by pure GSPN solvers such as **SHARPE** or **GreatSPN** [STP96, ABC+95]. However, **SPNP** is capable of handling such models. For further details we refer the reader to [IbTr90] and [ChTr92].

The SRN (or GSPN) models can be solved using **SPNP**. The **SPNP** input file contains the specification of timed transitions and corresponding firing rates, immediate transitions, the input and output arc for each transition, and the initial number of tokens in each place. From this information **SPNP** generates the reachability graph, eliminates all vanishing markings, and constructs the

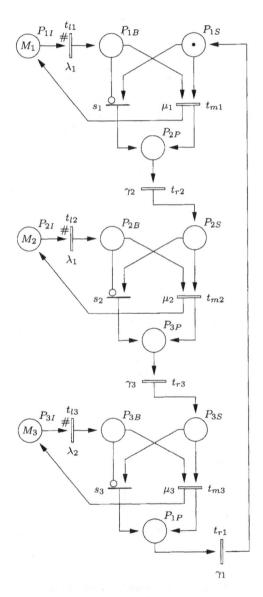

Fig. 13.41 GSPN model for the single service polling system.

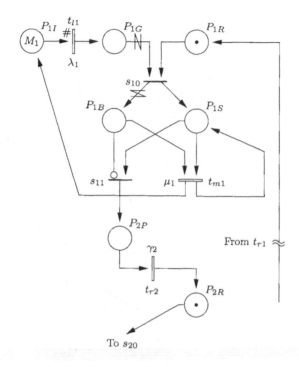

Fig. 13.42 SRN model of gated service polling system.

underlying CTMC. In order to obtain the steady-state probability of each marking, a combination of SOR and the Gauss–Seidel method is used. It is possible to obtain the mean number of tokens in each place as well as more general reward-based measures. Furthermore, transient and sensitivity analysis can also be carried out.

Problem 13.9   Specify and solve the GSPN model of the single server polling system using **SHARPE**. Determine the size of the underlying CTMC as a function of the number of customers $M_1 = M_2 = M_3 = M$. Determine the steady-state and the transient performance measures.

Problem 13.10   Specify and solve the SRN model of the gated service polling system using **SPNP**. Determine the steady-state and transient performance measures.

Problem 13.11   Extend the polling model of Fig. 13.41 to an Erlang-3 distributed polling time. Run the resulting model on both **SHARPE** and **SPNP**.

13.2.3   Client–Server Systems

Consider a distributed computing system with one file server and N workstations, interconnected by a LAN. Because the communication medium is shared by several workstations, the server and each workstation should have access to the network before it starts transmitting request/reply messages. It is assumed that a workstation does not start generating a new request until it has received the reply of its previous request. This assumption is based on the fact that in many situations future operations at a client workstation depend on the outcome of the current operation (current service request). As long as a workstation's request is outstanding, it can continue its local processing. Our model deals only with the aspect of a user's processing that requires access to the server. Our exposition here is based on [ICT93].

In our system model, we assume that each time the server captures the access right to the network it transmits at most one reply message. The order of service at the server is FCFS.

It is important to distinguish between the client server models here and the single buffer polling system presented in Section 13.2.2. The difference is that in a single buffer polling system the station does not wait for reply to its message. Thus, in a polling system, as soon as a station transmits its message, it is ready to generate a new one after the reply to the last transmitted message has been received. The interdependencies of the access control of the network and workload between the client workstations and the server also make the analysis of client server systems different from the traditional analysis of plain LANs. We consider client server systems based on a token ring network. For another model with CSMA/CD, see [ICT93].

The first step of the analysis is to find an appropriate SRN model for the system. Since there might be a confusion between the word token in a Petri net or in a token ring network, we refer to token in a ring network as *network token* and to the token associated with the SRN as the PN token. We assume that the time until a client generates a request is exponentially distributed with mean $1/\lambda$. The time for the network to move from one station to the next station (polling time) is also assumed to be exponentially distributed with mean $1/\gamma$. The processing time of a request at a server is exponentially distributed with mean $1/\mu$, and the time required for the server to process a request is assumed to be exponentially distributed with mean $1/\beta$. The system consists of N clients and one server. We note that the assumption of exponential distribution can be realized by resorting to phase-type expansions.

As can be seen in Fig. 13.43, we consider a tagged client and lump the remaining clients into one *superclient*. This superclient subsystem enables us to closely approximate $(N-1)$ multiples of the single client subsystem so that we reduce the number of states of the underlying CTMC.

The part of the SRN model that deals with the tagged client is described in Fig. 13.44. The place P_{TI} contains one PN token and represents the condition that the client is idle (in the sense the client does not generate a network-based request). The firing rate of the timed transition t_{ta} is λ. Thus, the firing of t_{ta} represents the event that the client has generated a request. The place P_{TA} represents the condition that the network token has arrived. Suppose that there is a PN token in P_{TS}. Then if there is a PN token in P_{TA}, the

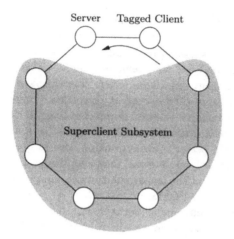

Fig. 13.43 A possible configuration of network relative to the tagged client.

timed transition t_{ts} whose mean time to fire is the mean request transmission time is enabled; otherwise the immediate transition s_1 fires. The place P_{SI} represents the condition that the client's request has arrived at the server, and the place P_{TW} represents the condition that the client is waiting to receive a reply from the server. The place P_{SP} represents the condition that the client has finished transmitting its request, if it is reached through t_{ts}, or the client has no request to transmit, if it is reached via s_1. The timed transition t_{sp} represents the event that the network token is being passed to its neighbor, the server in this configuration. The place P_{SS} represents the condition that the server has received the network token and can commence transmitting a reply, if one is ready.

It should be noted that in this model, a network token is released from a station as soon as it finishes transmitting its packet. This operation of the token ring network is called the *single-token* operation and is used in IEEE 802.5 token ring networks [Ibe87]. Single-token operation requires a station that has completed its packet transmission to release a network token only after its packet header has returned to the station. In the case that the round-trip delay of the ring is lower than the transmission time, the two schemes become identical.

Let us consider now the superclient subsystem. The corresponding SRN for this subsystem is shown in Fig. 13.45 and can be interpreted as follows: Place P_{OI} initially contains $N - 1$ PN tokens and represents the condition that no member of the superclient subsystem has generated a request. The number of tokens can generally be interpreted as the number of idle members of the subsystem. The firing rate of the timed transition t_{oa} is marking-dependent so that if there are l PN tokens in place P_{OI}, the firing rate of t_{oa} is given by $l \cdot \lambda (0 \leq l \leq N - 1)$. The places P_{OW}, P_{OS}, P_{TP}, and P_{OA}

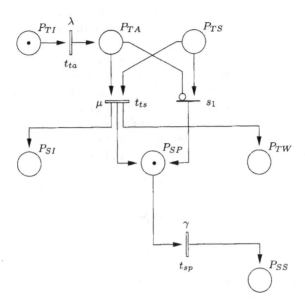

Fig. 13.44 SRN for tagged client subsystem.

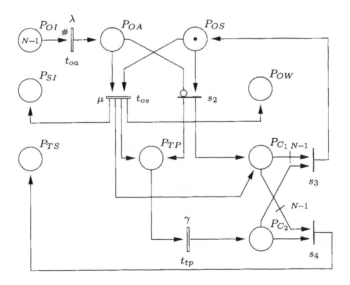

Fig. 13.45 SRN for the superclient subsystem in token ring network.

play the same role respectively as P_{TW}, P_{TS}, P_{SP}, and P_{TA} of Fig. 13.44. Similarly, the transitions s_2, t_{tp}, and t_{os} correspond respectively to s_1, t_{sp}, and t_{ts} of Fig. 13.44. Furthermore, the two additional places P_{C2} and P_{C1} are introduced and used to keep a count of the number of times the PN token has gone through P_{OS}. As soon as each member of the superclient subsystem has been polled, a PN token is deposited in both places P_{C1} and P_{C2}. Note that P_{C2} receives a PN token only after a walk time has been completed while P_{C1} receives the PN token immediately after a request, if one is transmitted, or prior to the commencement of the polling of the next member, if no transmission takes place. A regular arc of multiplicity $N-1$ from place P_{C1} is input to transition s_4 and an inhibitor arc of multiplicity $N-1$ is input from P_{C1} to the immediate transition s_3. Transition s_3 is enabled if a PN token is in place P_{C2} and fewer than $N-1$ PN tokens are in place P_{C1}. This represents the condition that the network token leaves the superclient subsystem if fewer than $N-1$ clients have been polled. If the number of PN tokens in place P_{C1} reaches $N-1$, implying all members of the subsystem have been polled, then with a PN token in P_{C2}, s_4 fires immediately and a PN token is deposited in P_{TS}. In this situation the tagged client now has access to the network after all members of the superclient subsystem had their chance to transmit their requests.

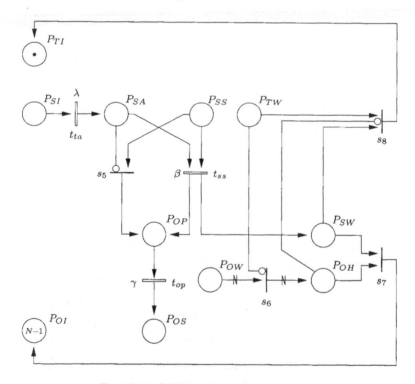

Fig. 13.46 SRN for the server subsystem.

The SRN for the subsystem is shown in Fig. 13.46. The places P_{OP}, P_{SS}, and P_{SA} serve the same purpose as P_{SP}, P_{TS}, and P_{TA} in Fig. 13.44. A token in place P_{SI} represents the condition that the server has received a request, while the place P_{TW} (P_{OW}) represents the condition that a client from the superclient subsystem (the tagged client) is waiting for a reply to its request. The firing rates of the timed transitions t_{op}, t_{ss}, and t_{sa} are given by γ, β, and λ, respectively. The condition that the server has completed serving one request is represented by the place P_{SW} and a determination is to be made regarding whose request was just serviced. In order to separate the requests from the superclient subsystem the server receives before that of the tagged client from those the server receives after that of the tagged client, the immediate transition s_6 is used. To denote that the input arc from P_{OW} to s_6 has a variable multiplicity, the zigzag sign on that arc is used.

Let us assume P_{OW} contains k PN tokens, $0 < k \leq N-1$, and P_{TW} has no token. Then the immediate transition s_6 is enabled and fires immediately by removing all k PN tokens from P_{OW} and deposits them in P_{OH}. The number of clients from the superclient subsystem whose requests were received by the server before that of the tagged client is therefore given by the marking of P_{OH}. The tagged client's request cannot be replied to until requests from the superclient subsystem that arrived before it have been replied to, or stated differently. Due to the FIFO queue at the server, the immediate transition s_8 cannot fire as long as there is a PN token in P_{OH}. As soon as the immediate transition s_7 fires, a PN token is deposited in P_{OI}, which means that a waiting member of the superclient subsystem becomes idle again. The complete SRN model for the system is shown in Fig. 13.47.

Problem 13.12   Specify and run the client server SRN using the SPNP package. Compute the average response time as a function of the number of stations N. In each case, keep track of the number of states and the number of nonzero transitions in the underlying CTMC.

Problem 13.13   Extend the previously described model to allow for an Erlang-2 distributed polling time. Solve the new SRN model using SPNP.

13.2.4   ISDN Channel

13.2.4.1   The Baseline Model   In this section we present a CTMC modeling a simplified version of an ISDN channel. Typically, heterogeneous data are transported across an ISDN channel. In particular, discrete and continuous media are served at the same time, although different qualities of service requirements are imposed by different types of media on the transport system. It is well known that voice data is very sensitive to delay and delay variation (jitter) effects. Therefore, guarantees must be given for a maximum end-to-end delay bound as far as voice data is concerned. On the other hand, continuous media streams such as voice can tolerate some fraction of data loss without suffering from perceivable quality degradation. Discrete data, in

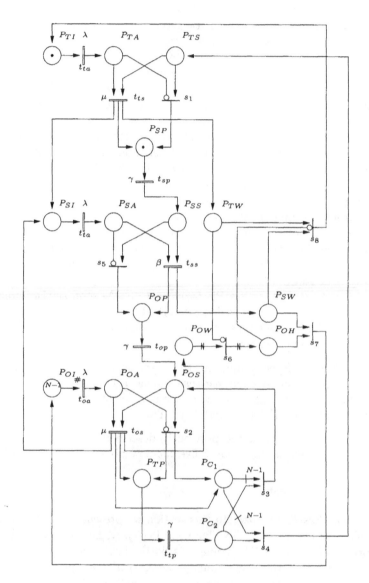

Fig. 13.47 SRN for the client–server system $(N > 1)$.

contrast, are very loss sensitive but require only weak constraints as far as delay is concerned. Trade-off analysis is therefore necessary to make effective use of limited available channel bandwidth.

In our simplified scenario, a single buffer scheme is assumed that is shared between a voice stream and discrete data units. For the sake of simplicity, the delay budget available for voice packets to be spent at the channel is assumed to be just enough for a voice data unit transmission. Any queueing delay would lead to a violation of the given delay guarantee. Therefore, voice packets arriving at a nonempty system are simply rejected and discarded. Rejecting voice data, on the other hand, is expected to have a positive impact on the loss rate of data packets due to sharing of limited buffer resources. Of course, the question arises to what extent the voice data loss must be tolerated under the given scenario.

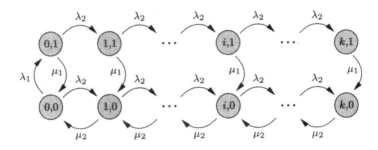

Fig. 13.48 CTMC modeling an ISDN channel with voice and discrete data.

Table 13.39 Parameter values for the ISDN channel in Fig. 13.48

Parameter	Value
λ_1	1.0
λ_2	2.0
μ_1	5.0
μ_2	10.0
k	10

For the time being, we assume Poisson arrivals for discrete data packets with parameter λ_2 and for voice data units with parameter λ_1. The transmission times of data packets are exponentially distributed with mean $1/\mu_2$ and for voice data with mean $1/\mu_1$. Buffer capacity is limited to $k = 10$. The resulting CTMC is depicted in Fig. 13.48 and the parameters summarized in Table 13.39. States are represented by pairs (i, j), where i, $0 \leq i \leq k$ denotes the number of discrete data packets in the channel and $j, 0 \leq j \leq 1$ indicates the presence of a voice data unit in the channel.

Under these assumptions and the model shown in Fig. 13.48, some numerical computations are carried out using the SHARPE software package. First, transient and steady-state results for the rejection probability of data packets are presented in Fig. 13.49 and for the rejection of voice data in Fig. 13.50. Comparing the results, it can be seen that the rejection probability of voice data very quickly approaches the steady-state value of a little more than 33%. Rejection probability of discrete data is much smaller, on the order of 10^{-6}, and steady-state is also reached in coarser time scale. It was assumed that the system was initially empty.

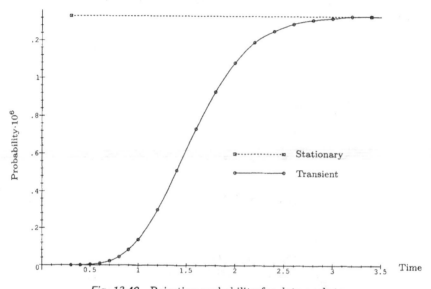

Fig. 13.49   Rejection probability for data packets.

The channel-idle probability of 66% is approached in a similar time scale as the rejection probability reaches its steady state for voice data, as can be seen in Fig. 13.51. Often, a GSPN model specification and automated generation of the underlying CTMC is much more convenient and less error prone due to possible state-space explosion. The GSPN in Fig. 13.52 is equivalent to the CTMC in Fig. 13.48 [STP96]. The GSPN was also solved using the SHARPE package.

The results of Fig. 13.53 illustrate the qualitative and quantitative difference in rejection probabilities of voice vs. discrete data. Note that rejection probabilities increase for voice data as channel capacity is increased. This effect is due to the reduced loss rate of discrete data data packets as a function of increased buffer size.

Figure 13.54 shows the impact that the arrivals of discrete data packets have on voice rejection probabilities as compared to rejection of discrete data

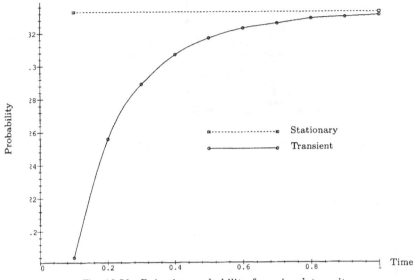

Fig. 13.50 Rejection probability for voice data units.

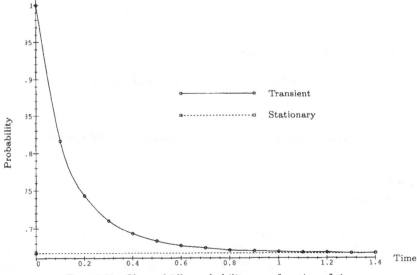

Fig. 13.51 Channel idle probability as a function of time.

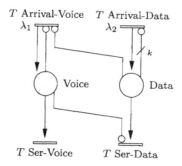

Fig. 13.52   SPN representation of the CTMC in Fig. 13.48.

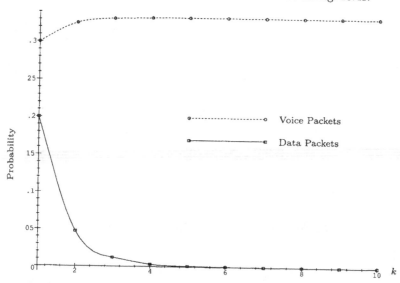

Fig. 13.53   Rejection probabilities as a function of channel capacity k: voice vs. data.

packets themselves. Both steady-state and transient probabilities are presented. Note that voice data are much more sensitive to an increase in λ_2.

13.2.4.2 Markov Modulated Poisson Processes   Markov modulated Poisson processes (MMPP) are introduced to represent correlated arrival streams and bursty traffic more precisely. An MMPP (see Section 6.8.2) is a stochastic arrival process with an arrival rate governed by an m-state CTMC. The arrival rate of an MMPP is assumed to be modulated according to an independent CTMC. The GSPN model of Fig. 13.52 is now modified so as to have two-state MMPP data arrivals as shown in Fig. 13.55 [STP96]. A token moves between the two places MMPP$_1$ and MMPP$_2$. When transition $T_{\mathrm{arrival-data1}}$ is enabled and there are fewer than k packets transmitted across the channel,

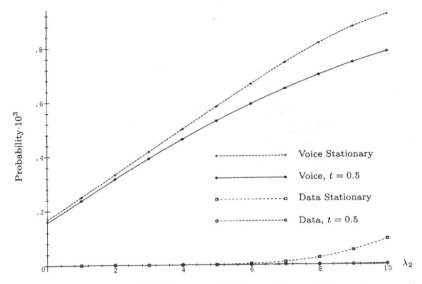

Fig. 13.54 Rejection probabilities as a function of λ_2 and time: voice vs. discrete data.

then discrete data is arriving with rate λ_{21}. Alternatively, the arrival rate is given by λ_{22}. If λ_{21} differs significantly from λ_{22}, bursty arrivals can be modeled this way. Note that the MMPP process is governed by the firing rates of T_{21} and T_{12}, which are given by rates a and b, respectively.

Table   13.40 Parameters to the MMPP-based ISDN model in Fig. 13.55

Parameter	Value
λ_1	1.0
λ_{21}	9.615
λ_{22}	0.09615
μ_1	5.0
μ_2	10.0
a	8.0
b	2.0
k	10

The results in Fig. 13.56 indicate that for small buffer sizes a Poisson traffic model assumption leads to more pessimistic rejection probability predictions than those based on MMPP assumptions. This effect is quickly reversed for discrete data if channel capacity is increased. For larger capacities there

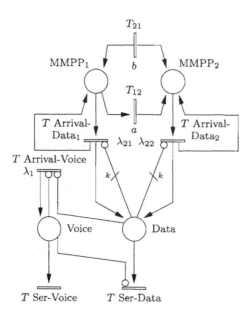

Fig. 13.55   A GSPN including an MMPP modulating data packet arrival rates.

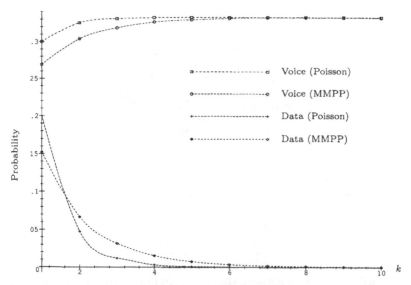

Fig. 13.56   Rejection probabilities as a function of channel capacity k: Poisson vs. MMPP.

is hardly any difference in rejection probability for voice data, regardless of Poisson- or MMPP-type traffic model. In general, though, correlated or bursty traffic can make a significant difference in performance measures as opposed to merely Poisson-type traffic. But our modeling methods based on GSPN (and SRN), which allow us to compactly specify and automatically generate and solve large CTMCs and MRMs for steady-state and transient behavior, are capable of handling such non-Poisson traffic scenarios.

13.2.5 ATM Network Under Overload

Consider a variant of the ATM network model studied in [WLTV96], consisting of three nodes, N_1, N_2, and N_3. Two links, $N_1 N_3$ and $N_2 N_3$, have been established as shown in Fig. 13.57. Now suppose that the network connection

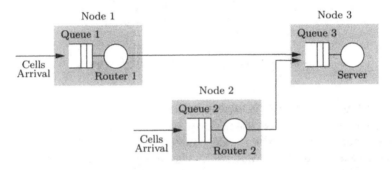

Fig. 13.57 The [WLTV96] queueing model before rerouting.

between N_1 and N_3 is down. Before the global routing table is updated, the cells destined to node N_3 arriving at node N_1 will be rerouted to node N_2 (see Fig. 13.58). Redirected cells from node N_1 may be rejected and thus lost if the input queue of node N_2 is full. Right after the rerouting starts, the cell loss probability of N_1 will overshoot because node N_2 is overloaded.

In the original paper, the burstiness of the ATM traffic is modeled by a two-state MMPP arrival process (see Section 6.8.2). The cell transmission time is modeled by an Erlang-5 distribution. Here we approximate the cell arrival process by a Poisson process and the cell service time by an exponential distribution. We can easily allow MMPP arrivals and Erlang service time, but the underlying CTMC will become larger. First, we analyze the simple case before the link $N_1 N_3$ breaks down. Here we assume that node N_1 and node N_2 have the same structure. That is, their buffer sizes and the service rates are the same. Furthermore, we assume that the job arrival rates are the same for the two nodes and that the buffer size at node N_3 is large enough to handle all the jobs coming from node N_1 and node N_2. Thus, node N_1 and node N_2 can be modeled by a finite-buffer M/M/1/N queue, where N is the buffer size.

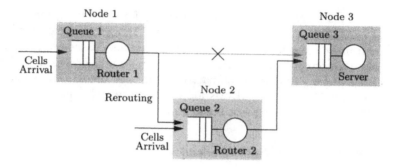

Fig. 13.58 The [WLTV96] queueing model after rerouting.

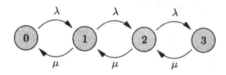

Fig. 13.59 The CTMC for the $M/M/1/3$ queue.

The CTMC and the GSPN models for the node are shown in Fig. 13.59 and Fig. 13.60, respectively. We use **SHARPE** to find the steady-state probabilities for each state as well as the steady-state cell loss probability before the link N_1N_3 breaks down. Fig. 13.61 shows the **SHARPE** input file of the CTMC and the GSPN models of the M/M/1/3 queue together with their outputs. Both models give the same results for the cell loss probability.

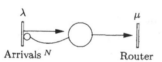

Fig. 13.60 The GSPN model for the M/M/1/N queue.

When the link N_1N_3 breaks down, the state transition diagram of the subnet can be given as shown in Fig. 13.62. Each state is identified by a pair (i,j) with $i,j \in \{0,\ldots,3\}$, representing the number of cells in the buffers of node N_1 and node N_2. The cell arrival rates are λ_1 and λ_2 and the cell transmission rates are μ_1 and μ_2. States 30, 31, 32, and 33 represent the buffer full states for node N_1, while states 03, 13, 23, and 33 represent the buffer full states for node N_2.

The loss probability (L_1) of cells destined to node N_3 going through node N_1 is determined by two elements:

1. The probability that node N_1 is full, denoted by p_1.

```
bind                                    gspn  noReroutingG
lam 0.1                                      * places
mu        1/2.73                        que   0
N    3                                  end
end
                                        * timed trans
format 6                                arr       ind lam
                                        router    ind mu
markov noReroutingM                     end
0   1   lam
1   2   lam                             * immediate trans
2   3   lam                             end
3   2   mu
2   1   mu                              * input arcs
1   0   mu                              que router 1
end                                     end

* steady-state probability state 0     * output arcs
var   p0 prob(noReroutingM,0)           arr que 1
expr p0                                 end

* steady-state probability state 1     * inhibitor arcs
var   p1 prob(noReroutingM,1)           que arr N
expr p1                                 end

* steady-state probability state 2     * steady-state cell loss probability
var   p2 prob(noReroutingM,2)           var  Lg 1- util(noReroutingG,arr)
expr p2                                 expr Lg
                                        end
* steady-state probability state 3
var   p3 prob(noReroutingM,3)           * Result:
expr p3                                 * p0: 7.310607e-01
                                        * p1: 1.995796e-01
* steady-state cell loss probability    * p2: 5.448523e-02
var   Lm prob(noReroutingM,3)           * p3: 1.487447e-02
expr Lm                                 * Lm: 1.487447e-02
                                        * Lg: 1.487447e-02
```

Fig. 13.61 SHARPE input file for the CTMC and GSPN models of the M/M/1/3 queue together with the results.

2. The probability that the rerouted cells are dropped because node N_2 is full, denoted by d.

In short: $L_1 = p_1 + (1 - p_1)d$.

The loss probability (L_2) of cells destined to node N_3 going directly through node N_2 is simply the probability that node N_2 is full, which is denoted by p_2.

In Figs. 13.63, 13.64, and 13.65 the SHARPE input file for the irreducible Markov reward model from Fig. 13.62 is given. From line 1 to line 6 we assign the values to the input parameters, and from lines 10 through 62 the Markov reward model is defined. Lines 10 through 26 describe the CTMC. By using loops in the specification of Markov chain transitions, we can generate the Markov chain automatically. Without the loop functionality, it would be cumbersome to specify the Markov chain manually, especially when the state space gets large. From lines 29 through 45, the reward rates $r_{i,j}$ are assigned

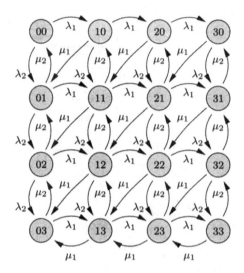

Fig. 13.62 The CTMC for Fig. 13.59 after rerouting.

to the states (i,j). By default, any state not assigned a reward rate is assumed to have the reward rate 0. In order to analyze the transient behavior of the system, the initial state probabilities are assigned from the values for the steady-state probabilities before the link $N_1 N_3$ breaks down. In other words:

$$\pi_{i,j} = \pi_{s,i} \times \pi_{s,j},$$

where $\pi_{i,j}$ is the initial state probability for state (i,j) in the transient model. The probability $\pi_{s,i}$ is the steady-state probability of node N_1 in state i before the link breaks down. Similarly, the probability $\pi_{s,j}$ is the steady-state probability of node N_2 in state j before the link breaks down. The fact we wish to carry out transient analysis of an irreducible CTMC and hence need to assign initial state probabilities is indicated to **SHARPE** with the keyword **readprobs** in line 10. The assignment of the initial probabilities is done from line 46 to line 62.

Reward rates for computing the average queue length of node N_1 are written in a separate file called **reward.quellength**. In this file, we assign the reward rates to indicate the number of jobs in the first node. The contents of the file are as follows:

```
bind
r00 0        r01 0        r02 0        r03 0
r10 1        r11 1        r12 1        r13 1
r20 2        r21 2        r22 2        r23 2
r30 3        r31 3        r32 3        r33 3
end
```

In line 65, the keyword **include** is used to read the file. From lines 67 through 69, we ask for the steady-state expected queue length of N_1. This

```
 1  bind                                       32  3_0      r30
 2  lam1     0.1                               33  0_1      r01
 3  lam2     0.1                               34  1_1      r11
 4  mu1      1/2.73                            35  2_1      r21
 5  mu2      1/2.73                            36  3_1      r31
 6  end                                        37  0_2      r02
 7                                             38  1_2      r12
 8  format 5                                   39  2_2      r22
 9                                             40  3_2      r32
10  markov rerouting readprobs                 41  0_3      r03
11  loop i, 0, 2                               42  1_3      r13
12    loop j, 0, 2                             43  2_3      r23
13      $(j)_$(i) $(j+1)_$(i)    lam1          44  3_3      r33
14      $(j+1)_$(i) $(j)_$(i+1)  mu1           45  end
15    end                                      46  0_0      0.53444974708449
16  end                                        47  0_1      0.14590480208172
17  loop i, 0, 2                               48  0_2      0.03983201038346
18    loop j, 0, 3                             49  0_3      0.01087414045033
19      $(j)_$(i) $(j)_$(i+1)    lam2          50  1_0      0.14590480208172
20      $(j)_$(i+1) $(j)_$(i)    mu2           51  1_1      0.03983201673616
21    end                                      52  1_2      0.01087414040931
22  end                                        53  1_3      0.00296864077281
23  loop i, 0, 2                               54  2_0      0.03983201038346
24    $(i)_3 $(i+1)_3 lam1                     55  2_1      0.01087414040931
25    $(i+1)_3 $(i)_3 mu1                      56  2_2      0.00296864028815
26  end                                        57  2_3      0.00081043891908
27                                             58  3_0      0.01087414045033
28  reward                                     59  3_1      0.00296864077281
29  0_0      r00                               60  3_2      0.00081043891908
30  1_0      r10                               61  3_3      0.00022124985778
31  2_0      r20                               62  end
```

Fig. 13.63   First part of input file for the Markov reward model.

```
63                                       91
64  * reward rates length queue1         92  var p2 exrss (rerouting)
65  include reward.que1length            93  echo steady-state probability
66                                                  that queue 2 is full
67  var que1length exrss (rerouting)     94  expr p2
68  echo steady-state queue1 length      95
69  expr que1length                      96  * reward rates throughput queue1
70                                       97  include reward.tput1
71  * reward rates queue1 to be full     98
72  include reward.que1full              99  var tputa11 exrss (rerouting)
73                                      100
74  var p1 exrss (rerouting)            101  * throughput cell arrivals queue 1
75  echo steady-state probability       102  bind
            that queue1 is full         103  tputa1   tputa11
76  expr p1                             104  end
77                                      105
78  bind                                106  * reward rates throughput queue2
79  p p1                                107  include reward.tput2
80  end                                 108
81                                      109  var tputa22 exrss (rerouting)
82  * reward rates for length queue2    110
83  include reward.que2length           111  * throughput cell arrival queue 2
84                                      112  bind
85  var que2length exrss (rerouting)    113  tputa2   tputa22
86  echo steady-state queue2 length     114  end
87  expr que2length                     115
88                                      116
89  * reward rates queue2 to be full    117  * reward rates throughput router2
90  include reward.que2full             118  include reward.tputr2
```

Fig. 13.64   Second part of input file for the Markov reward model.

```
119
120  * find throughput for router 2
121  var tputr2 exrss (rerouting)
122
123  * find steady-state drop probability for rerouted cells
124  var d  (tputa2+tputa1-tputr2)/tputa1
125
126  var L1 p+(1-p)*d
127  echo steady-state loss probability of cells destined to
128  echo node N3 going through node N1
129  expr L1
130
131  bind k 3
132
133  * start the transient analysis
134  loop t , 3, 90 , 3
135
136  bind
137  * pt is transient probability that queue1 is full.
138      pt      sum(i,0,k, tvalue(t;rerouting,3_$(i)))
139
140  * p2t is transient probability that queue2 is full.
141      p2t     sum(i,0,k, tvalue(t;rerouting,$(i)_3))
142
143  * tputa2t is transient throughput queue2
144      tputa2t lam2 * ( 1 - sum(i,0,k, tvalue(t;rerouting,$(i)_3)))
145
146  * tputa1t is transient throughput for queue1
147      tputa1t lam1 * ( 1 - sum(i,0,k, tvalue(t;rerouting,3_$(i))))
148
149  * tputr2t is transient throughput for router2
150      tputr2t mu2 * (1 - sum(i,0,k, tvalue(t;rerouting,$(i)_0)))
151
152  * dt is transient drop probability for rerouted cells
153      dt      (tputa2t + tputa1t - tputr2t)/tputa1t
154
155  * L1t is loss probability of cells destined to node N3
156  * going through node N1.
157      L1t     pt +(1-pt)* dt
158
159  end
160
161  end
162
163  end
```

Fig. 13.65 Third part of input file for the Markov reward model.

result is achieved by asking for the expected steady-state reward rate using the keyword exrss. In line 72, we assign the new reward rates so that the

steady-state probability that node N_1 is full can be calculated. Thereafter we bind the value of $p1$ to p. Similarly, from lines 83 through 121, we assign different reward rates and ask for the steady-state average queue length of node N_2, the steady-state probability for node N_2 to be full, the throughput of node N_1, node N_2, and router 2. Finally, we get the cell drop probability for the rerouted cells and the loss probability of cells destined to node N_3 going through node N_1. With these steps, we can see the power of the Markov reward model: By just changing the reward rates, we can get different performance measures, such as throughput, average queue length, and rejection probability for the arriving cells.

From line 133 to line 163, we perform transient analysis of the model. This result is achieved by using the keyword **tvalue**. With **tvalue**, we can get the probability that the system is in state (i, j) at time t. We use the nested sum functions to add up the transient probabilities. From lines 138 through 150, we use the **evaluated word** notation. An evaluated word is made up of one or more occurrences of $\$(n)$. Whenever the evaluated word is used, the string $\$(n)$ is expanded into an ASCII string whose digits are the number n. For example, in line 138, the state name $3\_\$(i)$ includes the states: 3_0, 3_1, 3_2, and 3_3. In Figs. 13.66 and 13.67 the results from the analysis are shown.

```
* steady-state queue1 length
  que1length:     3.53173e-01

* the steady-state probability that queue1 is full
  p1:     1.48745e-02

* steady-state queue2 length
  que2length:     8.06567e-01

* the steady-state probability that queue2 is full
  p2:     7.95456e-02

* The steady-state loss probability of cells destined to
* node N3 going through node N1
  L1:     9.00124e-02
```

Fig. 13.66   Steady-state results.

From the results, we can see that although both router 1 and router 2 have the same cell arrival rate, the steady-state probability for node N_2 to be full is more than four times bigger than that of node N_1. This condition occurs because router 2 has two cell sources, one coming from the original cell source, the other from the rerouted cells coming from router 1. However, the loss probability of cells destined to node N_3 going directly through N_2 is smaller than that going through N_1 first. From the transient results, we can answer questions such as:

- How long does it take to reach steady state after the line breakdown?

- What is the loss probability at a particular time before the system is in steady state?

These are important questions that need to be considered when we make performance and reliability analysis. Fig. 13.68 is derived from the transient analysis.

Problem 13.14   Modify the model of overload described in Section 13.2.5 to allow for two-state MMPP (see Section 6.8.2) sources and three-stage Erlang service. You may find it convenient to use GSPN rather than a direct use of CTMC.

```
t=3.000000                     t=33.000000                    t=63.000000
  pt        <- 0.014874          pt        <- 0.014874          pt        <- 0.014874
  p2t       <- 0.033728          p2t       <- 0.079335          p2t       <- 0.079545
  tputa2t <- 0.096627            tputa2t <- 0.092067            tputa2t <- 0.092046
  tputa1t <- 0.098513            tputa1t <- 0.098513            tputa1t <- 0.098513
  tputr2t <- 0.144398            tputr2t <- 0.182903            tputr2t <- 0.183044
  dt        <- 0.515083          dt        <- 0.077924          dt        <- 0.076279
  L1t       <- 0.522296          L1t       <- 0.091639          L1t       <- 0.090019

t=6.000000                     t=36.000000                    t=66.000000
  pt        <- 0.014874          pt        <- 0.014874          pt        <- 0.014874
  p2t       <- 0.051051          p2t       <- 0.079424          p2t       <- 0.079545
  tputa2t <- 0.094895            tputa2t <- 0.092058            tputa2t <- 0.092045
  tputa1t <- 0.098513            tputa1t <- 0.098513            tputa1t <- 0.098513
  tputr2t <- 0.162341            tputr2t <- 0.182962            tputr2t <- 0.183044
  dt        <- 0.315357          dt        <- 0.077225          dt        <- 0.076276
  L1t       <- 0.325541          L1t       <- 0.090951          L1t       <- 0.090016

t=9.000000                     t=39.000000                    t=69.000000
  pt        <- 0.014874          pt        <- 0.014874          pt        <- 0.014874
  p2t       <- 0.062675          p2t       <- 0.079475          p2t       <- 0.079545
  tputa2t <- 0.093732            tputa2t <- 0.092052            tputa2t <- 0.092045
  .                              .                              .
  .                              .                              .
  .                              .                              .
  L1t       <- 0.092832          L1t       <- 0.090024          L1t       <- 0.090012
```

Fig. 13.67   Results of transient analysis.

13.2.6   UMTS Cell with Virtual Zones

The aim of this section is to model a modern system from mobile communications. The traditional 2G mobile networks are similar in their behavior to the traditional telephone switching system with a fixed number of channels. This system can be modeled by an M/M/c/c loss system. The situation changes substantially when we consider third generation mobile systems since these have a dynamic capacity that depends on the interference levels in the covered area and the number of active users. Moreover, the coding system used in the Universal Mobile Telecommunication System (UMTS), the 3G system in Europe, implies that the received power at the base station is required to be equal for all active users. This implies that the distance of the mobile user from the base station is also an important factor because of the signal fading [Jake94]. In [SBE+00], capacity bounds have been derived based on interference levels for both maximum number of active users and maximum distance covered by the base station. The most important formulae are given in the following.

13.2.6.1   Capacity bounds   Assuming an ideal free space propagation of the radio signal, the following equation holds [Jake94]:

$$P_R = P_S \cdot \left(\frac{\lambda}{4\pi d} \right)^2 \cdot g_b \cdot g_m \qquad (13.21)$$

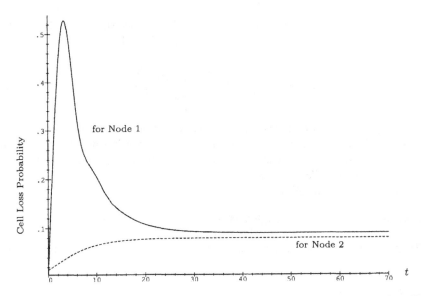

Fig. 13.68 Cell loss probability for the two nodes in the network discussed in Section 13.2.5.

where

P_R = received power,

P_S = transmitted power,

λ = the wave length,

d = distance between user endpoint (UE) and base station B,

g_m = antenna gain of the mobile equipment,

g_b = antenna gain of the base station B.

Given the maximum transmission power of the mobile equipment, $P_{S\,max}$, the maximum distance of a mobile UE from Node B is then given by [BeZr02]

$$d_{\max} = \frac{\lambda}{4\pi} \cdot \sqrt{\frac{P_{S\,max}}{P_R}} \cdot g_b \cdot g_m . \qquad (13.22)$$

For the single service case the received power P_R depends on the number of active users n and the interference levels in the cell. Assuming the antenna gains g_b and g_m to be 1, the maximum distance between UE and node B has been derived in [BeZr02] as

$$d_{\max}(n) = \frac{\lambda}{4\pi} \cdot \sqrt{\frac{P_{S\,max}}{N_0} \cdot (S - \varepsilon \cdot (n-1))} . \qquad (13.23)$$

where

S = service factor, ε = interference factor, N_0 = basic noise.

If we assume that the cell covers an ideal circle with the cell radius r, the maximum number of active users within a single UMTS cell can be calculated as [BeZr02]

$$n_{\max}(r) = \left(\frac{S}{\varepsilon} - \frac{N_0 \cdot (4\pi r)^2}{\varepsilon \cdot P_{S\max} \cdot \lambda^2} \right) + 1 . \tag{13.24}$$

Call Admission Control   Based on the bounds defined above, a call admission control (CAC) algorithm is developed. Given n existing active connections in the cell and a new call at a distance d_{new} from the node B (the Base Station in the UMTS terminology) and assuming N codes (channels), the CAC algorithm is as follows:

STEP 1   Assume that a $(n+1)$-st new user is asking for a service.

STEP 2   If $n + 1 > \min(N, n_{\max})$ then reject the call, else continue with step 3.

STEP 3   Calculate $d_{\max}(n+1)$ from Eq. (13.23).

STEP 4   If $d_{\mathrm{new}} > d_{\max}$ then reject the call, else continue with step 5.

STEP 5   Check for all existing active connections $i = 1 \ldots n$, if for any connection i the distance $d_i > d_{\max}$ then reject the new call, otherwise accept it.

Note that step 5 aims to protect existing connections from interruption because of the admission of a new call.

The Queueing Model   In order to evaluate the performance of the UMTS cell using the CAC algorithm described above, we propose a queueing model for the system. The main modeling difficulty herein stems from the fact that the distance of the UE from the base station needs to be included as part of the state information of each connection. As the distance is a continuous quantity the resulting model possesses a continuous state space which is intractable for the discrete-state methods. Even if we abstract from the continuous nature of the distance parameter by discretizing its domain, the need to store the value for each connection significantly increases the complexity of the system description and solution. Therefore, we choose the following approximation: We assume that the cell area is divided into Z concentric zones of equal area and that the radius of the inner border of Zone z, d_z, represents the distance of all connections in that zone. Given that the overall arrival rate over the cell area is λ, we can probabilistically split the traffic over the cell area by the vector $(\alpha_1, \alpha_2, \ldots, \alpha_Z)$ where $\sum_{i=1}^{Z} \alpha_i = 1$. Then the arrival rate to zone i can is as $\lambda_i = \alpha_i \cdot \lambda$.

With these assumptions, the UMTS cell can be described by a multiple queueing system as shown in Fig. 13.69. These queues are not independent

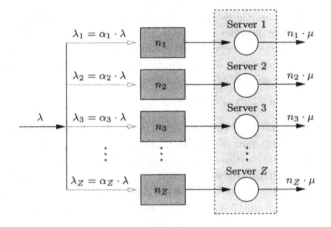

Fig. 13.69 Queueing model of one UMTS cell with Z virtual zones.

Table 13.41 System parameters

Parameter	Symbol	Values
Number of codes	N	64
Radius of the cell	r	4000 m
Spreading factor	SF	$64 \frac{\text{chips}}{\text{symbol}}$
Service factor	S	32
Signal-to-noise ratio	SNR	2 dB
Maximum transmission rate	$P_{S\,\text{max}}$	125 mW
Basic noise	N_0	-80 dBm
Interference factor	ε	0.5
Burst time	$1/\mu$	100s
Input traffic load	λ/μ	$0.25, \ldots, 0.70$

since the total number of connections in all zones cannot exceed n_{max}, where n_{max} is bounded by the maximum number of codes or the maximum number of users based on the interference as expressed in Eq. (13.24). That is, in any permitted system state, $\sum_{i=1}^{Z} n_i \leq n_{\text{max}}$, where n_i is the number of active connections in zone i. Assuming that the call arrival process is a Poisson process with rate λ, and an exponentially distributed burst size within each active connection with mean burst time $1/\mu$. Therefore, the model will result into a homogeneous CTMC. The example that we give in this section does not consider mobility of the user endpoints.

MOSEL-2 Model of the UMTS Cell   In order to express the call admission protocol in the UMTS cell adequately in the MOSEL-2-specification, we use the FUNC-construct of the MOSEL-2 formal description technique, which was not needed for the simpler examples in Section 12.3. Another feature of MOSEL-2 which proves useful for the construction of a larger model are the MOSEL-2 loop construts. As we see in the MOSEL-2-model illustrated in Figs. 13.70, 13.71, and 13.72, loop constructs such as

```
<1..Z>{ CONST N# := 2 * alpha# * N; }
```

allow a compact and easy to read declaration of a set of expressions. Macros can be used in every part of a MOSEL-2-model and substantially reduce the number of code lines without obfuscating the readability of the description. A loop construct definiton always starts with the character @ followed by a parameter range list and a loop body. The parameter range list is enclosed by the characters < and > and defines a range of values which are inserted in the code of the loop body during expansion. In the loop body the character # marks the places at which the values from the parameter range list are inserted. Loop expansion takes place in a preprocessing step, which the MOSEL-2 evaluation environment performs before checking the syntactic correctness of the (expanded) MOSEL-2-model. In case of our UMTS-model, the loop declaration above would be expanded into a set of four expressions, each of them defining a constant for one of the four zones in system. The system parameters are listed in Table 13.41.

Using the steady-state probabilities of the CTMC, performance measures like the burst blocking probabilities and system utilization are obtained. The MOSEL-2 results are compared with simulation results.

13.2.6.2 Numerical Results Figures 13.73 and 13.74 compare analytic-numerical results with those obtained by the detailed simulation for two different values of interference. The curves for both blocking probabilities and bandwidth utilization show a high degree of conformity.

Figures 13.75 and 13.76 show the influence of inter-code interference factors on the blocking probability and utilization. The results show clearly that the UMTS system is very sensitive to the value of the interference factor.

13.2.7 Handoff Schemes in Cellular Mobile Networks

Wireless cellular networks experience the handoff phenomenon, in which a call already in progress in a cell due to user mobility is "handed-over" (switched) to another cell. Handoff traffic into a cell consists of calls already in progress in neighboring cells that attempt to enter the cell under study and compete for the same resources as with the new calls that may be initiated in the same cell. Due to the widespread deployment of cellular networks and services in recent years and bandwidth scarcity over the air, cellular network dimensioning is a topic of paramount importance to wireless service providers.

One very popular scheme that is seen as a good compromise in terms of performance and complexity is the so called, "guard channel scheme" [HMPT01,

```
// UMTS cell with Z = 4 zones

// Constants and parameters

CONST Z := 4;   // number of zones
CONST SF := 32; // service factor
// interference factor
PARAMETER eps := 0.10, 0.30, 0.40, 0.45, 0.50;
PARAMETER lambda := 0.25, 0.30, 0.35, 0.40, 0.45, 0.50, 0.55, 0.60, 0.65,
                    0.70;

CONST alpha1 := 0.25; CONST alpha2 := 0.25; CONST alpha3 := 0.25;
CONST alpha4 := 0.25;

CONST r := 4000; // radius
CONST ps := 125.0;
CONST no := 0.00000001; // basic noise

CONST pi := 3.14159265358979;
CONST l := 0.15;

CONST N   := SF * 2; // spreading factor
@<1..Z>{ CONST N# := 2 * alpha# * N; }
CONST mue := 1.0 / 100;

@<1..Z>{CONST
  dd# := (SQRT(#)-SQRT(#-1))/SQRT(Z)*r;
}
CONST d1 := @<1..1>"+"{dd#};   CONST d2 := @<1..2>"+"{dd#};
CONST d3 := @<1..3>"+"{dd#};   CONST d4 := @<1..4>"+"{dd#};
```

Fig. 13.70   Constant and parameter declarations of the **MOSEL-2** UMTS model.

HoRa86, NaKa96]. In this scheme, a fixed number of the channels that are available in a given cell are reserved for handoff calls. For a cell with (total) capacity of C cells, g channels are reserved for handoff calls. Therefore, when a new call arrives and there are g or fewer channels available the call will be rejected. Note that this scheme provides a minimum for the number of handoff calls g that will exist in the cell at any given time.

Two important Quality of Service (QoS) measures have been defined in cellular networks; the first one is the *new call blocking probability*, a measure similar to the one that we see in telephone trunk systems. The second one, is the *handoff call dropping probability*. Dropping a handoff call is highly undesirable as it leads to a forced call termination.

Significant evidence [ChLu95, RaTa00, ZeCh99] suggests that handoff traffic is not Poisson. Hence we consider non-Poisson handoff traffic. More specifically, we consider handoff traffic to form a renewal process with r-stage Erlang and hyperexponential distributions interarrival times.

13.2.7.1   Model Description   We assume that the cellular network under study is homogeneous, i.e., all cells are identical and experience the same traffic patterns. This allows us to consider only one cell for our performance study and capture all interactions with neighboring cells through a handoff call arrival

```
// Nodes
@<1..Z>{ NODE Zone#[N#]; }

// Functions and conditions

FUNC value(z) := (SF - eps * (z - 1));

// function to find the maximum distance
FUNC max_d(z) := IF value(z) < 0 THEN 0
  ELSE 1 / (4 * pi) * SQRT (ps / no * value(z));
// the smallest index # for which max_d(x+1) < d<#+1>
FUNC find_index (x) := IF max_d(x+1) < d1 THEN 0
  @<2..Z>{ ELIF max_d(x+1) < d# THEN <#-1> }
  ELSE  Z;

// sum of all zones of index > i
FUNC tsum(i) := @<1..Z>"+"{(IF #>i THEN Zone# ELSE 0)};

COND accept(w) := tsum(find_index(w)) = 0;

// sum of all zones
FUNC zone_sum := @<1..Z>"+"{Zone#};

@<1..Z>{ COND blocks_# := zone_sum >= N
                          OR # > find_index (zone_sum)
                          OR NOT accept (zone_sum); }

// Assertions
ASSERT zone_sum <= N;

// Rules
@<1..Z>{IF NOT blocks_# TO Zone# WITH lambda*alpha#;}
@<1..Z>{IF Zone# > 0 FROM Zone# WITH Zone# * mue; }
```

Fig. 13.71 Node, function, and condition part of the MOSEL-2 UMTS model.

process. We further assume that the number of channels, C, that are allocated to the "cell under study" are fixed over time (i.e., the system employs the fixed channel allocation scheme).

We assume that ongoing call (new or handoff) completion times are exponentially distributed with parameter μ_d and the time at which the mobile station engaged in the call departs the cell is also exponentially distributed with parameter μ_h. We also assume that the interarrival times of new calls are exponentially distributed with parameter λ_n and of handoff call interarrivals are generally distributed with (cumulative) distribution function $P(T \leq t) = G(t)$ and density function $g(t)$ having finite mean $1/\lambda_h$ which is independent of new call arrival times. Note that new calls that find all $C - g$ channels busy will be blocked as also handoff calls which find all C channels busy will be dropped. The state transition diagram for this model is shown in Fig. 13.77. In this figure, a dotted arc denotes a system transition triggered by the arrival of a handoff call and $\mu = \mu_d + \mu_h$. Because of a general dis-

```
// Results
@<1..Z>{ RESULT MZ# := MEAN (Zone#); }
@<1..Z>{ RESULT blk# := PROB (blocks_#); }
PRINT MT := @<1..Z>"+"{MZ#};
PRINT util := MEAN (zone_sum);
PRINT blk := @<1..Z>"+"{alpha# * blk#};

// Pictures

PICTURE "Blocking"        PICTURE "Utilization"
PARAMETER lambda          PARAMETER lambda
CURVE blk                 CURVE util
XLABEL "call rate"        XLABEL "call rate"
YLABEL "blocking";        YLABEL "utilization";
```

Fig. 13.72   Result and picture part of the MOSEL-2 UMTS model.

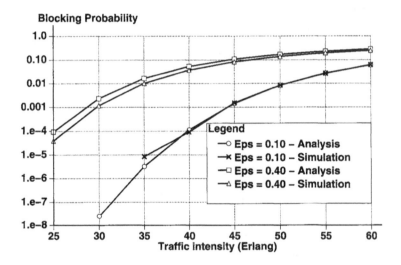

Fig. 13.73   Blocking probability — validation against simulation.

tribution, the stochastic process under consideration is not a CTMC but is a Markov regenerative process. Such processes can be solved numerically using packages such as SHARPE (see Section 12.4) or TimeNET [ZFGH00]. But we use a simulative solution here. In the particular case where handoff call interarrival times follow the exponential distribution with parameter λ_h, Haring et al. [HMPT01] provided closed-form formulae for the above-mentioned performance indices.

Since the two types of calls, new and handoff, are treated differently, performance measures of interest are the steady-state *new call blocking probability* and *handoff call dropping probability*. The new call blocking probability is simply the probability that an incoming new call finds at least $C - g$ channels

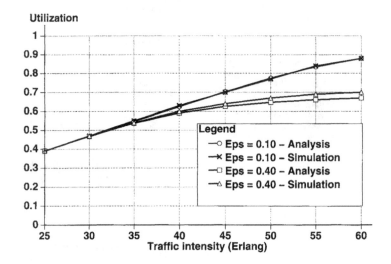

Fig. 13.74 Utilization — validation against simulation.

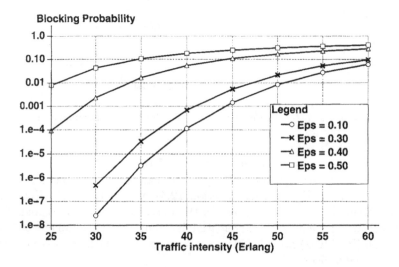

Fig. 13.75 Blocking probability versus call rate in single cell scenario — uniform traffic.

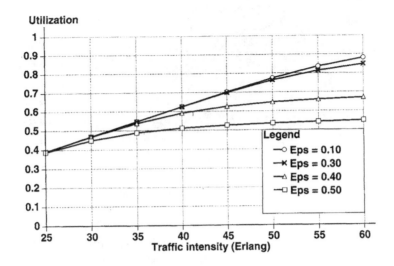

Fig. 13.76   Utilization versus call rate in single cell scenario — uniform traffic.

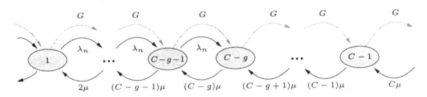

Fig. 13.77   State transition diagram for the performance model of the cellular system.

busy in the system, while the handoff dropping probability is the probability that an incoming handoff call finds all the channels busy in the system.

13.2.7.2 Simulation with **CSIM-19**   Simulation was done in CSIM-19 under different exponentially and nonexponentially distributed handoff interarrival times. We set $\lambda_n = 0.5$ new calls per minute, $\mu_d = 0.05$, $\mu_h = 1/3$ calls per minute. We choose the number of guard channels $g = 3$. For the purpose of comparison we consider different types of distributions for handoff interarrival times with fixed mean 2.5 minutes per call.

We choose the parameters $r = 3$ and each stage arrival rate is 1.2 per minute for 3-stage Erlang distribution and $\alpha_1 = 0.4, \alpha_2 = 0.6, \lambda_1 = 1, \lambda_2 = 0.2857$ for 2-stage hyperexponential distribution. We plot the blocking probability and dropping probability versus the number of channels for $C = 6$ to $C = 10$ in Figs. 13.78 and 13.79, respectively. The simulation results compare well with the numerical solution of the MRGP obtained by Dharma et al. [DTL03]. Note that 3-stage Erlang distribution gives the lowest blocking probability whereas hyperexponential distributed handoff interarrivals give the highest

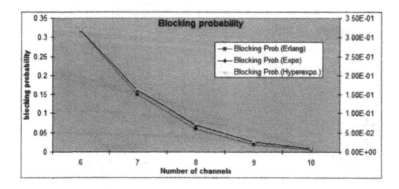

Fig. 13.78 Blocking probability for fixed mean handoff interarrival time.

value. Also, we observe that, with exponentially distributed handoff interarrivals, blocking probability lies in between that of hyperexponential and 3-stage Erlang cases. Similarly, in Fig. 13.79 we observe that the 3-stage Erlang distribution gives the lowest probability, hyperexponential distribution gives the highest probability and exponential distribution stays in between 3-stage Erlang and hyperexponential distributions. Also note that when the number of channels increases, the dropping probabilities tend to approach each other for various distributions as we expect. To verify the results of the CSIM-19 simulation, a simulative solution was also calculated with the tool SPNP.

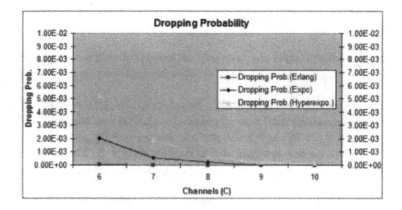

Fig. 13.79 Dropping probability for fixed mean handoff interarrival time.

SRN model constructed for wireless handoff case with C as number of channels and g as number of guard channel is depicted in Fig. 13.80.

In this 90% confidence interval is derived and the simulation is run 20 times. Table 13.42 provides the comparison of results obtained from CSIM-19

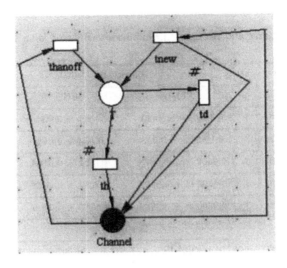

Fig. 13.80   Wireless handoff model in SPNP when $C = 6$ and $g = 3$.

and SPNP when the handoff arrival distribution is 3-stage Erlang. The value after $\pm$ represents half-width interval when 90% confidence interval is desired.

13.3   CASE STUDIES OF HIERARCHICAL MODELS

We discuss two examples of hierarchical models in this section. Several other hierarchical models are discussed in Section 10.5. For further exposition on hierarchical models, see [MaTr93].

13.3.1   A Multiprocessor with Different Cache Strategies

Section 13.1.1 introduces different models for loosely and tightly coupled multiprocessor systems. Now we model multiprocessor systems with different cache coherence protocols. Our treatment is based on [YBL89]. The outer model is a product-form queueing network, some of whose parameters are obtained from several lower-level CTMC models. The outer model shown in Fig. 13.81 (where $n = 2$ CPUs are assumed).

The model consists of open and closed job classes where closed job classes capture the flow of normal requests through the system and open job classes model the additional load imposed by the cache. For the queueing network (outer model), the following notation is used:

Table 13.42 Blocking and dropping probabilites calculated with the CSIM-19 and SPNP tools

		Blocking	
C	g	CSIM-19	SPNP
6	3	0.317016 ± 0.023	0.321 ± 0.034
7	3	0.150243 ± 0.013	0.159 ± 0.018
8	3	0.060049 ± 0.0078	0.0631 ± 0.0109
9	3	0.019583 ± 0.0009	0.0184 ± 0.0018
10	3	0.006009 ± 0.0009	0.0057 ± 0.0016
		Dropping	
C	g	CSIM-19	SPNP
6	3	0.000037 ± 0.00002	0.000042 ± 0.000028
7	3	0.000012 ± 0.00001	0.000016 ± 0.000017
8	3	0	0.000002 ± 0.000003
9	3	0	0
10	3	0	0

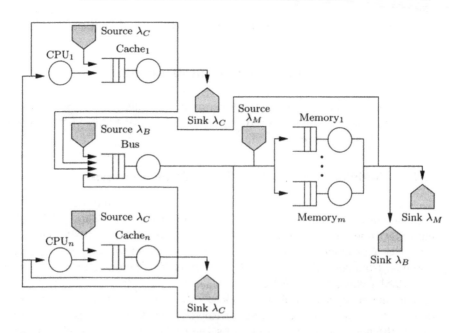

Fig. 13.81 The outer model for a multiprocessor system with cache protocols.

m Number of memory modules

n Number of CPUs

μ_{cpu} Service rate at the CPU

μ_{cache} Service rate at the cache

μ_{bus} Service rate at the bus

μ_{mem} Service rate at the main memory modules

For the CTMC cache models (inner models), we use

C Overall number of blocks in each cache

C_{sc} Number of shared cache blocks (blocks shared by all caches)

C_{pc} Number of private cache blocks

C_{ic} Number of instruction cache blocks

S Degree of sharing

f_w Probability that data/instructions are written

$1 - f_w$ Probability that data/instructions are read

h Probability of a given request to a private cache being a hit

u Probability that a previously accessed block has not been modified

d Probability that a private block, selected for replacement, is modified and needs to be written back

p_{ic} Probability that instructions are accessed during a memory request

$1 - p_{\text{ic}}$ Probability that a data block is accessed during a memory request; these blocks can be either shared or private

The cache behavior is modeled by the jobs of the open job classes. These additional customers induce additional traffic:

- The traffic at the cache due to requests from other caches (loading of missed blocks).

- The requests from other processors due to state updates.

- Additional bus traffic due to invalidation signals.

Fig. 13.82 CTMC for an instruction cache block (inner model 1).

If we access blocks in a cache, we have to differentiate between accessing private blocks, shared blocks, and instructions. Private blocks can be owned only by one cache at a time. No other cache is allowed to have a copy of these blocks at the same time. Since it is possible to read and modify private blocks, they have to be written back to the main memory after being modified. In order to keep shared block consistent, a cache coherence protocol needs to be used because shared blocks can be modified by any other processor that also owns this block. Items in the instruction cache can only be read but never be modified. Therefore, no cache coherence protocol is necessary for instruction blocks. Now we develop and solve three inner models: For an instruction cache block, for a private cache block, and for a shared cache block. A block in the *instruction cache* can either be not in the cache or valid. If a block is selected for replacement, it is overwritten (no write back is necessary) since these blocks can not be modified. This assumption leads to the CTMC, shown in Fig. 13.82.

The transition rates are given by

$$a = \frac{(1 - f_w)p_{ic}}{C_{ic}}\mu_{cpu}, \quad b = \frac{1 - h}{C}\mu_{cpu},$$

with $C_{ic} = p_{ic}C$. Steady-state probabilities of this simple CTMC are easily obtained:

$$\pi_0^{ic} = \frac{b}{a + b}, \qquad \pi_1^{ic} = \frac{a}{a + b}. \tag{13.25}$$

Since *private cache blocks* can also be modified, they have to be written back

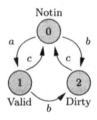

Fig. 13.83 CTMC for a private cache block (inner model 2).

after being modified. Modified blocks are called *dirty*. The corresponding state diagram is shown in Fig. 13.83. The transition rates are given by

$$a = (1 - p_{\text{ic}})(1 - S)\frac{1 - f_w}{C_{\text{pc}}}\mu_{\text{cpu}},$$

$$b = (1 - p_{\text{ic}})(1 - S)\frac{f_w}{C_{\text{pc}}}\mu_{\text{cpu}},$$

$$c = \frac{1 - h}{C}\mu_{\text{cpu}},$$

with $C_{\text{pc}} = (1 - p_{\text{ic}}(1 - S)C$. Steady-state probabilities of this CTMC are

$$\pi_0^{\text{pc}} = \frac{c}{a + b + c}, \qquad \pi_1^{\text{pc}} = \frac{ac}{(a + b + c)(b + c)}, \qquad \pi_2^{\text{pc}} = \frac{b}{b + c}. \qquad (13.26)$$

For private and instruction cache blocks no coherence protocol is necessary, while for *shared blocks* a *cache coherence protocol* must be used since these blocks can become inconsistent in different caches. One of the first cache coherence protocols found in the literature is Goodman's write-once scheme [Good83]. When using this protocol, the cache blocks can be in one of five states:

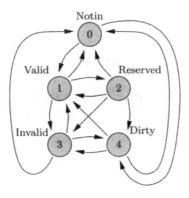

Fig. 13.84 CTMC for shared cache blocks (inner model 3).

1. Notin: The block is not in the local cache.

2. Invalid: The block is in the local cache but not consistent with other caches.

3. Valid: The block is in the local cache and consistent with the main memory.

4. Reserved: The block is in the local cache, consistent with the main memory, and not in any other cache.

5. Dirty: The block is in the local cache, inconsistent with the main memory, and not in any other cache.

If a read miss to a shared cache block occurs and a block in another cache exists that is in state dirty, this cache supplies the block as well as writing it back to the main memory. Otherwise (no cache has a dirty copy), each cache with a copy of that block sets the block state to valid. In case of a write miss the block is loaded from memory or, if the block is dirty, the block is loaded from the cache that has a dirty copy (this cache then invalidates its copy). When a write hit to an already dirty or reserved cache block occurs, the write proceeds locally, and if the state is valid, the block is written. In case of a write hit to an invalid cache block, the local state of the cache block is changed. Finally, if a block is loaded into the cache, it can happen that there is not enough space in the cache and a block in the cache is to be selected for replacement. Blocks are replaced in the order invalid, dirty, reserved, and valid.

The state diagram of the CTMC for shared cache blocks is shown in Fig. 13.84. Transition rates for this CTMC are as follows:

- The processor under consideration generates a read request to a shared block:

$$\text{notin} \rightarrow \text{valid} : \frac{S(1 - f_w)(1 - p_{\text{ic}})}{C_{\text{sc}}} \mu_{\text{cpu}} ,$$

$$\text{invalid} \rightarrow \text{valid} : \frac{S(1 - f_w)(1 - p_{\text{ic}})}{C_{\text{sc}}} \mu_{\text{cpu}} .$$

If the cache block is in state dirty, reserved, or valid, the read request can proceed without any state change and bus transaction.

- Another processor generates a read request to a shared block:

$$\text{reserved} \rightarrow \text{valid} : \frac{(n - 1)S(1 - f_w)(1 - p_{\text{ic}})}{C_{\text{sc}}} \mu_{\text{cpu}} ,$$

$$\text{dirty} \rightarrow \text{valid} : \frac{(n - 1)S(1 - f_w)(1 - p_{\text{ic}})}{C_{\text{sc}}} \mu_{\text{cpu}} .$$

If the cache block is in the states valid, invalid, or notin, no state change occurs.

- The processor under consideration generates a write request to a shared block:

$$\text{valid} \rightarrow \text{reserved} : \frac{(n - 1)S f_w(1 - p_{\text{ic}})}{C_{\text{sc}}} \mu_{\text{cpu}} ,$$

$$\text{reserved} \rightarrow \text{dirty} : \frac{(n - 1)S f_w(1 - p_{\text{ic}})}{C_{\text{sc}}} \mu_{\text{cpu}} ,$$

$$\text{invalid} \rightarrow \text{dirty} : \frac{(n-1)Sf_w(1-p_{ic})}{C_{sc}}\mu_{cpu},$$

$$\text{notin} \rightarrow \text{dirty} : \frac{Sf_w(1-p_{ic})}{C_{sc}}\mu_{cpu}.$$

If a block is already dirty, the write request proceeds locally without any state change and bus transaction.

- Another processor generates a read request to a shared block:

$$\text{dirty} \rightarrow \text{invalid} : \frac{(n-1)Sf_w(1-p_{ic})}{C_{sc}}\mu_{cpu},$$

$$\text{reserved} \rightarrow \text{invalid} : \frac{(n-1)Sf_w(1-p_{ic})}{C_{sc}}\mu_{cpu},$$

$$\text{valid} \rightarrow \text{invalid} : \frac{(n-1)Sf_w(1-p_{ic})}{C_{sc}}\mu_{cpu}.$$

- A write or read miss in the local cache can result in a page replacement, which leads to the following state transitions:

$$\text{reserved} \rightarrow \text{notin} : \frac{1-h}{C}\mu_{cpu},$$

$$\text{dirty} \rightarrow \text{notin} : \frac{1-h}{C}\mu_{cpu},$$

$$\text{valid} \rightarrow \text{notin} : \frac{1-h}{C}\mu_{cpu},$$

$$\text{invalid} \rightarrow \text{notin} : \frac{1-h}{C}\mu_{cpu}.$$

Let

$$a = \frac{1-h}{C}\mu_{cpu}, \quad b = \frac{S(1-f_w)(1-p_{ic})}{C_{sc}}\mu_{cpu}, \quad c = \frac{Sf_w(1-p_{ic})}{C_{sc}}\mu_{cpu},$$

with $C_{sc} = (1-p_{ic})S \cdot C$. Then the steady-state probabilities of the CTMC are given by

$$\pi_0^{sc} = \frac{c}{a+b+c},$$

$$\pi_1^{sc} = \frac{a(\pi_0^{sc} + \pi_3^{sc}) + (n-1)a(1 - \pi_0^{sc} - \pi_3^{sc})}{(n-1)a + nb + c},$$

$$\pi_2^{sc} = \frac{b\pi_1^{sc}}{(n-1)a + nb + c},$$

$$\pi_3^{\text{sc}} = \frac{b(n-1)(1-\pi_0^{\text{sc}})}{a+nb+c},$$

$$\pi_4^{\text{sc}} = 1 - \pi_0^{\text{sc}} - \pi_1^{\text{sc}} - \pi_2^{\text{sc}} - \pi_3^{\text{sc}}. \tag{13.27}$$

Now we are ready to parameterize the outer model. To analyze the outer model given in Fig. 13.81, we need the visit ratios for each station and the arrival rates of the open job classes. We see from the following that these parameters are functions of the state probabilities of the three inner models:

Visit Ratio for a Remote Cache i: A cache miss is either supplied by a remote cache that has a dirty copy or by one of the main memory modules if no dirty copy exists. Because the probability of a dirty and valid copy are $1 - (1 - \pi_4^{\text{sc}})^{n-1}$ and $1 - (1 - \pi_1^{\text{ic}})^{n-1}$, respectively, the visit ratio of the remote cache i is given by

$$e_{C_i} = \frac{(1-p_{\text{ic}})S(\pi_0^{\text{sc}} + \pi_3^{\text{sc}})(1-(1-\pi_4^{\text{sc}})^{n-1})}{n-1}$$
$$+ \frac{p_{\text{ic}}\pi_0^{\text{ic}}(1-(1-\pi_1^{\text{ic}})^{n-1})}{n-1}.$$

Visit Ratio to Main Memory Module i: The main memory is involved when a cache miss to a private block occurs, a cache miss to a shared block occurs and no other cache has a dirty copy, and a cache miss to instructions occurs and no other cache contains these instructions. Hence the visit ratio is

$$e_{M_i} = \frac{(1-p_{\text{ic}})(1-S)(1-h) + (1-p_{\text{ic}})S(\pi_0^{\text{sc}} + \pi_3^{\text{sc}})(1-\pi_4^{\text{sc}})^{n-1}}{m}$$
$$+ \frac{p_{\text{ic}}\pi_0^{\text{ic}}(1-\pi_1^{\text{ic}})^{n-1}}{m}.$$

Visit Ratio to the Bus: The bus is used to broadcast invalidations and to transmit missed blocks. Hence the visit ratio is

$$e_B = 2(1-p_{\text{ic}})S(\pi_0^{\text{sc}} + \pi_3^{\text{sc}}) + 2(1-p_{\text{sc}})(1-S)(1-h)$$
$$+ (1-p_{\text{sc}})Sf_w\pi_1^{\text{sc}} + (1-p_{\text{sc}})(1-S)f_w u + p_{\text{ic}}\pi_0^{\text{ic}}.$$

Once we know the visit ratios and the structure of the network, we can compute the routing probabilities, using Eq. (7.5). As a result, we get for the routing probabilities

$$p_{\text{cache}_i \to \text{proc}_i} = \frac{(e_{C_i} - e_B) + \sum_{j=1}^{m} e_{M_j}}{e_{C_i}}, \quad \forall i = 1, \ldots, n,$$

$$p_{\text{cache}_i \to \text{bus}} = \frac{e_B - \sum_{j=1}^{m} e_{M_j}}{e_{C_i}}, \quad \forall i = 1, \ldots, n,$$

$$p_{\text{bus}\to\text{proc}_i} = \frac{e_B - \sum\limits_{j=1}^{m} e_{M_j}}{e_B}, \quad \forall i = 1, \ldots, n,$$

$$p_{\text{bus}\to\text{mem}_j} = \frac{e_{M_j}}{e_B} \quad \forall j = 1, \ldots, m.$$

Next we obtain the arrival rates for each open customer class:

Derivation of λ_C: The possible customers of a local cache are the requests from its owner processor $(1 - p_{\text{ic}})(1 - S)(1 - h)$, the loading of missed blocks $(1 - p_{\text{ic}})S(\pi_0^{\text{sc}} + \pi_3^{\text{sc}}) + p_{\text{ic}}\pi_0^{\text{ic}}$, the requests from other processors for state updating due to a write request to a shared block $(n-1)(1-p_{\text{ic}})Sf_w(\pi_1^{\text{sc}} + \pi_2^{\text{sc}})$, and requests from other processors for state updating due to a read request for a reserved block $(n-1)(1-p_{\text{ic}})S(1-f_w)\pi_2^{\text{sc}}$. Therefore, the open customer arrival rate, denoted by λ_C, is given by

$$\begin{aligned}
\lambda_C = \rho_i \mu_{cpu_i} \big(&(1 - p_{\text{ic}})(1 - S)(1 - h) + (1 - p_{\text{ic}})S(\pi_0^{\text{sc}} + \pi_3^{\text{sc}}) \\
&+ p_{\text{ic}}\pi_0^{\text{ic}} + (n - 1)(1 - p_{\text{ic}})Sf_w(\pi_1^{\text{sc}} + \pi_2^{\text{sc}}) \\
&+ (n - 1)(1 - p_{\text{ic}})S(1 - f_w)\pi_2^{\text{sc}} \big).
\end{aligned} \tag{13.28}$$

Derivation of λ_B: A selected block needs to be written back to main memory only if it is private and modified or shared and dirty. The probability for a shared block to be selected is given by $C_{\text{sc}}(1 - \pi_0^{\text{sc}})/C$ and the probability for a private block to be selected is given by $C_{\text{pc}}(1 - \pi_0^{\text{pc}})/C$. The selected block is written back if it is private and modified or shared and dirty. The arrival rate at the bus is, therefore, given by

$$\begin{aligned}
\lambda_B = \rho_i \mu_{cpu_i} n \big(&(1 - p_{\text{ic}})(1 - S)(1 - h) + (1 - p_{\text{ic}})S\pi_0^{\text{sc}} \big) \\
&\cdot \left(\frac{C_{\text{pc}}(1 - \pi_0^{\text{pc}})}{C}d + \frac{C_{\text{sc}}(1 - \pi_0^{\text{sc}})}{C}\pi_4^{\text{sc}} \right).
\end{aligned} \tag{13.29}$$

Derivation of λ_M: There are two possibilities: A shared read results in a miss and one of the remote caches has a dirty copy $(1 - p_{\text{ic}})S(1 - f_w)(\pi_0^{\text{sc}} + \pi_3^{\text{sc}})(1 - (1 - \pi_4^{\text{sc}})^{n-1})$, or a write hit to a valid shared block occurs and a write through operation is performed $(1 - p_{\text{ic}})Sf_w + (1 - p_{\text{ic}})(1 - S)f_w u$. The arrival rate at the memory modules is, therefore, given by

$$\begin{aligned}
\lambda_M = \rho_i \mu_{cpu_i} n \big(&(1 - p_{\text{ic}})S(1 - f_w)(\pi_0^{\text{sc}} + \pi_3^{\text{sc}})(1 - (1 - \pi_4^{\text{sc}})^{n-1}) \\
&+ (1 - p_{\text{ic}})Sf_w + (1 - p_{\text{ic}})(1 - S)f_w u \big).
\end{aligned} \tag{13.30}$$

Because the arrival rates λ_C, λ_B, and λ_M are functions of server utilizations that are computed from the solution of the outer model, the outer model needs to be solved using fixed-point iteration. The solution algorithm is sketched in the following:

STEP 1 Solve the inner model for shared/private/instruction cache blocks to obtain the steady-state probabilities (see Eqs. (13.25), (13.26), and (13.27)).

STEP 2 Determine the arrival rates λ_C (Eq. (13.28)), λ_B (Eq. (13.29)), and λ_M (Eq. (13.30)) of the open customer classes, assuming the initial value $\rho_i = 1$ and using the steady-state probabilities of the inner model.

STEP 3 Solve the outer model using the computed arrival rates to obtain new values for the utilizations ρ_i.

STEP 4 The computed utilizations are used to determine the new arrival rates.

STEP 5 The last two steps are repeated until the results of two successive iteration steps differ less than ϵ.

Table 13.43   Performance measures for the cache model with $\mu_{\text{bus}} = 1.00$

No. of Processors	ρ_{cpu}	ρ_{bus}	System Power
2	0.475	0.356	0.950
3	0.454	0.426	1.278
4	0.438	0.495	1.754
5	0.425	0.562	2.127
6	0.413	0.619	2.478
7	0.402	0.679	2.817
8	0.391	0.734	3.131
9	0.377	0.785	3.395
10	0.363	0.821	3.634
15	0.279	0.902	4.180
20	0.219	0.902	4.374

For our example we assume the following parameter values (all time units are msec):

$$C = 2048 \qquad \mu_{\text{cpu}} = 0.25 \qquad h = 0.98$$
$$S = 0.1 \qquad \mu_{\text{cache}} = 1.0 \qquad f_w = 0.3$$
$$u = 0.1 \qquad \mu_{\text{bus}} = 1.0 \qquad p_{\text{ic}} = 0.1$$
$$d = 0.4 \qquad \mu_{\text{mem}} = 0.25$$

In Tables 13.43 and 13.44 the performance measures for different mean bus service times are given as a function of the number of processors. Here, the measure *system power* is the sum of all processor utilizations. In Table 13.43 we see that by increasing the number of processors from 2 to 20, the system power increases by a factor of 4.6. Because the system bottleneck is the bus, it makes more sense to use a faster bus or to use a dual bus instead of increasing the number of processors. In Table 13.44 the results are shown for a faster bus with $\mu_{\text{bus}} = 1.5$. As a result, the system power is increased by a factor of 6.5.

Table 13.44 Performance measures for the cache model with $\mu_{bus} = 1.50$

No. of Processors	ρ_{cpu}	ρ_{bus}	System Power
2	0.511	0.255	1.023
3	0.486	0.303	1.458
4	0.470	0.354	1.882
5	0.459	0.404	2.297
6	0.451	0.451	2.707
7	0.445	0.501	3.115
8	0.439	0.549	3.515
9	0.432	0.601	3.892
10	0.427	0.643	4.269
15	0.379	0.817	5.688
20	0.331	0.909	6.614

13.3.2 Performability of a Multiprocessor System

We consider a model that was developed by Trivedi, Sathaye, et al.[TSI+96] to help determine the optimal number of processors needed in a multiprocessor system. A Markov reward model is used for this purpose. In this model, it is assumed that all failure events are mutually independent and that a single repair facility is shared by all the processors. Assume that the failure rate of each processor is α. A processor fault is covered with probability c and is not covered with probability $1 - c$. Subsequent to a covered fault, the system comes up in a degraded mode after a brief reconfiguration delay, while after an uncovered fault, a longer reboot action is required. The reconfiguration times are assumed to be exponentially distributed with mean $1/\gamma$, the reboot times are exponentially distributed with mean $1/\delta$, and the repair times are exponentially distributed with mean $1/\beta$. It is also assumed that no other event can take place during a reconfiguration or a reboot phase. The CTMC modeling the failure/repair behavior of this system is shown in Fig. 13.85. In state i, $1 \leq i \leq n$, the system is up with i processors functioning, and $n - i$ processors are waiting for repair. In states (c_{n-i}), $i = 0, \cdots, n-2$, the system is undergoing a reconfiguration. In states (b_{n-i}), $i = 0, \cdots, n-2$, the system is being rebooted. In state 0, the system is down waiting for all processors to be repaired.

The reward rate in a state with i processors functioning properly correspond to some measure of performance in that configuration. Here the loss probability is used as a performance measure. An M/M/i/b queueing model is used to describe the performance of the multiprocessor system, which is represented by the PFQN shown in Fig. 13.86. This model contains two stations. Station mp is the processor station with i processors; it has type ms for multiple servers, each server having service rate μ. The other station is *source*, which represents the job source with rate λ. Because there is a limited

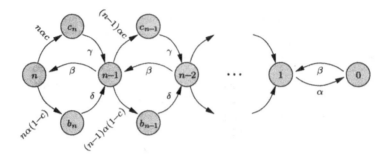

Fig. 13.85 CTMC for computing the performability of a multiprocessor system.

number b of buffers available for queueing the jobs, the closed product-form network with a fixed number b of jobs is chosen.

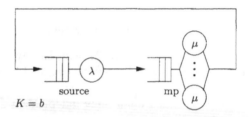

Fig. 13.86 PFQN model of the M/M/i/b queue.

The loss probability can be obtained via the throughput *tput* of station *mp*. Note that we have used a two-level hierarchical model. The lower level (inner model) is the PFQN (Fig. 13.86), while the outer model is the CTMC (Fig. 13.85). For state i of the CTMC, the PFQN with i processors is evaluated, and the resulting loss probability is used as a reward rate in state i. Then the expected steady-state reward rate is the overall performability measure of the system, given by

$$\text{EXRSS} = \sum_{i=1}^{n} r_i \pi_i,$$

where π_i is the steady-state probability of state i, and r_i is the loss probability of station *mp*. We consider just the operational state i because in the down states c_i, b_i and 0, the loss probability (reward rate) is 1.

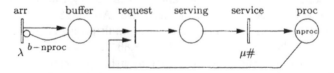

Fig. 13.87 GSPN model of the M/M/i/b queue.

It is also possible to model the lower level model using the GSPN shown in Fig. 13.87. The initial number *nproc* of tokens in place *proc* means that there are *nproc* processors available. When a new job arrives in place *buffer*, a token is taken from place *proc*. Jobs arrive at the system when transition *arr* fires. There is a limitation for new jobs entering the system caused by the inhibitor arc from place *buffer* to transition *arr*. Thus, *arr* can only fire when the system is not already full. There can be only *b* jobs in the system altogether, *nproc* being served (in place *serving*) and $b - nproc$ in place *buffer*. The firing rates are λ for transition *arr* and $k\mu$ for transition *service*. Here k is the number of tokens in place *serving*; the notation for this marking-dependent firing rate and in Fig.13.87 is $\mu\#$. The expected reward rate in steady state is called the *total loss probability* (TLP). It is defined as the fraction of jobs rejected either because the buffer is full or the system is down.

```
 1   format 8                          39  func R(nproc,L) \
 2   bind                              40     1-util(queue,arr;nproc,L)
 3   gamma 1/6000                      41
 4   beta 12                           42  markov one(L)
 5   tau 1.0                           43  1 0 gamma
 6   delta 360                         44  0 1 tau
 7   mu 100                            45  reward
 8   C 0.98                            46  1 R(1,L)
 9   b 10                              47  0 1
10   end                              48  end
11                                     49
12   gspn queue (nproc,lambda)         50  markov two(L)
13   * places                          51  2 x2 2* gamma*C
14   buffer 0                          52  2 y2 2* gamma*(1-C)
15   serving 0                         53  x2 1 delta
16   proc nproc                        54  y2 1 beta
17   end                               55  1 2 tau
18   * timed transitions               56  1 0 gamma
19   arr ind lambda                    57  0 1 tau
20   service dep serving mu            58  reward
21   end                               59  2 R(2,L)
22   * immediate transitions           60  x2 1
23   request ind 1                     61  1 R(1,L)
24   end                               62  0 1
25   * input arcs                      63  end
26   buffer request 1                  64
27   serving service 1                 65  include markov.3
28   proc request 1                    66  include markov.4
29   end                               67  * ...
30   * output arcs                     68
31   arr buffer 1                      69  loop L,50,200,50
32   request serving 1                 70     expr exrss(one;L)
33   service  proc 1                   71     expr exrss(two;L)
34   end                               72     expr exrss(three;L)
35   * inhibitor arcs                  73     expr exrss(four;L)
36   buffer arr b-nproc                74  * ...
37   end                               75  end
38                                     76  end
```

Fig. 13.88 Input file for multiprocessor performability model.

A possible SHARPE input file for this hierarchical performability model is shown in Fig. 13.88. An explanation of the lines of the SHARPE input file is given in Table 13.45.

Table 13.45 SHARPE code description

Line	Description
3–9	Assign the values to the input parameters.
12–37	Define the GSPN model; the model has two parameters, the number of processors *nproc* and the job arrival rate *lambda*.
39–40	Define a function to use for reward rates.
42–48	Specification of a Markov model for *nproc* = 1 processor.
46–47	Assign reward rates to all states.
50–63	Specification of a Markov model for *nproc* = 2 processors.
59–64	Assign reward rates to all states.
65	Input file with specification of a Markov model for 3 processors.
66	Input file with specification of a Markov model for 4 processors.
67	Define models for more processors.
69	Let the arrival rate L vary from 50 to 200 by increments of 50.
70–75	Get the TLP for each number of processors.

In Fig. 13.89 the total loss probability and system unavailability (TLP with $\lambda = 0$) are plotted as functions of the number of processors. The difference between the unavailability and the total loss probability is caused only by limited buffer size b.

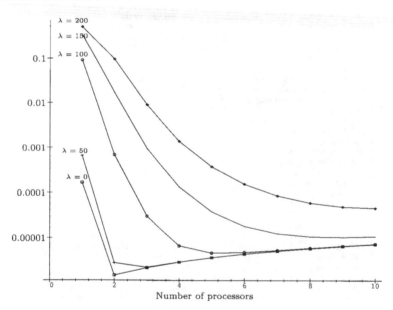

Fig. 13.89 TLP and unavailability.

Glossary

$\#(P_i, m)$	number of tokens at a particular place P_i in marking m
$1/\mu_i$	mean service time of the jobs at the ith node
$\alpha(r)$	factor for computing the shadow service time
argmin	function that returns the index of the minimum value
A–C	Allen–Cunneen
ABA	asymptotic bounds analysis
AIX	advanced IBM Unix
APU	associate processing unit
ASCII	American standard code for information interchange
ASYM-MVA	asymmetric MVA
ASYM-SCAT	asymmetric SCAT
ASPA	average sub-system population algorithm
ASYM-SUM	asymmetric SUM
ATM	asynchronous transfer mode
AVL	Adelson-Velskii and Landis
$b_{i,j}$	probability that a job at node i leaves the node after the jth phase
B	normalized blocking factor
B_N	blocking factor

BA	behavior aggregates
BCH	Boxma, Cohen, and Huffels (formula)
BCMP network	queueing network with several job classes, different queueing strategies, and generally distributed service times, named after Baskett, Chandy, Muntz, and Palacios
BDD	binary decision diagram
BFS	Bolch, Fleischmann, and Schreppel method
BJB	balanced job bounds analysis
BMAP	batch Markovian arrival process
BOTT	bottapprox method
BRI	basic rate interface
c_{Ai}	coefficient of variation of the interarrival time
$\hat{c}_A$	coefficient of variation of the batch interarrival time
c_{Bi}	coefficient of variation of the service time
$c_{B\infty}$	coefficient of variation of the service time when applying the closing method
$c_i(j)$	capacity function of node i; the number of jobs completed by station i if there are j jobs in the queue
$c_X = \frac{\sigma_X}{\bar{X}}$	normalized standard deviation or coefficient of variation
ceiling	function that rounds up to the next integer
cl	index for closed job classes
corr	correction factor
C_k	Cox distribution with k phases
$C(\boldsymbol{\mu})$	linear cost function
CAC	call admission control
CBQ	class-based queueing
CBR	constant bit rate
CCNC	coalesce computation of normalizing constants
CDF	cumulative distribution function
CDMA	code division multiple access
CPH	continuous phase-type distribution
CPU	central processing unit
CSMA/CD	carrier sense multiple access with collision detection
CSPL	C-based stochastic Petri net language
CTMC	continuous-time Markov chain
$d(\mathbf{S})$	function of the jobs in the network
driv **context**	a job executes driver code to set up or perform an I/O operation

D	deterministic distribution
$D_{irs}(\mathbf{K})$	change of contribution in the SCAT algorithm if a class-s job is removed from the network
D^+	output arcs
D^-	input arcs
DA	diffusion approximation
DAC	distribution analysis by chain
DBS	database server
DDP	delay differentiation parameter
DES	discrete-event simulation
DiffServ	differentiated services
DPH	discrete phase-type distribution
DS	differentiated services
DSCP	differentiated services code point
DTMC	discrete-time Markov chain
ϵ	stopping condition
$\mathbf{e}$	vector of the visit ratios $\mathbf{e} = (e_0, e_1, \ldots, e_N)$
e_i	visit ratio; the mean number of visits of a job to the ith node, also known as the relative arrival rate
e_∞	visit ratio of the closing node when applying the closing method
e_{iq}^*	number of visits at node i in chain q, also called visit ratio
$ec_i(k)$	effective capacity function
$\mathcal{ERG}$	extended reachability graph
E_k	Erlang distribution with k phases
E_r	set of pairs (j, s) that can be reached from pair (i, r) in a finite number of steps
ECN	explicit congestion notification
EXRSS	expected steady-state reward rate
$f(c_A, c_B, \rho)$	Kulbatzki approximation
f_{ol}	fraction of all I/Os that overlap with a primary task
$f_{\mathbf{X}}(\mathbf{s}; \mathbf{t})$	probability density function
$f_X(x) = \frac{dF_X(x)}{dx}$	probability density function
$\mathbf{fix}_{ir}$	functions needed for the fixed-point iteration
floor	function that rounds to the next lower integer
F_i	miss ratio function $F_i = 1 - H_i$

$F_i(k_i)$	functions corresponding to the state probabilities $\pi_i(k_i)$ of the ith node
$F_{ir}(\mathbf{K})$	contribution of class-r jobs at the ith node for a given population $\mathbf{K}$
$F_X(x) = P(X \leq x)$	cumulative distribution function
FB	full batch policy
FCFS	first-come-first-served
FDDI	fiber distributed data interface
FES	flow equivalent server method
FFS	fastest free server
FIFO	first-in-first-out (= FCFS)
FSM	finite state machine
FTCS	symposium on fault tolerant computing
FTP	file transfer protocol
$g(\lambda)$	core function of the summation method
$\Gamma(\alpha)$	gamma distribution
G	general distribution, also used for the normalization constant
$G(K)$	normalization constant of a single class queueing network
$G(\mathbf{K})$	normalization constant of a multiclass queueing network
$G_{\mathbf{KLB}}$	correction factor Krämer–Langenbach-Belz
$G_n(k)$	auxilary function for computing $G(K)$
$G_n(\mathbf{k})$	auxiliary function for computing $G(\mathbf{K})$
$G_N^{(i)}(k)$	normalization constant of the network with node i removed from the network
$G_N^{(i)}(\mathbf{k})$	normalization constant of the multiclass network with node i removed from the network
GE	Gaussian elimination algorithm
GI	general distribution with independent interarrival times
GPPL	general-purpose programming language
GPRS	general packet radio service
GreatSPN	analysis package for analyzing stochastic Petri nets
GREEDY	special case of the minimum batch policy
GSM	global system for mobile communication
GSPN	generalized SPN
GSSMC	general state-space Markov chain

GTH	Grassmann, Taksar, and Heyman algorithm
GUI	graphical user interface
h	function used in bottleneck analysis
h_i	hit ratio of the cache
$h(\rho, m)$	factor for computing $\overline{W}_{GI/D/m}$ in the case of the Cosmetatos approximation
H_k	hyperexponential distribution with k phases
$H(\pi)$	entropy function
HOL	non-preemptive head-of-line queueing strategy
HTML	hypertext markup language
HTTP	hypertext transfer protocol (protocol of WWW)
I/O	input/output
I	$= [i^-, i^+]$ interval
I_{μ_n}	$= [\mu_n^-, \mu_n^+]$ interval of the service rates
IBP	interrupted Bernoulli process
IGL	intermediate graphical language
IMMD IV	institute for mathematical machines and data processing IV
IntServ	integrated services
IS	infinite server
ISDN	integrated services digital network
ISP	Internet service provider
iSPN	integrated environment for modeling using stochastic Petri nets
IP	Internet protocol
IPP	interrupted Poisson process
k_i	number of jobs at the ith node
$(k_1, k_2, \ldots, k_N)$	state of the network
k_{ir}	number of jobs in the rth class at the ith node
kern **context**	job executes kern code
K	constant number of jobs in a closed network, also used for the number of jobs in the system
K_r	number of jobs of the rth class in the network
K^*	overall number of jobs in the fictious network
$\mathbf{K} = (K_1, \ldots, K_R)$	number of jobs in the various classes, known as the population vector
$\mathbf{K}_s$	mean number of class-s jobs at node i that an arriving class-r job sees given the network population $\mathbf{k}$

$\overline{K}$	mean number of jobs in the system
$\overline{K}_{FJ}$	mean number of parallel-sequential jobs with fork–join synchronization
$\overline{K}_i$	mean number of jobs at node i
$\overline{K}_{ir}$	mean number of class-r jobs at node i
$\overline{K}_{iR^*}^*$	overall number of class-R^* jobs at node i in the fictious network
$\overline{K}_i^*$	mean number of jobs at the ith node in the pseudo open network
KLB	Krämer–Langenbach-Belz
λ	throughput
$\lambda^{(0)}$	initial value for the throughput (used in bottleneck analysis)
λ_{0i}	arrival rate of jobs from outside to the ith node
$\lambda_1(\lambda_{02})$	throughput of the primary task at node 0
λ_2^{max}	maximum throughput of the open class in case of parallel processing with asynchronous tasks
λ_i	overall arrival rate of jobs at the ith node
$\hat{\lambda}$	arrival rate of batches
$\lambda_{ij,r}$	arrival rate of class-r jobs from node i to node j
λ_{ir}	throughput, the rate at which jobs of the rth class are serviced and leave the ith node
$\lambda_{iR^*}^*$	throughput of the fictious network at node i, class -R^*
λ_{opt}	optimistic bound, largest possible throughput
λ_{pes}	pessimistic bound, lowest possible throughput
λ_r	overall throughput
$L(\theta)$	likelihood estimation
$L(\lambda, \mu_1, \ldots, \mu_N, y_1, y_2)$	Lagrange function with objective function λ and two Lagrange multipliers y_1 and y_2
L_i	load factor of open classes regarding to station i
$L_X(s)$	Laplace transform of the pdf: Laplace–Stieltjes transform of the CDF
LAN	local area network
LB	local balance property
LBANC	local balance algorithm for normalizing constants
LBPS	last batch processor sharing
LCFS	last-come-last-served
Little's theorem	Little's theorem, also known as Little's law
LST	Laplace–Stieltjes transform

LT	Laplace transform
μ_i	service rate of the jobs at the ith node
μ_∞	service rate of the closing node when applying the closing method
μ_{ir}	service rate of the ith node for jobs of the rth class
$\boldsymbol{\mu}^m = (\mu_1, \mu_2, \cdots, \mu_m)$	vector of service rates
m_0	$m_0 \in \mathcal{M}$ denotes the initial marking of a PN
m_i	number of parallel servers at the ith node
maxval	function that delivers the maximum value of a set of numbers
M	exponential distribution
$\overline{M}_{ir}$	mean number of customers of class-i who arrive during the waiting time of the tagged customer and receive service before him
$M \Rightarrow M$	Markov implies Markov property
$\mathbf{M}(t)$	time-averaged behavior of the CTMC
MAC	media access control
MAM	matrix-analytic method
MAP	Markovian arrival process
MB	minimum batch policy
MEM	maximum entropy method
MGM	matrix-geometric method
MMBP	Markov-modulated Bernoulli process
MMFS	Markov-modulated fluid source
MMPP	Markov-modulated Poisson process
MOSEL	modeling, specification, and evaluation language
MOSEL-2	modeling, specification, and evaluation language, 2nd revision
MOSES	modeling, specification, and evaluation system
MRGP	Markov regenerative process
MRM	Markov reward model
MTBDD	multi terminal binary decision diagram
MTTA	mean time to absorption
MTTF	mean time to failure
MTTR	mean time to repair
MVA	mean value analysis
MVALDMX	mean value analysis load-dependent mixed
$\boldsymbol{\nu}(n)$	vector of the state probabilities at time n

$\nu_i(k)$	conditional throughput of node i with k jobs
ν	vector of the state probabilities of a discrete time Markov chain (DTMC)
$\tilde{\nu}$	limiting state probabilities
nc_{Dm}	factor of the Cosmetatos approximation
$\mathcal{N}$	product-form network
$N_{i,r}$	number of class-r jobs at node i in the shadow node
$\overline{N}_{ir}$	mean number of customers of class-i found in the queue by the tagged (priority r) customer who receive service before him
NAM	network animator
NHPP	nonhomogeneous Poisson process
ns	network simulator
NPFQN	non-product-form queueing network
op	index for open job classes
$\hat{\theta}$	estimate of θ
ODE	ordinary differential equation
One-limited service	station transmits at most one frame when the token arrives
$\pi^{(L)}$	probability of loss
$\pi_m^{(L)}$	probability that all servers are occupied
π_g	individual state for all $g \subseteq G$
$\pi_i(k)$	marginal probability that the ith node contains exactly $k_i = k$ jobs
$\pi_i^{(W)}$	state probability of node i of a non-lossy system
$\pi_j(v)$	unconditional state probabilities
π_k	probability of the number of jobs in the system
$\pi(\mathbf{S}_1, \ldots, \mathbf{S}_N)$	state probability of a network with multiple job classes
$\boldsymbol{\pi}(u)$	state probabilitiy vector at any instant of time u
$\dot{\boldsymbol{\pi}}$	derivative of the transition probabilities
$p_{0,js}$	probability in an open network that a job from outside enters the jth node as a job of the sth class
p_{0j}	probability that a job entering the network from outside first enters the jth node
p_{i0}	probability that a job leaves the network just after completing service at node i
p_{ij}	probability that a job is transferred to the jth node after service completion at the ith node (routing probability)

$p_{ij}(t)$	simplified transition probabilities for time-homogeneous CTMCs
$p_{ij}(u, v)$	transition probability of the CTMC to travel from state i to state j during $[u, v)$
$p_{ij}^{(n)}(k, l)$	probability that the Markov chain transits from state i at time k to state j at time l in exactly $n = l - k$ steps
$p_{ir,0} = \sum_{j=1}^{N} \sum_{s=1}^{R} p_{ir,js}$	probability in an open network that a job of the rth class leaves the network after having been serviced at the ith node
$p_{ir,js}$	probability that a job of the rth class at node i is transferred to the sth class at node j (routing probability)
$\phi(y, t)$	time-average accumulated reward
$\psi(y, t)$	distribution of the accumulated reward over a finite time $[0, t)$, also called performability
$\psi_{i,r}$	correction factor for computing the shadow service time in the case of a network with class switching and mixed priorities
pdf	probability density function
pmf	probability mass function
pri_r	priority of a class-r job with respect to a quota node
$\Phi(x)$	CDF of the standard normal distribution
P	places of a Petri net
P_m	waiting probability
$\dot{\mathbf{P}}$	derivative of the matrix of the transition probabilities
$\mathbf{P}$	stochastic transition matrix
$\mathbf{P}(u, v)$	matrix of the transition probabilities
$\mathbf{P}^{(n)}$	matrix of the n-step transition probabilities
PADS	parallel and distributed simulation
PANDA	Petri net analysis and design assistant
PASTA	Poisson arrivals see time averages theorem
PEPSY	performance evaluation and prediction system
PFQN	product-form queueing network
PH	phase-type distribution
PHB	per-hop behavior
PN	Petri net
$PR(n)$	scaling function of the extended SCAT algorithm
PRIOMVA	extended MVA for the analysis of priority queueing networks

PRIOSUM	extension of the summation method to priority networks
PRS	preemptive resume queueing strategy
PS	processor sharing
PSL	plain simulation language
$q_r(t)$	priority of class-r at time t
Q	queue length
$\mathbf{Q}$	infinitesimal generator matrix
$\hat{Q}$	mean queue length for batches
$\overline{Q}$	mean queue length
$\overline{Q}_i$	mean number of tasks waiting for the join synchronization in queue Q_i
$\overline{Q}_{ir}$	mean queue length of class-r jobs at the ith node
$\overline{Q}_{\text{batch}}$	mean queue length for individual customers in a batch system
QBD	quasi-birth–death process
QNAP	queueing network analysis package
QoS	quality of service
ρ	utilization
ρ_{bott}	utilization of the bottleneck node
ρ_i	utilization of node i
$\rho_i r$	utilization of node i by class-r jobs
$\rho_k^{(W)}$	specific utilization needed for calculating the utilization of an asymmetric non-lossy system
ρ_k^L	utilization of the kth server of a lossy system
r_i	reward rate assigned to state i
r_r	starting priority of class-r
$\mathcal{RG}$	reduced reachability graph
$\mathcal{RS}$	reachability set
R	number of priority classes
$\overline{R}$	mean remaining service time of a busy server, also called mean residual life
RECAL	recursion by chain algorithm
RED	random early detection
RESQ	research queueing network package
RESTART	repetitive simulation trials after reaching threshold
ROMTBDD	reduced and ordered MTBDD
RPS	rotational position sensing system

RR	round robin
RSVP	resource reservation protocol
$\sigma_X^2 = \mathbf{var}(X)$	variance of X or second central moment
$s_{(i,r)}^{virt}$	virtual service time of class-r jobs at quota node i
s_{ci}	mean contention time
$s_{i,r}$	service time of class-r jobs at node i
$\tilde{s}_{i,r}$	service time of class-r jobs at shadow node i
s_{iq}^*	service rate in chain q at node i
$s_{ir}(j)$	load-dependent mean service time or normal service time
s_{li}	mean latency time
s_{si}	mean seek time
s_{ti}	mean transfer time
$\mathbf{S}_i = (k_{i1}, \ldots, k_{iR})$	state of the ith node
$\mathbf{S} = (\mathbf{S}_1, \ldots, \mathbf{S}_N)$	overall state of the network with multiple classes
S_N	synchronization overhead
SB	station balance property
SBP	switched Bernoulli process
SCAT	self correcting approximation technique
SHARPE	symbolic hierarchical automated reliability performance evaluator
SIRO	service in random order
SLA	service level agreement
SMP	semi-Markov process
SNR	signal-to-noise ratio
SOR	successive over relaxation
SP	simulation package
SPN	stochastic Petri net
SPNP	stochastic Petri net package
SPP	switched Poisson process
SQL	structured query language
SRN	stochastic reward net
SUM	summation method
$\mathcal{T}$	tangible markings
T	response time
$\overline{T}$	mean response time

$\overline{T}_{FJ}$	mean response time of parallel-sequential jobs with fork–join synchronization
$\overline{T}_i$	mean response time at node i
$\overline{T}_{ir}$	mean response time of class-r jobs at node i
$\overline{T}_J$	mean join time
$\overline{T}_{J,i}$	mean time spent by task i waiting for the join synchronization in queue Q_i
$\overline{T}_{\mathrm{opt}}$	optimistic bound, lowest possible mean response time
$\overline{T}_{\mathrm{pes}}$	pessimistic bound, greatest possible mean response time
$\overline{T}_r$	mean target rotation time
$\overline{T}_S$	mean response time of a purely sequential job
$\overline{T}_{S,i}$	mean response time of task i belonging to a purely sequential job
T_d	delay time of a frame
T_t	transfer time of a frame
Tcl/Tk	tool command language/tool kit
TCP	transmission control protocol
TLP	total loss probability
TLP(t)	transient expected total loss probability
TPM	transition probability matrix
Type 1 blocking	blocking after service, transfer blocking, production blocking, or nonimmediate blocking
Type 2 blocking	blocking before service, service blocking, communication blocking, or immediate blocking
Type 3 blocking	repetitive service, or rejection blocking
user **context**	job executes user code
U	number of chains
UMTS	universal mobile telecommunication system
Upper time limit	time limit within which jobs need to be served in a real-time system
URL	uniform resource locator (WWW address)
$\mathcal{V}$	vanishing markings
VRT	variance reduction technique
ω	relaxation parameter
ω_0	optimal relaxation parameter
$\omega_0^{\approx}$	approximation of the optimal relaxation parameter
$\omega_{r,k}$	factor for computing the shadow service time
W	waiting time

$W(i,j)$	weighting function of the extended SCAT algorithm
$\overline{W}$	mean waiting time
$\overline{W}_0$	mean remaining service time of a job in service
$\overline{W}_i$	mean waiting time at node i
$\overline{W}_{ir}$	mean waiting time of class-r jobs at node i
$\overline{W}_r$	mean waiting time of an arriving customer of priority class-r
WAN	wide area network
WEIRDP	queueing strategy where the first job in the queue is assigned to the pth part of the processor and the rest of the jobs are assigned to the $1-p$th part of the processor
WinPEPSY	Windows version of the queueing network package PEPSY
WIP	mean number of workpieces in the system (work in progress)
WLS	WebLogic servers
WWW	World Wide Web
ξ_r	set of jobs that cannot be preempted by class-r jobs
x_{ave}	average relative utilization of the individual nodes in the network
x_i	relative utilization
x_{max}	maximum relative utilization
x_{sum}	sum of the relative utilizations
$X_t : t \in T$	family of random variables
$\overline{X}$	$=E[X]$, mean value or expected value
$\overline{X^n}$	$= E[X^n]$, nth moment
XPEPSY	X-version of the queueing network package PEPSY
$Y(t)$	accumulated reward in the finite time horizon $[0,t)$
$z_i(k)$	number of possibilities to distribute k tasks to $n_i + 1$ nodes (including the join queue Q_i)
$Z(t)$	instantaneous reward rate at time t
$Z_{FJ}(k)$	number of states in the fork–join sub-network

Bibliography

The numbers following each entry indicate backreferences to the pages on which the entry is mentioned.

ABB+89. M. Ajmone Marsan, G. Balbo, A. Bobbio, G. Chiola, and A. Conte, G. Cumani. The Effect of Execution Policies on the Semantics and Analysis of Stochastic Petri Nets. *IEEE Transactions on Software Engineering*, 15:832–846, 1989. 113, 114

AbBo97. M. Abu El-Qomsan and G. Bolch. Analyse von Warteschlangennetzen mit asymmetrischen Knoten und mehreren Auftragsklassen. Technical report TR-I4-97-07, Universität Erlangen-Nürnberg, IMMD IV, 1997. 580

ABC84. M. Ajmone Marsan, G. Balbo, and G. Conte. A Class of Generalized Stochastic Petri Nets for the Performance Evaluation of Multiprocessor Systems. *ACM Transactions on Computer Systems*, 2(2):93–122, May 1984. 96

ABC+95. M. Ajmone Marsan, G. Balbo, G. Conte, S. Donatelli, and G. Franceschinis. *Modelling with Generalized Stochastic Petri Nets*. John Wiley, New York, 1995. 700, 760

Abde82. A. Abdel-Moneim. Weak Lumpability in Finite Markov Chains. *Journal of Applied Probability*, 19:685–691, 1982. 222

Aber94.   R. Abernethy. The New Weibull Handbook. 536 Oyster Road,
           North Palm Beach, Florida, 1994. 40, 45

Aber00.   R. B. Abernethy. *The New Weibull Handbook*. Barringer & Asso-
           ciates, Humble, TX, fourth edition, November 2000. 615

AbMa93.   H. Abdalah and R. Marie. The Uniformized Power Method for
           Transient Solutions of Markov Processes. *Computers and Opera-
           tions Research*, 20(5):515–526, June 1993. 218

ABP85.   I. Akyildiz, G. Bolch, and M. Paterok.   Die erweiterte
           parametrische Analyse für geschlossene Warteschlangennetze. In
           H. Beilner, editor, *Proc. 3rd GI/NTG Conf. on Measurement,
           Modeling and Performance Evaluation of Computer and Com-
           munication Systems (MMB '85)*, volume 110 of *Informatik-
           Fachberichte*, pages 170–185, Dortmund, Germany, Berlin, Octo-
           ber 1985. Springer. 369, 414, 415

ABS04.   B. Almási, G. Bolch, and J. Sztrik. Heterogeneous Finite-Source
           Retrial Queues. *Journal of Mathematical Sciences*, 121:2590–2596,
           2004. 302

AgBu83.   S. Agrawal and J. Buzen. The Aggregate Server Method for Ana-
           lyzing Serialization Delays in Computer Systems. *ACM Transac-
           tions on Computer Systems*, 1(2):116–143, May 1983. 549

AgTr82.   J. Agre and S. Tripathi.   Modeling Reentrant and Nonreen-
           trant Software. *ACM Sigmetrics Performance Evaluation Review*,
           11(4):163–178, 1982. 549

AkBo83.   I. Akyildiz and G. Bolch.   Erweiterung der Mittelwertanalyse
           zur Berechnung der Zustandswahrscheinlichkeiten für geschlossene
           und gemischte Netze. In P. J. Kühn and K. M. Schulz, editors,
           *Proc. 2nd GI/NTG Conf. on Measurement, Modeling and Per-
           formance Evaluation of Computer and Communication Systems
           (MMB '83)*, volume 61 of *Informatik-Fachberichte*, pages 267–276,
           Stuttgart, Germany, Berlin, February 1983. Springer. 388, 394,
           403

AkBo88a.   I. Akyildiz and G. Bolch. Mean Value Analysis Approximation
           for Multiple Server Queueing Networks. *Performance Evaluation*,
           8(2):77–91, April 1988. 435, 437

AkBo88b.   I. Akyildiz and G. Bolch. Throughput and Response Time Opti-
           mization in Queueing Network Models of Computer Systems. In
           T. Hasegawa, H. Takagi, and Y. Takahashi, editors, *Proc. Int.
           Seminar on Performance of Distributed and Parallel Systems*,

pages 251–269, Kyoto, Japan, Amsterdam, December 1988. Elsevier. 443

AKH97. S. C. Allmaier, M. Kowarschik, and G. Horton. State Space Construction and Steady-State Solution of GSPNs on a Shared Memory Multiprocessor. In *Proc. 7th Int. Workshop on Petri Nets and Performance Models (PNPM '97)*, pages 112–121, St. Malo, France, Los Alamitos, CA, June 1997. IEEE Computer Society Press. 103

AkSo97. N. Akar and K. Sohraby. Finite and Infinite QBD Chains: A Simple and Unifying Algorithmic Approach. *Proc. Conference on Computer Communications (IEEE Infocom)*, pages 1105–1113, 1997. 135

Akvo89. I. Akyildiz and H. von Brand. Exact Solutions for Open, Closed and Mixed Queueing Networks with Rejection Blocking. *Theoretical Computer Science*, 64(2):203–219, 1989. 593

Akyi87. I. Akyildiz. Exact Product Form Solution for Queueing Networks with Blocking. *IEEE Transactions on Computers*, 36(1):122–125, January 1987. 592, 593, 594

Akyi88a. I. Akyildiz. Mean Value Analysis for Blocking Queueing Networks. *IEEE Transactions on Software Engineering*, 14(4):418–428, April 1988. 592

Akyi88b. I. Akyildiz. On the Exact and Approximate Throughput Analysis of Closed Queueing Networks with Blocking. *IEEE Transactions on Software Engineering*, 14(1):62–70, January 1988. 592, 593

Alle90. A. Allen. *Probability, Statistics and Queueing Theory with Computer Science Applications*. Academic Press, New York, 2nd edition, 1990. 15, 51, 252, 265, 266, 270, 483

AlPe86. T. Altiok and H. Perros. Approximate Analysis of Open Networks of Queues with Blocking: Tandem Configurations. *AIIE Transactions*, 12(3):450–461, March 1986. 592

ALRL04. A. Avižienis, J.-C. Laprie, B. Randell, and C. Landwehr. Basic Concepts and Taxonomy of Dependable and Secure Computing. *IEEE Transactions on Dependable and Secure Computing*, 1(1):11–33, January–March 2004. 7

Alti82. T. Altiok. Approximate Analysis of Exponential Tandem Queues with Blocking. *European Journal of Operational Research*, 11:390–398, October 1982. 592

AMM65.   B. Avi-Itzhak, W. Maxwell, and L. Miller. Queueing with alternating priorities. *Operations Research*, 13(2):306–318, March–April 1965. 147

ANO96.   S. Asmussen, O. Nerman, and M. Olsson. Fitting Phase-type Distributions via the EM Algorithm. *Scandinavian Journal of Statistics*, 23:419–441, 1996. 119

Are.   Arena Simulator: http://www.arenasimulation.com/. Last verified on February 08, 2005. 612

Arta99a. J. R. Artalejo. Accessible bibliography on retrial queues. *Math. and Comp. Modeling*, 30:1–6, 1999. 302

Arta99b. J. R. Artalejo. Classified bibliography of research on retrial queues. In *Special issue of Trabajos de Investigación Operativa (TOP)*, 7(2), pages 187–211, December 1999. 302

ASP72.   ASPOL Reference Manual. Control Data Corporation, 1972. Pub. No. 17314200. 640

AUT05.   AUTOMOD Simulator: http://www.autosim.com/. Last verified on February 08, 2005, 2005. 609

Baer85.   M. Baer. Verlustsysteme mit unterschiedlichen mittleren Bedienungszeiten der Kanäle. *Wissenschaftliche Zeitschrift der Hochschule für Verkehrswesen–Friedrich List*, 1985. Sonderheft 15. 284, 286, 287, 292, 294

BaIa83.   S. Balsamo and G. Iazeolla. Some Equivalence Properties for Queueing Networks with and without Blocking. In A. Agrawala and S. Tripathi, editors, *Proc. Performance '83*, pages 351–360, Amsterdam, 1983. North-Holland. 593

BaKe94.   F. Bause and P. Kemper. *QPN-Tool for the qualitative and quantitative analysis of Queueing Petri Nets*. Number 794 in LNCS. Springer, Berlin, 1994. 700

Bard79.   Y. Bard. Some Extensions to Multiclass Queueing Network Analysis. In *Proc. 4th Int. Symp. on Modelling and Performance Evaluation of Computer Systems*, volume 1, pages 51–62, Vienna, New York, February 1979. North-Holland. 422, 512

Bard80.   Y. Bard. A Model of Shared DASD and Multipathing. *Communications of the ACM*, 23(10):564–572, October 1980. 545

Bard82.   Y. Bard. Modeling I/O Systems with Dynamic Path Selection and General Transmission Networks. *ACM Sigmetrics Performance Evaluation Review*, 11(4):118–129, 1982. 545

bas.   baseSim Simulator: http://www.ibrightsolutions.co.uk/. Last verified on February 08, 2005. 612

BaTh94.   J. Barner and K.-C. Thielking. Entwicklung, Implementierung und Validierung von analytischen Verfahren zur Analyse von Prioritätsnetzen. Studienarbeit, Universität Erlangen-Nürnberg, IMMD IV, 1994. 446

BBA84.   S. Bruell, G. Balbo, and P. Afshari. Mean Value Analysis of Mixed, Multiple Class BCMP Networks with Load Dependent Service Stations. *Performance Evaluation*, 4:241–260, 1984. 405, 409

BBBZ03.   K. Begain, J. Barner, G. Bolch, and A. Zreikat. The Performance and Reliability Modelling Language MOSEL and Its Application. *International Journal of Simulation*, 3(3–4):66–80, 2003. 679, 700

BBC77.   R. Brown, J. Browne, and K. Chandy. Memory Management and Response Time. *Communications of the ACM*, 20(3):153–165, March 1977. 544

BBC+98.   D. Black, S. Blake, M. Carlson, E. Davies, Z. Wang, and W. Weiss. An Architecture for Differentiated Services. IETF RFC 2475, December 1998. 721

BBG04.   P. Bazan, G. Bolch, and R. German. WinPEPSY-QNS Performance Evaluation and Prediction System for Queueing Networks. In K. Al-Begain and G. Bolch, editors, *Proc. of the 11th International Conference on Analytical and Stochastic Modelling Techniques and Applications (ASMTA'04)*, pages 147–150, Magdeburg, Germany. SCS-European Publishing House, Erlangen, Germany, June 2004. 665, 700

BBH01.   K. Begain, G. Bolch, and H. Herold. *Practical Performance Modeling — Application of the MOSEL Language*. Kluwer Academic Publishers, 2001. 677, 678, 679

BBQH03.   O. Boxma, S. Borst, R. N. Queija, and M. Harchol-Balter. Heavy Tails: Performance Models and Scheduling Disciplines. In *Tutorial of ITC 18*, Berlin, 2003. 30, 257

BBQZ03.   S. C. Borst, O. J. Boxma, R. N. Queija, and A. P. Zwart. The Impact of the Service Discipline on Delay Asymptotics. *Performance Evaluation*, 54(2):175–206, 2003. 257

BBS77.   G. Balbo, S. Bruell, and H. Schwetmann. Customer Classes and Closed Network Models–A Solution Technique. In B. Gilchrist, editor, *Proc. IFIP 7th World Computer Congress*, pages 559–564, Amsterdam, August 1977. North-Holland. 374, 384

BCH79.   O. Boxma, J. Cohen, and N. Huffels. Approximations of the Mean Waiting Time in a M/G/s-Queueing System. *Operations Research*, 27:1115–1127, 1979.   269, 272

BCMP75.   F. Baskett, K. Chandy, R. Muntz, and F. Palacios. Open, Closed, and Mixed Networks of Queues with Different Classes of Customers. *Journal of the ACM*, 22(2):248–260, April 1975.   341, 353, 357, 552

BCNN01.   J. Banks, J. S. Carson, B. L. Nelson, and D. M. Nicol. *Discrete-Event System Simulation*. Prentice Hall, Upper Saddle River, NJ, third edition, 2001.   610, 612, 614, 634

BEA99.   J. Barceló, L. Escudero, and J. Artelejo, editors. *1st International Workshop on Retrial Queues*, Special issue of Trabajos de Investigación Operativa (TOP), 7(2), Madrid, Spain, December 1999. Sociedad de Estadística e Investigación Operativa.   299, 302

Beau77.   M. D. Beaudry. Performance related Reliability Measures for Computing Systems. In *Proc. 7th Annual Int. Conf. on Fault Tolerant Computing Systems*, pages 16–21. IEEE, June 1977.   8

Beau78.   M. Beaudry. Performance-Related Reliability Measures for Computing Systems. *IEEE Transactions on Computers*, 27(6):540–547, 1978.   80, 91

Bega93.   K. Begain. Using Subclasses of PH Distributions for the Modeling of Empirical Distributions. In *Proc. 18th ICSCS*, pages 141–152, Cairo, Egypt, 1993.   25

BeGo81.   P. Bernstein and N. Goodman. Concurrency Control in Distributed Database Systems. *Computing Surveys*, 13(2):185–221, June 1981.   549

BEM01.   G. Bolch, L. Essafi, and H. de Meer. Dynamic Priority Scheduling for Proportional Delay Differentiated Services. In *Proc. 2001 Aachen International Multiconference on Measurement, Modeling and Evaluation of Computer and Communication Systems, Aachen, Germany, September 11–14, 2001*, 2001.   723

BeMe78.   F. Beutler and B. Melamed. Decomposition and Customer Streams of Feedback Networks of Queues in Equilibrium. *Operations Research*, 26(6):1059–1072, 1978.   345

BeZr02.   K. Begain and A. Zreikat. Interference Based CAC for Uplink Traffic in UMTS Networks. In *Proc. World Wireless Congress*, pages 298–303, June 2002.   783, 784

BFH88.   M. Bär, K. Fischer, and G. Hertel. *Leistungsfähigkeit–Qualität–Zuverlässigkeit*, chapter 4. Transpress Verlagsgesellschaft mbH, Berlin, 1988. 284, 286, 288, 291, 294

BFKT01.   P. Buchholz, M. Fischer, P. Kemper, and C. Tepper. New features in the APNN toolbox. In P. Kemper, editor, *Tools of Aachen 2001 Int. Multi-conference on Measurement, Modelling and Evaluation of Computer-Communication Systems*, pages 62–68, 2001. 700

BFS87.   G. Bolch, G. Fleischmann, and R. Schreppel. Ein funktionales Konzept zur Analyse von Warteschlangennetzen und Optimierung von Leistungsgrößen. In U. Herzog and M. Paterok, editors, *Proc. 4th GI/ITG Conf. on Measurement, Modeling and Performance Evaluation of Computer and Communication Systems (MMB '87)*, volume 154 of *Informatik-Fachberichte*, pages 327–342, Erlangen, Germany, Berlin, September 1987. Springer. 440, 443

BGJ92.   G. Bolch, M. Gaebell, and H. Jung. Analyse offener Warteschlangennetze mit Methoden für geschlossene Warteschlangennetze. In *Operations Research Proc. 1992*, pages 324–332, Aachen, Germany, Berlin, September 1992. DGOR–Deutsche Gesellschaft für Operations Research, Springer. 507, 508

BGMT98.   G. Bolch, S. Greiner, H. de Meer, and K. Trivedi. *Queueing Networks and Markov Chains*. John Wiley & Sons, New York, 1998. 678

BHST02.   A. Bobbio, A. Horváth, M. Scarpa, and M. Telek. Acyclic Discrete Phase Type Distributions: Properties and a Parameter Estimation Algorithm, October 2002. 118, 119

BKLC84.   R. Bryant, A. Krzesinski, M. Lakshmi, and K. Chandy. The MVA Priority Approximation. *ACM Transactions on Computer Systems*, 2(4):335–359, November 1984. 514, 516, 517

BMW94.   H. Beilner, J. Mäter, and C. Wysocki. The Hierarchical Evaluation Tool HIT. In *Short Papers and Tool Descriptions of the 7th International Conference on Modelling Techniques and Tools for Computer Performance Evaluation*, pages 62–68, 1994. 700

BNO01.   S. Balsamo, V. de Nitto Persone, and R. Onvural. *Analysis of Queueing Networks with Blocking*. Kluwer Academic Publishers, Boston, Dordrecht, London, 2001. 600

BoBr84.   G. Bolch and W. Bruchner. Analytische Modelle symmetrischer Mehrprozessoranlagen mit dynamischen Prioritäten. *Elektronische Rechenanlagen*, 26(1):12–19, 1984. 280, 281, 282

BoCu92.   A. Bobbio and A. Cumani. ML Estimation of the Parameters of a PH Distribution in Triangular Canonical Form. *Computer Performance Evaluation*, pages 33–46, 1992. 118

BoFi93.   G. Bolch and M. Fischer. Bottapprox: Eine Engpaßanalyse für geschlossene Warteschlangennetze auf der Basis der Summationsmethode. In H. Dyckhoff, U. Derigs, M. Salomon, and H. Tijms, editors, *Operations Research Proc. 1993*, pages 511–517, Amsterdam, Berlin, August 1993. DGOR–Deutsche Gesellschaft für Operations Research, Springer. 447, 449

BoHe95.   G. Bolch and H. Herold. MOSEL — MOdeling Specification and Evaluation Language. Technical Report TR-I4-95-02, Universität Erlangen-Nürnberg, IMMD IV, 1995. 679

BoHe96.   G. Bolch and H. Herold. MOSEL, A new Language for the Markov Analyzer MOSES. Technical report TR-I4-96-02, Universität Erlangen-Nürnberg, IMMD IV, 1996. 533

BoKi92.   G. Bolch and M. Kirschnick. PEPSY-QNS - Performance Evaluation and Prediction SYstem for Queueing NetworkS. Technical report TR-I4-21-92, Universität Erlangen-Nürnberg, IMMD IV, October 1992. 658

BoKo81.   O. Boxma and A. Konheim. Approximate Analysis of Exponential Queueing Systems with Blocking. *Acta Informatica*, 15:19–66, 1981. 592

Bolc83.   G. Bolch. Approximation von Leistungsgrößen symmetrischer Mehrprozessorsysteme. *Computing*, 31:305–315, 1983. 267

Bolc89.   G. Bolch. *Leistungsbewertung von Rechensystemen*. Leitfäden und Monographien der Informatik. B.G.Teubner Verlagsgesellschaft, Stuttgart, 1989. 479, 482

Bolc91.   G. Bolch. Analytische Leistungsbewertung–Methoden und Erfolge. *PIK–Praxis der Informationsverarbeitung und Kommunikation*, 4:231–241, 1991. 705

BoRo94.   G. Bolch and K. Roggenkamp. Analysis of Queueing Networks with Asymmetric Nodes. In M. Woodward, S. Datta, and S. Szumko, editors, *Proc. Computer and Telecommunication Systems Performance Engineering Conf.*, pages 115–128, London, 1994. Pentech Press. 536, 577

BoSc91.   G. Bolch and A. Scheuerer. Analytische Untersuchungen Asymmetrischer Prioritätsgesteuerter Wartesysteme. In W. Gaul, A. Bachem, W. Habenicht, W. Runge, and W. Stahl, editors,

Operations Research Proc. 1991, pages 514–521, Stuttgart, Berlin, September 1991. DGOR–Deutsche Gesellschaft für Operations Research, Springer. 283, 284, 285, 290, 294

BoTe97.   A. Bobbio and M. Telek. Non-Exponential Stochastic Petri Nets: an Overview of Methods and Techniques, 1997. 119

BoTr86.   A. Bobbio and K. S. Trivedi. An Aggregation Technique for the Transient Analysis of Stiff Markov Chains. *IEEE Transactions on Computers*, 35(9):803–814, September 1986. 222, 228

BPST98.   A. Bobbio, A. Puliafito, M. Scarpa, and M. Telek. WebSPN: Non-Markovian Stochastic Petri Net Tool. In *Int. Conf on WEB-based Modeling and Simulation*, January 1998. 678

Bran82.   A. Brandwajn. Fast Approximate Solution of Multiprogramming Models. *ACM Sigmetrics Performance Evaluation Review*, 11(4):141–149, 1982. 544

BrBa80.   S. Bruell and G. Balbo. *Computational Algorithms for Closed Queueing Networks*. Operating and programming systems series. North-Holland, Amsterdam, 1980. 330, 353, 356, 357, 380, 384

BRH92.   A. D. B. R. Haverkort, A. P. A. van Moorsel. MGMtool: A Performance Analysis Tool Based on Matrix Geometric Methods. In J. H. R. Pooley, editor, *Proc. 6th International Conference on Computer Performance Evaluation, Modelling Techniques and Tools*, pages 312–316. Edinburgh University Press, 1992. 135

BRH94.   D.-J. S. B. R. Haverkort, A. P. A. van Moorsel. Xmgm: Performance Modelling using Matrix Geometric Techniques. In V. K. Madisetti, E. Gelenbe, and J. C. Walrand, editors, *MASCOTS '94, Proc. 2nd International Workshop on Modeling, Analysis, and Simulation On Computer and Telecommunication Systems, January 31 - February 2, 1994, Durham, North Carolina, USA*, pages 152–157. IEEE Computer Society, 1994. 135

BrSe91.   I. Bronstein and K. Semendjajew. *Taschenbuch der Mathematik*. B.G.Teubner Verlagsgesellschaft, Stuttgart, 1991. 567

BRT88.   J. Blake, A. Reibman, and K. S. Trivedi. Sensitivity analysis of reliability and performability measures for multiprocessor systems. In *Proc. 1988 ACM SIGMETRICS Int. Conf. on Measurement and Modeling of Computer Systems*, pages 177–186, Santa Fee, NM, May 1988. 91

BrTa97.   L. Bright and P. G. Taylor. Equilibrium Distributions for Level-Dependent Quasi-Birth-and-Death Processes. In R. R.

Chakravarthy and A. S. Alfa, editors, *Matrix-Analytic Methods in Stochastic Models*, volume 183 of *Lecture Notes in Pure and Applied Mathematics*, pages 359–375. Marcel Dekker, 1997. 135

Brum71.   S. Brumelle. Some Inequalities for Parallel-Server Queues. *Operations Research*, 19:402–413, 1971. 269

Brya90.   R. E. Bryant. Symbolic Simulation — Techniques and Applications. In *DAC '90: Proc. 27th ACM/IEEE conference on Design automation*, pages 517–521. ACM Press, 1990. 611

BuHe77.   W. Bux and U. Herzog. The Phase Concept: Approximation of Measured Data and Performance Analysis. In *Computer Performance*, pages 23–38, 1977. 119

Bux81.   W. Bux. Local-Area Subnetworks: A Performance Comparison. *IEEE Transactions on Communication*, 29(10):1465–1473, October 1981. 147

Buze71.   J. Buzen. *Queueing Network Models of Multiprogramming*. PhD thesis, Div. of Engineering and Applied Physics, Harvard University, 1971. 321, 369, 371, 372, 373, 375

Buze73.   J. Buzen. Computational Algorithms for Closed Queueing Networks with Exponential Servers. *Communications of the ACM*, 16(9):527–531, September 1973. 372, 373

BVDT88.   M. Boyd, M. Veeraraghavan, J. Dugan, and K. S. Trivedi. An Approach to Solving Large Reliability Models. In *Proc. IEEE/AIAA DASC Symp.*, San Diego, 1988. 12

C++.   C++SIM Simulator: http://cxxsim.ncl.ac.uk/. Last verified on February 08, 2005. 612

CBC⁺93.   G. Ciardo, A. Blakemore, P. Chimento, J. Muppala, and K. S. Trivedi. Automated Generation and Analysis of Markov Reward Models using Stochastic Reward Nets. In C. D. Meyer and R. J. Plemmons, editors, *Linear Algebra, Markov Chains and Queueing Models*, volume 48 of *IMA Volumes in Mathematics and Its Applications*, pages 145–191. Springer, New York, 1993. 99, 219

CFGR95.   G. Chiola, G. Franceschinis, R. Gaeta, and M. Ribaudo. GreatSPN 1.7: GRaphical Editor and Analyzer for Timed and Stochastic Petri Nets. *Performance Evaluation*, 24(1,2):137–159, November 1995. 678

CFMT94.   G. Ciardo, R. Fricks, J. Muppala, and K. S. Trivedi. *SPNP Users Manual Version 4.0*. Duke University, Department of Electrical Engineering, Durham, NC, March 1994. 101

CGL94. G. Ciardo, R. German, and C. Lindemann. A Characterization of the Stochastic Process Underlying a Stochastic Petri Net. *IEEE Transactions on Software Engineering*, 20:506–515, 1994. 114

CGM83. A. Chesnais, E. Gelenbe, and I. Mitrani. On the Modeling of Parallel Access to Shared Data. *Communications of the ACM*, 26(3):196–202, March 1983. 549

Chan72. K. Chandy. The Analysis and Solutions for General Queuing Networks. In *Proc. 6th Annual Princeton Conf. on Information Sciences and Systems*, pages 224–228, Princeton, NJ, March 1972. Princeton University Press. 336, 340, 360

ChCl03. J. Chung and M. Claypool. Analysis of Active Queue Management. In *Proc. 2nd IEEE International Symposium on Network Computing and Applications (NCA)*, pages 359–366, Cambridge, Massachusetts, USA, April 2003. 648

ChCl05. J. Cheung and Claypool. NS by example, http://nile.wpi.edu/NS/. Last verified on February 08, 2005, 2005. 651, 653

ChLa74. A. Chang and S. Lavenberg. Work Rates in Closed Queueing Networks with General Independent Servers. *Operations Research*, 22:838–847, 1974. 325

ChLa83. K. Chandy and M. Lakshmi. An Approximation Technique for Queueing Networks with Preemptive Priority Queues. Technical report, Department of Computer Sciences, University of Texas, February 1983. 517

ChLu95. E. Chlebus and W. Ludwin. Is handoff traffic really Poissonian? *IEEE ICUPC '95 Conf. Record*, pages 348–353, Nov. 1995. 787

ChMa83. K. Chandy and A. Martin. A Characterization of Product–Form Queuing Networks. *Journal of the ACM*, 30(2):286–299, April 1983. 10, 364

ChMi81. K. M. Chandy and J. Mishra. Asynchronous Distributed Simulation via a Sequence of Parallel Computations. *Communications of the ACM*, 24(4), 1981. 610, 611

ChNe82. K. Chandy and D. Neuse. Linearizer: A Heuristic Algorithm for Queueing Network Models of Computing Systems. *Communications of the ACM*, 25(2):126–134, February 1982. 427

ChSa80. K. Chandy and C. Sauer. Computational Algorithms for Product Form Queueing Networks. *Communications of the ACM*, 23(10):573–583, October 1980. 369, 370

CHT77.   K. Chandy, J. Howard, and D. Towsley. Product Form and Local Balance in Queueing Networks. *Journal of the ACM*, 24(2):250–263, April 1977. 339, 341, 364

CHT02.   D. Chen, Y. Hong, and K. S. Trivedi. Second-Order Stochastic Fluid Models with Fluid Dependent Flow Rates. *Performance Evaluation*, 49(1/4):341–358, 2002. 115

ChTr92.   H. Choi and K. S. Trivedi. Approximate Performance Models of Polling Systems Using Stochastic Petri Nets. In *Proc. IEEE INFO-COM '92*, volume 3, pages 1–9, Florence, Italy, Los Alamitos, CA, May 4–8 1992. IEEE Computer Society Press. 760

CHW75a.   K. Chandy, U. Herzog, and L. Woo. Approximate Analysis of General Queueing Networks. *IBM Journal of Research and Development*, 19(1):43–49, January 1975. 502

CHW75b.   K. Chandy, U. Herzog, and L. Woo. Parametric Analysis of Queueing Networks. *IBM Journal of Research and Development*, 19(1):36–42, January 1975. 369, 410, 414

Chyl86.   P. Chylla. *Zur Modellierung und approximativen Leistungsanalyse von Vielteilnehmer-Rechensystemen.* Dissertation, Faculty for Mathematics and Computer Science, Technical University Munich, 1986. 479, 481

ChYu83.   W. Chow and P. Yu. An Approximation Technique for Central Server Queueing Models with a Priority Dispatching Rule. *Performance Evaluation*, 3:55–62, 1983. 421

Ciar95.   G. Ciardo. Discrete-Time Markovian Stochastic Petri Nets. In *Proc. 2nd. International Workshop on Numerical Solutions for Markov Chains*, pages 339–358, 1995. 119

CiGr87.   B. Ciciani and V. Grassi. Performability Evaluation of Fault-Tolerant Satellite Systems. *IEEE Transactions on Computers*, 35(4):403–409, 1987. 91, 93

CiLi98.   G. Ciardo and G. Li. Efficient Approximate Transient Analysis for a Class of Deterministic and Stochastic Petri Nets, 1998. 117

Cinl75.   E. Cinlar. *Introduction to Stochastic Processes.* Prentice-Hall, Englewood Cliffs, NJ, 1975. 51, 142, 303

CiTr93.   G. Ciardo and K. S. Trivedi. A Decomposition Approach for Stochastic Reward Net Models. *Performance Evaluation*, 18(1):37–59, July 1993. 12

CKT93.   H. Choi, V. Kulkarni, and K. Trivedi. Transient Analysis of Deterministic and Stochastic Petri Nets. In M. A. Marsan, editor,

Application and Theory of Petri Nets, Lecture Notes in Computer Science, pages 166–185. Springer, 1993. 117

CKT94.   H. Choi, V. G. Kulkarni, and K. S. Trivedi. Markov regenerative stochastic Petri nets. In *Proceedings of the 16th IFIP Working Group 7.3 international symposium on Computer performance modeling measurement and evaluation*, volume 20, pages 337–357. Elsevier Science Publishers B. V., 1994. 114

CMP99.   X. Chao, M. Miyazawa, and M. Pinedo. *Queueing Networks - Customers, Signals and Product Form Solutions*. John Wiley, New York, 1999. 299

CMST90.   G. Ciardo, R. Marie, B. Sericola, and K. S. Trivedi. Performability Analysis Using Semi-Markov Reward Processes. *IEEE Transactions on Computers*, 39(10):1251–1264, 1990. 80, 91

CMT90.   M. Calzarossa, R. Marie, and K. S. Trivedi. System Performance with User Behavior Graphs. *Performance Evaluation*, 11(3):155–165, July 1990. 105, 106

CMT91.   G. Ciardo, J. Muppala, and K. S. Trivedi. On the Solution of GSPN Reward Models. *Performance Evaluation*, 12(4):237–254, July 1991. 103, 105

CNT99.   C. Ciardo, D. Nicol, and K. S. Trivedi. Discrete-Event Simulation of Fluid Stochastic Petri Nets. *IEEE Transactions on Software Engineering*, 25(2):207–217, 1999. 115, 669

CoGe86.   A. Conway and N. Georganas. RECAL–A New Efficient Algorithm for the Exact Analysis of Multiple-Chain Closed Queuing Networks. *Journal of the ACM*, 33(4):768–791, October 1986. 369, 400, 418

CoHo86.   E. J. Coffman and M. Hofri. Queueing models of secondary storage devices. *Queueing Systems*, 1(2):129–168, September 1986. 147

CoSe84.   P. Courtois and P. Semal. Bounds for positive eigenvectors of nonnegative matrices and for their approximations by decomposition. *Journal of the ACM*, 31(4):804–825, 1984. 197

Cosm76.   G. Cosmetatos. Some Approximate Equilibrium Results for the Multiserver Queue (M/G/r). *Operations Research Quarterly, USA*, pages 615–620, 1976. 268, 271

Cour75.   P. Courtois. Decomposability, Instabilities and Saturation in Multiprogramming Systems. *Communications of the ACM*, 18(7):371–377, July 1975. 185

Cour77.   P. Courtois. *Decomposability: Queueing and Computer System Applications.* Academic Press, New York, 1977. 185, 193

Cox55.   D. Cox. A Use of Complex Probabilities in the Theory of Stochastic Processes. In *Proc. Cambridge Philosophical Society*, volume 51, pages 313–319, 1955. 26, 118, 120

CPN05.   CPN Tools (http://www.daimi.au.dk/CPnets/), 2005. 700

CrBe96.   M. Crovella and A. Bestavros. Self-Similarity in World Wide Web Traffic: Evidence and Possible Causes. In *Proc. SIGMETRICS'96: The ACM International Conference on Measurement and Modeling of Computer Systems.*, Philadelphia, Pennsylvania, May 1996. 616

Crom34.   C. Crommelin. Delay Probability Formulae. *Post Off. Electronic Engineer*, 26:266–274, 1934. 269

CSI05.   CSIM 19 Simulator: http://www.mesquite.com/. Last verified on February 08, 2005, 2005. 612, 639, 640, 700

CTM89.   G. Ciardo, K. S. Trivedi, and J. Muppala. SPNP: Stochastic Petri Net Package. In *Proc. 3rd Int. Workshop on Petri Nets and Performance Models (PNPM '89)*, pages 142–151, Kyoto, Japan, Los Alamitos, CA, December 1989. IEEE Computer Society Press. 666

Cuma82.   A. Cumani. On the Canonical Representation of Homogeneous Markov Processes Modelling Failure-Time Distributions. *Microelectronics and Reliability*, 22:583–602, 1982. 118

Cuma85.   A. Cumani. ESP — A Package for the Evaluation of Stochastic Petri Nets with Phase-Type Distributed Transition Times. In *International Workshop on Timed Petri Nets*, pages 144–151. IEEE Computer Society, 1985. 119

DAB00.   V. R. D. A. Bini, B. Meini. Analyzing M/G/1 Paradigms Through QBDs: the Role of the Block Structure in Computing the Matrix G. In G. Latouche and P. Taylor, editors, *Proc. 3rd Conference on Matrix Analytic Methods*, pages 73–85, 2000. 135

DaBh85.   C. Das and L. Bhuyan. Bandwith Availability of Multiple-Bus Multiprocessors. *IEEE Transactions on Computers*, 34(10):918–926, 1985. 91, 92, 93

Dadu96.   H. Daduna. Discrete Time Queueing Networks: Recent Developments. In *Proc. Performance '96*, Lausanne, October 8 1996. Tutorial Lecture Notes, Revised Version. 242

DCC+02. D. Deavours, G. Clark, T. Courtney, D. Daly, S. Derisavi, J. Doyle, W. Sanders, and P. Webster. The Moebius Framework and Its Implementation. *IEEE Trans. on Soft. Eng.*, 28(10):956–969, October 2002. 700

DeBu78. P. Denning and J. Buzen. The Operational Analysis of Queueing Network Models. *Computing Surveys*, 10(3):225–261, September 1978. 453

Deng96. S. Deng. Empirical Model of WWW Document Arrivals at Access Link. In *Proc. 1996 IEEE International Conference on Communications*, pages 1797–1802, June 1996. 616

Devr86. L. Devroye. *Non-Uniform Random Variate Generation*. Springer-Verlag, New York, 1986. 614, 622

Dijk59. E. W. Dijkstra. A Note on Two Problems in Connection with Graphs. *Numerische Math*, 1:269–271, 1959. 651

DoIy87. L. Donatiello and B. Iyer. Analysis of a Composite Performance Reliability Measure for Fault-Tolerant Systems. *Journal of the ACM*, 34(4):179–199, 1987. 91, 94

DoRa99. C. Dovrolis and P. Ramanathan. A Case for Relative Differentiated Sevices and Proportional Differentiation Model. In *Proc. IEEE Network '99*, 1999. 721, 724

Dosh90. B. Doshi. Single Server Queues With Vacations–A Survey. *Queueing Systems*, 1, 1(1):29–66, 1990. 146

Down01. A. Downey. Evidence for Long-Tailed Distributions in the Internet, 2001. 30

DPL02. M.-S. Do, Y. Park, and J.-Y. Lee. Channel Assignment with QoS Guarantees for a Multiclass Multicode CDMA System. *IEEE/ACM Transactions on Vehicular Technology*, 51(5):935–948, September 2002. 135

DSR99. C. Dovrolis, D. Stiliadis, and P. Ramanathan. Proportional Differentiated Services: Delay Differentiation and Packet Scheduling. In *Proc. ACM SIGCOMM'99*, Cambridge, 1999. ACM Press. 721

DTL03. S. Dharmaraja, K. Trivedi, and D. Logothetis. Performance Modeling of Wireless Networks with Generally Distributed Hand-off Interarrival Times. *Computer Communications*, 26(15):1747–1755, 2003. 791

DuCz87. A. Duda and T. Czachórski. Performance Evaluation of Fork and Join Synchronization Primitives. *Acta Informatica*, 24:525–553, 1987. 565, 570

Duda87.   A. Duda. Approximate Performance Analysis of Parallel Systems. In *Proc. 2nd Int. Workshop on Applied Mathematics and Performance/Reliability Models and Computer/Communication Systems*, pages 189–202, 1987. 561, 570

EaSe83.   D. Eager and K. Sevcik. Performance Bound Hierarchies for Queueing Networks. *ACM Transactions on Computer Systems*, 1(2):99–115, May 1983. 458

EaSe86.   D. Eager and K. Sevcik. Bound Hierarchies for Multiple-Class Queueing Networks. *Journal of the ACM*, 33(1):179–206, January 1986. 452, 458

EBA01.   L. Essafi, G. Bolch, and A. Andres. An Adaptive Waiting Time Priority Scheduler for the Proportional Differentiation Model. In *Proc. ASTC, HPC'01*, Seattle, 2001. ev. 723

Ehre96.   D. Ehrenreich. Erweiterung eines Kommunikationsnetzes auf der Basis von Leistungsuntersuchungen. Diplomarbeit, Universität Erlangen-Nürnberg, IMMD IV, July 1996. 709

Erla17.   A. K. Erlang. Solution of Some Problems in the Theory of Probabilities of Significance in Automatic Telephone Exchanges. *Elektrotkeknikeren*, 13, 1917. 7, 10

EsBo03.   L. Essafi and G. Bolch. Performance Evaluation of Priority based Schedulers in the Internet. In D. Al-Dabass, editor, *Proc. 17th European Simulation Multiconference (ESM 2003)*, pages 25–33, Nottingham, UK, Erlangen, Germany, June 2003. SCS-European Publishing House. 282, 724

ESV00.   D. Eager, D. Sorin, and M. Vernon. AMVA Techniques for High Service Time Variability. In *Proc. ACM SIGMETRICS 2000, Santa Clara, California*, 2000. 504

FaTe97.   G. Falin and J. Templeton. *Retrial Queues*. Chapman & Hall, 1997. 299, 302

FeJo90.   R. Feix and M. R. Jobmann. MAOS — Model Analysis and Optimization System. Technical report, University of Munich, FB Informatik, 1990. 700

Fell68.   W. Feller. *An Introduction to Probability Theory and Its Applications*, volume 1. John Wiley, New York, 3rd edition, 1968. 15, 51, 63

FeWh96.   A. Feldmann and W. Whitt. Fitting Mixtures of Exponentials to Long-Tail Distributions to Analyze Network Performance Models, December 1996. 119

FiMe93.   W. Fischer and K. Meier-Hellstern. The Markov-Modulated
            Poisson Process (MMPP) Cookbook. *Performance Evaluation*,
            18(2):149–171, 1993. 303, 305, 306

Fish97.   G. S. Fishman. *Concepts Algorithms and Applications*. Springer-
            Verlag, London, UK, 1997. 610, 637

Flei89.   G. Fleischmann. *Modellierung und Bewertung paralleler Pro-
            gramme*. Dissertation, Universität Erlangen-Nürnberg, IMMD IV,
            1989. 577

FlJa93.   S. Floyd and V. Jacobson. Random Early Detection Gateways for
            Congestion Avoidance. *IEEE/ACM Transactions on Networking*,
            1(4):397–413, 1993. 648

FoGl88.   B. Fox and P. Glynn. Computing Poisson Probabilities. *Commu-
            nications of the ACM*, 31(4):440–445, April 1988. 218

Förs89.   C. Förster. Implementierung und Validierung von MWA–
            Approximationen. Studienarbeit, Universität Erlangen-Nürnberg,
            IMMD IV, 1989. 513

Fran03.   M. Frank. *Geschlossene Warteschlangennete und Kanban-
            Systeme*. PhD thesis, Department of Mathematic, Technical Uni-
            versity Clausthal, Germany, 2003. 601, 603, 754

FrBe83.   D. Freund and J. Bexfield. A New Aggregation Approximation
            Procedure for Solving Closed Queueing Networks with Simultane-
            ous Resource Possession. *ACM Sigmetrics Performance Evalua-
            tion Review*, pages 214–223, August 1983. Special Issue. 547

Fuji99.   R. M. Fujimoto. *Parallel and Distributed Simulation Systems*.
            Wiley-Interscience, 1999. 610, 611, 638, 639

GaKe79.   F. Gay and M. Ketelsen. Performance Evaluation for Gracefully
            Degrading Systems. In *Proc. IEEE Int. Symp. on Fault-Tolerant
            Computing, FTCS*, pages 51–58, Los Alamos, CA, Los Alamitos,
            CA, 1979. IEEE Computer Society Press. 91, 93

Gard86.   W. A. Gardner. *Introduction to Random Processes with Applica-
            tions to Signals and Systems*. Macmillan, New York, 1986. 303

GBB98.   S. Greiner, G. Bolch, and K. Begain. A Generalized Analysis
            Technique For Queueing Networks with Mixed Priority Strategy
            and Class Switching. *Computer Communications*, 21(9):819–832,
            July 1998. 526

GCS+86.   A. Goyal, W. Carter, E. de Souza e Silva, S. Lavenberg, and K. S.
            Trivedi. The System AVailability Estimator (SAVE). In *Proc.*

16th Int. Symp. on Fault-Tolerant Computing, FTCS, pages 84–89, Vienna, Austria, Los Alamitos, CA, July 1986. IEEE Computer Society Press. 12

Gele75.   E. Gelenbe. On Approximate Computer System Models. *Journal of the ACM*, 22(2):261–269, April 1975. 464, 465, 601

GeMi80.   E. Gelenbe and I. Mitrani. *Analysis and Synthesis of Computer Systems*. Academic Press, London, 1980. 465

GePu87.   E. Gelenbe and G. Pujolle. *Introduction to Queueing Networks*. John Wiley, Chichester, 1987. 479, 481

Germ00.   R. German. *Performance Analysis of Communication Systems: Modelling with Non-Markovian Stochastic Petri Nets*. John Wiley, West Sussex, England, 2000. 113, 114, 116, 117, 140, 303

GeTr83.   R. Geist and K. S. Trivedi. The Integration of User Perception in the Heterogeneous M/M/2 Queue. In A. Agrawala and S. Tripathi, editors, *Proc. Performance '83*, pages 203–216, Amsterdam, 1983. North-Holland. 284, 294

GHPS93.   J. D. Gee, M. D. Hill, D. N. Pnevmatikatos, and A. J. Smith. Cache Performance of the SPEC92 Benchmark Suite. *IEEE Micro*, 13(4):17–27, 1993. 611

GKZH94.   R. German, C. Kelling, A. Zimmermann, and G. Hommel. TimeNET-A Toolkit for Evaluating Non-Markovian Stochastic Petri Nets. Technical report, Technical University Berlin, 1994. 700

GlIg89.   P. W. Glynn and D. L. Iglehart. Importance Sampling for Stochastic Simulations. *Management Science*, 35(11):1367–1392, 1989. 610, 637

GLT87.   A. Goyal, S. Lavenberg, and K. S. Trivedi. Probabilistic Modeling of Computer System Availability. *Annals of Operations Research*, 8:285–306, March 1987. 124

GLT95.   R. German, D. Logothetis, and K. Trivedi. Transient Analysis of Markov Regenerative Stochastic Petri Nets: A Aomparison of Approaches. In *Proc. 6th International Workshop on Petri Nets and Performance Models*, pages 103–112, 1995. 117

Gold89.   D. Goldberg. *Genetic Algorithms in Search, Optimization & Machine Learning*. Addison-Wesley Longman, 1989. 41

GoNe67a.   W. Gordon and G. Newell. Closed Queuing Systems with Exponential Servers. *Operations Research*, 15(2):254–265, April 1967. 340, 346, 347

GoNe67b.   W. Gordon and G. Newell. Cyclic Queuing Systems with Restrict-
            ed Queues. *Operations Research*, 15(2):266–277, April 1967. 592

Good83.   J. Goodman. Using Cache Memory to Reduce Processor-Memory
            Traffic. In *Proc. 10th Int. Symp. on Computer Architecture*, pages
            124–131, New York, 1983. IEEE. 797

GöSc96.   C. Görg and F. Schreiber. The RESTART/LRE Method for Rare
            Event Simulation. In *Proc. 28th Conf. on Winter Simulation*,
            pages 390–397. ACM Press, 1996. 638

GoTa87.   A. Goyal and A. Tantawi. Evaluation of Performability in Acyclic
            Markov chains. *IEEE Transactions on Computers*, 36(6):738–744,
            1987. 91, 94

GoTr99.   S. Gokhale and K. S. Trivedi. A Time/Structure Based Soft-
            ware Reliability Model. *Annals of Software Engineering*, 8:85–121,
            1999. 617

GPS.   GPSS Simulator: http://www.minutemansoftware.com/. Last
            verified on February 08, 2005. 612

Gran76.   J. Grandell. *Doubly Stochastic Poisson Processes*, volume 529 of
            Lecture Notes in Mathematics. Springer, 1976. 303

GrBo96a.   S. Greiner and G. Bolch.   Approximative analytische Leis-
            tungsbewertung am Beispiel eines UNIX-basierten Multiprozes-
            sor Betriebssystems.   *Informatik, Forschung und Entwicklung*,
            11(3):111–124, 1996. 724

GrBo96b.   S. Greiner and G. Bolch. Modeling and Performance Evaluation
            of Production Lines Using the Modeling Language MOSEL. In
            *Proc. 2nd IEEE / ECLA / IFIP Int. Conf. on Architecture and
            Design Methods for Balanced Automation Systems (BASYS'96)*,
            Lisbon, Portugal, 1996. 679

GrHa85.   H. Gross and C. Harris. *Fundamentals of Queueing Theory*. John
            Wiley, New York, 2nd edition, 1985. 51, 86, 233, 251, 252, 265

GSS94.   D. Goldsman, L. W. Schruben, and J. J. Swain. Tests for Tran-
            sient Means in Simulated Time Series. *Naval Research Logistics*,
            41:171–187, 1994. 635

GTH85.   W. Grassmann, M. Taksar, and D. Heyman. Regenerative Analy-
            sis and Steady State Distribution for Markov Chains. *Operations
            Research*, 33:1107–1116, September 1985. 152, 159

Gün89.   L. Gün. Experimental Results on Matrix-Analytical Solution
            Techniques — Extensions and Comparisons. *Communications in
            Statistics — Stochastic Models*, 5:669–682, 1989. 135

Hahn88. T. Hahn. Implementierung und Validierung der Mittelwertanalyse für höhere Momente und Verteilungen. Studienarbeit, Universität Erlangen-Nürnberg, IMMD IV, 1988. 513

Hahn90. T. Hahn. Erweiterung und Validierung der Summationsmethode. Diplomarbeit, Universität Erlangen-Nürnberg, IMMD IV, 1990. 446

HaJa03. M. Hassan and R. Jain. *High Performance TCP/IP Networking: Concepts, Issues, and Solutions.* Prentice-Hall, 2003. 647

Hans03. T. Hanschke. Approximate Analysis of the $GI^X/G^{[b,b]}/c$ Queue. *Journal Operations Research*, 2003. 299

HaOs97. B. R. Haverkort and A. Ost. Steady-State Analysis of Infinite Stochastic Petri Nets: a Comparison Between the Spectral Expansion and the Matrix-Geometric Method. In *Proc. 7th International Workshop on Petri Nets and Performance Models*, pages 36–45, 1997. 135

HaSp95. T. Hanschke and T. Speck. AMS–An APL Based Modeling System for Production Planning. In *Operations Research Proceedings 1994*, pages 318–323, Berlin, 1995. Springer. 296

HaTo99. Z. Haraszti and J. K. Townsend. The Theory of Direct Probability Redistribution and Its Application to Rare Event Simulation. *ACM Transactions on Modeling and Computer Simulation (TOMACS)*, 9(2), 1999. 610, 638

HaTr93. R. Haverkort and K. S. Trivedi. Specification and Generation of Markov Reward Models. *Discrete-Event Dynamic Systems: Theory and Applications*, 3:219–247, 1993. 12, 91

Have96. B. R. Haverkort. SPN2MGM: Tool Support for Matrix-Geometric Stochastic Petri Nets. In V. K. Madisetti, E. Gelenbe, and J. C. Walrand, editors, *Proc. 2nd International Computer Performance and Dependability Symposium*, pages 219–228. IEEE Computer Society, 1996. 135

Have98. B. R. Haverkort. *Performance of Computer Communication Systems: a Model-Based Approach.* John Wiley & Sons, 1998. 135

HaYo81. L. Hagemann and D. Young. *Applied Iterative Methods.* Academic Press, New York, 1981. 175

HBAK86. K. Hoyme, S. Bruell, P. Afshari, and R. Kain. A Tree-Structured Mean Value Analysis Algorithm. *ACM Transactions on Computer Systems*, 4(2):178–185, November 1986. 409

HeGe97.   A. Heindl and R. German. A Fourth-Order Algorithm with Automatic Stepsize Control for the Transient Analysis of DSPNs. In *Proc. 7th International Conference on Petri Nets and Performance Models — PNPM97*, pages 60–69. IEEE Computer Society, 1997. 117

Heid86.   P. Heidelberger. Statistical Analysis of Parallel Simulations. In *WSC '86: Proc. 18th conference on Winter simulation*, pages 290–295. ACM Press, 1986. 639

Heid95.   P. Heidelberger. Fast Simulation of Rare Events in Queueing and Reliability Models. *ACM Transactions on Modeling and Computer Simulation (TOMACS)*, 5(1), 1995. 610, 638

HeKi97.   K. J. Healy and R. A. Kilgore. Silk: A Java-Based Process Simulation Language. In *Proc. 1997 Winter Simulation Conference*, pages 475–482, 1997. 612

HeLu03.   D. P. Heyman and D. Lucantoni. Modeling Multiple IP Traffic Streams with Rate Limits. *IEEE/ACM Transactions on Networking*, 11(6):948–958, 2003. 303

Hend72.   W. Henderson. Alternative Approaches to the Analysis of the M/G/1 and G/M/1 Queues. *Operations Research*, 15:92–101, 1972. 140

HeSo84.   D. Heyman and M. Sobel. *Stochastic Models in Operations Research*, volume 1 and 2. McGraw-Hill, New York, 1984. 278, 279

HeTr82.   P. Heidelberger and K. S. Trivedi. Queueing Network Models for Parallel Processing with Asynchronous Tasks. *IEEE Transactions on Computers*, 31(11):1099–1109, November 1982. 551, 552, 555, 558

HeTr83.   P. Heidelberger and K. S. Trivedi. Analytic Queueing Models for Programs with Internal Concurrency. *IEEE Transactions on Computers*, 32(1):73–82, January 1983. 549

HKNT98.   G. Horton, V. Kulkarni, D. Nicol, and K. Trivedi. Fluid Stochastic Petri Nets: Theory, Applications, and Solution Techniques. *European Journal of Operational Research*, 105:184–201, 1998. 114, 115

HLM96.   G. Haring, J. Lüthi, and S. Majumdar. Mean Value Analysis for Computer Systems with Variabilities in Workload. In *Proc. IEEE Int. Computer Performance and Dependability Symp. (IPDS '96)*, pages 32–41, Urbana-Champaign, September 1996. 708

HMPT01.   G. Haring, R. Marie, R. Puigjaner, and K. Trivedi. Loss Formulae and Their Application to Optimization for Cellular Networks. *IEEE Transaction on Vehicular Techology*, VT-50:664–673, 2001. 787, 789

HMS99.   H. Hermanns, J. Meyer-Kayser, and M. Siegle. Multi Terminal Binary Decision Diagrams to Represent and Analyse Continuous Time Markov Chains. In *Proc. 3rd Int. Workshop on the Numerical Solution of Markov Chains*, pages 188–207, 1999. 120, 122

HMT91.   D. Heimann, N. Mittal, and K. S. Trivedi. Dependability Modeling for Computer Systems. In *Proc. Annual Reliability and Maintainability Symposium*, pages 120–128, Orlando, FL, January 1991. 14, 71, 80, 91, 92, 93, 94

HoKl79.   A. C. M. Hopmans and J. P. C. Kleijnen. Importance Sampling in Systems Simulation: A Practical Failure? *Mathematics and Computers in Simulation XXI*, pages 209–220, 1979. 638

HoRa86.   D. Hong and S. Rappaport. Traffic Model and Performance Analysis for Cellular Mobile Radio Telephone Systems with Prioritized and Non-Prioritized Handoff Procedures. *IEEE Transaction on Vehicular Techology*, VT-35:72–92, 1986. 787

HoSe83.   M. H. van Hoorn and L. P. Seelen. The SPP/G/1 queue: Single Server Queue with a Switched Poisson Process as Input Process. *Operations Research-Spektrum*, 5:205–218, 1983. 304

Hova81.   A. Hordijk and N. van Dijk. Networks of Queues with Blocking. In F. Kylstra, editor, *Proc. Performance '81*, pages 51–65, Amsterdam, 1981. North-Holland. 593

Howa71.   R. Howard. *Dynamic Probabilistic Systems*, volume 2: Semi-Markov and Decision Processes. John Wiley, New York, 1971. 51, 76

HsLa87.   C. Hsieh and S. Lam. Two Classes of Performance Bounds for Closed Queueing Networks. *Performance Evaluation*, 7(1):3–30, February 1987. 458

HsLa89.   C. Hsieh and S. Lam. PAM–A Noniterative Approximate Solution Method for Closed Multichain Queueing Networks. *Performance Evaluation*, 9(2):119–133, April 1989. 421

HSZT00.   C. Hirel, R. A. Sahner, X. Zang, and K. S. Trivedi. Reliability and Performability Modeling Using SHARPE 2000. In *Proc. 11th International Conference on Computer Performance Evaluation: Modelling Techniques and Tools*, pages 345–349. Springer-Verlag, London, UK, 2000. 11, 543, 547, 549, 550, 678, 688, 689, 695, 700

HTT00.   C. Hirel, B. Tuffin, and K. S. Trivedi. SPNP: Stochastic Petri Nets.
           Version 6.0. In *Proc. 11th International Conference on Computer
           Performance Evaluation: Modelling Techniques and Tools*, pages
           354–357. Springer-Verlag, London, UK, 2000. 11, 612, 666, 700

Husl81.   R. Huslende. A Combined Evaluation of Performance and Relia-
           bility for Degradable Systems. In *Proc. 1981 ACM SIGMETRICS
           Conf. on Measurements and Modeling of Computer Systems*, Per-
           formance Evaluation Review, pages 157–164, Las Vegas, Nevada,
           September 1981. 91, 94

HWFT97.   C. Hirel, S. Wells, R. Fricks, and K. S. Trivedi. iSPN: An Inte-
           grated Environment for Modeling Using Stochastic Petri Nets. In
           *Tool Descriptions 7th Int. Workshop on Petri Nets and Perfor-
           mance Models (PNPM '97)*, pages 17–19, St. Malo, France, Los
           Alamitos, CA, June 1997. IEEE Computer Society Press. 674

Ibe87.   O. Ibe. Performance Comparison of Explicit and Implicit Token-
           Passing Networks. *Computer Communications*, 10(2):59–69, 1987.
           764

IbTr90.   O. Ibe and K. S. Trivedi. Stochastic Petri Net Models of Polling
           Systems. *IEEE Journal on Selected Areas in Communications*,
           8(10):146–152, December 1990. 108, 112, 759, 760

ICT93.   O. Ibe, H. Choi, and K. S. Trivedi. Performance Evaluation of
           Client–Server Systems. *IEEE Transactions on Parallel and Dis-
           tributed Systems*, 4(11):1217–1229, November 1993. 763

IEEE90.   IEEE. IEEE Std. 610.12-1990 — IEEE Standard Glossary of Soft-
           ware Engineering Terminology, February 1990. 5, 7

Jack54.   R. R. P. Jackson. Queueing Systems with Phase Type Service.
           Operations Research Quarterly, 5(2):109–120, 1954. 7

Jack57.   J. Jackson. Networks of Waiting Lines. *Operations Research*,
           5(4):518–521, 1957. 341

Jack63.   J. Jackson. Jobshop-Like Queuing Systems. *Management Science*,
           10(1):131–142, October 1963. 340, 341, 342

Jaek91.   H. Jaekel. Analytische Untersuchung von Multiple Server Sys-
           temen mit Prioritäten. Diplomarbeit, Universität Erlangen-
           Nürnberg, 1991. 271, 279, 282

Jain91.   R. Jain. *The Art of Computer Systems Performance Analysis*.
           John Wiley, New York, 1991. 1

Jake94.   W. Jakes, editor. *Microwave Mobile Communications*. Wiley-
           IEEE Press, 1994. 782

JaLa82.   P. Jacobson and E. Lazowska. Analyzing Queueing Networks with Simultaneous Resource Possession. *Communications of the ACM*, 25(2):142–151, February 1982. 547, 548

JaLa83.   P. Jacobson and E. Lazowska. A Reduction Technique for Evaluating Queueing Networks with Serialization Delays. In A. Agrawala and S. Tripathi, editors, *Proc. Performance '83*, pages 45–59, Amsterdam, 1983. North-Holland. 544, 548, 549

Jav.   JavaSim Simulator: http://javasim.ncl.ac.uk/. Last verified on February 08, 2005. 612

Jens53.   A. Jensen. Markoff Chains as an Aid in the Study of Markoff Processes. *Skandinavian Aktuarietidskr.*, 36:87–91, 1953. 216

Jobm91.   M. R. Jobmann. *Leistungsanalyse von Rechen- und Kommunikationssystemen — Konzepte der Modellauswertung und Definition einer Modellierungssprache.* PhD thesis, Universität Hamburg, Hamburg, Feb 1991. 700

JSI.   JSIM Simulator: http://chief.cs.uga.edu/~jam/jsim/. Last verified on February 08, 2005. 612

Kauf84.   J. Kaufmann. Approximation Methods for Networks of Queues with Priorities. *Performance Evaluation*, 4:183–198, 1984. 523, 525

Kelt86.   W. D. Kelton. Replication Splitting and Variance for simulating Discrete Parameter Stochastic Process. *Operations Research Letters*, 4:275–279, 1986. 635

Kelt89.   W. D. Kelton. Random initialization methods in simulation. *IIE Transactions*, 21:355–367, 1989. 635

KeRi78.   B. W. Kernighan and D. M. Ritchie. *The C Programming Language*. Prentice-Hall, Englewood Cliffs, NJ, 1978. 670

Kero86.   T. Kerola. The Composite Bound Method for Computing Throughput Bounds in Multiple Class Environments. *Performance Evaluation*, 6(1):1–9, March 1986. 458

KeSn78.   J. Kemeny and J. Snell. *Finite Markov Chains*. Springer, New York, 2nd edition, 1978. 192, 222

KGB87.   S. Kumar, W. Grassmann, and R. Billington. A Stable Algorithm to Calculate Steady-State Probability and Frequency of a Markov System. *IEEE Transactions on Reliability*, R-36(1), April 1987. 152, 159

KGTA88. D. Kouvatsos, H. Georgatsos, and N. Tabet-Aouel. A Universal Maximum Entropy Algorithm for General Multiple Class Open Networks with Mixed Service Disciplines. Technical report DDK/PHG-1. N, Computing Systems Modelling Research Group, Bradford University, England, 1988. 471

KiLi97. Y. Y. Kim and S. Q. Li. Performance Evaluation of Packet Data Services over Cellular Voice Networks. *Proc. Conference on Computer Communications (IEEE Infocom)*, pages 1022–1029, 1997. 135

Kimu85. T. Kimura. Heuristic Approximations for the Mean Waiting Time in the GI/G/s Queue. Technical report B55, Tokyo Institute of Technology, 1985. 267, 272, 483

King70. J. Kingman. Inequalities in the Theory of Queues. *Journal of the Royal Statistical Society*, Series B 32:102–110, 1970. 269

King90. J. King. *Computer and Communication Systems Performance Modeling*. Prentice-Hall, Englewood Cliffs, NJ, 1990. 51, 88, 148, 303

Kirs91. M. Kirschnick. Approximative Analyse von WS-Netzwerken mit Batchverarbeitung. Diplomarbeit, Universität Erlangen-Nürnberg, IMMD IV, 1991. 296

Kirs93. M. Kirschnick. XPEPSY Manual. Technical report TR-I4-18-93, Universität Erlangen-Nürnberg, IMMD IV, September 1993. 659

Kirs94. M. Kirschnick. The Performance Evaluation and Prediction SYstem for Queueing NetworkS (PEPSY-QNS). Technical report TR-I4-18-94, Universität Erlangen-Nürnberg, IMMD IV, June 1994. 660, 678, 700

Klei65. L. Kleinrock. A Conservation law for a Wide Class of Queueing Disciplines. *Naval Research Logistic Quart.*, 12:181–192, 1965. 278

Klei75. L. Kleinrock. *Queueing Systems*, volume 1: Theory. John Wiley, New York, 1975. 38, 252, 255, 257, 265

Klei76. L. Kleinrock. *Queueing Systems*, volume 2: Computer Applications. John Wiley, New York, 1976. 266, 453

Klei95. J. P. C. Kleijnen. *Reliability and Maintenance of Complex Systems, ed. by Süleyman Özekici*, volume 154, chapter Simulation: Runlength Selection and Variance Reduction Techniques, pages 411–428. Springer, June 1995. 614

KMS90.　U. R. Krieger, B. Müller-Clostermann, and M. Sczittnick. Modeling and Analysis of Communication Systems Based on Computational Methods for Markov Chains. *Journal on Selected Areas in Communication*, 8(9):1630–1648, 1990. 700

KoAl88.　D. Kouvatsos and J. Almond. Maximum Entropy Two-Station Cyclic Queues with Multiple General Servers. *Acta Informatica*, 26:241–267, 1988. 471

Koba74.　H. Kobayashi. Application of the Diffusion Approximation to Queueing Networks, Part 1: Equilibrium Queue Distributions. *Journal of the ACM*, 21(2):316–328, April 1974. 463, 464, 465

Koba78.　H. Kobayashi. *Modeling and Analysis: An Introduction to System Performance Evaluation Methodology*. Addison-Wesley, Reading, MA, 1978. 15, 370

Koba79.　H. Kobayashi. A Computational Algorithm for Queue Distributions via Polya Theory of Enumeration. In M. Arato and M. Butrimenko, editors, *Proc. 4th Int. Symp. on Modelling and Performance Evaluation of Computer Systems*, volume 1, pages 79–88, Vienna, Austria, February 1979. North-Holland. 370

KoBu03.　S. Kounev and A. Buchmann. Performance Modeling and Evaluation-of Large-Scale J2EE Applications. In *Proc. 29th International Conference of the Computer Measurement Group (CMG) on Resource Management and Performance Evaluation of Enterprise Computing Systems (CMG-2003), Dallas, Texas, December 7-12*, 2003. 735, 742

Köll74.　J. Köllerström. Heavy Traffic Theory for Queues with Several Servers. *Journal of Applied Probability*, 11:544–552, 1974. 270

KoRe76.　A. Konheim and M. Reiser. A Queueing Model with Finite Waiting Room and Blocking. *Journal of the ACM*, 23(2):328–341, April 1976. 592

KoRe78.　A. Konheim and M. Reiser. Finite Capacity Queuing Systems with Applications in Computer Modeling. *SIAM Journal on Computing*, 7(2):210–299, May 1978. 592

KöSc92.　D. König and V. Schmidt. Zufällige Punktprozesse. In J. Lehn, N. Schmitz, and W. Weil, editors, *Teubner Skripten zur Mathematischen Stochastik*, 1992. 303

Kouv85.　D. Kouvatsos. Maximum Entropy Methods for General Queueing Networks. In *Proc. Int. Conf. on Modelling Techniques and Tools for Performance Analysis*, pages 589–608, 1985. 471, 472

KrLa76. W. Krämer and M. Langenbach-Belz. Approximate Formulae for General Single systems with Single and Bulk Arrivals. In *Proc. 8th Int. Teletraffic Congress (ITC)*, pages 235–243, Melbourne, 1976. 270, 483

KrNa99. U. R. Krieger and V. Naumov. Analysis of a Delay-Loss System with a Superimposed Markovian Arrival Process and State-Dependent Service Times. In W. S. et al., editor, *Numerical Solution of Markov Chains (NSMC99), Zaragoza, Spain*, pages 261–279. Prensas Universitarias de Zaragoza, 1999. 135

Krze87. A. Krzesinski. Multiclass Queueing Networks with State-Dependent Routing. *Performance Evaluation*, 7(2):125–143, June 1987. 364

KTK81. A. Krzesinski, P. Teunissen, and P. Kritzinger. Mean Value Analysis for Queue Dependent Servers in Mixed Multiclass Queueing Networks. Technical report, University of Stellenbosch, South Africa, July 1981. 405

Kuba86. P. Kubat. Reliability Analysis for Integrated Networks with Application to Burst Switching. *IEEE Transactions on Communication*, 34(6):564–568, 1986. 91

Kühn79. P. Kühn. Approximate Analysis of General Queuing Networks by Decomposition. *IEEE Transactions on Communication*, 27(1):113–126, January 1979. 479, 482

Kulb89. J. Kulbatzki. Das Programmsystem PRIORI: Erweiterung und Validierung mit Hilfe von Simulationen. Diplomarbeit, Universität Erlangen-Nürnberg, IMMD IV, 1989. 266, 271

Kulk96. V. Kulkarni. *Modeling and Analysis of Stochastic Systems*. Chapman & Hall, London, 1996. 140, 303

KuMi01. K. Kumaran and D. Mitra. Performance and Fluid Simulations of a Novel Shared Buffer Management System. *ACM Transactions of Modeling and Computer Simulation*, 11(1):43–75, 2001. 610

LaCa79. A. M. Law and J. M. Carlson. A Sequential Procedure for Determining the Length of Steady State Simulation. *Operations Research*, 27:131–143, 1979. 634

LaKe00. A. M. Law and W. D. Kelton. *Simulation Modeling and Analysis*. McGraw-Hill Higher Education, third edition, 2000. 609, 612, 614, 622, 625, 632, 635, 636, 637, 646

LaLi83. S. Lam and Y. Lien. A Tree Convolution Algorithm for the Solution of Queueing Networks. *Communications of the ACM*, 26(3):203–215, March 1983. 384

Lam81.   S. Lam. A Simple Derivation of the MVA and LBANC Algorithms from the Convolution Algorithm. Technical report 184, University of Texas at Austin, November 1981. 370

LaRa93.   G. Latouche and V. Ramaswami. A logarithmic Reduction Algorithm for Quasi-Birth–Death Processes. *Journal of Applied Probability*, 30:605–675, 1993. 134, 135

LaRa99.   G. Latouche and V. Ramaswami. *Introduction to Matrix Analytic Methods in Stochastic Modeling*. Series on Statistics and Applied Probability. ASA-SIAM, Philadelphia, PA, 1999. 128, 132, 134, 135

LaRe80.   S. Lavenberg and M. Reiser. Stationary State Probabilities at Arrival Instants for Closed Queueing Networks with Multiple Types of Customers. *Journal of Applied Probability*, 17:1048–1061, 1980. 384

LaSl75.   S. S. Lavenberg and D. Slutz. Introduction to Regenerative Simulation. *IBM Journal of Research and Development*, 19:458–462, 1975. 610

Lato93.   G. Latouche. Algorithms for Infinite Markov Chains with Repeating Columns. In C. D. Meyer and R. J. Plemmons, editors, *Linear Algebra, Markov Chains, and Queueing Models*, volume 48 of *The IMA Volumes in Mathematics and Its Applications*, pages 231–265. Springer, 1993. 135

Lave83.   S. Lavenberg. *Computer Performance Modeling Handbook*. Academic Press, New York, 1983. 247, 493, 625

Lawl82.   J. Lawless. *Statistical Models and Methods for Lifetime Data*. John Wiley, 1982. 45

LaZa82.   E. Lazowska and J. Zahorjan. Multiple Class Memory Constrained Queueing Networks. *ACM Sigmetrics Performance Evaluation Review*, 11(4):130–140, 1982. 544

LeVe90.   T.-J. Lee and G. de Veciana. Model and Performance Evaluation for Multiservice Network Link Supporting ABR and CBR Services. *IEEE Communications Letters*, 4(11):375–377, 1990. 135

LeWi88.   Y. Levy and P. Wirth. A Unifying Approach to Performance and Reliability Objectives. In *Proc. 12th International Teletraffic Conference*, pages 1173–1179, Amsterdam, 1988. Elsevier. 91

LeWi04.   J. S. H. Leeuwaarden and E. E. M. Winands. Quasi-Birth–Death Processes with an Explicit Rate Matrix. *SPOR — Reports in Statistics, Probability and Operations Research*, 2004. 133, 135

LiHw97.   S. Li and C. Hwang. On the Convergence of Traffic Measurement and Queueing Analysis: A Statistical-Matching and Queueing (SMAQ) tool. *IEEE/ACM Transactions on Networking*, 5(1):95–110, 1997. 135

LiSh96.   C. Lindemann and G. Shedler. Numerical Analysis of Deterministic and Stochastic Petri Nets with Concurrent Deterministic Transitions. *Performance Evaluation*, 27–28:565–582, 1996. 117

LiTh99.   C. Lindemann and A. Thümmler. Transient Analysis of Determinsitic and Stochastic Petri Nets with Concurrent Deterministic Transitions. *Performance Evaluation*, 36–37:35–54, 1999. 117

Litt61.   J. Little. A Proof of the Queuing Formula $L = \lambda W$. *Operations Research*, 9(3):383–387, May 1961. 88, 89, 111, 245, 578

LLY00.   M. K. H. Leung, J. C. S. Lui, and D. K. Y. Yau. Characterization and Performance Evaluation for Proportional Delay Differentiated Services. In *Proc. of IEEE International Conference on Network Protocols*, Osaka, 2000. 723

LMN90.   D. M. Lucantoni, K. S. Meier-Hellstern, and M. F. Neuts. A Single-Server Queue with Server Vacations and a Class of Non-Renewal Arrival Processes. *Advances in Applied Probability*, 22:676–705, 1990. 303, 304, 306, 307

LMR01.   A. Luchetta, S. Manetti, and A. Reatti. SAPWIN — A Symbolic Simulator as a Support in Electrical Engineering Education. *IEEE Transactions on Education*, 44(2):213, 2001. 611

Los.   Los Alamos National Laboratory Cache Trace: http://public.lanl.gov/radiant/research/measurement/traces.html. Last verified on February 08, 2005. 611

LPA98.   S. Li, S. Park, and D. Arifler. SMAQ: A Measurement-Based Tool for Traffic Modeling and Queuing Analysis Part I: Design Methodologies and Software Architecture. *IEEE Communications Magazine*, pages 56–65, August 1998. 135

LTK+00.   C. Lindemann, A. Thümmler, A. Klemm, M. Lohmann, and O. Waldhorst. Quantitative System Evaluation with DSPNexpress 2000. In *Proc. 2nd Int. Workshop on Software and Performance (WOSP), Ottawa, Canada*, pages 12–17, 2000. 700

LTWW94.   W. Leland, M. Taqqu, W. Willinger, and D. Wilson. On the Self-Similar Nature of Ethernet Traffic (Extended Version). *IEEE/ACM Transactions on Networking*, 2(1), February 1994. 29, 30, 303

Luca91.   D. Lucantoni. New Results on the Single Server Queue with a
            Batch Markovian Arrival Process. *Stochastic Models*, 7(1):1–46,
            1991. 307, 309, 311

Luca93.   D. Lucantoni. The BMAP/G/1 Queue: A Tutorial. In L. Donatiel-
            lo and R. Nelson, editors, *Models and Techniques for Performance
            Evaluation of Computer and Communications Systems*. Springer,
            New York, 1993. 309, 310, 311

LZGS84.   E. Lazowska, J. Zahorjan, G. Graham, and K. Sevcik. *Quanti-
            tative System Performance — Computer System Analysis Using
            Queueing Network Models*. Prentice-Hall, Englewood Cliffs, NJ,
            1984. 385, 458, 544, 545, 546, 547, 708

LüLl00.   J. Lüthi and C. Lladó. Splitting Techniques for Interval Parame-
            ters in Performance Models. Technical report 2000-07, Universität
            der Bundeswehr, München, 2000. 41

MAD94.   D. Menascé, V. Almeida, and L. Dowdy. *Capacity Planning and
            Performance Modeling–From Mainframes to Client–Server Sys-
            tems*. Prentice-Hall, Englewood Cliffs, NJ, 1994. 706, 707, 708

MaMc73.   M. MacDougall and J. McAlpine. Computer System Simulation
            with ASPOL. In *IEEE Proc. 1973 Symposium on Simulation of
            Computer Systems*, pages 92–103, 1973. 640

MaNi98.   M. Matsumoto and T. Nishimura. Mersenne Twister: A 632-
            Dimensionally Equidistributed Uniform Pseudo-Random Number
            Generator. *ACM Transactions on Modeling and Computer Simu-
            lation*, 8:3–30, 1998. 614

Marc74.   W. Marchal. *Simple Bounds and Approximations in Queuing Sys-
            tems*. D.sc. dissertation, George Washington University, Depart-
            ment of Operations Research, Washington, D.C., 1974. 269

Marc78.   W. Marchal. Some Simpler Bounds on the Mean Queueing Time.
            Operations Research, 26(6), 1978. 265, 266, 483

MaRe93.   M. Malhotra and A. Reibman. Selecting and Implementing Phase
            Approximations for Semi-Markov models. *Stochastic Models*,
            9(4):473–506, 1993. 119

Mari78.   R. Marie. Méthodes itératives de résolution de modèles
            mathématiques de systèmes informatiques. *R.A.I.R.O. Informa-
            tique/Computer Science*, 12(2):107–122, 1978. 497

Mari79.   R. Marie. An Approximate Analytical Method for General
            Queueing Networks. *IEEE Transactions on Software Engineer-
            ing*, 5(5):530–538, September 1979. 492

Mari80.   R. Marie. Calculating Equilibrium Probabilities for $\lambda(n)/C_k/1/N$ Queues. *ACM Sigmetrics Performance Evaluation Review*, 9(2):117–125, 1980. 492, 494, 497, 498

Mart72.   J. Martin. *System Analysis for Data Transmission*. Prentice-Hall, Englewood Cliffs, NJ, 1972. 267

MaSt77.   R. Marie and W. Stewart. A Hybrid Iterative-Numerical Method for the Solution of a General Queueing Network. In *Proc. 3rd Symp. on Measuring, Modelling and Evaluating Computer Systems*, pages 173–188, Amsterdam, 1977. North-Holland. 492

Mat05.   Mathematica: http://www.wolfram.com/. Last verified on February 08, 2005, 2005. 10, 11

MaTr93.   M. Malhotra and K. S. Trivedi. A Methodology for Formal Specification of Hierarchy in Model Solution. In *Proc. 5th Int. Workshop on Petri Nets and Performance Models (PNPM '93)*, pages 258–267, Toulouse, France, October 1993. 101, 793

McMi84.   J. McKenna and D. Mitra. Asymptotic Expansions and Integral Representations of Moments of Queue Lengths in Closed Markovian Networks. *Journal of the ACM*, 31(2):346–360, April 1984. 458

McQi98.   D. McDonald and K. Qian. An Approximation Method for Complete Solutions of Markov-Modulated Fluid Models. *Queueing System Theory Applied*, 30(3-4):365–384, 1998. 629

MCT94.   J. Muppala, G. Ciardo, and K. S. Trivedi. Stochastic Reward Nets for Reliability Prediction. *Communications in Reliability, Maintainability and Serviceability (An nternational journal published by SAE international)*, 1(2):9–20, July 1994. 92, 93, 101

MeAl98.   D. A. Menasce and V. A. F. Almeida. *Capacity Planning for Web Performance: Metrics, Models and Methods*. Prentice Hall, Upper Saddle River, NJ, 1998. 666

MeAl00.   D. Menasce and V. Almeida. *Scaling for E-Business — Technologies, Models, Performance and Capacity Planning*. Prentice Hall, Upper Saddle River, NJ, 2000. 743

Medh03.   J. Medhi. *Stochastic Models in Queueing Theory*. Academic Press, San Diego, London, 2003. 299, 300, 302, 303

MeEs98.   W. Meeker and L. Escobar. Maximum Likelihood for Location-Scale Based Distributions, 1998. 43

Mein98.   B. Meini. Solving M/G/1 Type Markov Chains: Recent Advances and Applications. *Communications in Statistics — Stochastic Models*, 14(1&2):479–496, 1998. 261

MeNa82.   D. Menasce and T. Nakanishi. Optimistic Versus Pessimistic Concurrency Control Mechanisms in Database Management Systems. *Information Systems*, 7(1):13–27, 1982. 549

Mena02.   D. A. Menasce. *Capacity Planning for Web Services: Metrics, Models and Methods*. Prentice Hall, Upper Saddle River, NJ, 2002. 708

MeSe97.   H. de Meer and H. Sevcikova. PENELOPE: Dependability Evaluation and the Optimization of Performability. In R. Marie, R. Plateau, and G. Rubino, editors, *Proc. 9th Int. Conf. on Modeling Techniques and Tools for Computer Performance Evaluation*, Lecture Notes in Computer Science 1245, pages 19–31, St. Malo, France, Berlin, 1997. Springer. 94

Meye78.   J. F. Meyer. On Evaluating the Performability of Degradable Computing Systems. In *Proc. 8th Int. Symp. on Fault-Tolerant Computing*, pages 44–49, June 1978. 8

Meye80.   J. Meyer. On Evaluating the Performability of Degradable Computing Systems. *IEEE Transactions on Computers*, 29(8):720–731, August 1980. 76, 77, 86, 93

Meye82.   J. Meyer. Closed-Form Solutions of Performability. *IEEE Transactions on Computers*, 31(7):648–657, July 1982. 91, 93

MFV99.   B. Mueller-Clostermann, C. Flues, and M. Vilents. VITO: A Tool for Capacity Planning of Computer Systems. In D. Baum, N. Mueller, and R. Roedler, editors, *Proc. of the Conference on Measurement, Modelling and Evaluation of Computer-Communication Systems, Trier 1999, Description of Tools*, pages 115–119, 1999. 700

Mitz97.   U. Mitzlaff. *Diffusionsapproximation von Warteschlangensystemen*. Dissertation, Technical University Clausthal, Germany, July 1997. 463, 464

MMKT94.   J. Muppala, V. Mainkar, V. Kulkarni, and K. S. Trivedi. Numerical Computation of Response Time Distributions Using Stochastic Reward Nets. *Annals of Operations Research*, 48:155–184, 1994. 83, 87, 101

MMT94.   M. Malhotra, J. Muppala, and K. Trivedi. Stiffness-Tolerant Methods for Transient Analysis of Stiff Markov Chains. *International Journal on Microelectronics and Reliability*, 34(11):1825–1841, 1994. 216, 218, 219, 221

MMT96. J. Muppala, M. Malhotra, and K. Trivedi. Markov Dependability Models of Complex Systems: Analysis Techniques. In S. Ozekici, editor, *Reliability and Maintenance of Complex Systems*, pages 442–486. Springer, Berlin, 1996. 216, 219

MMW57. C. Mack, T. Murphy, and N. Webb. The Efficiency of N Machines Unidirectionally Patrolled by One Operative when Walking Time and Repair Time are Constant. *Journal of the Royal Statistical Society*, Series B 19(1):166–172, 1957. 147

Moor72. F. Moore. Computational Model of a Closed Queuing Network with Exponential Servers. *IBM Journal of Research and Development*, 16(6):567–572, November 1972. 370

Morr87. J. Morris. A Theoretical Basis for Stepwise Refinement and the Programming Calculus. *Science of Computer Programming*, 9(3):287–306, December 1987. 14

MSA96. M. Meo, E. de Souza e Silva, and M. Ajmone Marsan. Efficient Solution for a Class of Markov Chain Models of Telecommunication Systems. *Performance Evaluation*, 27/28(4):603–625, October 1996. 159

MSW88. P. Martini, O. Spaniol, and T. Welzel. File Transfer in High Speed Token Ring Networks: Performance Evaluation by Approximate Analysis and Simulation. *IEEE Journal on Selected Areas in Communications*, 6(6):987–996, July 1988. 714, 716

MTBH93. H. de Meer, K. S. Trivedi, G. Bolch, and F. Hofmann. Optimal Transient Service Strategies for Adaptive Heterogeneous Queueing Systems. In O. Spaniol, editor, *Proc. 7th GI/ITG Conf. on Measurement, Modeling and Performance Evaluation of Computer and Communication Systems (MMB '93)*, Informatik aktuell, pages 166–170, Aachen, Germany, Berlin, September 1993. Springer. 90, 91, 94

MTD94. H. de Meer, K. S. Trivedi, and M. Dal Cin. Guarded Repair of Dependable Systems. *Theoretical Computer Science*, 128:179–210, July 1994. Special Issue on Dependable Parallel Computing. 85, 91, 92, 93, 94, 106

Munt72. R. Muntz. Network of Queues. Technical report, University of Los Angeles, Department of Computer Science, 1972. Notes for Engineering 226 C. 360, 384

Munt73. R. Muntz. Poisson Departure Process and Queueing Networks. In *Proc. 7th Annual Princeton Conf. on Information Sciences and*

Systems, pages 435–440, Princeton, NJ, March 1973. Princeton University Press. 339, 341, 354

MüRo87.   B. Müller-Clostermann and G. Rosentreter. Synchronized Queueing Networks: Concepts, Examples and Evaluation Techniques. *Informatik-Fachberichte*, 154:176–191, 1987. 549, 577

MuTr91.   J. Muppala and K. S. Trivedi. Composite Performance and Availability Analysis Using a Hierarchy of Stochastic Reward Nets. In G. Balbo, editor, *Proc. 5th Int. Conf. on Modeling Techniques and Tools for Computer Performance Evaluation*, pages 322–336, Torino, Amsterdam, 1991. Elsevier Science (North-Holland). 87, 91, 93, 101

MuTr92.   J. Muppala and K. S. Trivedi. Numerical Transient Solution of Finite Markovian Queueing Systems. In U. Bhat and I. Basawa, editors, *Queueing and Related Models*, pages 262–284. Oxford University Press, Oxford, 1992. 90, 101, 216, 218, 244

MuWo74a.   R. Muntz and J. Wong. Asymptotic Properties of Closed Queueing Network Models. In *Proc. 8th Annual Princeton Conf. on Information Sciences and Systems*, pages 348–352, Princeton, NJ, March 1974. Princeton University Press. 453

MuWo74b.   R. Muntz and J. Wong. Efficient Computational Procedures for Closed Queueing Network Models. In *Proc. 7th Hawaii Int. Conf. on System Science*, pages 33–36, Hawaii, January 1974. 378, 379

NaFi67.   T. Naylor and J. Finger. Verification of Computer Simulation Models. *Management Science*, 14:92–101, October 1967. 14

NaGa88.   W. Najjar and J. Gaudiot. Reliability and Performance Modeling of Hypercube-Based Multiprocessors. In G. Iazeolla, P. Courtois, and O. Boxma, editors, *Computer Performance and Reliability*, pages 305–320. North-Holland, Amsterdam, 1988. 91

NaKa96.   M. Naghshineh and I. Katzela. Channel Assignment Schemes for Cellular Mobile Telecommunication Systems: A Comprehensive Survey. *IEEE Personal Communications*, 3 (3):10–31, 1996. 787

Nam.   Network Animator: http://www.isi.edu/nsnam/nam/. Last verified on February 08, 2005. 654

Naou97.   V. Naoumov. Matrix-Multiplicative Approach to Quasi-Birth-and-Death Processes Analysis. In R. R. Chakravarthy and A. S. Alfa, editors, *Matrix-analytic methods in stochastic models*, volume 183 of *Lecture notes in pure and applied mathematics*, pages 87–106. Marcel Dekker, 1997. 135

NBBB98.   K. Nichols, S. Blake, F. Baker, and D. Black. Definition of the
          Differentiated Services Field (DS Field) in IPv4 and IPv6 Headers.
          IETF RFC 2474, December 1998. 721

NeCh81.   D. Neuse and K. Chandy. SCAT: A Heuristic Algorithm for
          Queueing Network Models of Computing Systems. *ACM Sigmet-
          rics Performance Evaluation Review*, 10(3):59–79, 1981. 427, 428,
          431, 435

Nels82.   W. Nelson. *Applied Life Data Analysis*. John Wiley and Sons,
          New York, 1982. 631

Nels90.   R. D. Nelson. A Performance Evaluation of a General Parallel
          Processing Model. *ACM SIGMETRICS Performance Evaluation
          Review*, 18(1):13–26, 1990. 135

NeOs69.   G. Newell and E. Osuna. Properties of Vehicle-Actuated Signals:
          II. One-Way Streets. *Transportation Science*, 3:99–125, 1969. 147

NeSq94.   R. D. Nelson and M. F. Squillante. The MAtrix-Geometric qUeue-
          ing model Solution package (MAGUS) User Manual. Technical
          report rc, IBM Research Division, 1994. 135

NeTa88.   R. Nelson and A. Tantawi. Approximate Analysis of Fork/Join
          Synchronization in Parallel Queues. *IEEE Transactions on Com-
          puters*, 37(6):739–743, June 1988. 577

Neut79.   M. F. Neuts. A Versatile Markovian Point Process. *Journal of
          Applied Probability*, 16:764–779, 1979. 128, 309

Neut81.   M. Neuts. *Matrix-Geometric Solutions in Stochastic Models: An
          Algorithmic Approach*. The Johns Hopkins University Press, Bal-
          timore, MD, 1981. 3, 127, 132, 133, 134, 264

Neut89.   M. F. Neuts. *Structured Stochastic Matrices of M/G/1 Type and
          Their Applications*, volume 5 of *Probability: Pure and Applied —
          A Series of Textbooks and Reference Books*. Marcel Dekker, 1989.
          259, 260, 314, 316

Newe69.   G. Newell. Properties of Vehicle-Actuated Signals: I. One-Way
          Streets. *Transportation Science*, 3:30–52, 1969. 147

Nico90.   V. Nicola. Lumpability of Markov Reward Models. Technical
          report, IBM T.J. Watson Research Center, Yorktown Heights,
          New York, 1990. 12, 202, 221

NKW97.   V. Naoumov, U. R. Krieger, and D. Wagner. Analysis of a Multi-
          server Delay-Loss System with a General Markovian Arrival Pro-
          cess. In R. R. Chakravarthy and A. S. Alfa, editors, *Matrix-
          Analytic Methods in Stochastic Models*, volume 183 of *Lecture*

Notes in Pure and Applied Mathematics, pages 43–66. Marcel Dekker, 1997. 135

Noet79.   A. Noetzel. A Generalized Queueing Discipline for Product Form Network Solutions. *Journal of the ACM*, 26(4):779–793, October 1979. 364

Ns05.   Network Simulator: http://www.isi.edu/nsnam/ns/. Last verified on February 08, 2005, 2005. 612, 639, 640, 651, 700

NSH93.   V. F. Nicola, P. Shahabuddin, and P. Heidelberger. Techniques for Fast Simulation of Highly Dependable Systems. In *Proc. 2nd International Workshop on Performability Modelling of Computer and Communication Systems*, 1993. 638

Oeze97.   S. Oezekici. Markov Modulated Bernoulli Process. *Mathematical Methods of Operations Research*, 45(3):311–324, 1997. 628

OMN.   OMNeT++ Discrete Event Simulation System: http://www.omnetpp.org. Last verified on February 08, 2005. 612

OnPe86.   R. Onvural and H. Perros. On Equivalences of Blocking Mechanisms in Queueing Networks with Blocking. *Operations Research Letters*, 5(6):293–298, December 1986. 593

OnPe89.   R. Onvural and H. Perros. Some Equivalences Between Closed Queueing Networks with Blocking. *Performance Evaluation*, 9(2):111–118, April 1989. 592

OPN05.   OPNET Technologies Inc. : http://www.opnet.com/. Last verified on February 08, 2005, 2005. 612, 639, 700

Ost01.   A. Ost. *Performance of Communication Systems: A Model-Based Approach with Matrix-Geometric Methods*. Springer, 2001. 135, 303

OTC05.   OTCL: Object TCL extensions: http://bmrc.berkeley.edu/research/cmt/cmtdoc/otcl/. Last verified on February 08, 2005, 2005. 651

PaFl95.   V. Paxson and S. Floyd. Wide Area Traffic: The Failure of Poisson Modeling. *IEEE/ACM Transactions on Networking*, 3(3):226–244, 1995. 303

Park02.   D. Parker. *Implementation of Symbolic Model Checking for Probabilistic Systems*. PhD thesis, Faculty of Science, University of Birmingham, 2002. 120, 122

PaWi00.   K. Park and W. Willinger. *Self-Similar Network Traffic and Performance Evaluation.* John Wiley & Sons, New York, 2000. 625

PeAl86.   H. Perros and T. Altiok. Approximate Analysis of Open Networks of Queues with Blocking: Tandem Configurations. *IEEE Transactions on Software Engineering,* 12(3):450–461, March 1986. 592

Perr81.   H. Perros. A Symmetrical Exponential Open Queue Network with Blocking and Feedback. *IEEE Transactions on Software Engineering,* 7(4):395–402, July 1981. 592

Perr94.   H. Perros. *Queueing Networks with Blocking.* Oxford University Press, Oxford, 1994. 592, 600

PeSn89.   H. Perros and P. Snyder. A Computationally Efficient Approximation Algorithm for Feed-Forward Open Queueing Networks with Blocking. *Performance Evaluation,* 9(9):217–224, June 1989. 592, 600

Petr62.   C. A. Petri. *Kommunikation mit Automaten.* Dissertation, University of Bonn, Bonn, Germany, 1962. 94

Pitt79.   B. Pittel. Closed Exponential Networks of Queues with Saturation: The Jackson Type Stationary Distribution and Its Asymptotic Analysis. *Mathematics of Operations Research,* 4:367–378, 1979. 593

PJL02.   K. Pawlikowski, H. D. Jeong, and J. S. Lee. On Credibility of Simulation Studies of Telecommunication Networks. *IEEE Communication Magazine,* 4:132–139, Jan 2002. 609

PSSP95.   C. D. Pegden, R. E. Shannon, R. P. Sadowski, and C. D. Pegden. *Introduction to Simulation Using Siman.* McGraw-Hill, New York, second edition, 1995. 612

PuAi86.   G. Pujolle and W. Ai. A Solution for Multiserver and Multiclass Open Queueing Networks. *INFOR,* 24(3):221–230, 1986. 479, 481, 505, 601

Qua.   Scalable Network Technology, QualNet Simulator: http://www.scalable-networks.com. Last verified on February 08, 2005. 612

RaBe74.   E. Rainville and P. Bedient. *Elementary Differential Equations.* Macmillan, New York, 1974. 214

RaLa86.   V. Ramaswami and G. Latouche. A General Class of Markov Processes with Explicit Matrix-Geometric Solutions. *Operations Research-Spektrum,* 8:209–218, 1986. 133

Rama80.   V. Ramaswami. The N/G/1 Queue and Its Detailed Analysis. *Advances in Applied Probability*, 12:222–261, 1980. 309

Rama88.   V. Ramaswami. A Stable Recursion for the Steady State Vector in Markov Chains of M/G/1 Type. *Communications in Statistics — Stochastic Models*, 4(1):183–188, 1988. 260

Rama98.   V. Ramaswami. The Generality of Quasi Birth-and-Death Processes. In A. S. Alfa and R. R. Chakravarthy, editors, *Advances in Matrix-Analytic Methods for Stochastic Models*, pages 93–113. Notable Publications, Inc, 1998. 135

Rand86.   S. Randhawa. Simulation of a Wafer Fabrication Facility Using Network Modeling. *Journal of the Paper Society of Manufacturing Engineers*, 1986. 756

RaTa00.   M. Rajaratnam and F. Takawira. Nonclassical Traffic Modeling and Performance Analysis of Cellular Mobile Networks with and without Channel Reservation. *IEEE Transaction on Vehicular Techology*, VT-49:817–834, 2000. 787

RaTr95.   A. Ramesh and K. S. Trivedi. Semi-Numerical Transient Analysis of Markov Models. In R. Geist, editor, *Proc. 33rd ACM Southeast Conf.*, pages 13–23, 1995. 216

REA.   REAL network simulator: http://www.cs.cornell.edu/skeshav/real/overview.html. Last verified on February 08, 2005. 651

Reis81.   M. Reiser. Mean-Value Analysis and Convolution Method for Queue-Dependent Servers in Closed Queueing Networks. *Performance Evaluation*, 1, 1981. 405

ReKo74.   M. Reiser and H. Kobayashi. Accuracy of the Diffusion Approximation for Some Queueing Systems. *IBM Journal of Research and Development*, 18(2):110–124, March 1974. 463, 464, 465, 467, 470

ReKo75.   M. Reiser and H. Kobayashi. Queueing Networks with Multiple Closed Chains: Theory and Computational Algorithms. *IBM Journal of Research and Development*, 19(3):283–294, May 1975. 278, 384

ReLa80.   M. Reiser and S. Lavenberg. Mean-Value Analysis of Closed Multichain Queuing Networks. *Journal of the ACM*, 27(2):313–322, April 1980. 369, 384, 385, 393, 401, 456, 578

ReTr88.   A. Reibman and K. Trivedi. Numerical Transient Analysis of Markov Models. *Computers and Operations Research*, 15(1):19–36, 1988. 218, 221

ReTr89.   A. Reibman and K. Trivedi. Transient Analysis of Cumulative Measures of Markov Model Behavior. *Stochastic Models*, 5(4):683–710, 1989. 216, 219

ReWa84.   R. Redner and H. Walker. Mixture Densities, Maximum Likelihood, and the EM Algorithm, 1984. SIAM Rev., vol. 26, no. 2, pp. 195–239. 41

RiSm02a.   A. Riska and E. Smirni. MAMSolver: A Matrix Analytic Methods Tool. In *Proc. 12th International Conference on Computer Performance Evaluation, Modelling Techniques and Tools (TOOLS 2002)*, volume 2324 of *Lecture Notes in Computer Science*, pages 205–211. Springer, 2002. 135

RiSm02b.   A. Riska and E. Smirni. M/G/1-Type Markov Processes: A Tutorial. In M. C. Calzarossa and S. Tucci, editors, *Performance Evaluation of Complex Systems: Techniques and Tools, Performance 2002, Tutorial Lectures*, volume 2459 of *Lecture Notes in Computer Science*, pages 36–63. Springer, Berlin, 2002. 259

RKS05.   J. Roszik, C. Kim, and J. Sztrik. Retrial Queues in the Performance Modeling of Cellular Mobile Networks using MOSEL. *International Journal of Simulation: Systems, Science and Technology*, 6:38–47, 2005. 302

RST89.   A. Reibman, R. Smith, and K. S. Trivedi. Markov and Markov Reward Model Transient Analysis: An Overview of Numerical Approaches. *European Journal of Operational Research*, 40:257–267, 1989. 75

SaCh81.   C. Sauer and K. Chandy. *Computer Systems Performance Modelling*. Prentice-Hall, Englewood Cliffs, NJ, 1981. 26, 339, 369, 542, 544

SaMa85.   C. Sauer and E. MacNair. The Evolution of the Research Queueing Package. In D. Potier, editor, *Proc. Int. Conf. on Modelling Techniques and Tools for Performance Analysis*, pages 5–24, Paris, Amsterdam, 1985. North-Holland. 719

Sarg94.   R. G. Sargent. A Historical View of Hybrid Simulation/Analytic Models. In *Proc. 26th Conference on Winter Simulation*, December 1994. 12

SaTr87.   R. Sahner and K. S. Trivedi. Performance and Reliability Analysis Using Directed Acyclic Graphs. *IEEE Transactions on Software Engineering*, 13(10):1105–1114, October 1987. 3, 577

Saue81.   C. Sauer.   Approximate Solution of Queueing Networks with
           Simultaneous Resource Possession. *IBM Journal of Research and
           Development*, 25(6):894–903, November 1981. 544

Saue83.   C. Sauer. Computational Algorithms for State-Dependent Queue-
           ing Networks. *ACM Transactions on Computer Systems*, 1(1):67–
           92, February 1983. 384, 409

SBE+00.   J. Schüler, K. Begain, M. Ermel, T. Müller, and M. Schweigel.
           Performance Analysis of a Single UMTS Cell. In *Proc. Europ.
           Wireless Conference*, page 6, 2000. 782

Schr70.   L. Schrage. An Alternative Proof of a Conservation Law for the
           Queue G/G/1. *Operations Research*, 18:185–187, 1970. 278

Schw79.   P. Schweitzer. Approximate Analysis of Multiclass Closed Net-
           works of Queues. In *Proc. Int. Conf. on Stochastic Control and
           Optimization*, pages 25–29, Amsterdam, June 1979. 422

Schw84.   P. Schweitzer. Aggregation Methods for Large Markov Chains. In
           G. e. a. Iazeolla, editor, *Mathematical Computer Performance and
           Reliability*. North-Holland, Amsterdam, 1984. 199

SeMi81.   K. Sevcik and I. Mitrani.   The Distribution of Queuing Net-
           work States at Input and Output Instants. *Journal of the ACM*,
           28(2):358–371, April 1981. 384, 385, 401, 578

Sevc77.   K. Sevcik. Priority Scheduling Disciplines in Queueing Network
           Models of Computer Systems. In *Proc. IFIP 7th World Comput-
           er Congress*, pages 565–570, Toronto, Amsterdam, 1977. North-
           Holland. 522, 526

Shah94.   P. Shahabuddin. Importance Sampling for the Simulation of High-
           ly Reliable Markovian Systems. *Management Science*, 40(3):333–
           352, 1994. 637

ShBu77.   A. Shum and J. Buzen.   The EPF Technique: A Method for
           Obtaining Approximate Solutions to Closed Queueing Networks
           with General Service Times. In *Proc. 3rd Symp. on Measur-
           ing, Modelling and Evaluating Computer Systems*, pages 201–220,
           1977. 350

Shru81.   L. W. Shruben. Control of Initialization Bias in Multivariate Sim-
           ulation Response. *Communications of ACM*, 24(4):246–252, 1981.
           634

ShSa83.   J. G. Shanthikumar and R. G. Sargent. A Unifying View of Hybrid
           Simulation/Analytic Models and Modeling. *Operations Research*,
           30:1030–1052, 1983. 12

ShYa88.   J. Shanthikumar and D. Yao. Throughput Bounds for Closed Queueing Networks with Queue-Dependent Service Rates. *Performance Evaluation*, 9(1):69–78, November 1988. 458

Sim05a.   Simscript II.5 Simulator: http://www.caciasl.com/. Last verified on February 08, 2005, 2005. 609, 612

Sim05b.   A Collection of Simulation Resources on the Internet, http://www.idsia.ch/~andrea/simtools.html. Last verified on February 08, 2005, 2005. 613

Sim05c.   Simula Simulator: http://www.isima.fr/asu/. Last verified on February 08, 2005, 2005. 609, 612

SLM86.   E. de Souza de Silva, S. Lavenberg, and R. Muntz. A Clustering Approximation Technique for Queueing Networks with a Large Number of Chains. *IEEE Transactions on Computers*, 35(5):419–430, May 1986. 421

SmBr80.   C. Smith and J. Browne. Aspects of Software Design Analysis: Concurrency and Blocking. *ACM Sigmetrics Performance Evaluation Review*, 9(2):245–253, 1980. 549

SMK82.   C. Sauer, E. MacNair, and J. Kurose. The Research Queueing Package: Past, Present and Future. *Proc. National Computer Conf.*, pages 273–280, 1982. 541

SmSa95.   A. J. Smith and R. H. Saavedra. Measuring Cache and TLB Performance and Their Effect on Benchmark Runtimes. *IEEE Transaction on Computer*, 44(10):1223–1235, 1995. 611

SmTr90.   R. Smith and K. S. Trivedi. The Analysis of Computer Systems Using Markov Reward Processes. In H. Takagi, editor, *Stochastic Analysis of Computer and Communication Systems*, pages 589–629. Elsevier Science (North-Holland), Amsterdam, 1990. 75, 77

SnSc85.   M. Snell and L. W. Schruben. Weighting Simulation Data to Reduce Initialization Effects. *IIE Transactions*, 17(4):354–363, 1985. 635

SoGa89.   E. de Souza e Silva and H. Gail. Calculating Availability and Performability Measures of Repairable Computer Systems Using Randomization. *Journal of the ACM*, 36(1):171–193, 1989. 91, 216

SOL.   S. Spatz, N. C. Oguz, and R. Lingambhotla. TELPACK application: http://www.sice.umkc.edu/telpack/. Last verified on March 15, 2005. 135

SoLa89.   E. de Souza e Silva and S. Lavenberg. Calculating Joint Queue-Length Distributions in Product-Form Queuing Networks. *Journal of the ACM*, 36(1):194–207, January 1989. 370

SOQW95.   W. Sanders, W. Obal, A. Qureshi, and F. Widjanarko. The Ultra-SAN modeling environment. *Performance Evaluation*, 24(1):89–115, October 1995. 700

Spir79.   J. Spirn. Queuing Networks with Random Selection for Service. *IEEE Transactions on Software Engineering*, 5(3):287–289, May 1979. 364

Squi91.   M. F. Squillante. MAGIC: A Computer Performance Modelling Tool Based on Matrix-Geometric Methods. In G. Balbo and G. Serazzi, editors, *Proc. 5th International Conference on Computer Performance Evaluation, Modelling Techniques and Tools*, pages 411–425. North-Holland, Amsterdam, 1991. 135

Stan02.   Standard Performance Evaluation Corporation (SPEC). SPEC-jAppServer2002 Documentation. Specifications, SPEC, November 2002. http://www.spec.org/osg/jAppServer/. 735

Stew90.   W. Stewart. *MARCA: MARCOF CHAIN ANALYZER, A Software Package for Makrov Modelling, Version 2.0*. North Carolina State University, Department of Computer Science, Raleigh, NC, 1990. 700

Stew94.   W. Stewart. *Introduction to Numerical Solution of Markov Chains*. Princeton University Press, Princeton, NJ, 1994. 123, 199, 201, 219, 700

StGo85.   W. Stewart and A. Goyal. Matrix Methods in Large Dependability Models. Research report RC 11485, IBM T.J. Watson Research Center, Yorktown Heights, NY, November 1985. 174, 175, 176, 219

STP96.   R. Sahner, K. S. Trivedi, and A. Puliafito. *Performance and Reliability Analysis of Computer Systems — An Example-Based Approach Using the SHARPE Software Package*. Kluwer Academic Publishers, Boston, MA, 1996. 3, 11, 12, 18, 76, 93, 213, 216, 543, 547, 549, 550, 678, 688, 689, 695, 697, 760, 770, 772

STR88.   R. Smith, K. S. Trivedi, and A. Ramesh. Performability Analysis: Measures, an Algorithm and a Case Study. *IEEE Transactions on Computers*, 37(4):406–417, April 1988. 91

Stre86.   J. Strelen. A Generalization of Mean Value Analysis to Higher Moments: Moment Analysis. *ACM Sigmetrics Performance Evaluation Review*, 14(1):129–140, May 1986. 370

SuDi84.   R. Suri and G. Diehl. A New Building Block for Performance Evaluation of Queueing Networks with Finite Buffers. *ACM Sigmetrics Performance Evaluation Review*, 12(3):134–142, August 1984. 600

SuDi86.   R. Suri and G. Diehl. A Variable Buffer-Size Model and Its Use in Analyzing Closed Queueing Networks with Blocking. *Management Science*, 32(2):206–225, February 1986. 592, 600

SuHi84.   R. Suri and R. Hildebrant. Modeling Flexible Manufacturing Systems Using Mean Value Analysis. *Journal of Manufacturing Systems*, 3(1):27–38, 1984. 749

SZT99.   H. Sun, X. Zang, and K. S. Trivedi. The Effect of Web Caching on Network Planning. *Computer Communications*, 22(14):1343–1350, September 1999. 641

Sztr05.   J. Sztrik. Tool Supported Performance Modelling of Finite-Source Retrial Queues with Breakdowns. *Publicationes Mathematicae*, 66:197–211, 2005. 302

Taka75.   Y. Takahashi. A Lumping Method for Numerical Calculations of Stationary Distributions of Markov Chains. Research report B 18, Tokyo Institute of Technology, Department of Information Sciences, Tokyo, June 1975. 198

Taka90.   H. Takagi. *Stochastic Analysis of Computer and Communication Systems*. North-Holland, Amsterdam, 1990. 759

Taka93.   H. Takagi. *Queueing Analysis: A Foundation of Performance Evaluation*, volume 1-3. North-Holland, Amsterdam, 1991–1993. 108

Tane95.   A. Tanenbaum. *Distributed Operating Systems*. Prentice-Hall, Englewood Cliffs, NJ, 1995. 261

Tay87.   Y. Tay. *Locking Performance in Centralized Databases*. Academic Press, New York, 1987. 549

TBJ+95.   M. Telek, A. Bobbio, L. Jereb, A. Puliafito, and K. Trivedi. Steady State Analysis of Markov Regenerative SPN with Age Memory Policy. In *Quantitative Evaluation of Computing and Communication Systems*, volume 977 of *LNCS*, pages 165–179. Springer, 1995. 113, 114

ThBa86.   A. Thomasian and P. Bay. Analytic Queueing Network Models for Parallel Processing of Task Systems. *IEEE Transactions on Computers*, 35(12):1045–1054, December 1986. 577

Thom83.   A. Thomasian. Queueing Network Models to Estimate Serialization Delays in Computer Systems. In A. Agrawala and S. Tripathi, editors, *Proc. Performance '83*, pages 61–81, Amsterdam, 1983. North-Holland. 549

ThRy85.   A. Thomasian and I. Ryu. Analysis of Some Optimistic Concurrency Control Schemes Based on Certification. *ACM Sigmetrics Performance Evaluation Review*, 13(2):192–203, August 1985. 549

Tijm86.   H. Tijms. *Stochastic Modelling and Analysis: A Computational Approach*. John Wiley, New York, 1986. 269, 272

TMWH92.   K. S. Trivedi, J. Muppala, S. Woolet, and B. Haverkort. Composite Performance and Dependability Analysis. *Performance Evaluation*, 14:197–215, 1992. 79, 85, 90, 91, 93

Tom.   TomasWeb Simulator: http://www.tomasweb.com/. 612

Tows80.   D. Towsley. Queuing Network Models with State-Dependent Routing. *Journal of the ACM*, 27(2):323–337, April 1980. 364

TrDo00.   H. T. Tran and T. V. Do. Computational Aspects for Steady State Analysis of QBD Processes. *Periodica Polytechnica — Electrical Engineering*, 44(2):179–200, 2000. 135

Triv99.   K. Trivedi. SPNP User's Manual. Version 6.0. Technical report, Center for Advanced Computing and Communication, Department of Electical and Computer Engineering, Duke University, Durham, NC, September 1999. 669, 678

Triv01.   K. S. Trivedi. *Probability and Statistics with Reliability, Queuing, and Computer Science Applications*. John Wiley and Sons, New York, second edition, 2001. 1, 15, 38, 46, 51, 53, 125, 126, 142, 216, 241, 247, 252, 284, 289, 290, 294, 303, 350, 543, 614, 621, 628, 631, 632, 633, 635, 644

Triv02.   K. S. Trivedi. SHARPE 2002: Symbolic Hierarchical Automated Reliability and Performance Evaluator. *International Conference on Dependable Systems and Networks (DSN)*, 2002. 689

TrKu93.   K. Trivedi and V. Kulkarni. FSPNs: Fluid Stochastic Petri Nets. In *Proc. 14th International Conference on Applications and Theory of Petri Nets*, volume 691 of *LNCS*, pages 24–31. Springer, 1993. 114, 115

TSI+96.   K. S. Trivedi, A. S. Sathaye, O. C. Ibe, R. C. Howe, and A. Aggarwal. Availability and Performance-Based Sizing of Multiprocessor Systems. *Communications in Reliability, Maintainability and Serviceability*, 1996. 803

TSIH90.   K. S. Trivedi, A. Sathaye, O. Ibe, and R. Howe. Should I Add a Processor? In *Proc. 23rd Annual Hawaii Int. Conf. on System Sciences*, pages 214–221, Los Alamitos, CA, January 1990. IEEE Computer Society Press. 85, 91, 93

TuSa85.   S. Tucci and C. Sauer. The Tree MVA Algorithm. *Performance Evaluation*, 5(3):187–196, August 1985. 409

TuTr00.   B. Tuffin and K. S. Trivedi. Implementation of Importance Splitting Techniques in Stochastic Petri Net Package. In *LNCS 1786: Computer Performance Evaluation, Modeling Techniques and Tools — 11th International Conference, TOOLS 2000, Schaumburg, IL, March 2000. Proceedings / Boudewijn R. Haverkort, Henrik C. Bohnenkamp, Connie U. Smith (Eds.)*, pages 216–229. Springer-Verlag, London, UK, 2000. 610, 638

TuTr01.   B. Tuffin and K. S. Trivedi. Importance Sampling for the Simulation of Stochastic Petri Nets and Fluid Stochastic Petri Nets. In *Proc. High-Performance Computing Symposium (HPC), Advanced Simulation Technologies Conference (ASTC 2001), Seattle*, pages 228–235, April 2001. 610, 637

VePo85.   M. Veran and D. Potier. QNAP 2: A Portable Environment for Queueing Systems Modelling. In D. Potier, editor, *Proc. Int. Conf. on Modelling Techniques and Tools for Performance Analysis*, pages 25–63, Paris, Amsterdam, 1985. North-Holland. 678, 700, 719

Vesi04.   R. A. Vesilo. Long-Range Dependence of Markov Renewal Processes. *Australian & New Zealand Journal of Statistics*, 46(1):155–171, 2004. 303

Wals85.   R. Walstra. Nonexponential Networks of Queues: A Maximum Entropy Analysis. *ACM Sigmetrics Performance Evaluation Review*, 13(2):27–37, August 1985. 471

Welc83.   P. P. Welch. Statistical Analysis of Simulation Results. In S. S. Lavenberg, editor, *Computer Performance Modeling Handbook*, pages 268–328. Academic Press, New York, 1983. 614, 634

Welc95.   B. Welch. *Practical Programming in Tcl and Tk*. Prentice-Hall, Upper Saddle River, NJ, 1995. 674

Whit83a.   W. Whitt. Performance of the Queueing Network Analyzer. *Bell System Technical Journal*, 62(9):2817–2843, November 1983. 479, 481

Whit83b.   W. Whitt. The Queueing Network Analyzer. *Bell System Technical Journal*, 62(9):2779–2815, November 1983. 479, 481, 601

Whit97.   P. P. White. RSVP and Integrated Services in the Internet: A Tutorial. IEEE Communiction Magazin, May 1997. 720

Wilh77.   N. Wilhelm. A General Model for the Performance of Disk Systems. *Journal of the ACM*, 24(1):14–31, January 1977. 545

Wing01.   J. M. Wing. Formal Methods. In J. J. Marciniak, editor, *Encyclopedia of Software Engineering*, pages 547–559. John Wiley & Sons, New York, 2nd edition, 2001. 10

Wirt71.   N. Wirth. Program Development by Stepwise Refinement. *Communications of the ACM*, 14(4):221–227, April 1971. 14

WLTV96.   C. Wang, D. Logothetis, K. Trivedi, and I. Viniotis. Transient Behavior of ATM Networks under Overloads. In *Proc. IEEE INFOCOM '96*, pages 978–985, San Francisco, CA, Los Alamitos, CA, March 1996. IEEE Computer Society Press. 101, 775, 776

Wolf82.   R. Wolff. Poisson Arrivals See Time Averages. *Operations Research*, 30:223–231, 1982. 141, 145, 401

Wolt99.   K. Wolter. *Performance and Dependability Modeling with Second Order FSPNs*. PhD thesis, TU Berlin, Germany, 1999. 115

Wong75.   J. Wong. *Queueing Network Models for Computer Systems*. PhD thesis, Berkley University, School of Engineering and Applied Science, October 1975. 378, 379, 380

WTE96.   W. Willinger, M. S. Taqqu, and A. Erramilli. A Bibliographical Guide to Self-Similar Traffic and Performance Modeling for Modern High-Speed Networks. In F. P. Kelly, S. Zachary, and I. Ziedins, editors, *Stochastic Networks: Theory and Applications*, pages 339–366. Clarendon Press, 1996. 303

Wu82.   L. Wu. Operational Models for the Evaluation of Degradable Computing Systems. In *Proc. 1982 ACM SIGMETRICS Conf. on Measurements and Modeling of Computer Systems*, Performance Evaluation Review, pages 179–185, College Park, Maryland, New York, April 1982. Springer. 91

XF03.   D. G. X. Fang. Performance Modeling and QoS Evaluation of MAC/RLC Layer in GSM/GPRS Networks. *Proc. ICC 2003 — IEEE International Conference on Communications*, 26(1):271–275, May 2003. 135

Xgr05.   Xgraph application: http://www.isi.edu/nsnam/xgraph/. Last verified on February 08, 2005, 2005. 654

YaBu86.   D. Yao and J. Buzacott. The Exponentialization Approach to Flexible Manufacturing System Models with General Processing Times. *European Journal of Operational Research*, 24:410–416, 1986. 593

YBL89.   Q. Yang, L. Bhuyan, and B. Liu. Analysis and Comparison of Cache Coherence Protocols for a Packet-Switched Multiprocessor. *IEEE Transactions on Computers*, 38(8):1143–1153, August 1989. 793

Ye01.   Q. Ye. On Latouche–Ramaswami's Logarithmic Reduction Algorithm for Quasi-Birth-and-Death Processes. Research Report 2001-05, Department of Mathematics, University of Kentucky, March 2001. 135

YST01.   L. Yin, M. A. J. Smith, and K. S. Trivedi. Uncertainty Analysis in Reliability Modeling. In *Proc. Annual Reliability and Maintainability Symposium, (RAMS)*, pages 229–234, Philadelphia, PA, January 2001. 631

ZaWo81.   J. Zahorjan and E. Wong. The Solution of Separable Queueing Network Models Using Mean Value Analysis. *ACM Sigmetrics Performance Evaluation Review*, 10(3):80–85, 1981. 400, 401

ZeCh99.   H. Zeng and I. Chlamtac. Handoff Traffic Distribution in Cellular Networks. In *Proc. IEEE Wireless Communications and Networking Conference*, volume 1, pages 413–417, 1999. 787

ZES88.   J. Zahorjan, D. Eager, and H. Sweillam. Accuracy, Speed, and Convergence of Approximate Mean Value Analysis. *Performance Evaluation*, 8(4):255–270, August 1988. 427

ZFGH00.   A. Zimmermann, J. Freiheit, R. German, and G. Hommel. Petri Net Modelling and Performability Evaluation with TimeNET 3.0. In *Proc. 11th Int. Conf. on Modelling Techniques and Tools for Computer Performance Evaluation (TOOLS'2000)*, volume 1786 of *LNCS*, pages 188–202. Springer, 2000. 11, 679, 789

Zisg99.   H. Zisgen. *Warteschlangennetzwerke mit Gruppenbedienung*. PhD thesis, Department of Mathematic, Technical University Clausthal, Germany, 1999. 298, 601

ZKP00.   B. Zeigler, T. G. Kim, and H. Praehofer. *Theory of Modeling and Simulation*. Academic Press, New York, second edition, 2000. 609

Zoma96.   A. Y. H. Zomaya. *Parallel and Distributed Computing Handbook*. McGraw-Hill, New York, 1996. 638

ZSEG82. J. Zahorjan, K. Sevcik, D. Eager, and B. Galler. Balanced Job Bound Analysis of Queueing Networks. *Communications of the ACM*, 25(2):134–141, February 1982. 456

Index